Fundamentals of Legume Breeding

Girish Prasad Dixit • Harsh Kumar Dikshit
Gyan Prakash Mishra • Muraleedhar S. Aski
Editors

Fundamentals of Legume Breeding

A Text for Students and Practitioners

Springer

Editors
Girish Prasad Dixit
Indian Institute of Pulses Research
Kanpur, Uttar Pradesh, India

Harsh Kumar Dikshit
Division of Genetics
Indian Agricultural Research Institute
New Delhi, Delhi, India

Gyan Prakash Mishra
Division of Seed Science and Technology
Indian Agricultural Research Institute
New Delhi, Delhi, India

Muraleedhar S. Aski
Division of Genetics
Indian Agricultural Research Institute
New Delhi, India

ISBN 978-981-96-7837-2 ISBN 978-981-96-7838-9 (eBook)
https://doi.org/10.1007/978-981-96-7838-9

This Springer imprint is published by the registered company Springer Nature Singapore Pte Ltd.
The registered company address is: 152 Beach Road, #21-01/04 Gateway East, Singapore 189721, Singapore

Foreword

Legumes are essential for global food and nutrition security and sustainable agriculture. As important crops, they necessitate ongoing innovation to address issues including malnutrition, climate change, and the requirement for resilient agricultural systems. The book *Fundamentals of Legume Breeding* describes fundamental concepts and advances in legume breeding. The book covers importance, origin, reproduction biology, breeding procedures, techniques, and tools used in genetic improvement of legumes. The chapters are systematically structured to describe both conventional and modern tools including recent developments. Information on current status of pre-breeding, legume improvement, and conservation has added value to the publication. The book bridges knowledge gaps with regard to modern breeding tools for development of high yielding, micronutrient dense, and climate resilient varieties of different grain legumes. I congratulate the editors Drs. Girish Prasad Dixit, Harsh Kumar Dikshit, Gyan Prakash Mishra, and Muraleedhar S. Aski along with all other contributors of different chapters for bringing out this publication. I hope this book will greatly benefit students, research scholars, scientists, and others having interest in legume breeding.

T. R. Sharma
Deputy Director General (Crop Science)
Indian Council of Agricultural Research
Krishi Bhawan, New Delhi

Preface

The world population is expected to reach 9.3 billion by 2050. To ensure food and nutritional security both production and productivity have to be enhanced. It is estimated that nearly 840 million people will be undernourished by 2030. Post-COVID approximately 270 million people are facing food insecurity. The grain legumes have pivotal role in addressing food insecurity and healthy and balanced diets for all. These crops have potential of accelerating progress toward the Sustainable Development Goals (SDGs) of the 2030 Agenda. Climate change, micronutrient deficiencies, and biotic and abiotic stresses are the major challenges before the legume breeders.

The book entitled *Fundamentals of Legume Breeding* compiles the crop-based breeding techniques/procedures for major legume crops and is a textbook for graduate and postgraduate students and researchers working on grain legumes. Seventeen chapters covering different aspects of legume breeding have been included in the book. The book covers summer and winter season grain legumes. The book covers origin and distribution, floral biology and plant morphology, cytogenetics, genetics of qualitative and quantitative traits, breeding objectives, biotic and abiotic stresses, nutrient use efficiency, biofortification, wide hybridization, recombination breeding, and marker-assisted selection for each grain legume. The overall breeding progress made in different aspects of different crops, breeding using molecular tools, pre-breeding, and maintenance breeding are suitably covered. The book compiles fundamental concepts and latest breeding advances for each legume crop. The developments in conventional and molecular approaches including genomics are included for each legume crop. The book will expose students, researchers, and teachers to modern breeding tools and concepts in the legume breeding.

The editors are hopeful that the compiled information will serve as basic resource material for students, researchers, and teachers. Students trained in legume breeding techniques will serve as human resources for enhancing the production and productivity of the legume crops. The crop-wise information has been compiled by the experienced plant breeders. The editors are extremely thankful to all the authors for their contribution.

Kanpur, Uttar Pradesh, India — Girish Prasad Dixit
New Delhi, Delhi, India — Harsh Kumar Dikshit
New Delhi, Delhi, India — Gyan Prakash Mishra
New Delhi, Delhi, India — Muraleedhar S. Aski

Contents

Editors and Contributors

About the Editors

Girish Prasad Dixit is the Director of the ICAR-Indian Institute of Pulses Research (IIPR), Kanpur. He had served as Project Coordinator of AICRP on Chickpea, AICRP on MULLaRP, and AICRP on Linseed for more than 12 years. He has more than three decades of experience in pulses research and development, and has more than 24 varieties of different pulses including field pea, chickpea, and urdbean to his credit. He has coordinated the release of more than 81 chickpea varieties including the first of its kind, marker-assisted selection-derived varieties for Fusarium wilt resistance and drought tolerance. He has published more than 100 research papers in high-impact journals and edited 6 books. He is recipient of several awards of repute and has traveled to many countries under research and management activities.

Harsh Kumar Dikshit is a Principal Scientist working at the ICAR-Indian Agricultural Research Institute, Delhi. He obtained his doctorate on plant breeding and pathology from G. B. Pant University of Agriculture and Technology, Uttarakhand. He got his master's in Plant Breeding from G. B. Pant University and bachelor's from Kanpur University. He has more than three decades of experience working as a scientist in the field of Plant Breeding, first at Indian Institute of Pulses Research, Kanpur, and then till date at ICAR-IARI. He is a recipient of multiple fellowships and awards and has published numerous articles in high-impact peer-reviewed journals.

Gyan Prakash Mishra is working as a Senior Scientist at ICAR-Indian Agricultural Research Institute, PUSA, New Delhi. He has obtained his PhD and master's in Genetics from ICAR-IARI and his bachelor's in Agriculture from Banaras Hindu University. He has also briefly worked at Purdue University, Indiana, USA, and University of California, USA. He holds 2 patents and 40 research articles in high-impact peer-reviewed journals. He has got more than 15 years of experience working on genetics and plant breeding and has been honored with multiple fellowships and awards.

Muraleedhar S. Aski is a Senior Scientist specializing in Lentil and Mungbean Breeding at the Division of Genetics, ICAR-Indian Agricultural Research Institute, New Delhi. Previously, he served as a Chickpea Breeder at ICAR-IIPR Kanpur. His research focuses on crop improvement using conventional, molecular and transgenic methods. Dr Aski earned his bachelor's degree from UAS Dharwad, a Master's from ANGRAU, Bapatla, and a PhD in Genetics from IARI, New Delhi. He has played a key role in developing 15 legume crop varieties, including high-protein chickpea IPC 2005-62 and India's first biofortified lentil, L 4717 (Pusa Agethi). As a faculty member, he teaches and supervises graduate, post-graduate and PhD students. His work on genetic resources, stress evaluations and nutrient gene mapping has earned him the GRISAAS 2019 Young Scientist Award and ISGPB Fellow 2025. His research is well-cited, with 1392 Google Scholar citations, an H-index of 20 and an i10-index of 42.

Contributors

Muraleedhar S. Aski Division of Genetics, ICAR-Indian Agricultural Research Institute, New Delhi, India

Surendra Barpete International Centre for Agricultural Research in the Dry Areas, Amlaha, Madhya Pradesh, India

C. Bharadwaj ICAR-Indian Agricultural Research Institute, Pusa, New Delhi, India

Abhishek Bohra ICAR-Indian Institute of Pulses Research (IIPR), Kanpur, Uttar Pradesh, India

Arpita Das Bidhan Chandra Krishi Viswavidyalaya, Mohanpur, West Bengal, India

P. Dhanasekar Nuclear Agriculture and Biotechnology Division, Bhabha Atomic Research Centre, Trombay, Mumbai, Maharashtra, India

Harsh Kumar Dikshit Division of Genetics, ICAR-Indian Agricultural Research Institute, New Delhi, India

Girish Prasad Dixit ICAR-Indian Institute of Pulses Research (IIPR), Kanpur, Uttar Pradesh, India

Padmavati G. Gore ICAR-National Bureau of Plant Genetic Resources, New Delhi, India

Sanjay Gupta ICAR-Indian Institute of Soybean Research Institute, Indore, Madhya Pradesh, India

Soma Gupta Division of Genetics, ICAR-Indian Agricultural Research Institute, New Delhi, India

Sonali Habde Nuclear Agriculture and Biotechnology Division, Bhabha Atomic Research Centre, Trombay, Mumbai, Maharashtra, India

Esha Jaiswal ICAR-Indian Institute of Pulses Research, Kanpur, India

Souframanien Jegadeesan Nuclear Agriculture and Biotechnology Division, Bhabha Atomic Research Centre, Trombay, Mumbai, Maharashtra, India

Sudhir Kumar Jha Indian Institute of Pulses Research, Kanpur, Uttar Pradesh, India

Rahul Kaldate Assam Agricultural University, Jorhat, Assam, India

P. K. Katiyar ICAR-Indian Institute of Pulses Research, Kanpur, India

Manju Kohli Division of Genetics, ICAR-Indian Agricultural Research Institute, New Delhi, India

D. Kumar Central Arid Zone Research Institute, Jodhpur, Rajasthan, India

Kuldeep Kumar Indian Institute of Pulses Research, Kanpur, Uttar Pradesh, India

Neeraj Kumar ICAR-Indian Agricultural Research Institute, Pusa, New Delhi, India

Prema Kumar Division of Genetics, ICAR-Indian Agricultural Research Institute, New Delhi, India

Shiv Kumar South Asia and China Program, International Center for Agricultural Research in the Dry Areas, National Agriculture Science Complex, New Delhi, India

Manisha Kumari ICAR-National Bureau of Plant Genetic Resource Regional Station, Shimla, Himachal Pradesh, India

Pummi Kumari ICAR-National Bureau of Plant Genetic Resource Regional Station, Shimla, Himachal Pradesh, India

H. R. Mahala Central Arid Zone Research Institute, Jodhpur, Rajasthan, India

Chirag Maheshwari ICAR-Indian Agricultural Research Institute, Pusa, New Delhi, India

Gyan Prakash Mishra Division of Genetics, ICAR-Indian Agricultural Research Institute, New Delhi, India

Suraj Mishra Division of Genetics, Indian Agricultural Research Institute, New Delhi, India

Ramakrishnan M. Nair World Vegetable Center, South Asia/Central Asia, Patancheru, Hyderabad, Telangana, India

V. Nataraj ICAR-National Soybean Research Institute, Indore, Madhya Pradesh, India

Ravi Pamarthi ICAR-National Bureau of Plant Genetic Resources, New Delhi, India

Rinky Resma Panda The Graduate School, ICAR-Indian Agricultural Research Institute, New Delhi, India

Ashok K. Parihar ICAR-Indian Institute of Pulses Research (IIPR), Kanpur, Uttar Pradesh, India

Dhanasekar Punniyamoorthy Nuclear Agriculture and Biotechnology Division, Bhabha Atomic Research Centre, Trombay, Mumbai, Maharashtra, India

K. R. Ramya ICAR-National Bureau of Plant Genetic Resources, New Delhi, India

J. C. Rana The Alliance of Bioversity International & CIAT—India Office, New Delhi, India

Meenal Rathore Indian Institute of Pulses Research, Kanpur, Uttar Pradesh, India

Jenia Roy Division of Genetics, ICAR-Indian Agricultural Research Institute, New Delhi, India

V. P. Sahi Department of Genetics and Plant Breeding, Sam Higginbottom University of Agriculture Technology and Sciences, Prayagraj, Uttar Pradesh, India

S. J. Satheesh Naik ICAR-Indian Institute of Pulses Research, Kanpur, Uttar Pradesh, India

K. B. Saxena International Crops Research Institute for the Semi-Arid Tropics (ICRISAT), Patancheru, Telangana, India

Al AinCity, Abu Dhabi, United Arab Emirates

Rachit K. Saxena ICAR - Indian Agricultural Research Institute (IARI), Dhemaji, Assam, India

Manish Sharma Sardarkrushinagar Dantiwada Agricultural University, Dantiwada, Gujarat, India

M. Shivakumar ICAR-National Soybean Research Institute, Indore, Madhya Pradesh, India

G. P. Singh ICAR-National Bureau of Plant Genetic Resources, New Delhi, India

K. H. Singh ICAR-Indian Institute of Soybean Research, Indore, Madhya Pradesh, India

Mohar Singh ICAR-National Bureau of Plant Genetic Resource Regional Station, Shimla, Himachal Pradesh, India

Raghav Sood ICAR-National Bureau of Plant Genetic Resource Regional Station, Shimla, Himachal Pradesh, India

J. Souframanien Nuclear Agriculture and Biotechnology Division, Bhabha Atomic Research Centre, Trombay, Mumbai, Maharashtra, India

Pradeep Kumar Thakur ICAR-National Bureau of Plant Genetic Resource Regional Station, Shimla, Himachal Pradesh, India

U. S. Thendral The Graduate School, ICAR-Indian Agricultural Research Institute, New Delhi, India

Kuldeep Tripathi Germplasm Evaluation Division, ICAR-National Bureau of Plant Genetic Resources, New Delhi, India

Shailesh Tripathi ICAR-Indian Institute of Pulses Research, Kanpur, Uttar Pradesh, India

T. Vinutha ICAR-Indian Agricultural Research Institute, Pusa, New Delhi, India

Prachi S. Yadav Division of Genetics, ICAR-Indian Agricultural Research Institute, New Delhi, India

Global Research Status of Grain Legumes

Girish Prasad Dixit and S. J. Satheesh Naik

Abstract

Looking into the multifaceted importance of grain legumes/pulses, the year 2016 was declared as International Year of Pulses by 68th United Nations General Assembly. The United Nations General Assembly (UNGA) program supports policies that promote the health, environmental, and economic benefits of pulses. As a highly nutritious, protein-dense food source, pulses serve as an alternative to animal-based proteins, offering unmatched environmental sustainability. As a result, the global demand for pulses has grown in both pulse-producing and non-pulse-producing countries. Pulses belong to the Leguminosae (Fabaceae) family, specifically the Papilionoideae subfamily. Although there are over 1300 edible legume species, only a small number are widely cultivated for human consumption. From 1961 to 2022, a significant shift in area (36%), yield (43%), and production (59%) has been recorded under different grain legumes. However, the cultivation of legumes faces several challenges—including genetic and developmental issues, agricultural conditions, and socio-economic factors—which often lead to lower yields compared to cereals. Climate change conditions adding new dimension for lowering the yield, requiring urgent achievement of new science tools. The advancing science of molecular genetics, phenomics, and speed breeding facilities is needed to be deployed judiciously for developing future varieties having new level of productivity. Scientific innovations, combined with policy initiatives, are promoting the cultivation and consumption of legumes due to their multidimensional benefits—for agriculture, human and livestock health, and climate change mitigation. Pulses are remarkable seeds that help protect the environment and sustain nutritional security.

G. P. Dixit (✉) · S. J. Satheesh Naik
ICAR-Indian Institute of Pulses Research,
Kanpur, Uttar Pradesh, India

Keywords
Grain legumes · Genetic gain · Biological nitrogen fixation · Papilionoideae

1 Introduction

Global food security is vulnerable due to monoculturing of low-value crops, resulting in unbalanced diets worldwide (Raza et al. 2019). The World Health Organization (WHO) estimates that approximately 9.9% of the global population is undernourished—a figure that may rise by 2030. In this context, grain legumes play a vital role in improving both dietary quality and ecosystem sustainability. Also known as dry pulses, grain legumes belong to the Fabaceae (or Leguminosae) family and are cultivated not only for human and livestock consumption but also for enhancing soil health (FAO 2023). When consumed alongside staple cereals, grain legumes contribute significantly to nutritional security. Moreover, they offer substantial benefits to the planet through their role in sustainable agriculture and environmental protection (Mudryj et al. 2014; Martens et al. 2012; Koivunen et al. 2016). Recognizing the importance of pulses, the 68th United Nations General Assembly (UNGA) stated 2016 as the International Year of Pulses. The health and environmental benefits and policy backup by the UNGA initiative increased the demand for pulses in several pulse-growing and non-pulse-growing countries for proteinaceous seeds and for enviro-economic impacts (Reckling et al. 2016). Symbiotic nitrogen fixation increases soil nitrogen (Williams et al. 2014). Grain legumes play a critical role in promoting global nutritional well-being and enhancing ecosystem resilience.

The Fabaceae (Leguminosae) family includes over 19,500 species that are cultivated across a wide variety of climatic regions worldwide (Yahara et al. 2013; Smýkal et al. 2015). Grain legumes, also known as dry pulses, are generally classified into two groups—major and minor—based on dietary importance, research and development priorities, and the

G. Prasad Dixit et al. (eds.), *Fundamentals of Legume Breeding*, https://doi.org/10.1007/978-981-96-7838-9_1

scale of cultivation. The list of major grain legumes encompasses common bean (*Phasiolous vulgaries*), chickpea (*Cicer arietinum*), lentil (*Lens culinaris*), cowpea (*Vigna unguiculata*), and pigeonpea (*Cajanus cajan*). There are more than a dozen minor grain legumes such as lupins, grass peas, mesquite, carob, alfalfa, and clover, green gram, black gram, lathyrus pea, jackbean, winged bean, common vetch, lablab, hyacinth bean, velvet bean, cowitch, yam bean, etc. (Table 1) for sustainable agricultural development and nutritional security worldwide (Jose et al. 2020, Asik et al. 2022, Tendai et al. 2017; Lenka and Lal 2012). Pulses are cultivated in more than 50 countries, with India leading in production, consumption, and import. They are rich in digestible proteins, complex carbohydrates, micronutrients, vitamins, and minerals. Their high fibre content and low fat make them beneficial for managing cholesterol levels and promoting bowel health. Additionally, integrating pulses into various cropping systems enhances soil health, improves resource-use efficiency, reduces the emission of climate-harming gases, and increases overall agricultural biodiversity. The ability of pulses to withstand water stresses makes their cultivation easy in dryland ecosystem.

Table 1 Grain legumes along with the botanical names and cultivated countries

Common name	Botanical name	Growing/ consuming regions/ countries
Major grain legumes		
Beans (Kidney bean, navy bean, pinto bean, black turtle bean, haricot bean)	*Phaseolus vulgaris*	Asia, Tropical Africa, West Indies
Chick peas	*Cicer arietinum*	India, Pakistan, Afghanistan
Peas	*Pisum sativum*	Asia, Australia, Europe
Cowpeas	*Vigna unguiculata*	Asia, Tropical Africa, West Indies
Broad beans	*Vicia faba*	Asia, Tropical Africa
Lentils	*Lens culinaris*	Asia, Canada, Australia,
Pigeonpeas	*Cajanus cajan*	India, Tropical Africa
Bambara beans	*Vigna subterranea*	Tropical Africa
Minor grain legumes		
Green gram	*Vigna radiata*	India, Myanmar, Bangladesh
Black gram	*Vigna mungo*	India, Myanmar, Bangladesh
Lathyrus pea	*Lathyrus latifolius*	India,
Jackbean	*Canavalia species*	Tropical Africa
Winged bean	*Psophocarpus tetragonolobus*	Tropical Africa
common vetch	*Vicia sativa*	Tropical Africa
Lupins	*Lupinus* spp	Australia
Lablab, hyacinth bean	*Lablab purpureus*	India, Africa
Velvet bean, cowitch	*Mucuna pruriens var. utilis*	
Yam bean	*Pachyrhizus erosus*	

Between 1961 and 2022, the area under grain legumes increased by 36%, while yield and total production rose by 43% and 59%, respectively (Fig. 1). During this period, six legume species accounted for 90% of the total cultivation area and 88% of total production. These dominant species include various types of beans (Phaseolus spp. and Vigna spp.), chickpea (*Cicer arietinum*), garden pea (*Pisum sativum*), cowpea (*Vigna unguiculata*), lentil (*Lens culinaris*) and pigeonpea (*Cajanus cajan*) worldwide. Pulses production in India registered a remarkable increase during the past decade reaching 27.3 million tonnes in 2021–22. Between 2015-16 and 2021–22, production rose by 67%, from 16.4 million tonnes to 27.3 million tonnes. Hence, imports declined >50%, suggesting India is gradually heading towards self-sufficiency in pulses. According to Crop Husbandry, Agriculture Inputs, Demand and Supply group (NITI Aayog 2023), the pulses demand would be 35–38 MT in 2030–31 and 49–57 MT in 2047–48 (Fig. 2). Consumption of pulses was 27 grams/day in 2011–12, against the requirement of 80–97 grams/day.

2 Grain Legumes Origin, Taxonomy, Domestication, and Constraints in Production

2.1 Origin and Centres of Domestication

The domestication of grain legumes was independent at different locations, known as 'centres of domestication'; later on, the crops have dispersed to elsewhere via trade migration. Mesoamericans introduced common bean to Europe from via trade (Gepts 2003; Angioi et al. 2010). Approximately 10,000 years ago, key Neolithic legumes—including chickpea (*Cicer arietinum*), pea (*Pisum sativum*), lentil (*Lens culinaris*), and bitter vetch (*Vicia ervilia*)—were first domesticated in the region between the upper Euphrates and Tigris rivers, an area known as the Fertile Crescent in present-day southeastern Turkey and northern Syria. These crops were subsequently disseminated through trade routes to the Indian subcontinent and various Southeast Asian countries (Lev-Yadun et al. 2000; Abbo and Gopher 2017; Abbo et al. 2008; Dönmez and Belli 2007; Kerem et al. 2007; Simon and Muehlbauer 1997; Singh 1997; Varshney et al. 2013; Willcox et al. 2009).

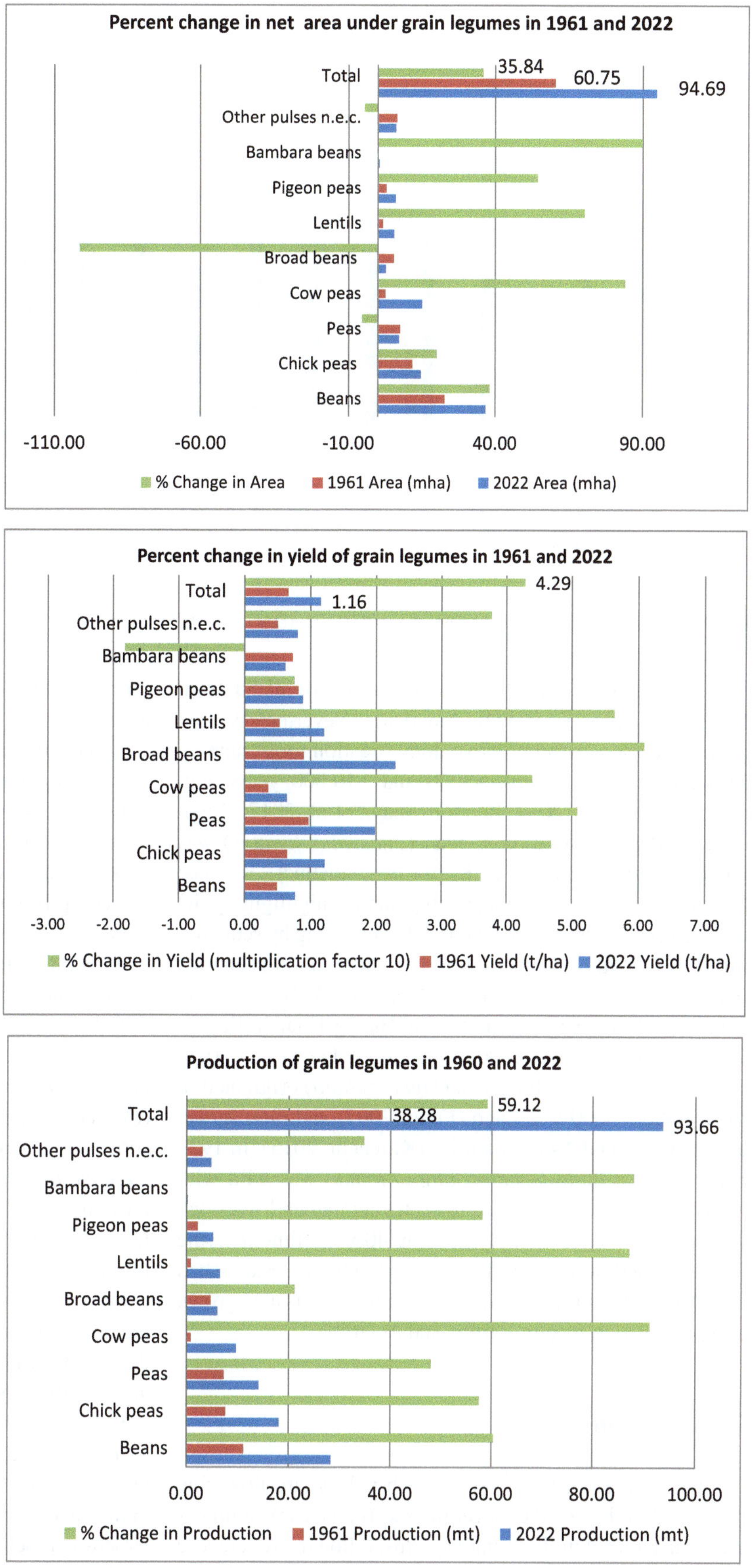

Fig. 1 Global level comparison between net cultivable area (mha), yield (t/ha), and production (mt) of grain legumes from 1960 to 2022. (Source: FAOSTAT 2022)

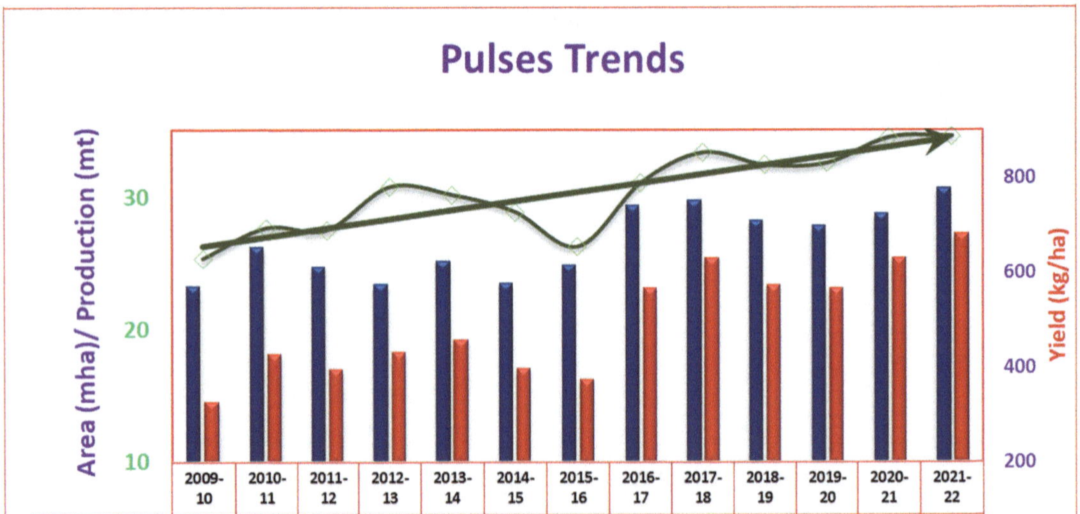

	2009-10	2010-11	2011-12	2012-13	2013-14	2014-15	2015-16	2016-17	2017-18	2018-19	2019-20	2020-21	2021-22
Area (m.ha)	23.28	26.28	24.78	23.47	25.21	23.55	24.91	29.44	29.82	28.28	27.86	28.78	30.73
Production (m.ton)	14.66	18.24	17.09	18.34	19.25	17.15	16.32	23.13	25.41	23.4	23.15	25.46	27.30
Yield (Kg/ha)	630	694	690	781	764	728	655	789	853	827	831	885	888

Fig. 2 Pulse trends in India since 2009–10 to 2021–22. (DAC GoI 2022)

It was thought that the common bean originated from Peru and Ecuador (South America). Nevertheless, the genomic evidences support two domestication roots, namely Mesoamerica and Andes, ~8000 and ~10,000 years ago, respectively (Gaut 2014; Rendón-Anaya et al. 2017; Singh et al. 2019). Likewise, lima bean (*Phaseolus lunatus*) was independently domesticated in two distinct regions: Mesoamerica and the Andean highlands, highlighting its dual centres of origin. Application of genotyping by sequencing (GBS) method supported both domestication scenarios (Chacón-Sánchez and Martínez-Castillo 2017). Genomic data traced soybean origin to the Huang-Huai Valley in Central China ~5000–9000 years ago (Han et al. 2016; Sedivy et al. 2017). Mungbean (*Vigna radiata*) originated in India ~4000–6000 years ago (Fuller 2007) and then spread across Africa and Asian countries (Kim et al. 2015). In 1929 narrow-leafed lupin (*Lupinus angustifolius*) was wild with high levels of seed-alkaloids toxic for human and livestock consumption. Fascinatingly, within 40-year timeline, the narrow-leafed lupin was transformed into an established crop (Hondelmann 1984; Wang et al. 2021) and is now grown as a protein-rich feed in Europe and Australia.

2.2 Taxonomy

Legumes, also known as protein-rich pulses, are flowering plants belonging to the Leguminosae (Fabaceae) family—the third-largest family of angiosperms, following Asteraceae and Orchidaceae, among seed-bearing plants on Earth. Leguminosae/Fabaceae has the global distribution including forming ecologically and economically important Mediterranean part to rainforest, temperate, tropical, arid, seasonally dry, and savanna ecosystems (Schrire et al. 2005). In Latin, Legume = legere which means to gather. The term pulse = puls or porridge, a cooked bean dish in ancient Romans (Albala 2007). The family names *Leguminosae* and *Fabaceae* are used interchangeably to refer to a plant family that comprises approximately 751 genera and 19,500 species (Lewis et al. 2005; Morris 2003).

Leguminosae/Fabaceae family has three sub-families: (1) Caesalpinioideae (2) Mimosoideae, and (3) Papilionoideae, each has distinct morphological traits. The majority of edible, nutritious legume crops belongs to the subfamily *Papilionoideae*. Notable examples include common beans (*Phaseolus* spp.), chickpea (*Cicer arietinum*), pea (*Pisum sativum*), pigeonpea (*Cajanus cajan*), cowpea (*Vigna unguiculata*), various *Vigna* species, and lentil (*Lens culinaris*), among others (Lewis et al. 2005; Morris 2003). Other lesser-known members of the legume family include clover, liquorice, lupin, etc. Legume taxonomy is relatively simple compared to cereals and brassicas. Further, the availability of the inter-generic hybrids in nature is difficult and isolation of desired plant type is a challenge (Smartt 1979, 1981). There are 28 tribes under sub-family Papilionoideae together with the seed and forage legumes.

Important grain legumes belong to two tribes, viz. Vicieae and Phaseoleae. Both groups have epigeal/hypogeal germination and herbaceous growth with variable seed texture, structure, biochemical compositions, and genetic behaviours (de Vogel 1980). The Leguminosae (Fabaceae) family plays a vital role in biological nitrogen fixation (BNF) through symbiotic associations with rhizobial bacteria, a process that is several times more efficient than nitrogen fixation by free-living bacteria.

2.3 Domestication of Grain Legumes

Domesticated plants are different from their ancestors in several agro-morphological, biochemical, and economic traits (Fig. 3), collectively referred to as the domestication syndrome (Harlan et al. 1973; Hammer 1984). Domestication is a continuous evolutionary process. Therefore, the differences observed between wild progenitors and domesticated types are not solely the result of human selection; natural selection also plays a significant role. Following initial domestication events, successive modifications in the genome can occur as crops are introduced to new environments. These changes, often driven by adaptation to biotic and abiotic stresses, can give rise to novel phenotypes. In some cases, these phenotypic differences are so pronounced that the domesticated forms may appear to belong to entirely different species, even though they do not. Crucial domestication traits related to economically valuable improvements show clear distinctions between the progenitor and domesticated types. (Fig. 3) (Abbo et al. 2014).

Historically grain legumes have also undergone several rounds of domestication and diversification. The important economic traits in legumes, viz. pod size, pod shattering, seed dormancy, and size are domesticated characters. Pod shattering and seed dormancy are key domestication traits in cool-season legumes, as their modification was essential for improving harvestability and uniform germination under cultivation (Abbo et al. 2014) and warm-season legumes (Parker et al. 2021a, b). Therefore, the present crop improvement researcher focusses on consumers'/farmers' preferences after domestication. Other major improved trait of grain legumes includes plant architecture/phenology, shape of grains, nutritional and anti-nutritional quality, and flavour of the grains (Vaz Patto 2016; Vaz Patto et al. 2015).

2.4 Constraints for Cultivation of Grain Legumes

Grain legumes are the second largest consumed food crop after cereals. Pulses farming, consumption, and trade are steadily growing worldwide. They are especially valuable in dry regions, where they complement staple foods and offer a low-cost, plant-based alternative to animal proteins. In many low-income countries, pulses make up a significant share of daily protein intake—ranging from 10% to 35% in Africa

Fig. 3 Pigeonpea proginator (*Cajanus cajanifolius*) (top two images) and the post-domesticated improved *Cajanus cajan* cv. IPA15–2(Sharada) (bottom two images) at flowering and podding stage

and 25–35% in India—making them essential for food and nutritional security (Joshi and Rao 2016). Therefore, legumes could help diversification of food for rural livelihood in low per capita income countries, if consumed with staple cereal diets (Satheesh et al. 2019; Abberton 2010; Singh 2017). Ecologically, legumes have positive impact on biodiversity and ecosystem processes through symbiotic biological nitrogen fixation, enhancing soil fertility (Duchene et al. 2017).

Despite their importance, legume cultivation and productivity face several challenges worldwide. Key issues include low-realized farm yields, instability in production, and a significant decline in pulse cultivation area—particularly in India's Indo-Gangetic Plains (IGP). Other major constraints are inadequate seed replacement rates, substantial post-harvest losses, wide fluctuations in market prices, and limited access to quality inputs. Additionally, poor technology transfer, insufficient seed storage infrastructure, low levels of mechanization, suboptimal crop management practices, and cultivation on marginal or degraded soils further hinder productivity. These challenges are compounded by the often-vulnerable socio-economic status of pulse-growing farmers. Climate constraints are additional layer of existing problems with the farmers. Nevertheless, legumes exhibit considerable genetic diversity for various yield traits, but harnessing this variation effectively remains a challenge. Breeding for improved resistance/tolerance to diseases and drought requires extensive research and development. Changing climatic conditions, combined with a growing global population and low-capital investment in legume cultivation, are intensifying the challenges related to food insecurity (Erokhin and Gao 2020). In addition, several socio-economic constraints continue to hinder pulse production worldwide. These include limited access to high-yielding, disease-resistant varieties, inadequate availability of quality or certified seeds, a lack of well-defined agronomic packages of practices for legumes, and insufficient dissemination and extension tools (Kebede 2020). There are other constraints such as post-harvest losses, limited value addition and inadequate output, and international markets. Pulses are often overlooked for one or other reasons by the farmers and the government institutions in comparison to the cereals, commercial/cash crops, and horticultural crops.

3 Breeding and Biotechnological Advancement for Improving the Economic Traits in Grain Legumes

3.1 Genetic Enhancement for Yield and Its Attributing Traits

While grain legume seeds in gene banks harbour substantial genetic variation, this diversity has not been fully harnessed in active breeding operations. Efforts to enhance the nutritional quality of grain legumes are limited. Factors affecting their nutritional quality, such as protein amino acid profiles and net protein utilization, require systematic exploration. Breeding nutri-dense cultivars in food legumes is essential. Presently, more focus has been diverted to improve the productivity and breeding for biotic and abiotic stresses. Further the crop management have paved the way to realize the good yield under optimal conditions. Pulses, when combined with cereals, can supply adequate calories and essential nutrients for human body.

3.1.1 Pod Shattering

During the domestication, selection against pod shattering is one of the key trait humans selected in legumes for breeding. It is an important seed dispersal mechanism in wild legumes but hinders the economic harvest and net yield in farming. Apart from the genetic constitution, several environmental indicators, viz. low-relative humidity and high temperatures, influence pod shattering (Zhang et al. 2018). Understanding the genetic architecture governing pod shattering was well-dispensed target across the world. Evidentially, pod shattering is a domestication syndrome (Table 2). For instance, Parker et al. 2020, elucidated a major QTL *PvPdh1* (ortholog of soybean *Pdh1*) associated with shattering resistance in common bean/dry beans and DNA markers flanking to the QTL region for introgression of the QTL in common bean (Parker et al. 2021a). In the recent past, the QTL *PvMYB26* locus has been precise to 22.5 kb region using transcriptome sequencing for pod shattering in common bean (Di Vittori et al. 2020). In cowpea, Lo et al. 2018 identified QTLs *CPshat3* and in adzuki bean (*Vigna angularis*), Takahashi et al. 2020 identified QTLs *CPshat5* and *VaMYB26*. Having

Table 2 Pod shattering genes/QTLs available in grain legumes

Legumes	Chromosome	Gene/QTL	References
Common bean	Pv03	*PvPdh1*	Parker et al. (2020)
	Pv05	*PvMYB26*	Rau et al. (2019)
Cowpea	Vu03 and Vu05	*CPshat3 (VuNAC007) and CPshat5 (VuMYB26)*	Lo et al. (2018)
Adzuki bean	Va07	*VaMYB26*	Takahashi et al. (2020)
Pea	LG III	*Dpo or Dpo1*	Weeden et al. (2002)
Lentil	LG IV	*Pi*	Ladizinsky (1979)
Chickpea	LG6	*PDH1*	Aguilar-Benitez et al. (2020)
Grass pea	LTr VII	*pod1 and pod2*	Talukdar (2011)

reference genome is crucial to discover QTL and candidate genes for pod shattering in grain legumes.

3.2 Advancement in High Yielding, Resilient, and Nutrient Dense Varieties/Hybrid Development

The Consultative Group on International Agricultural Research (CGIAR) institutes, namely International Crops Research Institute for Semi-Arid Tropics (ICRISAT), International Institute of Tropical Agriculture (IITA), International Centre for Agricultural Research in the Dry Areas (ICARDA), and the Alliance of Bioversity, are working in close connection with the national agricultural system towards the legumes enhancement. Apart from the international research organizations, different national agricultural research systems are also working for genetic improvement of legumes, viz. pre-breeding, line development for target traits, and breeding and hybrid breeding. As a result of collective research and developmental activities, >1300 improved varieties have been released in different crops for the cultivation of the farmers worldwide in the past one decade (https://www.cgiar.org/, https://iipr.icar.gov.in/).

In India, a number of area- and season-specific varieties have been developed in different pulses, having a positive impact of 15–20% in increasing pulse production. The significant achievements comprise reduction in ~20–25 days maturity duration in chickpea, lentil, green gram, and pigeonpea. Cytoplasmic Male Sterility (CMS) based hybrid cultivars in pigeonpea and marker-assisted breeding varieties in chickpea. India has released ~775 improved varieties having resistant to biotic/abiotic stresses (https://iipr.icar.gov.in/).

3.2.1 Common Bean

In Tanzania, through bean value chain initiatives, several improved bean varieties have developed. However, farmers are unwilling to adopt improved beans because of missing organoleptic traits. In Southern Ethiopia, a study evaluated and selected common bean varieties based on high yield and other agronomic traits. Notable varieties included Nasir, Hawassa-dume, and Angar. In Zambia, six improved common bean varieties are known and grown, including Chambeshi, Lukupa, Kapisha, Luangeni, Lyambai, and improved Kabulangeti. In other regions, there are additional improved varieties such as Dark Red Kidney (DRK) Bean, Omo-95, Awash-1, Granscope, Roba, Argene, Chore, Ebaya, Awash Melka, Dinkinesh, SNNPR-1-20, Melka Dima, and Nasir. Muroki et al. 2023, in Burundi, investigated that MAC70 and RWR1272 biofortified common bean varieties are excellent sources of iron and zinc and also retained significant bioavailable portion after cooking. Similarly, Kenya in 2017 released Angaza, Faida, and Nyota biofortified varieties. In Ethiopia, more than 60 improved bean varieties have been released since 1970, including a biofortified variety introduced in 2019. These high-yielding, resilient varieties are either lines or crosses with materials from the International Centre for Tropical Agriculture (CIAT).

3.2.2 Chickpea

Chickpea, called chana in India or garbanzo bean, has a greater role in sustaining global agriculture and nutrition. Amongst pulses, chickpea cultivated 14.80 mha globally, contributing 14.24 mt to the global food supply annually. In India, it has a 50% share of the total pulse basket of the country. It contributed significantly in 'Pulse Revolution' in India from a level of 7.3 million tonnes in 2014–15 to 13.5 mt in 2021–22 (Department of Agriculture Cooperation, GoI 2022). India released >250 improved varieties specific to different agro-niches. Pusa Chickpea 4005 is a drought-tolerant high-yielding chickpea variety, which was developed through marker-assisted selection. It's an excellent choice for regions with water scarcity. IPCMB 19–3 (Samriddhi) is resistant to Fusarium wilt and boasts a high protein content of 22.9%. It was also developed using Marker-Assisted Selection (MAS).

3.2.3 Peas, Dry (Field Pea Dry)

An important cool/rabi season pulse grown around the world and China is top producer of green peas, with an impressive output of 12.2 million tonnes. India follows closely, producing 4.8 million tonnes of peas. The USA contributes approximately 0.31 million tonnes to global pea production. France is another significant producer, contributing around 0.23 million tonnes. Egypt produces approximately 0.15 million tonnes of peas (Döring 2022). Other notable pea-producing countries include Canada, Russia, the United Kingdom, Kenya, Algeria, Morocco, and Peru. In India, the average productivity is 1.4 tonnes/ha. The enhanced yield in pea was due to development of several improved varieties. The key varieties of field pea in India are Early Bejar; this variety is known for its early maturity and good yield. It's suitable for regions with shorter growing seasons. Asauji peas are well adapted to various climates and offer good yields. They are popular among farmers. Jawahar Matar is a reliable variety with consistent performance. It's widely cultivated in different states of India. Arket peas are versatile and can be used both as green vegetables and dried pulses.

3.2.4 Cowpeas, Dry

Globally cowpea is cultivated in 14.5 mha with 6.2 mt production (FAOSTAT 2021). In India, cowpea is also grown as rice fallows in Chhattisgarh, Andhra Pradesh, Orissa, Madhya Pradesh, Karnataka, and Tamil Nadu. Due to climate smart nature of the cowpea, it's a part of food security programme in African countries. Recently Ghana developed seven varieties, viz. Asare-Moya, Kum-Zoya, Saka-Buro,

Aluba-Kpole, Yor-Kpitio, UCC-Early, and the Aduapa. Similarly, some of the improved varieties of cowpea bred in India are Anaswara, Kairali, Varun, C-152, Pusa Phalguni, Amba (V-16) (Medium duration), Ramba (V240) (M), Swarna (V-38) (M), GC-3, Pusa Sampada (V-585), and Shreshtha (V-37) (M). Furthermore, the researchers have been working on developing biofortified cowpea varieties. For instance, Pant Lobia 1 to 5 series are popular iron and zinc biofortified cowpea varieties developed through conventional breeding (PC report, AICRP on Kharif Pulses 2022).

3.2.5 Lentils

Lentil is a crop of >50 countries. The total cultivated area globally is around 6.1 million hectares, producing 6.33 million tonnes of seeds. Leading producers include Canada, India, Australia, Turkey, and Nepal. Canada and Australia export lentils to South Asia. In India, it's a rabi crop and grown on 1.4 mha with 1.26 mt of production (2021–22). It is grown as rainfed crop after rice, maize, pearl millet, or kharif fallow. Lentil is susceptible to Fusarium wilt, rust, low biomass, and lack of early vigour and varieties amenable to machine harvesting.

Nevertheless, several improved varieties have been released in India and elsewhere. Recently the biofortified lentil varieties are available in India, Bangladesh, and Nepal. They play a crucial role in addressing malnutrition and enhancing food security. The variety IPL_220 was released in India; it contains up to 113 ppm of iron (Yadava et al. 2022). This biofortified variety contributes to improving nutritional intake. BARI Masur-9 is a super-early lentil variety boasts an impressive 78 ppm of iron content. It's a significant step toward combating hidden hunger in Bangladesh. A lentil variety named Khajurah Masuro-4 was introduced in Nepal and has an iron content exceeding 75 ppm. Such biofortified cultivars are essential for achieving nutritional security in the third world countries.

3.2.6 Pigeonpea

Pigeonpea, known as red gram or arhar, is a rainy season/ Kharif crop. It is the second most leading crop after chickpea in terms of both area and production. It contributes 15% of the total pulses in India. Due to shift in the crop growing conditions, the extra-early (<130 days maturity) and early (up to 150 days maturity) crops are in high demand in the north-western plains as an alternative to the rice-wheat cropping system. The mid-early and medium groups (151 to 180 days maturity) occupy major pigeonpea growing areas in south and central India. During 2021–22, 4.2 m tonnes production was recorded in the country, which is much below the demand. Morphologically, the pigeonpea flower is suitable for self-pollination like other pulses. Pigeonpea is an often cross-pollinated crop (Saxena et al. 1990) with 20–70% outcrossing (Saxen et al. 2010). Often cross pollination offers both boom (exploitation of heterosis) and burst (varietal purity maintenance) for pigeonpea improvement. Solomon et al. (1957), reported that the first-generation hybrids are 20–25% better parent heterosis (Saxena and Sharma 1990).

By 1974, male sterility systems are better known to play a greater role in productivity enhancement of many crops, including pigeonpea, through enabling exploitation of heterosis/hybrid vigour. The 17 years of research efforts on GMS hybrids (1974) helped in the development of 6 GMS hybrids, namely ICPH 8 (1991) with 31–40% yield superiority (Saxena et al. 1992), and this was the first hybrid in pigeonpea. This was followed by five additional hybrids, viz. PPH 4 in 1994, CoH 1 in 1994, CoH 2 in 1997, AKPH 4104 in 1997, and AKPH 2022 in 1998 (Saxena et al. 2010). Nevertheless, these hybrids couldn't reach the farmers' fields because of seed parent segregation (female into sterile (50%) and fertile (50%)) and seed quality law.

It took 28 years to develop the first CGMS hybrid (1978) GTH 1, released by Central Varietal Release Committee, Govt. of India, for cultivation in the central zone of India in 2006. Currently, there are 42 registered CMS lines with ICAR-NBPGR, and more than 50 new stable CMS lines and > 100 restorers have been developed by ICAR and ICRISAT for use in hybrid breeding programmes. A total of five CGMS hybrids are available for cultivation, viz. ICPH 2671 (2014), ICPH 3762 (2014), ICPH 2740 (2015), IPH 15-03 (2020), and IPH 09-5 (2021) by ICAR and ICRISAT (Bohra et al. 2020).

3.2.7 Bambara Beans, Dry

It is native to Africa and also identified as earth pea, jugo bean, nyimo bean or ditloo, Bambara nut, Bambara bean, Congo goober, earth pea, ground bean, and hog peanut (Majola et al. 2021). The plant resembles peanuts, bearing the pods underground. Globally, it is produced 0.2 mt from a 0.25 mha area. Sub-Saharan Africa, Southeast Asia, the United States, Australia, and Brazil are the major growers. West African countries, namely Burkina Faso, Niger, and Cameroon, produce >74%. Bambara groundnut is climate resilient and capable of thriving under low-fertility soils (Olayide et al. 2018; Mayes et al. 2019; Pasipanodya et al. 2022). Therefore, the not significant progress towards varietal development has gone into this crop. However, a few landraces which are popular in South Africa are SB11-1A, SB19-1A, SB12-3B, and Bambara-12 (Unigwe et al. 2016).

3.2.8 Mungbean and Urdbean

Mung (*Vigna radiata* (L.) Wilczek) and urd (*V. mungo* (L.) Hepper) have a distinct role in the diversification of Indian agriculture due to across-seasons and location cultivation. They are mainly grown during kharif season either as sole crop or intercropped with cereals and cash crops. Short-duration and photo-thermos-insensitive genotypes have expanded their cultivation during spring or summer season in the North and during rabi season in the southern peninsular zone of India. Several improved varieties were released worldwide for cultivation: Samrat, Virat, IPM2–3, SML 668 and Meha in India; Crystal, Jade-AU, Regur in Australia; and Zhonglv No. 1 and Bao 8824-17 in China. Similarly, a number of varieties are released in Urdbean also; WBU 109 (Sulata), UH-1, LU 391, and KUG 479 are resistant to Mungbean Yellow Mosaic Virus. IPU02-43, NUL 7, and VBG 04-008 are resistant to MYMV and powdery mildew. TU 40 has the durable resistant to powdery mildew across the country.

3.2.9 Lathyrus

The presence of a neurotoxin Beta N-oxalyl-L-alpha-beta diaminopropionic acid (ODAP) is constraint for Lathyrus cultivation. However, it was grown in 0.28 mha in 2020–21 in India (Chhattisgarh, Bihar, Madhya Pradesh, and West Bengal). Ratan, Prateek, and Mahateora are candidate varieties with low ODAP levels available for cultivation in India.

4 Genomic Resources in Grain Legumes

Legumes are often referred to as ‘orphan crops’ because they typically receive less attention in advanced research techniques, agricultural training, extension services, and international trade. Nevertheless, the advanced genomic tools and techniques have significantly impacted agricultural research, enabling scientists to enhance crop yield, improve crop traits, and ensure food security. Researchers are exploring genomic tools to improve legume breeding. These tools hold promise for enhancing crop quality and yield. The quest for better grain legumes continues, driven by the need to address malnutrition-related challenges and sustain pulse production worldwide. Grain legumes are not only vital for global food security but also offer untapped potential for nutritional enhancement through strategic breeding efforts.

4.1 Next-Generation Sequencing (NGS)

NGS is widely used. Post-sanger sequencing technologies have revolutionized genomic research by allowing the sequencing of thousands of genomes to study diversity within or between germplasm pools of numerous agricultural crops and livestock. *Arabidopsis thaliana*, the first multicellular organism sequenced completely by a consortium of multinations using bacterial artificial chromosome (BAC) 2000. Presently, the genome assembly of *Arabidopsis thaliana* having the highest accuracy in terms of sequence information, which is popularly called ‘gold standard’ (Kersey 2019; Bohra et al. 2020). Similar efforts are going on to sequence different warm and cool season grain legumes (Table 3). Efforts are being made to improve the situation for orphan crops, including grain legumes. This initiative could lead to the breeding of more nutritious and resilient plants, enhancing food security (Ramsay et al. 2021; Kreplak et al. 2019).

4.2 Comparative Genomics

It aids in plant breeding by comparing the genomes of other species for gene identification. Comparative genomics in grain legumes is a fascinating field that involves analysing the genomes of different legume species to understand their evolution, domestication, and genetics. Comparative studies help in understanding the impact of domestication on traits that have economic importance, both above and below ground. Research focusing on cool-season legumes, with a particular emphasis on chickpea, explores the developmental dynamics that contribute to yield and nutritional value (Basu and Parida 2023). Further, the translational genomics in agriculture (TGA) for leading legumes uncovers advances in gene expression analysis for precise cloning and gene editing targets (Rajeev et al. 2015).

4.3 Genotyping and Marker-Assisted Breeding

These methods involve identifying and selecting for specific genes that confer economic values, viz. disease resistance or drought tolerance. Varshney et al. (2014) confirmed the QTL hotspot on linkage group 04 of chickpeas (CaLG04); having QTLs for drought tolerance component traits can explain up to 58.2% of phenotypic variation. Similarly, chickpea introgression lines possessing superior haplotypes from an ABCC3-type ABC transporter gene (CaABCC3) of a CaqSW2.4 QTL, which enhances seed weight and yield under drought stress, have also been developed (Basu et al. 2019).

4.4 Genome Editing Tools

Techniques, viz. CRISPR/Cas9, allow modifications in plant genomes, leading to improved crop varieties in agriculture. Utilizing CRISPR/Cas9 for gene editing offers a straightforward, adaptable, and accurate approach to genetic modification in plants. CRISPR/Cas9 utilizes two essential molecules: a Cas9 endonuclease and a guide RNA (gRNA). The gRNA consists of two small RNA molecules, CRISPR RNA

Table 3 Availability of genomic resources in legumes

Name	Botanical name	Genome size (n)	Sequencing technologies	Assembly size (Mb)	Ploidy level	References
Adzuki bean	*V. angularis*	612 Mb	Illumina + 454 XLR	443	$2n = 2 \times = 22$	Kang et al. (2015)
			454XLR, Pacbio	523		Sakai et al. (2015)
Andean lupin, pearl lupin	*Lupinus mutabilis*	1 Gb			$2n = 2 \times = 48$	
Bambara groundnut/ ground-bean	*Vigna subterranea*	550 Mb	Illumina	512	$2n = 2 \times = 22$	Chang et al. (2019)
Beach pea/ beach cowpea	*Vigna marina*	365 Mb	Illumina	365.6	$2n = 2 \times = 22$	Singh et al. (2019)
Bitter vetch	*V. ervilia*	4.2 Gb			$2n = 2 \times = 14$	
Black gram	*Vigna mungo*	574 Mb	10X Genomics + HiC	499	$2n = 2 \times =22$	Pootakham et al. (2020)
Blue lupin (narrow-leafed)	*L. angustifolius*	924 Mb	Illumina	598	$2n = 2 \times = 40$	Yang et al. (2013)
			Illumina	609		Hane et al. (2016)
			Pacbio	615.7		Wang et al. (2021)
Chickpea	*C. arietinum* (cultivated)	740 Mb	Illumina	532.3	$2n = 2 \times = 16$	Varshney et al. (2013)
			Illumina + 454 XLR	519.8		Jain et al. (2013)
			Illumina + 454 XLR	510.9		Parween et al. (2015)
	C. reticulatum (wild)		Illumina	416		Gupta et al. (2017)
Common bean/ dry beans	*P. vulgaris*	587 Mb	454 XLR + Illumina + Sanger	474.3	$2n = 2 \times = 22$	Schmutz et al. (2014)
			454 XLR + SOLID + Sanger	549.6		Vlasova et al. (2016)
Common vetch	*Vicia sativa*	1.8 Gb	Illumina + Pacbio	1541	$2n = 2 \times = 14$	Shirasawa et al. (2021)
			ONT PromethION + HiC	1651		Xi et al. (2022)
Cowpea/ yardlong bean	*V. unguiculata*	613 Mb	Illumina	568–610	$2n = 2 \times = 22$	Spriggs et al. (2018)
			Pacbio + BioNano	519.4		Lonardi et al. (2019)
			Illumina	632.8		Xia et al. (2019)
Faba beans/ broad bean	*V. faba*	13 Gb			$2n = 2 \times = 12$	
Grass pea	*L. sativus*	6.2 Gb	ONT PromethION	6237	$2n = 2 \times = 14$	Emmrich et al. (2020)
Horse gram	*Macrotyloma uniflorum*	343.6 Mb	Illumina	259.2	$2n = 2 \times = 20$	Shirasawa et al. (2021)
Hyacinth bean	*L. purpureus*	367 Mb	Illumina	395	$2n = 2 \times = 22$	Chang et al. (2019)
Jack bean	*Canavalia gladiata*				$2n = 2 \times = 22$	
Lentil	*L. culinaris* (cultivated)	4 Gb	Pacbio + ONT PromethION + BioNano+ HiC	3760	$2n = 2 \times = 14$	Ramsay et al. (2021)
	Lens ervoides (wild)			2870		
Lima bean	*Phaseolus lunatus*	622 Mb	Illumina	623	$2n = 2 \times = 22$	Garcia et al. (2021)
Moth bean	*Vigna aconitifolia*	1.1 Gb			$2n = 2 \times = 22$	
Mungbean/Green gram	*V. radiata*	579 Mb	Illumina + 454 XLR	431	$2n = 2 \times = 22$	Kang et al. (2014)
			Pacbio	475		Ha et al. (2021)
Pea/dry pea	*P. sativum*	4.4 Gb	Illumina + Pacbio + BioNano	3920	$2n = 2 \times = 14$	Kreplak et al. (2019)
Pigeonpea	*Cajanus cajan*	852 Mb	Illumina	605.7	$2n = 2 \times = 22$	Varshney et al. (2012)
			Illumina	648.2		Mahato et al. (2018)
			None	548		Marla et al. (2020)
Rice bean	*V. umbellata*	562 Mb	Illumina + Pacbio	414	$2n = 2x =22$	Kaul et al. (2019)

Name	Botanical name	Genome size (n)	Sequencing technologies	Assembly size (Mb)	Ploidy level	References
Tepary bean	*P. acutifolius*	682 Mb	Pacbio + HiC	512.6–661.8	$2n = 2\times = 22$	Moghaddam et al. (2021)
White lupin	*Lupinus albus*	451 Mb	Pacbio + BioNano	451	$2n = 2\times = 50$	Hufnagel et al. (2020)
			Pacbio + HiC	558.9		Xu et al. (2020)
Winged beans	*Psophocarpus tetragonolobus*	1.2 Gb			$2n = 2\times = 18$	
Yellow lupin	*L. luteus*				$2n = 2\times = 52$	

(crRNA) and transactivating crRNA (tracrRNA). The crRNA is a 20-bp that is matching to target DNA, while the tracrRNA is binding scaffold for the Cas9 endonuclease. Recognition of the target site by Cas9 hinges on the existence of a particular protospacer-adjacent motif (PAM) located right next to the target site. One significant limitation of the CRISPR/Cas9 system is the strict PAM requirement, which restricts the flexibility of target site selection (Jinek et al. 2012).

4.4.1 Base Editing

Base Editing is a cutting-edge technique that allows direct permanent alteration of a specific base into another in a controlled way. An essential tool in genetic engineering, a base editor combines a catalytically inactive CRISPR/Cas9 domain with a deaminase domain to facilitate accurate editing of genomes' specific point mutations.

4.4.2 CRISPR Interference (CRISPRi) and Activation (CRISPRa)

CRISPR Interference (CRISPRi) and Activation (CRISPRa) are sophisticated genetic engineering methods originating from the CRISPR/Cas system. CRISPRi was developed by Stanley Qi in collaboration with Wendell Lim, Adam Arkin, Jonathan Weissman, and Jennifer Doudna. CRISPRi utilizes catalytically inactive/dead Cas9 (dCas9) fused with repressor domains to inhibit gene expression at targeted genomic locations. This method achieves gene repression by disrupting transcription initiation or elongation. On the other hand, CRISPRa involves using dCas9 combined with transcriptional activator domains to boost gene expression at specific genomic sites. It enhances transcription by attracting activators to the specific site. CRISPRi and CRISPRa allow for precise manipulation of gene expression without changing the DNA sequence, providing a wide range of uses in genomics, gene regulation research, biotechnology, and potential medical treatments (Qi et al. 2013).

4.4.3 CRISPR/Cas Multiplex Editing

CRISPR/Cas Multiplex Editing involves targeting and modifying multiple genomic loci simultaneously through the CRISPR/Cas system. This method allows researchers to modify multiple genes or regulatory elements in the genome simultaneously, providing a potent tool for enhancing trait development in different organisms, including pulse crops. When working with CRISPR/Cas multiplex editing, the key is to create several guide RNAs that each target specific areas of the genome. These gRNAs are utilized along with the Cas nuclease for site-specific double-strand breaks (DSBs) at target sites. Researchers can induce simultaneous DNA cleavage at multiple genomic locations by introducing multiple gRNAs along with the Cas protein into the target cells. After the DSBs are created, the cell's internal repair system can mend the breaks using either non-homologous end joining (NHEJ) or homology-directed repair (HDR). During this repair process, researchers can introduce desired genetic modifications, such as insertions, deletions, or replacements, at each of the targeted sites (Xing et al. 2014).

4.5 Epigenetic Modifications

It involves methylation of DNA and modifications in histone molecules for regulating gene expression patterns and cellular identity. Through precise modulation of epigenetic marks, researchers can control gene expression levels and potentially create specific traits in organisms, such as pulse crops.

5 Precision Phenotyping and Speed Breeding

Yield is a sum total of physiological traits, viz. root engineering, input acquisition, photosynthetic efficiency, etc. are some of the key factors which contribute to yield. Translating component traits quantified under controlled environments to field conditions is useful in enhancing the breeding value of the trait. These traits can be used either directly in selection or as surrogates to identify superior genotypes. Therefore, advanced phenotyping techniques help in removing the bottlenecks in phenotyping physiological traits.

Speed breeding technique leverages controlled environmental conditions to accelerate plant development. By optimizing light, temperature, and other growth parameters, researchers can significantly shorten the life cycle, allowing rapid advancement of generations and the swift development of novel plant varieties. Standardized protocols are now available in chickpea and pigeonpea for expediting breeding and pre-breeding process. At the International Crops Research Institute for the Semi-Arid Tropics (ICRISAT) in Hyderabad, India, the facility and protocol of speed breeding are accessible for wide use by the national and international agricultural researchers. The goal was to overcome the challenges posed by chickpea's long growth cycle and sensitivity to day length. The optimized photoperiod for chickpea involves 13 h of light during vegetative growth, followed by 8 h of darkness, and then 13 h of light during flowering and pod filling stages.

Broad-spectrum light (5700 K LED) accelerates early vegetative growth and pod formation. Far-red light (735 nm) promotes early flowering. With this protocol, chickpea breeding teams can now produce 6–7 generations per year under semi-controlled greenhouse conditions. Flowering occurs 20–25 days after sowing, compared to the normal flowering at 40–50 days after sowing (Sajja et al. 2024). Similarly, the World's First Pigeonpea Speed Breeding Protocol has been developed by ICRISAT. By doing so, the breeding cycle can be shortened to just 2–4 years, compared to the conventional period of 7 years. Pigeonpea is crucial for food security and soil health globally. The speed breeding protocol enables researchers to develop climate-resilient, nutritionally superior pigeonpea varieties at an unprecedented pace. The shortened breeding cycle allows for faster development of climate-resilient, nutritionally superior, and higher-yielding grain legume varieties (Gangashetty et al. 2024).

6 Agronomic Interventions for Accelerated Yield

The optimum crop husbandry and crop management are the prerequisite for realizing the potential yield of any crop. Nevertheless, the large areas under grain legumes cultivation are still under self-subsistent farming in Asia and Africa. However, the agronomy of legumes helped significant increase in farm yields. The novel techniques of seed enhancements (agronomic ways, seed coating/pelleting, seed priming, and bio-priming) to ensure proper germination followed by optimum crop stands to avoid suboptimal conditions have paved the way in extra-large kabuli chickpea. New agro-techniques of pigeonpea, and vigna crops to overcome initial stress and water-logging problems in low-laying areas.

Integrated management of soil and foliar nutrients, including biochar, green manuring, residue retention through conservation agriculture, helped pulses to popularize in resource-poor conditions. Some of the advanced agronomic practices for management of weeds are through application of pendimethaline (30 EC) @ 1.0 Kg a.i./ha as pre-emergence application and Imazethapyr @ 70–80 g/ha as post-emergence application at 20–25 DAS during kharif season after assuring the germination of weeds. Adoption of life-saving irrigations for realizing economic yield levels. Seed priming by soaking the seed in water for 6–8 h ensures better germination. Ridge planting of pigeonpea in high rainfall and low-lying plain areas. Use of critical nutrients such as sulphur @ 20–25 Kg/ha. Adoption of efficient and remunerative cropping systems viz. rice-wheat-mungbean, pigeonpea-wheat and maize/sorghum / pearl millet-chickpea/ lentil in India. Raised bed planting in mungbean and urdbean and ridge and furrow system in pigeonpea help in optimum population management.

7 Advancement in Biotic Stress Management

All leguminous crops are prone to a number of biotic stresses during their growth and development. They exhibit a spectrum of insect pests and disease symptoms and often succumb to severe effects on the different parts of the plants after being infected. The major fungi are rust (Uromyces spp., Phakopsora spp., and Puccinia spp.): Rust fungi attack leguminous plants, causing characteristic reddish-brown pustules on leaves, stems, and pods. Powdery Mildews (Erysiphe spp. and Podosphaera spp.): These fungi create powdery white patches on the plant surface, affecting photosynthesis and overall plant health. Downy Mildews (Peronospora spp.): Downy mildews cause yellowish lesions on leaves, leading to reduced vigour and yield. Ascochyta Blight (Ascochyta spp. and Phoma spp.): it results in necrotic lesions on leaves and stems, impacting crop productivity. Botrytis Gray Mold (Botrytis spp.): Botrytis spp. cause grayish mold on various plant parts. Anthracnose (Colletotrichum spp.): Anthracnose fungi lead to dark lesions on leaves and pods, affecting yield. Root Rot (Aphanomyces euteiches, Rhizoctonia solani, Fusarium spp., etc.): Root rot pathogens attack the root system, impairing nutrient uptake and plant stability (Punja 2006).

After fungal diseases, viruses are one of the major disease-causing agents and are the most problematic to manage simply because they make use of host machinery for their proliferation. The total number of viruses known to date exceeds 2000 (http://www.ictvonline.org). Some important groups of viruses that infect grain legumes are Geminiviruses, Potyviruses, Tospoviruses etc. In pulse crops, the majority of the virus's trespass is by thrips, whitefly, aphids, and mites and even via. Seed from the genera Begomovirus, Becurtovirus, and Curtovirus. Sweet potato leaf curl virus and Dolichos yellow mosaic in lablab beans are seed transmittable (Renukadevi et al. 2022).

Srivastava and Joshi (2011), estimated that at least 150–250 different species of insect pests cause different levels of economic damage to legumes. The infestation at or after flowering and podding stage by pod borer complex causes serious problems for crop productivity (Shanower et al. 1999). The chewing pests, viz. *Maruca vitrata, Helicoverpa armigera, Maruca testulalis,* and *Melanogromyza obtuse*, and pod-sucking pests, viz. *Calvigralla gibbosa* and *C. scutellarius*, are major threats in India and other pulse-growing countries (Saxena et al. 2018).

The advent of science led to the effective control of diseases post-infection. Some of the promising methods are *Biological Control*: Biofumigation (Biological Soil Disinfection): Utilizing plant materials (such as Brassica species) that release volatile compounds upon decomposition to suppress soil-borne pathogens. Biocontrol Agents such as Trichoderma, Coniothyrium minitans, and Paenibacillus jamilae HS-26 are employed to inhibit pathogenic fungi, viz. Fusarium, Rhizoctonia, and Sclerotinia sclerotiorum. These agents can be applied directly to the soil or plant surfaces. *Agronanotechnology:* Nanoscale materials (e.g., silver nanoparticles) exhibit antifungal properties and can be used to control fungal diseases. *Ghost Techniques*: These involve using non-pathogenic fungal strains to outcompete and displace virulent pathogens. *Genetic Approaches*: RNA Interference (RNAi): Silencing specific fungal genes using RNA molecules to hinder pathogenicity. *Genome Editing Tools:* Techniques, viz. CRISPR-Cas9, enable targeted modifications in fungal genomes, potentially leading to disease-resistant traits in crops. PCR-based protocols for the detection of viruses causing MYMIV have been confirmed in many wild Vignas (Naimuddin and Gupta 2011; Naimuddin and Pratap 2011). The begomovirus causing YMD in mung and urd can be detected by immunosorbent electron microscopy (ISEM), Enzyme immunoassays, PCR-tests, and nucleic acid hybridization.

Several field-tested standardized protocols were developed for effective management of pests and diseases, viz. Seed treatment@5 ml Imidacloprid (17.8 SL)/kg followed by usual plant growth promoting agent (PGPA) to manage soil-borne insect damages. For sucking pests Imidacloprid (17.8 SL)@3 ml/10 L water @ 600 litres solution for one hectare standing crop. Biocontrol agents namely Trichoderma + carboxin (4 + 1 g/kg seed) have been recommended in grain legumes for managing soil-borne diseases.

8 Policy Initiatives

The Food and Agriculture Organization (FAO) dedicated 2016 as the international year of pulses to promote utilization and value addition of pulses in food system, create awareness about the benefits of legumes in agro-ecosystems and human health, foster enhanced research in grain legumes and pulses in crop rotations, and overcome difficulties of pulse trade in international market. Further, to boost the Whole Grain Initiative, the World Health Organization (WHO) and the Food and Agricultural Organization (FAO) have joined forces with the Whole Grain Initiative to promote whole grains in healthy and sustainable diets across the globe. The UNGA is advocating for the national governments to encourage reviewing the dietary guidance and supporting policies to increase whole grain consumption.

In this regard, to achieve pulse self-sufficiency in India, several pulse-pro initiatives have launched, for instance, a seed-hub programme to ensure quality seed availability. DAC&FW, GoI, supported 150 seed hubs across the country in ICAR Institutes/State/Central Agricultural Universities/ Krishi Vigyan Kendra's along with 'Enhancing breeder seed for increasing Indigenous Production of Pulses in India' for making quality seeds available locally for the most suitable varieties of different pulse crops in 2016–17. Since the inception of the seed hub project >4.0 lakh quintals of quality seeds have been produced and distributed to the farmers (https://iipr.icar.gov.in/iiprseedportal/).

9 Conclusion

Plant proteins are ideal source in place of meat proteins for maintaining global food & climate sustainability. Achieving local production of pulses is a key to success. Genetic enhancement coupled with better management of crop husbandry is essential for higher productivity. To make necessary changes in the legume plant, it is important to understand that legumes' botany, inheritance of economic traits, better crop management practices and encouraging policy supports are indispensable. It is well known that the pulses had received less research donation from the international and national funding agencies in comparison to cereals. However, the progress achieved through conventional breeding, management and fluctuating policy support was sufficient to meet the targets set so far. But the future demand is challenging, and accordingly, we need to tailor our plants to yield more per unit input under changing climate. It has been observed that improvement for yield has taken place at the cost of narrowing the genetic variability, which ultimately leads to the loss of resistance-related genes in legumes. This was confirmed by estimating genetic variation between wild and cultivated populations at genomic levels.

Nevertheless, the recent genomic tools and techniques are amicable to research advancement in grain legumes. We still need to go a long run to understand the epigenetic changes associated in them. More scientific dedications are needed to unveil the epigenetic changes when crop experiences stress condition. Transferring beneficial genes from wild to desirable agronomic base for developing potential introgression lines is still difficult. The modern tools, viz. omics, gene editing, and optimization of rapid generation turnover techniques for expediting the target breeding programme, are still inadequate in most of the important pulses, including chickpea, pigeonpea, and lentil. Nevertheless, it is anticipated that the growing phonemic facilities and rapid generation advancement techniques will help to overcome the bottleneck of breeding cycles. Therefore, a well-balanced blending of genomics, phenomics, metabolomics, and genome editing tools provides a unique opportunity to decipher productivity-limiting factors in grain legumes and translate wisdom into varieties.

References

Abberton M (2010) Enhancing the role of legumes: potential and obstacles. In: Abberton M, Conant R, Batello C (eds) Integrated crop management, vol. 11: Grassland, carbon sequestration: management, policy and economics. Proceedings of the workshop on the role of grassland carbon sequestration in the mitigation of climate change, Rome, pp 177–187

Abbo S, Zezak I, Schwartz E, Lev-Yadun S, Kerem Z, Gopher A (2008) Wild lentil and chickpea harvest in Israel: bearing on the origins of Near Eastern farming. J Archaeol Sci 35(12):3172–3177. https://doi.org/10.1016/j.jas.2008.07.004

Abbo S, Lev-Yadun S, Gopher A (2014) Plant domestication in the Near East. In: Encyclopaedia of the history of science, technology, and medicine in Non-Western cultures. https://doi.org/10.1007/978-94-007-3934-5_10006-1

Abbo H, Gopher A (2017) Near Eastern plant domestication: a history of thought. Trends Plant Sci 22(6):491–511. https://doi.org/10.1016/j.tplants.2017.03.010

Aguilar-Benitez D, Rubio J, Millán T, Gil J, Die JV, Castro P (2020) Genetic analysis reveals PDH1 as a candidate gene for control of pod dehiscence in chickpea. Mol Breed 40:40

Albala K (2007) Beans: a history. Berg, New York

Angioi SA, Rau D, Attene G, Nanni L, Bellucci E, Logozzo G et al (2010) Beans in Europe: origin and structure of the European landraces of Phaseolus vulgaris L. Theor Appl Genet 121:829–843

Asik D, Ankita T, Nath CP, Gupta DS, Hazra KK (2022) A comprehensive review on grain legumes as climate -smart crops: challenges and prospects. Environm Challenges 7:100479. https://doi.org/10.1016/j.envc.2022.100479

Basu U, Upadhyaya HD, Srivastava R, Daware A, Malik N, Sharma A, Bajaj D, Narnoliya L, Thakro V, Kujur A, Tripathi S, Bharadwaj C, Hegde VS, Pandey AK, Singh AK, Tyagi AK, Parida SK (2019) ABC transporter-mediated transport of Glutathione conjugates enhances seed yield and quality in Chickpea. Plant Physiol 180(1):253–275. https://doi.org/10.1104/pp.18.00934. Epub 2019 Feb 8. PMID: 30737266; PMCID: PMC6501113

Basu U, Parida SK (2023) The developmental dynamics in cool season legumes with focus on chickpea. Plant Mol Biol 111:473–491. https://doi.org/10.1007/s11103-023-01340-w

Bohra A, Chand Jha U, Godwin ID, Kumar VR (2020) Genomic interventions for sustainable agriculture. Plant Biotechnol J 18(12):2388–2405. https://doi.org/10.1111/pbi.13472. Epub 2020 Sep 22. PMID: 32875704; PMCID: PMC7680532

Chacón-Sánchez MI, Martínez-Castillo J (2017) Testing domestication scenarios of lima bean (Phaseolus lunatus L.) in Mesoamerica: insights from genome-wide genetic markers. Front Plant Sci 8:1551

Chang Y, Liu H, Liu M, Liao X, Sahu SK, Fu Y et al (2019) The draft genomes of five agriculturally important African orphan crops. Giga Sci 8(3). https://doi.org/10.1093/gigascience/giy152

de Vogel EF (1980) Seedlings of dicotyledons. Centre for Agricultural Publishing and Documentation, Wageningen

Department of Agriculture Cooperation, GoI (2022) https://desagri.gov.in/wp-content/uploads/2023/05/Agricultural-Statistics-at-a-Glance-2022.pdf

Di Vittori V, Bitocchi E, Rodriguez M, Alseekh S, Bellucci E, Nanni L et al (2020) Pod indehiscence in common bean is associated to the fine regulation of PvMYB26 and a non-functional abscission layer. J Exp Bot 72:1617–1633

Dönmez E, Belli O (2007) Urartian plant cultivation at Yoncatepe (Van), eastern Turkey. Econ Bot 61(3):290–298. https://doi.org/10.1663/0013-0001(2007)61[290:upcayv]2.0.co;2

Döring M (2022) Pisum sativum L. in English Wikipedia – Species Pages. Wikimedia Foundation. Checklist dataset https://doi.org/10.15468/c3kkgh accessed via GBIF.org on 2024-04-13

Duchene O, Vian JF, Celette F (2017) Intercropping with legume for agroecological cropping systems: Complementarity and facilitation processes and the importance of soil microorganisms. A review. Agric Ecosyst Environ 240:148–161

Emmrich PMF, Sarkar A, Njaci I, Kaithakottil GG, Ellis N, Moore C, Edwards A, Heavens D, Waite D, Cheema J, Trick M, Moore J, Webb A, Caiazzo R, Thomas J, Higgins J, Swarbreck D, Kumar S, Mundree S, Loose M, Yant L, Martin C, Wang TL (2020) A draft genome of grass pea (Lathyrus sativus), a resilient diploid legume. bioRxiv. https://doi.org/10.1101/2020.04.24.058164

Erokhin V, Gao T (2020) Impacts of COVID–19 on trade and economic aspects of food security: Evidence from 45 developing countries. Int J Environ Res Public Health 17(16):5775

FAO (2022) World food and agriculture – statistical yearbook 2022. FAO, Rome. https://doi.org/10.4060/cc2211en

FAO (2023) World Pulses Day 2023 highlights how pulses are at the core of sustainability. Food and Agriculture Organization of the United Nations, Rome. https://www.fao.org/fao-stories/article/en/c/1624661/

Food and Agriculture Organization of the United Nations (2023) FAO statistical yearbook 2023 – world food and agriculture. https://reliefweb.int/report/world/fao-statistical-yearbook-2023-world-food-and-agriculture

FAOSTAT (2021) Food and Agriculture Organization corporate statistical database. http://www.fao.org/faostat/en/#data/QC

Fuller DQ (2007) Contrasting patterns in crop domestication and domestication rates: recent archaeobotanical insights from the Old World. Ann Bot 100:903–924

Gangashetty PI, Belliappa SH, Bomma N et al (2024) Optimizing speed breeding and seed/pod chip based genotyping techniques in pigeonpea: a way forward for high throughput line development. Plant Methods 20:27. https://doi.org/10.1186/s13007-024-01155-w

García MN, Martinez CJ, Perucini-Avendaño M, Camacho-Díaz B, Jiménez-Aparicio A, Ortiz GD (2021) Phenolic compounds in legumes: composition, processing and gut health. In: Legumes. https://doi.org/10.5772/intechopen.98202

Gaut BS (2014) Thecomplexdomesticationhistoryofthecommonbean. Nat Genet 46:663–664

Gepts P (2003) Crop domestication as a long-term selection experiment. Plant Breed Rev 24:1–44

Gupta S, Nawaz K, Parween S, Roy R, Sahu K, Kumar Pole A, Khandal H, Srivastava R, Kumar Parida S, Chattopadhyay D (2017) Draft genome sequence of Cicer reticulatum L., the wild progenitor of chickpea provides a resource for agronomic trait improvement. DNA Res 24(1):1–10. https://doi.org/10.1093/dnares/dsw042. PMID: 27567261; PMCID: PMC5381347

Ha J, Satyawan D, Jeong H, Lee E, Cho KH, Kim MY, Lee SH (2021) A near-complete genome sequence of mungbean (Vigna radiata L.) provides key insights into the modern breeding program. Plant Genom 14(3):e20121

Hammer K (1984) Das Domestikationssyndrom. Kult 32:11–34

Han Y, Zhao X, Liu D, Li Y, Lightfoot DA, Yang Z et al (2016) Domestication footprints anchor genomic regions of agronomic importance in soybeans. New Phytol 209:871–884

Hane JK, Ming Y, Kamphuis LG, Nelson MN, Garg G, Atkins CA, Bayer PE, Bravo A, Bringans SD, Cannon S, Edwards D, Foley R, Gao LL, Harrison MJ, Huang W, Hurgobin B, Li S, Liu CW, McGrath A, Morahan G (2016) A comprehensive draft genome sequence for lupin (Lupinus angustifolius), an emerging health food: insights into plant–microbe interactions and legume evolution. Plant Biotechnol. https://doi.org/10.1111/pbi.12615

Hufnagel B, Marques A, Soriano A, Marquès L, Divol F, Doumas P, Sallet E, Mancinotti D, Carrere S, Marande W, Arribat S, Keller J, Huneau C, Blein T, Aimé D, Laguerre M, Taylor J, Schubert V, Nelson M et al (2020) High-quality genome sequence of white lupin provides insight into soil exploration and seed quality. Nat Commun 11(1):1–12. https://doi.org/10.1038/s41467-019-14197-9

Harlan JR, de Wet JMJ, Price EG (1973) Comparative Evolution of Cereals.545. Evolution 27:311–325

Hondelmann W (1984) Thelupin—ancientandmoderncropplant. Theor Appl Genet 68:1–9

Jain M, Misra G, Patel RK, Priya P, Jhanwar S, Khan AW, Shah N, Singh VK, Garg R, Jeena G, Yadav M, Kant C, Sharma P, Yadav G, Bhatia S, Tyagi AK, Chattopadhyay D (2013) A draft genome sequence of the pulse crop chickpea (Cicer arietinum L.). Plant J 74:715–729

Jinek M, Chylinski K, Fonfara I, Hauer M, Doudna JA, Charpentier E (2012) A programmable dual-RNA-guided DNA endonuclease in adaptive bacterial immunity. Science 337(6096):816–821. https://doi.org/10.1126/science.1225829

Jose CJ-L, Singh KB, Clemente A, Nelson MN, Ochatt S, Smith PMC (2020) Editorial: legumes for global food security. Front Plant Sci 11. https://doi.org/10.3389/fpls.2020.00926

Joshi PK, Rao PP (2016) Global and regional pulse economies–current trends and outlook, IFPRI discussion paper. International Food Policy Research Institute (IFPRI), Washington, DC, p 149

Kang YJ et al (2014) Genome sequence of mungbean and insights into evolution within Vigna species. Nat Commun 5:5443

Kang YJ, Satyawan D, Shim S, Lee T, Lee J, Hwang WJ, Kim SK, Lestari P, Laosatit K, Kim KH, Ha TJ, Chitikineni A, Kim MY, Ko J-M, Gwag J-G, Moon J-K, Lee Y-H, Park B-S, Varshney RK, Lee S-H (2015) Draft genome sequence of adzuki bean, Vigna angularis. Sci Rep 5:8069. https://doi.org/10.1038/srep08069

Kaul T, Eswaran M, Thangaraj A et al (2019) Rice Bean (iVigna umbellata/i) draft genome sequence: unravelling the late flowering and unpalatability related genomic resources for efficient domestication of this underutilized crop. bioRxiv. https://doi.org/10.1101/816595

Kebede E (2020) Grain legumes production and productivity in Ethiopian smallholder agricultural system, contribution to livelihoods and the way forward. Cogent Food Agric 6(1):1722353

Kerem Z, Lev-Yadun S, Gopher A, Weinberg P, Abbo S (2007) Chickpea domestication in the Neolithic Levant through the nutritional perspective. J Archaeol Sci 34(8):1289–1293. https://doi.org/10.1016/j.jas.2006.10.025

Kersey PJ (2019) Plant genome sequences: past, present, future. Curr Opin Plant Biol 48:1–8

Kim SK, Nair RM, Lee J, Lee S-H (2015) Genomic resources in mungbean for future breeding programs. Front Plant Sci 6:626

Koivunen E et al (2016) Digestibility and energy value of pea (Pisum sativum L.), faba bean (Vicia faba L.) and blue lupin (narrow-leaf) (Lupinus angustifolius) seeds in broilers. Anim Feed Sci Technol 218:120–127

Kreplak J, Madoui M-A, Cápal P, Novák P, Labadie K, Aubert G et al (2019) A reference genome for pea provides insight into legume genome evolution. Nat Genet 51:1411–1422

Ladizinsky G (1979) The genetics of several morphological traits in the lentil. J Hered 70:135–137

Lenka NK, Lal R (2012) Soil-related constraints to the carbon dioxide fertilization effect. Crit Rev Plant Sci 31(4):342–357. https://doi.org/10.1080/07352689.2012.674461

Lev-Yadun S, Gopher A, Abbo S (2000) The cradle of agriculture. Science 288:1602–1603

Lewis G, Schrire B, Mackinder B, Lock M (eds) (2005) Legumes of the world. Royal Botanic Gardens, Kew, Richmond

Lo S, Muñoz-Amatriaín M, Boukar O, Herniter I, Cisse N, Guo YN et al (2018) Identification of QTL controlling domestication-related traits in cowpea (Vigna unguiculata L. Walp). Sci Rep 8:6261

Lonardi S, Muñoz-Amatriaín M, Liang Q, Shu S, Wanamaker SI, Lo S, Tanskanen J, Schulman AH, Zhu T, Luo C, Alhakami H, Ounit R, Hasan AM, Verdier J, Roberts PA, Santos RP, Ndeve A, Doležel J, Vrána J et al (2019) The genome of cowpea (Vigna unguiculata [L.] Walp.). Plant J 98(5):767–782. https://doi.org/10.1111/tpj.14349

Mahato AK, Sharma AK, Sharma TR, Singh NK (2018) An improved draft of the pigeonpea (Cajanus cajan (L.) Millsp.) genome. Data Brief 16:376–380. https://doi.org/10.1016/j.dib.2017.11.066

Majola NG, Gerrano AS, Shimelis H (2021) Bambara groundnut (Vigna subterranea [L.] Verdc.) production, utilisation and genetic improvement in Sub-Saharan Africa. Agronomy 11:1345

Marla SS, Mishra P, Maurya R, Singh M, Wankhede DP, Kumar A, Yadav MC, Subbarao N, Singh SK, Kumar R (2020) Refinement of draft genome assemblies of Pigeonpea (Cajanus cajan). Front Genet 11:607432. https://doi.org/10.3389/fgene.2020.607432

Martens SD, Tiemann TT, Bindelle J, Peters M, Lascano CE (2012) Alternative plant protein sources for pigs and chickens in the tropics—nutritional value and constraints: a review. J Agric Rural Dev Trop 113:101–123

Mayes S, Ho WK, Chai HH, Gao X, Kundy AC, Mateva KI, Zahrulakmal M, Hahiree MKIM, Kendabie P, Licea LCS, Massawe F, Mabhaudhi T, Modi AT, Berchie JN, Amoah S, Faloye B, Abberton M, Olaniyi O, Azam-Ali SN (2019) Bambara groundnut: an exemplar underutilised legume for resilience under climate change. Planta 250:803–820

Moghaddam SM, Oladzad A, Koh C, Ramsay L, Hart JP, Mamidi S, Hoopes G, Sreedasyam A, Wiersma A, Zhao D, Grimwood J, Hamilton JP, Jenkins J, Vaillancourt B, Wood JC, Schmutz J, Kagale S, Porch T, Bett KE, Buell CR, McClean PE (2021) The tepary bean genome provides insight into evolution and domestication under heat stress. Nat Commun 12(1):2638. https://doi.org/10.1038/s41467-021-22858-x. PMID: 33976152; PMCID: PMC8113540

Morris B (2003) Legumes. In: Katz SH (ed) Encyclopedia of food and culture. 3 vols

Mudryj AN, Yu N, Aukema HM (2014) Nutritional and health benefits of pulses. Appl Physiol Nutr Metab 39:1197–1204

Muroki MW, Lydiah MW, Robert F, Andrew K, Symon MM (2023) Cooking properties and nutrient retention of biofortified common bean (Phaseolus vulgaris) varieties: the case of Burundi. Front Sustain Food Syst (Sec Nutrition and Sustainable Diets) 7. https://doi.org/10.3389/fsufs.2023.1062978

Naimuddin AM, Gupta S (2011) Identification of Mungbean yellow mosaic India virus infecting Vigna mungo var. silvestris L. Phytopathologia Mediterranea 50:94–100

Naimuddin AM, Pratap A (2011) First report of natural infection of Mungbean yellow mosaic India virus in two wild species of Vigna. New Dis Rep 23:21

NITI Aayog (2023) Annual report October 2023, p 154

Olayide OE, Donkoh SA, Ansah IGK, Adzawla W, O'Reilly PJ, Mayes S, Feldman A, Halimi RA, Nyarko G, Ilori CO, Alabi T (2018) Assessing socioeconomic factors influencing production and commercialization of Bambara groundnut as an indigenous climate resilient crop in Nigeria. In: Leal Filho W (ed) Handbook of climate change resilience. Springer, Cham, pp 1–19

Parker TA, Berny Mier Y, Teran JC, Palkovic A, Jernstedt J, Gepts P (2020) Pod indehiscence is a domestication and aridity resilience trait in common bean. New Phytol 225:558–570

Parker TA, de Sousa LL, de Oliveira Floriani T, Palkovic A, Gepts P (2021a) Toward the introgression of PvPdh1 for increased resistance to pod shattering in common bean. Theor Appl Genet 134:313–325

Parker TA, Lo S, Gepts P (2021b) Pod shattering in grain legumes: emerging genetic and environment-related patterns. Plant Cell 33:179–199

Parween S, Nawaz K, Roy R, Pole AK, Venkata Suresh B, Misra G et al (2015) An advanced draft genome assembly of a desi type chickpea (Cicer arietinum L.). Sci Rep 5:12806. https://doi.org/10.1038/srep12806

Pasipanodya JT, Horn LN, Achigan-Dako EG, Musango R, Sibiya J (2022) Utilization of plant genetic resources of Bambara groundnut conserved ex situ and genetic diversification of its primary genepool for semi-arid production. Agriculture 12(4):492

PC report, AICRP on Kharif Pulses (2022) p 85

Pootakham W, Nawae W, Naktang C, Sonthirod C, Yoocha T, Kongkachana W, Sangsrakru D, Jomchai N, Somta P, Laosatit K, Tangphatsornruang S (2020) A chromosome-scale assembly of the black gram (Vigna mungo) genome. Mol Ecol Resour 21(1):238–250. https://doi.org/10.1111/1755-0998.13243

Punja ZK (2006) Recent developments toward achieving fungal disease resistance in transgenic plants. Can J Plant Pathol 28:S298–S308

Qi LS, Larson MH, Gilbert LA, Doudna JA, Weissman JS, Arkin AP, Lim WA (2013) Repurposing CRISPR as an RNA-guided platform for sequence-specific control of gene expression. Cell 152(5):1173–1183. https://doi.org/10.1016/j.cell.2013.02.022

Rajeev V, Kudapa HB, Lekha P, Chitikineni A, Thudi M, Bohra A, Gaur P, Janila P, Fikre A, Kimurto P, Ellis N (2015) Translational Genomics in Agriculture: Some Examples in Grain Legumes. Crit Rev Plant Sci 34(1–3):169–194

Ramsay L, Koh CS, Kagale S, Gao D, Kaur S, Haile T et al (2021) Genomic rearrangements have consequences for introgression breeding as revealed by genome assemblies of wild and cultivated lentil species. bioRxiv:453237

Rau D, Murgia ML, Rodriguez M, Bitocchi E, Bellucci E, Fois D et al (2019) Genomic dissection of pod shattering in common bean: mutations at non-orthologous loci at the basis of convergent phenotypic evolution under domestication of leguminous species. Plant J 97:693–714

Raza A, Razzaq A, Mehmood SS, Zou X, Zhang X, Lv Y, Xu J (2019) Impact of climate change on crops adaptation and strategies to tackle its outcome: a review. Plants 8:34. https://doi.org/10.3390/plants8020034

Reckling M et al (2016) Trade-offs between economic and environmental impacts of introducing legumes into cropping systems. Front Plant Sci 7:669

Rendón-Anaya M, Montero-Vargas JM, Saburido-Alvarez S, Vlasova A, Capella-Gutiérrez S, Ordaz-Ortiz JJ et al (2017) Genomic history of the origin and domestication of common bean unveils its closest sister species. Genome Biol 18:60

Renukadevi P, Sangeetha B, Malathi VG, Nakkeeran S, Satya VK (2022) Geminivirus: detection, diagnosis and management. Academic Press, Cambridge, MA. Satya. Enigmatic emergence of seed transmission of geminiviruses; pp. 285–306

Sajja SB, Mathew A, Pasupuleti P, Radhakrishnan T (2024) Speed breeding to accelerate crop improvement. In: Digital agriculture. Springer, Cham, pp 425–443. ISBN 978-3-031-43548-5

Sakai H, Naito K, Yu T, Sato T, Yamamoto T, Muto I, Itoh T, Tomooka N (2015) The Vigna Genome Server, ' Vig GS': a genomic knowledge base of the Genus Vigna based on high-quality, annotated genome sequence of the Azuki Bean, Vigna angularis (Willd.) Ohwi & Ohashi. Plant Cell Physiol 57(1):2015. https://doi.org/10.1093/pcp/pcv189

Satheesh Naik SJ, Lamichaney A, Bohra A, Mishra RK, Singh F, Datta D, Singh IP, Singh NP (2019) Identification of tolerant genotypes against pulse beetle as a source to reduce post harvest losses in pigeonpea [Cajanus cajan (L.) Millisp.]. Legume Res Int J 44(4):480–485. https://doi.org/10.18805/LR-4112

Saxena KB, Sharma D (1990) Pigeonpea genetics. In: Nene YL, Hall SD, Sheila VK (eds) The Pigeonpea. CAB International, Wallingford, pp 137–158

Saxena KB, Singh L, Gupta MD (1990) Variation for natural out-crossing in pigeonpea. Euphytica 39:143–148

Saxena KB, Chauhan YS, Johansen C, Singh L (1992) Recent developments in hybrid pigeonpea research. In: New frontiers in pulses research and development. Proceedings of national symposium. 10-12 November 1989: Directorate of Pulses Research. Kanpur 208024, India, pp 58–69

Saxena KB, Sultana R, Mallikarjuna N, Saxena RK, Kumar RV, Sawargaonkar SL, Varshney RK (2010) Male sterility systems in pigeonpea and their role in enhancing yield. Plant Breed 129:125–134

Saxena KB, Chauhan YS, Kumar CVS, Hingane AJ, Kumar RV, Saxena RK, Rao GVR (2018) Developing improved varieties of pigeonpea. Burleigh Dodds Science Publishing Limited, pp 1–30. https://doi.org/10.19103/AS.2017.0023.33

Schmutz J, McClean PE, Mamidi S, Wu GA, Cannon SB, Grimwood J, Jenkins J, Shu S, Song Q, Chavarro C, Geffroy V, Moghaddam SM, Gao D, Abernathy B, Barry K, Blair M, Brick MA, Chovatia M, Gepts P et al (2014) A reference genome for common bean and genome-wide analysis of dual domestications. Nat Genet 46(7):707–713. https://doi.org/10.1038/ng.3008

Shanower TG, Romeis J, Minja EM (1999) Insect pests of pigeonpea and their management. Annu Rev Entomol 44:77–96

Shirasawa K, Chahota R, Hirakawa H, Nagano S, Nagasaki H, Sharma T, Isobe S (2021) A chromosome-scale draft genome sequence of horsegram (Macrotyloma uniflorum). GigaByte 2021:gigabyte30. https://doi.org/10.46471/gigabyte.30. PMID: 36824333; PMCID: PMC9650294

Schrire BD, Lavin M, Lewis GP (2005) Global distribution patterns of the Leguminosae: Insights from recent phylogenies. Biol Skr 55:375–422

Sedivy EJ, Wu F, Hanzawa Y (2017) Soybean domestication: the origin, genetic architecture and molecular bases. New Phytol 214:539–553

Simon CJ, Muehlbauer FJ (1997) Construction of a chickpea linkage map and its comparison with maps of pea and lentil. J Hered 38:115–119

Singh KB (1997) Chickpea (Cicer arietinum L.). Field Crop Res 53:161–170

Singh N (2017) Pulses: an overview. J Food Sci Technol 54(4):853–857

Singh AK, Velmurugan A, Gupta DS, Kumar J, Kesari R, Konda A et al (2019) Draft genome sequence of a less-known wild Vigna: Beach pea (V. marina cv. ANBp-14-03). Crop J 7:660–666

Smartt J (1979) Econ Bot 33:329–337

Smartt J (1981) Euphytica 30:445–449

Spriggs A, Henderson ST, Hand ML, Johnson SD, Taylor JM, Koltunow A (2018) Assembled genomic and tissue-specific transcriptomic data resources for two genetically distinct lines of Cowpea (Vigna unguiculata (L.) Walp). Gates Open Res 2:7. https://doi.org/10.12688/gatesopenres.12777.2. PMID: 29528046; PMCID: PMC5841572

Srivastava CP, Joshi N (2011) Insect pest management in pigeonpea in Indianscenario: a critical review. Indian J Entomol 73(1):63–75

Smýkal P, Coyne CJ, Ambrose MJ, Maxted N, Schaefer H, Blair MW et al (2015) Legume crops phylogeny and genetic diversity for science and breeding. Crit Rev Plant Sci 34:43–104. https://doi.org/10.1080/07352689.2014.897904

Solomon S, Argikar GP, Solanki MS, Morbad R (1957) A study of heterosis in Cajanus cajan (L.) Millsp. Indian J Genet 17:90–95

Takahashi Y, Kongjaimun A, Muto C, Kobayashi Y, Kumagai M, Sakai H et al (2020) Same Locus for Non-shattering Seed Pod in Two Independently Domesticated Legumes, Vigna angularis and Vigna unguiculata. Front Genet 11:748

Talukdar D (2011) Genetics of pod indehiscence in grass pea (Lathyrus sativus L.). J Crop Improv 25:161–175

Tendai PC, Modi AT, Mabhaudhi T (2017) Expounding the value of grain legumes in the semi- and arid tropics. Sustain For 2017(9):60. https://doi.org/10.3390/su9010060

Unigwe AE, Gerrano AS, Adebola P, Pillay M (2016) Morphological variation in selected accessions of Bambara groundnut (Vigna subterranea L. Verdc) in South Africa. J Agric Sci 8:69

Varshney RK, Chen W, Li Y, Bharti AK, Saxena RK, Schlueter JA et al (2012) Draft genome sequence of pigeonpea (Cajanus cajan), an orphan legume crop of resource-poor farmers. Nat Biotechnol 30:83–89. https://doi.org/10.1038/nbt.2022

Varshney RK, Song C, Saxena RK, Azam S, Yu S, Sharpe AG, Cannon S, Baek J, Rosen BD, Tar'an B et al (2013) Draft genome sequence of chickpea (Cicer arietinum) provides a resource for trait improvement. Nat Biotechnol 31(3):240–246

Varshney RK, Terauchi R, McCouch SR (2014) Harvesting the promising fruits of genomics: applying genome sequencing technologies to crop breeding. PLoS Biol 12:e1001883. https://doi.org/10.1371/journal.pbio.1001883

Vaz Patto MC (2016) Grain legume protein quality: a hot subject. Arbor 192:a314

Vaz Patto MC, Amarowicz R, Aryee ANA, Boye JI, Chung HJ, Martín-Cabrejas MA et al (2015) Achievements and challenges in improving the nutritional quality of food legumes. Crit Rev Plant Sci 34:105–143

Vlasova A, Capella-Gutiérrez S, Rendón-Anaya M et al (2016) Genome and transcriptome analysis of the Mesoamerican common bean and the role of gene duplications in establishing tissue and temporal specialization of genes. Genome Biol 17:32. https://doi.org/10.1186/s13059-016-0883-6

Wang P, Zhou G, Jian J, Yang H, Renshaw D, Aubert MK et al (2021) Whole-genome assembly and resequencing reveal genomic imprint and key genes of rapid domestication in narrow-leafed lupin. Plant J 105:1192–1210

Weeden NF, Brauner S, Przyborowski JA (2002) Genetic analysis of pod dehiscence in pea (*Pisum sativum* L.). Cell Mol Biol Lett 7:657–663

Willcox G, Buxo R, Herveux L (2009) Late Pleistocene and early Holocene climate and the beginnings of cultivation in northern Syria. The Holocene 19(1):151–158

Williams CM et al (2014) Effects of three pulse crops on subsequent barley, canola, and wheat. Agron J 106:343–350

Xi H, Nguyen V, Ward C, Liu Z, Searle IR (2022) Chromosome-level assembly of the common vetch (Vicia sativa) reference genome. GigaByte 2022:gigabyte38. https://doi.org/10.46471/gigabyte.38

Xing HL, Dong L, Wang ZP et al (2014) A CRISPR/Cas9 toolkit for multiplex genome editing in plants. BMC Plant Biol 14:327. https://doi.org/10.1186/s12870-014-0327-y

Xia Q, Pan L, Zhang R, Ni X, Wang Y, Dong X, Gao Y, Zhang Z, Kui L, Li Y, Wang W (2019) The genome assembly of asparagus bean, Vigna unguiculata ssp. sesquipedialis. Sci Data 6(1):124

Xu W, Zhang Q, Yuan W, Xu F, Muhammad Aslam M, Miao R, Li Y, Wang Q, Li X, Zhang X, Zhang K, Xia T, Cheng F (2020) The genome evolution and low-phosphorus adaptation in white lupin. Nat Commun 11(1):1069. https://doi.org/10.1038/s41467-020-14891-z. PMID: 32103018; PMCID: PMC7044338

Yadava DK, Choudhury PR, Hossain F, Kumar D, Sharma TR, Mohapatra T (2022) Biofortified varieties: sustainable way to alleviate malnutrition, 4th edn. Indian Council of Agricultural Research, New Delhi, p 106

Yahara T, Javadi F, Onoda Y, de Queiroz LP, Faith DP, Prado DE, Akasaka M et al (2013) Global legume diversity assessment: concepts, key indicators, and strategies. Taxon 62:249–266. https://doi.org/10.1002/tax.621012

Yang HA, Tao Y, Zheng ZQ, Zhang QS, Zhou GF et al (2013) Draft genome sequence, and a sequence-defined genetic linkage map of the legume crop species Lupinus angustifolius L. PloS One 8:e64799

Zhang Q, Tu B, Liu C, Liu X (2018) Pod anatomy, morphology and dehiscing forces in pod dehiscence of soybean (Glycine max (L.) Merrill). Flora 248:48–53

Common Bean (*Phaseolus vulgaris* L.) Breeding

Manisha Kumari, Pradeep Kumar Thakur, Pummi Kumari, Raghav Sood, Mohar Singh, and J. C. Rana

Abstract

Phaseolus vulgaris L., commonly known as the common bean, is a globally important crop, providing essential nutrition and serving as a significant source of income for millions of people worldwide. Rich in dietary fiber, minerals, amino acids, and phytochemicals, it offers numerous health benefits. This chapter offers an in-depth overview of the origin, distribution, genetic diversity, gene flow assessment, genetic resources, conservation strategies, crop hybridization, molecular markers, and mass-assisted breeding. Efforts in germplasm conservation, facilitated by gene banks worldwide, ensure the preservation and accessibility of diverse genetic resources for research and breeding purposes. These collections, comprising thousands of accessions from numerous countries, provide a valuable reservoir of traits for enhancing common bean cultivars adaptability and resilience to evolving challenges like climate change and pests. Understanding the genetic diversity of common beans is vital for developing improved varieties capable of meeting the increasing demand for food in a changing world. Research endeavors utilizing advanced molecular tools and genomic resources offer insights the species evolutionary dynamics, helping to develop resilient, high-yielding cultivars that are well-suited to various agricultural environments. Over the last two decades, the crop improvement programme has adopted marker-assisted selection method after identifying numerous important genes and the molecular markers associated with them. Furthermore, bean transformation protocols have been established to enable the introgression of foreign genes.

Keywords

Common bean · Genetic resources · Conservation strategies · Hybridization · Breeding

1 Introduction

Phaseolus is a diverse genus in the Fabaceae family, containing about seventy species distributed globally (Ulloa Ulloa et al. 2018; Chacón-Sánchez et al. 2021). This species is known with various common names like kidney bean, haricot bean, snap bean, common bean, navy bean, string bean, and garden bean. This is arguably the most important edible grain legume globally, having a commercial worth higher than all other legume crops combined, making it the most important leguminous crop grown for direct human use (Broughton et al. 2003; Porch et al. 2013; Graham and Vance 2003). It is a self-pollinating diploid species with $2n = 2x = 22$ chromosomes, and its haploid genome size is estimated to be between 587 and 637 Mbp (Arumuganathan and Earle 1991; Bennett and Leitch 1995, 2010). The common bean is grown worldwide due to its versatility in cultivation methods, uses, phenotypic diversity, and ability to adapt to a wide range of environments (De Ron et al. 2016). In several European countries, *Phaseolus vulgaris* plays a significant role in traditional cuisine and lifestyle. *Phaseolus* species are extensively diverse crops that can be cultivated under various kinds of environments and elevations ranging from sea level to 3000 m height, and for a variety of purposes, including green as vegetable and dry as pulses. Beans are part of one of the world's most diversified production systems due to their incredible range in plant types and growth behavior (Broughton et al. 2003). In addition to a wide variety of major and minor gene pools, races, and intermediate forms, their genetic resources occasionally contain introgressions

M. Kumari · P. K. Thakur · P. Kumari · R. Sood · M. Singh (✉)
ICAR-National Bureau of Plant Genetic Resource Regional Station, Shimla, Himachal Pradesh, India
e-mail: Mohar.Singh2@icar.gov.in

J. C. Rana (✉)
The Alliance of Bioversity International & CIAT—India Office, New Delhi, India
e-mail: j.rana@cgiar.org

G. Prasad Dixit et al. (eds.), *Fundamentals of Legume Breeding*, https://doi.org/10.1007/978-981-96-7838-9_2

from both wild and domesticated species. Thus, beans are a crop that may be grown in response to several kinds of agronomic and consumer preference. While beans are largely produced on small holdings, they are cultivated in a number of methods, bush beans grown in a monoculture to a complex mixture of climbing or indeterminate beans with maize and sugarcane, other cereals, or fruit crops.

Phaseolus comprises the most numerous and diverse genera of pulse crops, which consist about 70 species from Central and North America (Freytag and Debouck 2002). Of these, five species—*P. vulgaris* (Common bean), *P. dumosus* (Year bean), *P. coccineus* (Runner or Scarlet bean), *P. acutifolius* (Tapery bean), and *P. lunatus* (Lima bean)—have been domesticated, while a few other species show early signs of domestication (Delgado-Salinas et al. 2006). The best-known among them is the common bean (*P. vulgaris*), which is cultivated across tropical, subtropical, and temperate regions of the world today (Table 1).

All the mentioned species are self-pollinating, except for *Phaseolus coccineus*, which undergoes cross-pollination. It is a significant vegetable crop grown during the winter season. It is cultivated for both, the dry seeds are consumed as pulses and the soft green pods are consumed fresh vegetable. It is an inexpensive source of protein, calories, and micronutrients in the diets of people across several countries, such as Mexico and Brazil. Dried beans are rich in protein and are often considered comparable to meat in their nutritional content. It is primarily grown in India for its tender green pods, although it is cultivated in large numbers in the United States for processing purposes. This legume not only serves an important role in human nutrition but also enhances soil fertility to a great extent because of its efficient nitrogen fixation. Nutritionists consider the common bean an almost ideal food because of its high protein content, along with its rich levels of fiber, complex carbohydrates, and other essential dietary components. The 100 g of green common bean pods contains 1.7 g protein, 4.5 g carbohydrates, 50 mg calcium, 11 mg vitamin C, and 221 IU vitamin A (Gopalkrishnan 2007).

Table 1 Species closely associated with *Phaseolus vulgaris*

Species	Geographic location
P. acutifolius	Mexico, Southwestern United States
P. albescens	Western Mexico
P. coccineus	Guatemala, Honduras, Mexico
P. costaricensis	Eastern Costa Rica, Western Panama
P. dumosus	Western Guatemala, Mexico
P. parvifolius	Southwestern United States, Guatemala, Pacific coast of Mexico and Central America
P. persistentus	Guatemala

Sources: Porch et al. (2013), Bellucci et al. (2014)

Beans are a rich source of essential minerals like iron, copper, and zinc; amino acids such as lysine and tryptophan; and beneficial phytochemicals like flavonoids and antioxidants. It possesses medicinal properties that help control diabetes and heart issues, reduce the risk of obesity, and serve as a natural remedy for bladder infections. Additionally, it has carminative and restorative qualities that aid in relieving constipation and diarrhea (Duke 1981; Hangen and Bennik 2003; Thompson et al. 2009). These health benefits may be due to the grain's fiber content, along with antioxidant compounds such as phenolics. Domestic bean consumption has recently grown because of the increasing demand for ethnic foods, their rich content of essential minerals and vitamins, and the believed health benefits, including their cholesterol-lowering effects (Andersen et al. 1984; Andersen and Gustafson 1989). The most widely consumed form of beans is as dry seeds, although varieties for alternative uses, such as snap bean or shell bean, have also been developed. Snap bean cultivars have a fleshy, succulent mesocarp and contain little to no fiber in the green pod walls and sutures, while shell beans are harvested when the seeds are still immature, before the pod has completely dried.

Common beans represent a major source of daily protein in numerous countries, especially in Latin America, Africa, and parts of Asia, as well as in the traditional diets of the Middle East and Mediterranean areas (Broughton et al. 2003; Casquero et al. 2006). This legume is a key component of the Mediterranean diet and is becoming more and more popular in the United States as people become more interested in "ethnic" and healthy foods (Blair and Izquierdo 2012). Beans are an important commercial food legume and vegetable crop in Canada, the USA, and Europe. Their consumption is particularly high in African countries, with per capita intake ranging from 50 to 60 kg annually in Rwanda, Kenya, and Uganda (Broughton et al. 2003; Buruchara et al. 2011).

Its rich nutritional profile, diverse forms (including fresh, canned, frozen pods or seeds, precooked/dehydrated seeds, and packaged dry seeds), and cooking versatility make it a highly attractive and profitable crop. Consumption habits differ substantially between geographic locations and cultures. In fact, it is grown extensively across all five continents, with cultivation ranging from 52°N to 32°S latitude, and from near sea level in the United States and Europe to altitudes exceeding 3000 meters above sea level in the Andean region of South America. Nearly half of the world's supply of dry beans is produced in American countries; the leading producers are Brazil, the USA, Mexico, and Central America. India, China, and Myanmar are the top producers of beans in Asia. In Europe, bean cultivation is primarily focused in areas around the Mediterranean, including the Iberian Peninsula, Italy, and the Balkan states, though production does not fully meet the total demand.

In India, it is an important commercial pulse crop in the North-Western Himalayan region, renowned for its cooking quality and unique organoleptic characteristics. In India, it got evolved in two ecotypes; the hilly rajmash type grows in temperate climate, while the tropical type thrives well in the central and southern part of India. In India, pulses account for almost one-fifth of total food grain area and contribute to roughly one-fifth of total food grain production, and common bean is cultivated in 288 thousand ha area with a production of 2727 thousand metric tonnes and an average productivity of 9.46 t/ha (Anonymous 2023). In India, it is mainly grown in Himachal Pradesh, Jammu and Kashmir, Punjab, Uttar Pradesh, Haryana, Bihar, Madhya Pradesh, Gujarat, Maharashtra, Karnataka, Andhra Pradesh, and Tamil Nadu. Farmers in the mountainous regions of the Himalayas cultivate indigenous bean varieties through a combination of monocropping to control diseases and pests, while also enhancing production stability, household consumption, and income generation (Rana et al. 2015). The economic significance of the common bean justifies ongoing efforts to develop new varieties that are suitable for mechanical harvesting, resistant to pests and diseases, and possess exceptional nutritional value. With the ongoing population growth and rising bean consumption, the demand for *P. vulgaris* is anticipated to increase (Akibode and Maredia 2011), and crucial issues related to this species must be addressed in the future. To meet the challenges of climate change and the rapidly growing demand for food, future efforts will focus on enhancing bean productivity, food quality, and resilience to both biotic and abiotic stresses.

2 Botanical Description of *Phaseolus vulgaris*

Phaseolus vulgaris is a herbaceous plant that can be either vining (indeterminate) or bush-type (pseudo-determinate). Vining varieties are perennial and grow 2–5 meters long, while bush-types are annuals and typically reach heights of 20–60 cm. The leaves are trifoliate, arranged alternately along the stem, and vary in color from green to purple. Most species have five petals on their flowers, which resemble papilionaceous or butterfly shapes. Flowers have a single large upright petal, flanked by two horizontal "wing" petals, and supported by two petals at the bottom that form a boat-like "keel." Generally flower have ten stamens, out of which nine form a tube around the ovary and one is independent and lies above the ovary, though some species have different stamen arrangements. The legume, which bears one-to-many seeds, is the fruit of the Fabaceae family. It is a single-carpelled pod that appears in a variety of sizes and forms. The pods, either green or yellow, grow up to 20 cm in length, may be straight or curved, and contain 4–12 seeds each. Their surface is either smooth or slightly hairy, with a prominent beak. The seeds are kidney-shaped, elongated, nearly globular, or slightly compressed, and come in a variety of colors, including white, red, brown, purple or black.

3 Origin and Distribution

The majority of the *Phaseolus* genus species are thought to have originated in Mesoamerica based on their geographic spread (Freytag and Debouck 1996, 2002; Delgado-Salinas et al. 1999, 2006). Archaeological evidence shows that the species may have originated and been domesticated as early as 5000 B.C. (Purseglove 1968; Bitocchi et al. 2012, 2013); however, evidence suggests a more recent origin in Mesoamerica (Kaplan and Lynch 1999). According to data from multiple loci sequence analyses, the common bean was domesticated approximately 8000 years ago (Mamidi et al. 2011). Currently, approximately 400 species have been identified within the genus *Phaseolus* (Freytag and Debouck 2002). Five species, *P. vulgaris*, *P. dumosus*, *P. coccineus*, *P. acutifolius*, and *P. lunatus* have been domesticated and are currently under cultivation. *P. dumosus* and *P. coccineus*, both Mesoamerican species, are the nearest relatives of *P. vulgaris*, and all three are capable of partial intercrossing. The other domesticated species, *P. lunatus* and *P. acutifolius*, are more distantly related. Wild relatives of the common bean, including *P. coccineus*, *P. acutifolius*, and *P. lunatus*, serve as valuable sources of anthracnose resistance and can help broaden the genetic diversity of cultivated beans (Kumar et al. 2021). As per the sequence data of the α-amylase inhibitor gene, *P. vulgaris* diverged from *P. dumosus* and *P. coccineus* across two million year ago (Gepts et al. 1999). Among the five domesticated *Phaseolus* species, *P. vulgaris* holds the greatest economic significance, serving as the main grain legume consumed directly by humans. These species exhibit genetic and phenotypic variations, particularly in their reproductive systems, which can be either self-pollinating or outcrossing.

Domestication is the outcome of a selection process that makes plants better suited for cultivation and human utilization. Plants and animals undergo both natural and human selection during domestication. The characteristics which separate domesticated species out from wild species are known as the "domestication syndrome" (Hammer 1984). These traits include the absence of seed dispersal and dormancy, a more compact plant structure, improved yield, synchronized and earlier flowering, sensitivity to photoperiod, harvest index, and changes in seed pigmentation (Gepts and Debouck 1991; Koinange et al. 1996). The domesticated common bean developed from wild gene pools through separate domestication events (Gepts and Debouck 1991; Kwak and Gepts 2009; Kwak et al. 2009) in two primary domestication centers: Mesoamerica (western Mexico) and the Southern Andes of South America. Two major cultivar groups emerged from two primary centers: Mesoamerica and the Southern Andes. Mesoamerican cultivars comprised mainly small-seeded beans with a predominantly "S" phaseolin type, while Andean beans contained large-seeded "T" phaseolin-type cultivars (Gepts et al. 1986; Gepts and Bliss 1988; Sonnante et al. 1994; Chacon et al. 2005). However, a smaller domestication center in Colombia has been identified, where small-seeded cultivars with the "B" phaseolin-type seed protein are found (Gepts and Debouck 1991; Chacon et al. 1996; Tohme et al. 1996).

The wild *P. vulgaris* is broadly distributed from northern Mexico to north-western Argentina (Toro et al. 1990) and is distinguished by two primary eco-geographical gene pools: Mesoamerica and the Andes. Numerous studies based on various datasets, including plant morphology, have revealed that these two gene pools have analogous geographical topologies in the wild and domesticated (Singh et al. 1991b), seed proteins (Gepts et al. 1986), allozymes (Koenig and Gepts 1989), restriction fragment length polymorphism (Becerra-Velasquez and Gepts 1994), random amplified polymorphic DNA (RAPD, Freyre et al. 1996), amplified fragment length polymorphism (AFLP; Rossi et al. 2009), and simple sequence repeats (microsatellites, SSRs; Kwak and Gepts 2009). Similar results have also shown genetic diversity utilizing amplified fragment length polymorphism (AFLP) markers (Beebe et al. 2001; Papa and Gepts 2003; Zizumbo-Villarreal et al. 2005), SSR markers (Gaitan-Solis et al. 2002; Blair et al. 2006), DNA sequencing (Gepts et al. 2008), and single nucleotide polymorphism (SNP) markers (Galeano et al. 2009, 2012; Blair et al. 2013). These tools assist in providing answers to several concerns about the domestication, evolution, and diversity of common beans that cannot be provided by phenotypic approaches alone (Arif et al. 2010). For example, genes related to the domestication events in the Andean and Mesoamerican regions, as well as evolutionary traits such as seed shattering, have been discovered (Bellucci et al. 2014; Gaut 2014). Additionally, polymorphisms in genes associated with drought tolerance have also been identified (Cortes et al. 2012a, b).

SNP, SSR, and AFLP markers are the most commonly used markers in investigations on the genetic variation of common beans due to their low cost, effectiveness, and simplicity. These two gene pools are further differentiated by geographic factors and partial reproductive barriers (Gepts and Bliss 1985; Gioia et al. 2013). Additionally, within each gene pool, the species is divided into various eco-geographic races and seed types. Crosses between different gene pools or races may result in negative combining ability and lethality issues, complicating the breeding of common beans (Singh et al. 1991; Kelly and Miklas 1998; Blair et al. 2007). For genotypic data, several projects have generated large SNP datasets. The U.S. Bean Coordinated Agricultural Project (Bean CAP) has developed SNP resources for two diversity panels: the Mesoamerican Diversity Panel (MDP) and the Andean Diversity Panel (ADP) (Moghaddam et al. 2014).

Global distribution of common bean began with their primary American origins and domestication sites (Zeven 1997). The common bean developed in a climate that ranged from subtropical to temperate, with rainy and dry seasons. The bean favors areas with adequate rainfall over those with excessive or dry precipitation (Beebe et al. 2013). It is believed that the common bean was introduced to Europe on two separate occasions: Mesoamerican beans brought by the Spanish and Portuguese during their colonization of the Americas in 1506, and Andean beans introduced later in 1528, during Pizarro's expedition in Peru (Gioia et al. 2013). Different agricultural practices, cropping patterns, and farmer preferences have led to the development of common bean cultivars and landraces in Europe (Zeven 1997; Pipan and Meglic 2019). Moreover, the crossing between Andean and Mesoamerican genotypes played a key role in creating considerable genotypic and phenotypic diversity within the European common bean (Rodino et al. 2007; Gioia et al. 2013).

According to Papa and Gepts (2003), certain wild populations in Columbia belong to a transect area. Wild forms of the common bean create a third gene pool situated in a confined area between Northern Peru and Ecuador (Debouck et al. 1993). These populations are commonly regarded as the likely ancestors of *P. vulgaris*, with the unique type I phaseolin, the main seed storage protein, being considered a key evolutionary marker (Kami et al. 1995).

Kami et al. (1995) suggested that type I phaseolin is the ancestral form of the other *P. vulgaris* phaseolin sequences. This is due to the fact that duplications can happen anywhere within a sequence, while deletions are restricted to the tandem direct repeat sites. The common bean is said to have originated in the western slopes of the Andes, specifically in northern Peru and Ecuador. The wild bean spread northward to Colombia, Central America, and Mexico, then southward to southern Peru, Bolivia, and Argentina, resulting in the Mesoamerican and Andean gene pools (Kami et al. 1995). Phaseolin, the primary storage protein in common beans, has served as a key marker for identifying gene pool affiliation (Gepts and Bliss 1988; Sustar-Vozlic et al. 2006; Logozzo et al. 2007). Phaseolin types S (Sanilac), M (Middle America), and B (Boyaca) are linked to the Mesoamerican origin of genotypes of common beans, whereas genotypes with phaseolin types T (Tendergreen), C (Contender), and H (Huevo de Huanchaco) are related with the Andean gene pool. However, since Nani et al. (2011) developed three indel spanning markers SHP1-A, SHP1-B, and SHP1-C, subsequent research has increasingly utilized these markers for identifying gene pools (Pipan and Meglic 2019).

4 Diversity

The genus *Phaseolus* is most diverse in central and southern Mexico, which shares a border with Guatemala in Mesoamerica. There are 70–80 wild species along with the progenitors of five domesticated species (Freytag and Debouck 2002). It is debatable whether secondary centers of diversity exist in Ethiopia, Turkey, and Spain (Harlan 1971; Santalla et al. 2002; Ocampo et al. 2005; Kami et al. 1995). Recent studies on Indian and Chinese beans have expanded Asia's secondary centers of diversity for common bean (Zhang et al. 2008; Sharma et al. 2013). Using morphological, biochemical, and molecular methods (Jaaska 1996; Pueyo and Delgado-Salinas 1997), the evolutionary relationships between domesticated *Phaseolus* species and with combined analysis suggested that the *Phaseolus* is monophyletic (Delgado-Salinas et al. 1999). In addition to comprehending the kind and extent of genetic diversity derived from various features, it is preferable to find accessions that exhibit resistance to the most common pests and illnesses (Rana et al. 2015). India is home to a rich diversity of common beans, with numerous landraces and cultivated varieties adapted to various agroclimatic conditions across the country. In India, common beans are grown in diverse regions, including the hills of Himachal Pradesh, Uttarakhand and Jammu and Kashmir along with the plains of Haryana, Punjab, Madhya Pradesh, Uttar Pradesh, Karnataka, and Maharashtra. Each region may have its own unique bean varieties selected over generations by local farmers to suit local growing conditions and culinary preferences. The common bean's interspecies diversity is divided into four gene pools: primary, secondary, tertiary, and quaternary.

4.1 Primary Gene Pool

Phaseolus vulgaris cultivars and wild populations make up the main gene pool. Wild variants of *Phaseolus vulgaris*, which range from north Mexico to northwest Argentina (Toro et al. 1990; Singh 2001), are the direct ancestors of common bean cultivars (Berglund-Brucher and Brucher 1976). Two main gene pools may be identified within both wild and cultivated populations: the Andean and the Mesoamerican gene pools (Gepts and Bliss 1985; Koenig and Gepts 1989; Kairaillah et al. 1990). *Phaseolus vulgaris* is only sexually compatible in a few domesticated and wild varieties. F1 hybrid weakening has been observed in various parents from the Andes and Mesoamerica (Pathania et al. 2014).

Incompatibility genes in domesticated and wild populations in Mesoamerica and the Andes create hybridization barriers (van Reheenen et al. 1979; Shii et al. 1980; Singh and Gutierrez 1984; Vieira et al. 1989). These genes lead to hybrid weakness and phenotypic irregularities in later progenies (Koinange and Gepts 1992; Debouck et al. 1993; Freyre et al. 1996; Singh and Molina 1996), along with inviable F_1 hybrids (Shii et al. 1980; Gepts and Bliss 1985). Consequently, in certain wild and cultivated populations with varied gene pools, there is little gene exchange and introgression (Kornegay et al. 1992; Mumba and Galwey 1999).

4.2 Secondary Gene Pool

The secondary gene pool of the common bean encompasses *P. coccineus*, *P. dumosus* (*P. polyanthus* Greenman), and *P. costaricensis* Freytag and Debouck. Under natural and controlled conditions, these three species readily cross with domesticated common beans, primarily when the common bean serves as the female parent (Baggett 1956; Camarena and Baudoin 1987; Singh et al. 1997). Making a cross can be challenging when the common bean acts as the pollen donor because the offspring often revert to the maternal parent (Smartt 1970; Ockendon et al. 1982; Hucl and Scoles 1985; Debouck 1999). Certain crosses can produce partially sterile offspring (Ibrahim and Coyne 1975; Manshardt and Bassett 1984). *Phaseolus coccineus* cannot reproduce with common beans (Smartt 1979, 1990; Manshardt and Bassett 1984; Hoover et al. 1985; Debouck and Smartt 1995).

4.3 Tertiary Gene Pool

Phaseolus parvifolius and *Phaseolus acutifolius* are two species included in the tertiary gene pool. (Debouck 1999). By means of embryo rescue, it is possible to perform crosses between these two species and the common bean. Interspecific F_1 progeny that survive are sterile (Menjia-Jimenez et al. 1994; Singh et al. 1998). Hybrid fertility can be restored by either chromosomal doubling or multiple generations of backcrossing with recurrent common bean parents (Prendota et al. 1982; Pratt et al. 1985; Thomas and Waines 1984). Fertile hybrids are produced in rare situations (Honma 1956). The genotype plays a crucial role in the level of cross compatibility between *P. vulgaris* and *P. acutifolius* due to the presence of two complementary dominant genes found in certain populations (Parker and Michaels 1986). Choosing appropriate genotypes of common beans as parents for the hybrids encourages interspecific hybridization and enhances hybrid viability (Federici and Waines 1988).

4.4 Quaternary Gene Pool

A group of genes known as the quaternary gene pool is found in common beans and comprises *Phaseolus lunatus*, *P. carteri* Freytag and Debouck, *P. filiformis* Benth, and *P. angustissimus* A. Grey. There have been many attempts to cross *P. filiformis*, *P. angustissimus*, and *P. lunatus* with the common bean have failed to produce fertile hybrids. As a result, these species ought to be included in the common bean's quaternary gene pool to allow for gene introgression.

4.5 Landraces

Landraces are unique but diverse populations that frequently go by local names, lack "formal" crop improvement, and are characterized by a unique ability to adapt to the local biotic and abiotic stresses of farming. They are also closely associated with the customs, knowledge, habits, dialects, and festivals of the people who created and continue to use them. Despite being a vital part of agro biodiversity, landraces are in danger of going extinct because they are farmed by traditional farmers and require both in situ and ex-situ protection. In the Northern Himalayas, a number of popular landraces of rajmash are grown in particular niches and are identified by their local names.

These includes Shopian, Bandipora, Gurez, Anantnag, Kupwara, Baramula, etc. (Kashmir), Poonch, Kistwar, Gool, Bani, Bhadarwah, Madwaha etc. (Jammu), Chamba, Barot, Kullu, Kinnauri (Himachal Pradesh), Harsar, Joshimath, Chamoli, Dhakna, Munsayari etc. (Uttarakhand). These include both determinate (bushy), indeterminate (pole) types and distinguished by grain size and contrasting colors. Molecular markers have been extensively employed in the common bean to evaluate genetic variation and genetic structure of several important collections and to develop maps (Zhang et al. 2008; Kwak and Gepts 2009). Landraces can offer genetic diversity for abiotic stress tolerance, disease and pest resistance. Another argument for landrace protection is the growing replacement of local, indigenous, or traditional agriculture by industrial agriculture. A parallel tendency is the substitution of genetic monoculture for native crop varieties.

5 Status of Genetic Resources

Germplasm conservation for crops such as common beans and their wild relatives can be accomplished by preserving seeds at low temperatures and moisture levels for significant span of time. However, certain conservation barriers may threaten the genetic integrity of preserved accessions. Periodic renewal of materials preserved as seeds is necessary to counteract the declining viability of the seed. Protocols are being implemented all around the world to reduce the likelihood that the genetic composition of preserved samples will be altered by selection, mutations, random drift, or accidental contamination. Large *Phaseolus* germplasm collections were created to help scientists improve the quality and production of the crop through the collection, preservation, assessment, recording, and dissemination of germplasm. These collections, which contain genotypes of both cultivated and wild *Phaseolus* species, are kept in storage facilities all over the world. Seed samples are often available for breeding or research purposes upon request; moreover, a paper trail for material transfer agreements is provided (Lioi and Piergiovanni 2013). Genetic resources constitute an abundant supply of such "new" genes.

Besides the common bean (*P. vulgaris*) and several wild *Phaseolus* species, these collections also include four other domesticated *Phaseolus* species: year-long bean (*Phaseolus dumosus*), runner bean (*Phaseolus coccineus*), tepary bean (*Phaseolus acutifolius*), and lima bean (*Phaseolus lunatus*). Most of these collections came from their center of origin, primarily in the Andes and Mesoamerica. There are smaller collections of common bean accessions available at breeding stations of national, sub-national, or multi-country regional programme (ECABREN and SABRN bean networks in East and Southern Africa, respectively), as well as through non-governmental organizations like Seed Savers Exchange in Decorah, Iowa. For genetic research and bean breeding, germplasm accessibility usually does not pose a barrier. The collection of germplasm from throughout the world can be classified according to several factors: phenotype (like disease responses, photoperiod reaction, growth habit, seed characteristics, etc.), genotype (i.e., sequence- or marker-sequence-based classification), pedigree, gene pool, race, or geographic origin. Ideally, all germplasm accessions would retain these qualities together. In reality, though, the characterizations are insufficient and unrelated. The GRIN system maintains substantial phenotypic data for the US germplasm collection. Except for a few populations, pedigree information is typically limited and is typically found at institutions with extensive breeding programmes (e.g., CIAT). For information on geographic origin, Legume Info geographical information system map of germplasm origin is a helpful resource (https://legumeinfo.org). This interactive viewer shows the phenotypic information stored in GRIN together with the geo-coordinates for the bean collection. Phenotypic categories or values can be displayed, and the viewer's data can be examined by geographic location or by phenotype (e.g., seed size or photoperiod).

6 Conservation

6.1 Ex Situ Conservation of *Phaseolus vulgaris* Genetic Resources

Ex situ conservation can be done for cultivated common beans as well as their wild relatives by preserving the seeds for extended periods of time at low humidity and temperature. Periodic rejuvenation is necessary for materials kept as seeds in order to balance the deteriorating viability of the seeds. Globally accepted protocols are intended to reduce the likelihood that mutations, selection, random drift, or unintentional contamination might alter the genetic makeup of samples that are being preserved. Across the world, these collections stored contain genotypes of both domesticated and wild *Phaseolus* species. Gene banks maintain significant ex situ collections of common bean germplasm. Information regarding *P. vulgaris* accessions kept in gene banks can be found on the Genesys (http://www.genesys-pgr.org). According to this platform, 63 institutions from 138 countries globally hold about 83,000 accessions. Landraces (about 61,000), improved cultivars (>8000), breeding materials (>2000), and wild forms (>1500) are primary accessions (Table 2).

The collection of particularly interesting wild Phaseoleae (also known as Phaseolinae) is kept at the National Botanic Garden of Meise, Belgium. There are now 1886 accessions in this collection, which includes a wide range of genetic diversity and includes 225 species from the Phaseoleae tribe, mostly from the Phaseolinae subtribe. *Phaseolus* and *Vigna* are the two most well-represented genera, with 41 species (712 accessions) and 67 species (978 accessions), respectively (http://www.br.fgov.be/research/collections/living/phaseolus). Furthermore, smaller collections are dispersed across the worldwide. These collections collectively provide a significant and usually unrestricted supply of genetic variation for studies in plant breeding and genetics. The PHASELIEU project partners of the European Union assessed the condition of the minor European germplasm collections in 2001 and produced the Catalogue of Bean Genetic Resources (Amurrio and Santalla 2001). The European Cooperative Programme for Plant Genetic Resources (ECPGR), which supports the sensible and efficient preservation of plant genetic resources, is now coordinated by Bioversity International, a member of the CGIAR Consortium and a UN FAO collaborator. The European Cooperative Programme for Plant Genetic Resources (ECPGR) built a site that allows users to access passport data of common bean accessions stored at over 20 global gene banks (http://www.ecpgr.cgiar.org/germplasm_databases.html).

Table 2 Some gene banks holding genetic resources of common bean

Institution, location	Number of accessions	Number of taxa
International Centre for Tropical Agriculture (CIAT; Cali, Colombia)	>36,000	44
United States Department of Agriculture (USDA-ARS) at Washington State University	>15,000	50
INIFAP, Texcoco, Mexico	>13,000	3
Botanic Garden, Meise, Belgium	>700	46
University UACh, Chapingo, Mexico	>1400	5

Source: FAO-WIEWS

6.2 In Situ Conservation of *Phaseolus vulgaris* Genetic Resources

The use of in situ conservation for crop species themselves as well as for their wild relatives has gained popularity in recent decades. This alternative method of conservation aims to save crop wild relatives by sustaining the ecosystem in its whole. The basis for in situ conservation of crop plants, which was created especially for on-farm conservation, is the genetic resources preserved by custodian farmers who keep up and employ traditional landraces or varieties, enabling the plants evolution to continue in the habitat in which they are traditionally grown. The ICAR-NBPGR has 4597 total accessions, which comprise 2152 indigenous and 2445 exotic germplasm collections. Such collections represent an abundance of genetic diversity for studying species-wide variability and using novel genetic marker methods to discover novel sources of disease resistance.

Breeding efforts are undergoing ICAR NBPGR for the introgression of desired characters such as resistant to bean common mosaic virus (BCMV) and anthracnose to improve crop yield, nutritional quality and adaptation to different environmental conditions. Breeders utilize sources of resistance found in wild bean relatives or other related species to incorporate genes conferring resistance into commercial varieties. These efforts are essential for ensuring food security and coping with the challenges posed by climate change and evolving agricultural needs. Pre-breeding often begins with the collection and characterization of diverse germplasm representing different landraces, wild relatives, and varieties of the common bean.

7 Gene Flow for Crop Improvement

Common beans may be found in both wild and cultivated forms in most of their geographical zones (Zizumbo-Villarreal et al. 2005). Gene flow between the wild and domesticated bean populations has been shown to result in significant introgression of alleles from the domesticated gene pool to the wild gene pool (Freyre et al. 1996; Beebe et al. 2001). Despite the common bean's high degree of self-pollination, 2–3% outcrossing has been documented. Genes from wild populations are introgressed into domesticated populations and vice versa at regions of sympatric distribution, creating wild-weedy-domesticated hybrid complexes due to genetic compatibility between the two populations. They are a valuable source of genes for farmers and plant breeders because the resultant of this introgression usually has morphological traits similar to one or both parents, such as larger seeds than the wild parent or colors or patterns of seed like those found in wild beans (Debouck and Smartt 1995; Beebe et al. 2001). Nevertheless, they also offer a risk of the enormous introduction of genes from cultivated to wild populations (Gepts et al. 1999; Papa and Gepts 2003). Recent research on wild populations in Mesoamerica, where traditional agriculture is still practiced, shows that potential for wild-domesticated hybridization exist despite a primarily autogamous reproductive system. They propose an unequal gene flow that is at least three times higher between domesticated and wild species (Papa and Gepts 2003; Chacon et al. 2005; Zizumbo-Villarreal et al. 2005). The prevalence of wild characteristics, the size differences between domesticated and wild populations, or human selection favoring hybrids within domesticated populations might all be contributing factors to this asymmetric gene flow (Papa and Gepts 2003). The gene flow from domesticated populations can therefore cause a displacement of genetic variability in wild populations (Papa and Gepts 2003). Early cultivars may have spread from a domesticated region to neighboring regions, where they interbred with native populations to produce modern landraces (Beebe et al. 2001).

7.1 Hybridization

Several hybridization techniques are used in common bean breeding programs to add genetic variety and produce new cultivars with desired traits like increased yield, resistance to disease, and better nutritional value. Some hybridization methods are mentioned below:

Hand emasculation and pollination: In this method, remove the male reproductive organs (anthers) from the female parent flower to prevent self-pollination. The stigma is then delicately dusted with pollen from the selected male parent. Hand emasculation and pollination ensure controlled crosses and are often employed to preserve genetic purity in breeding lines.

Artificial hybridization: This technique involves controlled pollination conducted in a laboratory or controlled environment. It allows breeders to control environmental factors such as temperature, humidity, and light, ensuring successful crosses even between genetically distant parents. Artificial hybridization is useful for overcoming barriers to natural pollination and is commonly employed to introduce specific traits into common bean cultivars.

Intraspecific hybridization: Breeding within the same species, it is a common practice to enhance traits such as quality, yield, and adaptability. The aim of intraspecific hybridization is to create superior cultivars that satisfy the demands of both farmers and consumers by combining advantageous genes from several varieties or breeding lines.

Interspecific hybridization: Crossing between different species within the same genus, such as *Phaseolus coccineus* or *Phaseolus lunatus*, is another approach to introduce novel genetic variation into common bean breeding programme. Through interspecific hybridization, the genetic base of cultivated beans can be expanded, resilience to biotic and abiotic stresses can be improved, and the variety of traits accessible for selection can be increased.

Wide hybridization or distant hybridization: This method involves crossing genetically distant species or even genera to introduce desirable traits not present in common bean germplasm. Novel genes for characteristics like disease resistance, abiotic stress tolerance, or nutritional quality may be introduced as a result of widespread hybridization. However, it often requires specialized techniques and may involve overcoming barriers to hybridization such as genetic incompatibility or sterility.

Among these methods, hand emasculation and pollination are typically the most frequently used in common bean breeding programme due to their simplicity, effectiveness, and applicability to a broad range of breeding objectives. However, the choice of method depends on various factors, including the breeding goals, available resources, and the genetic diversity of the breeding population.

7.2 Interspecific Hybridization

Interspecific hybridization among *Phaseolus* species: The genus *Phaseolus* originated in the Neotropics. Though the definition of a bean is well known, the number of *Phaseolus* species is less certain. The number of species in Central America is estimated to be 50–60, although further research is needed (Debouck 1999). To investigate the variability issue, it is practically significant to comprehend the relationships among the different species. Recent phylogenetic investigations involving both wild and cultivated species of the plant and utilizing morphological, biochemical, and molecular data (seed proteins, isozymes, nuclear, chloroplastic, and mitochondrial DNA, among other things) have revealed that the genus *Phaseolus* is monophyletic (Debouck 1999). At the subgeneric level, there could be two to nine subclades (Baudoin et al. 1998; Delgado-Salinas et al. 1999). *P. lunatus* is included in one lineage; contrarily the common bean is included in different (Maquet et al. 1999). *P. vulgaris*, *P. polyanthus*, and *P. coccineus* are three species that are members of the same branch of the evolution (Schmit et al. 1993). However, variations appear in the quantity, sort, and nature of the DNA studied. Few systematic studies have been conducted to check for disease resistance, insect resistance, and other critical traits in wild common beans and also in wild and domesticated germplasm of alien species. *P. polyanthus* is renowned for its ability to withstand both BGYMV and ascochyta blight. *P. coccineus* confers resistance to anthracnose, root rots, white mould, BYMV, and BGYMV. *P. acutifolius* has leafhopper tolerance, and some tepary bean accessions show substantial levels of resistance to CBB (Common Bacterial Blight) and bruchids (Baudoin et al. 2001; Debouck 1999; Schmit and Baudoin 1992; Singh 1999).

Therefore, the absence of biotic resistance is one of the main production restrictions; intraspecific hybridizations among *Phaseolus* species have been adopted due to the sluggish progress in identifying beneficial genes in related species. In contrast, between 1940 and 1985, *P. vulgaris* and *P. coccineus* often intercrossed. It was discovered that segregants in reciprocal crosses with *P. coccineus* as the female parent spontaneously returned back to the donor parent cytoplasm after a few generations (Baudoin et al. 1998). In an attempt to bridge *P. vulgaris* with wild *P. coccineus* variations, the possibility that domestication is the cause of reproductive isolation was considered. However, few commercial cultivars have been created in this way. When *P. vulgaris* is the pollen donor, P. *polyanthus* crosses more readily with *P. coccineus* and allied forms than with the latter (Baudoin et al. 2001). Although *P. polyanthus* is a member of the *P. vulgaris* clade, its ability to participate in interspecific hybridizations is restricted due to the introduction of *P. coccineus* genes into its nuclear genome. It is easy to perform crossings between *P. vulgaris* and *P. costaricensis* without embryo rescue, especially when *P. vulgaris* is utilized as the female parent. However, it is uncertain if *P. coccineus* genes have altered *P. costaricensis* nuclear genome (Debouck 1999). After the *P. vulgaris* phylum and the primary gene pool have been fully utilized, *P. coccineus* and its associates may be the diversity researcher with the most potential.

8 Molecular Marker and Marker-Assisted Breeding

Molecular techniques like genetic mapping and marker-assisted selection (MAS) are used to identify genomic regions associated with important characteristics like disease resistance, drought tolerance, yield, and nutritional quality. This allows breeders to introgress beneficial alleles into elite breeding lines more efficiently. With climate change leading to increased environmental stresses such as drought, heat, and soil salinity, there's a growing emphasis on producing common bean varieties with improved tolerance to these abiotic stresses. Breeding for traits like water use efficiency, heat tolerance, and salinity tolerance is essential for ensuring stable yields under adverse conditions. Genetic enhancement involves the deliberate manipulation of an organism's genetic makeup to introduce or enhance desirable traits. This can be accomplished through traditional breeding methods, such as crossing, selection, and backcrossing, as well as through modern biotechnological techniques like genetic engineering and genome editing. Marker-assisted selection (MAS), transgenic strategies, genome editing (e.g., CRISPR-Cas9), and other biotechnological techniques are examples of improvement techniques.

P. vulgaris target traits can be achieved indirectly by using molecular markers such as amplified fragment length polymorphism (AFLP), sequence characterized amplified region (SCAR), restriction fragment length polymorphism (RFLP), and simple sequence repeats (SSR) and that are closely related to underlying genes or have been developed from actual gene sequences (Xu and Crouch 2008). It has been determined that several common bean diseases have resistance genes associated with distinct sequence characterized amplified region (SCAR) markers. For example, Pse-1 gene which is resistance to halo blight a severe bacterial seed borne disease caused by *Pseudomonas syringae* pv. *phaseolicola* (Burkh.) in common bean was tagged and mapped by Miklas et al. 2009. To assist breeders, a major effort to identify RAPD markers tightly linked to different genes was undertaken. To date, 23 RAPD and five SCAR markers associated with 15 distinct resistance genes have been discovered, along with QTL conditioning resistance to seven important common bean diseases (Kelly and Miklas 1998). Codominant marker pairs, flanking markers, repulsion-phase linkages, recombination-facilitated MAS, and SCAR markers have all been shown to increase selection efficiency. Breeders now have chances that were not possible with previous breeding methods thanks to marker-assisted selection for disease resistance in common beans.

9 Conclusions

The common bean is an important crop globally, contributing significantly to food security and nutrition. Its importance is underscored by its widespread cultivation, diverse uses, and adaptability to various environmental conditions. This leguminous crop holds a prominent place in traditional diets across many cultures and regions. Efforts to conserve common bean genetic resources are crucial for maintaining the crop's resilience and ensuring future breeding endeavors. Various gene banks worldwide house extensive collections of common bean germplasm, providing researchers and breeders with valuable genetic diversity to support crop improvement initiatives. Advancements in molecular genetics have facilitated the characterization and utilization of common bean germplasm, enabling targeted breeding for desired traits. Marker-assisted selection and genomic technologies offer efficient tools for accelerated development of enhanced bean varieties with improved agronomic and nutritional attributes. All things considered, pre-breeding, genetic enhancement, and hybrid development are crucial elements of modern plant breeding programs that aid in the creation of better crop varieties that tackle the issues of climate change, food security, and sustainable agriculture. To fully utilize this adaptable crop and solve upcoming agricultural issues, more research, conservation efforts, and stakeholder involvement are essential.

10 Future Prospects

Nutritional Enhancement There's a growing interest in enhancing the nutritional profile of common beans, such as increasing protein content, iron, zinc, and other essential nutrients. Breeding programmes aim to develop beans with higher nutritional value to address malnutrition and improve overall health outcomes.

Climate Resilience Agriculture faces serious challenges as a result of climate change, including common bean production. Breeding programme are focusing on developing varieties that can thrive under changing climatic conditions, including extreme temperatures, drought, and increased pest and disease pressure.

Sustainable Agriculture Sustainable agricultural practices are becoming increasingly important. Bean breeding programme are integrating traits that promote sustainable farming practices, such as reduced input requirements (water, fertilizers, pesticides) and improved soil health.

Consumer Preferences Consumer preferences are evolving, with an increasing demand for organic, non-GMO, and specialty bean varieties. Breeding programmes are responding to these preferences by developing beans tailored to specific market niches, such as beans suitable for organic production.

Technological Advancements Advances in breeding technologies, such as genomic selection, high-throughput phenotyping, and CRISPR-based gene editing, are accelerating the breeding process. These technologies enable breeders to more efficiently identify and incorporate desirable traits into new varieties, reducing the time and resources required for traditional breeding methods.

Market Diversification Common beans have a huge range of culinary uses beyond traditional dishes, like soups, salads, snacks, and baked goods. Breeding programmes may target specific market segments by developing beans with unique flavor profiles, textures, and colors to meet consumer demand for diverse culinary experiences.

Value-Added Traits In addition to agronomic traits like yield and disease resistance, breeding programme may focus on incorporating value-added traits like enhanced shelf life, cooking quality, and processing characteristics. These traits can improve the marketability and profitability of common bean varieties for both farmers and consumers.

Biofortification Biofortification involves increasing the concentration of micronutrients in crops through breeding or biotechnology. Common bean breeding programmes may focus on biofortification to address micronutrient deficiencies, such as iron and zinc, especially in areas where beans are a primary food. By addressing these various aspects, common bean breeding programme can contribute to sustainable agriculture, food security, and nutrition while meeting the diverse needs of farmers and consumers worldwide.

References

Akibode S, Maredia M (2011) Global and regional trends in production, trade and consumption of food legume crops. SPIA Report department of agricultural, food and resource economics, Michigan State University, East Lansing, MI

Amurrio M, Santalla M, De Ron, A. M. (Eds.) (2001) Catalogue of Bean genetic resources. PHASELIEU – FAIR-PL97-3463. Spain: Mision Biologica de Galicia (CSIC)

Andersen JW, Gustafson NJ (1989) Hypocholesterolemic effect of oat and bean products. Michigan Dry Bean Digest 13:2–5

Andersen JW, Story L, Sieling B, Chen W-JL, Petro MS (1984) Hypocholesterolemic effects of oat-bran or bean intake for Hypercholesterolemic men. Am J Clin Nutr 40:1146–1155

Anonymous (2023) Ministry of Agriculture and Farmers Welfare. https://agriwelfare.gov.in/en/Agricultural_Statistics_at_a_Glance

Arif IA, Bakir MA, Khan AH, Farhan AH, Al Homaidan AA, Bahkali AH, Al Sadoon M, Shbrak M (2010) A brief review of molecular techniques to assess plant diversity. Int J Mol Sci 11:2079–2096

Arumuganathan K, Earle E (1991) Nuclear DNA content of some important plant species. Plant Mol Biol Rep 9:208–218

Baggett JR (1956) The inheritance of resistance to strains of bean yellow mosaic virus in the interspecific cross *Phaseolus vulgaris* × *P. coccineus*. Plant Dis Rep 40:702–707

Baudoin JP, Fofana B, du Jardin P, Vekemans X (1998) Diversity and genetic organization of the genus *Phaseolus*. Analysis of the global and chloroplastic genome. In: International symposium on breeding of protein and oil crops. EUCARPIA, Fundacion Pedro Barrie De La Maza, pp 75–76

Baudoin JP, Camarena F, Lobo M, Mergeai G (2001) Breeding *Phaseolus* for intercrop combinations in Andean highlands. In: Cooper HD, Spillane C, Hodgkin T (eds) Broadening the genetic basis of the crop. CAB International, Wallingford, pp 373–384

Becerra-Velasquez VL, Gepts P (1994) RFLP diversity in common bean (*Phaseolus vulgaris* L.). Genome 37:256–263

Beebe S, Rengifo J, Gaitan E, Duque MC, Tohme J (2001) Diversity and origin of Andean landraces of common bean. Crop Sci 41(3):854–862

Beebe SE, Rao IM, Blair MW, Acosta-Gallegos JA (2013) Phenotyping common beans for adaptation to drought. Front Physiol 4. https://doi.org/10.3389/fphys.2013.00035

Bellucci E, Bitocchi E, Rau D, Rodriguez M, Biagetti E, Giardini A, Attene G, Nanni L, Papa R (2014) Genomics of origin, domestication and evolution of Phaseolus vulgaris. In: Genomics of plant genetic resources, vol 1, pp 483–507

Bennett MD, Leitch IJ (1995) Nuclear DNA amounts in angiosperms. Ann Bot 76:113–116

Bennett MD, Leitch IJ (2010) Angiosperm DNA C-values database (release 7.0, Dec. 2010)

Berglund-Brucher O, Brucher H (1976) The South American wild bean (*Phaseolus aborigineus* Burk.) an ancestor of common bean. Econ Bot 30:257–272

Bitocchi E, Nanni L, Bellucci E, Rossi M, Giardini A, Zeuli PS, Logozzo G, Stougaard J, McClean P, Attene G, Papa R (2012) Mesoamerican origin of the common bean (Phaseolus vulgaris L.) is revealed by sequence data. Proc Natl Acad Sci 109(14):788–796

Bitocchi E, Bellucci E, Giardini A, Rau D, Rodriguez M, Biagetti E, Santilocchi R, Spagnoletti Zeuli P, Gioia T, Logozzo G, Attene G (2013) Molecular analysis of the parallel domestication of the common bean (*Phaseolus vulgaris*) in Mesoamerica and the Andes. New Phytol 197(1):300–313

Blair MW, Izquierdo P (2012) Use of the advanced backcross-QTL method to transfer seed mineral accumulation nutrition traits from wild to Andean cultivated common beans. Theor Appl Genet 125:1015–1031

Blair MW, Giraldo MC, Buendia HF, Tovar E, Duque MC, Beebe SE (2006) Microsatellite marker diversity in common bean (*Phaseolus vulgaris* L.). Theor Appl Genet 113:100–109

Blair MW, Iriarte G, Beebe S (2007) QTL analysis of yield traits in an advanced backcross population derived from a cultivated Andean 9 wild common bean (*Phaseolus vulgaris* L.) cross. Theor Appl Genet 112:1149–1163

Blair MW, Cortes AJ, Penmetsa RV, Farmer A, Carrasquilla-Garcia N, Cook DR (2013) A high-throughput SNP marker system for parental polymorphism screening, and diversity analysis in common bean (*Phaseolus vulgaris* L.). Theor Appl Genet 126:535–548

Broughton WJ, Hernandez G, Blair M, Beebe S, Gepts P, Vanderleyden J (2003) Beans (*Phaseolus* spp.) – model food legumes. Plant Soil 252:55–128

Buruchara R, Chirwa R, Sperling L, Mukankusi C, Rubyogo JC, Mutonhi R (2011) Development and delivery of bean varieties in Africa: the Pan-Africa Bean Research Alliance (PABRA) model. Afr Crop Sci J 19:227–245

Camarena F, Baudoin JP (1987) Obtention des premiers hybrids interspecifiques entre *Phaseolus vulgaris*et*Phaseolus polyanthus* avec le cytoplasme de cette derniere forme. Bull Rech Agron Gembloux 22:43–55

Casquero PA, Lema M, Santalla M, De Ron AM (2006) Performance of common bean landraces from Spain in the Atlantic and Mediterranean environments. Genet Resour Crop Evol 53:1021–1032

Chacon MI, Gonzalez AV, Gutierrez JP, Beebe S, Debouck DG (1996) Increased evidence for common bean (*Phaseolus vulgaris*) domestication in Colombia. Ann Rep Bean Improv Coop 39:201–202

Chacon MI, Pickersgill S, Debouck DG (2005) Domestication patterns in common bean (*Phaseolus vulgaris* L.) and the origin of Mesoamerican and Andean cultivated races. Theor Appl Genet 110:432–444

Chacón-Sánchez MI, Martínez-Castillo J, Duitama J, Debouck DG (2021) Gene flow in Phaseolus beans and its role as a plausible driver of ecological fitness and expansion of cultigenb. Front Ecol Evol 9:618709. https://doi.org/10.3389/fevo.2021.618709

Cortes AJ, Chavarro M, Madrinan S, This D, Blair MW (2012a) Molecular ecology and selection of drought related Asr gene polymorphisms in wild and cultivated common bean (*Phaseolus vulgaris* L.). BMC Genet 13:58. https://doi.org/10.1186/1471-2156-13-58

Cortes AJ, This D, Chavarro MC, Madrinan S, Blair MW (2012b) Nucleotide diversity patterns at the drought related DREB encoding genes in wild and cultivated common bean (*Phaseolus vulgaris* L.). Theor Appl Genet 125(5):1069–1085. https://doi.org/10.1007/s00122-012-1896-5

De Ron AM, Rodino AP, Santalla M, Gonzalez AM, Lema MJ, Martín I, Kigel J (2016) Seedling emergence and phenotypic response of common bean germplasm to different temperatures under controlled conditions and in open field. Front Plant Sci 2(7):205992

Debouck DG (1999) Diversity in *Phaseolus* species in relation to the common bean. In: Singh SR (ed) Common bean improvement in the 21st century. Kluwer Academic Publisher, Dordrecht, pp 25–52

Debouck DG, Smartt J (1995) Beans. In: Smartt J, Simmonds NW (eds) Evolution of crop plants, 2nd edn. Longman, Harlow, pp 287–294

Debouck DG, Toro O, Paredes OM, Johnson WC, Gepts P (1993) Genetic diversity and ecological distribution of *Phaseolus vulgaris* (Fabaceae) in northwestern South America. Econ Bot 47:408–423

Delgado-Salinas A, Turley T, Richman A (1999) Phylogenetic analysis of the cultivated and wild species of Phaseolus (Fabaceae). Syst Bot 24:438–460

Delgado-Salinas A, Bibler R, Lavin M (2006) Phylogeny of the genus *Phaseolus*(Leguminosae): a recent diversification in an ancient landscape. Syst Bot 31:779–791

Duke JA (1981) Handbook of legumes of world economic importance. Plenum Press, New York, USA/ London, UK, pp 195–200

Federici CT, Waines JG (1988) Interspecific hybrid compatibility of selected *Phaseolus vulgaris* L. lines with *Phaseolus acutifolius* A. Gray, *Phaseolus lunatus* L., *Phaseolus filiformis* Bentham. Ann Rep Bean Improv Coop 31:201–202

Freyre R, Ríos R, Guzmán L, Debouck DG, Gepts P (1996) Ecogeographic distribution of *Phaseolus spp*.(Fabaceae) in Bolivia. Econ Bot 1:195–215

Freytag GF, Debouck DG (1996) Phaseolus costaricensis, a new wild bean species (Phaseolinae, Leguminosae) from Costa Rica and Panama, CentralAmerica. Novon 6:157–163

Freytag GF, Debouck DG (2002) Taxonomy, distribution and ecology of genus *Phaseolus* (Leguminosae – Papilionoideae) in North America, Mexico and Central America. Sida, Botanical Miscellany 23. Botanical Research Institute of Texas [BRIT], Forth Worth, p 298

Gaitan-Solis E, Duque MC, Edwards KJ, Tohme J (2002) Microsatellite repeats in common bean (*Phaseolus vulgaris*): isolation, characterization, and cross-species amplification in Phaseolus ssp. Crop Sci 42:2128–2136

Galeano CH, Fernández AC, Gomez M, Blair MW (2009) Single strand conformation polymorphism based SNP and indel markers for genetic mapping and synteny analysis of common bean (*Phaseolus vulgaris* L.). BMC Genomics 10:1–4

Galeano CH, Cortés AJ, Fernández AC, Soler A, Franco-Herrera N, Makunde G, Vanderleyden J, Blair MW (2012) Gene-based single nucleotide polymorphism markers for genetic and association mapping in common bean. BMC Genet 13(1):1–11

Gaut BS (2014) The complex domestication history of the common bean. Nat Genet 46:663–664

Gepts P, Bliss FA (1985) F_1 hybrid weakness in the common bean: differential geographic origin suggests two gene pools in cultivated bean germplasm. J Hered 76:447–450

Gepts P, Bliss FA (1988) Dissemination pathways of common bean (*Phaseolus vulgaris*, Fabaceae) deduced from phaseolin electrophoretic variability. II Europe and Africa. Econ Bot 42:86–104

Gepts P, Debouck D (1991) Origin, domestication and evolution of the common bean (*Phaseolus vulgaris* L.). In: Van Shoonhoven A, Voysest O (eds) Common beans: research for crop improvement. Commonwealth Agricultural Bureau, Wallingford, pp 7–53

Gepts P, Osborn TC, Rashka K, Bliss FA (1986) Electrophoretic analysis of phaseolin protein variability in wild form and landraces of common bean, *Phaseolus vulgaris*: evidence for multiple centers of domestication. Econ Bot 40:451–468

Gepts P, Papa R, Coulibaly S, Mejia AG, Pasquet R (1999) Wild legume diversity and domestication—insights from molecular methods. In: Vaughan D (ed) Wild legumes, Proc. 7th MAFF International Workshop on Genetic Resources. National Institute of Agrobiological Resources, Tsukuba, Japan, pp 19–31

Gepts P, Aragao FJ, Barros ED, Blair MW, Brondani R, Broughton W, Galasso I, Hernández G, Kami J, Lariguet P, McClean P (2008) Genomics of Phaseolus beans, a major source of dietary protein and micronutrients in the Tropics. In: Moore PH, Ming R (eds) Genomics of Tropical Crop Plants. Springer, Berlin, pp 113–143

Gioia T, Logozzo G, Attene G, Bellucci E, Benedettelli S, Negri V, Papa R, Spagnoletti Zeuli P (2013) Evidence for introduction bottleneck and extensive inter-gene pool (Mesoamerica x Andes) hybridization in the European common bean (*Phaseolus vulgaris* L.) germplasm. PLoS One 8(10):e75974

Gopalkrishnan P (2007) Vegetable crops. 176. New India Publisher. Delhi India gram. Biosci Discovery 4(2):214–219

Graham PH, Vance CP (2003) Legumes: importance and constraints to greater use. Plant Physiol 131(3):872–877. https://doi.org/10.1104/pp.017004

Hammer K (1984) Das domestication syndrome. Kulturpfl anze 32:11–34

Hangen L, Bennik MR (2003) Consumption of black beans and navy beans (*Phaseolus vulgaris*) reduced azoxymethaneinduced colon cancer in rats. Nutr Cancer 44:60–65. http://www.kew.org/cvalues/

Harlan JR (1971) Agricultural origins. Centers and noncenters. Science 174:468–474

Honma S (1956) A bean interspecific hybrid. J Hered 47:217–220

Hoover EE, Brenner ML, Ascher PD (1985) Comparison of development of two bean crosses. Hortic Sci 20:84–86

Hucl P, Scoles GJ (1985) Interspecific hybridization in common bean. A review. Hortic Sci 20:352–357

Ibrahim AM, Coyne DP (1975) Genetics of stigma shape, cotyledon position and fl ower color in reciprocal crosses between *Phaseolus vulgaris* L. × *Phaseolus coccineus* (Lam.) and implications in breeding. J Am Soc Hortic Sci 100:622–626

Jaaska V (1996) Isozyme diversity and phylogenetic affinities among the *Phaseolus* beans (Fabaceae). Plant Syst Evol 200:233–252

Kairaillah MM, Adams MW, Sears BB (1990) Mitochondrial polymorphisms of Malawian bean lines: further evidence for two major gene pools. Theor Appl Genet 80:753–761

Kami J, Becerra V, Debouck DG, Gepts P (1995) Identification of presumed ancestral DNA sequences of phaseolin in *Phaseolus vulgaris*. Proc Natl Acad Sci 92:1101–1104

Kaplan L, Lynch TF (1999) *Phaseolus* (Fabaceae) in archaeology: AMS radiocarbon dates and their significance for pre-Colombian agriculture. Economic Botany 53(3):261–272. https://doi.org/10.1007/BF02866636

Kelly JD, Miklas PN (1998) The role of RAPD markers in breeding for disease resistance in common bean. Mol Breed 4:1–11

Koenig R, Gepts P (1989) Allozyme diversity in wild *Phaseolus vulgaris*: further evidence for two major centers of genetic diversity. Theor Appl Genet 78:809–817

Koinange EMK, Gepts P (1992) Hybrid weakness in wild *Phaseolus vulgaris* L. J Hered 83:135–139

Koinange EMK, Singh SP, Gepts P (1996) Genetic control of the domestication syndrome in common bean. Crop Sci 36:1037–1045

Kornegay J, White JW, Ortiz de la Cruz O (1992) Growth habit and gene pool effects on inheritance of yield in common bean. Euphtyica 62:171–180

Kumar S, Singh M, Malhotra N, Blair MW, Sharma JP, Gupta R (2021) Introgression of anthracnose resistance into the background of locally adapted common bean landraces. Euphytica 217(52). https://doi.org/10.1007/s10681-021-02784-1

Kwak M, Gepts P (2009) Structure of genetic diversity in the two major gene pools of common bean (*Phaseolus vulgaris* L., Fabaceae). Theor Appl Genet 118:979–992

Kwak M, Kami JA, Gepts P (2009) The putative Mesoamerican domestication center of *Phaseolus vulgaris* is located in the Lerma-Santiago basin of Mexico. Crop Sci 49:554–563

Lioi L, Piergiovanni AR (2013) Genetic and genomic resources of grain legume improvement; European Common Bean, 1st edn. Elsevier, pp 11–40

Logozzo G, Donnoli R, Macaluso L, Papa R, Knüpffer H, Zeuli PS (2007) Analysis of the contribution of Mesoamerican and Andean gene pools to European common bean (*Phaseolus vulgaris* L.) germplasm and strategies to establish a core collection. Genet Resour Crop Evol 54:1763–1779

Mamidi S, Rossi M, Annam D, Moghaddam S, Lee R, Papa R, McClean P (2011) Investigation of the domestication of common bean (*Phaseolus vulgaris*) using multilocus sequence data. Funct Plant Biol 38(12):953–967

Manshardt RM, Bassett MJ (1984) Inheritance of stigma position in *Phaseolus vulgaris*× *Phaseolus coccineus* hybrid populations. J Hered 75:45–50

Maquet A, Vekemans X, Baudoin JP (1999) Phylogenetic study on wild allies of Lima bean and implication on its origin. Plant Syst Evol 218:43–54

Menjia-Jimenez A, Munoz C, Jacobsen HJ, Roca WM, Singh SP (1994) Interspecific hybridization between common and tepary beans: increased hybrid embryo growth, fertility and efficiency of hybridization through recurrent and congruity back crossing. Theor Appl Genet 88:324–331

Miklas PN, Fourie D, Wagner J, Larsen RC, Mienie CMS (2009) Tagging and Mapping Pse-1 Gene for Resistance to Halo Blight in Common Bean Differential Cultivar UI-3. Crop Sci 49(1):41–48

Moghaddam SM, Song Q, Mamidi S, Schmutz J, Lee R, Cregan P, Osorno JM, McClean PE (2014) Developing market class specific In Del markers from next generation sequence data in *Phaseolus vulgaris* L. Front Plant Sci 5:185. https://doi.org/10.3389/fpls.2014.00185

Mumba LE, Galwey NW (1999) Compatibility between wild and cultivated common bean (Phaseolus *vulgaris* L.) genotypes of the Mesoamerican and Andean gene pools: evidence from the inheritance of quantitative characters. Euphytica 108:105–119

Ocampo CH, Martin JP, Sanchez- Yelamo MD, Ortiz JM, Toro O (2005) Tracing the origin of spanish common bean cultivars sing biochemical and molecular markers. Genet Resour Crop Evol 52:33–40

Ockendon DJ, Currah L, Taylor JD (1982) Transfer of resistance to halo blight (*Pseudomonas phaseolicola*) from*Phaseolus vulgaris*to *Phaseolus coccineus*. Ann Rep Bean Improv Coop 25:84–85

Papa R, Gepts P (2003) A symmetry of gene flow and differential geographical structure of molecular diversity in wild and domesticated common bean (*Phaseolus vulgaris* L.) from Mesoamerica. Theor Appl Genet 106:239–250

Parker JP, Michaels TE (1986) Simple genetic control of hybrid plant development in interspecific crosses between *Phaseolus vulgaris* and *Phaseolus acutifolius*A. Gray. Plant Breed 97:315–323

Pathania A, Sharma SK, Sharma SN (2014) Broadening the Genetic Base of Grain Legumes; Common Bean. Springer, New Delhi, pp 11–50

Pipan B, Meglic V (2019) Diversification and genetic structure of the western-to-eastern progression of European *Phaseolus vulgaris* L. germplasm. Plant Biol 19:442. https://doi.org/10.1186/s12870-019-2051-0

Porch TG, Beaver JS, Debouck DG, Jackson S, Kelly JD, Dempewolf H (2013) Use of wild relatives and closely related species to adapt common bean to climate change. Agronomy 3(2):433–461

Pratt RC, Bresssan RA, Hasegawa PM (1985) Genotypic diversity enhance recovery of hybrids and fertile back- crosses of *Phaseolus vulgaris*L. × *P. acutifolius* A. Gray. Euphytica 34:329–344

Prendota K, Baudoin JP, Marechal R (1982) Fertile allopolyploids from cross the *Phaseolus acutifolius* × *Phaseolus vulgaris*. Bull Rech Agron Gembloux 17:177–189

Pueyo JJ, Delgado-Salinas A (1997) Presence of alpha-amylase inhibitor in some members of the subtribe Phaseolinae (Phaseoleae = Fabaceae). Am J Bot 84:79–84

Purseglove JW (1968) Tropical Crops: Dicotyledons, Longmans. Longmans, London

Rana JC, Sharma TR, Tyagi RK, Chahota RK, Gautam NK, Singh M, Sharma PN, Ojha SN (2015) Characterisation of 4274 accessions of common bean (*Phaseolus vulgaris* L.) germplasm conserved in the Indian gene bank for phenological, morphological and agricultural traits. Euphytica 205:441–457

Rodino AP, Lema EM, Marlene PB, Santalla M, De Ron AM (2007) Assessment of runner bean (*Phaseolus coccineus* L.) germplasm for tolerance to low temperature during early seedling growth. Euphytica 155:63–70. https://doi.org/10.1007/s10681-006-9301-6

Rossi M, Bitocchi E, Bellucci E, Nanni L, Rau D, Attene G, Papa R (2009) Linkage disequilibrium and population structure in wild and domesticated populations of *Phaseolus vulgaris* L. Evol Appl 2(4):504–522

Santalla M, Rodino AP, de Ron AM (2002) Allozyme evidence supporting Southwestern Europe as a secondary center of genetic diversity for the common bean. Theor Appl Genet 104:34–944

Schmit V, Baudoin JP (1992) Screening for resistance to ascochyta blight in populations of *P. coccienus* L. and *P. polythanus* Greenman. Field Crop Res 30:155–165

Schmit V, du Jardin P, Baudoin JP, Debouck DG (1993) Use of chloroplast DNA polymorphism for the phylogenetic study of seven *Phaseolus* taxa including *P. vulgaris* and *P. coccineus*. Theor Appl Genet 87:506–516

Sharma PN, Diaz LM, Blair MW (2013) Genetic diversity of two Indian common bean germplasm collections based on morphological and microsatellite markers. Plant Genet Resour 11:121–130

Shii CT, Mok MC, Temple SR, Mok DWS (1980) Expression of developmental abnormalities in hybrid of *Phaseolus vulgaris* L. interaction between temperature and allelic dosage. J Hered 71(4):218–222

Singh SP (1999) Integrated genetic improvement. In: Singh SP (ed) Common bean improvement in the 21st century. Kluwer Academic Publishers, Dordrecst, pp 133–165

Singh SP (2001) Broadening the genetic base of common bean cultivars. Crop Sci 41:1659–1675

Singh SP, Gutierrez JA (1984) Geographical distribution of DL1 and DL2 genes causing hybrid dwarfism in *Phaseolus vulgaris* L. their association with seed size and their significance in breeding. Euphytica 33(2):337–345

Singh SP, Molina A (1996) Inheritance of crippled trifoliate leaves occurring in interracial crosses of common bean and its relationship with hybrid dwarfism. J Hered 87(6):464–469

Singh SP, Nodari R, Gepts P (1991) Genetic diversity in cultivated common bean. I. Allozymes. Crop Sci 31:19–23

Singh SP, Gutierrez JA, Molina A, Urrea C, Gepts P (1991b) Genetic diversity in cultivated common bean. II. Marker–based analysis of morphological and agronomic traits. Crop Sci 31:23–29

Singh SP, Debouck DG, Roca WM (1997) Successful interspecific hybridization between *Phaseolus vulgaris* L. and *Phaseolus costaricensis* Freytag & Debouck. Ann Rep Bean Improv Coop 40:40–41

Singh SP, Cardona C, Morales FJ, Pastor-Corrales MA, Voysest O (1998) Gamete selection for upright carioca bean with resistance to five diseases and a leafhopper. Crop Sci 38:666–672

Smartt J (1970) Interspecific hybridization between cultivated American species of the genus *Phaseolus*. Euphytica 19:480–488

Smartt J (1979) Interspecific hybridization in grain legumes: a review. Econ Bot 33:329–337

Smartt J (1990) Grain legumes. Evolution and genetic resources. Cambridge University Press, Cambridge, p p379

Sonnante G, Stockton T, Nodari RO, Becerra Velasquez VL, Gepts P (1994) Evolution of genetic diversity during the domestication of common-bean (*Phaseolus vulgaris* L.). Theor Appl Genet 89:629–635

Sustar-Vozlic J, Maras M, Kavornik B, Meglic M (2006) Genetic diversity and origin of Slovene common bean (*Phaseolus vulgaris* L.) germplasm as revealed by AFLP markers and phaseolin analysis. J Am Hortic Sci 131(2):242–249

Thomas CV, Waines JG (1984) Fertile backcross and allotetraploid plants from crosses between tepary beans and common beans (*Phaseolus acutifolius, Phaseolus vulgaris*). J Hered 75:93–98

Thompson MD, Brick MA, McGinley JN, Thompson HJ (2009) Chemical composition and mammary cancer inhibitory activity of dry bean. Crop Sci 49:179–186

Tohme J, Gonzalez O, Beebe S, Duque MC (1996) AFLP analysis of gene pools of a wild bean core collection. Crop Sci 36:1375–1384

Toro O, Tohme J, Debouck DG (1990) Wild bean (*Phaseolus vulgaris* L.): Description and distribution. Centro Internacional de Agricultura Tropical, Cali, Colombia. 252(1):55–128

Ulloa Ulloa C, Acevedo-Rodríguez P, Beck S, Belgrano MJ, Bernal R, Berry PE et al (2018) Vascular Plants of the Americas VPA Website. Available online at: http://www.tropicos.org/Project/VPA

van Reheenen HA, Muigai SGS, Kitivo DK (1979) Male sterility in beans (*Phaseolus vulgaris* L.). Euphytica 28:761–763

Vieira AL, Patto Ramalho MA, dos Santos JB (1989) Crossing incompatibility in some bean cultivars utilized in Brazil. Rev Brasil Genet 12:169–171

Xu Y, Crouch JH (2008) Marker assisted selection in plant breeding: from publication to practices: Review and interpretation. Crop Sci 48:391–407

Zeven AC (1997) The introduction of the common bean (*Phaseolus vulgaris* L.) into Western Europe and the phenotypic variation of dry beans collected in The Netherlands in 1946. Euphytica 94:319–328

Zhang X, Blair MW, Wang S (2008) Genetic diversity of Chinese common bean (*Phaseolus vulgaris* L.) landraces assessed with simple sequence repeat markers. Theor Appl Genet 117:629–640

Zizumbo-Villarreal D, Colunga-GarcíaMarín P, de la Cruz EP, Delgado-Valerio P, Gepts P (2005) Population structure and evolutionary dynamics of wild–weedy–domesticated complexes of common bean in a Mesoamerican region. Crop Sci 45(3):1073–1083

Chickpea (*Cicer arietinum* L.) Breeding

Shailesh Tripathi and Girish Prasad Dixit

Abstract

Chickpea (*Cicer arietinum* L.) is a vital cool-season legume cultivated in over 50 countries and consumed in more than 120 countries. Renowned as a rich source of dietary protein, it is also gaining popularity as a health food in developed nations. Besides its nutritional value, chickpea plays a crucial role in crop rotations with cereals, enhancing soil health and boosting the overall productivity of cropping systems. In India, scientists from National Agricultural Research Systems (NARS), in collaboration with CGIAR institutes such as the International Crops Research Institute for the Semi-Arid Tropics (ICRISAT) and the International Centre for Agricultural Research in Dry Areas (ICARDA), have been actively developing chickpea cultivars tailored to both water-limited conditions and irrigated environments. The collaborative efforts focus on traits valued by farmers, traders, and consumers. The major objectives of the chickpea improvement program include increasing yield potential, developing short-duration varieties, and enhancing tolerance to abiotic stresses like terminal drought and extreme temperatures during the reproductive stage. Additionally, the program targets resistance to major diseases such as fusarium wilt, dry root rot, ascochyta blight, and botrytis grey mold, as well as pests like *Helicoverpa* pod borer. Improvements in seed traits, including color, size, and shape, are also prioritized. The development of high-yielding, short-duration, and wilt-resistant varieties of both desi and kabuli types has significantly impacted chickpea cultivation and production. Notably, machine-harvestable chickpea varieties are now widely adopted by Indian farmers, transforming chickpea farming. Significant progress has been achieved in refining screening techniques and identifying diverse sources of resistance to major abiotic and biotic stresses. Intensive efforts have also been made to ensure the availability of quality seeds in India through seed hubs supported by the Department of Agriculture and Farmers Welfare, Government of India. As a result, the seed replacement rate for chickpea in India has risen from less than 10% in 2005 to nearly 30% in recent years. Furthermore, the share of new varieties (less than 10 years old) now exceeds 81%, marking a significant advancement in chickpea breeding and production.

Keywords

Breeding · Chickpea · Genetic improvement · Improved varieties · Research strategies

1 Introduction

Chickpea (*Cicer arietinum* L.) is an annual, self-pollinating diploid crop (2n = 2x = 16) cultivated primarily during the cool season in the Indian subcontinent, the Middle East, eastern Africa, and the Mediterranean regions. South and Southeast Asia account for the majority of the global chickpea-growing area. In addition to being a vital source of plant-based protein for millions in developing countries, chickpea is recognized as a healthy and nutritious food in developed nations. As a legume, it contributes to improving soil fertility, thereby promoting sustainable cropping systems. Currently, chickpea is grown on 14.8 million hectares globally, with an annual production of 18.1 million tonnes (FAOSTAT 2023). However, its potential yield remains unmet due to various biotic and abiotic stresses, negatively affecting total production. To ensure food security for millions of resource-poor farmers in developing regions, especially in the semi-arid tropics, increasing chickpea production and productivity remains a key focus for the research community. Among pulse crops, chickpea ranks second in global acreage and third in total production. Over the past six decades, the global chickpea cultivation area has expanded

S. Tripathi (✉) · G. P. Dixit
ICAR-Indian Institute of Pulses Research, Kanpur, India

G. Prasad Dixit et al. (eds.), *Fundamentals of Legume Breeding*, https://doi.org/10.1007/978-981-96-7838-9_3

from 11.9 million hectares to 14.15 million hectares, while production has risen from 7.3 million tonnes to 16.36 million tonnes (FAOSTAT 2023). Yield improvements, from 610 kg/ha (1961–1963) to 1153.2 kg/ha (2020–2022), have been achieved through the adoption of improved varieties and advanced crop production practices (Fig. 1).

Pulses production in India has witnessed a significant increase over the last decade, reaching 27.3 million tonnes (mt) in 2021–2022. From 2010–2011 to 2021–2022, pulses production grew by 50%, rising from 18.2 mt to 27.3 mt. Notably, chickpea production reached a record high of 13.5 mt in 2021–2022 (Fig. 2).

As per the recent Working Group report on Crop Husbandry, Agriculture Inputs, Demand, and Supply (NITI Aayog 2024, the demand for pulses in India is projected to reach 35–38 mt by 2030–2031 and 49–57 mt by 2047–2048. Chickpea has played a pivotal role in driving the "Pulse Revolution," contributing significantly to making India nearly self-sufficient in pulses.

Research efforts in chickpea have primarily focused on improving resistance and tolerance to various biotic and abiotic stresses that hinder its productivity. Significant progress has been achieved in developing early-maturing varieties that can endure end-of-season stresses, such as drought and heat, while exhibiting resistance to common diseases like Fusarium wilt and Ascochyta blight. However, breeding for resistance or tolerance to other biotic and abiotic stresses remains limited due to a lack of suitable donor germplasm. The low genetic diversity in cultivated chickpea germplasm necessitates intensified efforts to explore novel genes in wild relatives. To date, only a few annual wild relatives, including *Cicer reticulatum*, *Cicer echinospermum*, and *Cicer pinnatifidum*, have been successfully utilized in interspecific crosses. The presence of post-fertilization incompatibility barriers (Ahmad et al. 1988; Ahmad and Slinkard 2004) has restricted the use of other wild *Cicer* species from the secondary and tertiary gene pools.

Desi-kabuli introgression is another widely employed strategy for introducing desirable traits. Desi chickpea varieties have contributed genes for Fusarium wilt resistance and drought tolerance, while kabuli types have provided genes for larger seed size and resistance to Ascochyta blight. These targeted breeding approaches have played a crucial role in addressing key challenges in chickpea improvement. Genotype-independent regeneration and transformation protocols have been successfully standardized in chickpea, enabling the development of transgenics with enhanced tolerance to drought and resistance to pod borer. Molecular markers have been extensively utilized for association studies and QTL mapping, leading to the identification of numerous marker-trait associations and quantitative trait loci (QTLs) linked to drought tolerance, Fusarium wilt resistance, and Ascochyta blight resistance in chickpea. The growing demand for chickpea in the preparation of functional foods, nutraceuticals, dietary supplements, and cosmetics has emphasized the need to develop varieties with improved nutritional quality and specific end-use applications.

2 Advances in Chickpea Breeding and Genetics

2.1 Early Maturity

Phenology plays a critical role in the adaptation of crops to diverse environments (Gaur et al. 2008). Chickpea is a photo-thermo-sensitive crop with a maturity duration ranging from

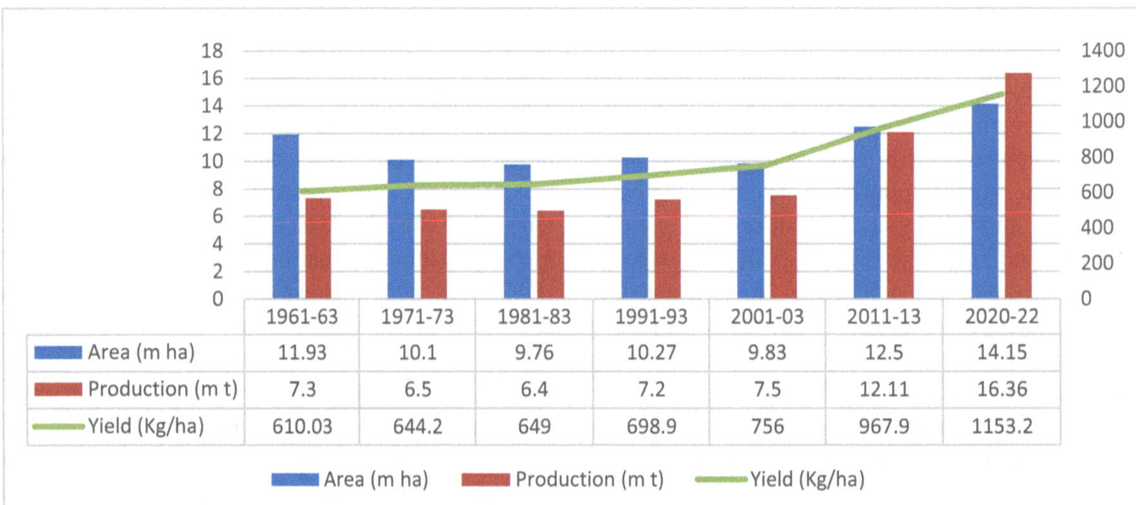

	1961-63	1971-73	1981-83	1991-93	2001-03	2011-13	2020-22
Area (m ha)	11.93	10.1	9.76	10.27	9.83	12.5	14.15
Production (m t)	7.3	6.5	6.4	7.2	7.5	12.11	16.36
Yield (Kg/ha)	610.03	644.2	649	698.9	756	967.9	1153.2

Fig. 1 Trends in global area, production and yield of chickpea (1961–2022)

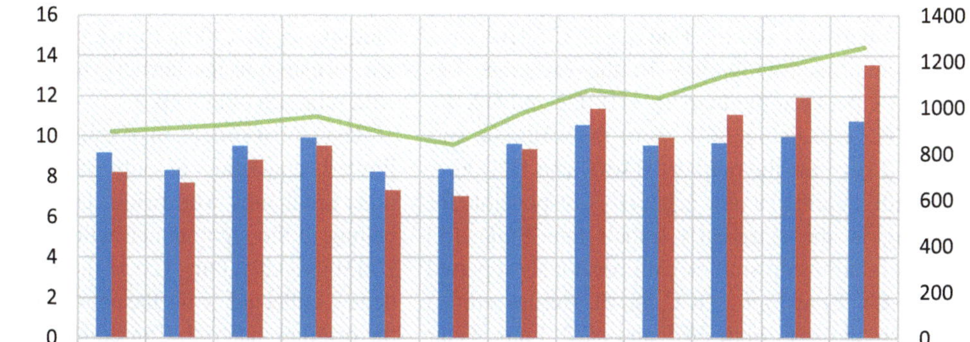

	2010-11	2011-12	2012-13	2013-14	2014-15	2015-16	2016-17	2017-18	2018-19	2019-20	2020-21	2021-22
Area (m ha)	9.19	8.32	9.51	9.93	8.25	8.39	9.63	10.56	9.55	9.68	9.99	10.74
Production (m t)	8.22	7.7	8.83	9.53	7.33	7.06	9.38	11.37	9.94	11.08	11.91	13.54
Yield (kg/ha)	895	912	929	960	889	840	974	1077	1041	1142	1192	1261

Fig. 2 Trends in area, production and yield of chickpea in India (2010–22)

Fig. 3 Chickpea cultivation in rice fallow

80 to 180 days, depending on the genotype and cultivation conditions (Ellis et al. 1994). Early-maturing chickpea varieties are particularly important in South and Southeast Asia, where the crop is predominantly grown under rainfed conditions during the post-rainy season but often encounters terminal drought during seed development and pod filling stages.

In the Mediterranean environments (e.g., Australia), early-maturing genotypes are vital for autumn-sown rainfed crops as they can escape terminal drought. Similarly, in temperate regions (e.g., Canada), they are crucial for summer-sown chickpea to avoid frost at the end of the season. Additionally, early-maturing varieties are essential for the horizontal expansion of chickpea cultivation in rice fallow regions of northeastern India (Fig. 3), where approximately 14 million hectares of land in South Asia remain fallow during the winter season after rice harvest (Subbarao et al. 2001). The development of short-duration cultivars played a significant role in expanding chickpea cultivation to central and southern India during the 1960s, regions where it was previously restricted to the northern areas. These efforts have enabled better utilization of available agricultural land and improved productivity under challenging environmental conditions.

Gumber and Sarvjeet (1996) performed inheritance analysis and found that early flowering in the cultivar ICCV 2 was controlled by two duplicate genes, operating in a homozygous recessive condition. Kumar and van Rheenen (2000) identified a major gene, efl-1, responsible for flowering time in ICCV 2 when crossed with the medium-duration genotype JG 62. The RILs from this cross exhibited a bimodal distribution with nearly equal peaks, indicating the segregation of other minor genes in the population. ICC 5810 (Harigantras), a black-seeded landrace from central India, is another source of early flowering, reported to be photoperiod-insensitive (Roberts et al. 1985). This variety flowered about 2 months earlier than the local cultivar 'Hadas' (Or et al. 1999). The flowering time difference between ICC 5810 and Hadas was attributed to a major gene named ppd. (photoperiod-dependent) (Or et al. 1999). Further experiments indicated that the efl-1 gene from ICCV 2, responsible for earliness, was not allelic to the ppd. gene in ICC 5810, a conclusion that was later supported by Hegde (2010). Similarly, Anbessa et al. (2006) reported two major genes with duplicate recessive epistasis controlling the days to flowering.

Significant progress has been made in mapping major genes associated with flowering time. Using the RIL mapping population derived from the cross ICCV 2 × JG 62, Cho et al. (2002) identified a major QTL for flowering time located between the simple sequence repeat (SSR) markers

TA 57 and TA 127 on linkage group 3 (LG 3). Similarly, Lichtenzveig et al. (2006) identified two QTLs for flowering time from RILs derived from the Hadas × ICC 5810 cross. One QTL, accounting for 53–56% of the variation, was mapped on LG 1 between the markers H1F022 and GA 40. The second QTL, explaining 18–22% of the variation, was located on LG 2 between the markers H4B09 and H1B06.

Hegde (2010) identified a new gene, *efl-3*, responsible for early flowering in the extra-early kabuli line BGD 132 through allelic relationship studies with previously reported genes *efl-1* (ICCV 2) and *efl-2* (ICC 5810). He determined that *efl-3* is non-allelic to both *efl-1* and *efl-2*. Later, Gaur et al. (2015a) discovered another early flowering gene, *efl-4*, in ICC 16641/ICC 16644, which was also found to be non-allelic to any of the previously reported early flowering genes. To date, four major early flowering genes have been reported in chickpea: *efl-1* (ICCV 96029, ICCV 2), *efl-2* (ICC 5810), *efl-3* (BGD 132), and *efl-4* (ICC 16641, ICC 16644). These discoveries hold significant value for developing effective breeding strategies to enhance chickpea adaptation to short-season environments. Researchers can now select specific early flowering genes based on their genetic background and linkage relationships with other traits (Gaur et al. 2015b).

Chickpea breeding programs in countries such as India, Myanmar, Bangladesh, Ethiopia, Australia, and Canada prioritize the development of early maturing varieties. Through collaborative efforts between national and international organizations, many early maturing, high-yielding, and Fusarium wilt-resistant cultivars have been successfully developed. ICRISAT has contributed significantly by developing two super-early, cold-tolerant lines, ICCV 96029 and ICCV 96030, which mature within 75–80 days in South India. These lines are particularly suitable for cultivation in North India, primarily for vegetable purposes, including green leaves and pods. Ongoing efforts aim to combine early maturity with resistance to Ascochyta blight and tolerance to chilling stress. These advancements are focused on creating chickpea cultivars adapted to the short growing seasons typical of Mediterranean and temperate regions, such as Australia and Canada.

2.2 Pre-Breeding for Introgression of Rare Alleles

The genus *Cicer* consists of 43 species, of which eight are annual. These wild species serve as a valuable reservoir of genes that can be utilized in wide hybridization programs by chickpea breeders. The limited genetic variability within cultivated chickpea poses a significant challenge to the development of varieties tolerant to biotic and abiotic stresses (Ladizinsky and Adler 1975; Kazan and Muehlbauer 1991; Ahmad and Slinkard 1992; Tayyar and Waines 1996; Siddique et al. 2000). To overcome this limitation, breeders are focusing on the introgression of beneficial alleles from wild relatives of chickpea.

The global collection of annual wild *Cicer* species is quite limited, consisting of only 593 entries maintained across nine genebanks worldwide. Of these, just 285 accessions belong to the eight annual wild *Cicer* species (Berger et al. 2003). Notably, only 120 of these accessions were independently collected, with the remainder being selections from the original material. Consequently, only a small fraction of the genetic diversity present in wild populations has been captured in ex situ collections. Additionally, the challenges of cultivating perennial wild *Cicer* species outside their natural habitats have further restricted their evaluation and utilization in chickpea improvement programs.

Among the annual wild relatives, only *C. reticulatum* and *C. echinospermum* are fully cross-compatible with *C. arietinum*, producing viable hybrids. While there are some reports of successful crosses involving *C. bijugum*, *C. pinnatifidum*, and *C. judaicum* with the cultivated species, these have not yet been utilized for developing pre-breeding lines. Ahmad and Slinkard (2003) demonstrated that no pre-fertilization barriers exist to prevent hybridization between *C. reticulatum*, *C. echinospermum*, and other wild *Cicer* species. These cross-compatible species could serve as bridge species, facilitating the introgression of desirable traits from wild species that are not directly cross-compatible with *C. arietinum*.

The use of wild *Cicer* species in breeding programs has largely been limited to developing resistance against cyst nematode (*Heterodera ciceri*) (Di Vito et al. 1996; Singh and Ocampo 1997; Erskine et al. 2001; Malhotra et al. 2002). However, selections derived from interspecific hybrids of *C. arietinum* × *C. echinospermum* have been developed, exhibiting resistance to root lesion nematode, enhanced resistance to *Phytophthora*, and moderate resistance to Ascochyta blight (Knights et al. 2002). Additionally, crosses involving *C. arietinum*, *C. reticulatum*, and *C. echinospermum* have shown higher heterotic potential for various traits, including yield (Jaiswal et al. 1987; Singh and Ocampo 1993, 1997). Recent advancements include the development of tall, erect, and machine-harvestable genotypes, such as Kundan (IPC 2015–132), from interspecific crosses (IIPR 2024).

Embryo rescue can be employed to overcome species barriers in *Cicer* hybridization (van Rheenen 1991; Muehlbauer et al. 1994; Singh and Ocampo 1997). Another approach to facilitate interspecific hybridization between distant *Cicer* species is protoplast fusion. However, the application of this technique is limited by the lack of established protocols for regenerating whole plants from single cells or protoplasts.

3 Efforts in Breeding for Abiotic Stresses

3.1 Drought Stress

Chickpea is frequently subjected to terminal drought in the semi-arid tropics (SAT) and the West Asia and North Africa (WANA) region. In these areas, chickpea is typically sown during the post-rainy season and cultivated as a rainfed crop, making it vulnerable to terminal drought during the pod-filling stage. The primary goal of chickpea improvement programs in these regions is to maximize yield under water-limited rainfed conditions. Currently, the development of early-maturing varieties that can escape drought is the most effective strategy. However, this approach often comes with a yield penalty due to a shorter crop duration (Saxena 1987). Therefore, it is essential to develop early-maturing chickpea lines with enhanced yield potential.

There are several strategies to mitigate the impact of drought stress in chickpea cultivation. One approach is to exploit physiological mechanisms of drought tolerance, which include traits such as deeper root systems, reduced leaf area, proline accumulation in the cytosol, and osmotic adjustment. For rainfed post-rainy season crops like chickpea, the ability to efficiently utilize water during the pod-filling stage under receding soil moisture is crucial for overcoming drought stress. Root traits, such as increased root length and vigor, play a vital role in enabling plants to extract water from deeper soil layers. For instance, ICC 4958, a drought-tolerant line, has demonstrated superior performance under drought conditions, attributed to its high root biomass (Saxena et al. 1993; Krishnamurthy et al. 1996; Ali et al. 2005). Further studies on root growth parameters have shown a positive correlation between root biomass and yield under drought stress (Kashiwagi et al. 2006). Additionally, efforts have been made to develop drought-tolerant, Fusarium wilt-resistant, and high-yielding lines through three-way crosses (Saxena 2003).

ICRISAT has developed a modified monolith method for large-scale screening of root traits in chickpea (Serraj et al. 2004). To address the challenges of estimating root profiles in shallow pot-grown plants, a PVC cylinder culture system was designed at ICRISAT (Kashiwagi et al. 2006). Significant genetic variability in root traits has been observed within the mini-core collection of chickpea germplasm (Kashiwagi et al. 2005). In addition to the drought-tolerant accession ICC 4958, another kabuli line, ICC 8261, is also noted for its prolific and deeper root system. The challenges of simultaneously selecting for root traits and yield led to the development of molecular markers for root traits in chickpea (Saxena 2003). The identification of molecular markers closely linked to QTLs regulating root traits has facilitated the introgression of these QTLs into adapted cultivars. A significant genomic region on linkage group 4 (LG 4), containing multiple QTLs for drought tolerance components, has been identified, explaining up to 58.2% of the phenotypic variation. This "QTL hotspot" has been successfully introgressed into elite chickpea cultivars using marker-assisted backcross breeding, leading to the development and release of several drought-tolerant varieties in India. Additionally, chickpea introgression lines with superior haplotypes from the ABCC3-type ABC transporter gene (*CaABCC3*) of the CaqSW2.4 QTL, which enhances seed weight and yield under drought stress, have also been reported (Basu et al. 2019). A list of drought-tolerant chickpea varieties recently developed and released through the All India Coordinated Research Project on Rabi Pulses is provided in Table 3.

Another key trait contributing to drought avoidance in chickpea is the reduction in leaf area, either through leaf shedding or smaller leaf morphology, which ultimately minimizes water loss. Saxena (2003) reported two chickpea accessions, ICC 5680 (with fewer leaflets) and ICC 10448 (with narrower leaflets), exhibiting this trait. In ICC 5680, transpirational water loss was reduced by 30% compared to ICC 4958 under controlled environmental conditions. To mitigate drought stress, breeding lines combining the large root traits of ICC 4958 with the fewer leaflets trait of ICC 5680 need to be developed.

The genetic transformation approach offers an alternative method for understanding the genetic control of drought resistance (Sharma and Ortiz 2000). Researchers have developed chickpea transgenics by introducing the dehydration response element DREB 1A from *Arabidopsis thaliana* using the stress-inducible promoter rd29A. DREB 1A is a transcription factor that regulates a cascade of stress-responsive genes, including those involved in drought, salinity, and cold tolerance. Additionally, chickpea has been transformed with the P5CSF-129A gene, which enhances tolerance to osmotic stress by increasing proline accumula-

Table 3 Chickpea varieties in India for drought tolerance developed through MAS

Chickpea varieties	Gene/QTL introgressed	Institute
BGM 10216 (2020)	QTL hotspot from ICC 4958	ICAR-IARI, N Delhi
Pusa Chickpea 4005 (2021)	QTL hotspot from ICC 4958	ICAR-IARI, N Delhi
IPC L4-14 (2021)	QTL hotspot from ICC 4958	ICAR-IIPR, N Delhi
Pusa JG 16 (2022)	QTL hotspot from ICC 4958	ICAR-IARI, N Delhi
NC 7 (2023)	ABC transporter gene (CaABCC3)	NIPGR, New Delhi
NC 9 (2024)	*bHLH* trans. factor gene *(CabHLH10)*	NIPGR, N Delhi

tion in the plant. Another gene from the same pathway has also been overexpressed. Some selected transgenic plants carrying the P5CSF gene have shown up to a fivefold increase in proline content.

3.2 Low Temperature Stress

Singh et al. (1990) highlighted several advantages of winter-sown chickpea compared to those planted in the spring. Winter-sown chickpea exhibits a higher germination rate, reduced incidence of Fusarium wilt (FW), improved nitrogen fixation, more efficient moisture utilization, increased crop height suitable for mechanized harvesting, enhanced protein content, and higher yield. However, in regions such as WANA and northern India, winter sowing faces challenges, particularly chilling or frost injury during the seedling and flowering stages. Most chickpea cultivars are sensitive to chilling temperatures during the flower induction stage (Croser et al. 2003). Freezing temperatures reduce growth vigor and vegetative biomass, while chilling temperatures during flowering lead to flower and pod abortion.

For winter-sown chickpea cultivars in the WANA region, cold tolerance at both the seedling and flowering stages is crucial. Research has identified several cold-tolerant lines from both cultivated (Singh et al. 1995) and wild (Robertson et al. 1995) gene pools at ICARDA. Notably, the line ILC 8262, the mutant ILC 8617, and the breeding line FLIP 87-82C have consistently exhibited strong cold tolerance (rated 3 on a 1–9 scale) across multiple years and locations (Singh et al. 1990, 1995). Collaborative efforts between ICARDA and national agricultural research systems (NARS) in the WANA region have resulted in the development of cold-tolerant cultivars resistant to ascochyta blight. These advancements have enabled the successful winter cultivation of kabuli chickpea in WANA, offering a 50–100% yield advantage compared to spring-sown chickpea.

In regions such as the Indian subcontinent and Australia, chickpea often encounters chilling temperatures during the early reproductive stage. To address this, ICRISAT developed several breeding lines (ICCV 88502, 88503, 88506, 88510, 88516) capable of setting pods under low-temperature conditions (ICRISAT 1994; Srinivasan et al. 1998). In southern Australia, average daily temperatures drop below 15 °C during winter and early spring, leading to pollen sterility and bud drop (Clarke and Siddique 2004). Research at the University of Western Australia identified two critical stages in chickpea development that are particularly sensitive to low temperatures: pollen formation, resulting in sterile pollen, and pollen tube development (Lawlor et al. 1998). At low temperatures, slow pollen tube growth hinders timely fertilization, leading to flower abortion.

The rate of pollen tube growth under low temperatures is strongly correlated with a plant's cold tolerance, making it a valuable trait for selecting cold-tolerant lines. In western Australia, the chickpea breeding program successfully developed two desi cultivars, WACPE 2075 (Sonali) and WACPE 2095 (Rupali), by combining early maturity, moderate resistance to Ascochyta blight (AB), and chilling tolerance. Using pollen selection methods, cold tolerance from ICCV 88516 (CTS 60543) was introduced into chill-sensitive Australian cultivars, such as Amethyst and Tyson (Clarke et al. 2004). Furthermore, researchers identified restriction fragment length polymorphism (RFLP) markers linked to chilling tolerance, which were later converted into sequence characterized amplified region (SCAR) markers. These markers were used to select chill-tolerant progeny from a cross between Amethyst and ICCV 88516 (Millan et al. 2006).

3.3 Salinity Stress

Soil salinity poses a major challenge to crop productivity, affecting nearly 80 million hectares of the world's arable land (Flowers et al. 2010). Globally, salinity impacts 20% (45 million hectares) of irrigated land and 2% (32 million hectares) of dryland. Regions such as India, Pakistan, Iran, WANA, and Australia experience significant reductions in chickpea production due to saline soils. In South Asia, salinity affects approximately 36 million hectares, while 36.8 million hectares in Southeast Asia and 30 million hectares in Central Asia are similarly impacted (Abrol et al. 1988).

Soil salinity hampers germination, resulting in poor plant establishment. In saline environments, desi chickpea varieties display high anthocyanin content in their foliage, while kabuli types exhibit yellowing leaves, along with a reduction in biomass, seed size, and grain yield. Earlier studies suggested limited genotypic variation in chickpea for salt tolerance (Saxena 1984). However, subsequent research has identified genotypes with low-to-moderate salt tolerance (Singh and Singh 1984; Ashraf and Waheed 1992; Kathiria et al. 1997; Maliro et al. 2004). In India, the desi chickpea variety Karnal Chana 1 (CSG 8963) has been released for cultivation under saline conditions. Additionally, in Australia, the ICRISAT breeding line ICCV 96836 (released as Genesis 836) has demonstrated high salt tolerance in pot trials (Maliro et al. 2004).

Screening of over 250 diverse genotypes at ICRISAT, including mini-core collections, breeding lines, wild relatives, and previously identified salinity-tolerant lines, revealed more than sixfold variation in seed yield per plot under saline conditions. Some genotypes showed a 20% higher yield compared to the best-performing check, CSG 8962 (Vadez et al. 2006). Seed yield under salinity was strongly correlated with seed yield under control conditions, indicating that the yield under saline conditions was partly determined by the genotype's inherent yield potential and a salinity tolerance component. There was no significant relationship between biomass at the late vegetative stage and seed yield at maturity. The seed yield per plot correlated strongly with the number of pods per plant under salinity but had a weak correlation with seed size, suggesting that salinity impacts pod setting and/or grain formation rather than seed development and filling. Moreover, maturity showed an inverse parabolic relationship with salinity tolerance, with extra-early genotypes displaying lower seed yield under salinity. Given the lack of correlation between biomass at 50 DAS and seed yield at maturity, screening for salinity tolerance in chickpea is more effective at the maturity stage.

A total of 188 RILs derived from the cross ICCV 2 × JG 11 were evaluated for yield and related traits under two conditions: a control soil with 0 mM NaCl and a salinity-treated soil with 80 mM NaCl, over a span of 2 years. The QTL analysis identified two genomic regions on CaLG05 (28.6 cM) and CaLG07 (19.4 cM), each harboring QTLs associated with various salinity tolerance traits. Two major QTLs for yield under salinity, explaining 12% and 17% of the phenotypic variation, were located within these regions. Pushpavalli et al. (2015) identified multiple QTLs for salinity tolerance on linkage groups (LGs) 3 and 6, including a major QTL on LG 6 that accounted for 37% of the variation in seed number. Similarly, Atieno et al. (2021) identified six QTLs on CaLG04, CaLG05, and CaLG06 associated with salt tolerance in an RIL population developed from the cross between Rupali (salt-sensitive) and Genesis836 (salt-tolerant). A total of 21 QTLs were mapped to two distinct genomic regions on CaLG04. One of these regions controlled traits such as the number of filled pods, leaf necrosis, seed number, and seed yield under saline conditions, co-localizing with four QTLs linked to salinity tolerance. The other region governed seed weight and growth-related traits independently of salinity.

4 Breeding for Resistance to Biotic Stresses

4.1 Resistance to Diseases

4.1.1 Fusarium Wilt

Fusarium wilt, caused by *Fusarium oxysporum* f. sp. *ciceri*, is a major disease affecting chickpea in regions between 30°N and 30°S, impacting 33 countries across Asia, Africa, and South America (Nene et al. 1996). Yield losses in susceptible genotypes can exceed 90%. Eight distinct pathogen races (0, 1A, 1B/C, 2, 3, 4, 5, and 6) have been identified, each with its own geographical distribution. Races 2, 3, and 4 are prevalent in India (Haware and Nene 1982), while races 0, 1B/C, 5, and 6 are commonly found in the Mediterranean region and the USA (Jimenez-Diaz et al. 1993; Halila and Strange 1996). Race 1, originally identified in India, was later designated as 1A and is also found in the USA and Mediterranean region (Jimenez-Diaz et al. 1993). Races 0 and 1B/C cause the yellowing syndrome (yellowing pathotype), while the other races result in the typical vascular wilt (wilting pathotype) (Trapero-Casas and Jimenez-Diaz 1985; Jimenez-Diaz et al. 1993).

Effective field, greenhouse, and laboratory techniques for resistance screening against Fusarium wilt (FW) have been developed (Nene et al. 1981) (Fig. 4). At ICRISAT, over 13,500 chickpea accessions were tested against race 1 of FW, resulting in the identification of 160 resistant accessions, mostly of the desi type (Haware et al. 1992). Desi × Kabuli crosses have been utilized to incorporate FW resistance into Kabuli cultivars, but many of the resulting varieties exhibit an undesirable brown tinge in seed color instead of the preferred cream or white seed coat color. Screening of 12 large-seeded Kabuli genotypes for FW resistance led to the identification of two resistant accessions from Mexico (ICC 14194 and ICC 17109), which are now being used in breeding extra-large Kabuli varieties with both FW resistance and Kabuli-type seed color (Gaur et al. 2006). Some germplasm lines exhibit resistance to multiple FW races. For example,

Fig. 4 Wilt sick plot screening at ICRISAT and vascular discoloration of xylem in vertical section

WR 315 is resistant to all races except race 3, while JG 74 is resistant to races 0, 1A, 3, 4, and 6 (Haware 1998).

Significant progress has been made in understanding the genetics of Fusarium wilt (FW) resistance and mapping FW resistance genes (Ahmad et al. 2005; Millan et al. 2006; Sharma and Muehlbauer 2007). Resistance to six FW races (0, 1A, 2, 3, 4, and 5) is controlled by 1 to 3 genes per race, while the genetics of races 1B/C and 6 remain unexplored. The genes responsible for wilt resistance delay the onset of wilting symptoms, a phenomenon known as "slow wilting." Early studies suggest a monogenic recessive inheritance of resistance to race 1A (Kumar and Haware 1982; Sindhu et al. 1983). For "late wilting" in race 1 susceptible genotypes, three independent genes—h1, h2, and H3—were identified, with resistance requiring combinations of any two (Upadhyaya et al. 1983; Singh et al. 1987). Molecular markers have been identified for at least one resistance gene for each of these six races. The first wilt resistance gene to be tagged in chickpea was H1 (foc-1) (Mayer et al. 1997). It has been found that wilt resistance genes are clustered into two groups on different linkage groups (Sharma and Muehlbauer 2007). A cluster of five wilt resistance genes (foc-1, foc-2, foc-3, foc-4, and foc-5) spanned 8.2 cM on linkage group 2 (LG 2) of chickpea (Winter et al. 2000). Pratap et al. (2017) developed highly resistant lines with the foc-2 gene in the background of Pusa 256 using marker-assisted backcrossing (MABC).

4.1.2 Ascochyta Blight

Ascochyta blight, caused by *Ascochyta rabiei* (Pass.) Labr., is a devastating foliar disease that impacts chickpea crops across West and Central Asia, North Africa, North America, and Australia. In the Indian subcontinent, it is particularly prevalent in northwestern India and Pakistan, where it thrives under cool, cloudy, and humid conditions. The pathogen shows considerable genetic variability, driven by random mating between pathotypes, which often undermines the durability of resistance in newly developed cultivars.

Significant efforts have been undertaken to identify chickpea genotypes resistant to Ascochyta blight. At ICARDA, over 13,000 germplasm accessions were screened, leading to the identification of 11 resistant *kabuli* genotypes (ILC 72, ILC 196, ILC 201, ILC 202, ILC 2506, ILC 2956, ILC 3274, ILC 3279, ILC 3346, ILC 3956, and ILC 4421) and 6 resistant *desi* genotypes (ICC 3634, ICC 4200, ICC 4248, ICC 4368, ICC 5124, and ICC 6981) (Reddy and Singh 1984). Further studies by Singh and Reddy (1993) identified 3 *desi* accessions (ICC 4475, ICC 6328, and ICC 12004) and 2 *kabuli* accessions (ILC 200 and ILC 6482) as resistant to six pathogen races after extensive screening from 1979 to 1991. Some resistant lines, such as ILC 72, ILC 195, ILC 482, and ILC 3279, have been directly released as commercial varieties in various countries.

Despite these efforts, the lack of highly resistant sources in both cultivated and wild chickpea species, combined with the pathogenic variability of the disease, has impeded the development of varieties with durable resistance to Ascochyta blight (Pande et al. 2005). To address this challenge, ICRISAT has focused on enhancing resistance in *desi* chickpea by incorporating genes from multiple sources. Genotypes such as ICCV 04512, ICCV 04514, and ICCV 04516 have demonstrated a high level of resistance, with a score of 2 on a 1–9 rating scale (where 1 = resistant and 9 = susceptible) against multiple Ascochyta blight isolates (Gaur and Gowda 2005).

Screening for Ascochyta blight resistance is carried out under artificial epiphytotic conditions at institutions such as Punjab Agricultural University (PAU), Ludhiana (Fig. 5). Prominent AB-resistant varieties released across various countries include Pusa 261, PBG 1, GNG 469, and Gaurav in India; Dasht, NIFA 88, CM 72, CM 88, CM 98, and CM 2000 in Pakistan; Dwelley, Sanford, Myles, Evans, and Sierra in the USA; CDC Frontier, CDC Anna, CDC Cabri, CDC Desiray, CDC Nika, and Amit in Canada; and Sonali, Rupali, Genesis 508, Genesis 090, Genesis 836, Yorker, Flipper, Nafice, and Almez in Australia.

Millan et al. (2006) reviewed advancements in the identification of markers and QTLs associated with Ascochyta blight resistance. At ICRISAT, a genetic linkage map comprising 84 markers (82 SSRs and 2 ESTs) was developed using an F2 population derived from an intra-specific cross between ICCV 04516 (AB-resistant) and Pb 7 (AB-susceptible). Three QTLs for AB resistance were identified: one on LG3 and two (QTL 2 and QTL 3) on LG4. The QTLs located on LG4 were further validated in another population (ICRISAT 2005).

4.1.3 Botrytis Gray Mould

Botrytis gray mold (BGM), caused by *Botrytis cinerea* Pres., is a major foliar disease affecting chickpea (Haware and McDonald 1992; Corbin 1975). This necrotrophic fungus exhibits a broad host range, significant variability, and wide adaptability. Screening efforts involving over 12,000 germplasm accessions and breeding lines have identified no strong resistance to BGM; however, moderate resistance has been observed in some genotypes, such as ICC 14344, which was released in India as Avrodhi (IIPR 1997) (Pande et al. 2002). Resistance to BGM is controlled either by a single dominant gene (Rewal and Grewal 1989) or through the complementary action of dominant genes (Rahul et al. 1995).

Erect chickpea accessions such as ICCL 87322 and ICCV 88510 exhibited reduced disease impact, likely due to improved air circulation that limits humidity and the spread of the disease. Wild chickpea species have been observed to possess higher levels of resistance. At ICRISAT, 36 germplasm accessions from seven annual wild *Cicer* species were

Fig. 5 Screening for Ascochyta blight resistance at PAU, Ludhiana

screened for their response to BGM under controlled growth room conditions. Among these, three accessions of *C. bijugum* (ICCW 41, 42, and 91) demonstrated promising resistance (Haware and McDonald 1992). However, this resistance has not yet been incorporated into cultivated chickpea varieties.

4.1.4 Root Rot Diseases

Three types of root rot diseases—dry root rot (DRR), Phytophthora root rot, and collar rot—are known to affect chickpea. Dry root rot, caused by *Rhizoctonia bataticola*, is particularly significant in dry regions where chickpea crops are often exposed to temperatures exceeding 30 °C. Limited efforts have been made to identify races of the DRR pathogen. Pundir et al. (1988) identified 47 tolerant accessions from ICRISAT's chickpea germplasm collection. Progress in breeding for root disease resistance has been achieved alongside efforts to improve resistance to Fusarium wilt (FW). At ICRISAT, breeding lines are routinely screened for FW and DRR simultaneously in a multiple root disease screening nursery. Although no immunity to DRR has been identified, several germplasm lines and cultivars exhibit low to moderate resistance, including SAKI 9516 (Dua et al. 2001) and breeding lines RSG 130, RSG 132, and RSG 191 (Chitale et al. 1990). To map genomic regions associated with DRR resistance, an RIL population derived from a cross between the susceptible line BG 212 and the moderately resistant line ICCV 08305 was screened against the Rb 6 isolate of *R. bataticola* using the paper towel method under controlled conditions (Karadi et al. 2021). Genotyping was performed using the SNP genotyping platform, Affymetrix® Axiom® CicerSNP array, which identified a minor QTL, *qDRR-8*, on CaLG08. This QTL explained 6.70% of the phenotypic variance (PVE) with a LOD score of 3.34, providing insights into DRR resistance.

Phytophthora root rot, caused by *Phytophthora medicaginis* Hansen, is a significant disease affecting chickpea crops in northern New South Wales and Queensland (Knights et al. 2003). Field resistance was first identified in a chickpea accession from Tajikistan (Brinsmead et al. 1985), which has been successfully incorporated into several well-adapted cultivars in northeastern Australia. This has resulted in a 56–112% yield increase compared to susceptible cultivars (Singh et al. 1994). Higher levels of resistance have also been identified in accessions of *Cicer echinospermum* (Knights et al. 2003), and efforts are underway to introgress this resistance into the genetic background of cultivated varieties.

Collar rot, caused by *Sclerotium rolfsii*, is commonly observed in regions with high soil moisture and warm temperatures (around 30 °C) during sowing. The disease is further aggravated by the presence of undecomposed organic matter near the soil surface and low soil pH. To date, no breeding programs specifically aimed at developing resistance to collar rot in chickpea have been reported.

4.1.5 Viral Diseases

Viral diseases in chickpea are sporadic but can cause significant yield losses in certain regions. Common symptoms include foliage discoloration, phloem browning, and stunted growth. Several viruses are associated with stunting, such as *Chickpea chlorotic dwarf monogeminivirus* (CCDV) and chickpea luteovirus (CpLV) in India and Pakistan; *Faba bean necrotic yellows nanovirus* (FBNYV) in Syria, Turkey, and Lebanon; and *Bean leaf roll luteovirus* (BLRV) in Algeria (Horn et al. 1995, 1996; Yahia et al. 1997). These viruses are transmitted by aphid vectors, primarily *Aphis craccivora* Koch and *Myzus persicae* Subzer (Kaiser and Danesh 1971).

4.2 Resistance to Insect Pest

4.2.1 Pod Borer

The pod borer (*Helicoverpa armigera* Hübner) is the most significant insect pest affecting chickpea worldwide. Its pest status is attributed to its biological traits, including high mobility, polyphagy, a high reproductive rate, diapause, and a wide host range (Fig. 6). Yield losses in chickpea due to pod borer are estimated to range from 18% to 24%, amounting to an annual economic loss of over $542 million (Ryan 1997).

Highly tolerant lines to pod borer are absent in cultivated chickpea germplasm. Screening of over 14,800 accessions identified ICC 506, which showed a mean pod damage of 8.6% compared to 29.9% in the susceptible variety Annigeri (Lateef and Pimbert 1990). Several other lines with low levels of resistance have also been reported (see Sharma et al. 2003 for a review). In early-maturing *desi* chickpea types, the predominance of additive genetic variance and high genetic variability make the pedigree selection method ideal for improving pod borer resistance. Conversely, medium- and long-duration *desi* and *kabuli* types exhibit both additive and dominance variance, indicating that selection should be delayed until the F5 generation (Gowda et al. 2005).

At ICRISAT, over 160 accessions of annual wild *Cicer* species were screened for resistance to pod borer. Slow larval growth was observed on 21 accessions (Sharma et al. 2005), with the majority of resistant accessions belonging to *C. bijugum*, a member of the tertiary gene pool of chickpea. These accessions are valuable sources of resistance to pod borer, provided crossability barriers are overcome using in vitro techniques. Preliminary screening of some perennial wild *Cicer* species identified *C. microphyllum* and *C. canariense* as promising sources of resistance, with pod borer ratings as low as 1.0. However, the presence of crossability barriers with cultivated chickpea limits their direct use in breeding for *Helicoverpa* resistance (Sharma et al. 2006).

The validation of markers for *Helicoverpa* resistance in chickpea has been limited due to the lack of sources with a high level of resistance. However, significant progress has been made in the transformation and regeneration of transgenic chickpea plants. Transgenics for pod borer resistance have been developed using *Bt* and proteinase inhibitor genes (Sharma and Ortiz 2000).

In 2004–2005, 20 transgenic chickpea events containing *cry1Ac* and *cry1Ab* genes were bioassayed for resistance to pod borer. These transgenic lines demonstrated a 35–40% reduction in the larval weight of *H. armigera* compared to control plants. The evaluation of these transgenic lines is ongoing in contained field trials (ICRISAT 2005).

4.2.2 Other Insect Pests

Leaf miner (*Liriomyza cicerina* Rondani) is a major insect pest affecting chickpea crops in WANA and southern Europe. Researchers have actively sought resistance sources in both cultivated and wild chickpea species. At ICARDA, a screening of 6800 *kabuli* chickpea germplasm lines under natural field infestation identified 31 tolerant lines. Among them, four lines—ILC 726, ILC 1776, ILC 3350, and ILC 5901—showed strong potential for resistance. Most tolerant lines exhibited smaller leaflets and seeds, while ILC 5901, the most resistant, was distinguished by its multi-pinnate leaves (Malhotra et al. 1996).

Some accessions of wild chickpea species have demonstrated high resistance to leaf miner, with scores of 2 on a 1–9 scale (where 1 = plants free from damage and 9 = highly susceptible). Wild accessions exhibiting strong resistance include *C. echinospermum* (ILWC 35), *C. bijugum* (ILWC 227), *C. judaicum* (ILWC 206, 207, 210, 211, and 223), *C. pinnatifidum* (ILWC 9, 97, 226, and 252), *C. cuneatum* (ILWC 40, 185, 187, and 232), and *C. chorassanicum* (ILWC 147) (Robertson et al. 1995).

The seed beetle, or bruchid (Callosobruchus species), is a significant storage pest of chickpea, with no known resistance in cultivated varieties. However, certain wild species have demonstrated immunity to bruchid damage. These include *C. reticulatum* (ILWC 104), *C. bijugum* (ILWC 65, 67, 68, 70, 73, 74, 75, 83, and 177), *C. echinospermum* (ILWC 39, 179, and 181), *C. judaicum* (ILWC 46, 54, 173, 174, 176, and 189), and *C. cuneatum* (ILWC 187). These wild species scored 1 on a 1–9 scale of bruchid susceptibility, where 1 indicates seeds free from damage and 9 represents high susceptibility (Robertson et al. 1995). Despite this, transferring bruchid resistance from wild to cultivated chickpea remains a significant challenge for breeders.

Fig. 6 Chickpea pod borer (*Helicoverpa armigera* Hubner)

Notably, transgenic chickpea carrying the *α-amylase inhibitor 1 (αAI 1)* gene has shown strong protection against two bruchid species, *C. maculatus* and *C. chinensis* (Higgins et al. 2004).

5 Efforts in Breeding for Resistance to Multiple Stresses

Cultivars with resistance to multiple stresses are essential for stabilizing chickpea production in regions where the crop is exposed to multiple challenges simultaneously. While some cultivars exhibit good resistance to individual stresses, those with combined resistance to multiple stresses are rare. To address this, breeders must focus on multiple or multi-parental crosses to incorporate resistance genes from diverse sources into a single genotype.

Pundir et al. (1988) documented phenotypic data from the evaluation of ICRISAT's chickpea germplasm, identifying several accessions with resistance to multiple diseases. For example, accession ICC 12237 demonstrated resistance to Fusarium wilt (FW), dry root rot (DRR), and black root rot, while ICC 11321 showed resistance to FW, Ascochyta blight (AB), and botrytis gray mold (BGM). To further explore multiple resistance in chickpea, over 200 accessions resistant to FW and root rot were evaluated at 24 locations across 11 countries over 5 years. Among these, three Indian accessions (ICC 10803, 11550, and 11551) and three Iranian accessions (ICC 2862, 9023, and 9032) exhibited resistance to FW and root rot in approximately two-thirds of the locations (Nene et al. 1989).

Wild relatives of chickpea serve as valuable sources of resistance to multiple stresses. For example, *C. bijugum* accessions ILWC 70 and ILWC 73 exhibit resistance to Ascochyta blight (score 3), Fusarium wilt (score 1), bruchid (score 1), and cyst nematode (score 2), along with tolerance to low-temperature stress (score 2–3) (Robertson et al. 1995). These wild species present significant potential for breeding cultivars with resistance to multiple stresses. However, developing such varieties through conventional breeding methods is time-intensive and challenging. Marker-assisted selection (MAS) offers a more efficient and effective approach for identifying, tracking, and combining stress-resistance genes in segregating generations.

6 Efforts in Breeding for Wide Adaptation

Most national breeding programs focus on developing varieties tailored to specific locations. However, breeding for widely adapted varieties is hindered by significant genotype × environment (G × E) interactions. Selecting genotypes with wide adaptation or low G × E interaction—those that demonstrate stable average yields across diverse locations and years—often results in the exclusion of lines that perform well in unfavorable environments.

Some widely adapted chickpea genotypes have been identified through international nurseries and successfully released in multiple countries. For instance, FLIP 84-92C and FLIP 84-79C were released in Algeria, Morocco, and Tunisia; ICCV 2 in India, Sudan, and Myanmar; and ICCV 88202 in Australia, Myanmar, and India. The genotype × environment (G × E) interactions of 46 chickpea genotypes—41 from India, 3 from Australia, and 2 from the Mediterranean basin—were evaluated across seven locations over 3 years (Berger et al. 2006). Overall, research indicates that varieties developed by state-based breeding programs tend to exhibit specific adaptation rather than wide adaptation.

7 Efforts in Breeding for Enhanced Nutritional Quality

Chickpea provides numerous health benefits and, when combined with other pulses and cereals, may help combat diseases such as cardiovascular disease (CVD), type 2 diabetes, digestive disorders, and certain cancers (Jukanti et al. 2012). The global demand for high-quality food legumes is increasing due to growing awareness of their health and nutritional value. As a critical source of protein for populations with cereal-dominated diets, chickpea contains 20–22% protein and is rich in fiber and essential minerals, including phosphorus (P), calcium (Ca), magnesium (Mg), iron (Fe), and zinc (Zn). Its lipid fraction is rich in unsaturated fatty acids, and it has high levels of carotenoids such as β-carotene, cryptoxanthin, lutein, and zeaxanthin (Abbo et al. 2005).

Zinc and iron content in chickpea germplasm range from 1.10 to 5.91 mg/100 g and 0.50 to 8.54 mg/100 g, respectively (Grewal et al. 2020). These levels are comparable to some of the best biofortified crops developed under the HarvestPlus program. Notably, cultivars such as Heera, H82-2, and H214 have been identified for high zinc content (>4 mg/100 g), while L550, KGD1168, PG114, JG74, and ICCV6 exhibit higher iron content (>6 mg/100 g). Chickpea lacks major anti-nutritional factors such as vicin in faba bean and trypsin inhibitors in soybean. However, a notable drawback of chickpea consumption is flatulence, caused by the high concentration of raffinose family oligosaccharides (RFOs). Since humans lack the enzyme needed to digest RFOs, these oligosaccharides pass undigested through the stomach and upper intestine. Despite this issue, limited efforts have been made to breed chickpea varieties with reduced RFO concentrations.

Research on genetic variability for quality traits in chickpea has shown significant potential for improvement. Protein content in both cultivated and wild species exhibits wide variability, ranging from 12.4% to 31.5%, offering opportunities to develop cultivars with higher protein levels. However, sulfur-containing amino acids, particularly methionine and cysteine, are limiting in chickpea. Ashur et al. (1973) found that chickpea meets the requirements for all essential amino acids except methionine and cysteine in humans. This has driven the use of gene transfer technology to enhance the levels of these amino acids. Transgenic chickpea lines carrying a seed-specific chimeric gene encoding sunflower seed albumin (SSA) have shown 24–94% higher methionine content but 10–15% lower cysteine levels compared to non-transgenic controls (Higgins et al. 2004).

A comprehensive assessment of genetic variability for nutritional quality traits and anti-nutritional factors in both cultivated and wild chickpea genepools is essential. Understanding the genetics and genotype × environment (G × E) interactions for these traits is critical for their effective utilization in breeding programs. Recently, interest in chickpea has been increasing, as it is being recognized as a functional food with significant potential health benefits.

8 Efforts in Breeding Large-Seeded Kabuli Chickpea

The market price and consumer preference for chickpea are largely influenced by grain appearance, including size, shape, and color. For kabuli chickpea, large seed size is a highly desirable quality trait, as it is primarily consumed as whole grain. Larger seeds fetch premium prices in the market, making this trait a breeding priority in India. Under the All India Coordinated Research Project (AICRP) on Chickpea, several kabuli varieties with seed sizes exceeding 30 g/100 seeds or 8–9 mm in diameter have been released, including JGK 1, JGK 2, JGK 3, KAK 2, LBeG 7, Vihar, and Virat. Additionally, extra-large seeded kabuli varieties with seed sizes over 45 g/100 seeds, such as Phule G 0517, PKV Kabuli 4, and MNK 1, have also been developed (Fig. 7).

9 Breeding for Labor-Saving Traits

Reducing cultivation costs and minimizing losses caused by weeds are key objectives in agriculture. With the increasing mechanization among Indian farmers to enhance efficiency and lower expenses, there is a rising demand for crop varieties compatible with combined harvesting. This necessitates a shift from traditional short and semi-spreading plant types to tall and erect ones. Several chickpea varieties suitable for machine harvesting, such as NBeG 47, JG 24, Phule Vikram, Pusa Parvati (BG 3062), RVG 204, and Kundan (IPC 2015-132), have been released for commercial cultivation in various zones across India.

Weeds pose a significant challenge, particularly in rabi pulses and rice-fallow ecologies, leading to yield losses of up to 70–80%. Manual weeding increases production costs and, given the current shortage of farm labor in India, is not a feasible solution (Fig. 8). Developing post-emergence herbicide (PEH) tolerant varieties of rabi pulses, including chickpea, is essential to address weed management challenges in the context of resource-efficient agriculture. Collaborative research projects involving ICAR-Indian Institute of Pulses Research (Kanpur), ICAR-Indian Agricultural Research Institute (New Delhi), the International Crops Research Institute for the Semi-Arid Tropics (ICRISAT), and other NARS partners have screened released varieties and germplasm accessions of chickpea across India, identifying genetic variation for PEH tolerance.

Fig. 7 Large and extra-large seeded kabuli chickpea

Fig. 8 Manual weeding in chickpea increases production cost

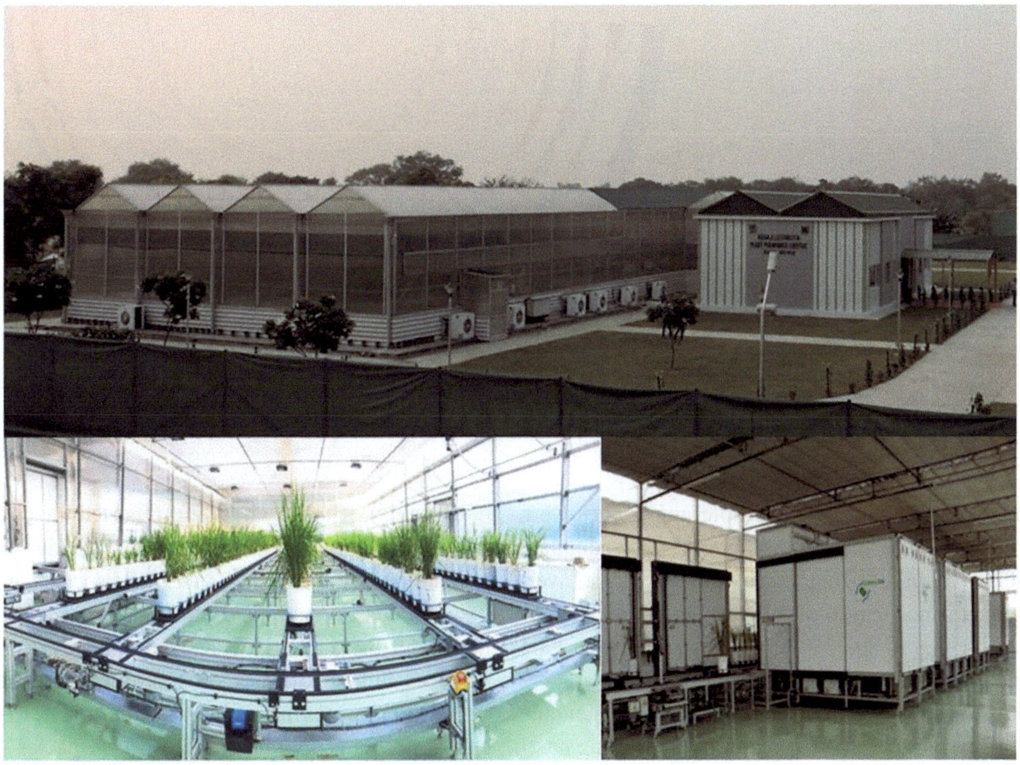

Fig. 9 Nanaji Deshmukh Plant Phenomics Centre at ICAR-IARI, New Delhi

10 Precision Phenotyping

Achieving higher yields in crops like chickpea requires the optimization of various physiological traits, including light, water, and nutrient use efficiency. Key factors contributing to high yields include root architecture, water and nutrient acquisition, and photosynthetic efficiency. Drought stress adds complexity to the process, making it challenging to phenotype traits associated with yield under such conditions. Translating drought-related traits screened in controlled environments to field conditions is crucial for their effective use in breeding programs. These traits can then be applied directly in selection or used as proxies to identify superior genotypes. Next-generation phenotyping techniques are set to play a vital role in overcoming the challenges associated with phenotyping physiological traits (Fig. 9).

11 Genome Sequence Information and Its Utilization

Chickpea faces significant challenges due to limited genetic variation and production bottlenecks, highlighting the need for its genome assembly. Varshney et al. (2013) published the first draft genome sequence of a Canadian kabuli chickpea variety, CDC Frontier (~738 Mb) (Fig. 10), utilizing Illumina HiSeq 2000 short reads and bacterial artificial chro-

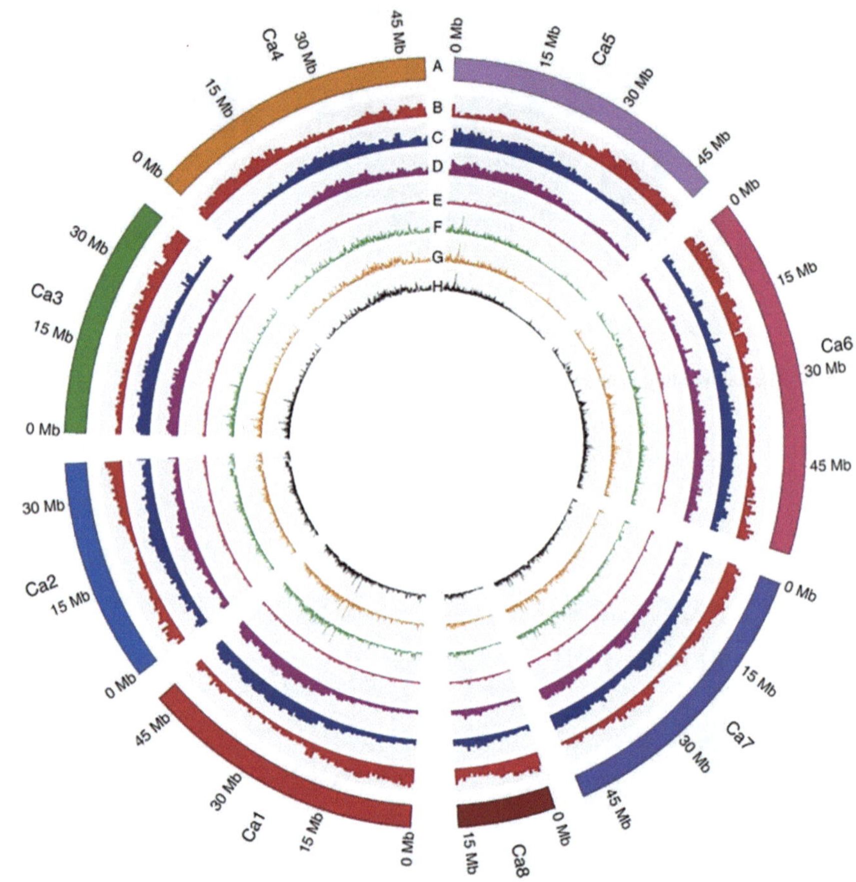

Fig. 10 Chickpea genome. (From Varshney et al. (2013) *Nature Biotechnology* 31, 240–246) Pseudomolecules (A), gene density (red, B); repeat density (blue, C); retrotransposon density (violet, D); transposon density (magenta, E); average pair-wise nucleotide diversity (theta pi) across 17 *desi* chickpea varieties (green, F), 12 *kabuli* chickpea varieties (orange, G), and 29 leading chickpea varieties (black, H)

mosome (BAC) end sequencing. Multiple approaches were applied for gene prediction, culminating in a non-redundant gene set (OGSv1.0) comprising 28,269 genes with an average transcript sequence length of 3055 bp. Functional annotations were assigned to 89.73% of the genes, leaving 2904 genes unannotated.

Similarly, Jain et al. (2013) generated the draft genome sequence of a desi chickpea genotype, ICC 4958, using next-generation sequencing platforms, BAC end sequences, and a genetic map. Their analysis predicted a genome length of 740 Mb, with 27,571 genes and 210 Mb of repeat elements. The study revealed that chickpea shares the largest number of gene families (864) with Medicago, compared to 118, 219, and 131 gene families shared with soybean, pigeonpea, and grapevine, respectively. This finding underscores the close evolutionary relationship between chickpea and Medicago.

Draft and reference genome assemblies serve as crucial benchmarks for genomic studies, providing valuable resources for identifying target genes associated with disease and insect pest resistance, as well as traits of agronomic importance. They also offer a foundation for designing markers used in advanced breeding experiments, such as QTL mapping and association mapping for key traits.

Recently, Khan et al. (2024) reported the development of the *Cicer* super-pangenome based on de novo genome assemblies of eight annual *Cicer* wild species. A graph-based genome was constructed to capture structural variations between cultivated and wild species, uncovering genetic differences linked to traits such as disease resistance. These genetic variations pave the way for transferring economically important traits from wild *Cicer* species to elite cultivated chickpea varieties through marker-assisted selection or gene editing.

References

Abbo S, Molina C, Jungmann R, Grusak MA, Berkovitch Z, Reifen R, Kahl G, Winter P, Reifen R (2005) Quantitative trait loci governing carotenoid concentration and weights in seeds of chickpea (*Cicerarietinum* L.). Theor Appl Genet 111:185–195

Abrol IP, Yadav JSP, Massoud FI (1988) Salt-affected soils and their management. FAO Soil Bulletin 39. Food and Agriculture Organization of the United Nations, Rome, Italy

Ahmad F, Slinkard AE (1992) Genetic relationships in the genus *Cicer* L. as revealed by polyacrylamide gel electrophoresis of seed storage proteins. Theor Appl Genet 84:668–692

Ahmad F, Slinkard AE (2003) Limitations to bridge-species facilitated alien gene transfers in chickpea: pre-fertilization events. J Genet Breed 57:69–74

Ahmad F, Slinkard AE (2004) The extent of embryo and endosperm growth following interspecific hybridization between *Cicer arietinum* L. and related annual species. Genet Resour Crop Evol 51:765–772

Ahmad F, Slinkard AE, Scoles GJ (1988) Investigations into barrier(s) to interspecific hybridization between *Cicerarietinum* L. and eight other annual Cicer species. Plant Breed 100:193–198

Ahmad F, Gaur PM, Croser J (2005) Chickpea (*Cicer arietinum* L.). In: Singh RJ, Jauhar PP (eds) *Grain legumes*, Vol 1: *genetic resources, chromosome engineering and crop improvement*. CRC Press, Boca Raton, pp 187–217

Ali MY, Johansen C, Krishnamurthy L, Hamid A (2005) Genotypic variation in root systems of chickpea (*Cicer arietinum* L) across environments. J Agron Crop Sci 191:464–472

Anbessa Y, Warkentin T, Vandenberg A, Ball R (2006) Inheritance of time to flowering in chickpea in a short-season temperate environment. J Hered 97:55–61

Ashraf M, Waheed A (1992) Screening chickpea (*Cicerarietinum* L.) for salt tolerance. Tropenlandwirt 93:45–55

Ashur SS, Clark HE, Moon WH, Malzer JL (1973) Nitrogen in adult human subjects who consumed wheat and rice supplemented with chickpea, sesame, milk or whey. Am J Clin Nutr 26:1195–1201

Atieno J, Colmer TD, Taylor J, Li Y, Quealy J, Kotula L, Nicol D, Nguyen DT, Brien C, Langridge P, Croser J, Hayes JE, Sutton T (2021) Novel salinity tolerance loci in chickpea identified in glasshouse and field environments. Front Plant Sci 12:667910

Basu U, Upadhyaya HD, Srivastava R, Daware A, Malik N, Sharma A, Bajaj D, Narnoliya L, VirevolThakro AK, Shailesh Tripathi C, Bharadwaj VSH, Pandey AK, Singh AK, Tyagi AK, Parida SK (2019) ABC transporter-mediated transport of glutathione conjugates enhances seed yield and quality in chickpea. Plant Physiol 180:253–275

Berger J, Abbo S, Turner NC (2003) Ecogeography of annual wild *Cicer* species: the parlous state of the world collection. Crop Sci 43:1076–1090

Berger JD, Ali M, Basu PS, Chaudhary BD, Chaturvedi SK, Deshmukh PS, Dharmaraj PS, Dwivedi SK, Gangadhar GC, Gaur PM, Kumar J, Pannu RK, Siddique KHM, Singh DN, Singh DP, Singh SJ, Turner NC, Yadava HS, Yadav SS (2006) Genotype by environment studies demonstrate the critical role of phenology in adaptation of chickpea (*Cicerarietinum* L.) to high and low yielding environments of India. Field Crop Res 98:230–244

Brinsmead RB, Rettke ML, Irwin JAG, Ryley MJ, Langdon PW (1985) Resistance in chickpea to *Phytophthoramegaperma* f. sp. Medicaginis. Plant Disease 69:504–506

Chitale K, Tyagi RNS, Singh RD (1990) Reaction of chickpea cultivars to collar rot disease induced by Sclerotium rolfsii. Indian J Mycol Plant Pathol 20:262

Cho S, Kumar J, Shultz JF, Anupama K, Tefera F, Muehlbauer FJ (2002) Mapping genes for double podding and other morphological traits in chickpea. Euphytica 125:285–292

Clarke H, Siddique KHM (2004) Response of chickpea genoptypes to low temperature stress during reproductive development. Field Crop Res 90:323–334

Clarke H, Khan TN, Siddique KHM (2004) Pollen selection for chilling tolerance at hybridization leads to improved chickpea cultivars. Euphytica 139:65–74

Corbin EJ (1975) Present status of chickpea research in Australia. In: International workshop on grain legumes, held during 13–16 January 1975 at ICRISAT, Patancheru, India, pp 87–97

Croser JS, Clarke HJ, Siddique KHM (2003) Low temperature stress: implications for chickpea (*Cicer arietinum* L.) improvement. Crit Rev Plant Sci 22:185–219

Di Vito M, Singh KB, Greco N, Saxena MC (1996) Sources of resistance to cyst nematode in the cultivated and wild Cicer species. Genet Resour Crop Evol 43:103–107

Dua RP, Chaturvedi SK, Sewak S (2001) Reference varieties of chickpea for IPR regime. Indian Institute of Pulses Research, Kanpur

Ellis RH, Lawn RJ, Summerfield RJ, Qi A, Roberts EH, Chays PM, Brouwer JB, Rose JI, Yeates SJ, Sandover S (1994) Towards the reliable prediction of time to flowering in six annual crops: V chickpea (*Cicer arietinum* L.). Exp Agric 30:271–282

Erskine W, Siddique K, Khan T, Cowling W (2001) Utilization of grain legume diversity. In: Maxted N, Bennett SJ (eds) Plant genetic resources of legumes in the Mediterranean. Kluwer Academic Publishers, Dordrecht, pp 311–326

FAOSTAT Data (2023). http://www.fao.org/statistics/en (last assessed on 20th May 2024)

Flowers TJ, Gaur PM, Gowda CL, Krishnamurthy L, Samineni S, Siddique KH, Turner NC, Vadez V, Varshney RK, Colmer TD (2010) Salt sensitivity in chickpea. Plant Cell Environ 33(4):490–509. https://doi.org/10.1111/j.1365-3040.2009.02051.x

Gaur PM, Gowda CLL (2005) Trends in world chickpea production, research and development. In: Knights EJ, Merrill R (eds) Proceeding of focus 2005: chickpea in the farming systems, held during 21-23 September 2005 at Goondiwindi, Qld, Australia, pp 8–15

Gaur PM, Pande S, Upadhyaya HD, Rao BV (2006) Extra-large kabuli chickpea with high resistance to fusarium wilt. Int J SAT Agric Res. Available online at http://www.icrisat.org/journal/cropimprovement/v2i1/v2i1extralargekabuli.pdf

Gaur PM, Kumar J, Gowda CLL, Pande S, Siddique KHM, Khan TN et al (2008) Breeding chickpea for early phenology: perspectives, progress and prospects. In: Kharkwal MC (ed) Proceedings of Fourth International Food legumes Research Conference. IARI, New Delhi, India, pp 39–48

Gaur PM, Samineni S, Tripathi S, Varshney RK, Gowda CLL (2015a) Allelic relationships of flowering time genes in chickpea. Euphytica 203:295–308. https://doi.org/10.1007/s10681-014-1261-7

Gaur PM, Samineni S, Krishnamurthy L, Kumar S, Ghanem ME, Beebe S, Rao I, Chaturvedi SK, Basu PS, Nayyar H, Jayalakshmi V, Babbar A, Varshney RK (2015b) High temperature tolerance in grain legumes. Legume Perspect 7:23–24. ISSN 2340-1559

Gowda CLL, Ramesh S, Chandra S, Upadhyaya HD (2005) Genetic basis of pod borer (*Helicoverpaarmigera*) resistance and grain yield in desi and kabuli chickpea (*Cicerarietinum* L.). Euphytica 145:199–214

Grewal SK, Sharma KP, Bharadwaj RD, Hegde V, Tripathi S, Sarvjeet Singh PK, Jain PK, Agrawal, Mondal B (2020) Understanding genotypic variation and identification of promising genotypes for iron and zinc in chickpea (*Cicer arietinum* L.). J Food Compos Anal 89:103458

Gumber RK, Sarvjeet S (1996) Genetics of flowering time in chickpea – a preliminary report. Crop Improv 23:295–296

Halila MH, Strange RN (1996) Identification of the causal agent of wilt of chickpea in Tunisia as *Fusarium oxysporum* f. sp. *ciceris* race 0. Phytopathol Mediterr 35:67–74

Haware MP (1998) Diseases of chickpea. In: Allen DJ, Lenne JM (eds) The pathology of food and pasture legumes. CAB International, Wallingford, UK, pp 473–516

Haware MP, McDonald D (1992) Integrated management of botrytis gray mold of chickpea. In: Haware MP, Faris DG, Gowda CLL (eds) Botrytis Gray Mold of Chickpea: Summary Proceedings of the BARI/ICRISAT Working Group Meeting to Discuss Collaborative Research on Botrytis Gray Mold of Chickpea. ICRISAT, Patancheru, India, pp 1–6

Haware MP, Nene YL (1982) Races of *Fusarium oxysporum* F. sp. *ciceris*. Plant Dis 66:809–810

Haware MP, Nene YL, Pundir RPS, Narayana RJ (1992) Screening of world chickpea germplasm for resistance to fusarium wilt. Field Crop Res 30:147–154

Hegde VS (2010) Genetics of flowering time in chickpea in a semi-arid environment. Plant Breed 29(6):683–687

Higgins TJ, Chiaiese P, Sarmah B, Molvig L, Tabe L (2004) Genetic improvement of chickpeas: seed composition and protection. In: Proceedings 5th European conference on grain legumes and 2nd international conference on legume genomics and genetics, held during 7-11 June 2004 at Dijon, France, pp 21–22

Horn NM, Makkouk KM, Kumari SG, vanden Heuvel JFJM, Reddy DVR (1995) Survey of chickpea (*Cicerarietinum* L.) for chickpea stunt disease and associated viruses in Syria, Turkey and Lebanon. Phytopathol Mediterr 34:192–198

Horn NM, Reddy SV, vanden Heuvel JFJM, Reddy DVR (1996) Survey of chickpea (*Cicerarietinum* L.) for chickpea stunt disease and associated viruses in India and Pakistan. Plant Dis 80:286–290

ICRISAT (1994) Cold-tolerant chickpea varieties ICCV 88503, ICCV 88506 and ICCV 88510. In: Plant material description no. 53. International Crops Research Institute for the Semi-Arid Tropics, Patancheru, India

ICRISAT (2005) Harnessing biotechnology for the poor. Archival Report. International Crops Research Institute for the Semi-Arid Tropics, Patancheru, India

IIPR (Indian Institute of Pulses Research) (1997) Package of practices for pulse crops in India. Indian Institute of Pulses Research, Kanpur

IIPR (Indian Institute of Pulses Research) (2024). https://iipr.icar.gov.in/varieties/

Jain M, Misra G, Patel RK, Priya P, Jhanwar S, Khan AW, Shah N, Singh VK, Garg R, Jeena G, Yadav M, Kant C, Sharma P, Yadav G, Bhatia S, Tyagi AK, Chattopadhyay D (2013) A draft genome sequence of the pulse crop chickpea (Cicer arietinum L.). Plant J 74(5):715–729

Jaiswal HK, Singh BD, Singh AK, Singh RM (1987) Improvement of chickpea (*Cicer arietinum* L.) through introgression of genes from *Cicer reticulatum*. Indian J Agric Sci 57:880–883

Jimenez-Diaz RM, Alcala-Jimenez AR, Hervas A, Trapero-Casas JL (1993) Pathogenic variability and host resistance in the *Fusarium oxysporum* f. sp. *ciceris*/*Cicer arietinum*pathosystem. In: Arseniuk E, Goral T (eds) Fusarium mycotoxins, taxonomy, pathogenicity and host resistance. Proceedings of the 3rd European Seminar, Radzikov, Poland: Plant breeding and Acclimatization Institute, pp 87–94

Jukanti AK, Gaur PM, Gowda CLL, Chibbar RN (2012) Nutritional quality and health benefits of chickpea (*Cicer arietinum* L.): a review. Br J Nutr 108:S11–S26

Kaiser WJ, Danesh D (1971) Etiology of virus-induced wilt of *Cicer arietinum*. Phytopathology 61:453–457

Karadi A, Samineni S, Sajja S, Sharma M, Thudi M, Mallikarjuna BP, Vishwanatha KP, Varshney RK, Gaur PM (2021) Molecular mapping of dry root rot resistance genes in chickpea (*Cicer arietinum* L.). Euphytica 217:123

Kashiwagi J, Krishnamurthy L, Upadhyaya HD, Krishna H, Chandra S, Vincent V, Serraj R (2005) Genetic variability for drought-avoidance root traits in the minicore germplasm collection of chickpea (*Cicer arietinum* L.). Euphytica 146:213–222

Kashiwagi J, Krishnamurthy L, Crouch JH, Serraj R (2006) Variability of root length density and its contribution to seed yield in chickpea (*Cicer arietinum* L) under terminal drought stress. Field Crop Res 95:171–181

Kathiria KB, Nayagapara DR, Vaddoria MA, Poshiya VK (1997) Screening of chickpea genotypes for salinity tolerance during germination and early seedling growth. Gujarat Agric Univ Res J 22:28–32

Kazan K, Muehlbauer FJ (1991) Allozyme variation and phylogeny in annual species of *Cicer* (*Leguminosae*). Plant Syst Evol 175:11–21

Khan AW, Garg V, Sun S et al (2024) *Cicer* super-pangenome provides insights into species evolution and agronomic trait loci for crop improvement in chickpea. Nat Genet 56:1225. https://doi.org/10.1038/s41588-024-01760-4

Knights EJ, Brinsmead RB, Fordyce M, Wood JA, Kelly A, Harden S (2002) Use of the wild relative *Cicerechinospermum* in chickpea improvement. In: McComb JA (ed) Plant breeding for the 11th Millenium: proceedings 12th Australasian plant breeding conference. Australian Plant Breeding Assoc. Inc., Perth, Australia, p 145

Knights EJ, Southwell RJ, Schwinghamer MW (2003) Evaluation of wild Cicer species for resistance to phytophthora root rot. In: Sharma RN, Shrivastava GK, Rathore AL, Sharma ML, Khan MA (eds) Chickpea Research for the Millennium: Proceedings of the International Chickpea Conference, held during 20–22 January, 2003 at Indira Gandhi Agricultural University, Raipur, India, pp 54–57

Krishnamurthy L, Ito O, Johansen C (1996) Genotypic differences in root growth dynamics and its implications for drought resistance in chickpea. In: Ito O, Johansen C, AduGyamfi JJ, Katayama K, Kumar Rao JVDK, Rego TJ (eds) Dynamics of roots and nitrogen in cropping Systems of the Semi-Arid Tropics, JIRCAS Agriculture Series No. 3. Japan International Research Center for Agriculture Sciences, Tsukuba, Japan, pp 235–250

Kumar J, Haware MP (1982) Inheritance of resistanceto *Fusarium* wilt in chickpea. Phytopathology 72:1035–1036

Kumar J, van Rheenen HA (2000) A major gene for time of flowering in chickpea. J Hered 91:67–68

Ladizinsky G, Adler A (1975) The origin of chickpea as indicated by seed protein electrophoresis. Israel J Bot 24:183–189

Lateef SS, Pimbert MP (1990) The search for host resistance to *Helicoverpaarmigera*. In: Chickpea and Pigeonpea at ICRISAT: Summary Proceedings of First Consultative Group Meeting on Host selection Behaviour of Helicoverpaarmigera, March 1990, ICRISAT, Patancheru, India, pp 14–18

Lawlor HJ, Siddique KHM, Sedgley RH, Thurling N (1998) Improving cold tolerance and insect resistance in chickpea and the use of AFLPs for the identification of molecular markers for these traits. In: Drew RA (ed) International symposium on biotechnology of tropical and subtropical species – part II. Acta Horticulturae, pp 185–192. https://doi.org/10.17660/ActaHortic.1998.461.18

Lichtenzveig J, Bonfil DJ, Zhang HB, Shtienberg D, Abbo S (2006) Mapping quantitative trait loci in chickpea associated with time of flowering and resistance to *Didymellarabiei* the causal agent of Ascochyta blight. Theor Appl Genet 113:1357–1369

Malhotra RS, Singh KB, van Rheenen HA, Pala M (1996) Genetic improvement and agronomic management of chickpea with emphasis on the Mediterranean region. In: Saxena NP, Saxena MC, Johansen C, Virmani SM, Harris H (eds) Adaptation of chickpea in the West Asia and North Africa region. ICRISAT, Hyderabad, India and ICARDA, Aleppo, Syria, pp 217–232

Malhotra RS, Singh KB, Di Vito M, Greco N, Saxena MC (2002) Registration of ILC 10765 and ILC 10766 chickpea germplasm lines resistant to cyst nematode. Crop Sci 42:1756

Maliro MFA, McNeil D, Kollmorgen J, Pittock C, Redden R (2004) Screening chickpea (*Cicerarietinum* L.) and wild relatives germplasm from diverse country sources for salt tolerance. In: Proceedings of international crop science congress, held during 26 Sept – 1 Oct 2004 at Brisbane, Australia. Available at: http://www.cropscience.org.au/icsc2004/poster/3/6/2/1280_malirom

Mayer M, Tullu A, Simon C, Kumar J, Kaiser W, Kraft J et al (1997) Development of a DNA marker for *Fusarium* wilt resistance in chickpea. Crop Sci 37:1625–1637

Millan T, Clarke HJ, Siddique KHM, Buhariwalla HK, Gaur PM, Kumar J, Gil J, Kahl G, Winter P (2006) Chickpea molecular breeding: new tools and concepts. Euphytica 147:81–103

Muehlbauer FJ, Kaiser WJ, Simon CJ (1994) Potential for wild species in cool season food legume breeding. Euphytica 73:109

Nene YL, Haware MP, Reddy MV (1981) Chickpea diseases: resistance-screening techniques, Information Bulletin No. 10. ICRISAT, Patancheru

Nene YL, Haware MP, Reddy MV, Phillips JC, Castro EL, Kotasthane SR, Gupta O, Singh G, Shukla P, Sah RP (1989) Identification of broad-based and stable resistance to wilt and root rots of chickpea. Indian Phytopathol 42:499–505

Nene YL, Sheila VK, Sharma SB (1996) A world list of chickpea and Pigeonpea pathogens, 5th edn. ICRISAT, Patancheru, p 27

NITI Aayog (2024) Working group report on crop husbandry, agriculture inputs, demand and supply, Government of India

Or E, Hovav R, Abbo S (1999) A major gene for flowering time in chickpea. Crop Sci 39:315–322

Pande S, Singh G, Narayana Rao J, Bakr MA, Chaurasia PCP, Joshi S, Johansen C, Singh SD, Kumar J, Rahman MM, Gowda CLL (2002) Integrated Management of Botrytis Gray Mold of Chikcpea, Information Bulletin No. 61. ICRISAT, Patancheru, p 26

Pande S, Siddique KHM, Kishore GK, Bayaa B, Gaur PM, Gowda CLL, Bretag TW, Crouch JH (2005) Ascochyta blight of chickpea (*Cicerarietinum* L.): a review of biology, pathogenicity and disease management. Aust J Agric Res 56:317–332

Pratap A, Chaturvedi SK, Tomar R, Rajan N, Malviya N, Thudi M, Saabale PR, Prajapati U, Varshney RK, Singh NP (2017) Marker-assisted introgression of resistance to fusarium wilt race 2 in Pusa 256, an elite cultivar of desi chickpea. Mol Gen Genomics 292(6):1237–1245

Pundir RPS, Reddy KN, Mengesha MH (1988) ICRISAT chickpea germplasm catalog: evaluation and analysis. ICRISAT, Patancheru, p 94

Pushpavalli R, Krishnamurthy L, Thudi M, Gaur PM, Rao MV, Siddique KH, Colmer TD, Turner NC, Varshney RK, Vadez V (2015) Two key genomic regions harbour QTLs for salinity tolerance in ICCV 2 × JG 11 derived chickpea (*Cicer arietinum* L.) recombinant inbred lines. BMC Plant Biol 15:124. https://doi.org/10.1186/s12870-015-0491-8

Rahul C, Singh IS, Gupta AK (1995) Inheritance of resistance to Botrytis grey mould in chickpea (*Cicerarietinum* L.). Legum Res 18:1–4

Reddy MV, Singh KB (1984) Evaluation of a world collection of chickpea germplasm accessions for resistance to Ascochyta blight. Plant Dis 68:900–901

Rewal N, Grewal JS (1989) Inheritance of resistance to *Botrytis cinerea* Pers. in *Cicer arietinum* L. Euphytica 44:61–63

Roberts EH, Hadley P, Summerfield RJ (1985) Effects of temperature and photoperiod on flowering in chickpea (*Cicerarietinum* L.). Ann Bot 55:881–892

Robertson LD, Singh KB, Ocampo B (1995) A catalog of annual wild Cicer species. Syria, ICARDA, Aleppo, p 171

Ryan JG (1997) A global perspective on pigeonpea and chickpea sustainable production systems: present status and future potential. In: Asthana AN, Ali M (eds) Recent advances in pulses research. Indian Society of Pulses Research and Development, Indian Institute of Pulses Research, Kanpur, pp 1–31

Saxena NP (1984) Chapter 12: chickpea. In: Goldsworthy PR, Fisher NM (eds) The physiology of tropical field crops. John Wiley & Sons Ltd, Chichester, pp 419–452

Saxena NP (1987) Screening for adaptation to drought: case studies withchickpea and pigeonpea. In: Consultants' workshop: adaptation of Chickpea and Pigeonpea to abiotic stresses, 19–21 December 1984, ICRI-SAT, India

Saxena NP (2003) Management of drought in chickpea – a holistic approach. In: Saxena NP (ed) Management of agricultural drought – agronomic and genetic options. Oxford & IBH Publishing Co Pvt. Ltd, New Delhi, pp 103–122

Saxena NP, Krishnamurthy L, Johansen C (1993) Registration of a drought-resistant chickpea germplasm. Crop Sci 33:1424

Serraj R, Krishnamurthy L, Kashiwagi J, Kumar J, Chandra S, Crouch JH (2004) Variation in root traits of chickpea (*Cicerarietinum* L.) grown under terminal drought. Field Crop Res 88:115–127

Sharma KD, Muehlbauer FJ (2007) Fusarium wilt of chickpea: physiological specialization, genetics of resistance and resistance gene tagging. Euphytica 157:1–14

Sharma KK, Ortiz R (2000) Program for the application of genetic transformation for crop improvement in the semi-arid tropics, *in vitro* cellular developmental biol. Plant 36:83

Sharma HC, Gowda CLL, Sharma KK, Gaur PM, Mallikarjuna N, Buhariwalla HK, Crouch JH (2003) Host plant resistance to pod borer, *Helicoverpaarmigera* in chickpea. In: Sharma RN, Shrivastava GK, Rathore AL, Sharma ML, Khan MA (eds) Chickpea Research for the Millennium: Proceedings of the International Chickpea Conference, held during 20–22 January, 2003 at Indira Gandhi Agricultural University, Raipur, India, pp 118–137

Sharma HC, Pampapathy G, Lanka SK, Ridsdill-Smith TJ (2005) Antibiosis mechanism of resistance to pod borer, *Helicoverpaarmigera*in wild relatives of chickpea. Euphytica 142:107–117

Sharma HC, Bhagwat M, Pampapathy G, Sharma J, Ridsdill-Smith TJ (2006) Perennial wild relatives of chickpea as potential sources of resistance to *Helicoverpaarmigera.* Genet Resour Crop Evol 53:131–138

Siddique KHM, Brinsmead RB, Knight R, Knights EJ, Paull JG, Rose IA (2000) Adaptation of chickpea (*Cicerarietinum* L.) and faba bean (*Vicia faba* L.) to Australia. In: Knight R (ed) Linking research and marketing opportunities for pulses in the 21st century. Kluwer Academic Publishers, Dordrecht, pp 289–303

Sindhu JS, Singh KP, Slinkard A (1983) Inheritance of resistance to Fusarium wilt in chickpeas. J Hered 74:68

Singh KB, Ocampo B (1993) Interspecific hybridization in annual *Cicer* species. J Genet Breed 47:199–204

Singh KB, Ocampo B (1997) Exploitation of wild Cicer species for yield improvement in chickpea. Theor Appl Genet 95:418–423

Singh KB, Reddy MV (1993) Resistance to six races of *Ascochyta rabiei* in the world germplasm collection of chickpea. Crop Sci 33:186–189

Singh NB, Singh RA (1984) Screening for alkalinity and salinity tolerance in chickpea. Int Chickpea Newsl 11:25–26

Singh H, Kumar J, Smithson JB, Haware MP (1987) Complementationbetween genes for resistance to race 1 of *Fusarium oxysporum* f. sp. *Ciceri*in chickpea. Plant Pathol 36:539–543

Singh KB, Malhotra RS, Saxena MC (1990) Sources for tolerance to cold in *Cicer* species. Crop Sci 30:1136–1138

Singh KB, Malhotra RS, Halila MH, Knights EJ, Verma MM (1994) Current status and future strategy in breeding chickpea for resistance to biotic and abiotic stresses. Euphytica 73:137–149

Singh KB, Malhotra RS, Saxena MC (1995) Additional sources of tolerance to cold in cultivated and wild *Cicer* species. Crop Sci 35:1491–1497

Srinivasan A, Johansen C, Saxena NP (1998) Cold tolerance during early reproductive growth of chickpea (*Cicerarietinum* L.): characterization of stress and genetic variation in pod set. Field Crop Res 57:181–193

Subbarao GV, Kumar Rao JVDK, Kumar J, Johansen C, Deb UK, Ahmed I, Krishna Rao MV, Venkataratnam L, Hebbar KR, Sai MVSR, Harris D (2001) Spatial distribution and quantification of rice-fallows in South Asia: potential for legumes. ICRISAT, Patencheru

Tayyar RI, Waines JG (1996) Genetics relationships among annual species of *Cicer* (Fabaceae) using isozyme variation. Theor Appl Genet 92:245–254

Trapero-Casas A, Jimenez-Diaz RM (1985) Fungal wilt and root rot diseases of chickpea in southern Spain. Phytopathology 75:1146–1151

Upadhyaya HD, Smithson JB, Haware MP, Kumar J (1983) Resistance to wilt in chickpea. II. Further evidence for two genes for resistance to race 1. Euphytica 32:749–755

Vadez V, Krishnamurthy L, Gaur PM, Upadhyaya HD, Hoisington DA, Varshney RK, Turner NC, Siddique (2006) Tapping the large genetic variability for salinity tolerance in chickpea. In: Proceedings of 13th Australian Society of Agronomy Conference, held during 10–14 September 2006 at Perth, Australia. Available online at http://www.regional.org.au/au/asa/2006/concurrent/environment/4561_vadez.htm

van Rheenen HA (1991) Chickpea breeding – progress and prospects. Plant Breed Abstr 61(9):997–1009

Varshney R, Song C, Saxena R et al (2013) Draft genome sequence of chickpea (*Cicer arietinum*) provides a resource for trait improvement. Nat Biotechnol 31:240–246. https://doi.org/10.1038/nbt.2491

Winter P, Benko-Iseppon AM, Hüttel B et al (2000) A linkage map of chickpea (*Cicer arietinum* L.) genome based on recombinant inbred lines from a *C. arietinum* × *C. reticulatum* cross: localization of resistance genes for fusarium wilt races 4 and 5. Theor Appl Genet 101:1155–1163

Yahia AA, Ouada MA, Hadjarab K, Belfendes R, Sarni K (1997) Identification of chickpea stunt viruses in Algeria. Bull OEPP 27:265–268

Pigeonpea [*Cajanus cajan* (L.) Millsp.] Breeding

K. B. Saxena and Rachit K. Saxena

Abstract

Pigeonpea [*Cajanus cajan* (L.) Millsp.] plays a significant role in meeting the challenges of global food and social security of the farming communities, especially under the looming threats of climate change, soil degradation, and escalating production costs. The grains of this pulse contain high protein and a range of mineral nutrients, and this makes pigeonpea a valuable food supplement, especially for vegetarian masses. The crop improvement activities in this crop are being carried out since the last three-quarters of a century and these efforts have yielded several significant accomplishments, but unfortunately, the same cannot be said about its on-farm productivity. In this chapter, besides summarizing the key breeding accomplishments, the major constraints encountered in this journey are also briefly outlined. Also, the authors have made some suggestions to attain a positive growth trajectory in the near future by developing high-yielding cultivars/hybrids using new plant breeding and genomics technologies.

Keywords

Breeding accomplishments · Hybrids · Maintenance breeding · Molecular breeding · New breeding methods · Productivity trends

K. B. Saxena (✉)
International Crops Research Institute for the Semi-Arid Tropics (ICRISAT), Patancheru, Telangana, India

Al AinCity, Abu Dhabi, United Arab Emirates

R. K. Saxena
ICAR - Indian Agricultural Research Institute (IARI), Dhemaji, Assam, India

1 Pigeonpea: A Crop for Nutritional and Social Security

The term "Pulse" is a derivative of a Latin word "puls" or "pultis," meaning thick soup; and this group embraces 1800 different plant species. The evidence related to the cultivation of lentil has been found in the Egyptian pyramids. Also, the traces of dry pea cultivation were discovered in a Switzerland village dating back to the stone age (AGT Food and Ingredients Inc.). Among pulses, pigeonpea [*Cajanus cajan* (L.) Millspaugh] is an important food crop of the tropical and subtropical regions of the world. It occupies a key position in subsistence agriculture because of its various farmer- and environment-friendly attributes. Globally, the crop is grown on 6–7 m ha (Fig. 1), and India is its largest (80–85%) producer. The other significant regions where pigeonpea grows are Nepal and Myanmar in Asia, Southeast Africa, South America, and the Caribbean. Pigeonpea is an ideal crop for sustainable agricultural systems due to its multiple benefits and usage. Besides providing high protein and vital nutrients, its multiple usage makes pigeonpea ideal crop for healthy rural livelihoods. These include its use as fuel, fodder, feed, fresh vegetable, and as growing-medium for commercial products like mushrooms and lac resin (Saxena et al. 2021). The cultivation of pigeonpea also adds to soil conservation and fertility enhancement.

It is pity to note that in spite of serious genetic enhancement research efforts spread over five decades, the on-farm productivity of pigeonpea has remained low at around 700–750 kg/ha (Fig. 1). But, on the other hand, it is also true that significant genetic advances have been made in improving some key traits like disease resistance, earliness, and some market-preferred seed traits. This research article, besides reviewing the important crop breeding accomplishments, also proposes new breeding technologies and discusses key inherent constraints encountered by the pigeonpea breeders.

G. Prasad Dixit et al. (eds.), *Fundamentals of Legume Breeding*, https://doi.org/10.1007/978-981-96-7838-9_4

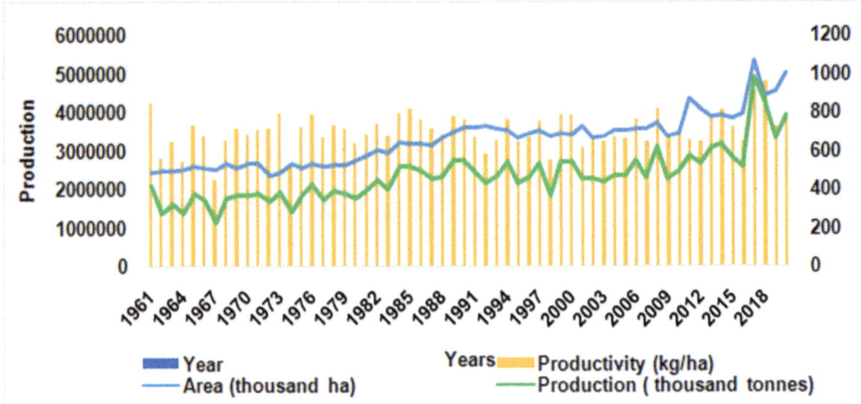

Fig. 1 Pigeonpea global area, production, and productivity

2 Origin, Taxonomy, and Nomenclature

Information on the origin of a given species is valuable as it tries to decipher the past natural events, which led to the evolution of the present-day domesticated biotypes. The cultivation of pigeonpea in the tropics and semi-arid tropics dates back to centuries, and it has not only played an important role in the nutritional security but also added to sustainability of rainfed agricultural systems. During the early periods, the cultivars adapted to the prevailing ecosystems were evolved following some major changes brought about by natural mutations, cross-pollinations, and elimination of unfit individuals. This was accompanied by deliberate selections by ancient growers for various visibly favorite traits.

The successful search of the center of origin of a species provides opportunities to look for additional variability in the vicinity that could not be fetched earlier. Such a valuable information is usually gathered through the revelations from excavations and their carbon-dating data and various biotechnological approaches. Now some modern genomics tools are also used in identifying the center of origin, key centers of diversity, and in pegging the putative routes of spreading the crop in new areas and eco-systems.

2.1 Origin of Domesticated Pigeonpea

Researchers of seventeenth to nineteenth centuries opined two contrasting views regarding the origin of pigeonpea. Based on a few direct and indirect evidence and speculations, one group of historians favored West African origin of pigeonpea, while the other opined that the domesticated form of pigeonpea evolved from a wild species in India. Hence, for a considerable period, this subject remained a matter of speculation and discussion.

2.1.1 The Theory of West African Origin

The West African theory of the origin of pigeonpea was put forward by de Candolle (1885) and later supported by Purseglove (1968) and others. There were two considerations which led to this belief. First, the scientists found *Atylosiaker stingii* (Haines) growing wild in Senegal, Ghana, Togo, and Nigeria. It is a perennial shrub and distantly related to the cultivated species *Cajanus cajan*. In comparison to the cultivated types ($2n = 22$), *A. kerstingii* has $2n = 32$ chromosomes, and this diversity did not permit hybridization between the two species. Besides this, an exclusive distribution of *A. kerstingii* in Africa and the discovery of a single pigeonpea seed in Egyptian tombs of 12th dynasty firmed that this crop was cultivated in Africa around 2200–2400 BC (Sturtevant 1972).

2.1.2 The Theory of Indian Origin

Considering the vast variability in the Indian pigeonpea germplasm and the existence of different related wild species in the forests, De Candolle (1813) and Vavilov (1951) believed that "India Orientalis" as the home of pigeonpea. Vavilov (1951) also postulated that pigeonpea originated in the "Ghats (hilly forests)" of eastern coast and central plateau, and that the early Asiatic name of the species *kayan* was linked to the present species name *cajan*. De (1974a, b) believed that the ecosystem of the Ghats is ideal for preserving the diversity of different tropical deciduous woodland species.

In archaeological findings and carbon dating, data carry a lot of weightage in speculating the age and the past events of any object. In the context of pigeonpea, the excavations carried out at Sanganaleallu and Tuljapu Garhi in Maharashtra state of India revealed the presence of small-sized pigeonpea seeds, and these were carbon-dated back between the second to third century BC to third century AD (Kajale 1974). Besides this, the other archaeological findings from two Neolithic sites of Gopalpur and Golbai Sassan in Odisha state also suggested that pigeonpea was cultivated there around 3400–3000 BC.

2.2 The Progenitor of Pigeonpea and Its Habitat

The detailed studies conducted by van der Maesen (1980) confirmed that *Cajanus cajanifolius* (Haines), a wild species of secondary gene pool of genus *Cajanus*, is the definite progenitor of pigeonpea; and the two species are critically linked through a single gene. The other synonyms of this progenitor species found in literature are *Atylosia cajanifolia* (Haines) and *Cantharosspermum cajanifolium* (Haines, Raizada). The detailed survey conducted in the eastern and southern forests revealed that *Cajanus cajanifolius* is endemic to these ecologies. In fact, the plants of this species can still be found in hills of East Godavari, Srikakulam, Visakhapatnam of Andhra Pradesh state, Nayagarh district of Orissa, and Baster district of Chhattisgarh. It is also believed that the spread of *C. cajanifolius* is restricted due to various reasons, such as pod shattering, palatable grazing material, and the consumption of its green and dry seeds by the local tribes.

In 1980, Dr. L. J. G. van der Maesen, the former Principal Germplasm Botanist at ICRISAT, combed the forests, and he located some pockets of the self-sown scattered plants of *C. cajanifolius*. He collected four specimens, which are still preserved at ICRISAT's Gene Bank. The genomic studies showed that the two species shared remakable molecular similarities between them (Saxena et al. 2014). Also, Varshney et al. (2017) identified the genomic regions of the wild species that had undergone selective sweeps during the process of domestication. These gene mutations were considered as "kick-off" points for trait diversification toward the cultivated pigeonpea. The *C. cajanifolius* (Fig. 2) resembles pigeonpea in all morphological traits except the presence of a prominent strophiole on its seeds. The local tribes still consume the seeds like its domesticated counterpart, and they call it as "Ban Arhar" meaning "wild pigeonpea."

Fig. 2 *Cajanus cajanifolius*, the progenitor of pigeonpea

2.3 Global Dispersal Routes of Pigeonpea

The theory of Indian origin of pigeonpea is now well established and accepted worldwide. According to Mula and Saxena (2010), the pigeonpea seeds reached across the continents to over 100 countries through various primary, secondary, and tertiary dispersion agencies. It is believed that the pigeonpea seeds were first taken from India to Africa and Madagascar at least two millennia BC by traders, and to the new world along with the slave trade during the post-Columbian period (van der Maesen 1986; De 1974a, b). Pigeonpea, then, moved away from India to Indo-China and Australia via the Malay Archipelago.

2.4 Taxonomy of Genus *Cajanus*

Earlier *Cajanus* was regarded as a monotypic genus, but it was always considered very close to genus *Atylosia.* Based on this kinship, Lackey (1977) concluded that "*Cajanus* is quite possibly nothing more than a cultigen of *Atylosia*." Later, van der Maesen (1986) extensively studied the two genera—*Cajanus* DC and *Atylosia* W.& A.; and based on their morphological, cytological, and biochemical properties and hybridization patterns, he merged the genus *Atylosia* W. & A. with genus *Cajanus*. Consequently, the sub-tribe *Cajaninae*, instead of 13, now contains 12 genera and 34 species; and among these, pigeonpea is the only cultivated food crop of this tribe.

Harlan and de Wet (1971) developed the famous germplasm classification that was based on primary criteria of genetic diversity and cross hybridization. Using this rule, the genus *Cajanus* of sub-tribe *Cajaninae* was sub-divided into primary (1 species), secondary (10 species), tertiary (24 species), and quaternary (10 species) gene pools. In this scheme, a wild species, *Cajanus platycarpus*, was placed in the tertiary gene pool. However, based on the genetic affinity of this species with the cultivated types, Saxena et al. (2018a) opined that instead of tertiary gene pool, the *Cajanus platycarpus* should find a place in a new subgroup within the secondary gene pool.

2.5 Scientific Nomenclature

The genus *Cajanus* was founded by de Candolle in 1813, and the first scientific name to pigeonpea, *Arbor trifolia indica*, was given by Bauhin and Cherla during 1650–1651 (cited by van der Maesen 1986). A century later, Linnaeus (1753) gave the first binomial name to pigeonpea as *Cytisuscajan* (L.). According to Watt (1889), the species name *cajan* evolved from Malaya language, while Vavilov (1951) opined that this species name is supposedly linked to the early Asiatic name *kayan*.

In the past literature, one can find binomials such as *Cytisus pseudo-cajan* Jacq., *Cajanus pseudu-cajan* Schi. & Gull., *C. indicus*, *C. flavus*, *C. bicolor*, *C. cajan* Huth., and *C. cajan* Druce; and these should now be considered synonyms of the presently accepted binomial "*Cajanus cajan* (Linnaeus) Millspaug." After merging the two related taxa *Cajanus* and *Atylosia*, the pigeonpea binomial has now been revised to "*Cajanus cajan* (Linnaeus) Millspaug van der Maesen."

2.6 Ethno-Botanical Nomenclature

Amarsimha (c. 200 BC), in his lexicon, "Amarkosha" mentioned *adhaki*, *kakshi*, and *tuvarika* as the names of pigeonpea. In the Marathi book *Gathasaptasati* (third to fourth century AD), pigeonpea has been referred to as *Tuvari*. It is believed that the popular Indian vernaculars of pigeonpeawere derived either from Sanskrit (*Adhuki*, *Adhuka*, *Arhar*) or Dravidian (*Tuvari*, *Tuvarai*, *Tuar*, or *Tur*) languages. The most common name used for pigeonpea in north India is *arhar*, and it is considered a corrupted form of *adhaki*.

The names for pigeonpea in the Americas are derived from African and European languages. It is believed that the Indian Telugu word *kandulu* led to the derivation of Portuguese name *guandu*, the Spanish *gandul*, and the Egyptian *gandoles*. According to van der Maesen (1986), common pigeonpea names across various countries include *Katjang* in Malaysia, *Cguando* in Brazil, *Gandolu* in Puerto Rico, *Guango* and *Congo pea* in Jamaica, *Quinchoncho* in Venezuela, *Pois de Congo* in Africa, *Thora parippu* in Sri Lanka, and *Mu Dau* in China. The globally recognized English name, "pigeonpea," was first used by Plukenet in 1692 in Barbados, where the crop was planted in barren land primarily to feed the local pigeon bird population.

3 Floral Morphology, Cytology, and Genetics

3.1 Floral Morphology

Reddy (1990) examined the floral morphology of pigeonpea and reported zygomorphic corolla with petals as its characteristic feature (Fig. 3). The outermost petal (standard) is clawed at the base, and the wing petals are asymmetrically biauriculate, obovate, with a clawed base. The innermost petals (keel) are boat-shaped, wrapping the androecium and gynoecium underneath. The stamens are 10 in number arranged in a diadelphous (9 + 1) pattern. The anthers are ellipsoidal and yellow in color. The style is long, filiform, glabrous, and connected to a thickened, capitate (swollen) stigma. The ovary is superior, sessile, with a marginal placenta and containing 2–9 ovules.

Pigeonpea flowers, despite of belonging to the leguminous family, are not truly cleistogamous. The buds remain cleistopetalous (closed) only for the first 2–3 days, during which 70–80% of them undergo self-pollination. After this period, both pollinated and unpollinated buds open and remain in this state for the next 24–48 h. During this time, pollinating insects visit the flowers searching for nectar and facilitate cross-pollination. The unpollinated flowers drop off the plant. Thus, pigeonpea experiences both self- and cross-pollination on the same branch or plant. The diadelphous (9 + 1) stamens are dimorphic in height. Four of the ten stamens have short filaments, while the remaining six, including the odd posterior (free) stamen, have long filaments. The anthers are connivent and dehisce through longitudinal slits. The short stamens' anthesis occurs before that of the taller ones. Pollen grains from both types of stamens are monads (single), yellow, powdery, binucleate, and equally viable. The active life of pigeonpea flowers stretches for a few days, and during this period, the nectar secretion in them continues. Various foragers visit open flowers, and when these insects depress the keel for harvesting nectar, the wet receptive stigma is exposed and rubs the pollinators' body to affect cross pollination. In fact, in pigeonpea, to date there is no comprehensive information available on the biology of nectary, nectar chemistry, nectar dynamics, and the mechanism of nectar robbing.

3.2 Cytology of Pigeonpea

Roy (1933) reported the presence of 11 chromosomes in female gametophytic tissues of pigeonpea. Krishnaswamy and Aiyengar (1935) examined the pollen mother cells and confirmed this chromosome count in the gametophyte. Naithani (1941) was the first to report the somatic chromosome number ($2n = 22$) in pigeonpea. Van der Maesen (1986) noted that all 32 wild relatives of pigeonpea had a chromosome count identical to that of *Cajanus cajan*. In a single odd report, Gill and Husaini (1986) observed the presence of 16 gametic chromosomes in *C. kertsingii*.

In pigeonpea, both natural and induced tetraploids have been reported. Pathak (1948), Pathak and Yadav (1951), and Saxena et al. (1982) reported the existence of naturally occurring tetraploids ($2n = 44$) in pigeonpea. Such plants characteristically have large and thick leaves, and their flowers are partial male or female sterile. The colchicine-induced tetraploids were reported by Bhattacharjee (1956), and the partial male-sterility and poor pod set were attrib-

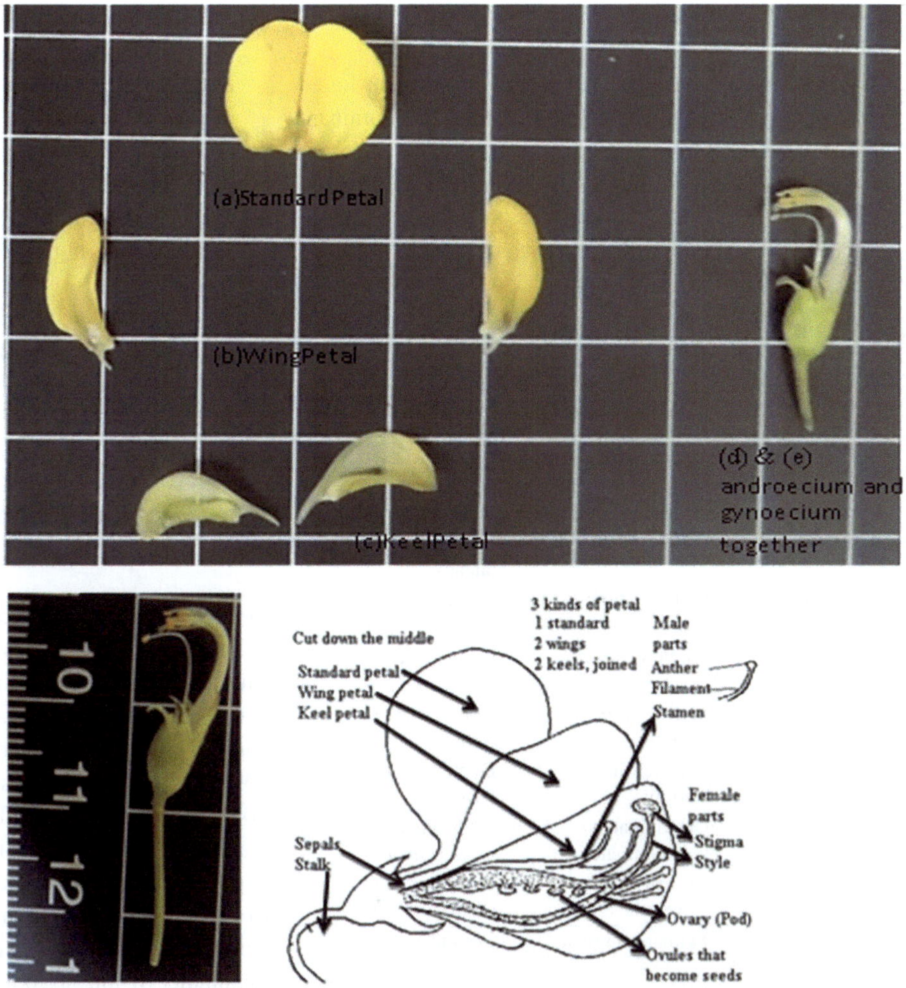

Fig. 3 Pigeonpea flower parts

uted to varying degrees of multivalent formation and irregular disjunction in the second division of meiosis. The first case of aneuploidy ($2n = 23$) in pigeonpea was reported by D'Cruz and Jadhav (1972). Also, a large variation in the chromosome number ($2n = 8$ to 28) was reported by Bajaj et al. (1980) in the anther-derived callus tissues of pigeonpea.

The somatic chromosomes of pigeonpea are small with the longest one measuring 2.7μ and the shortest 1.35μ (Naithani 1941). The first detailed karyotype analysis in pigeonpea was carried out by Deodikar and Thakar (1956). Considerable genotypic variation in total chromatin length was reported by Sinha and Kumar (1979), while Pundir and Singh (1985) observed variation in satellited chromosomes. Dundas et al. (1983) discovered that one chromosome bivalent was attached to the nucleolus, whereas Kumar et al. (1987) identified two sites for nucleolar organization. The first detailed pachytene analysis of pigeonpea chromosomes was conducted by Reddy (1981), who characterized the chromosomes based on their relative length, arm length, nucleolar association, and chromomere structure.

3.3 Genetic Control of Some Key Traits

Genes are known to carry information about a given trait from one generation to the other through sexual reproduction. The phenotypic expression of any particular gene could be prominent (major gene) or a little (minor gene). The action of minor genes, in general, is estimated using various quantitative genetics tools, while the inheritance of qualitative traits is studied using various genetic theories proposed by Mendel.

In case of pigeonpea, such exercises with respect to reproductive and growth traits have an unsatisfactory record. A summary of these traits along with the assigned gene symbols (Table 1) was published recently by Srivastava and Saxena (2023). To save on the space, the references are not given in this chapter, and the interested readers can go through the above-mentioned publication. Some of the interesting points noted from these genetic studies are listed below.

1. Photo-sensitivity in pigeonpea was controlled by three dominant genes. These genes act in hierarchical order,

i.e., $PS_3 > PS_2 > PS_1$, and the photo-insensitive genotype is $ps_1ps_1ps_2ps_2ps_3ps_3$.

2. In ICPL 87119, the resistance to fusarium wilt was found to be controlled by two independent genes, one dominant and another recessive.
3. Pod color in ICP 3773 was controlled by unstable (jumping) genes with random expression for the colored streaked on the pods within a branch or plant.

Table 1 Genetics of some key qualitative traits in pigeonpea

Trait	Variation	Genetics
Reproductive traits		
Cleistogamous	Normal vs cleistogamous	Dominance of normal
Photo-sensitivity	Sensitive vs insensitive	Dominance of sensitivity
Flowering	Early vs late	Dominance of lateness
Flower color	Red vs yellow	Dominance of red color
Flower color	Purple vs yellow	Dominance of purple
Genetic male sterility	Male fertility vs sterility	Dominance of fertility
Cytoplasmic male sterility	Male fertility vs sterility	Dominance of fertility
Growth habit	Determinate vs indeterminate	Dominance of indeterminate
Pod development	Normal vs open carpel	Dominance of normal
Pod color	Maroon vs green	Dominance of maroon
Seed color	White vs brown	Dominance of brown
Seed size	Bold vs small	Dominance of small
Seed strophiole	Present vs absent	Dominance of strophiole
Plant traits		
Plant type	Erect vs creeping	Dominance of erect type
Plant height	Tall vs dwarf	Dominance of tallness
Stem color	Purple vs green	Dominance of purple color
Branching habit	Spreading vs erect	Dominance of erect type
Leaflet shape	Lanceolate vs obcordate	Dominance of lanceolate
Disease resistance		
Fusarium wilt	Susceptible vs resistance	Dominance of susceptibility
SM virus	Susceptible vs resistance	Dominance of susceptibility
Phytophthora blight	Susceptible vs resistance	Dominance of susceptibility
Seed quality		
Protein content	Low vs high	Dominance of low protein

Derived from: Srivastava and Saxena (2023)

4. The fertility restoration of A_4 CMS system was controlled by two independent dominant genes; and for the stable fertility restoration in a genotype, the presence of both the genes was necessary.
5. Genetic studies showed that fertility restoration in A_1 and A_2 CMS systems was prone to temperature variations.
6. Transgressive segregations with significant gains in the seed size and high protein were reported in some interspecific crosses.
7. The quantitative genetics studies conducted so far in pigeonpea have confirmed the presence of both additive and dominance gene actions for yield, seed size, plant height, pods/plant, and seeds/pod.

4 Pollination and Hybridization Systems

Since pigeonpea is a perennial plant, its flowering and pod setting/growth continue simultaneously on the same plant. This provides opportunities to make hybridizations for extended periods. At Patancheru, it was observed that the anthesis in pigeonpea starts from 06.00 h and continues till 16.00 h. The peak anthesis takes place between 09.00 and 10.00 h. The duration of flower opening also depends on the weather conditions, and it may vary from 6 to 36 h. The stigma of pigeonpea remains receptive for extended period to allow the pollen to germinate and fertilize the egg cells. In a comprehensive study using a CMS line as recipient and its maintainer B-line as pollinator, Dalvi and Saxena (2009) demonstrated that in pigeonpea the stigma becomes receptive a day prior to flower opening; and it peaks during the next 3 days, including the day of flower opening. After this period, the petals start unfolding with about 35% of the buds remain unpollinated, but the stigma remained receptive to receive pollen grains. The flowers remained in this state for the next 2 days.

4.1 Emasculation and Cross-Pollination Methods

It has been observed that the hybridization success is high if the early formed buds on the plants are chosen for emasculation. At each floral site in a plant, a bunch of 5–6 buds of different ages are present, but for emasculation only two buds are selected and retained, and the rest are manually removed. The fresh tightly closed buds, approximately two thirds the size of mature buds should be selected on different branches of a plant. Such buds generally have bright yellow corolla without any greenish hue.

The selected floral bud is firmly held between thumb and the middle finger with the index finger used to provide sup-

Fig. 4 Step involved in manual emasculation and cross pollination

port (Fig. 4). The curved side of the standard petal is held toward the crosser, and the sepal covering the keel is removed first by a pair of scissors. Then, the corolla is opened by inserting one of the tips of a fine pointed forceps at the base of the keel, and the bud is open by moving the forceps upward to the tip of the standard. Such a bud will open with slight pressure of the supporting index finger and expose the staminal column. At this point, the anthers from the bud are carefully removed. The selected pollinator buds should be fully grown but still closed. From this bud, the staminal column is removed with forceps. Subsequently, its anthers are brushed on stigma of the emasculated buds to affect cross fertilization. The hand-pollinated buds are identified by tying a loose thread (Fig. 4). Different colored threads can be used to identify different crosses on the same plant.

The success of crossing is determined by both, the ambient temperature and humidity, besides the skill level of worker. An average pod set of 20–40% is often recorded. The perennial nature of pigeonpea allows for an extended flowering period. Saxena et al. (1976) successfully utilized ratooning of early and medium maturity types and their flowers on regenerated growth to cross divergent maturity types. Fertilization typically occurs 48–54 h after pollination. Pod development is completed approximately 20 days after blooming, with seeds reaching physiological maturity in 30 days and becoming ready for harvest in about 40 days.

4.2 Natural Cross-Pollination

The occurrence of natural out-crossing in pigeonpea was first recognized in 1909 when Howard et al. (1919) noted extensive bee activity on pigeonpea flowers (Fig. 5). They also observed significant variation in various traits among progenies selected from single plants in farmers' fields. Between 1912 and 1913, this research team conducted the first field experiment to assess the extent of natural out-crossing in pigeonpea. In this study, genotypes with yellow petals and a white seed coat (recessive traits) were used as the female parent, while those with red petals and a brown seed coat (dominant traits) served as the contaminant. By evaluating the flower and seed color of suspected hybrid plants, the estimated out-crossing in the progenies ranged from 2.25% to 12.0%. Subsequently, it was found that the natural cross-pollination in pigeonpea occurred in almost every field, but its extent varied considerably across the countries (Table 2) and growing environments within a country. The main reason for this variation is the density of the pollinator insects foraging in a field. It has been reported that the pigeonpea fields are visited by over a dozen insect species (Table 3), but the most common pollinators include *Apis mellifera*, *A. dorsata*, *Megachile bicolor*, *Megachilelanata*, and insects from *Xylocopa* (carpenter bee) and *Bombus* (bumble bee) species.

Fig. 5 A honeybee working on an open pigeonpea flower

Table 2 Natural out-crossing in pigeonpea reported from different countries

Country	Out-crossing (%)
Australia	2–40
Cameroon	15–39
China	0–60
India	0–42
Kenya	23–46
Myanmar	20–40
Nepal	20–25
Sri Lanka	1–20
Trinidad	2–40
Uganda	8–22

Derived from: Saxena et al. (2016)

Table 3 Major insect species visiting pigeonpea crop and help in pollen transfers

Family	Insect species
Apidae	*Apis mellifera* *Bombus terrestris auct*
Megachillidae	*Megachile kyotensis Alfken* *Megachile nipponica nipponica Cockerell*
Anthophoridae	*Xylocopa violacea L.* *Xylocopa valga Gerstaecker*

5 Domesticated and Wild Pigeonpea Genetic Resources

Plant genetic resources, including wild species and landraces, are considered a valuable reservoir of useful genes, and their potential use in crop improvement is significant. In this context, the role of "gene banks" in the collection, conservation, evaluation, and distribution of germplasm has become extremely critical. In pigeonpea, the major germplasm collection exists at International Crops Research Institute for the Semi-Arid Tropics (ICRISAT), Patancheru, India, and ICAR-National Bureau of Plant Genetic Resources (ICAR-NBPGR), New Delhi, India. ICRISAT genebank has 13,632 accessions from 74 countries. This collection has consideration variation for different traits (Table 4). The biological status of the collection indicates the presence of 8215 landraces, 4795 breeding materials, 67 cultivars, and 555 accessions from 67 wild species of genus *Cajanus*.

Similarly, the ICAR-NBPGR genebank holds a total of 11,229 pigeonpea accessions. These genetic resources represent a wide range of diversity for almost all the morpho-agronomic characters. In addition to these resources, 1366 accessions are also preserved at the ICAR-Indian Institute of Pulses Research (ICAR-IIPR), Kanpur, India. Several other countries have also secured the pigeonpea accessions, which include 4116 accessions at U.S. Department of Agriculture (USDA), USA; 1288 accessions at Kenya Agricultural Research Institute's National Genebank (KARI-NGBK) of Kenya; and 433 accessions at National Plant Genetic Resources Laboratory, Philippines. Of the wild species, some can be crossed (Table 5), mostly from the secondary gene pool.

Frankel and Brown (1984) suggested that small groups of diverse germplasm should be created and made available to the researchers. Following this approach, a core collection comprising of 1290 accessions capturing most of the available diversity in pigeonpea germplasm was developed (Reddy et al. 2005). Furthermore, a mini-core collection in pigeonpea, comprising 146 accessions, was also constituted with respect to different morpho-agronomic, biotic, or abiotic traits (Upadhyaya et al. 2006a). Another informative set of 300 accessions, referred to as "Reference Set," was also constituted using diversity indices based on Simple Sequence Repeat (SSR) markers data on a composite collection of pigeonpea (Upadhyaya et al. 2006b). It is important to mention that the reference set of pigeonpea has been used to establish the center of origin, dispersal route, and identification of genomic segments undergone natural/artificial selection sweeps using whole genome sequencing data (Varshney et al. 2017).

Table 4 Crossable wild Cajanus species

Species	Habitat	Major features
C. scarabaeoides	India, Sri Lanka, Australia, Africa	Creeper, antibiosis for pod borer, rounded grey seeds with black mosaic
C. sericeus	India	Erect shrub, densely branched, grey or black rectangular-rounded seeds
C. albicans	India, China	Perennial climber with woody base, rectangular dark colored seeds
C. cajanifolius	India	Erect, open branches, rectangular-rounded seeds with light orange color
C. acutifolius	Australia	Semi-spreading, leaves covered with short silvery hairs, seeds dark and oblong
C. lanceolatus	Australia	Erect slender shrub, seeds globose to compressed and dark mottled
C. latisepalus	Australia	Erect shrub, pubescent branches, seeds flattened-globose
C. lineatus	India, Sri Lanka	Erect, open branches, seeds flattened
C. reticulates	Australia	–
C. trinervius	India	Erect, branches pubescent, dark, flat seed

Table 5 Phenotypic variability for different traits recorded in the conserved pigeonpea germplasm grown at ICRISAT, Patancheru

Trait	Minimum	Maximum
Days to 50% flowering	45.0	237.0
Days to 75% maturity	85.0	299.0
Plant height (cm)	39.0	385.0
Primary branches (number)	2.0	66.0
Secondary branches (number)	0.3	146.0
Racemes (number)	6.0	915.0
Seed pod^{-1} (number)	1.6	9.0
100—seed mass (g)	2.8	25.8
Seed protein (%)	12.4	32.0

Source: ICRISAT

Furthermore, the systematic evaluation of wild *Cajanus* species has identified the accessions with resistance to biotic stresses, such as Alternaria blight (Sharma et al. 1987), Phytophthora blight (Rao et al. 2003), sterility mosaic virus (Kulkarni et al. 2003; Rao et al. 2003), pod borer, *Helicoverpa armigera* (Rao et al. 2003; Sujana et al. 2008; Sharma et al. 2009), pod fly (Rao et al. 2003), pod fly and wasps (Sharma et al. 2003), and nematodes (Rao et al. 2003). Besides this, such groups with respect to tolerance to abiotic stress including salinity (Srivastava et al. 2006) and drought (Rao et al. 2003) tolerance, and photoperiod insensitivity (Rao et al. 2003). Few of the above-mentioned sources have been used to develop pre-breeding populations with the aim to transfer favorable alleles from wild relatives to cultivated types (Sharma and Upadhyaya, 2016). Two such back-cross populations (BP) namely ICPL 87119 (*C. cajan*) × ICPW 15613 (*C. acutifolius*) and ICPL 87119 × ICPW 29 (*C. cajanifolius*) were used to identify quantitative trait loci (QTLs) for yield-related traits and to discover novel alleles from wild species, which can be used for improvement of traits through genomics-assisted breeding (Saxena et al. 2020a).

Additionally, to effectively utilize wild germplasm in pigeonpea improvement, systematic genome-wide surveys of accessions are needed, along with the identification of adaptive alleles for coping with environmental extremes. The incorporation of diverse wild alleles into cultivated backgrounds will be essential for trait analysis and breeding. Availability of huge amount of genomic resources coupled with advances in next-generation sequencing provides unprecedented opportunities to catalogue the genome-wide diversity across different accessions at a very cost effective and timely manner and mining the superior alleles but also for systematic and precise use of wild species, without introducing linkage drag, for pigeonpea improvement (Bohra et al. 2020a, b). Introgression libraries (ILs) developed after introducing smaller and useful genomic segments from wild species in the genetic background of cultivated species are valuable resources. High-density genotyping information on ILs is required to locate the short chromosome segments and further its use in breeding program (Saxena et al. 2020b). Additionally, several other approaches are being developed for the effective utilization of wealth of germplasm to the crop improvement programs, e.g., haplotype mining for the preferred traits, genome-wide characterization, etc.

6 Cropping Systems and Production Trends

The cropping system is defined as "a combination of crops in time and space that provides farmers high levels of stable returns." The technology of growing these crops for their optimum productivity in an ecological system is influenced by various social, economic, and edaphic factors, besides the volume and distribution of precipitation.

6.1 The Major Pigeonpea Cropping Systems

The traditional pigeonpea cropping systems involve growing medium (160–180 days) and late (>200 days) maturing varieties with a variety of short-aged crops like sorghum, pearl millet, maize, groundnut, cotton, sunflower, or soybean. These pigeonpea cultivars are sown after the longest day, and their harvesting is done well after the shortest day, and these also display a strong photo-sensitive reaction. In this system, the intercrops are generally harvested in about 100–120 days and that leaves sufficient time for pigeonpea to grow and establish an optimum canopy for reasonable yields. The pigeonpea plants also develop deep and voluminous roots, and these help in exploiting residual moisture after the removal of the companion crop. In contrast, the early cultivars mature in 100–140 days, and these are cultivated as a sole crop; and their harvesting is completed before the sowing of a post-rainy season wheat crop. The emergence of the early types has helped in extending the adaptation of pigeonpea beyond 35° latitude and high (>1000 m) altitudes, and to the arid areas where pigeonpea was never grown earlier (Saxena et al. 2019a).

The most profitable and widespread crop rotation involve rice in the rainy season followed by wheat crop in the post-rainy months. This cropping system is popular in the western and northern areas of north India, including the Indo-Gangetic plains. This crop combination not only has brought food security in the country but also earning valuable foreign currency. The extensive adoption of this cereal-cereal crop rotation over the decades has shown clear signs of soil fatigue in terms of deteriorating soil structure, poor drainage, increased salinity, and finally, a visible decline in the productivity. Dahiya et al. (2002) opined that due to heavy use of chemical fertilizers and liberal use of irrigation over the years are drifting the cropping environment toward unsustainability. In this context, the pigeonpea-wheat rotation has come for a good to rejuvenate this eco-system.

6.2 Production and Productivity Trends

Nigam et al. (2021) presented a comprehensive analysis of the production and productivity trends of pigeonpea, and the following information is primarily credited to them. Between the period of 1961 and 1963, the pigeonpea production at global level recorded an increase by 147.6%, which is largely due to 96% (2.6 m ha) increase in area and of 26% (176 kg/ha) in yield. The annual growth rates of area, production, and yield were 1.4%, 1.8%, and 0.4%, respectively. Asia is the largest (83.6%) contributor to global pigeonpea area and 78.2% to the production. During the same period, the yield increased by 17% (116.4 kg/ha). Despite various challenges, both the area and production of pigeonpea in India have shown consistent growth over the past five and a half decades. The contribution of Africa is 13% to the global pigeonpea area and 19% to its production. The average yield of pigeonpea in Africa is 1196 kg/ha. The leading pigeonpea-producing countries in Africa are Kenya, Tanzania, Uganda, Mozambique, and Malawi, with the extent of their annual cultivated area largely influenced by export demand and market prices.

Between 1961 and 1963, pigeonpea production in Africa grew by 821% (0.76 million tons), area by 311% (0.55 million hectares), and yield by 126% (666 kg/ha). Notable pigeonpea-producing countries in the Caribbean include small island nations such as Haiti, the Dominican Republic, and Puerto Rico, where the crop is primarily grown as a vegetable (green peas) for both domestic consumption and export markets. Collectively, the Caribbean islands cultivate pigeonpea on 0.15 million hectares, producing a total of 0.13 million tons, with an average yield of 874 kg/ha. This region contributes 3% to both the global area and production. Pigeonpea production here has increased by 309%, primarily driven by a 409% expansion in area.

6.3 Utilization Pattern

The projected global demand for pigeonpea is expected to reach 5.58 million tons by 2030, increasing to 6.42 million tons by 2040. However, global supply is anticipated to meet demand during both of these periods. Over 98% of the world's pigeonpea production is concentrated in developing countries, with supply from developed nations remaining almost negligible. South Asia is the largest producer, accounting for about 72% of global production, followed by sub-Saharan Africa and East Asia Pacific (Nigam et al. 2021). The two leading pigeonpea-producing countries, India and Myanmar, are projected to become self-sufficient and slightly surplus in production by 2030, maintaining this status until 2040. In contrast, countries in sub-Saharan Africa, including Kenya, Tanzania, Uganda, and Malawi, may experience a small gap between projected demand and supply. Meanwhile, domestic supply in Venezuela and Nepal is expected to significantly decline and fail to meet demand. Future research efforts should focus on introducing high-yielding technologies and exploring value-added opportunities in these regions.

7 Breeding Objectives and Constraints

7.1 Breeding Objectives

Pigeonpea is an important high protein pulse that is grown extensively under subsistence agriculture. In spite of intensive R&D efforts, it has been tagged as a low-yielding crop. There may be a number of agronomic, biological, and social factors responsible for this scenario. Over the years, this issue is continuously haunting the research managers, scientists, and political leaders. To overcome this undesirable scenario, the pigeonpea breeders all over the country are engaged in developing disease resistant widely adapted high-yielding cultivars, but the issues till persists. Hence, the prime sole objective of pigeonpea breeding program should be to enhance the productivity on a sustainable basis. In this context, the use of two emerging technologies, i.e., CMS-based hybrid breeding and use of genomics tools, should be given the highest priority. The other objective should be to transform pigeonpea from partially cross-pollinated to self-pollinated crop. The efforts made in different research areas, and their accomplishments are highlighted in the following text. Prior to this, the constraints often encountered in this endeavor are also briefly discussed.

7.2 Inherent Constraints Encountered in Breeding Pigeonpea

In pigeonpea, a number of genetic and biological processes and their complex interactions among themselves and with local environment adversely affect breeding value of the selections. Some of the key natural and plant breeding constraints are briefly described herewith.

7.2.1 Limited Genetic Diversity

Kumar et al. (2003) studied the pedigrees of all the pigeonpea cultivars released in India and concluded that only a limited number of parental lines from primary gene pool were used in the past breeding programs. They further postulate that this is one of the key factors for low productivity of new cultivars. The primary gene pool of pigeonpea germplasm includes over 15,000 accessions, and these exhibit tremendous phenotypic variability. The same, however, is not true about their molecular diversity. Since the molecular diversity in secondary gene pool is greater, it deserves to be utilized in the pigeonpea breeding programs. Unfortunately, pigeonpea breeders did not exploit this resource in the past, perhaps due to strong linkage drag and paucity of resources.

7.2.2 Long-Generation Turnover Time

Breeding efforts in pigeonpea are limited by the long time it takes to complete one seed-to-seed generation. It is because all the medium and long duration germplasm are highly sensitive to photo-period, and the plants start flowering only under long periods of darkness, and this limits only a single generation within a year. Therefore, breeding of new cultivars becomes a long-term (10–12 years) endeavor. So far, no photo-insensitive germplasm is available in pigeonpea to overcome this constraint.

7.2.3 Harvest Index

Harvest index is an indicator of grain productivity of plants in relation to their biomass production. This parameter is closely linked to the efficiency of plants to transfer their dry matter to the developing grains. Being a perennial species, the pigeonpea plants transfer a greater proportion of their carbohydrates as food reserve in roots, stem, and branches, and only a small portion is available for fulfilling the grain needs. This results in poor (0.2–0.3) harvest indices of pigeonpea that leads to low yields.

7.2.4 Biotic and Abiotic Stresses

Pigeonpea encounters various biotic (diseases and insects) and abiotic (water-logging, drought, salinity, etc.) stresses during its life cycle. These stresses result in total or partial yield losses. So far, the resistance breeding in pigeonpea has been limited to wilt and sterility mosaic diseases only. The main reasons for the failure to overcome the other stresses are (i) non-availability of stable resistant sources (particularly insects), (ii) inadequate knowledge of stresses (e.g., drought), (iii) lack of screening appropriate technologies, and (iv) intra-accession variability due to inadequate maintenance breeding programs.

7.2.5 Poor Response to Selection

A perusal of national coordinated trials shows that there have been tremendous genetic gains through crop breeding for simply inherited traits such as earliness, seed size, and resistance to wilt and sterility mosaic diseases. But the gains in grain yields were far from the expectations. Swaminathan (1973) opined this failure to poor selection efficiency and various physiological and management limitations. Saxena and Nigam (2022, 2023) postulated that for seed yield the genotype-environment interactions in pigeonpea are extremely large even at micro level; which induce tremendous non-heritable phenotypic variability among individual plants; and this results in poor heritability and breeding value.

7.2.6 Genetic Contamination of Selections

Besides self-pollination, the pigeonpea plants are also prone to insect-aided cross-pollination. Surprisingly, in spite of knowing the ill effects of out-crossing, the pigeonpea breeders always resorted to pure line breeding while developing new cultivars. Since the breeding activities are invariably performed in open fields, the crop is exposed to random insect visitations. This leads to some undetected cross-hybridization of individual plants. In the following season, such out-crossed (natural hybrids) may be selected by breeders due to their hybrid vigor. Such selections do not breed true in succeeding generations and, consequently, adversely affect their breeding values. Besides this, the natural cross-pollination renders considerable difficulties in maintaining genetic purity of breeding stocks and parental lines identified for use in breeding and academic research activities.

8 Conventional Breeding Methods and Accomplishments

Like most autogamous crops, the commonly used breeding methods have also been applied for genetic enhancement of pigeonpea. These include direct introduction, pure line selection from open-pollinated germplasm, hybridization followed by pedigree selection, and mutation breeding. The major plant breeding accomplishments resulted in the release of 87 pure line cultivars for various agro-ecological regions (Table 6). Of these, respectively 32 and 50 were developed through germplasm and pedigree breeding, besides these 5 cultivars were derived from mutant populations.

Table 6 Pigeonpea inbred cultivars developed through different breeding methods

Maturity group	Total releases	Germplasm selection	Hybridization	Mutation
Early	34	6	24	4
Medium	37	17	19	1
Late	16	9	7	0
Total	**87**	**32**	**50**	**5**

Fig. 6 A favorite vegetable type pigeonpea landrace ICP 7035

8.1 Germplasm Introduction and Selection

The earliest breeding activities in pigeonpea started with the collection of landraces from farmers' fields. In this endeavor, some breeders directly released the elite germplasm as cultivars, while the others exploited the intra-accession variability of the heterogeneous landraces. They selected promising individual plants and advanced them through pedigree breeding. Using this approach, a number of the medium and long duration cultivars were developed, and some of them are still preferred by farmers. The most popular cultivars developed from germplasm selection were C 11, T 7, BDN 1, HYD 3C, NA 1, etc. Beside this, some important sources of diseases resistance were also identified from the germplasm. These include 15-3-3 (Fusarium wilt), ICP 366 (sterility mosaic virus), ICP 7105 (Alternaria blight), PB 9 (Phytophthora blight), ICP 7035 (wilt and sterility mosaic virus), and many more. ICP 7035, a collection from Madhya Pradesh (Fig. 6), has proved to a source of highest (8.8%) sugar content, and this trait made ICP 7035 the most popular vegetable type in a number of countries. Fiji was the first to release it as national pigeonpea variety and named it as 'Kamika'.

8.2 Pedigree Breeding

In the subsequent breeding programs, the emphasis was given to enhance some specific traits through hybridization and pedigree selection. These included earliness, seed size, pod size, plant type, disease resistance, and yield (Choudhary and Nadarajan, 2011). At present, there are a number of pigeonpea varieties at farmers' disposal in each maturity group.

8.2.1 Early Maturity

The program for breeding early maturing cultivars started with the selection of an early flowering spontaneous mutant (T-1) within a late maturing landrace. Subsequently, this source was used as donor to breed early maturing cultivars through pedigree breeding. The popular early maturing (120–140 days) cultivars, such as T. 21, Prabhat, UPAS 120, Manak, Pant A2, Pusa 992, and ICPL 88039, were also bred. The new early maturing cultivars developed through pedigree breeding are potentially high yielding (2000–2500 kg/ha) and are invariably cultivated as a sole crop under good crop management. This crop culture also provides stability to the production system through enrichment of soil (Saxena et al. 2019). The area under early maturing pigeonpea, however, is <10% and, hence, do not significantly influence the production statistics at the national level. Recently, some super early pigeonpea lines maturing in <90 days were also bred through pedigree breeding (Srivastava and Saxena 2019). These early maturing cultivars have helped farmers in diversifying their cropping systems and exploring some new niches (Saxena et al. 2019a).

In the longer maturity groups also, a number of high yielding cultivars have been bred through pedigree breeding, but the productivity increases at farmers' level have been elusive. In fact, in this maturity group, there exists a huge gap between the potential and realized yield. This was evident from the demonstrations of variety IPA 203 conducted in Uttar Pradesh. This variety produced >4000 kg/ha under optimum crop management practices, whereas under farmers' conditions its yields were 1500 kg/ha or less. The realized on-farm low yields were attributed to poor crop management practices and various stress factors (A. K. Choudhary, pers. com). Some of the key accomplishments recorded through pedigree breeding are briefly discussed in the following text.

8.2.2 Biotic and Abiotic Stresses

The host-plant resistance breeding in pigeonpea has been attempted for two major diseases—fusarium wilt and sterility mosaic virus; and notable successes have been accomplished in this endeavor. All these cultivars were bred using the standard pedigree system (Choudhary et al. 2013). The base populations for implementing pedigree selections were either heterogeneous germplasm or breeding populations derived from single, double, or backcrosses. In pigeonpea, the pure line selections made within landraces were very effective in developing disease-resistant cultivars both in India and Africa. For example, in India, variety Maruti, a pure line selection from the landrace 15-3-3 has proved a boon to the farmers of wilt-prone areas of peninsular India (Bantilan and Joshi 1996). Similarly, in Africa, the most popular pigeonpea variety 'Nandolowanswana' is a pure line selection from a Tanzanian landrace ICP 9145 (Silim 2001). Besides these, a number of disease-resistant varieties were also developed for different regions using hybridization and pedigree breeding (Singh et al. 2016). Similarly, pure-line selections from landraces have also resulted in the identification of a pod borer (*Helicoverpa armigera*) and Maruca (*Marucavitrata*) tolerant genotypes (Lateef 1992; Saxena et al. 2002). Besides these, a few water-logging tolerant pure-line selections have also been derived from landraces and breeding populations (Sultana et al. 2013).

8.2.3 Biofortification for Nutritional Sustainability

Protein content is the first and foremost quality trait in pulses. Its content in the primary gene pool of *Cajanus* is around 20–22%, but some of its wild relatives such as *C. scarabaeoides*, *C. sericeus*, and *C. albicans* have up to 32% protein (Saxena et al. 1987). The high-protein trait was successfully transferred from wild to cultivated types through pedigree breeding. In this program, segregants with 26–28% seed protein were recovered within inter-specific populations. Besides high protein, the two selections HPL 40 and HPL 8 also produced 2100 and 1660 kg/ha grain yield, respectively (Saxena et al. 2023). This simply means that, in comparison traditional cultivars, the cultivation of high protein lines in one hectare plot can provide additional 100 kg of digestible protein (Table 7) to the farming family. The biological evaluation of such lines using rat-feeding trials showed that high-protein genotypes were nutritionally superior to the standard pigeonpea cultivars (Singh et al. 1990). Recently, a transgressive selection with the highest-ever protein (35.6%) content was recovered from an inter-specific cross Baigani x *Cajanus scarabaeoides* (Saxena et al. 2023a) has further enhanced the scope of increasing the protein content of pigeonpea.

Table 7 Seed and protein yields harvested from high protein lines

Genotype	Maturity (days)	100-seed wt. (g)	Yield (kg/ha)	Protein (%)	Protein yield (kg/ha)
HPL 40-5	169	9.6	2100	26.9	452
HPL 40- 17	169	8.5	2070	26.5	440
BDN 1 (C)	168	9.6	2020	23.2	373
SEm±	0.9	0.18	160	0.46	–
CV (%)	0.9	3.4	17.3	3.0	–

Source: Singh et al. (1990)

8.3 Mutation Breeding

Early mutation breeding research in pigeonpea primarily focused on determining effective doses of various chemical and physical mutagens and documenting the induced genetic variation for different morphological traits. Among these, treatments using ethyl methane sulfonate ($C_3H_8SO_3$), fast neutrons, and gamma rays successfully created valuable genetic variation in pigeonpea, although their practical application in breeding high-yielding cultivars was limited. Genetic improvements through mutation breeding in pigeonpea have been achieved for traits such as yield, earliness, seed size, and disease resistance (Pawar et al. 1991). To date, only five commercial pigeonpea cultivars (Co 3, Co 5, TT 5, TT 6, TAT 10) have been developed through mutagenesis (Singh et al. 2016).

8.4 Variety Development and Testing System

The national pigeonpea varietal development and testing programs was launched soon after the establishment of All India Coordinated Pulses Improvement Programme in 1966. Since then, some changes were made with respect to the formulation of trials related to different maturity durations and their agronomy, but no major change was witnessed in the procedures for selecting cultivars for release. Overall, this system has served quite well in identifying new cultivars but the productivity issues are still haunting, it seems that some new approaches, if introduced may help in enhancing the realized yields. The authors feel that in the present system the performance of test materials is highly influenced by genotype-environment interactions, and this leads to instability in the varietal performance. Therefore, one option could be to give more emphasis on breeding cultivars with specific adaptation. This will reduce g x e interactions and allow better expression of yield under relatively uniform environmental conditions and soil type. Perhaps, it is not easy option as it would require major changes at a large scale.

9 Evolution of Extra Early Maturing Pigeonpea

Earliness in crop plants has always fascinated both researchers and farmers equally. All the traditional pigeonpea cultivars and landraces are of long (6–9 months) duration. The emergence of early maturing genotypes is fairly recent, and this subject has been reviewed by Saxena et al. (2019a). It was in 1953, when an early maturing mutant, identified within the population of a long duration landrace that was grown in a farmer's field in Gorakhpur district of Uttar Pradesh, was selected and preserved. This mutant was designated as "T-1," and it was used in hybridization with a long duration genotype T-190 to breed the first early maturing cultivar "T-21" at Kanpur in 1961. This inbred was adopted for cultivation in rotation with wheat, but, at this point, a necessity of breeding even earlier pigeonpea cultivars was felt so that this crop rotation could be extended to other wheat-growing areas.

The first extra-early maturing variety, "Co-1," maturing in 120–130 days, was developed in 1970 from a spontaneous mutation in a local landrace grown in Perambalur, Tamil Nadu. Concomitantly, another early maturing variety, "Prabhat," was bred in Kanpur from a spontaneous mutation of the T-21 variety. In 1974, the widely popular high-yielding early maturing variety "UPAS 120" was developed at Pantnagar from a spontaneous mutant of the P-4758 germplasm. Later, three additional extra-early maturing cultivars were bred at Pantnagar by selecting spontaneous mutants from the UPAS 120 population. More recently, Srivastava et al. (2012) demonstrated transgressive segregation for extremely early types within a segregating population from two early-flowering pigeonpea parents. These "super early" selections flowered significantly earlier than both parent varieties, making them the earliest maturing pigeonpea lines ever reported in the Cajanus genus.

In India, over thirty early and extra-early maturing, pigeonpea cultivars have been released for cultivation. The successful adaptation of these cultivars to various soil and climatic conditions has prompted scientists to explore the potential for growing them in regions where pigeonpea was previously not cultivated. According to estimates from the Directorate of Economics and Statistics (DES, 2017), the adoption of these cultivars by farmers has expanded the area under pigeonpea by over 100,000 hectares in India. The successful breeding of extra-early maturing varieties, such as Manak and ICPL 88039, has contributed to stabilizing the pigeonpea-wheat rotation, which is now widely adopted in Punjab, Haryana, and the Indo-Gangetic plains. Additionally, extra-early maturing pigeonpea cultivars have shown promise when introduced to new areas. One such successful application has been in the rainfed hilly regions of the country, where pigeonpea has made a significant impact. Despite low productivity in these areas (300–400 kg/ha), on-farm trials have demonstrated that pigeonpea, particularly ICPL 88039, can be successfully grown at elevations up to 1580 meters (Champawat district), with grain yields ranging from 1250 to 1878 kg/ha.

10 Evolution of Hybrid Technology and Present Position

Unlike other pulses, pigeonpea offers a unique opportunity to breeders for exploiting hybrid vigor for its yield enhancement, and this is primarily due to the availability of natural out-crossing which can allow large-scale seed production of its hybrids. The development of this path-breaking technology was painstaking with research of over three decades. The entire hybrid pigeonpea breeding scenario, starting from its concept to the release of first hybrid, has been captured in an article published by Saxena et al. (2018a, b, c) in *Plant Breeding Reviews*. A brief overview is reproduced here.

To launch this ambitious project, as a first step, plans were made to breed a stable cytoplasmic-nuclear male sterility (CMS) system using the cytoplasm of wild species and nuclear genome of pigeonpea. The successes were achieved when Tikka et al. (1997) selected a natural hybrid within the open pollinated population of *C. scarabaeoides* (A_2 cytoplasm) and by crossing *C. cajanifolius* (A_4 cytoplasm) with a cultivated type (Saxena et al. 2005). These two CMS sources were used to (i) identify their fertility restorers; (ii) develop experimental hybrids in early, medium, and late maturing groups; and (iii) test them under diverse environments to assess their yield and stability potential.

In this program at ICRISAT, some high-performing hybrids were identified after their multi-location testing (Table 8). Among the early maturing hybrids ICPH 2433, ICPH 2438, and ICPH 2363 were outstanding with, respectively, 54%, 42%, and 36% heterosis over the best control. Similarly, ICPH 3491 (57%), ICPH 3497 (44%), and ICPH 3481 (41%) in medium; and ICPH 2307 (53%), ICPH 2306 (39%), ICPH 2896 (38%) in late maturing groups were found outstanding. This was followed by the development and release of three A_4 CMS-based hybrids. These included ICPH 2671, ICPH 2740, and ICPH 3762 by ICRISAT.

Since this technology was new, the on-farm validation of hybrid yields was essential. For this purpose, a series of on-farm trials were organized in different states of India. The

Table 8 Yield (kg/ha) and standard heterosis of hybrids of different maturity groups in multi-location trials

Maturity	Hybrid	Loc.	Mean yield (kg/ha)	Heterosis (%)
Early	ICPH 2433	25	2306[a]	54
	ICPH 2438	25	2127[a]	42
	ICPH 2363	25	2048[a]	36
Medium	ICPH 3491	18	2919[a]	57
	ICPH 3497	18	2686[a]	44
	ICPH 3481	18	2637[a]	41
Late	ICPH 2307	5	2855[a]	53
	ICPH 2306	5	2600[a]	39
	ICPH 2896	5	2579[a]	38

[a]Significantly different from the corresponding control variety at $p < 0.01\%$

Table 9 Yield of ICPH 2671 in the on-farm trials in four states

State	Farmers (no.)	Mean yield (kg/ha)		Standard Het. (%)
		Hybrid	control	
Maharashtra	782	969	717	35
Andhra Pradesh	399	1411	907	56
Jharkhand	288	1460	864	69
Madhya Pradesh	360	1940	1326	46
Total/Mean	1829	1445	954	52

Fig. 7 High-yielding hybrid ICPH 2740 in an on-farm trial in Maharashtra state

medium duration wilt and sterility mosaic resistant hybrid ICPH 2671 was evaluated in 1829 on-farm trials using farmers' own cultural practices. The highest number (782) of tests were conducted in Maharashtra state, and on average, the hybrid demonstrated 35% superiority over the best local control (Table 9). The performance of this hybrid in the three other states was equally good. Similarly, the on-farm productivity data of ICPH 2740 (220 trials) and ICPH 3762 (144 trials) were also outstanding with respectively 45% and 112% standard heterosis (Saxena et al. 2018a, b, c). The studies conducted on hybrid seed production technology showed that using a row ratio of 1 male: 4 female row can produce high economic yields. In this context, the on-farm trials of the seed production technology, conducted in collaboration with seed producers, confirmed the high harvests of hybrid seeds.

The large-scale on-farm testing of hybrids and their seed production demonstrated that the hybrids technology has a future, and it has a potential to yield 30–50% more grains than the best available varieties in the respective maturity groups (Fig. 7). This development also gave confidence that the hybrid technology has a potential to smash the low yield plateau in pigeonpea to benefit Indian farmers. In spite of such positives of the hybrid technology, at present the farmers are unable to reap its benefits. The main reason for this bottleneck is non-availability of the adequate seed quality control mechanism. The routing grow-out testing of the hybrids and their parents was not feasible in pigeonpea due to photo-sensitivity of the species. This limits to only one seed-to-seed crop per year. This single factor made the seed producers hesitant in adopting the hybrid technology, and ultimately, it restricted the exploitation of hybrid vigor at commercial level.

11 Maintenance Breeding in Pigeonpea

To retain the genetic potential of new as well as popular cultivars for fairly long periods, it is necessary that their genetic purity is maintained through the generations at the highest level. In the same way, for a sustainable breeding program, there is a need to preserve the purity of genetic stocks and parental lines. The maintenance breeding in the self-pollinated crops is relatively easier and cost effective than those of the cross- pollinated species. This group of food legumes maintain their genetic purity with a little effort under natural growing conditions, but it is not entirely true as some level of natural out-crossing occur in crops like faba bean, pigeonpea, and grass pea. Among these, pigeonpea is the toughest to deal because of the insect-aided cross-pollination. For producing genetically pure seed of pigeonpea in different volumes, various selfing techniques are used. These include small muslin cloth bag, large muslin cloth bag, nylon mosquito net, and isolation plots (Fig. 8).

Fig. 8 The pure seed production tools include selfing bags (1), nylon net cages (2), and isolation plot (3)

11.1 Natural Cross-Pollination and the Factors Responsible

Green et al. (1981) reported that with an average insect-aided out-crossing of 25–30%; it will take only two to three generations of open-pollination to destroy the potential of cultivars and selections. Hence, in pigeonpea, the maintenance breeding programs should be an inseparable part of crop breeding endeavor. This activity primarily deals with the principles and methods of producing high-quality products in various approved seed grades. The factors responsible for encouraging out-crossing in pigeonpea are briefly discussed herein.

11.1.1 Nectar Production

In pigeonpea, the nectar is secreted by floral nectaries, located at the base of perianth; and this attracts a variety of pollinating insects. The production of nectar is consistent across the flowering duration, and its amount is dependent on the balance between plant frugality and requirement of the plants to attract the pollinators. The nectarivore insects extend their proboscis, and during the process of nectar sucking, they brush anthers to collect pollen grains or pistil for pollination.

11.1.2 Extended Flowering Duration

The perennial and non-determinate growth habit of pigeonpea enables the plants to sustain both vegetative and reproductive stages simultaneously. This condition persists for extended periods, either until the desired pod load is reached or the growing conditions become unfavorable. The extended reproductive phase allows for the formation of more number of out-crossed pods on the plants.

11.1.3 Presence of Alternate Hosts for Pollinating Insects

Physical factors such as temperature, humidity, wind direction and velocity, along with the local habitat, influence natural out-crossing by facilitating the buildup of insect populations. It has also been observed that pigeonpea fields located near alternative host plants, such as fruit trees, grasses, and legumes, contribute to a sufficient population of pollinating insects, which in turn aids in cross-pollination.

11.2 Seed Classes of Hybrids and Their Parents

In India, there are five approved seed classes namely, nucleus seed, breeder seed, foundation seed, certified seed, and truthfully labelled seed. In this context, the standard seed production procedures, and their quality parameters for inbred cultivars are well established and documented. In pigeonpea, the following seed production grades have also been established for hybrid breeding.

11.2.1 Nucleus Seed Production

The nucleus seed of hybrid parents must be produced to the highest standards of genetic purity. To multiply the nucleus seed of the male sterile (A) line, it is essential to obtain the available nucleus seed of both the A and B lines from a trusted source. These seed lots should be inspected for any physical impurities before proceeding with multiplication. The seeds should be sown under an appropriately sized insect-proof net.

During flowering, each plant of the A line should be carefully inspected for any trace of pollen, and any suspicious plants should be removed. Similarly, each plant of the B line should be checked for its characteristic morphological traits, and rouging should be carried out to eliminate off-types. Seed multiplication of the nucleus seed should be done by hand-pollinating individual A line plants with single B line plants. The harvested seeds should be labeled, dried to approximately 9–10% moisture, and stored in a cool, dry place. Since the maintainers (B line) and restorers (R line) are male fertile, their nucleus seed production should follow the same process as other cultivars but under insectproof nets. In this case as well, individual plants should be inspected for their specific morphological traits, and strict rouging standards should be applied. After careful harvesting, threshing, and labeling, the seeds can be stored.

11.2.2 Breeder Seed Production

In this category of seed production, extremely high purity standards must be maintained. The seed is multiplied in isolation, at least 1000 meters away from other pigeonpea crops. For male sterile (A) line seed production, the nucleus seeds of the A and B lines should be sown in isolation at a 2 female: 1 male row ratio. During flowering, intensive rouging of both A and B lines should be carried out. At maturity, the male rows should be harvested first by cutting the plants at ground level and removing them from the field, followed by the harvesting of the A line plants. The breeder seed production of B and R lines should take place in separate isolations, with a minimum distance of 1000 meters between them.

11.2.3 Foundation Seed Production

This is a requirement for certified seed grade, and as such, relatively large quantities of parental seed must be produced. The seeds of all three parents are grown in separate isolations, with a minimum distance of 500 meters between them. For A line seed multiplication, a female to male row ratio of 4:1 is recommended. Each plant in the female rows should be examined for male sterility. At maturity, the male rows should be harvested first and removed from the field.

11.2.4 Certified Hybrid Seed Production

Hybrid seed production for distribution to farmers is classified as "Certified Seed" grade. A 4 female to 1 male row ratio typically yields good results. To eliminate off-type plants, at least two rounds of rouging are necessary: one before flowering and another during flowering. To prevent seed mixture, the male rows should be manually cut and removed from the field.

11.3 Important Features of Isolation Plots

A universal isolation distance for pigeonpea seed production is challenging to recommend due to various physical factors, as mentioned earlier. Therefore, local breeders or seed producers should rely on their judgment and past experience. Based on data from multiple locations over the years, ICRISAT suggests an isolation distance of 500 meters for certified hybrid seed production and 1000 meters for breeder and foundation seed production of hybrid parents. Another crucial factor is the selection of the isolation site. The isolation plot should be favorable for attracting pollinating insects, which is more likely if the area has natural habitats with bushes, fruit trees, and other vegetation. Such sites typically support high bee activity and improved pod set. Additionally, it is advantageous to locate production sites near temporary or permanent water bodies.

The planting design is crucial for optimizing hybrid yields. The primary goal should be to keep the pollen parent plants in a flowering state for as long as possible, which will extend insect visits and maximize cross-pollination. Additionally, the net seed production plot, sown in a 4 female: 1 male configuration, should be surrounded by a two-meter-wide belt of pollen parent plants, ideally sown 2–3 weeks later than the main plot (Fig. 9). This approach is effective in small seed production plots of about an acre. For larger seed production plots, it is recommended to divide the field into 2–3 subplots, each planted with the 4 female to 1 male combination. Sowing of each subplot should be staggered by 2 weeks. This hybrid seed production method has been successfully tested at multiple locations, with an average hybrid yield of 1019 kg/ha across 94 locations.

Seed quality control is a top priority in every commercial hybrid breeding program. In short-duration crops, it is relatively easy to assess the quality of hybrids and their parents by evaluating their progenies for specific morphological markers controlled by one or two dominant genes. However, pigeonpea is a long-duration crop (around 200 days), making it impossible to grow two crops per year. To address this challenge, seed quality testing methods based on molecular markers can be employed. For this purpose, a set of two simple sequence repeat (SSR) markers for hybridity testing has been developed (RK Saxena et al. 2010). Additionally, gene-based or gene-linked markers can also be utilized for hybridity testing.

11.4 Make Pigeonpea an Autogamous Species Through Breeding

The logical approach to address the issue of cross-pollination is to find out a stable genetic solution. In this context, a cleistogamous flower mutant (Fig. 10), identified within an F_7 inter-specific population (Saxena et al. 1994), was found

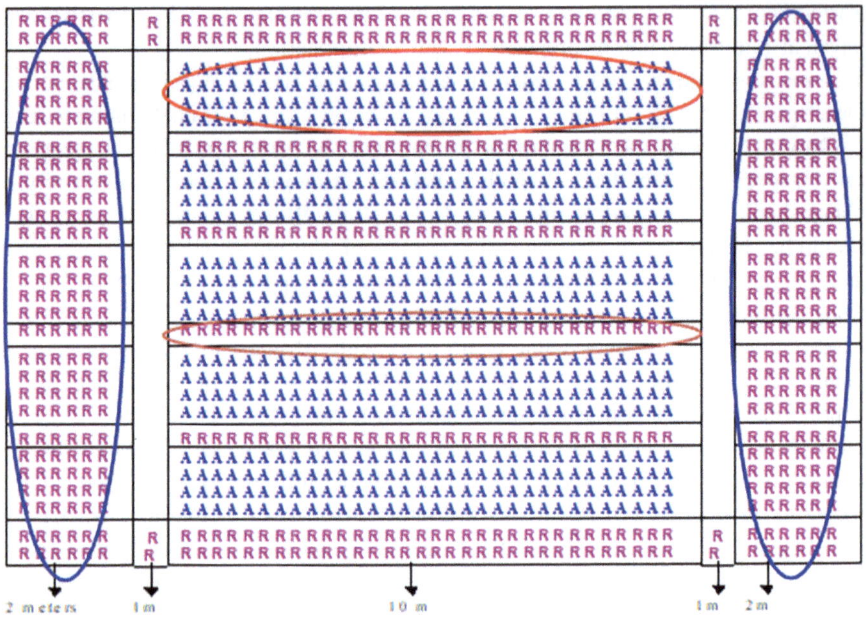

Fig. 9 An outline of a hybrid seed production plot. (A male sterile; R restorer line)

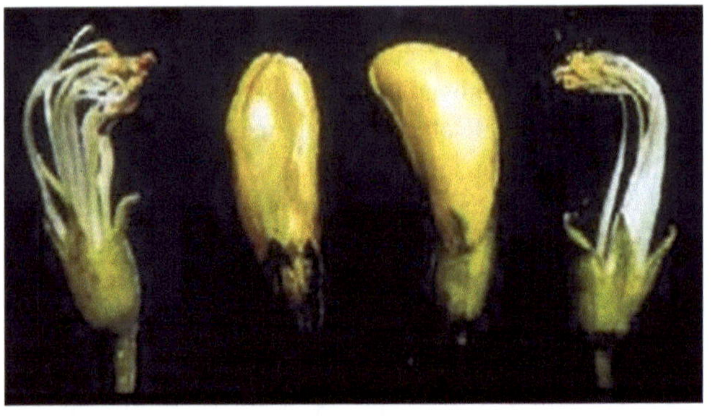

Fig. 10 A modified pigeonpea flower (left) with all 10 free stamen, which restricts out-crossing to <2% and a normal (9 + 1) flower which registers >30% out-crossing

most promising. The natural out-crossing in this genotype is <2% as compared to 25–30% in the normal flower type cultivars. Since this floral mutant is easy to identify and controlled by a single recessive gene pair, it could easily be incorporated into elite germplasm and cultivars through breeding. Recently, Yadav et al. (2019) have also identified this gene at molecular level, and thus, the breeding an autogamous pigeonpea cultivars could be accelerated using specific molecular tools. At ICRISAT, this trait has been incorporated in some known high-yielding cultivars. The authors feel that the incorporation of this trait in breeding programs could provide a long-lasting solution of the purity maintenance.

12 Some Unexplored Pigeonpea Breeding Technologies

12.1 Enhance Genetic Diversity Through Pre-breeding

The wild relatives of cultivated species are a well-established source of new genes, and their potential in crop improvement was recognized early in plant breeding. Since the genetic diversity within the primary gene pool of pigeonpea is limited, expanding its genetic diversity can be achieved by utilizing the secondary gene pool. However, the process of transferring genes from wild species to cultivated species is both resource-intensive and time-consuming. Therefore, this process can be divided into two main phases. The first phase involves developing unselected advanced-generation (F5/F6) populations for further breeding activities, a process referred to as "pre-breeding." In the second phase, the pre-breeding materials are used for further selection or as parents in variety development programs.

12.2 Adopt Speed Breeding Approach

Speed breeding provides an ideal platform to accelerate the process of cultivar development. Pure-line breeding takes about a decade to develop a cultivar. However, the off-season nurseries can reduce the breeding time. In pigeonpea also, the technology of "speed breeding" was developed. This technology, however, can be applied only in the early maturing germplasm due to its photoperiod-insensitivity (Saxena et al. 2019b) The speed breeding technology integrates the germination of immature (30 day from flowering) seed for rapid generation advancement and a "single pod descent" method of breeding for conserving the intra-cross variability. This allowed four generations within a year. This approach can fast-track the field evaluation of resulting homozygous inbred lines. This breeding approach can deliver new early maturing cultivars within a timeframe of 4–5 years.

12.3 Select Genotypes with Rapid Seed Filling

In contrast to control (48 days), the new selections took only 29 days from flowering to maturity (Srivastava et al. 2012). Such differences may arise from the presence of various genetic regulatory mechanisms that control the photoperiod response. When these genes are present, they induce indeterminacy in pigeonpea plants, prolonging their reproductive phase and causing significant delays in pod setting and maturity. Pazhamala et al. (2017), through RNA sequencing, also found that these genes remain active only during the period between flowering and pod setting. This "rapid seed filing" is a useful trait but has not been exploited so far by pigeonpea breeders. The selection of rapid pod filling genotypes will help in breeding cultivars with synchronous flowering and maturity, and these will allow easy and effective mechanization of insect management and harvesting.

12.4 Use of Heterotic Pools in Breeding

The concept of "heterotic group" was developed by breeders to identify genetically diverse parents which would produce heterotic hybrids when mated. It is considered an important genetic tool to facilitate breeding programs. This process involves grouping potential parental lines based on their combining ability, origin, or genetic distance. The genetic divergence among the test entries is assessed using various statistical or genomic tools. In pigeonpea, Saxena and Sawargaonkar (2014) were the first to establish seven heterotic groups based on multi-location specific combining ability data. They also found that heterosis for seed yield was significantly higher when parental lines from two distinct heterotic groups were crossed (Fig. 11). The use of SSR markers to select diverse parents in breeding is considered more effective as it eliminates environmental and genotype × environment interaction effects while forming the heterotic groups (Mudaraddi and Saxena 2015). This breeding approach can be very effective in developing high-yielding hybrids as well as pure-line cultivars. Sawargaonkar and Saxena (2020) demonstrated that in pigeonpea both high-yielding inbred lines and F1 hybrids can be bred by using the parents selected from diverse heterotic groups.

12.5 Select Inbred Lines from High-Yielding Hybrids

Crossing of two diverse inbred lines (A x R) integrates two different nuclear genomes into one, and this determines the phenotype of F_1 hybrid. Different types of gene actions such as additive, dominance, epistasis, etc. operate and determine the performance of a given hybrid. Of these, the additive genetic component of variation can be fixed in the inbred lines through pedigree selection. Saxena and Sharma (1990) while reviewing the gene action in pigeonpea concluded that in most hybrids the additive genetic variation played an important role in the expression of seed yield. This information suggested that the high-yielding hybrids are capable of producing high-yielding inbred lines also through pedigree breeding. Such inbred lines accumulate positive alleles in a single genotype.

In a similar exercise, Saxena et al. (1992) demonstrated that some of the hybrid-derived inbred lines achieved 70–75% of the productivity expressed by the hybrid ICPH 8. These inbreds performed significantly better that the control UPAS 120 and male parent ICPH 161. Since in pigeonpea a number of high-yielding hybrids are already available, this breeding approach can be fruitful in breeding high yielding inbred cultivars.

12.6 Exploit Temperature-Sensitive Male Sterility in Hybrid Breeding

This is a unique phenomenon where the expression of male sterility and fertility is controlled by changes in ambient temperature. The first temperature-sensitive male sterility was reported in rice (Shi 1981). Subsequently, Yuan (1987) outlined a methodology for its use in hybrid rice breeding. Since the temperature-sensitive male sterility system eliminates the need of male sterility maintainer (B-) line; it is called as a "two-parent hybrid" breeding system. Levings et al. (1980) opined that the reversal of male fertility to sterility is controlled by certain genetic factors housed in the cytoplasm, and their loss is correlated with the reversion of male sterility to pollen fertility.

In pigeonpea, the first temperature-sensitive male sterility was reported by Saxena (2014). The selections sown in mid-

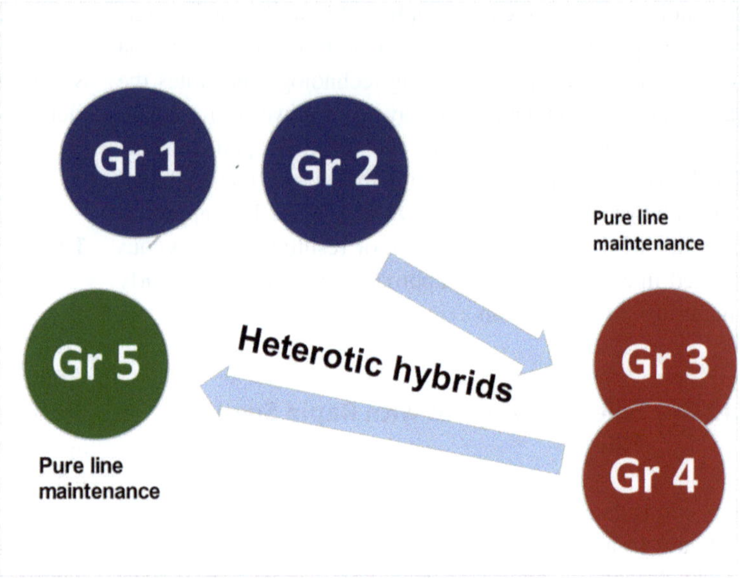

Fig. 11 The diverse heterotic group crosses(Gr 5 x Gr 4; and Gr2 x Gr3), which produced high-yielding hybrids

June were completely male sterile in September. In the month of November, these plants became fully fertile and produced self-pollinated pods. In the month of March, when temperature rose, these plants again reverted back to male-sterility. A perusal of temperature and pollen sterility/fertility data revealed that under the temperature regime of >25 °C that prevailed during the months of September and March, the plants were completely male-sterile. On the contrary, the winter months with temperatures of <20 °C, all the plants produced fertile pollen grains. This unique male sterility system has potential to produce quality seeds of the hybrids and their parents. It needs to be exploited for quantum jumps in enhancing the production and productivity of pigeonpea grains.

12.7 Exploiting Heterosis Through Breeding Sybrid Population

The conventional pure-line breeding procedures not only restrict recombination but also retain the undesirable linkages. Pigeonpea being a partially cross-pollinated crop offers scope to overcome these limitations by breeding composite cultivars. Saxena (2020) proposed a new population breeding method called as "sybrid breeding." It aims to exploit a part of heterosis in a single use population without using any male-sterility system. The sybrid, a mixture of inbred and hybrid, is cost effective and simple. At genetic level, this population will be benefitted from additive, dominance, and epistatic gene actions; and unlike synthetic populations, there will be no yield loss on account of inbreeding depression. In comparison to inbred cultivars, the sybrid population houses relatively broader spectrum of variability. Considering these advantages, a sybrid population, due to their hybrid component, is expected to express relatively greater buffering and stability.

12.8 Early Generation Testing of Crosses

The concept of early generation testing is not new, and it is recommended to conserve resources in breeding cultivars. In this approach, the unproductive crosses are discarded early in the breeding program, and only potential crosses are advanced for pedigree breeding. In most breeding programs, the shrinking resources may not allow the luxury of screening a large number of crosses, and therefore, a breeding approach such as early generation testing is likely to give good results. A study conducted in pigeonpea showed that the low-yielding crosses can be rejected on the basis of their F_1 performance alone. Further, to confirm their cross performance and final selection, the selected hybrids should be tested again using their F_2 bulks.

13 Genomics Advances for Pigeonpea R&D

A comprehensive review in this field has been published by Bohra et al. (2020a, b), it is recommended to gain further insight, readers should read this article. In the interest of this book chapter, we have summarized few highlights of available genomics resources in pigeonpea and provided information on recent advances.

13.1 Molecular Markers and Genotyping Platforms

Early molecular markers-based studies on pigeonpea employed restriction fragment length polymorphism (RFLP), random amplified polymorphic DNA (RAPD), and amplified fragment length polymorphism (AFLP) for genetic diversity assessments and trait-specific mapping. The advent of simple sequence repeat (SSR) markers revolutionized pigeonpea molecular analyses. Initially, SSR marker development involved labor-intensive procedures such as mining expressed sequenced tags (ESTs) and creating SSR-enriched genomic libraries (Saxena et al. 2010). The application of SSRs faced challenges in cultivated pigeonpea due to limited genetic diversity, impeding their use in marker-trait association studies. To address these limitations, researchers turned to high-throughput, automated, and cost-efficient DNA marker systems. Diversity arrays technology (DArT), a hybridization-based genotyping protocol, and next-generation sequencing (NGS) technologies emerged as viable alternatives. DArT facilitated genetic diversity analysis and linkage mapping, while NGS technologies, particularly single nucleotide polymorphism (SNP) markers, became the preferred choice for their automation feasibility and genome-wide distribution.

Subsequently, the development of genotyping platforms like BeadXpress and GoldenGate assay in pigeonpea allowed medium- to high-throughput genotyping of SNP markers (Roorkiwal et al. 2013). Cost-effective KASP technology was employed for the development of 1616 SNPs, referred to as PKAMs (pigeonpea KASP assay markers) (Saxena et al. 2012). Genome-wide SNP analysis using diverse pigeonpea accessions provided insights into crop domestication and demographic history (Saxena et al. 2014). In recent years, next-generation sequencing (NGS)-based genotyping methods, such as genotyping-by-sequencing (GBS), have gained prominence due to their ease in library preparation and greater multiplexing capacity, offering a cost- and time-efficient means to discover and genotype thousands of SNPs simultaneously (Saxena et al. 2017).

Whole-genome resequencing (WGRS) data from three studies including 20 *Cajanus* accessions (Kumar et al. 2016), 292 accessions (Varshney et al. 2017) and 104 accessions (Saxena et al. 2021a), have provided millions of sequence variations (SNPs) and first information on structural variations, such as copy number variations (CNVs) and presence and absence variations (PAVs) in breeding lines, landraces, and wild species accessions. There is a growing interest in CNVs and PAVs due to their association with agronomic traits, including flowering time, stress response, and evolutionary transitions. The re-sequencing data led to the development of the Axiom *b* array, containing 56,512 unique and informative sequence variations (Saxena et al. 2018a, b, c). Noteworthy is the incorporation of 1554 SNPs and 385 InDels associated with key agronomic traits, enhancing the array's suitability for identifying new haplotypes linked to these traits. To deploy genotyping in routine breeding practices, a mid-density genotyping array of 2000 markers has also been developed (Saxena et al. https://excellenceinbreeding.org/toolbox/services/pigeonpea-mid-density-genotyping-services).

13.2 Improved Genome Assembly

A major advancement after the publication of Bohra et al. (2020a, b) has been done in form of improved genome assembly of pigeonpea (Garg et al. 2022). Using Hi-C data, improved genome assembly for pigeonpea (Cajca.Asha_v2.0) has been developed. The comparison of draft genome assembly, i.e., C.cajan_V1.0 (Varshney et al. 2012) and improved genome assembly (Cajca.Asha_v2.0) has been nicely presented in Garg et al. (2022).The refined assembly boasts notable improvements, including a substantial increase in genome sequences anchored to chromosomes from 40.86% to an impressive 91.35%. Furthermore, the total number of genes has been streamlined to 29,482, a significant reduction from the earlier prediction of 48,680 genes. This upgraded chromosome-length reference genome stands out for its high quality, rendering it invaluable for genetic studies. Moreover, it promises enhanced precision in identifying genomic regions and genes associated with crucial traits in pigeonpea.

13.3 Diagnostic Markers for Selection and Quality Control

Using several mapping approaches, 20–30 traits have been mapped in pigeonpea (Bohra et al. 2020a, b). However, the initially identified marker trait associations (MTAs) cannot be deployed in selection program. Therefore, validation of MTAs has been considered as one of the most important steps in markers-based selection program. These validated markers have been termed as diagnostic markers. A brief account has been given below for the diagnostic markers developed in pigeonpea:

13.3.1 Diagnostic Markers for Fusarium Wilt (FW) and Sterility Mosaic Diseases (SMD)

A novel approach based on multiple datasets and through comprehensive data analysis successfully identified nine diagnostic markers for Fusarium wilt (FW) resistance and ten for Sterility Mosaic Disease (SMD) resistance in pigeonpea (Saxena et al. 2021b). These markers have been transformed into a diagnostic kit for regular use in crop improvement programs, specifically targeting the development of FW and SMD-resistant genotypes. Additionally, this study has also discovered two genes associated with FW resistance and four genes linked to SMD resistance. Further, customized and cost-effective Kompetitive allele-specific PCR genotyping assays have been developed for the identified genes, intending to facilitate their application in pigeonpea breeding programs. The utilization of this diagnostic marker kit holds significant promise for advancing pigeonpea varietal development programs, with a focus on ensuring resistance to FW and SMD, a key prerequisite for the inclusion of improved lines in the varietal release pipeline.

13.3.2 Diagnostic Markers for Cytoplasmic Male Sterility (CMS) and Fertility Restoration (Rf)

In the study on mitochondrial genome sequencing in pigeonpea by Tuteja et al. (2013), candidate chimeric open reading frames (ORFs) and genes were discovered in both the A-line (ICPA 2039) and its maintainer line (ICPB 2039). Through a comparative analysis of 34 mitochondrial genes between ICPA 2039 and ICPB 2039, a notable 10 bp deletion in *nad7a* was identified in the cytoplasmic male sterility (CMS) line. This deletion was subsequently utilized to formulate a PCR-based InDel marker named nad7a_del, as reported by Sinha et al. (2015). The *nad7a*_del marker proved effective in validating the genetic purity of A4-derived CMS lines.

In a parallel development, Saxena et al. (2018a, b, c) introduced two markers, namely "CcLG08_RFQI1" and "CcLG08_RFQI4," designed to distinguish restorer lines from non-restorer lines in the A4 hybrid system of pigeonpea. These markers were derived from QTL regions identified in a segregating F2 population originating from the male-sterile line (ICPA 2039) and fertility restorer line (ICPL 87119). The implementation of these markers for CMS and restorer lines significantly streamlines phenotyping activities, offering a more efficient approach to identifying CMS and restorer lines.

The significance of these DNA markers extends beyond simplifying phenotyping tasks; their applicability in low-cost genotyping addresses challenges associated with advanced genotyping methodologies. This dual-marker approach not only enhances the accuracy and efficiency of identifying CMS and restorer lines in pigeonpea but also contributes to overcoming technical challenges in the genotyping process.

13.3.3 Diagnostic Markers for Hybrid Seed Testing

To evaluate the purity of hybrid seeds, researchers initially developed SSR markers for specific hybrids (Saxena et al. 2010; reviewed in Bohra et al. 2020a, b). However, employing SSRs for assessing the purity of a large seed quantity in practical hybrid breeding can incur high costs. Consequently, low-density SNP assays have been optimized for numerous hybrids to reduce genotyping expenses for marker-based hybrid purity evaluation (https://excellenceinbreeding.org/module3/kasp). These SNPs are currently utilized in routine hybrid breeding programs (Saxena et al. 2020a, b).

Additionally, several other advancements in pigeonpea genomics have been achieved, such as initiation of haplotype-based breeding (Sinha et al. 2020), defining heterotic groups based on genomic prediction (Saxena et al. 2021a), deployment of early generation selection through markers, etc. Above-mentioned advances are proving to be useful in realizing the genetic gains in pigeonpea.

14 Emerging Challenges and Future Thrust Areas

In the context of nutritional food security in a changing climate, highly nutritious and stress-tolerant crops like pigeonpea play a vital role. Enhancing breeding efficiency through innovative designs and genomic technologies is essential for achieving stable increases in pigeonpea productivity. The integration of breeding approaches, particularly genomics, holds significant promise for developing climate-resilient crop varieties with enhanced nutrition. While genome sequencing has been undertaken for economically important crops like pigeonpea, the translation of this information into crop improvement programs is limited. Utilizing genome sequence data alongside well-characterized germplasm sets is crucial for developing superior crop varieties. The alarming susceptibility of high-yielding pigeonpea varieties to the major and upcoming diseases emphasizes the need for immediate attention to prevent negative impacts on yields.

Advances in pigeonpea crop improvement, especially in hybrid breeding, offer significant benefits for addressing issues like yield plateau. Shortening the crop breeding cycle, especially in crops with photoperiod sensitivity like pigeonpea, is key to improving genetic gains in breeding programs. Genomic selection models and marker-assisted breeding are expected to contribute significantly to this goal. Additionally, investments in strengthening germplasm collection programs, especially in characterization and their systematic use in crop improvement programs along with the quality seed delivery systems, are equally important for the overall success.

References

Bajaj YPS, Singh H, Gosal SS (1980) Haploid embryogenesis in anther cultures of pigeonpea (*Cajanus cajan*). Theor Appl Genet 58:157–159

Bantilan MCS, Joshi PK (1996) Returns to research and diffusion investments on wilt resistance in pigeonpea, Impact series, 1. International Crops Research Institute for the semi-Arid Tropics, Patancheru

Bhattacharjee SK (1956) Study of autotetraploid *Cajanus cajan* (Linn.) Millsp. Caryologia 9:149–159

Bohra A, Saxena KB, Varshney RK, Saxena RK (2020a) Genomics assisted breeding for pigeonpea improvement. Theor Appl Genet 133:1721–1737

Bohra A, Saxena KB, Varshney RK, Saxena RK (2020b) Genomics assisted breeding for pigeonpea improvement. Theor Appl Genet 133:1721–1737. https://doi.org/10.1007/s00122-020-03563-7

Choudhary AK, Nadarajan N (2011) Breeding improved cultivars of pigeonpea in India, Research bulletin. Indian Pulses Research Institute, Kanpur

Choudhary AK, Raje RS, Datta, Sultana R, Ontagodi T (2013) Conventional and molecular approaches towards genetic improvement in pigeonpea for insect resistance. Am J Plant Sci 4:372–385

D'Cruz R, Jadav AS (1972) Aneuploidy in tur (*Cajanus cajan* L. Millsp.). Res J Mahatma Phule Krishi Vidyapeeth 3:61–62

Dahiya SS, Chauhan YS, Johansen C, Waldia RS, Sekhon HS, Nandal JK (2002) Extra short duration pigeonpea for diversifying wheat-based cropping systems in sub-tropics. Exp Agric 38:1–11

Dalvi V, Saxena KB (2009) Stigma receptivity in pigeonpea [*Cajanus cajan* (L.) Millspaugh]. Indian J Genet Plant Breed 69:247–249

De DN (1974a) Pigeonpea. In: Hutchinson J (ed) Evolutionary studies in world crops: diversity and change in the Indian subcontinent. Cambridge University Press, London, pp 79–87

De C (1974b) The history and evolution of Pigeonpea (Cajanus cajan). Econ Bot 28(3):215–222

de Candolle AP (1813) Catalogus plantarum horticulture. In: Botanici monspeliensis. Paris/Strassbourg/Montpellier, pp 85–86

de Candolle AP (1885) Origin of cultivated plants. pp 332–333

Deodikar GB, Thakar CV (1956) Cyto-taxonomic evidence for the affinity between *C. indicus* Spreng. and certain erect species of *Atylosia* W. and A. Proc Indian Acad Sci (Sect B) 43:37–45

Dundas IS, Britten EJ, Byth DE (1983) Pachytene chromosome identification by a key based chromomeres in the pigeonpea. J Hered 74:461–464

Frankel OH, Brown AHD (1984) In: Holden JHW, Williams JT (eds) Plant genetic resources today: a critical appraisal in crop genetic resources, conservation and evaluation. George Allen and Unwin Ltd, London, pp 249–257

Garg V, Dudchenko O, Wang J, Khan A, Gupta S, Kaur P, Han K, Saxena RK, Kale S, Pham M, Yu J, Chitikineni A, Zhang Z, Fan G, Lui C, Valluri V, Meng F, Bhandari A, Liu X, Yang T, Chen H, Valliyodan B, Roorkiwal M, Shi C, Yang HB, Durand NC, Pandey MK, Li G, Barmukh R, Wnag X, Chen X, Lam HM, Jiang H, Zong X, Liang X, Liu X, Liao B, Guo B, Jackson S, Nguyen HT, Zhuang W, Shubo W, Wang X, Aiden EL, Bennetzen JL, Varshney RK (2022) Chromosome-length genome assemblies of six legume species provide insights into genome organization, evolution, and agronomic traits for crop improvement. J Adv Res 42:315–329. https://doi.org/10.1016/j.jare.2021.10.009

Gill LS, Husaini SWH (1986) Cytological observations in *Leguminosae* from Southern Nigeria. Willdenowia 15:521–527

Green JM, Sharma D, Reddy LJ, Saxena KB, Gupta SC, Jain KC, Reddy BVS, Rao MR (1981) Methdology and progress in the ICRISAT pigeonpea breeding program. In: Proceedings of the international workshop on Pigeonpeas, volume 1, 15–19 December 1980. ICRISAT, Patancheru, pp 437–449

Harlan JR, de Wet JMJ (1971) Towards a rational classification of cultivated plants. Taxon 20:507–517

Howard A, Howard GLC, Khan AR (1919) Studying the pollination of Indian crops I. Mem Dep Agric India (Bot Ser) 10:195–200

Kajale MD (1974) Plant economy at Bhokardan. In: Dev SB, Gupta S (eds) Excavations at Bhokardan (Bhogavardana). Nagpur University and Maharashtra Marathwada University, India, p 7

Krishnaswamy N, Ayyangar GNR (1935) Chromosome number in *Cajanus indicus* Spreng. Curr Sci 3:614–615

Kulkarni NK, Reddy AS, Lava Kumar P, Vijaynarasimha J, Rangaswamy KT, Muniyappa V, Reddy LJ, Saxena KB, Jones AT, Reddy DVR (2003) Broad-based resistance to pigeonpea sterility mosaic disease in accessions of *Cajanus scarabaeoides* (L.) Benth. Indian J Plant Prot 31:6–11

Kumar PS, Subrahmanyam NC, Faris DG (1987) Nucleolar behaviour in pollen mother cells in pigeonpea. J Hered 78:366–368

Kumar S, Gupta S, Chandra S, Singh BB (2003) How wide is the genetic base of pulse crops? In: Ali M, Singh BB, Kumar S, Viswadhar (eds) Pulses in new perspectives, pp 211–221

Kumar V, Khan AW, Saxena RK, Garg V, Varshney RK (2016) First generation hapmap in *Cajanus* spp. reveals untapped variations in parental lines of mapping populations. Plant Biotechnol J 14:1673–1681. https://doi.org/10.1111/pbi.12528

Lackey JA (1977) A synopsis of *Phaseoleae* (*Leguminosae: Papilionoidease*). PhD thesis, Iowa State University, Ames, Iowa USA, p 293

Lateef SS (1992) Scope and limitations of host plant resistance in pulses for the control of *Helicoverpaarmigera*. In: Sachan JN (ed) *Helicoverpa* management – current status and future strategies in pigeonpea. IIPR, Kanpur, pp 31–37

Levings C, Kim B, Pring D, Conde M, Mans R, Laughnan J, Gabay-Laughnan S (1980) Cytoplasmic reversion of CMS-S in maize: association with a tRNA positional event. Science 209:1021–1023

Linnaeus C (1753) Species Plantarum Holmiae 2:739

Mudaraddi HB, Saxena KB (2015) Utilization of SSR markers for identifying diverse parents and forming heterotic groups in Pigeonpea breeding. J Plant Breed Crop Sci 7(10):291–299

Mula MG, Saxena KB (2010) Lifting the level of awareness on pigeonpea – a global perspective. Monograph. International Crops Research Institute for the Semi-Arid Tropics, Patancheru. ISBN 9789290665359

Naithani SP (1941) Cytological studies on Indian pulses, Part 1, The somatic chromosomes, and the pro-chromosomes of *Cajanus*. Proc Indian Acad Sci 11:67–73

Nigam SN, Chaudhary S, Deevei KC, Saxena KB, Janila P (2021) Trends in legume production and future outlook. In: Saxena KB et al (eds) Genetic enhancement in major food legumes. Springer Nature, pp 7–48. https://doi.org/10.1007/978-3-030-64500-7_2

Pathak GN (1948) Cytological studies of a spontaneously-originated tetraploid *Cajanus cajan* Millsp. Indian J Genet 8:68–71

Pathak GN, Yadav RS (1951) Spontaneously-originated hexaploid and tetraploid plants in *Cajanus cajan* Millsp. Curr Sci 20:304

Pawar SE, Thakre RG, Reddy KS, Bhatia C (1991) Use of induced mutations in the breeding of pulse crops. In: Proceedings of the IAEA symposium plant mutation breeding for crop improvement, vol 1. IAEA, Vienna, pp 413–418

Pazhamala LT, Purohit S, Saxena RK, Garg V, Krishnamurthy L, Verdier J, Varshney RK (2017) Gene expression atlas of pigeonpea and its application to gain insights into genes associated with pollen fertility implicated in seed formation. J Exp Bot 68:2037–2054. https://doi.org/10.1093/jxb/erx010

Plukenet L (1692) Table 213, Fig 3, Phytographia. 3, London

Pundir RPS, Singh RB (1985) Crossability relationships among *Cajanus, Atylosia and Rhynchosia* species and detection of crossing barriers. Euphytica 34:303–308

Purseglove JW (1968) Tropical crops. Dicotyledons. Longmans Green, London, pp 236–241

Rao NK, Reddy LJ, Bramel PJ (2003) Potential of wild species for genetic enhancement of some semi-arid food crops. Genet Resour Crop Evol 50:707–721

Reddy LJ (1981) Pachytene analyses in *Cajanus cajan, Atylosia lineata* and their hybrid. Cytologia 46:397–412

Reddy LJ (1990) Pigeonpea morphology. In: The Pigeonpea. CAB International, Wallingford/Oxon, pp 47–88

Reddy LJ, Upadhyaya HD, Gowda CLL, Singh S (2005) Development of core collection in pigeonpea using geographic and qualitative morphological descriptors. Genet Resour Crop Evol 52:10499–11056

Roorkiwal M, Sawargaonkar SL, Chitikineni A, Thudi M, Saxena RK, Upadhyaya HD, Vales MI, Riera-Lizarazu O, Varshney RK (2013) Single nucleotide polymorphism genotyping for breeding and genetics applications in chickpea and pigeonpea using the BeadXpress platform. Plant Genome 6. https://doi.org/10.3835/plantgenome2013.05.0017

Roy B (1933) Studies in the development of the female gametophyte in some leguminous crop plants of India. Indian J Agric Sci 3:1098–1107

Sawargaonkar SL, Saxena KB (2020) Application of heterotic grouping, standard heterosis and performance in breeding high yielding inbred and hybrid cultivars. Indian J Genet Plant Breed 79:704–712

Saxena KB (2020) SyBrid – a new breeding method for food legumes. Int J Sci Res 9:199–200

Saxena KB, Nigam SN (2022) Genotype-environment interactions and their impact on seed yield in Pigeonpea (Cajanus cajan). J Agric Sci Technol 22(4):673–685

Saxena KB, Nigam SN (2023) Is micro-environment a hurdle in the genetic enhancement of pigeonpea? International. J Sci Res. https://doi.org/10.36106/ijsr/9327134

Saxena KB, Sawargaonkar SL (2014) First information on heterotic groups in pigeonpea *Cajanus cajan* (L.) Millsp. Euphytica 200:187–196

Saxena KB, Sharma D (1990) Pigeonpea genetics. In: Nene YL, Hall SD, Sheila VK (eds) The Pigeonpea. CAB International, Wallingford, pp 137–158

Saxena KB, Sharma D, Green JM (1976) Pigeonpea ratooning – an aid to breeders. Trop Grain Legume Bull 4:21. ISSN 0304-5765

Saxena KB, Dundas IS, Byth DE, Wallis ES (1982) Effect of chromosome doubling on some vegetative and reproductive characteristics of pigeonpea (*Cajanus cajan* (L.) Millsp.). Legum Res 5:83–86

Saxena KB, Faris DG, Singh U, Kumar RV (1987) Relationship between seed size and protein content in newly developed high protein lines of pigeonpea. Plant Foods Hum Nutr 36:335–340. ISSN 0921-9668

Saxena KB, Chauhan YS, Johansen C, Singh L (1992) Recent developments in hybrid pigeonpea research. In: New frontiers in pulses research and development. Directorate of Pulses Research, Kanpur, pp 58–69

Saxena KB, Jayasekera SJBA, Ariyaratne HP, Ariyanayagam RP, Fonseka HHD (1994) Frequency of natural out-crossing in partially cleistogamous pigeonpea lines in diverse environments. Crop Sci 34:660–662

Saxena KB, Chandrasena GDSN, Hettiarachchi K, Iqbal YB, Fonseka HHD, Jayasekera SJBA (2002) Evaluation of pigeonpea accessions and selected lines for reaction to *Maruca*. Crop Sci 42:615–618

Saxena KB, Kumar RV, Srivastava N, Bao S (2005) A cytoplasmic-nuclear male-sterility system derived from a cross between *Cajanus cajanifolius* and *Cajanus cajan*. Euphytica 145:289–294

Saxena RK, Saxena KB, Varshney RK (2010) Application of SSR markers for molecular characterization of hybrid parents and purity assessment of ICPH 2438 hybrid of pigeonpea [*Cajanus cajan* (L.) Millspaugh]. Mol Breed 26:371–380

Saxena RK, Penmetsa RV, Upadhyaya HD, Kumar A, Carrasquilla-Garcia N, Schlueter JA, Farmer A, Whaley AM, Sarma BK, May GD, Cook DR, Varshney RK (2012) Large-scale development of cost-effective SNP marker assays for genetic mapping in pigeonpea and comparative mapping in legumes. DNA Res 19:449–461. https://doi.org/10.1093/dnares/dss025

Saxena RK, von Wettberg E, Upadhyaya HD, Sanchez V, Songok S, Saxena KB, Kimurto P, Varshney RK (2014) Genetic diversity and demographic history of *Cajanus* spp. illustrated from genome-wide SNPs. PLoS One 9:e88568

Saxena RK, Kale SM, Kumar V, Parupalli S, Joshi S, Singh VK, Garg V, Das RR, Sharma M, Yamini KN, Ghanta A, Rathore A, Sameer Kumar CV, Saxena KB, Varshney RK (2017) Genotyping-by-sequencing of three mapping populations for identification of candidate genomic regions for resistance to sterility mosaic disease in pigeonpea. Sci Rep 7:1813. https://doi.org/10.1038/s41598-017-01535-4

Saxena KB, Sharma D, Vales MI (2018a) Development and commercialization of CMS pigeonpea hybrids. Plant Breed Rev 41:103–167

Saxena RK, Patel K, Sameer Kumar CV, Tyagi K, Saxena KB, Varshney RK (2018b) Molecular mapping and inheritance of restoration of fertility (Rf) in A4 hybrid system in pigeonpea (*Cajanus cajan* (L.) Millsp.). Theor Appl Genet 131:1605–1614. https://doi.org/10.1007/s00122-018-3101-y

Saxena RK, Rathore A, Bohra A, Yadav P, Das RR, Khan AW, Singh VK, Chitikineni A, Singh IP, Sameer Kumar CV, Saxena KB, Varshney RK (2018c) Development and application of high density Axiom® *Cajanus* SNP Array with 56K SNPs to understand the genome architecture of released cultivars and founder genotypes for redefining future pigeonpea breeding programs. Plant Genome 11:180005. https://doi.org/10.3835/plantgenome2018.01.0005

Saxena KB, Reddy KM, Patel SS (2019) Role of Pigeonpea in sustainable agriculture: enhancing soil fertility and crop production. Field Crop Res 231:1–11

Saxena KB, Choudhary AK, Srivastava RK, Bohra A, Saxena RK, Varshney RK (2019a) Origin of early maturing pigeonpea germplasm and its impact on adaptation and cropping systems. Plant Breed 138:243–251

Saxena KB, Saxena RK, Hickey LT, Varshney RK (2019b) Can a speed breeding approach accelerate genetic gain in pigeonpea? Euphytica 215:1–7

Saxena RK, Kale S, Mir RR, Mallikarjuna N, Yadav P, Das RR, Molla J, Sonnappa M, Ghanta A, Narasimhan Y, Rathore A, Sameer Kumar CV, Varshney RK (2020a) Genotyping-by-sequencing and multilocation evaluation of two interspecific backcross populations identify QTLs for yield-related traits in pigeonpea. Theor Appl Genet 133:737–749. https://doi.org/10.1007/s00122-019-03504-z

Saxena KB, Varma A, Kumar S (2020b) Optimization of low-density SNP assays for cost-effective marker-based hybrid purity evaluation in Pigeonpea. Mol Breed 40(2):12–21

Saxena KB, Choudhary AK, Dalvi VA, Saxena RK, Ghosh J, Singh S, Verma P, Kumar S (2021) Pigeonpea is significantly more than just a delicious pulse. J AgricSearch 8:177–187

Saxena RK, Jiang Y, Khan A, Zhao Y, Singh VK, Bohra A, Sonappa M, Rathore A, Sameer Kumar CV, Saxena KB, Reif J, Varshney RK (2021a) Characterization of heterosis and genomic prediction based establishment of heterotic pattern for developing better hybrids in pigeonpea. Plant Genome 14:e20125. https://doi.org/10.1002/tpg2.20125

Saxena RK, Hake A, Bohra A, Khan AW, Hingane A, Sultana R, Singh IP, Satheesh Naik SJ, Varshney RK (2021b) A diagnostic marker kit for fusarium wilt and sterility mosaic diseases resistance in pigeonpea. Theor Appl Genet 134:367–379. https://doi.org/10.1007/s00122-020-03702-0

Saxena KB, Reddy LJ, Saxena RK (2023) Breeding high protein pigeonpea genotypes and their agronomic and biological assessments. Plant Breed 142:129–139

Saxena KB, Srivastava N, Kumar RV (2023a) Transgressive segregation for seed protein in an inter-specific Cajanus cross. Indian J Plant Genet Resour 36:313–316

Sharma S, Upadhyaya HD (2016) Interspecific hybridization to introduce useful genetic variability for pigeonpea improvement. Indian J Genet 76:496–503

Sharma D, Kannaiyan J, Saxena KB (1987) Sources of resistance to alternaria blight in pigeonpea. SABRAO J 19:109–114

Sharma HC, Pampapathy G, Reddy LJ (2003) Wild relatives of pigeonpea as a source of resistance to the pod fly (*Melalagromyza* obtuse Malloch) and pod wasp (*Tranaostigmodes cajaninae* La Salle). Genet Resour Crop Evol 50:817–824

Sharma HC, Sujana G, Rao DM (2009) Morphological and chemical components of resistance to pod borer, *Helicoverpaarmigera* in wild relatives of pigeonpea. In: Arthropod – plant interactions, vol 3. Springer SBM, Dordrecht, pp 151–161

Shi MS (1981) Preliminary report of breeding and utilization of late japonica natural double purpose line. J Hubei Agric Sci 7:1–3

Silim SN (2001) Pigeonpea landraces and improved varieties in Africa: development and use. Euphytica 119(2):269–277

Singh U, Jambunathan R, Saxena KB, Subrahmanyam N (1990) Nutritional quality evaluation of newly developed high protein genotypes of pigeonpea (Cajanus cajan). J Sci Food Agric 50:201–209

Singh I, Bohra A, Singh F (2016) An overview of varietal development program of pigeonpea in India. Legume Perspect 11:39–42

Sinha SSN, Kumar P (1979) Mitotic analysis in thirteen varieties of *Cajanus cajan* (L.) Millsp. Cytologia 44:571–580

Sinha P, Saxena KB, Saxena RK, Singh VK, Suryanarayana V, Sameer Kumar CV, Mohan AVSK, Khan AW, Varshney RK (2015) Association of *nad7a* gene with cytoplasmic male sterility in pigeonpea (*Cajanus cajan*). Plant. Genome 8:2014.11.0084. https://doi.org/10.3835/plantgenome2014.11.0084

Sinha P, Singh VK, Saxena RK, Khan AW, Abbai R, Chitikineni A, Desai A, Molla J, Upadhyaya HD, Kumar A, Varshney RK (2020) Superior haplotypes for haplotype-based breeding for drought tolerance in pigeonpea (*Cajanus cajan* L.). Plant Biotechnol J 18:2482–2490. https://doi.org/10.1111/pbi.13422

Srivastava R, Saxena K (2019) The earliest maturing pigeonpea [Cajanus cajan (L.) Millspaugh] germplasm bred at ICRISAT. Genet Resour Crop Evol 66. https://doi.org/10.1007/s10722-019-00743-3.

Srivastava N, Saxena KB (2023) Annotated bibliography of hybrid pigeonpea, pp 1–180

Srivastava N, Vadez V, Upadhyaya HD, Saxena KB (2006) Screening for intra and inter specific variability for salinity tolerance in Pigeonpea (*Cajanus cajan*) and its related wild species. J SAT Agric Res. https://www.researchgate.net/publication/26507941_Screening_for_intra_and_inter_specific_variability_for_salinity_tolerance_in_Pigeonpea_Cajanus_cajan_and_its_related_wild_species

Srivastava RK, Vales MI, Sultana R, Saxena KB, Kumar RV, Thanki HP, Choudhary KN (2012) Development of super early pigeonpea with good yield potential from early x early crosses. J SAT Agric Res 10:1–6

Sturtevant EL (1972) In: Hedrick UP (ed) Edible plants of the world. Dover Publication, New York

Sujana G, Sharma HC, Rao DM (2008) Antixenosis and antibiosis components of resistance to pod borer *Helicoverpa armigera* in wild relatives of pigeonpea. Int J Trop Insect Sci 28:191–200

Sultana R, Vales MI, Saxena KB, Rathore A, Rao S, Rao SK, Mula MG, Kumar RV (2013) Water-logging tolerance in pigeonpea (*Cajanus cajan* (L.) Millsp.): genotypic variability and identification of tolerant genotypes. J Agric Sci 151:659–671

Swaminathan MS (1973) Pigeonpea breeding and improvement. Indian J Genet Plant Breed 33(2):125–134

Tikka SBS, Pawar LD, Chauhan RM (1997) First record of cytoplasmic-genic male sterility system in pigeonpea [*Cajanus cajan* (L.) Millsp.]. Gujarat Agric Univ Res J 22:160–162

Tuteja R, Saxena RK, Davila J, Shah T, Chen W, Xiao YL, Fan G, Saxena KB, Alverson AJ, Spillane C, Town CD, Varshney RK (2013) Cytoplasmic male sterility-associated chimeric open reading frames identified by mitochondrial genome sequencing of four *Cajanus* genotypes. DNA Res 20:485–495. https://doi.org/10.1093/dnares/dst025

Upadhyaya HD, Reddy LJ, Gowda CLL, Reddy KN, Singh S (2006a) Development of a mini core subset for enhanced and diversified utilization of pigeon pea germplasm resources. Crop Sci 46:2127–2132

Upadhyaya HD, Gowda CLL, Buhariwalla HK, Crouch JH (2006b) Efficient use of crop germplasm resources: identifying useful germplasm for crop improvement through core and mini collections and molecular marker approaches. Plant Genet Resour 4:25–35

van der Maesen LJG (1980) India is the native home of pigeonpea. In: de Wit HCD, Arends JC, Boelema G, de Groot CT, Leeuwenberg AJM (eds) Libergratulatorius in honorem, Landbouwhoge School, Miscellaneous paper no. 19. Veenman H and Zonen BV, Wageningen, pp 257–262

van der Maesen LJG (1986) *Cajanus* DC. and *Atylosia* W. & A. (*Leguminosae*), Agricultural University Wageningen papers 85-4 (1985). Agricultural University, Wageningen

Varshney RK, Chen W, Li Y, Bharti AK, Saxena RK, Schlueter JA, Donoghue MTA, Azam S, Fan G, Whaley AM, Farmer AD, Sheridan J, Iwata A, Tuteja R, Penmetsa RV, Wu W, Upadhyaya HD, Yang SP, Shah T, Saxena KB, Michael T, McCombie WR, Yang B, Zhang G, Yang H, Wang J, Spillane C, Cook DR, May GD, Xu X, Jackson SA (2012) Draft genome sequence of pigeonpea (*Cajanus cajan*), an orphan legume crop of resource-poor farmers. Nat Biotechnol 30:83–89. https://doi.org/10.1038/nbt.2022

Varshney RK, Saxena RK, Upadhyaya HD, Khan AW, Yu Y, Kim C, Rathore A, Kim D, Kim J, An S, Kumar V, Anuradha G, Yamini KN, Zhang W, Muniswamy S, Kim JS, Penmetsa RV, von Wettberg E, Datta SK (2017) Whole-genome re-sequencing of 292 pigeonpea accessions identifies genomic regions associated with domestication and agronomic traits. Nat Genet 49:1082–1088

Vavilov NI (1951) The origin, variation, immunity and breeding of cultivated plants. Chron Bot 13:1–366

Watt G (1889) A directory of economic products of India 2. Calcutta, India, pp 12–15

Yadav P, Saxena KB, Hingane A, Sameer Kumar CV, Kandalkar VS, Varshney RK, Saxena RK (2019) An "Axiom *Cajanus* SNP Array" based high density genetic map and QTL mapping for high-selfing flower and seed quality traits in pigeonpea. BMC Genomics 20:1–10. https://doi.org/10.1186/s12864-019-5595-3

Yuan LP (1987) Strategy concepts of hybrid rice breeding. Hybrid Rice 1:1–4

Mungbean [*Vigna radiata* (L.) R. Wilczek] Breeding

Gyan Prakash Mishra, Muraleedhar S. Aski, Harsh Kumar Dikshit, Manju Kohli, Soma Gupta, Prachi S. Yadav, Kuldeep Tripathi, Prema Kumar, Jenia Roy, and Ramakrishnan M. Nair

Abstract

The mungbean [*Vigna radiata* (L.) R. Wilczek] is mainly grown in East Asia, Southeast Asia, and the Indian subcontinent. This legume is known for its unique characteristics, including short growing period, low input requirements, wide adaptability, and resilience to various environmental stresses. Mungbean grains are highly nutritious, while sprouts and microgreens possess enhanced nutritional values. Breeding efforts are mainly aimed at improving its ability to withstand both biotic and abiotic stresses, including nutritional profile. However, mungbean breeders face the challenge of developing varieties with improved economic traits such as protein content, amino acid quality, and micronutrient contents.

Additionally, efforts are underway to develop varieties with increased resistance to pests and diseases, as well as enhanced adaptability to diverse agroclimatic conditions. Recent breeding strategies involve integrating desired genes from related and wild *Vigna* species through distant hybridization, marker-assisted selection, genomics approaches, etc. This chapter provides a comprehensive overview of the mungbean crop, including breeding advancements, variety development, stress resilience, and nutritional enhancement prospects, and addresses the ongoing challenges and future improvement opportunities on a global scale.

Keywords

Mungbean · Green gram · *Vigna radiata* · Breeding · Biofortification · Nutrition · Cultivation

1 Introduction

Mungbean [*Vigna radiata* (L.) Wilczek] is an economically important pulse crop, cultivated in India since ancient times. Its cultivation extends across Asia, Australia, the West Indies, the Americas, and tropical and subtropical regions of Africa. India accounts for nearly 65% of the global acreage and 54% of total mungbean production. This short-duration grain legume is highly valued for its wide adaptability and low input requirements, making it a key crop in various agricultural systems. Mungbean can fix atmospheric nitrogen (58–109 kg/ha) through symbiotic association with *Rhizobium*, meeting its own nitrogen needs while enriching the soil for subsequent crops too (Ali 1992). Its versatility makes it a wonder crop for its easy incorporation into various cropping systems, making it a valuable component of sustainable agriculture. In addition to its agronomic benefits, mungbean plays a vital role in human diets. Germinated seeds, sprouts, and microgreens offer enhanced nutritional value, with increased levels of some key vitamins like thiamine, niacin, and ascorbic acid. Mungbean is particularly known for its high protein content, which is easily digestible. On a dry weight basis, its seeds typically contain 24–28% protein, 1.0–1.5% oil, 3.5–4.5% fiber, 4.5-5.5% ash, and 62-65% carbohydrates (Tsou 1979).

Globally, mungbean is cultivated on approximately 7.2 million hectares, yielding over 750 kg per hectare (Nair and Schreinemachers 2020). It is primarily grown in tropical and subtropical regions and adapts well to various cropping systems (http://avrdc.org/intl-mungbean-network/). The crop thrives in various soil types, including well-drained loamy or sandy loams, with a pH between 5 and 8. Mungbean can grow up to an altitude of 1850 meters above mean sea level (AMSL) but is sensitive to temperature and photoperiod (Lambrides

G. P. Mishra (✉) · M. S. Aski · H. K. Dikshit · M. Kohli · S. Gupta · P. S. Yadav · P. Kumar · J. Roy
Division of Genetics, ICAR-Indian Agricultural Research Institute, New Delhi, India

K. Tripathi
Germplasm Evaluation Division, ICAR-National Bureau of Plant Genetic Resources, New Delhi, India

R. M. Nair
World Vegetable Center, South Asia/Central Asia, Patancheru, Hyderabad, India

G. Prasad Dixit et al. (eds.), *Fundamentals of Legume Breeding*, https://doi.org/10.1007/978-981-96-7838-9_5

and Godwin 2007). Major producers include Asian countries such as India, Myanmar, China, Thailand, Indonesia, and Bangladesh, which together account for about 90% of global production (Schafleitner et al. 2015; Alam et al. 2014b). India is the largest producer, consumer, and exporter of mungbeans, with cultivation covering around 4.32 million hectares area, yielding 2.17 million tons, with an average productivity of 502 kg/ha (IIPR 2018). Mungbean is valued for its low input requirements, adaptability, resilience to heat and drought, and high nutritional content, making it an ideal choice for smallholder farmers (Pratap et al. 2020).

Due to its short life cycle, broad adaptability, and resilience to diverse soil and climatic conditions, mungbean is cultivated throughout the year, across India. Over the past three decades, the development of disease- and pest-resistant varieties with shorter growth durations and photoperiod tolerance has significantly expanded its cultivation across various agroecological niches. A promising breeding approach involves the incorporation of beneficial genes from wild relatives through distant hybridization, paving the way for further advancements in mungbean varieties (Mishra et al. 2022a). The genus *Vigna*, which includes mungbean, comprises over 150 species, of which only seven are cultivated. These species are classified into two subgenera: the Asian and African subgenera. The Asian subgenus includes *Vigna radiata* (L.) Wilczek (mungbean), *Vigna mungo* (L.) Hepper (black gram), *Vigna aconitifolia* (Jacq.) Marechal (moth bean), *Vigna angularis* (Willd.) Ohwi & Ohashi (adzuki bean), and *Vigna umbellata* (Thunb.) Ohwi & Ohashi (rice bean). In contrast, the African subgenus consists of *Vigna unguiculata* (L.) Walp. (cowpea) and *Vigna subterranea* (L.) Verdc. (Bambara groundnut) (Dikshit et al. 2022). *Vigna* species are well-known for their ability to thrive under a range of agroclimatic conditions, making them prevalent across tropical and subtropical regions. Their adaptability and resilience have allowed them to become essential components of diverse agricultural systems in these areas (Fig. 1).

Fig. 1 A lush green mungbean crop at pod-development stage

2 Origin and Distribution

Mungbean is thought to have originated in the Indian subcontinent (Zukovskij 1962), where a rich diversity of both cultivated and wild *Vigna* can be recorded, indicating this region as its primary center of domestication (Baudoin and Marechal 1988). Himachal Pradesh and the Western Ghats are recognized as the key regions for wild mungbean diversity (Chandel 1981), with the upper Western Ghats and Deccan hills exhibiting the greatest variation among related *Vigna* species. Additionally, Bihar (a state in India) serves as the secondary center of diversity for mungbean in India. References to mungbean in ancient Vedic texts, such as the *Charak Samhita*, indicate its presence in India well before the Christian era (Jain and Mehra 1980). Some archaeological evidences, like presence of charred mungbean grains at Chalcolithic Navdatoli (1500–1000 BC) and Neolithic Chirand in Bihar (1800 BC–200 AD), also indicate its early cultivation in India (Kajale 1974). Additionally, carbonized grains of wild mungbean were also discovered at Daimabad in Maharashtra (Kajale 1977). The wild ancestor of mungbean, *Vigna radiata* var. *sublobata*, is commonly found as a weed in cultivated fields and wastelands across India. It also thrives in subtropical wetlands of northern and eastern Australia (Chandel et al. 1984; Lawna and Cottell 1988).

3 Crop Classification, Gene Pools, and Germplasm Availability in *Vigna* Species

The mungbean, which was previously classified under the genus *Phaseolus*, has been reclassified into the genus *Vigna*. Thus, it is sometimes also referred to as *Phaseolus aureus* or *Phaseolus radiatus*. These names have often been used interchangeably to describe the same plant species. The classification of mungbean is as follows:

Kingdom *Plantae* – Plants
Subkingdom *Tracheobionta* (Vascular plants)
Superdivision *Spermatophyta* (Seed plants)
Division *Magnoliophyta* (Flowering plants)
Class *Magnoliopsida* (Dicotyledons)
Subclass *Rosidae*
Order *Fabales*
Family *Fabaceae* (Pea family)
Genus *Vigna* Savi (cowpea)
Species *Vigna radiata* (L.) R. Wilczek

The genetic resources of *Vigna* species are categorized into primary, secondary, and tertiary gene pools. The GP-1 (primary gene pool) includes *V. radiata* var. *radiata*, *V. radiata* var. *sublobata*, and *V. radiata* var. *setulose*. The GP-2 (secondary gene pool) consists of *V. mungo* var. *mungo*, *V. mungo* var. *silvestris*, *V. trilobata*, *V. subramaniana*, *V. grandiflora*, *V. stipulacea*, *V. tenuicaulis*, *V. trinervia*, and *V. umbellata*. The GP-3 (tertiary gene pool) includes *V. angularis*, *V. dalzelliana*, *V. glabrescens*, *V. grandis*, *V. vexillate*, and *V. aconitifolia* (Smartt 1981, 1985). However, variations in gene pool classification have been recorded, like inclusion of *V. umbellata* in GP-3 and *V. aconitifolia* in GP-2. Hybridization studies have shown that *V. glabrescens* can produce viable offspring when crossed with *V. radiata*, with no fertility or crossover barriers (Pratap et al. 2014a; Kumar et al. 2004). A more accurate classification of *Vigna* species requires a reassessment of the gene pool classification by using both molecular techniques and traditional crossing methods (Mishra et al. 2022a; Chandel et al. 1984). The success of mungbean improvement depends on the availability of diverse genetic resources for targeted genetic improvement through introgression of desirable traits into cultivated genotypes, often supported by hormonal manipulations and tissue culture. Therefore, the collection, evaluation, documentation, and utilization of germplasm are critical for advancing mungbean breeding. The genetic reservoir for mungbean is relatively limited, with an international base collection of approximately 5600 accessions maintained at AVRDC, Taiwan.

In India, the National Bureau of Plant Genetic Resources (NBPGR) in New Delhi houses nearly 3500 mungbean accessions. Both cultivated varieties and wild species play a vital role as valuable sources of resistance to both biotic and abiotic stresses (Harlan 1984). Wild *Vigna* species are predominantly found in the Western and Eastern Ghats, northeastern hills, northwestern plains, peninsular region, and northern Himalayas. A diverse collection of these species is preserved as a wild garden at the Indian Institute of Pulses Research in Kanpur, India. NBPGR also maintains collections of wild *Vigna* species, including wild rice bean (*V. umbellata*) from the Khasi and Jaintia hills of Meghalaya, and various other species such as *V. aconitifolia* (as a weedy form), *V. minima*, *V. dalzelliana*, *V. mungo* var. *sylvestris*, *V. khandalensis*, *V. radiata* var. *sublobata*, and *V. vexillata* from the Western Ghats of Maharashtra and Rajasthan. Additionally, Dana (1998) collected different wild *Vigna* species, including *V. aconitifolia* var. *sylvestris*, *V. dalzelliana*, *V. hainiana*, *V. khandalensis*, *V. mungo* var. *sylvestris*, *V. radiata* var. *setulosa*, *V. radiata* var. *sublobata*, and *V. trilobata* from various Indian states such as Gujarat, Rajasthan, Maharashtra, Madhya Pradesh, Bihar, Orissa, and West Bengal (Padhi et al. 2022a; Mishra et al. 2022a).

4 Floral Biology and Plant Morphology Traits

Mungbean is an annual herb with an erect to sub-erect growth habit, having 30 to 130 cm height, deep roots, and abundant branching with occasional twinning tendency. Cultivated varieties typically exhibit a branched, erect posture, while wild types may have a prostrate form. The root system is taproot, while the stem is hollow, furrowed, and squarish, covered with hairs, and may display purple pigmentation. Root nodules facilitate nitrogen fixation through a symbiotic relationship with rhizobium. The leaves are alternate, compound, and typically trifoliate; sometimes quadra- or pentafoliate, and covered with hairs. The stipules are broad and ovate, while the petioles and rachises are grooved and pubescent (Mishra et al. 2022a). The lower two leaflets are opposite and asymmetrical, while the terminal one is large, symmetrical, ovate, entire, tapering toward the base, and pointed at the tip. Flowers appear in axillary or terminal racemes, with peduncles up to 13 cm long, bearing clusters of 10–20 flowers; while corolla is yellow, sometimes curved, and 5–10 cm long (Fig. 2) (Dikshit et al. 2020). The flower structure consists of one standard, two wings, two keels, a diadelphous androecium (9 + 1), and a monocarpellary gynoecium with a superior, unilocular ovary. The style is twisted below the stigmatic surface, which is covered with hairs, and the placentation is marginal. The calyx comprises five sepals: three large, free sepals and two small, fused sepals. The keel encloses the reproductive organs, and each pod contains about 10–12 seeds, which may vary in color

Fig. 2 Close-up of mungbean flowers and developing pods

from green or olive green to yellow, brown, or even blackish (Dikshit et al. 2020).

5 Cytogenetics of Mungbean

Cytological studies determine the number, size, and type of chromosomes in any crop species (Aski et al. 2024). The mungbean genome ($2n = 22$) is relatively small (579 Mb), and it is considered a model plant among *Vigna* species (Arumuganathan and Earle 1991). Bhatnagar et al. (1974) proposed the karyotype formula for mungbean as 4Lsm + 4Msm + 3 Mm, where L represents long chromosomes (2.7–3.5 μm), M denotes medium chromosomes (1.9–2.6 μm), sm stands for submedian centromeres, and m refers to median centromeres (Table 1).

6 Genetics of Economically Important Traits in Mungbean

Numerous studies have been conducted to investigate the genetics of both quantitative and qualitative traits in mungbean. Kumar et al. (2004) provided details about the inheritance of various economically significant traits. Regarding the twining habit, Khattak et al. (1999) identified a single dominant gene (T), while Pathak and Singh (1963) reported a single recessive gene controlling this trait. The semi-spreading habit was recorded as dominant over erect habit and is governed by a single dominant gene (Pathak and Singh 1963), while indeterminate growth habit is also reported to be controlled by a single dominant gene (Talukdar and Talukdar 2003). The pentafoliate leaf expression is reported to be controlled by a single gene (Chhabra 1990), while the trifoliate leaf shape is dominant over the entire leaf shape and is also controlled by a single dominant gene (Talukdar and Talukdar 2003). Several studies indicated two dominant genes, *Tlb1* and *Tlb2*, exhibiting duplicated gene action for the development of trilobed leaves (Sareen 1982). The narrow lanceolate leaf shape is reported to be controlled by two recessive genes, *nl1* and *nl2* (Dwivedi and Singh 1986).

Mukherjee and Pradhan (2002) reported that anthocyanin pigmentation in the hypocotyls is controlled by two supplementary genes (*Sh* and *Ph*), with recessive epistatic interaction. Purple hypocotyls show dominance over purple-spotted and green hypocotyls, while purple-spotted hypocotyls are dominant over green hypocotyls. The purple pigmentation on the stem, petiole, and veins of the leaves is governed by a single dominant gene, *Ppp1*, exhibiting pleiotropic effects (Dwivedi and Singh 1986). Stem fasciculation is controlled by a single recessive gene (*fs1*), with a pleiotropic effect on the number of floral organs. Pod pubescence is dominant over non-pubescence and is governed by duplicate gene action (Khadilkar 1963). Seed coat color in mungbean range from green, yellow, brown, tan, and black, to mosaics of green, black, and yellow. Khattak et al. (1999) reported monogenic inheritance for seed coat color, whereas black, black-spotted, and dull-green were found dominant over green, non-spotted, and shiny green seed coat colors, respectively. Chen and Liu (2001) suggested black and green seed colors controlled by a single gene (B), with black being dominant over green. Lambrides et al. (2004) also reported a single gene controlling seed coat color. In case of cotyledon color, green cotyledons are controlled by a single recessive gene (Thakare et al. 1980). Humphry et al. (2005) identified four loci for hard-seededness through QTL analysis, two of which were co-localized with loci controlling seed weight (Table 2).

7 Breeding Objectives

Mungbean breeders have now started seriously focusing on improving traits like protein content, amino acid quality, and micronutrient levels (Mishra et al. 2023). Furthermore, developing extra-early and photo-thermally insensitive gen-

Table 1 Cytological features of mungbean chromosomes

Chromosome number	Length of long arm (μm)	Length of short arm (μm)	Length of the chromosome (μm)	Type of chromosome
1	0.95	0.65	2.1	Metacentric
2	0.72	0.61	1.77	Submetacentric
3	1.00	0.62	1.62	Metacentric
4	0.84	0.69	1.53	Metacentric
5	0.77	0.63	1.40	Metacentric
6	0.75	0.60	1.35	Metacentric
7	0.70	0.57	1.27	Metacentric
8	0.76	0.44	1.20	Submetacentric
9	0.63	0.52	1.15	Metacentric
10	0.58	0.50	1.08	Metacentric
11	0.52	0.44	0.96	Metacentric

Derived from Vinora (1974), Vinora et al. (1999)

Table 2 Inheritance of economically important traits in mungbean

S. No.	Trait	Inheritance	Reference
1.	Maturity duration	Additive, non-additive, and epistatic gene action	Singh and Singh (1974), Khattak et al. (2001)
2.	Cotyledon color	Single recessive gene (green color)	Thakare et al. (1980)
3.	Flower color	Single dominant gene	Bose (1939)
4.	Hard seededness	One or few dominant genes	Lambrides (1996), Humphry et al. (2005)
5.	Inflorescence	Two dominant genes; compound types by two recessive genes; cluster types by one dominant gene	Sen and Ghosh (1959), Singh and Singh (1970)
6.	Nodulation	Additive action and non-additive gene action	Singh et al. (1985)
7.	Plant type and growth habit	Single dominant/recessive gene, semi-spreading is dominant over erect habit	Sen and Ghosh (1959), Pathak and Singh (1963) and Khattak et al. (1999)
8.	Pigmentation	Single dominant/recessive gene, anthocyanin in hypocotyl by supplementary gene action	Pathak and Singh (1963), Virk and Verma (1977) and Mukherjee and Pradhan (2002)
9.	Pubescence	Single dominant gene	Murty and Patel (1973), Sen and Ghosh (1959)
10.	Pod color	Single dominant gene	Bose (1939), Sen and Ghosh (1959), Murty and Patel (1973)
11.	Pod shattering	Single dominant gene	Verma and Krishi (1969)
12.	Photoperiod response	Insensitivity dominant over sensitivity and monogenic	Verma (1970)
13.	Stem fasciation	Single recessive gene	Dwivedi and Singh (1986)
14.	Seed coat color	One or few genes (mottling by a single gene)	Watt (1976), Khattak et al. (1999), Chen and Liu (2001) and Lambrides et al. (2004)
15.	Yield components	Additive and non-additive action	Singh and Singh (1971), Yohe and Poehlman (1972), Dasgupta et al. (1998) and Khattak et al. (2002)
16.	YMD resistance	Monogenic recessive	Malik et al. (1986), Reddy and Singh (1995), Saleem et al. (1998), Khattak et al. (1999), Khan et al. (2007), Reddy (2009a), Dhole and Reddy (2013), Sudha et al. (2013)
17.	YMD resistance	Complementary recessive genes	Shukla and Pandya (1985), Alam et al. (2014a)
18.	YMD resistance	Dominant and complementary recessive genes	Sandhu et al. (1985)
19.	YMD resistance	Three recessive genes	Mishra and Asthana (1996)
20.	YMD resistance	Modifying genes	Khattak et al. (2000)
21.	YMD resistance	Two recessive genes	Dhole and Reddy (2012), Singh et al. (2013b)
22.	YMD resistance	Three gene (2 dominant +1 recessive)	Markam et al. (2018)
23.	YMD resistance	Two dominant genes	Mahalingam et al. (2018)
24.	Powdery mildew (PM) resistance	One dominant gene	Chaitieng et al. (2002)
25.	PM resistance	Two dominant genes (*Pm1*, *Pm2*)	Reddy et al. (1994)
26.	PM resistance	One dominant gene	Khajudparn et al. (2007)
27.	PM resistance	One dominant gene (*Pm3*)	Reddy (2009b)
28.	PM resistance	Gene action: Quantitative & additive	Kasettranan et al. (2009)
29.	PM resistance	QTL (PMR1; PVE 65%)	Chaitieng et al. (2002)
30.	PM resistance	QTL (PVE 86%)	Humphry et al. (2003)
31.	PM resistance	Gene action: Additive & dominance	Gawande and Patil (2003)
32.	*Cercospora* leaf spot (CLS) resistance	One dominant gene	Chankaew et al. (2011), Singh et al. (2017c)
33.	CLS resistance	One recessive gene	Mishra et al. (1988)
34.	CLS resistance	One major QTL (LG03)	Chankaew et al. (2011)
35.	Bruchid resistance	One gene	Miyagi et al. (2004), Sun et al. (2009)
36.	Bruchid resistance	*Br1* locus (LG11)	Wang et al. (2016)
37.	Seed hardness	One dominant gene (*Br*); a few dominant genes	Kitamura et al. (1988), Young et al. (1992) and Humphry et al. (2005)
38.	Seed hardness	Gene on LG8	Kaga and Ishimoto (1998)
39.	Calcareous soil (Fe deficiency chlorosis)	Two genes with inhibitory action	Nopparat et al. (1997)
40.	Calcareous soil (Fe deficiency chlorosis)	One dominant gene (*IR*) + a few modifying genes	Srinives et al. (2010)
41.	Pre-harvest sprouting	Additive and non-additive gene action; high G × E interaction	Durga and Kumar (1997)

otypes could be beneficial for diverse cropping systems. Mungbean protein lacks methionine and cystine, two sulfur-containing amino acids, making breeding efforts necessary to increase their levels. Enhancing root length and lateral root development can improve phosphorus scavenging potential, so optimizing phosphorus acquisition and mobilization should be a key focus for mungbean breeders (Vinutha et al. 2022).

Mungbean grown during the kharif should exhibit semi-determinate maturity (90–120 d), medium plant height (70–75 cm), large inflorescences, numerous clusters per plant, 3–5 branches per plant, 8–10 seeds per pod, 3–4 grams per 100 seeds, tolerance to shattering, and moderate seed dormancy (Dikshit et al. 2020). For genotypes cultivated during spring or summer season, the ideal plant type includes a high pod set in the first reproductive flush, long pods (>10 seeds per pod), non-shattering pods at the top, short maturity duration (50–60 d), determinate growth habit, high harvest index, rapid initial growth, reduced photoperiod sensitivity, and a transition from vegetative growth to flowering. High-yielding, nutritionally rich cultivars are being developed using various breeding techniques like introduction, hybridization, early generation selection, and mutation. Additionally, the use of molecular markers in the genomics era enhances the accuracy of trait selection for the breeding of new varieties (Mishra et al. 2022a).

7.1 Breeding for Higher Yield

Systematic efforts to develop mungbean varieties in India began during early twentieth century, primarily focusing on improving locally adapted populations through pureline and mass selection methods. The hybridization, selection, and distant hybridization have led to the development of various mungbean varieties in India. Currently, breeding efforts prioritize the creation of short-duration, photoperiod- and thermo-sensitive varieties with resistance to biotic stresses including yellow mosaic virus and powdery mildew (Dikshit et al. 2020; Mishra et al. 2022a). This has led to the development of varieties suitable for cultivation in spring, summer, and even winter seasons in various regions, including rice fallow areas in peninsular India. Among the key breeding objectives, increasing yield continues to be a top priority. Since the establishment of the All India Coordinated Research Project on Pulses in 1967 (AICRP on MULLARP, now AICRP on Kharif Pulses), several high-yielding mungbean varieties having resistance/tolerance to various biotic and abiotic stresses, suitable for different agro-climatic zones of India have been released.

Pre-harvest sprouting during rainy season limits its potential for intercropping, and therefore, developing varieties with limited or no pre-harvest sprouting is essential. With the growing demand for mungbean as sprouts in breakfasts and microgreens in food, there is a need to breed specialized varieties that sprout early and with nutritious microgreens. Over the years, numerous short-duration, disease- and insect-pest-resistant, high-yielding mungbean varieties have been developed to suit various climatic conditions (Table 3).

7.2 Breeding for Abiotic Stress Tolerance

Tolerance to major abiotic stresses is key in the breeding of mungbean (Aski et al. 2023b). In the wake of climate change, mungbean productivity is mainly threatened by abiotic stresses like water-deficit stress and heat stress (HS). To mitigate these, plants employ different defense strategies, including escape, avoidance, or tolerance mechanisms (Aski et al. 2021). However, threshold temperatures for mungbean plants and their specific responses remain inconsistent and are not yet clearly defined. Traits such as an upright growth habit, lower canopy temperature, and small seed size have shown potential for better adaptation under heat stress (Devi et al. 2023). A simple and effective method for identifying stress-tolerant genotypes is screening of mungbean gene pools for abiotic stress tolerance in specific environments (Singh et al. 2022). Thus, precise phenotyping with modern phenomics tools can enhance breeding efficiency in mungbean (Naik et al. 2024).

Next-generation sequencing (NGS) data can be used to identify candidate genes for abiotic stress tolerance in mungbean. Furthermore, genomic selection, development of transgenic varieties (such as *mtlD* and *DREB*), genome-wide association studies (GWAS), and marker-assisted selection (MAS) are valuable tools for creating genotypes that are tolerant to biotic and abiotic stresses (Patel et al. 2017; Sarkar et al. 2016). The development of transgenics could also serve as an alternative approach to developing heat stress-tolerant mungbean genotypes (Devi et al. 2023). Identifying variations in root traits in mungbean is essential for developing breeding lines with improved root characteristics, which enhance water uptake and help mitigate heat and drought stress (Mishra et al. 2009d). Screening of AVRDC mungbean mini-core collection (296 genotypes) for 14 root-related traits showed significant variation, with genotypes possessing favorable root traits identified as potential donors for breeding cultivars better adapted to water-deficit stress (Aski et al. 2021).

7.3 Breeding for Biotic Stresses

Resistance to major biotic stresses including yellow mosaic disease (YMD) (Fig. 3), powdery mildew (Fig. 4), and *Cercospora* leaf spot (CLS) are key breeding objective in

Table 3 Short-duration cultivars of mungbean suitable for Spring/Summer/Rice Fallow cultivation

S. No.	Variety	Developed by	Seasons/ conditions	Year of release	Average yield (q/ha)	Days to maturity	Area of adaptation	Reaction to major diseases
1.	ADT 3	TNAU, Aduthurai	Rabi	1991	10.7	65–70	Tamil Nadu	Tolerant to MYMV and powdery mildew
2.	LGG 410	RARS-Lam, Guntur	Rice fallow	1994	14–16	65–70	Andhra Pradesh	Resistant to YMV
3.	TARM 2	BARC, Akola	Rabi	1994	9.5	65	Maharashtra	Tolerant to powdery mildew
4.	LGG 407	RARS-Lam, Guntur	Rabi	1995	14	70–75	Andhra Pradesh	Tolerant to YMV and powdery mildew
5.	Pusa 9072	IARI, New Delhi	Rabi	1995	9.0	65–75	SZ (Rabi)	Tolerant to powdery mildew
6.	LGG 450	Lam, Guntur	Rabi	1995	13.0	70–75	Andhra Pradesh	Tolerant to MYMV and powdery mildew
7.	SML 134	PAU, Ludhiana	Spring, Summer	1996	11.0	68	Punjab	Tolerant to MYMV
8.	TARM 1	BARC, Akola	Rabi (shining green seeds)	1996	12.4 6.7	79	Maharashtra	Resistant to powdery mildew
9.	TARM 18	BARC, Akola	Shining green seed	1996	12.0	68	Maharashtra	Resistant to powdery mildew
10.	LGG 460 (Lam 460)	RARS-Lam, Guntur	Rice fallow	1997	15–16	65–70	Andhra Pradesh	Resistant to YMV
11.	RMG 268	RAU, Durgapura	Kharif, Summer	1997	8.9	60	Rajasthan	Tolerant to *Cercospora* leaf spot
12.	CO 6	TNAU, Coimbatore	All seasons	1999	10.0	65	Tamil Nadu	Resistant to MYMV
13.	Pratap (SG1)	Shillongani, Assam	Summer, Kharif	1999	12–14	65–70	Assam	Resistant to *Cercospora* leaf spot and tolerant to YMV
14.	Pusa 9531	IARI, New Delhi	Summer	2000	9.0	60	CZ	Resistant to MYMV
15.	Pusa Vishal	IARI, New Delhi	Summer (large seed: 6 g/100 seed)	2000	11.0	62	NWPZ	Resistant to MYMV
16.	HUM 2 (Malaviya Jagriti)	BHU, Varanasi	Spring, Summer	2000	10.5	67	Uttar Pradesh and Uttaranchal	Moderately resistant to MYMV
17.	Samrat	IIPR, Kanpur	Summer	2001	10–12	60–65	Uttar Pradesh	Resistant to YMV
18.	HUM 6 (Malaviya Janpriya)	BHU, Varanasi	Spring	2001	10.0	68	Uttar Pradesh	Moderately resistant to MYMV
19.	SML 668	PAU, Ludhiana	Spring, Summer	2002	11.3	60	Punjab	Tolerant to MYMV
20.	Pant Mung 5	GBPUA&T, Pantnagar	Kharif, Summer	2002	14–15		Uttar Pradesh, Uttarakhand	Resistant to MYMV
21.	HUM 12 (Malaviya Janchetna)	BHU, Varanasi	Summer	2003	11.2	60–62	NEPZ	Moderately resistant to MYMV and *Cercospora* leaf spot
22.	IPM 02-3	IIPR, Kanpur	Large seeded (Kharif, Spring)	2003	11.0	62–68	NWPZ	Resistant to MYMV
23.	Meha (IPM 99-125)	IIPR, Kanpur	Spring	2004	9.8	66	NEPZ	Resistant to MYMV
24.	TMB 37	BARC, Mumbai	Summer	2005	11.0	65	NEPZ	Moderately resistant to YMV

(continued)

Table 3 (continued)

S. No.	Variety	Developed by	Seasons/ conditions	Year of release	Average yield (q/ha)	Days to maturity	Area of adaptation	Reaction to major diseases
25.	HUM 16 (Malaviya Jankalyani	BHU, Varanasi	Summer	2006	10.9	55–58	NEPZ	Resistant to MYMV
26.	TM-96-2	BARC and ANGRAU, LAM	Rice fallow	2007	9	65	Andhra Pradesh	Resistant to powdery mildew
27.	IPM 02-14	IIPR, Kanpur	Spring, Summer	2010	11.0	60–65	SZ	Resistant to MYMV
28.	SML 832	PAU, Ludhiana	Summer	2010	10–11	60–65	Punjab	Tolerant to MYMV and thrips
29.	Pairy mung	IGKV, Raipur	Rabi	2010	12 q/ha	65–70	Chhattisgarh	Tolerant to YMV and resistant to powdery mildew
30.	MH 421	CCS HAU, Hisar	Summer	2012	12.0	60	NWPZ	Resistant to MYMV
31.	SML 1115	PAU, Ludhiana	Summer	2016	10–11	65–70	NEHZ	Resistant to MYMV
32.	Virat (IPM 205-7)	IIPR, Kanpur	Summer	2016	10–11	56–58	Irrigated tract of IGP, Madhya Pradesh, Tamil Nadu	Resistant to MYMV
33.	MH 318	CCS HAU, Hisar	Spring (Rice-Wheat rotation)	2016	14 (Kharif) 10 (Spring, Summer)	60–61	Rainfed and irrigated areas of Haryana	Resistant to MYMV
34.	Yadadri WGG 42)	PJTSAU, Hyderabad	Kharif, Rabi, Summer	2016	10–12	55–60	Telangana	Resistant to MYMV
35.	Sri Rama (MGG 351)	PJTSAU, Hyderabad	Rabi, Summer, Rice fallow	2016	12–14	60–65	Telangana	Moderately resistant to MYMV
36.	GBM-1	NAU, Gujarat	Rabi	2016	11–12	102–105	Gujarat	–
37.	IPM 410-3 (Shikha)	IIPR, Kanpur	Summer, Spring	2016	11–12	65–70	NWPZ/ CZ	Resistant to MYMV
38.	MSJ 118 (Keshvanand mung 2)	RARI, Durgapura	Kharif, Spring	2016	7–8	60–65	Rajasthan	Moderately resistant to MYMV
39.	Pusa 1431	IARI, New Delhi	Spring	2018	12–14	56–66	NCR region, Delhi state	Resistant to MYMV, *Cercospora* leaf spot, anthracnose, web blight, and ULCV
40.	Tripura Mung 1 (TRCM 131)	ICAR RC for NEH region	Summer, Kharif	2018	10–11	55–60	Tripura	MYMV resistant
41.	SGC 16 (Rupohi)	AAU, Jorhat	Summer, Kharif	2018	12–13	65–70	Assam	Resistant to *Cercospora* leaf spot and YMV, moderately resistant to web blight
42.	GAM 5	AAU, Anand	Summer, Kharif	2018	8–11	60–65	Gujarat	Resistant to YMV
43.	Gujarat Mung-7 (GM-7)	Navsari	Summer, Kharif	2018	10–11	75–80	Gujarat	YMV resistant
44.	Kanika (IPM 302-2)	IIPR, Kanpur	Spring, Kharif	2018	10–12	65–75	Uttar Pradesh	MYMV resistant, green, shining, attractive, and medium-large seed (3.4 g/100 seed)
45.	Heera (IPM 409-4)	IIPR, Kanpur	Spring	2019	10–11	52–55	Uttar Pradesh	Resistant to MYMV

(continued)

Table 3 (continued)

S. No.	Variety	Developed by	Seasons/ conditions	Year of release	Average yield (q/ha)	Days to maturity	Area of adaptation	Reaction to major diseases
46.	Vasudha (IPM 312-20)	IIPR, Kanpur	Spring	2019	12–13	65–83	Uttar Pradesh	Resistant to MYMV and *Cercospora* leaf spot
47.	VBN 4 (VGG 10-008)	NPRC, Vamban	All seasons	2019	10.11	65–70	Tamil Nadu	Moderately resistant to MYMV, ULCV, and powdery mildew
48.	IPM 512-1 (Soorya)	IIPR, Kanpur	Spring	2020	12–13	60–65	NEPZ	Resistant to MYMV, *Cercospora* leaf spot, and anthracnose
49.	Pusa 1641	IARI, New Delhi	Summer	2020	12–13 q/ha	62–64	NCR of Delhi	Resistant to MYMV, low incidence of major insect pests
50.	MGG 385 (Madhira Pesara)	PJTSAU, Telangana	Kharif, Summer, Rabi, Rice fallow	2021	12–13	70–75	Telangana	Resistant to leaf crinkle, leaf curl, and MR to YMV
51.	GAM 8 (Hara Moti)	AAU, Anand	Spring, Summer	2022	11–12	70–80	Gujarat	Resistant to MYMV, moderately resistant to powdery mildew
52.	VBN 5 (VGG 15013)	TANU, Vamban	All seasons	2022	8–9	70–75	Tamil Nadu (except Nilgiris and Kanyakumari districts)	Resistant to MYMV
53.	Lam Pesara 574 (LGG 574)	RARS, Lam	Rabi, rice fallow	2022	15–16	65–70	Andhra Pradesh	Tolerant to MYMV, ULCV, web blight, and *Cercospora* leaf spot
54.	Lam Pesara 607 (LGG 607)	RARS, Lam	Kharif, Rabi, Rice- fallow	2022	15–17	60–65	Andhra Pradesh	Resistant to MYMV; tolerant to ULCV, *Cercospora* leaf spot, and powdery mildew
55.	TRCRM-147	ARS, Bidar	Summer	2022	8–9	63–65	Karnataka	Resistant to MYMV

Source: Project Coordinators Report (2022) AICRP-MULLaRP. CZ Central Zone; NEPZ North Eastern Plain Zone; NWPZ North Western Plain Zone; SZ Southern Zone; AAU Assam Agricultural University; ARS Agriculture Research Station; BARC Bhabha Atomic Research Center; BHU Banaras Hindu University; CCS HAU Chaudhary Charan Singh Haryana Agricultural University; GBPUA&T G.B. Pant University of Agriculture and Technology; HAU Haryana Agricultural University; IARI Indian Agricultural Research Institute; IIPR Indian Institute of Pulses Research; PAU Punjab Agricultural University; RAU Rajasthan Agricultural University; TNAU Tamil Nadu Agricultural University

Fig. 3 Yellow mosaic disease (YMD) infection under field conditions

mungbean (Devi et al. 2022). However, short-duration genotypes play an important role by offering multiple crop cycles per year, allowing the plants to escape terminal heat stress. Various innovative approaches, such as genomics, GBS-based GWAS, MAS, and genome editing, can be employed to enhance resistance to biotic stresses in plants (Dodia et al. 2019; Mishra et al. 2022c). Different sources of resistance to *Cercospora* leaf spot (CLS) and Mungbean Yellow Mosaic

Fig. 4 Powdery mildew infection on the mungbean leaf

Virus (MYMV) have been identified in several countries. In India, special emphasis has been placed on developing resistance to YMV, resulting in the release of several YMV-resistant varieties (Mishra et al. 2020; Mishra et al. 2017). Additionally, some commercially released varieties have demonstrated resistance to powdery mildew, Macrophomina blight, and Leaf Crinkle Virus. Certain AVRDC accessions have shown resistance to bruchids. Resistance to both MYMV and bruchids has been achieved through wide crosses, particularly between *V. radiata* and *V. radiata* var. *sublobata*. Disease-resistant genes have also been identified from amphidiploids resulting from mungbean×ricebean crosses. Mutation breeding has played a key role in developing several large-seeded, high-yielding, and MYMV-resistant lines. Additionally, some studies have reported resistance to Urdbean Leaf Crinkle virus (ULCV) (Mishra et al. 2022a).

Mungbean cultivation faces another major constraint due to insect pest infestations, especially during *Kharif* and spring seasons. Defoliators like hairy caterpillars, semiloopers, and til caterpillars are commonly infesting the crop, especially in vegetative stage. Thrips activity peaks during bud formation stage and becomes problematic at the time of peak flowering; leading to significant flower drop. Diseases also cause serious threats to mungbean crop, with YMV very often affecting the crop during *Kharif* and spring seasons in the northern regions, and powdery mildew mostly occurs during winter season in Southern regions of India. Additionally, mungbean is highly susceptible to bruchids, with limited resistance against this pest. However, resistance to bruchids has been observed in wild mungbean, wild black gram, and certain genotypes of *V. nepalensis* and *V. umbellata*. This highlights the need for breeding mungbean varieties with better biotic stress resistance to improve productivity and sustainability (Singh et al. 2023).

A GBS-based GWAS analysis conducted on 132 diverse mungbean genotypes, focusing on traits such as flowering time, YMD resistance, soil plant analysis development (SPAD) value, trichome density, and leaf area, identified 31,953 SNPs. Among these, 44 shared SNPs (from MLM and BLINK methods) were reported linked to various candidate genes, including flowering time (*VRADI10G01470*), MYMIV resistance (*VRADI09G06940*), SPAD value (*VRADI07G28100*), leaf area (*VRADI03G06560*), and trichomes (*VRADI06G04290*) (Kohli et al. 2024). Additionally, an RNA-seq analysis compared a resistant and a susceptible mungbean genotype under MYMIV-infected and control conditions and identified a complex gene network associated with MYMIV resistance. This network begins with the production of pathogen-associated molecular patterns (PAMPs), followed by the activation of various signaling cascades such as kinases, jasmonic acid, and brassinosteroids. This ultimately led to the expression of specific genes, including PR-proteins, virus resistance factors, and R-gene proteins, resulting in a resistance response (Dasgupta et al. 2021). Also, two major QTLs regulating SPAD value (qSPAD-7-1) and trichome density (qTric-7-2) were identified using an RIL (PMR1 × Pusa Baisakhi) on chromosome 7. The candidate genes regulating trichome density on the dorsal leaf surface include *VRADI07G24840, VRADI07G17780*, and *VRADI07G15650*, which encode for ZFP6, a TF bHLH DNA-binding superfamily protein, and MYB102, respectively. Thus, there is potential for developing customized genotypes with high SPAD value and high trichome density, which may play a role in yield and MYMIV resistance in mungbean (Kumari et al. 2024).

7.4 Breeding for Nutrient Use Efficiency

Phosphorus (P) is a key nutrient limiting crop productivity, particularly in the weathered soils of subtropical and tropical regions. Its low mobility and limited availability in soil pose challenges for plant uptake. Efforts are underway to develop mungbean varieties with improved phosphorus-use efficiency (PUE) by enhancing both P uptake and utilization. Kothari et al. (2023) evaluated 327 accessions across 18 *Vigna* species and observed significant variations in inter- and intra-species traits associated with P-uptake and utilization efficiency. Among these, seven mungbean genotypes demonstrated superior PUE (IC251950, IC585931, V1002532AG, IC371653, IC331615, V1001400AG, V1000532BG) showing promise for both P uptake efficiency (PupE) and P utilization efficiency (PutiE).

Interestingly, cultivated species demonstrated higher PUE over wild types in both *Vigna* (Kothari et al. 2023) and *Lens* species (Ramtekey et al. 2021). Similarly, Aski et al. (2022) examined root architectural traits in 110 diverse lentil genotypes and identified five highly efficient genotypes based on the Comprehensive Phosphorus Efficiency Measure (CPEM) for potential use in mungbean breeding. Studies on root system architecture (RSA) in mungbean and urdbean genotypes highlighted their ability to support crops under suboptimal nutrient and water conditions (Reddy et al. 2023).

Under low phosphorus (P) conditions, mungbean genotypes exhibited significantly higher antioxidant activity compared to normal P conditions. Traits such as root dry weight (RDW) and total dry weight (TDW) emerged as key contributors and can serve as effective indicators for screening P deficiency in mungbean (Reddy et al. 2022). Additionally, Reddy et al. (2020a) observed substantial variation in root traits among 153 mungbean genotypes under both optimal and low P conditions. Notably, released Indian mungbean varieties (IPM-288, TM 96–25, TM 96–2, M1477, PUSA1342) outperformed other genotypic groups in root traits (Reddy et al. 2020c, 2021b).

GBS-based GWAS conducted on 144 diverse mungbean genotypes for 12 PUE traits under optimal and low-P conditions identified 55,634 SNPs, including 136 shared SNPs (detected by GLM and MLM approaches). Three candidate genes (*VRADI11G08340, VRADI01G05520,* and *VRADI04G10750*) had missense SNPs in their coding sequences, causing significant variations in their tertiary protein structure (Reddy et al. 2020b). Additionally, association mapping (AM) using GBS-based GWAS on 120 mungbean genotypes identified 116 SNPs within 61 protein-coding genes. Of these, six genes (*VRADI01G04370, VRADI05G20860, VRADI06G12490, VRADI08G20910, VRADI08G00070, VRADI09G09030*) showed high expression in roots, shoot apical meristem, and leaves, suggesting their critical roles in regulating PupE and PutiE in mungbean (Reddy et al. 2021a).

7.5 Breeding for Bio-fortification in Mungbean

Micronutrient deficiencies impact nearly one-third of the global population, and the biofortification of staple crops, including mungbean, is recognized as a sustainable, long-term solution to address this issue (Devi et al. 2019; Gupta et al. 2021b). Scaling up the cultivation of biofortified mungbean will require strengthening existing partnerships and forging new collaborations (Nawade et al. 2018). In India, various national programs are actively engaged in developing biofortified varieties of multiple staple crops (Mishra et al. 2022b). Additionally, food processors and retailers must integrate biofortified crops into their product offerings to enhance their accessibility and impact (Kumar et al. 2022). Despite its nutritional benefits, the presence of anti-nutritional factors poses a significant challenge to improving its nutritional quality through biofortification (Sinha et al. 2023; Gupta et al. 2021a). Mineral biofortification, a process aimed at enhancing the levels of essential micronutrients like iron (Fe) and zinc (Zn) in mungbean, has emerged as a promising approach to address these challenges (Dikshit et al. 2022). Identifying germplasm with high micronutrient content is a vital step in developing mungbean varieties rich in these critical nutrients (Singh et al. 2013a; Aski et al. 2023a).

Research has shown significant variability in the Fe and Zn content in mungbean genotypes. Aneja et al. (2012) reported Fe content ranging from 29.95 to 100.97 mg/kg and Zn content between 20.13 and 35.70 mg/kg. Similarly, Taunk et al. (2012) observed Fe levels between 46.31 and 106.15 mg/kg and Zn levels from 23.31 to 40.46 mg/kg. Singh et al. (2013a) documented wide variations in a RIL population, with Fe ranging from 1.6 to 9.3 mg/100 g and Zn from 1.5 to 3.9 mg/100 g. Interestingly, Wu et al. (2020) conducted GWAS studies for seven minerals using inductively coupled plasma (ICP) spectroscopy in 95 cultivated mungbean genotypes and identified markers linked to QTLs regulating micronutrient content, offering valuable insights for future breeding programs (Table 4).

Mungbean seeds contain anti-nutritional factors (phytic acid), which bind with minerals like iron, zinc, calcium, and magnesium; thereby reducing their absorption in the small intestine (Weinberger et al. 2002). Processing techniques like fermentation, germination, dehulling, soaking, and cooking effectively lower these anti-nutritional components (Hemalatha et al. 2007). Sprouting of mungbean reduces phytic acid content by 76%, thereby enhancing zinc bioavailability by 3.0-fold and iron by 2.4-fold (Nair et al. 2012; El-Adawy 2002). Trypsin inhibitors are low-molecular-weight proteins that interfere with protein digestion by inhibiting proteolytic enzymes are also present in mungbean seeds. Their activity can be reduced by heat treatment, soaking, and sprouting (Chandrashekar et al. 1989; Gupta et al. 2023) which improve the nutritional quality and digestibility of mungbeans.

The genetic architecture of grain iron, zinc, and anti-nutritional factors (phytic acid, tannins) was analyzed in an AM panel of 145 genotypes using GWAS which identified 14,447 SNPs (20 shared SNPs via GLM and Blink) associated with grain micronutrients and anti-nutritional factors. Key candidate genes identified were *VRADI07G30190, VRADI01G09630,* and *VRADI09G05450* for Fe; *VRADI10G04830* for Zn; *VRADI08G09870* and *VRADI01G11110* for phytic acid; and *VRADI04G11580* and *VRADI06G15090* for tannins. Among the genotypes, PUSA 1333 and IPM 02-19 exhibited high grain Fe and Zn levels with low phytic acid and tannins, making them promising for nutritional enhancement (Sinha et al. 2023). Similarly, nutritional analysis of 120 cowpea (*Vigna unguiculata*) genotypes identified EC170579 and EC201086 as having high protein

Table 4 Markers, QTLs for grain micronutrient concentration in mungbean

Trait	LG/markers/ population	QTLs/MTAs	References
Zn content	LG4 (qZn-4-3), LG6, LG7 & LG11 (qZn-11-2)	15 QTLs	Singh et al. (2017b)
Fe content	LG4 (qFe-4-1), LG6 (qFe-11-1)	2 QTLs	Singh et al. (2017b)
Ca, Fe, K, Mn, P, S, Zn	6486 SNPs (22% PVE) by GBS	43 MTAs	Wu et al. (2020)
K content	Chromosome 1	*Vradi01g00820*, *Vradi01g00830*, *Vradi01g00840*	Wu et al. (2020)
P content	Chromosome 5	*Vradi05g16350*	Wu et al. (2020)
P content	Chromosome 7	*Vradi07g26320*, *Vradi07g26340*	Wu et al. (2020)
K content	Chromosome 7	*Vradi07g14180*	Wu et al. (2020)
K content	Chromosome 8	*Vradi08g22740*, *Vradi08g17100*	Wu et al. (2020)
Fe content	Chromosome 6	*Vradi06g09900*, *Vradi06g10020*, *Vradi06g10060*, *Vradi06g10120*, *Vradi06g10210*	Wu et al. (2020)
Seed Fe and Zn content	RIL (ML776 x Sattya)	QTLs	Singh et al. (2017b)
Seed Fe and Zn content	RIL from ML446 (high Fe) x Sattya (low Fe)	Fe: 01 QTL (qFe-11-1) on LG11; Zn: 04 QTLs (qZn-11-4, qZn-11-5 on LG11 & qZn-4-1, qZn-4-2) on LG4	Singh (2013)
Phytic acid, total P, and inorganic P	–	QTLs	Sompong et al. (2012)

content (>27%) along with elevated TSS, amylose, and TDF levels, highlighting their potential for developing high-value food and feed varieties (Padhi et al. 2022b). Similar findings have been reported in other legumes, including lentils (Kumar et al. 2019) and peanuts (Sanyal et al. 2023).

7.6 Breeding for Nutritionally Rich Mungbean for Microgreens

Microgreens are a category of functional foods that provide essential nutritional elements and deliver various health benefits when included as dietary supplements. Their global consumption is increasing due to the higher concentration of health-promoting components compared to their mature plant counterparts (Dhaka et al. 2023). Mungbean microgreens are particularly notable as they are easy to cultivate, cost-effective, and represent a novel nutritional resource (Fig. 5). However, the detailed diversity of their phytonutrient profile remains poorly understood (Priti et al. 2022; Mishra et al. 2022d).

Priti et al. (2021) evaluated 20 diverse mungbean genotypes grown as microgreens under plain-altitude (Delhi) and high-altitude (Leh) conditions. The study revealed significant genotypic variations for several nutritional and bioactive compounds, including ascorbic acid, tocopherol, carotenoids, flavonoids, total phenolics, antioxidant activities (DPPH and FRAP), peroxide activity, proteins, enzymes (peroxidase and catalase), as well as micronutrient and macronutrient content. The genotype MH810 emerged as superior across most of the studied parameters. Interestingly, microgreens grown in Leh outperformed those grown in Delhi, likely due to Leh's unique environmental conditions, including a wide temperature amplitude, high photosynthetically active radiation, and elevated UV-B levels (Priti et al. 2021).

Dhaka et al. (2023) investigated the growth conditions and nutritional profiles of six microgreens (mungbean, lentil, red radish, pearl millet, mustard, red cabbage) under a 16-h light/8-h dark cycle. The optimum growth temperature for these microgreens ranged from 24 to 28 °C, with the ideal harvesting stage varying from the 6th to 13th day, depending on the species. Specifically, mungbean microgreens can be harvested within 6–7 days after sowing (Dhaka et al. 2023). Further research is needed to evaluate different mungbean genotypes under diverse light conditions to identify the most nutrient-rich or biofortified microgreens.

8 Crop Wild Relatives (CWR) and Genetic Improvement

The limited genetic diversity of most mungbean cultivars necessitates immediate corrective actions, including the incorporation of unadapted accessions, exotic varieties, and wild relatives into hybridization programs. Wild relatives of *Vigna* species serve as valuable reservoirs of genes for addressing various biotic and abiotic stresses, as well as for enhancing yield and quality traits (Baudoin and Marechal 1988). The challenge lies in the effective utilization of the available *Vigna* genetic resources through pre-breeding efforts to develop more adaptable lines for plant breeders (Fig. 6). Significant diversity was reported in 206 accessions representing 14 wild *Vigna* species (*V. aconitifolia*, *V. bourneae*, *V. dalzelliana*, *V. glabrescens*, *V. hainiana*, *V. minima*, *V. mungo var. silvestris*, *V. pilosa*, *V. radiata var. setulosa*, *V. radiata var. sublobata*, *V. trilobata*, *V. umbellata*, *V. vexillata*,

Fig. 5 Mungbean (**a**) sprouts and (**b**) microgreens

Fig. 6 Various plant parts of different *Vigna* species

Vigna khandalensis) for 45 morphological traits (Bisht et al. 2005). In the improvement of mungbean and urdbean, the sub-genepools of wild types have shown significant potential. Accession PLN5 of *Vigna radiata var. sublobata* and IW3390 of *V. mungo var. silvestris* have been identified as valuable sources of resistance to YMV (Singh and Ahuja 1977). Additionally, TC1966 of *V. radiata var. sublobata* has been found to possess a gene for bruchid tolerance (Tomooka et al. 1992). These discoveries underscore the critical role of genetic diversity in wild *Vigna* species for advancing breeding programs in mungbean and related crops.

9 Conventional and Innovative Breeding Technologies

9.1 Pureline Breeding

In pureline breeding, plant breeders identify and select superior plant types from heterogeneous populations which arise due to natural selection pressures, resulting in new plant types with enhanced genetic potential for yield and economic traits. Breeders stabilize these desirable traits through continuous self-pollination and repeated selection cycles (Gupta and Kumar 2006; Tickoo et al. 2006).

9.2 Recombination Breeding

Recombination breeding involves creation of superior recombinants by using existing intra- or inter-specific genetic variability through hybridization (Figs. 7 and 8). After hybridization, various selection methods, such as pedigree, bulk, recurrent, backcross, or single-seed descent (SSD), are applied to guide phenotypic selection in subsequent generations. Among these, the pedigree method is most commonly employed for improving yield and related traits. Additionally, recurrent selection and population improvement strategies are suggested for accumulating desirable alleles and breaking unfavorable linkages, thereby increasing the efficiency of recombination breeding. Early-generation testing (Burton 1997) is often employed to eliminate undesirable progenies through selection in early generations (F2, F3, F4).

Fig. 7 Crossing in mungbean

9.3 Wide Hybridization

Wide hybridization, or interspecific crossing, has been reported between *Vigna unguiculata* and *V. vexillata* (Gomathinayagam et al. 1998). Similarly, Tyagi and Chawla (1999) documented crosses between *Vigna radiata* and *V. unguiculata* using in vitro culture techniques. However, subsequent growth and development of plantlets were not reported. Continued research is needed to explore interspecific crosses involving *V. vexillata* and other *Vigna* species using advanced techniques. Notably, successful crosses between *V. radiata* and *V. mungo* have led to the development of four mungbean varieties (Pant M 4, HUM 1, Meha, PM 6) and one urdbean variety (Mash 1008) with improved traits (Singh and Singh 1998; Singh and Dikshit 2002).

Most reports on crossbreeding among various *Vigna* species (Singh 1990; Dana and Karmakar 1990) indicate that *V. radiata* is typically used as the seed parent in hybrids with *V. mungo*, *V. umbellata*, and *V. angularis*, as reciprocal crosses are not viable. Sequential embryo rescue techniques are used to generate reciprocal hybrids between *V. mungo* and *V. radiata* (Gosal and Bajaj 1983a; Verma and Singh 1986). Successful crosses have also been made between *V. mungo* and *V. delzelliana* (Chavan et al. 1966), *V. glabrescens* (Dana 1968; Krishnan and De 1968), and *V. trilobata* (Dana 1966). In case of *V. mungo* × *V. umbellata*, hybrids were obtained only through embryo rescue (Biswas and Dana 1975; Chen et al. 1983). Crosses between mungbean and rice bean have been attempted to introduce resistance to MYMV and other beneficial traits into mungbean (Verma and Brar 1996). Additionally, four amphidiploids resulting from *V. mungo* (ML 267 and K 851) × rice bean (RBL 33 and RBL 140) crosses were successfully produced and evaluated (Dar et al. 1991). Singh et al. (2003) also successfully created hybrids between *V. radiata* × *V. umbellata*, which displayed intermediate morphology and resistance to MYMV.

Similarly, successful interspecific crosses have been achieved between *V. mungo* × *V. umbellata* (Pal et al. 2005). Hybridization between cultivated cowpea (*V. unguiculata* ssp. *unguiculata* and *V. unguiculata* ssp. *biflora*) and wild cowpea (*V. unguiculata* var. *spontanea*, *V. unguiculata* ssp. *alba*, *V. unguiculata* ssp. *stenophylla*, *V. unguiculata* ssp. *pawekiae*, and *V. unguiculata* ssp. *baoulensis*) have also been

Fig. 8 Pod setting after hybridization in mungbean

reported (Kouadio et al. 2007). The highest success was observed in the crosses between cultivated and annual inbred forms, while intermediate success was noted in different crosses between cultivated and wild allogamous forms. Crosses involving *V. unguiculata* ssp. *baoulensis* and cultivated forms showed a lower success rate. Additionally, Reddy and Singh (1990) identified a wild accession of *V. radiata* var. *sublobata*, PLN15, as a potential donor for traits such as pods per plant and seeds per pod. Resistance to MYMV was also reported in *V. umbellata*, *V. trilobata*, and *V. mungo* (Nagaraj et al. 1981; Singh and Dikshit 2002). Slow pollen tube growth, along with abnormalities in the stigmatic and stylar regions, may contribute to the low pod set as observed in *V. radiata* × *V. umbellata* and *V. mungo* × *V. umbellata* crosses (Thiyagu et al. 2008). However, differences in ploidy level and style length may not present significant barriers for *Vigna* species, as shown by the successful cross between the long-styled female parent *V. radiata* and the short-styled male parent *V. trilobata*. Crosses between diploid × tetraploid (*V. radiata* × *V. glabrescens*) (Chen et al. 1989; Krishnan and De 1968) and tetraploid × diploid (*V. glabrescens* × *V. umbellata*) have also been successful. In many cases, crossability is genotype-dependent (Rashid et al. 1988).

Strong pre-fertilization barriers have been observed in *V. radiata* × *V. umbellata* crosses, and interspecific hybrid seedling lethality was influenced by the parental genotypes (Kumar et al. 2007). Meiotic irregularities, such as unequal separation of tetrads, led to male sterility in F1 and subsequent generations of interspecific crosses. Female sterility was associated with the degeneration of megaspores during megasporogenesis (Pandiyan et al. 2008). Biswas and Dana (1975) reported a single fertile pod with two hybrid seeds when *V. angularis* was used as the male parent, resulting in a partially fertile interspecific hybrid. The barriers encountered in interspecific crosses varied in intensity across different studies (Biswas and Dana 1975; Chen et al. 1983). Sidhu (2003) successfully produced interspecific hybrids of *V. radiata* × *V. mungo* and *V. trilobata*. While the crosses between *V. radiata* × *V. trilobata* were successful, seeds from the *V. mungo* × *V. trilobata* crosses exhibited poor germination, and the germinated seedlings did not survive. In some interspecific crosses of *Vigna*, hybrid sterility was found to be of a segregational type (Biswas and Dana 1975; Karmakar and Dana 1987).

10 Overcoming Crossability Barriers in *Vigna*

Overcoming crossability barriers in *Vigna* involves a multi-faceted approach including a deeper understanding of the processes like pollen-stigma interaction, pollen germination, pollen tube growth, and fertilization (Sharma and Satija 1996). The methods to overcome crossability barriers in *Vigna* are listed in Table 5.

10.1 Reciprocal Crossing and Irradiation Treatment

Pandiyan et al. (2008) noted an increased pod set in a *V. radiata* × *V. umbellata* cross developed from gamma ray-irradiated parental lines. If fertilization barriers are caused by disharmony between the genomes of one species and the cytoplasm of the other, reciprocal crosses can help recover successful hybrids. For example, the *V. mungo* × *V. radiata* cross was unsuccessful, whereas the reciprocal cross (*V. radiata* × *V. mungo*) produced viable hybrids (Verma and Singh 1986; Ravi et al. 1987). Interspecific hybridization between *V. nakashimae* × *V. angularis* was successful in both directions. Generally, using a female parent with a higher chromosome number yields more successful outcomes than the reciprocal cross.

Table 5 Ways to overcome crossability barriers in various *Vigna* species

Method	Cross combination	References
Use of bridge species	(*V. mungo* × *V. radiata*) × *V. angularis*	Gupta et al. (2002)
Use of growth regulators	*V. radiata* × *V. umbellate*	Gupta et al. (2002)
Use of colchicine for chromosome doubling	*V. radiata* × *V. mungo*	Pande et al. (1990)
	V. radiata × *V. trilobata*	Dana (1966)
Attempting reciprocal crosses	*V. radiata* × *V. mungo*	Verma and Singh (1986), Ravi et al. (1987)
Embryo rescue technique	*V. marina* × *V. luteola*	Palmer et al. (2002)
	V. radiata × *V. unguiculata*	Tyagi and Chawla (1999)
	V. vexillata × *V. unguiculata*	Gomathinayagam et al. (1998)
	V. radiata × *V. trilobata*	Sharma and Satija (1996)
	V. radiata × *V. radiata* var. *Sublobata*	Sharma and Satija (1996)
	V. glabrescens × *V. radiata*	Chen et al. (1990)
	V. mungo × *V. radiata*	Gosal and Bajaj (1983a, b)

10.2 Bridge Species and Their Use

Transferring bruchid resistance from wild *Vigna* species is challenging due to cross-incompatibility. However, the use of bridge species (*V. nakashimae) resulted in* successful transfer of bruchid resistance from *V. umbellata* to azuki bean (Tomooka et al. 1992, 2000).

10.3 Use of Growth Hormones

In vivo hormonal treatments have proven highly effective in facilitating the recovery of interspecific hybrids in *Vigna*. An interspecific derivative (*V. mungo* × *V. radiata)* was reciprocally crossed with *V. angularis*, and pollinated pistils were treated with GA3 at 24 and 78 h post-pollination, resulting in a higher seed set. This demonstrated the positive impact of hormonal intervention on enhancing seed development in interspecific crosses (Pratap et al. 2014b).

11 Alien Gene Introgressions in *Vigna*

The successful introgression of yield-enhancing genes from urdbean into mungbean has resulted in a yield increase of 15–60% (Singh and Dixit 2002). Reddy and Singh (1990) reported that mungbean × urdbean derivatives exhibit several desirable traits, including lodging resistance, synchronous podding, and non-shattering. Verma and Brar (1996) noted that some hybrids, such as mungbean × ricebean and mungbean × *V. radiata* var. *sublobata*, showed a high degree of resistance to MYMV. From mungbean × urdbean crosses, three mungbean cultivars (HUM 1, Pant Moong 4, IPM99-125) and one urdbean cultivar (Mash 1008) were developed. These cultivars feature improved plant types, enhanced resistance to MYMV, and synchronous maturity (Fujii et al. 1989; Dongre et al. 1996). *Vigna mungo* var. *silvestris* has been reported to be immune to bruchids, while rice bean (*V. umbellata*) also shows immunity to bruchids. At AVRDC, efforts are underway to utilize *V. radiata* var. *sublobata* for bruchid resistance. Related species of mungbean, such as *V. umbellata* and *V. angularis*, demonstrate higher productivity levels. Additionally, *V. mungo* and the wild progenitor of mungbean, *V. radiata* var. *sublobata*, may contribute productivity alleles to elite mungbean lines through the Advanced Backcross QTL Analysis (AB-QTL) strategy (Mishra et al. 2009a, 2022a).

12 Molecular Cytogenetic Techniques

Khattak et al. (2007) first detected 25S and 5S rDNA sites in mungbean using fluorescent in situ hybridization (FISH) and identified active nucleolar organizer regions (NORs) through silver staining. Interestingly, one pair of chromosomes containing 5S rDNA showed significantly lower fluorescent signal intensity compared to other pairs with same type of sites. Additionally, the silver staining technique revealed active NORs on two chromosome pairs in mungbean.

13 Double Haploid (DH) Production and Micropropagation

DH breeding enables breeders to obtain fully homozygous genotypes from heterozygous parents in a single generation (Pratap et al. 2006). Bajaj and Singh (1980) successfully obtained callus and immature embryos from three genotypes, initially dominated by haploid callus cells. However, significant variations in chromosome complements were observed over time, and no mature embryos or haploid plants were recovered. Similarly, in urdbean (*V. mungo*; $2n = 22$), Gosal and Bajaj (1988) reported haploid plant regeneration at a low frequency. Micropropagation, a direct method for plant regeneration from individual somatic cells (Pratap et al. 2010), is used to accelerate breeding in mungbean. Efforts in mungbean have led to the development of a reliable micropropagation protocol, with shoot regeneration achieved from cotyledons, shoot tips, and cotyledonary nodes, bypassing the intermediate callus phase (Gulati and Jaiwal 1992). Limited attempts have been made to establish anther and microspore culture systems for *Vigna* species (Bajaj and Singh 1980; Gosal and Bajaj 1988).

14 Possibility of Hybrids in Mungbean

Mungbean hybrids have been developed exclusively through manual cross-pollination for agronomic performance evaluations of F1 plants. Khattak et al. (2004) studied 15 hybrid combinations, of which 11 resulted in reduced seed yield, while 4 showed higher seed yield, with a maximum heterosis of 27%. Based on a pot experiment, heterosis in seed yield was assessed for four parental pairs, showing an increase ranging from 52% to 96% (Soehendi and Srinives 2005). The major limitation is the absence of any sterility system in mungbean, which complicates the large-scale production of true hybrid seeds for agronomic evaluations. Despite these efforts, mungbean hybrids have not yet achieved commercial success.

15 Emerging Molecular Breeding Techniques

In mungbean, various molecular markers have been developed to map traits associated with both biotic and abiotic factors. SSR markers have been used to assess genetic diversity, although they show a low level of polymorphism (Tangphatsornruang et al. 2009). SSRs from cowpea and adzuki bean have also been employed to identify QTLs in mungbean (Kitsanachandee et al. 2013; Gupta et al. 2013). Additionally, several SNPs have been identified in mungbean, but only a small number have been validated as kompetitive allele-specific polymorphism (KASP) markers (Van et al. 2013; Islam and Blair 2018). Research using GBS-based GWAS to map the traits and evaluate diversity in mungbean remains limited (Reddy et al. 2020a, b, c).

Previous studies on genetic diversity in mungbean germplasm, including both cultivated and wild varieties, have highlighted the lower diversity in cultivated mungbean compared to its wild counterparts (Pratap et al. 2015). Although a few linkage maps have been developed, a high-density map using whole-genome sequences shows great potential (Sokolkova et al. 2020). The recent release of a reference genome for mungbean opens new possibilities for genomic research (Dasgupta et al. 2021). Therefore, QTL analyses using mapping populations or GWAS, along with molecular markers and high-throughput sequencing techniques, provide innovative approaches to uncover the genes that control various traits (Mishra et al. 2009b, c, 2020 Singh et al. 2009).

15.1 Marker-Assisted Breeding

The genetic basis of two phenological traits (days to heading and maturity) and eight agronomic traits (leaf nitrogen, plant height, primary branches, pod length, pods per plant, seeds per pod, 100-seed weight, and yield) was investigated using GBS-based GWAS in 153 mungbean genotypes. A total of 55,634 SNPs were identified through GBS, with 15,926 SNPs used for GWAS. The BLINK model identified 50 SNPs significantly associated with the traits, while BLAST analysis revealed 19 candidate genes that regulate various traits. These genes serve as a valuable resource for marker-assisted breeding of these traits in mungbean (Manjunatha et al. 2023).

15.2 Genetic Engineering and Genome Editing Tools

In mungbean, genetic transformation systems have been successfully established, leading to the development of several transgenic plants. However, these transgenics are primarily in the proof-of-concept phase (Bhajan et al. 2019). Genome editing in mungbean is still in its early stages, with researchers aiming to apply CRISPR/Cas9 technology to develop virus-resistant mungbean plants (Khatodia et al. 2018). The availability of a fully sequenced reference genome for Asian mungbean, along with freely accessible SNP data and germplasm linked to specific traits, coupled with favorable regulatory policies, suggests that SNP-trait associations will play a pivotal role in the future of mungbean breeding.

16 Significant Role of Research Centers for Improvement of Mungbean

Mungbean is primarily native to South Asia, where most of the research efforts are focused. Key international organizations involved in mungbean improvement include the World Vegetable Centre (AVRDC) in Taiwan and the Japan International Research Council for Agricultural Studies (JIRCAS) in Japan. In India, an extensive mungbean crop improvement program is being led by the Indian Agricultural Research Institute, New Delhi and Indian Institute of Pulses Research (IIPR) in Kanpur, in collaboration with various partners from the National Agricultural Research System (NARS). Several State Agricultural Universities (SAUs) and their research stations, such as UAS Dharwad, ANGRAU Lam Guntur, HAU Hissar, GBPUAT Pantnagar, CSAUAT Kanpur, TNAU Coimbatore, and PAU Ludhiana, are also actively involved in mungbean improvement. Additionally, research stations like ARS-Durgapura, ARS-Dholi, ARS-Berhampore (WB), and ARS-Berhampore (Orissa) play a significant role in advancing mungbean research programs (Mishra et al. 2022a; Pratap et al. 2022).

17 Key Thrust Areas for Mungbean Improvement

Ongoing focus on incorporating resistance to pre-harvest sprouting, bruchids, and key insect pests (such as thrips, jassids, and pod borers), as well as diseases like yellow mosaic virus, powdery mildew, and *Cercospora* leaf spot, remains crucial. This requires not only identifying resistant germplasm but also pyramiding multiple resistance genes to ensure long-term protection against pests and diseases. Efforts must be strengthened to explore and utilize diverse germplasm sources, including wild species, to access novel genes for key economic traits and adaptive features. This will expand the genetic base of mungbean cultivars and improve their resilience to changing environmental conditions. Additionally, developing mungbean varieties suited to different agroecological zones and cropping systems is critical. This includes breeding varieties with distinct vegetative and reproductive phases to integrate into cereal-cereal cropping systems and adapt to varying temperature regimes across regions. Given India's diverse climatic conditions, there is an urgent need to develop mungbean varieties that are less sensitive to temperature and photoperiod. This will ensure stable performance across various seasons and regions, enhancing the reliability of mungbean cultivation.

The ongoing goal of developing short-duration and high-yielding mungbean cultivars suitable for intensive cropping systems should be continued. Effective postharvest storage is essential to maintain the nutritional and physical quality of mungbean. Bruchids pose a significant threat to stored grains, causing considerable losses in quality and nutrition. Therefore, there is an urgent need to identify bruchid-resistant donor plants and associated molecular markers. Significant variability exists among different mungbean cultivars and germplasm accessions for nutrition-related traits, indicating that improvements in these traits are possible through straightforward breeding methods. The sprouting segment, in particular, holds high market value, but the grains must meet stringent quality standards to succeed in this market. By focusing on these key areas, ongoing research and breeding efforts in mungbean can greatly enhance productivity, resilience, and sustainability in its cultivation.

18 Conclusions

Mungbean is valued for its adaptability, low input requirements, and nutritional benefits, making it an essential component of vegetarian diets. In India, breeding programs are focused on developing varieties that are less sensitive to temperature and photoperiod, ensuring stable performance across seasons and regions. Efforts are also aimed at creating short-duration, high-yielding cultivars resistant to pests, diseases, and environmental stresses. Postharvest storage remains a challenge due to pests like bruchids, highlighting the need for breeding programs to identify resistant varieties. Given its importance for food, nutrition, and environmental security, ongoing research and breeding efforts are vital to improving mungbean productivity, resilience, and sustainability. By addressing emerging challenges and prioritizing key research areas, mungbean can continue to play a critical role in ensuring food security and sustainability.

Acknowledgments The technical support received from Mr. Dilip Kumar is duly acknowledged. This research was funded by the Indian Council of Agricultural Research (ICAR), Indian Agricultural Research Institute (IARI), New Delhi, and SERB, New Delhi (CRG/2019/002024).

References

Alam AKMM, Somta P, Srinives P (2014a) Generation mean and path analyses of reaction to mungbean yellow mosaic virus (MYMV) and yield-related traits in mungbean (*Vigna radiata* (L.) Wilczek). SABRAO J Breed Genet 46(1):150–159

Alam AKMM, Somta P, Srinives P (2014b) Identification and confirmation of quantitative trait loci controlling resistance to mungbean yellow mosaic disease in mungbean [*Vigna radiata* (L.) Wilczekl]. Mol Breed 34:1497–1506

Ali M (1992) Weeds are a great threat to kharif pulses. Indian Farming 42(27):29–30

Aneja B, Yadav NR, Chawla V et al (2012) Sequence-related amplified polymorphism (SRAP) molecular marker system and its applications in crop improvement. Mol Breed 30:1635–1648. https://doi.org/10.1007/s11032-012-9747-2

Arumuganathan K, Earle ED (1991) Nuclear DNA content of some important plant species. Plant Mol Biol Rep 9:211–215

Aski MS, Rai N, Reddy VRP, Gayacharan, Dikshit HK et al (2021) Assessment of root phenotypes in mungbean mini-core collection (MMC) from the World Vegetable Center (AVRDC) Taiwan. PLoS One 16(3):e0247810

Aski M, Mehra R, Mishra GP, Singh D, Yadav P et al (2022) Genotypic variation in root architectural traits under contrasting phosphorus levels in Mediterranean and Indian origin lentil genotypes. PeerJ 10:e12766

Aski MS, Mishra GP, Tokkas JP, Yadav PS, Rai N et al (2023a) Strategies for identifying stable lentil cultivars (*Lens culinaris* Medik) for combating hidden hunger, malnourishment, and climate variability. Front Plant Sci 14:1102879

Aski MS et al (2023b) Current and future strategies in breeding lentil for abiotic stresses. In: Muthu Arjuna Samy P, Ramasamy A, Chinnusamy V, Sunil Kumar B (eds) Legumes: physiology and molecular biology of abiotic stress tolerance. Springer, Singapore, pp 235–257. https://doi.org/10.1007/978-981-19-5817-5_9

Aski MS, Mishra GP, Yadav PS, Langyan S, Kumari J et al (2024) Cytological studies of lentil genome. In: Kumar J, Gupta DS, Kumar S (eds) The lentil genome: genetics, genomics and breeding. Elsevier, pp 131–158. https://doi.org/10.1016/B978-0-443-19409-2.00002-8

Bajaj YPS, Singh H (1980) In vitro androgenesis in mung bean (*Phaseolus aureus* L.). Ind J Experimental Biol 18:1316–1318

Baudoin JP, Marechal R (1988) Taxonomy and evolution of the genus *Vigna*. In: Mungbean (ed) Proceedings of second international symposium, November 16–20, 1987, Bangkok. AVRDC, Shanhua

Bhajan SK, Begum S, Islam MN et al (2019) In vitro regeneration and *Agrobacterium*-mediated genetic transformation of local varieties of mungbean (*Vigna radiata* (L). Wilczek). Plant Tissue Cult Biotechnol 29(1):81–97

Bhatnagar CP, Chandola RP, Saxena DK, Sethi S (1974) Cytotaxonomic studies on the genus Phaseolus. Indian J Genet Pl Breed 34A:800–804

Bisht IS, Bhat KV, Lakhanpaul S (2005) Diversity and genetic resources of wild *Vigna* species in India. Genet Res Crop Evol 52:53–68

Biswas MR, Dana S (1975) Black gram × rice bean cross. Cytologia 40:787–795

Bose RD (1939) Studies in Indian Pulses. 1X. Contribution to the genetics of mung (*Phaseolus radiaus* L.). Indian J Agric 9:575–594

Burton JW (1997) Soybean (*Glycine max* (L.) Merr.). Field Crop Res 53:171–186

Chaitieng B, Kaga A, Han OK et al (2002) Mapping a new source of resistance to powdery mildew in mungbean. Plant Breed 121:521–525

Chandel KPS (1981) Wild *Vigna* species in Himalyas. Plant Genet Resour Newsl 45:17–19

Chandel KPS, Lester RN, Starling RJ (1984) The wild ancestors of urd and mungbeans (*Vigna mungo* (L.) Hepper and *V. radiata* (L.) Wilczek). Bot J Linnean Soc 89:85–96

Chandrashekar S, Hunshal S, Malik DS (1989) Effect of soaking and germination temperatures on selected nutrients and anti-nutrients of mungbean. Food Chem 34:111–120

Chankaew A, Somta P, Sorajjapinun W et al (2011) Quantitative trait loci mapping of Cercospora leaf spot resistance in mungbean, *Vigna radiata* (L.) Wilczek. Mol Breed 28:255–264

Chavan VM, Patil GD, Bhapkar DG (1966) Improvement of cultivated *Phaseolus* species-need for interspecific hybridization. Indian J Genet Plant Breed 26:152–154

Chen H, Liu X (2001) Inheritance of seed color and lustre in mungbean (*Vigna radiata*). Agric Sci Technol 2:8–12

Chen NC, Baker RL, Honma S (1983) Interspecific crossability among four species of *Vigna* food legumes. Euphytica 32:925–937

Chen HK, Mok MC, Shanmugasundaram S et al (1989) Interspecific hybridization between *Vigna radiata* (L.) Wilczek and *V. glabrescens*. Theor Appl Genet 78:641–647

Chen HK, Mok MC, Mok DWS (1990) Somatic embryogenesis and shoot organogenesis from interspecific hybrid embryos of *Vigna glabrescens* and *V. radiata*. Plant Cell Rep 9:77–79

Chhabra AK (1990) Inheritance of lobed and pentafoliate leaf in mungbean. Indian J Pulses Res 3(1):69–72

Dana S (1966) Cross between *Phaseolus aureus* and *P. mungo*. Genetica 37:259–274

Dana S (1968) Hybrid between *Phaseolus mungo* and tetraploid *Phaseolus species*. Jpn J Genet 43:153–155

Dana S (1998) Collection of wild *Vigna* in seven states in India. Green J 1:9–12

Dana S, Karmakar PG (1990) Species relation in *Vigna* subgenus *Ceratotropis* and its implications in breeding. Plant Breed Rev 8:19–42

Dar GM, Verma MM, Gosal SS, Brar JS (1991) Characterization of some interspecific hybrids and amphiploids in *Vigna*. In: Sharma B, Mehra RB (eds) Golden jubilee celebration symposium on grain legumes. Indian Society of Genetics and Plant Breeding, February 9–11, 1991. IARI, New Delhi, pp 73–78

Dasgupta T, Banik A, Das S (1998) Combining ability in mungbean. Indian J Pulses Res 11:28–32

Dasgupta U, Mishra GP, Dikshit HK, Mishra DC, Bosamia T et al (2021) Comparative RNA-Seq analysis unfolds a complex regulatory network imparting yellow mosaic disease resistance in mungbean [*Vigna radiata* (L.) R. Wilczek]. PLoS One 16(1):e0244593

Devi J, Sanwal SK, Koley TK, Mishra GP, Karmakar P et al (2019) Variations in the total phenolics and antioxidant activities among garden pea (*Pisum sativum* L.) genotypes differing for maturity duration, seed and flower traits and their association with the yield. Sci Hortic 244:141–150

Devi J, Mishra GP, Sagar V, Kaswan V, Dubey RK et al (2022) Gene-based resistance to *Erysiphe* species causing powdery mildew disease in peas (*Pisum sativum* L.). Genes 13(2):316

Devi J, Sagar V, Mishra GP, Jha PK, Gupta N et al (2023) Heat stress tolerance in peas (*Pisum sativum* L.): current status and way forward. Front Plant Sci 13:1108276

Dhaka AS, Dikshit HK, Mishra GP, Tontang MT, Meena NL et al (2023) Evaluation of growth conditions, antioxidant potential, and sensory attributes of six diverse microgreens species. Agriculture 13(3):676

Dhole VJ, Reddy KS (2012) Genetic analysis of resistance to mungbean yellow mosaic virus in mungbean (*Vigna radiata*). Plant Breed 131:414–417

Dhole VJ, Reddy KS (2013) Development of a SCAR marker linked with a MYMV resistance gene in mungbean (*Vigna radiata* L. Wilcek). Plant Breed 132:127–132

Dikshit HK, Mishra GP, Somta P, Shwe T, Alam AKMM et al (2020) Classical genetics and traditional breeding in mungbean. In: Nair R, Schafleitner R, Lee SH (eds) The Mungbean genome. Compendium of plant genomes. Springer, Cham, pp 43–54. https://doi.org/10.1007/978-3-030-20008-4_4

Dikshit HK, Mishra GP, Aski M, Singh A, Virk PS et al (2022) Lentil biofortification. In: Kumar S, Dikshit HK, Mishra GP, Singh A (eds) Biofortification of staple crops. Springer, Singapore, pp 271–293. https://doi.org/10.1007/978-981-16-3280-8_11

Dodia SM, Joshi B, Gangurde SG, Thirumalaisamy PP, Mishra GP et al (2019) Genotyping-by-sequencing based genetic mapping reveals large number of epistatic interactions for stem rot resistance in groundnut. Theor Appl Genet 132(4):1001–1016

Dongre TK, Pawar SE, Thakare RG, Harwalkar MR (1996) Identification of resistant source to cowpea weevil [*Callosobruchus maculatus* (F.)] in *Vigna* sp. and inheritance of their resistance in black gram (*Vigna mungo* var. *mungo*). J Stored Prod Res 32:201–204

Durga KK, Kumar SS (1997) Screening for pre-harvest sprouting in pulses. Legum Res 20:193–197

Dwivedi S, Singh DP (1986) Inheritance of people pigmentation and seed colour in green gram. Crop Improv 13:63–66

El-Adawy TA (2002) Nutritional composition and antinutritional factors of chickpeas (*Cicer arietinum L.*) undergoing different cooking methods and germination. Plant Foods Hum Nutr 57:83–87

Fujii K, Ishimoto M, Kitamura K (1989) Patterns of resistance to bean weevils (Bruchidae) in *Vigna radiata-sublobata* complex inform the breeding of new resistant varieties. Appl Entomol Zool 24:126–132

Gawande VL, Patil JV (2003) Genetics of powdery mildew (*Erysiphe polygoni* D.C.) resistance in mungbean (*Vigna radiata* (L.) Wilczek). Crop Prot 22(3):567–571

Gomathinayagam P, Ram SG, Rathnaswamy R, Ramaswamy NM (1998) Interspecific hybridization between *Vigna unguiculata* and *V. vexillata* through in vitro embryo culture. Euphytica 102:203–209

Gosal SS, Bajaj YPS (1983a) *In vitro* hybridization in an incompatible cross- black gram x green gram. Curr Sci 52:556–557

Gosal SS, Bajaj YPS (1983b) Interspecific hybridization between *Vigna mungo* and *Vigna radiata* through embryo culture. Euphytica 32:129–137

Gosal SS, Bajaj YPS (1988) Pollen embryogenesis and chromosomal variation in anther of three food legumes – *Cicer arietinum*, *Pisum sativum* and *Vigna mungo*. SABRAO J 20:51–58

Gulati A, Jaiwal PK (1992) *In vitro* induction of multiple shoots and plant regeneration from shoot tips of mungbean (*Vigna radiata* (L) Wilczek). Plant Cell Tissue Org Cult 29:199–205

Gupta S, Kumar S (2006) Urd bean breeding. In: Ali M, Kumar S (eds) Advances in mung bean and Urd bean. Indian Institute of Pulses Research, Kanpur, pp 149–168

Gupta PV, Plaha P, Rathore PK (2002) Partially fertile interspecific hybrids between a black gram X green gram derivative and adzuki bean. Plant Breed 121:182–183

Gupta S, Gupta DS, Anjum TK et al (2013) Inheritance and molecular tagging of MYMIV resistance gene in blackgram (*Vigna mungo* L. Hepper). Euphytica 193:27–37

Gupta S, Das S, Dikshit HK, Mishra GP, Aski MS et al (2021a) Genotype by environment interaction effect on grain iron and zinc content of Indian and Mediterranean lentil genotypes. Agronomy 11:1761

Gupta S, Kushwah A, Singh A, Dikshit HK, Mishra GP et al (2021b) Nutritional profiling and antioxidant assessment of Indian and exotic lentil accessions. Indian J Genet 81:440–449

Gupta S, Dikshit HK, Mishra GP, Aski M (2023) Association mapping of total flavonoid content and antioxidant capacities in lentil (*Lens culinaris* Medik.). Indian J Genet Plant Breed 83(03):326–334

Harlan JR (1984) Evaluation of wild relatives of crop plants. In: Holden JHW, Willium SJT (eds) Crop genetic resources: conservation and evaluation. George Allen and Unwin, London, pp 212–222

Hemalatha S, Platel K, Srinivasan K (2007) Influence of germination and fermentation on bioaccessibility of zinc and iron from food grains. Eur J Clin Nutr 61:342–348

Humphry SME, Magner T, McIntyre CL et al (2003) Identification of major locus conferring resistance to powdery mildew (*Erysiphe polygoni* D. C.) in mungbean (*Vigna radiata* L. Wiczek) by QTL analysis. Genome 46:738–744

Humphry ME, Lambrides CJ, Chapman SC, Aitken EAB, Imrie BCR et al (2005) Relationships between hard-seededness and seed weight in mungbean (*Vigna radiata)* assessed by QTL analysis. Plant Breed 124:292–298. https://doi.org/10.1111/j.1439-0523.2005.01084.x

IIPR (2018) Indian Institute of Pulses Research (Kanpur), Project Coordinator's report (Mungbean & Urdbean). All India Coordinated Research Project on MULLaRP group meet at Sardarkrushinagar, Dantiwada, pp 1–46

Islam A, Blair MW (2018) Molecular characterization of mung bean germplasm from the USDA core collection using newly developed KASP-based SNP markers. Crop Sci 58:1659–1670. https://doi.org/10.2135/cropsci2018.01.0044

Jain HK, Mehra KL (1980) Evolution, adaptation, relationships and uses of the species of *Vigna* cultivated in India. In: Summerfield RJ, Bunting AH (eds) Advances in legume science. Royal Botanic Gardens, Kew, pp 459–468

Kaga A, Ishimoto M (1998) Genetic localization of a bruchid resistance gene and its relationship to insecticidal cyclopeptide alkaloids, the vignatic acids in mungbean (*V. radiata* L. Wilczek). Mol Gen Genet 258:378–384

Kajale MD (1974) Plant economy at Bhokardan. Appendix A. In: Dev SB, Gupta RS (eds) *Excavations at Bhokardan (Bhogavardana)* 1973. Nagpur University and Maharashtra Marathwada University, Maharashtra, pp 117–224

Kajale MD (1977) On the botanical findings from excavations at Diamabad, a Chalecolithic site in Western Maharashtra, India. Curr Sci 46:818–819

Karmakar PG, Dana S (1987) Cytogenetic identification of a *Vigna sublobata* collection. Nucleus 30:47–50

Kasettranan W, Somta P, Srinives P (2009) Genetics of the resistance to powdery mildew disease in mungbean (*Vigna radiata* L. Wilczek). J Crop Sci Biotechnol 12(1):37–42

Khadilkar BT (1963) Inheritance of two characters in green gram (*Phaseolus aureus* Roxb. Piper). Akola Agric Res Extn 2:1–3

Khajudparn P, Wongkaew S, Thipyapong P (2007) Mungbean powdery mildew resistance identification of gene for resistance to powdery mildew in mungbean. Afr Crop Sci Conf Proc 8:743–745

Khan MG, Ahmad W, Khattak GSS et al (2007) Mode of inheritance of resistance to mungbean yellow mosaic virus (MYMV) in mungbean *(Vigna radiata* (L.) Wilczek). Sarhad J Agric 23(4):1071–1074

Khatodia S, Bhatotia K, Bhel R (2018) A novel approach of development of virus resistant transgene free mungbean plants using CRISPR/Cas9 genome editing system. In: Genetic and natural resources towards food, energy and environment. Agrobios (International)

Khattak GSS, Haq MA, Ashraf M (1999) Inheritance and joint segregation pattern of testa colour and plant growth habit in mungbean (*Vigna radiata* (L) Wilczek). Trop Agric Res Extn 2:1–3

Khattak GSS, Haq MA, Ashraf M et al (2000) Genetics of mungbean yellow mosaic virus (MYMV) in mungbean (*Vigna radiata* L. Wilczek). J Genet Breed 54(4):237–243

Khattak GSS, Haq MA, Ashraf M, Jabbar A, Zamir R (2001) Inheritance of some important agronomic traits in mungbean (*Vigna radiata* (L.) Wilczek). Breed Sci 51:157–161

Khattak GSS, Ashraf M, Haq MA, Srinives P (2002) Genetic architecture of seed yield and yield componenets in mungbean (*Vigna radiata* L. Wilczek). Trop Agric 179:260–264

Khattak GSS, Ashraf M, Khan MS (2004) Assessment of genetic variation for yield and yield components in mungbean (*Vigna radiata* (L.) Wilczek) using generation mean analysis. Pak J Bot 36:583–588

Khattak GSS, Ashraf M, Zamir R, Saeed I (2007) High yielding desi chickpea (*Cicer arietinum* L.) variety "NIFA-2005". Pak J Bot 39(1):93–102

Kitamura K, Ishimoto M, Sawa M (1988) Inheritance of resistance to infestation with azuki bean weevil in *Vigna sublobata* and successful incorporation to *V. radiata*. Jpn J Breed 38(4):459–464

Kitsanachandee R, Somta P, Chatchawankanphanich O et al (2013) Detection of quantitative trait loci for mungbean yellow mosaic India virus (MYMIV) resistance in mungbean (*Vigna radiata* (L.) Wilczek) in India and Pakistan. Breed Sci 63:367–373. https://doi.org/10.1270/jsbbs.63.367

Kohli M, Bansal H, Mishra GP, Dikshit HK, Reddappa SB et al (2024) Genome-wide association studies for earliness, MYMIV resistance, and other associated traits in mungbean (*Vigna radiata* L. Wilczek) using genotyping by sequencing approach. PeerJ 12:e16653

Kothari D, Pargaien N, Tewari LM, Dikshit HK, Mishra GP et al (2023) Genetic variation for traits related to phosphorus use efficiency in *Vigna* species. Agronomy 13:305

Kouadio D, Echikh N, Toussaint A, Pasquet RS, Baudoin JP (2007) Organisation of the gene pool of *Vigna unguiculata* (L.) Walp., crosses between the wild and cultivated forms of cowpea. Biotechnol Agron Soc Environ 11:47–57

Krishnan R, De DN (1968) Cytogenetical studies in *Phaseolus* II. *Phaseolus mungo* × tetraploid *phaseolus* species and the amphidiploid. Indian J Genet Plant Breed 28:23–30

Kumar S, Gupta S, Chandra S, Singh BB (2004) How wide is the genetic base of pulse crops? In: Ali M, Singh BB, Kumar S, Dhar V (eds) Pulses in new perspective. Indian Society of Pulses Research and Development, Indian Institute of Pulses Research, Kanpur, pp 211–221. http://eprints.icrisat.ac.in/9335/1/Howwide_211-221_20004.pdf

Kumar NP, Pandiyan M, Veerabadhiran P (2007) Prefertilization barriers in *Vigna radiata* × *Vigna umbillata*. Plant Arch 7:377–380

Kumar H, Singh A, Dikshit HK, Mishra GP, Aski M et al (2019) Genetic dissection of grain iron and zinc concentrations in lentil (*Lens culinaris* Medik.). J Genet 98(3):66

Kumar S, Dikshit HK, Mishra GP, Singh A, Aski M et al (2022) Biofortification of staple crops: present status and future strategies. In: Kumar S, Dikshit HK, Mishra GP, Singh A (eds) Biofortification of staple crops. Springer, Singapore, pp 1–30. https://doi.org/10.1007/978-981-16-3280-8_1

Kumari N, Mishra GP, Dikshit HK, Gupta S, Roy A et al (2024) Identification of quantitative trait loci (QTLs) regulating leaf SPAD value and trichome density in mungbean (*Vigna radiata* L.) using

genotyping-by-sequencing (GBS) approach. PeerJ 12:e16722. https://doi.org/10.7717/peerj.16722
Lambrides CJ (1996) Breeding for improved seed quality traits in mung bean (*Vigna radiata* (L) Wilczek) using DNA markers. PhD thesis, University of Queensland, Australia
Lambrides CJ, Godwin ID (2007) Mungbean. In: Kole C (ed) Genome mapping and molecular breeding in plants—pulses, sugar and tuber crops. Springer, Heidelberg, pp 69–90. https://doi.org/10.1007/978-3-540-34516-9_4
Lambrides CJ, Lawn RJ, Godwin ID, Manners J, Imrie BC (2004) Segregation distortion for seed testa color in mungbean (*Vigna radiata* (L.) Wilczek). J Hered 95(6):532–535. https://doi.org/10.1093/jhered/esh078
Lawna RJ, Cottell A (1988) Wild mungbean and its relative in Australia. Biologist 35:267–273
Mahalingam A, Satya VK, Manivannan N et al (2018) Inheritance of mungbean yellow mosaic virus disease resistance in greengram [*Vigna radiata* (L.) Wilczek]. Int J Curr Microbiol Appl Sci 7(01):880–885
Malik LA, Sarwar G, Ali Y (1986) Inheritance of tolerance to mungbean yellow mosaic virus and some morphological characters. Pak J Bot 18:189–198
Manjunatha PB, Aski MS, Mishra GP, Gupta S, Devate NB et al (2023) Genome-wide association studies for phenological and agronomic traits in mungbean (*Vigna radiata* L. Wilczek). Front Plant Sci 14:1209288
Markam NK, Nair SK, Nanda HC et al (2018) Studies on allelic relationship for resistance to mungbean yellow mosaic virus disease in mungbean genotypes. Int J Chem Stud 6(2):2401–2403
Mishra SP, Asthana AN (1996) Inheritance of yellow mosaic virus resistance in mungbean (*Vigna radiata* (L.) Wilczek). In: Asthana AN, Kim DH (eds) Recent advances in Mungbean research. Indian Society of Pulses Research, IIPR, Kanpur, pp 214–219
Mishra SP, Asthana AN, Yadav L (1988) Inheritance of *Cercospora* leaf spot resistance in mungbean, *Vigna radiata* (L.) Wilczek. Plant Breed 100:228–229
Mishra GP, Singh RK, Meetul K, Singh SB (2009a) QTL mapping in crop plants: principle and methodology. In: Singh RK et al (eds) Molecular plant breeding: principle methods and applications. Studium Press LLC, pp 161–188. ISBN:978-1933699493
Mishra GP, Singh RK, Naik PK, Singh R, Singh SB (2009b) Advances in molecular plant breeding: an introduction. In: Singh RK et al (eds) Molecular plant breeding: principle, method and application. Studium Press LLC, pp 1–36. ISBN:978-1933699493
Mishra GP, Tiwari SK, Singh R, Singh SB (2009c) Marker assisted selection for improvement of quality traits in crop plants. In: Singh RK et al (eds) Molecular plant breeding: principle, method, and application. Studium Press LLC, pp 343–366. ISBN:978-1933699493
Mishra GP, Singh RK, Singh R, Singh SB (2009d) Molecular and physiological approaches for drought management in crops. In: Singh RK et al (eds) Molecular plant breeding: principle, method and application. Studium Press LLC, pp 367–408. ISBN:978-1933699493
Mishra GP, Singh B, Seth T, Singh AK, Halder J et al (2017) Biotechnological advancements and begomovirus management in okra (*Abelmoschus esculentus* L.): status and perspectives. Front. Plant Sci 8:360
Mishra GP, Dikshit HK, Ramesh SV, Tripathi K, Kumar RR et al (2020) Yellow mosaic disease (YMD) of mungbean (*Vigna radiata* (L.) Wilczek): current status and management opportunities. Front Plant Sci 11:918
Mishra GP, Dikshit HK, Tripathi K, Aski M, Pratap A et al (2022a) Mungbean breeding. In: Yadava DK et al (eds) Fundamentals of field crop breeding. https://doi.org/10.1007/978-981-16-9257-4_22
Mishra GP, Dikshit HK, Priti, Kukreja B, Aski M et al (2022b) Historical overview of biofortification in crop plants and its implications. In: Kumar S, Dikshit HK, Mishra GP, Singh A (eds) Biofortification of staple crops. Springer, Singapore, pp 31–61. https://doi.org/10.1007/978-981-16-3280-8_2
Mishra GP, Aski M, Bosamia T, Chaurasia S, Mishra DC et al (2022c) Insights into the host-pathogen interaction pathways through RNA-Seq analysis of *Lens culinaris* Medik. in response to *Rhizoctonia bataticola* infection. Genes 13:90
Mishra GP, Priti DHK, Aski M, Sangwan S et al (2022d) Microgreens: A novel food for nutritional security. In: Ramesh SV, Praveen S (eds) Conceptualizing plant-based nutrition. Springer, Singapore. https://doi.org/10.1007/978-981-19-4590-8_7
Mishra GP, Dikshit HK, Devi J, Aski MS, Durgesh K (2023) Epigenetics of abiotic stress tolerance in legumes. In: Muthu Arjuna Samy P, Ramasamy A, Chinnusamy V, Sunil Kumar B (eds) Legumes: physiology and molecular biology of abiotic stress tolerance. Springer, Singapore, pp 89–114. https://doi.org/10.1007/978-981-19-5817-5_4
Miyagi M, Humphry ME, Ma ZY et al (2004) Construction of bacterial artificial chromosome libraries and their application in developing PCR-based markers closely linked to a major locus conditioning bruchid resistance in mungbean (*Vigna radiata* L. Wilczek). Theor Appl Genet 110:151–156
Mukherjee A, Pradhan K (2002) Genetics of mungbean: I. Anthocyanin pigment in hypocotyl. J Interacad 6:434–437
Murty BK, Patel GJ (1973) Inheritance of some morphological characters in mungbean. Barilal Amritlal Coll Agric Mag 25:1–9
Nagaraj NC, Muniyappa V, Satyan BA, Shanmugam N, Jayarajan R, Vidhyasekaran P (1981) Resistance source for mungbean yellow mosaic virus. In: Proceedings of the national seminar on disease resistance in crop plants, pp 69–72
Naik YD, Sharma VK, Aski MS, Rangari KS, Kumar R et al (2024) Phenotypic profiling of lentil (*Lens culinaris* Medikus) accessions enabled identification of promising lines for use in breeding for high yield, early flowering and desirable traits. Plant Genet Resour Charact Util:1–9. https://doi.org/10.1017/S1479262124000042
Nair R, Schreinemachers P (2020) Global status and economic importance of mungbean. In: Nair R, Schafleitner R, Lee SH (eds) The mungbean genome. Compendium of plant genomes. Springer, Berlin
Nair RM, Schalfleitner R, Kenyon L et al (2012) Genetic improvement of mungbean. SABRAO J Breed Genet 44:177–190
Nawade B, Mishra GP, Radhakrishnan T, Dodia SM, Ahmad S et al (2018) High oleic peanut breeding: achievements, perspectives, and prospects. Trends Food Sci Technol 78:107–119
Nopparat S, Srinives P, Kaveta R et al (1997) An inheritance of mungbean tolerance to microessential element denciency in Takhli soil series. In: Romkaew et al (eds) Proceedings of national mungbean research conference VII. Phitsanulok, Thailand, pp 78–82. (in Thai with English abstract)
Padhi SR, John R, Bartwal A, Tripathi K, Gupta K et al (2022a) Development and optimization of NIRS prediction models for simultaneous multi-trait assessment in diverse cowpea germplasm. Front Nutr 9:1001551
Padhi SR, Bartwal A, John R, Tripathi K, Gupta K et al (2022b) Evaluation and multivariate analysis of Cowpea (*Vigna unguiculata* (L) Walp) germplasm for selected nutrients-mining for nutri-dense accessions. Front Sustain Food Syst 6:888041
Pal SS, Sandhu JS, Singh I (2005) Exploitation of genetic variability in interspecific cross between *Vigna mungo* × *V. umbellata*. Indian J Pulses Res 18:9–11
Palmer JL, Lawn RJ, Atkins SW (2002) An embryo rescue protocol for *Vigna* interspecific hybrids. Aust J Bot 50:331–338
Pande K, Raghuvanshi SS, Prakash D (1990) Induced high yielding amphiploid of *Vigna radiata* × *V. mungo*. Cytologia 55:249–253
Pandiyan M, Ramamoorthi N, Ganesh SK, Jebaraj S, Pagarajan P et al (2008) Broadening the genetic base and introgression of MYMV

resistance and yield improvement through unexplored genes from wild relatives in mungbean. Plant Mutat Rep 2:33–38

Patel KG, Thankappan R, Mishra GP, Mandaliya VB, Kumar A, Dobaria JR (2017) Transgenic Peanut (*Arachis hypogaea* L.) overexpressing *mtlD* gene showed improved photosynthetic, physio- biochemical, and yield-parameters under soil-moisture deficit stress in Lysimeter system. Front Plant Sci 8:1881

Pathak JN, Singh B (1963) Inheritance studies in green gram. Indian J Genet Pl Breed 23:215–218

Pratap A, Sethi GS, Chaudhary HK (2006) Relative efficiency of anther culture and wheat x maize techniques for haploid induction in triticale ×wheat and triticale × triticale hybrids. Euphytica 150:339–345

Pratap A, Choudhary AK, Kumar J (2010) *In vitro* techniques towards genetic enhancement of food legumes – a review. J Food Legumes 23:169–185

Pratap A, Tomar R, Rajan N et al (2014a) Towards enriching genomic resources in legumes. In: Legumes in the Omic Era. Springer, New York, pp 221–248

Pratap A, Malviya N, Tomar R, Gupta DS, Kumar J (2014b) *Vigna*. In: Pratap A, Kumar J (eds) Alien gene transfer in crop plants, Vol. 2: achievements and impacts. Springer Verlag, New York. https://doi.org/10.1007/978-1-4614-9572-7_8

Pratap A, Gupta S, Malviya N (2015) Genome scanning of Asiatic *Vigna* species for discerning population genetic structure based on microsatellite variation. Mol Breed 35:178

Pratap A, Douglas C, Prajapati U et al (2020) Breeding progress and future challenges: biotic stresses. In: Nair et al (eds) The mungbean genome. Springer Nature, pp 55–80. https://doi.org/10.1007/978-3-030-20008-4_5

Pratap A et al (2022) Biofortification of Mungbean. In: Kumar S, Dikshit HK, Mishra GP, Singh A (eds) Biofortification of staple crops. Springer, Singapore. https://doi.org/10.1007/978-981-16-3280-8_12

Priti MGP, Dikshit HK, Vinutha T, Mechiya T et al (2021) Diversity in phytochemical composition, antioxidant capacities, and nutrient contents among mungbean and lentil microgreens when grown at plain-altitude region (Delhi) and high-altitude region (Leh-Ladakh), India. Front Plant Sci 12:710812

Priti, Sangwan S, Kukreja B, Mishra GP, Dikshit HK et al (2022) Yield optimization, microbial load analysis, and sensory evaluation of mungbean (*Vigna radiata* L.), lentil (*Lens culinaris* subsp. *culinaris*), and Indian mustard (*Brassica juncea* L.) microgreens grown under greenhouse conditions. PLoS One 17(5):e0268085

Ramtekey V, Bansal R, Aski MS, Kothari D, Singh A et al (2021) Genetic variation for traits related phosphorus use efficiency in Lens species at seedling stage. Plan Theory 10:2711

Rashid KA, Smartt J, Haq N (1988) Hybridization in the genus *Vigna*. In: Shanmugasundaram S, Mclean BT (eds) Mungbean, proceeding of the second international symposium. Asian Vegetable Research and Development Centre, Shanhua, pp 205–214

Ravi, Singh JP, Minocha JL (1987) Meiotic behaviour of interspecific hybrids of *Vigna radiata* × *V. mungo*. In: Proceedings of the first symposium on crop improvement. Tamil Nadu Agricultural University, Coimbatore, pp 58–59

Reddy KS (2009a) Identification and inheritance of a new gene for powdery mildew resistance in mungbean (*Vigna radiata* L. Wilczek). Plant Breed 128(5):521–523

Reddy KS (2009b) A new mutant for yellow mosaic virus resistance in mungbean (*Vigna radiata* L. Wilczek) variety SML-668 by recurrent gamma-ray irradiation. In: Shu GY (ed) Induced plant mutation in the genomics era. Food and agriculture Organization of the United Nations, Rome, pp 361–362

Reddy KR, Singh DP (1990) The variation and transgressive segregation in the wide and varietal crosses of mungbean. Madras Agric J 77:12–14

Reddy KR, Singh DP (1995) Inheritance of resistance to Mungbean yellow mosaic virus. Madras Agric J 88(2):199–201

Reddy KS, Pawar SE, Bhatia CR (1994) Inheritance of powdery mildew (*Erysiphe polygoni* D.C.) resistance in mungbean [*Vigna radiata* (L.) Wilczek]. Theor Appl Genet 88:945–948

Reddy VRP, Aski M, Mishra GP, Dikshit HK, Singh A et al (2020a) Genetic variation for root architectural traits in response to phosphorus deficiency in mungbean at the seedling stage. PLoS One 15(6):e0221008

Reddy VRP, Das S, Dikshit HK, Mishra GP, Aski M et al (2020b) Genome wide association analysis for phosphorus use efficiency traits in mungbean (*Vigna radiata* L. Wilczek) using genotyping by sequencing approach. Front. Plant Sci 11:537766

Reddy VRP, Dikshit HK, Mishra GP, Aski M, Pandey R et al (2020c) Unravelling the phosphorus use efficiency associated traits in mungbean (*Vigna radiata* L.) under low phosphorus condition. Indian. J Genet 80(4):412–418

Reddy VRP, Das S, Dikshit HK, Mishra GP, Aski M et al (2021a) Genetic dissection of phosphorous uptake and utilization efficiency traits using GWAS in mungbean. Agronomy 11(7):1401

Reddy VRP, Dikshit HK, Mishra GP, Aski M, Singh A et al (2021b) Comparison of different selection traits for identification of phosphorus use efficient lines in mungbean. PeerJ 9:e12156

Reddy BVRP, Aski MS, Dikshit HK, Mishra GP (2022) Phosphorus deficiency induced physiological and antioxidant response in Mungbean. Legume Res. https://doi.org/10.18805/LR-4586

Reddy BVRP, Dikshit HK, Aski MS, Mishra GP (2023) Comparative analysis of root system architectural traits in genotypes of *Vigna radiata* L. and *Vigna mungo* L. Bangladesh J Bot 52(1):61–69

Saleem M, Haris WAA, Malik A (1998) Inheritance of yellow mosaic virus in mungbean (*Vigna radiata* L. Wilczek). Pak J Phytopathol 10(1):30–32

Sandhu TS, Brar JS, Sandhu SS et al (1985) Inheritance of resistance to mungbean yellow mosaic virus in greengram. J Res Pun Agric Univ 22(1):607–611

Sanyal R, Pradhan B, Jawed DM, Tribhuvan KU, Dahuja A et al (2023) Spatio-temporal expression pattern of Raffinose synthase genes determine the levels of Raffinose family oligosaccharides in peanut (*Arachis hypogaea* L.) seed. Sci Rep 13:795

Sareen PK (1982) Genetic analysis of the trilobite leaf mutant in mungbean (*Vigna* radiatavar. aureus). Curr Sci 51:1028–1029

Sarkar T, Radhakrishnan T, Kumar A, Mishra GP, Dobaria JR (2016) Stress inducible expression of *AtDREB1A* transcription factor in transgenic peanut (*Arachis hypogaea* L.) crop conferred tolerance to soil-moisture deficit stress. *Front*. Plant Sci 7:935

Schafleitner R, Nair R, Rathore A et al (2015) The AVRDC–The World Vegetable Center mungbean (*Vigna radiata*) core and mini core collections. BMC Genomics 16:344

Sen NK, Ghosh AK (1959) Genetic studies in green gram. Indian J Genet Plant Breed 19:210–227

Sharma J, Satija CK (1996) *In vitro* hybridization in incompatible crosses of *Vigna* species. Crop Improv 23:29–32

Shukla GP, Pandya BP (1985) Resistance to yellow mosaic in greengram. SABRAO J 17(3):165–171

Sidhu MC (2003) Cytogenetic and isozyme studies in interspecific hybrids of *Vigna radiata* and *V. mungo* with *V. trilobata*. Crop Improv 30:140–145

Singh DP (1990) Distant hybridization in genus *Vigna* –a review. Indian J Genet Plant Breed 50:268–276

Singh V (2013) Genotypic response and QTL identification for micronutrient (iron and zinc) contents in mungbean [*Vigna radiata* (L.) Wilczek]. Department of Genetics and Plant Breeding, College of Agriculture, CCS Haryana Agricultural University, Hisar, p 96

Singh BV, Ahuja MR (1977) *Phaseolus sublobatus* Roxb. A source of resistance to yellow mosaic virus for cultivated mung. Indian J Genet 37:130–132

Singh BB, Dikshit HK (2002) Possibilities and limitations of interspecific hybridization involving greengram and blackgram. Indian J Agric Sci 72:676–678

Singh KB, Singh JK (1970) Inheritance of black spot on seed coat and shining seed surface in mung bean. (*Phaseolus aureous* L). Indian J Hered 2:61–62

Singh TP, Singh KB (1971) Mode of inheritance and gene action for yield and its components in Phaseolus aureus. Can J Genet Cytol 14:517–525

Singh TP, Singh KB (1974) Heterosis and combining ability in (Phaseolus aureus Roxb.). Theor Appl Genet 44:42–46. https://doi.org/10.1007/BF00277702

Singh BB, Singh DP (1998) Variation for yield and yield components in early segregating generation of a wide cross between mungbean and urdbean. Indian J Genet Plant Breed 58:113–115

Singh BD, Singh RM, Murty BK, Singh RB (1985) Combining ability analysis of nodulation characters in mungbean (*Vigna radiata* L. Wilczek). Indian J Genet Plant Breed 45:271–277

Singh KP, Monika SPK, Kumar A (2003) Interspecific hybridization studies in *Vigna radiata* (L.) Wilczek and *Vigna umbellate* L. Nat J Pl Improv 5:16–18

Singh RK, Mishra GP, Thakur AK, Singh SB (2009) Molecular markers in plants. In: Singh RK et al (eds) Molecular plant breeding: principle, method and application. Studium Press LLC, pp 37–80. ISBN: 978-1933699493

Singh R, van Heusden AW, Yadav RC (2013a) Mapping micronutrients using recombinant inbred lines (RILs) in mungbean (*Vigna radiata* L.). Chem Sci Rev Lett 6(24):2094–2099

Singh V, Yadav RK, Yadav R et al (2013b) Stability analysis in Mungbean [*Vigna radiata* (L.) Wilczek] for nutritional quality and seed yield. Legum Res 36(1):56–61

Singh V, Yadav RK, Yadav NR et al (2017b) Identification of genomic Regions/genes for high iron and zinc content and cross transferability of SSR markers in mungbean (*Vigna radiata* L.). Legum Res 40:1004–1011

Singh DK, Vaishampayan A, Singh MN et al (2017c) Inheritance studies on cercospora leaf spot (CLS) resistance in mung and urdbean. Vegetos 30:388–391

Singh A, Dikshit HK, Mishra GP, Aski M, Kumar S et al (2022) Breeding for abiotic stress tolerance in lentil in genomic era. In: Kole C (ed) Genomic designing for abiotic stress resistant pulse crops. Springer, Cham, pp 145–167. https://doi.org/10.1007/978-3-030-91039-6_5

Singh AK, Kushwaha C, Shikha K, Chand R, Mishra GP et al (2023) Rust (*Uromyces viciae-fabae* Pers. de-Bary) of Pea (*Pisum sativum* L.): present status and future resistance breeding opportunities. Genes 14(2):374

Sinha MK, Aski MS, Mishra GP, Kumar MBA, Yadav PS et al (2023) Genome wide association analysis for grain micronutrients and anti-nutritional traits in mungbean (*Vigna radiata* (L.) R. Wilczek) using SNP markers. Front Nutr 10:1099004

Smartt J (1981) Gene pools in *Phaseolus* and *Vigna* cultigens. Euphytica 30:445–459

Smartt J (1985) Evolution of grain legumes. III. Pulses in the genus *Vigna*. Exp Agric 21:87–100

Soehendi R, Srinives P (2005) Significance of heterosis and heterobeltiosis in an F1 hybrid of mungbean (*Vigna radiata* (L.) Wilczek) for hybrid seed production. SABRAO J Breed Genet 37:97–105

Sokolkova A, Burlyaeva M, Valiannikova T (2020) Genome-wide association study in accessions of the mini-core collection of mungbean (*Vigna radiata*) from the World Vegetable Gene Bank (Taiwan). BMC Plant Biol 20:363

Sompong U, Somta P, Raboy V et al (2012) Mapping of quantitative trait loci for phytic acid and phosphorus contents in seed and seedling of mungbean (*Vigna radiata* (L.) Wilczek). Breed Sci 62:87–92

Srinives P, Kitsanachandee R, Chalee T et al (2010) Inheritance of resistance to iron deficiency and identification of AFLP markers associated with the resistance in mungbean (*Vigna radiata* (L.) Wilczek). Plant Soil 335:423–437

Sudha M, Karthikeyan A, Anusuya P et al (2013) Inheritance of resistance to mungbean yellow mosaic virus (MYMV) in inter and intra specific crosses of mungbean (*Vigna radiata*). Am J Plant Sci 4:1924–1927

Sun L, Chen XZ, Wang SH et al (2009) Heredity analysis and gene mapping of bruchid resistance of a mungbean cultivar V2709. Agric Sci China 7:672–677

Talukdar T, Talukdar D (2003) Inheritance of growth habit and leaf shape in mungbean (*Vigna radiata* (L.) Wilczek). Indian J Genet Plant Breed 63:165–166

Tangphatsornruang S, Somta P, Uthaipaisanwong P et al (2009) Characterization of microsatellites and gene contents from genome shotgun sequences of mungbean (*Vigna radiata* (L.) *Wilczek)*. BMC Plant Biol 9:137

Taunk J, Yadav NR, Yadav RC et al (2012) Genetic diversity among green gram [*Vigna radiata* (L.) Wilczek] genotypes varying in micronutrient (Fe and Zn) content using RAPD markers. Indian J Biotechnol 11:48–53

Thakare RG, Pawar SE, Joshna DC (1980) Green cotyledon mutant in mung bean (*Vigna radiata* (L) Wilczek). Euphytica 27:325–333

Thiyagu K, Jayamani P, Nadarajan N (2008) Pollen pistil intraction in inter-specific crosses of *Vigna* species. Cytologia 73:251–257

Tickoo JL, Lal SK, Chandra N et al (2006) Mung bean breeding. In: Ali M, Kumar S (eds) Advances in Mung bean and Urd bean. Indian Institute of Pulses Research, Kanpur, pp 110–148

Tomooka N, Lairungruang C, Nakeeraks P, Egawa Y, Thavarasook C (1992) Development of bruchid resistant mungbean using wild mungbean germplasm in Thailand. Plant Breed 109:60–66

Tomooka N, Kashiwaba K, Vaughan D, Ishimoto M, Egawa Y (2000) The effectiveness of evaluating wild species, searching for sources of resistance to bruchid beetle in the genus *Vigna* subspecies *Caratotropis*. Euphytica 115:27–41

Tsou SCS (1979) The role of vegetables in Asian diet. In: Inglett GE, Charalambous G (eds) Tropical foods: chemistry and nutrition; proceedings of (volume 2). Academic, New York, pp 401–415

Tyagi DK, Chawla HS (1999) Effects of seasons and hormones on crossability barriers and *in vitro* hybrid development between *Vigna radiata* and *V. unguiculata*. Acta Agronomica Hungarica 47:147–154

Van K, Kang YJ, Han KS, Lee YH, Gwag JG, Moon JK et al (2013) Genome-wide SNP discovery in mungbean by Illumina HiSeq. Theor Appl Genet 126:2017–2027. https://doi.org/10.1007/s00122-013-2114-9

Verma SNP, Krishi JN (1969) Inheritance of some qualitative characters in greengram (Phaseolus aureus Roxb.). Indian J Hered 1:105–106

Verma SNP (1970) Inheritance of photosensitivity in mungbean (*Phaseolus aureors* Roxb.). Mysore J Agric Sci 5:477–480

Verma MM, Brar JS (1996) Breeding approaches for increasing yield potential of mungbean. In: Asthana AN, Kim DH (eds) Recent advances in Mungbean research. Indian Society of Pulses Research and Development, Kanpur, pp 102–123

Verma RPS, Singh DP (1986) Problems and prospects of interspecific hybridization involving green gram and black gram. Indian J Agric Sci 56:535–537

Vinora CP (1974) Cytotaxonomic studies in the genus *Phaseolus*. Indian J Genet 34A:800

Vinora G, Blangiforti S, Creminini R (1999) Karyotype analysis of twelve species belonging to genus *Vigna*. Cytologia 64:117

Vinutha T, Kumar RR, Bansal N, Prashat GR, Goswami S et al (2022) Legumes and pulses: ways and means to enhance the protein quality. In: Ramesh SV, Praveen S (eds) Conceptualizing plant-based nutrition. Springer, Singapore. https://doi.org/10.1007/978-981-19-4590-8_6

Virk DS, Verma MM (1977) A dominant mutation in *Vigna radiata* var. *radiata*. Crop Improv 4:115–116

Wang L, Wu C, Zhong M et al (2016) Construction of an integrated map and location of a bruchid resistance gene in mung bean. Crop J 4(5):360–366

Watt EE (1976) The composition and origin of the texture layer and the inheritance of colour and luster of mungbean seeds. Dissertation Abstract Int-B 36:3163 B

Weinberger K, Khan J, Mazhar-ul-Haq B (2002) Consumption of iron rich foods and productivity: on the indirect impact of pulses and vegetable research. Asian Vegetable Research and Development Center, Shanhua

Wu X, Islam ASMF, Limpot N et al (2020) Genome-wide SNP identification and association mapping for seed mineral concentration in mungbean (*Vigna radiata* L.). Front Genet 11:656

Yohe JM, Poehlman JM (1972) Genetic variability in the mungbean, *Vigna radiata* (L.) Wilczek. Crop Sci 12:461–464

Young ND, Kumar L, Menancio-Hautea D et al (1992) RFPL mapping of a major bruchid resistance gene in mungbean (*Vigna radiata* L. Wilczek). Theor Appl Genet 84:839–844

Zukovskij PM (1962) Cultivated plants and their wild relatives. Farnham Royal, Bucks, Eng.: Commonwealth Agricultural Bureaux, London, p 107

Urdbean [*Vigna mungo* (L.) Hepper] Breeding

Sonali Habde, P. Dhanasekar, and J. Souframanien

Abstract

Black gram is one of the major pulse crops cultivated in India during both kharif and rabi seasons. Its cultivation is continuously being challenged by several biotic and abiotic stresses, which has widened the gap between its potential and realized yield. The extreme sensitivity of black gram to abiotic stresses, including high temperature, moisture stress and salinity along with auxiliary challenges of climate change including pre-harvest sprouting, and increased severity of diseases and pests, are the key issues in black gram breeding. Further, certain modifications in plant architectural features are required while breeding for plant types suitable for mechanical harvesting. Nowadays, market preferences based on the product profile of black gram is well characterized by seed size, appearance traits, battering quality, and reduced level of anti-nutritional factors. It necessitates a re-orientation of black gram breeding with a major emphasis on all these traits. Breeding offers a sustainable alternative to develop an ideal plant type of black gram with different combinations of desirable traits. This chapter discusses different developments in black gram breeding for major biotic, abiotic stresses, and quality traits reflecting the transition of black gram from an orphan legume to a mainstream crop.

Keywords

Black gram · Salinity · Bruchid · Pre-harvest sprouting · Battering · Mechanical harvesting

S. Habde · P. Dhanasekar · J. Souframanien (✉)
Nuclear Agriculture and Biotechnology Division, Bhabha Atomic Research Centre, Trombay, Mumbai, Maharashtra, India
e-mail: souf@barc.gov.in

1 Introduction

Black gram [*Vigna mungo* (L.) Hepper], widely referred to as urdbean or mash, is a proteinaceous (25–28%) grain legume crop. It is extensively cultivated in the Indian subcontinent, but less popularly in other Asian and South Pacific countries like Thailand and Australia (Jegadeesan and Reddy 2021). India is the largest producer of black gram with an approximate 4.4 m ha area under cultivation, with a production of 2.7 million tonnes (Nair et al. 2024). However, the productivity of black gram in India faces significant challenges, due to both biotic factors like mungbean yellow mosaic disease (MYMD), powdery mildew, (*Erysiphe polygoni*), bruchids (*Callosobruchus maculatus* and *C. chinensis*), and abiotic factors including pre-harvest sprouting, drought, and salinity (Verma et al. 2024). Knowledge of key traits imparting heat tolerance for breeding heat-tolerant black gram varieties is of utmost importance in the context of continuously rising temperature. The present shift in cultivation practices through mechanization demands altered plant type and certain modifications in plant architecture.

Cultivars with erect and semi-determinate plant types are the varieties of choice owing to their amenability for sole cropping and machine harvest compliance (Nair et al. 2024). Among the post-harvest traits, seed appearance and color are highly variable traits and their demands are subject to consumer preferences. Large seed size is a major target trait for modern breeding, and large-seeded black gram genotypes are highly preferred in the processing industry. Currently, owing to the functional properties of legume flours, increased utilization of grain legumes in composite flours for various food formulations has been observed (Vanniarajan et al. 2022). Additionally, the significant demand of black gram for processed fermented food in the form of idli, dosa, and vada in Southern Indian states has emphasized the need for breeding varieties with improved battering quality.

In this context, black gram breeding needs simultaneous consideration of multiple factors and well-defined breeding objectives to develop an ideal plant type with key traits. In

G. Prasad Dixit et al. (eds.), *Fundamentals of Legume Breeding*, https://doi.org/10.1007/978-981-96-7838-9_6

the present chapter, the importance, impact, screening procedures, gene action, trait inheritance, genetic resources, varieties, and recent development in genomic and marker resources for different biotic and abiotic stresses and quality traits relevant to black gram are discussed.

2 Drought/Water Stress

Black gram cultivation in India is commonly practiced across different agro-ecologies and cropping systems in rainy, post-rainy as well as spring seasons. Prevalence of water stress is mainly observed in rainfed agro-ecology and post-rainy season cropping in several regions of North India, including Haryana, Punjab, Western Uttar Pradesh, and Bihar, where stress coincides with the flowering and pod filling, resulting in 20–30% yield losses (Nair et al. 2024). Major repercussions of moisture deficit stress perturbing different plant metabolisms include enzyme activity, cellular turgor potential, leaf morphogenesis, reactive oxygen species, electro-neutrality of ions, carbon fixation, and leaf gas exchange (Punniyamoorthy and Jegadeesan 2023). These are discernible as reduced rate of photosynthesis, transpiration, and stomatal conductance, negatively influencing the characteristics and sizeable proportion of plant growth and yield. Plants adopt one or more of the three mechanisms to manage water deficit stress conditions: drought escape, avoidance, and tolerance (Punniyamoorthy and Jegadeesan 2023; Pandiyan et al. 2023). By altering crop maturation, plants escape drought by completing their whole life cycle prior to onset of drought. Short duration varieties coupled with suitable agronomic management practices help plants to escape from moisture stress situations and complete crop failure, albeit with yield reduction. Drought avoidance mechanisms help plants tackle water deficits encountered at any growth stage of their life cycle. These are exhibited through a few modifications, including root production, deep-root penetration, reduction in stomatal conductance, small and thick leaf formation, reduced evapotranspiration, osmolytes production, and biosynthesis of cuticular wax on leaf surface.

A series of interrelated events establishing homeostatic conditions, adoptive responses through synthesis of osmolytes, osmoprotectants, and activation of stress-responsive mechanisms enable plant function even at lower water potentials by means of drought tolerance (Pandiyan et al. 2023). Hence, inclusion of comprehensive screening criteria encompassing morphological, physiological, and biochemical parameters are essential while breeding for drought tolerance. Various parameters such as plant height stress tolerance index (PHSI), leaf area index, leaf water status, rate of transpirational water loss, cell membrane stability (membrane stability index), chlorophyll-a, chlorophyll-b, total chlorophyll, chlorophyll stability index, chlorophyll degradation (%), leaf water potential, net photosynthetic rate, leaf temperature, stomatal conductance, intercellular CO_2, leaf proline content, total soluble sugar, starch, nitrate reductase activity, peroxidase activity, total flavonoids, anthocyanin content, seed yield, seed protein content, protein yield and dry matter stress tolerance index (DMSI) could be used for screening drought tolerance in plants including black gram (Baroowa and Gogoi 2012, 2014, 2015; Prakash et al. 2018; Gurumurthy et al. 2019). In addition to this traits, cooler canopy, slow canopy wilting, epidermal conductance, leaf pubescence density, and water use efficiency are also suggested (Pandiyan et al. 2023).

Further, traits such as early maturity, 50% flowering, and podding can be considered as mass screening traits for selecting black gram cultivars with drought escape mechanisms. Positive correlations of soil moisture content with leaf chlorophyll, chlorophyll stability index, plant height, leaf area, cell membrane stability, yield and its negative correlation with leaf proline concentrations have been reported (Baroowa and Gogoi 2012). Root and shoot parameters that can be studied for drought tolerance include root and shoot length, number of roots, root diameter, fresh weight of roots, fresh weight of shoots, dry weight of roots, dry weight of shoots, root volume, and gas exchange characters (Prakash et al. 2018). Since limited reports are available about drought resistance in black gram, screening process for drought-tolerant traits in other *Vigna* species needs to be contemplated. Sai and Chidambaranathan (2019) comprehensively studied the response of black gram to drought stress during germination, flowering, and post-stress/re-watering stages. They observed higher tolerance in two varieties VBN4 and K1 due to the protection of tissues from oxidative damage during drought stress, imparted by increased synthesis of ABA (five-fold), proline (4.5-fold), and lipid peroxidase activity (five-fold). They identified that traits like antioxidant accumulation protect membrane stability and contribute the major share of drought stress tolerance in *Vigna mungo*. This study signifies the role of antioxidants and the use of this trait as an index for understanding the molecular response of black gram during drought stress.

Enhancing drought-tolerance mechanisms and development of drought-tolerant cultivars are effective alternatives for improving black gram yield, production, and productivity under water-deficit circumstances (Pandiyan et al. 2023). A list of some of the important drought tolerant genotypes identified in various screening processes in previous studies has been provided in Table 1. These genotypes can be successfully deployed for breeding climate resilience in black gram.

Table 1 Important drought and heat-tolerant genotypes identified in black gram

S. No.	Genotype	Screening method	References
Drought tolerance			
1	T9	Morpho-physiological characters	Baroowa and Gogoi (2014)
2	VBN(Bg)4, VBN(Bg)6	Morpho, physio and biochemical attributes	Pandiyan et al. (2017)
3	CBG-09-13	Better root dynamics and gas exchange	Prakash et al. (2018)
4	VBN-4, K1	Increased synthesis of ABA, proline and lipid peroxidase	Sai and Chidambaranathan (2019)
5	BDU-1, PU-31, Azad-3, Pant Urd-35, Shekhar-2	Drought tolerance	Maheswari et al. (2019)
6	BDU-1, Pant U-31, Pant U-19,WBU-108 (Sharada),WBU-109 (Sulota)	Varieties suitable under delayed monsoon	Maheswari et al. (2019)
7	PGRU95016, COBG05, IPU99209, IPU941, IPU243	Morpho-physiological and biochemical characters	Gurumurthy et al. (2019)
8	UPU 85-86	Fast quenching of Fm, high antioxidant activity, high membrane stability, high ETR, leaf NBI	Sen Gupta et al. (2021)
Heat tolerance			
1	VBG-07-001, VBG-06-010	Cellular response post-TIR	Partheeban et al. (2017)
2	UPUPU-85-86, IPU94-2, IPU-98/36, No.5/31, PGRU-95014, PGRU-95016, PLU-1, BGP-247	Morphological and physiological traits	Sen Gupta et al. (2021)
3	PantU31, Mash114, UTTARA, IPU18-04	Physiological, reproductive, and yield-related traits	Chaudhary et al. (2022)
4	LBG-1004, TBG-129, PU-1804, LBG-995, GBG-1, TBG-104, TBG-141, LBG-1006, LBG-1010	Biochemical parameters and cluster analysis	Pavithra et al. (2023)

Modified from Punniyamoorthy and Jegadeesan (2023), Nair et al. (2024)

3 High-Temperature Stress Tolerance

Black gram is a warm season legume, which thrives well under 25–35 °C temperature with high humidity; however, drastic reduction in yield under high temperatures exceeding 35 °C has been reported (Anitha et al. 2016; Sen Gupta et al. 2021). Owing to its high photothermal-sensitivity, excessive heat stress above the threshold across the locations has limited area under black gram cultivation during the summer season. Thus, breeding heat tolerant cultivars can substantially contribute to increasing the area under urdbean cultivation (Sen Gupta et al. 2021). When the soil and air temperature rise for a minimum period of time beyond a threshold level, heat stress damages plant irreversibly and harms growth and development (Lamaoui et al. 2018). Plant growth and seed yield are negatively impacted due to heat stress due to increased sensitivity of chloroplast thylakoid membranes, electron transport activity, respiration, photosystems I and II, and biological processes of nitrogen fixation, microspore formation, pollen viability, stigma receptivity, embryogenesis, seed set, and seed development (Punniyamoorthy and Jegadeesan 2023).

Thus, continual rising temperature affect the overall morphology, anatomy, and physiology of plants. Various plant adaptive responses, reprogramming, and activated defense mechanisms in response to heat stress are reflected in changed crop phenology. This necessitates the study of association of crop phenology with temperature alterations for a better understanding of the impacts and defense strategies employed by plants for its adaptation (Chaudhary et al. 2022). Knowledge of genetics underlying key traits imparting heat tolerance helps the breeder to make genetic improvements more precisely (Sen Gupta et al. 2021).

A rapid and reliable method to assess the thermotolerance of black gram genotypes was proposed by Partheeban et al. (2017). They standardized the temperature induction response technique (TIR) (induction at 36–46 °C for 3 h and lethal temperature at 52 °C for 3 h) and studied the cellular responses. The main principle in the induction response technique is to expose plants to milder temperatures before they are subjected to severe temperatures. This acclimation resulted in increased thermoprotection and significantly higher leaf recovery at challenging temperatures over untreated, directly exposed plants. During this evaluation, thermotolerant genotypes showed cellular level tolerance in terms of higher survival percentage and lower growth reduction. In line with the above study in the above process, Pavithra et al. (2023) carried out screening of 100 genotypes for temperature induction response with slight modifications in the above protocol (induction at 36–46 °C for 3 h and lethal temperature at 54 °C for 3 h). Based on the percent reduction in the seedling growth of induced over control and survival percentage, 27 highly tolerant genotypes were identified.

A diverse array of traits must be screened for ascertaining potential heat-tolerant genotypes, which include physiological traits like membrane damage (electrolyte leakage), chlorophyll content, chlorophyll fluorescence, photosynthetic efficiency, stomatal conductance, leaf area, relative leaf water content; reproductive traits like pollen germination percentage, pollen viability percentage; yield-related traits like the total number of pods per plants, total seeds per plant, seed yield per plant, and single seed weight (Chaudhary et al. 2022). Sen Gupta et al. (2021) assessed a diverse panel of 97 genotypes for yield under stress and non-stress. Based on the heat susceptibility index, their study identified 8 highly heat tolerant and 35 highly heat-sensitive genotypes. Comparison of highly heat sensitive and heat-tolerant genotypes revealed significant genotypic variability for physiological and biochemical traits viz leaf nitrogen balance index (NBI), chlorophyll (SPAD), epidermal flavonols, and anthocyanin contents under 42/25 °C max/min temperature. They observed a high membrane stability index among heat-tolerant lines.

Additionally, heat-tolerant genotypes also exhibited high electron transport rate, fast quenching of Fm following fluorescence kinetics, and high antioxidation activity resulting in scavenging of ROS. Susceptible lines displayed distortion of PSII leading to reduced quantum yield of PSII and reduced photosynthetic efficiency. Chaudhary et al. (2022) evaluated 26 genotypes for heat tolerance under a natural environment as well as in controlled conditions in growth chambers. The assessment was carried out for various physiological, reproductive and yield related traits which led to the identification of four promising genotypes for heat tolerance viz PantU31, Mash114, UTTARA, and IPU18-04. Tolerant genotypes exhibited high chlorophyll content, photosynthetic efficiency, stomatal conductance, leaf area, pods plant^{-1}, total seeds plant^{-1}, and low reduction in pollen germination % and pollen viability under heat stress conditions. Further high heritability was observed for chlorophyll content, stomatal conductance, seed yield plant^{-1}, pods plant^{-1}, and single seed weight, which holds promise for their inclusion in screening criteria for heat tolerance.

Heat stress can lead to the production of reactive oxygen species, which degrades polyunsaturated fatty acids further leading to the deterioration of membranes. Malondialdehyde (MDA) is the resulting product of this oxidative degradation. Therefore, in addition to proline, enzymatic antioxidants, superoxide dismutase (SOD), peroxidase (POX), catalase (CAT), ascorbate peroxidase (APX), superoxide free radical, and malondialdehyde (MDA) can also be used as a reliable marker to measure oxidative damage (Pavithra et al. 2023). Positive accumulation of proline, SOD and POX activity and higher enzyme activity in heat-tolerant black gram genotypes, and more MDA accumulation in the susceptible genotypes has been reported (Partheeban et al. 2017; Pavithra et al. 2023).

In black gram, Sen Gupta et al. (2021) used a set of 21 genetic markers previously identified and designed in relevance to abiotic stress tolerance, for establishing genetic differences between the heat-tolerant and sensitive lines. It was observed that the highly heat tolerant genotype (UPU 85-86) was distinctly clustered from the highly heat-sensitive genotype (PKGU-1). Further for enhancing the sustainability of black gram production in the context of climate change, information generated through "omics" studies can be tapped for the identification of prime candidates imparting tolerance for heat and drought stresses (Souframanien and Dhanasekar 2014).

4 Salinity Tolerance

Legumes are highly salt-sensitive crops and negative consequences of salinity in legumes include, osmotic stress, different ranges of ion toxicity, and nutritional imbalances due to increased uptake of cations (Na^+, Mg^{2+}, Ca^{2+}) and reduced nodulation by rhizobia (Hanumantha Rao et al. 2016). Adverse effects of osmotic stress are envisioned as interruption of membranes, nutrient imbalance, impaired ability to detoxify reactive oxygen species (ROS), differences in antioxidant enzymes, and decreased photosynthetic activity (Pang et al. 2010). It also inhibits water uptake, transport, cell elongation, expansion, leaf development, and growth (Hanumantha Rao et al. 2016). One of the most damaging effects of soil salinity is discerned under toxic NaCl levels lead to higher Na^+ influx which blocks K^+ uptake, results in lower productivity, and even leads to cell death (Ahmad and Umar 2011). Besides, reports are available in chickpea, cowpea and mungbean where salinity interfered with nodulation, nitrogen-fixing efficiency of nodules and nitrogen metabolism which is evident sensitivity of the process of symbiosis to salinity in legumes (Hanumantha Rao et al. 2016). The plant responds to salinity stress by adaptive modifications at the physiological, metabolic, and molecular levels.

Physiological responses related to ion uptake, transport, and compartmentalization include favored storage of Na^+ in older leaves, Na^+ extrusion, high Na^+ transport to shoot, high Cl^- uptake, low P and Zn uptake, etc. Whereas biochemical responses are mediated through biosynthesis of osmoprotectants, scavenging enzymes, compatible solutes, activation and synthesis of antioxidant enzymes/compounds, polyamines, and hormonal modulation (Roy et al. 2014; Hanumantha Rao et al. 2016).

Vigna mungo is an economically important pulse crop that grows in marginally salt-affected soils (Hasan et al. 2019a). Further black gram cultivation in arid and semi-arid region is restricted partly due to salinity. Also, groundwater salinity is a serious concern for black gram cultivation especially in rabi season (Shanthi et al. 2014). There is more scope to utilize area affected by inland as well as coastal salinity for black gram cultivation by breeding salinity tolerant varieties (Shanthi et al. 2021b). Hence salinity tolerance is one of the mandatory objectives in black gram breeding.

Accumulation of soluble salts in the uppermost topsoil has a significant impact on germination and early seedling growth. Salinity inhibits seed water imbibition and absorption, and also it induces harmful effects due to ion accumulation. These effects lead to a reduction in enzyme hydrolysis of stored carbohydrates into simple sugars and thus, inhibit seed germination (Rajasekar et al. 2022). Therefore, screening of diverse germplasms for their capacity to tolerate salt stress during seed germination and seedling emergence is essential.

Priyadharshini et al. (2019) carried out an early seedling stage salinity tolerance screening at six EC levels (0,4,7,10,13 and 16dS/m). Based on seed germination and dry matter production parameters, they identified three saline tolerant genotypes viz VNBG 017, AUB 3, and AUB 20. Gandhi et al. (2021) evaluated twenty black gram genotypes for seedling stage salinity tolerance using seawater. They reported reductions of 85.5%, 75.68%, 61.73%, and 22.30% in germination, root length, shoot length, and seedling length, respectively, under 8 EC (dS/m) level. Genotypes IC-204869, TPU-94-2, IC-214845, and IC-214844 were found to be saline tolerant. Screening of thirteen black gram varieties for their salinity tolerance at three EC levels (4.0 dS/m, 11.0 dS/m, and 16.0 dS/m) was undertaken by Shanthi et al. (2021a). They recorded observations on mean germination percentage, plumule length, radicle length, and dry matter weight. A positive and significant correlation between radical length and dry matter weight at 11.0 EC was observed. Based on the grand mean performance, the genotypes, VBG-14-016, Vamban 4, Vamban 8, and VBG-12-062 were identified as tolerant. This study concluded radicle length was the most useful parameter for the selection of salinity tolerant genotype. Rajasekar et al. (2022) screened ten black gram genotypes for salt stress at the germination stage using a 130 mM NaCl solution. The highest values for the stress tolerance index were obtained for genotypes LBG 787 (0.831) and KU7720 (0.722). Genotypes KU 7720, LBG 787, PLU 621, and VBG 18028 exhibited greater tolerance toward salinity at germination and seedling growth stage in this study.

Since the flowering and maturity stages both are equally sensitive to salinity stress due to their direct impact on yield, screening for salinity tolerance at this stage is crucial. The effect of irrigation water salinity on germination, crop stand at different stages and yield attributes in nineteen released and pre-released varieties of black gram were studied by Shanthi et al. (2014). In their study germination was least impacted but considerable seedling mortality was observed at increased salinity levels. Two genotypes (LBG-402 and LBG-738) maintained lower mortality values at 6.0 E.C. dS/m during both the years of study. At all three salinity levels (2.0, 4.0, and 6.0 EC dS/m), the genotype LBG-738 was identified for its comparative better performance in terms of highest seed yield under stress.

A study was undertaken to assess the morphological and yield attributes at vegetative, flowering, and maturity stages under different salinity stress conditions (Shanthi et al. 2021b). The increasing salinity levels adversely impacted days to 50% flowering, chlorophyll content, and single plant yield. They noted that increasing salt concentration leads to a proportionate increase in the number of days for flower initiation and a high level of salinity (16.0 EC dS/m), which leads to drying of the plant before flowering. The yield reduction up to ¼ over control was observed at a salinity level of 4.0 EC dS/m. Based on the survival and performance under 4.0 EC dS/m level, the genotypes VBG-14-016 and VBN 8 were identified as tolerant. These genotypes could be utilized as a donor for the development of salt-tolerant varieties.

A case study in black gram conducted by Rather and Natarajan (2021) delineated the impact of salt stress on morphological and biochemical changes in pre-flowering stage. This study revealed reductions of 50%, 56.25%, and 61.54%, 46.69% in root and shoot length and fresh weight of root and shoot, respectively. Decreases in photosynthetic pigment including chlorophyll a, chlorophyll b, and carotenoid constituents were observed at 71.86%, 72.04%, and 67.73%, respectively. NaCl treatment led to a reduction of 59.43% in total protein content over control plants, and a 112.68% increase in proline concentration was observed.

Comparative studies have been conducted between black gram and green gram for their variable responses to salinity stress in terms of dry matter and mineral ions accumulation in different plant parts. Additional parameters including relative water content, water uptake, and retention capacity have also been studied. These studies have provided conclusive evidence proving better salt tolerance of black gram over green gram (Raptan et al. 2001a, b; Hasan et al. 2019b).

It is interesting to note that sensitivity to salt stress and severity of imposed salt stress is highly genotype-specific. Cultivar-specific differences for chlorophyll content, leaf water potential, solute potential, leaf proline, and turgor potential as well as the accumulation of various osmolytes including proline, glycine betaine, soluble sugars, free amino acids, and polyamines in response to salt stress have been reported (Ashraf 1989; Sunanda and Ranganayakulu 2019).

Despite of known recalcitrancy in legumes, Bhomkar et al. (2008) successfully established a reproducible and efficient transformation system utilizing the nodal regions of the embryonal axis of black gram. *Agrobacterium tumefaciens* mediated genetic transformation of *Vigna mungo* was carried out for a gene *glyoxalase I* driven by a novel constitutive

Cestrum yellow leaf curling viral promoter. This transgene conferred tolerance to salt stress and Glyoxalase I activity in transgenic lines was correlated with their ability to withstand salt stress. Win and Oo (2017) reported salt stress–induced changes in the protein profile of two black gram varieties differing in salinity tolerance. Better salt tolerance exhibited by "U-Taung-2" over "Mut Pe Khaing To" was attributed to capability of metabolism for adaptive response to salt stress through osmotic adjustment and accumulation of newly synthesized proteins. This study demonstrates the utility of protein band profile study as cost-effective tool for the selection of salt tolerant cultivars. In a molecular diversity analysis study using microsatellite markers, Priyadharshini et al. (2023) screened 21 markers and identified 13 polymorphic markers with a transferability rate of 83%. Their study revealed 47 alleles in forty-eight black gram genotypes with an average of 3 alleles per locus. Further based on evaluation for coastal salinity tolerance, three saline-tolerant genotypes, namely AUB 9, AUB 14, and AUB 20, were identified.

Seed priming treatments proved to be helpful in lessening salt-induced adversities during germination and seedling development. Pre-germination seed halopriming with a sublethal dose of NaCl (50 mM) for 2 h was found to be effective in vitiating NaCl stress adversities and favoring better plantlets growth (Biswas et al. 2020). Seed treatment with 6 g of KNO3 for 12 h was found to be an effective approach to enhance seed germination and seedling quality traits including vigor index (Spandana et al. 2021). A novel insight through investigating alterations in metabolic pathways for adaptive responses to salinity tolerance in black gram genotypes were provided by Biswas et al. (2020). The comparative study of metabolic responses under nonprimed and haloprimed treatments of seedlings led to discovery of 35 core candidates of fundamental metabolome induced in salinity tolerance.

Transgenic approaches in legumes although gained momentum with a slower pace owing to a recalcitrant nature for regeneration still can contribute immensely toward developing pulse cultivars with tolerance against abiotic stresses (Punniyamoorthy and Jegadeesan 2023). In black gram, overexpression of gene *ALDRXV4* through a transgenic approach showed increased protection against drought and salinity through sustained photosynthetic efficiency, maintaining increased relative water content and reduced photo-oxidative damage through the accumulation of more reactive oxygen species in transgenic lines (Singh et al. 2016).

5 Pre-harvest Sprouting Tolerance

Pre-harvest sprouting (PHS) is the inadvertent germination of physiologically mature grains just before harvest in the pod itself, triggered by humid and untimely rainfall conditions, resulting in undesirable seed starch degradation and substantial yield losses (Verma et al. 2024). In the present context of inconsistent and unforeseeable rainfall patterns coinciding with crop maturity and harvest, pre-harvest sprouting tolerance becomes a vital trait for breeding climate-resilient plant type of black gram. A series of morpho physiological characteristics of pods and seeds are studied for screening for pre-harvest sprouting tolerance. These include water imbibition by pods, seed germination in pods, water imbibition by seeds, fresh-seed germination, pod width, pod length and thickness of pod wall, seed length and seed width and seed hardness (Verma et al. 2024). The hard seediness trait was identified as a key determinant of PHS tolerance in urdbean exhibiting inverse relationship with PHS; further, positive influences of seed weight, seed length, 100 seed weight, pod water imbibition, seed water imbibition, mean germination time, and fresh seed germination on PHS were also reported (Lamichaney et al. 2023).

Since α-amylase plays a crucial role in breaking down starch into simpler sugars during germination, its increased activity has a relevance with high germination rate and susceptibility to PHS; consequently, amylase content serves as a potential biochemical marker for evaluation of PHS tolerance and positive correlation of PHS with an increase in alpha-amylase activity has been reported (Verma et al. 2024). The absence of seed dormancy is an important key determinant of PHS; additionally, pod-wax content and thicker pod walls decide the fate of water imbibition by pod and can serve as targets for improving PHS tolerance. Significant differences in pre-harvest sprouting tolerance among the diverse urdbean genotypes have been reported (Singh et al. 2012; Lamichaney et al. 2023). The PHS-tolerant genotypes PLU 302, CN 35–15, and PLU 65, possessing hard seededness of 70–88% after harvesting, were suggested as potential donors conferring PHS tolerance (Lamichaney et al. 2023). Based on fresh seed germination criteria, wild accessions of *Vigna trilobata* (IC276983, IC349701, Trichy Local 1, LRM/13-33 and LRM/13-26) and *Vigna sublobata* (IC251440) have been identified for PHS tolerance (Lamichaney et al. 2021). Based on α-amylase activity and fresh-seed germination, IC485641 accession was identified as highly tolerant to PHS (Verma et al. 2024). Utilizing these novel sources in high-yielding black gram varieties can impart PHS tolerance without loss of yield.

Non-dormant seed is a favorable trait for timely cultivation and uniform germination, but the loss of seed dormancy results in pre-harvest sprouting (Lomlek et al. 2023). Hence, seed dormancy can be explored as one of the alternative traits in the reduction of pre-harvest sprouting through genetic improvement. Information about the genetic basis of seed dormancy is limited, but seed dormancy period of 3–4 months and dormant seed count ranging from 0 to 90% have been reported in harvested seed lot of black gram depending on cultivar type and environment (Lomlek et al. 2023).

QTL mapping study by Somta et al. (2020) investigated seed dormancy using seed water absorption as an index in recombinant inbred line (RIL) population derived from a cross between cultivated black gram (var. *mungo*; accession JP219132) and wild black gram (var. *silvestris*; accession TC2210). They reported a single major quantitative trait locus (QTL), *qSdwa6.1-* located on LG6 and accounting for 20.58% of seed dormancy variation in the RIL population. A narrow sense heritability value of 58.9% was observed for seed dormancy. Further, comparison of their findings with earlier reports inferred absence of any common QTLs among *Vigna* species, indicating different mechanisms for controlling seed dormancy in *Vigna* species. This obliged dedicated efforts in each species to unravel the genetics of seed dormancy.

In another study, quantitative trait loci (QTLs) controlling seed dormancy were mapped by Lomlek et al. (2023) using two F_2 populations (F2BGA and F2BGB) developed from a cross between cultivated (Chai Nat 80) and wild black gram PI213017 and TVNu1076 (Chai Nat 80 × PI213017 and Chai Nat 80 × TVNu1076), respectively. The broad-sense heritability (H^2) value for seed dormancy was reported as 81.53–99.80%. They reported a total of three QTLs: *qSdwa5.1+*, *qSdwa6.1-*, and *qSdwa6.2-* on two linkage groups with the percentage of variance explained by QTL ranging from 0.18% to 43.81%. Genes encoding the nitrate regulatory gene 2 protein (NRG2), calcium-dependent protein kinase 26, NDR1/HIN1-like (NHL) 13, HVA22-like protein, ethylene-responsive transcription factor (ERF20 and ERF22), bZIP16 TF, and a gene-producing xyloglucan endotransglucosylase/hydrolase protein 30 were identified as candidates for QTL regions mapped for seed dormancy. There is still immense scope for trait mapping and deciphering the inheritance of traits contributing to pre-harvest sprouting tolerance and seed dormancy for imparting climate readiness to black gram.

6 Mungbean Yellow Mosaic Virus (MYMV)

MYMD disease caused by mungbean yellow mosaic virus (MYMV) and mungbean yellow mosaic India virus (MYMIV) is a major threat to black gram production in South and Southeast Asia regions and has been reported to cause more than 50% reduction in yield (Nair et al. 2024). Mungbean yellow mosaic virus (MYMV) is a gemini virus belonging to the begomovirus having monopartite (1~2.9 kb DNA) or bipartite genomes (2~2.6 kb DNAs) referred to as "DNA-A" and "DNA-B," respectively, and transmitted by a whitefly (*Bemisia tabaci*) (Souframanien and Dhanasekar 2014). It causes irregular yellow-green patches on older leaves and complete yellowing of young leaves of susceptible varieties (Souframanien and Gopalakrishna 2006). Symptoms caused by YMV are largely dependent on host species susceptibility and vary from yellow mosaic spot (includes yellow mottle and necrotic mottle), necrosis, less flowering and pod development, and small and distorted pods with immature small seeds (Raizada 2020).

Disease control measures are based on the prevention of whitefly vectors through chemical measures. However, it suffers from undesirable consequences on environmental safety and human health. Hence exploring host plant resistance is the only sustainable alternative available. However, recessive resistance gene control, transmission through the white fly, lack of uniform screening procedures, and inconsistencies in scoring infection phenotypes have hindered the introgression of MYMV resistance into elite black gram genotypes (Souframanien and Gopalakrishna 2006). This has obliged indirect selection criteria using molecular markers linked to resistance genes. Based on resistance gene analogs (RGAs), Basak et al. (2005) identified a marker RGA 1F-CG/RGA 1R linked to MYMV tolerance in black gram. The 445 bp DNA fragment amplified was sequenced and named "VMYR1." The predicted amino acid sequence showed highly significant homology with aNB-ARC domain present in several gene products involved in plant disease resistance, nematode cell death, and human apoptosis. Souframanien and Gopalakrishna (2006) identified an ISSR marker $ISSR811_{1357}$ linked with the MYMD resistance gene at 6.8-cM distance. Sequence characterized amplified region (SCAR) primers designed (YMV1-F and YMV1-R) from this ISSR marker distinguished the MYMV resistant and susceptible plants in the RIL population, which was validated using diverse black gram genotypes differing in their MYMV reaction. Narasimhan et al. (2010) conducted a molecular diversity study among MYMD-resistant and MYMD-susceptible genotypes using 24 cowpea resistance gene analog (RGAs) markers, which grouped two resistant (IPU 02-33 and IPU 6-02) and two susceptible (LBG 20 and T9) black gram genotypes into distinct clusters. Maiti et al. (2011) identified two markers linked to MYMV resistance, which were amplified by using RGA primer pairs (RGASF1/RGASR1 and RGA22F2/RGA24R2) referred as YR4 and CYR1. Various QTL mapping studies for MYMD led to the identification of highly reliable and diagnostic set of SSR markers linked to MYMD resistance, which includes CEDG180 (Gupta et al. 2013), VR9 (Naik et al. 2017), and CEDG185 (Rambabu et al. 2018). RAPD (OPN-1 and OPF-1) primers and RAPD derived SCAR marker 20F/20R (developed from OPQ1 RAPD) have been utilized in the screening of diverse germplasm for their disease reaction, and this has validated the usefulness of resistant source PU-31 for the development of marker linked to MYMV (Prasanthi et al. 2013; Devi et al. 2017; Vishalakshi et al. 2017).

Recently, using a QTL-seq approach, Dhaliwal et al. (2022) identified large-effect QTL (*qMYMIV6.1.1*) on chromosome 6 spanning 3.4 Mb and explained 70% of total phenotypic variation for resistance to mungbean yellow mosaic India virus (MYMIV). They also developed three KASP markers for three possible candidate genes (serine- threonine kinase, UBE2D2, and BAK1/BRI1-ASSOCIATED RECEPTOR KINASE) underlying the identified genomic region. In addition to this, recently developed next-generation sequencing-based transcriptome resources are useful to decipher molecular responses underlying MYMIV-resistance. Ganguli et al. (2016) analyzed MYMIV-induced transcriptome of mock-inoculated (control) and MYMIV-infected black gram VM84. A total of 77 and 107 million of processed reads were obtained for the control and MYMIV-inoculated cultivars, respectively. Rise in approximately 30 million reads for the infected genotype was attributed to the expression of stress and defense pathway associated genes. Another study for comparative RNA-Seq transcriptome profiling was carried out using the same MYMIV-resistant line VM84 and a susceptible plant T9 by Kundu et al. (2019). They observed distinct gene expression landscapes (DEGs) with 2158 and 1679 differentially expressed genes in VM84 and T9, respectively. Prompt intense immune reaction through activation of basal and induced immune responses was reported in resistant plant VM84. Functional analysis of upregulated DEGs revealed the presence of NB-LRR, WRKY33, ankyrin, argonaute, and NAC transcription factors, which indicated their potential roles in MYMIV resistance.

Raizada (2020) carried out transcriptome sequencing of wild black gram and generated a huge set of genic SSR and SNP primers from the transcriptome dataset. This study demonstrated a co-segregation pattern for three genic-SSR markers (TWSSR10, VmGSSR86, VmGSSR547) in BSA for the YMD trait. SSR marker pair CEDG180 and CEDG116 delimiting most robust and validated QTL *qMYMVD_60* on LG10 for its association with MYMD resistance has also been identified (Subramaniyan et al. 2022). In a recent investigation, Raizada and Jegadeesan (2023) utilized mungbean-derived resistance gene-SSR markers for diversity analysis in a set of 44 black gram genotypes. This cross-species amplification led to the identification of three key SSRs (MRGSSR 12, MRGSSR 56, MRGSSR 77) effective in differentiating YMD-resistant and YMD-susceptible genotypes.

Vanniarajan et al. (2019) developed gamma rays induced mutants in the background of MDU 1 and VBN (Bg) 4. Their study identified four mutant lines viz ACM-16-30, ACM-16-29, ACM-6-17, and ACM-16-14 with YMV resistance. In another study, Tamilzharasi et al. (2020) evaluated 48 genotypes for mungbean yellow mosaic virus resistance through combined phenotypic and genotypic analysis. They identified total fourteen genotypes (VBG 11-053, LBG 808, CO-6, VBN-4, R 15-009, R 15-006, VBG-10-019, R 15-011, VBG-11-010, KU-52, KU-24, ACM 14001, ACM-015-29, and ACM-015-30) exhibiting resistance reaction with a disease incidence of 0.54–0.96%. Additionally, resistant genotypes were tagged by using molecular markers CEDG180, ISSR8111357, and YMV1 FR. Subramaniyan and Narayana (2023) deployed marker-assisted selection in a population developed from the cross combinations of MDU 1 (susceptible female parent) and TU 68 (resistant male parent) and isolated introgression lines viz IL-9-1, IL-12-1, IL-12-2, IL-12-7, IL-12-9, and IL-12-10 with higher seed yield under disease prevalent environment.

7 Powdery Mildew

Powdery mildew is one of the major constraints in black gram resulting in considerable loss annually. It is a foliar disease caused by fungus *Erysiphe polygoni*, which causes quantitative as well as qualitative loss of grains. Its occurrence is observed throughout the year, but increased severity in late-sown kharif crops and under favorable weather conditions has been reported (Priyanka et al. 2018). Disease infection during post-rainy and rice fallow cropping seasons in various popular black gram cultivars resulting in severe yield loss has been also reported (Tamilzharasi et al. 2023). Powdery mildew pathogen is an obligate parasite with a broad host range, and its infection spreads through dissemination of spores by wind (Deshmukh et al. 2022). Humid weather, the temperature of 27.2–30.3 °C, and relative humidity of 90% are reported to be conducive factors for disease infection. It affects all the growth stages of crop. Disease symptoms include white powdery patches on leaves, petioles, stems, and pods, which later become dull in color. These patches gradually increase in size and become circular covering the lower leaf surface and resulting in a reduction in photosynthesis and yield of up to 40–90%, depending upon the stage and time of infection reported (Priyanka et al. 2018; Deshmukh et al. 2022). Severe infection may also lead to yellowing of foliage, premature defoliation, and forced maturity of the infected plants. Genetic resistance is a reliable, efficient, durable, and cost-effective strategy over chemical measures.

Screening of parents, F_1, F_2, F_3, F_4, and BC_1F_2 of cross between powdery mildew susceptible VBN 11 and resistant LBG 17 black gram genotypes revealed monogenic recessive gene action governing resistance disease reaction (Punithavathy et al. 2023a). Variability and association analyses have reported high heritability coupled with high genetic advance as a percentage of the mean for percent disease index (PDI) for powdery mildew (Punithavathy et al. 2023b). Identification of potential resistance sources is crucial in disease resistance breeding. Screening for powdery mildew resistance under natural as well as artificial epiphy-

Table 2 List of black gram genotypes for powdery mildew resistance identified in different studies

S. no.	Genotypes/varieties	Disease reaction	References
1	BDU 3-22, OBG-647, BDU 3-23, BDU 3-2, KU-5-527, BDU 3-21, BDU-5, BDU-7, BDU-9, BDU-12, LBG-645, LBG-465, LBG-685 and LBG-20	Moderately resistant	Hadimani et al. (2015)
2	LBG-645	Highly resistant	Priyanka et al. (2018)
3	ADT 6	Moderately resistant	Shanthi et al. (2019)
4	VBN 8	Moderately resistant	Pandiyan et al. (2018)
5	LBG-17	Highly resistant	Khandappagol and Rangaiah (2019)
6	VBN-11	Moderately resistant	Geetha et al. (2020)
7	LBG 752, KPU 1720-140, RU 03-22, NUL 7, VBG 12-111, COG 13-08 and KUG 791.		Ahiladevi et al. (2020)
8	Pusa B-43, RVSTU 21-1 and Pant U-30 DKU 87 and PU 31	Immune Resistant	Deshmukh et al. (2022)
9	OBG-31, Ujala Prasad, Nayagarh-I local	Moderately resistant Resistant	Samal et al. (2022)

totic conditions by several investigators lead to the identification of promising genotypes with variable disease reactions including moderate resistance to high resistance. Table 2 contains the list of some promising genotypes identified for powdery mildew resistance in different studies.

Priyanka et al. (2018) screened 116 genotypes for powdery mildew resistance under artificially inoculated conditions. Genotype LBG-645 was found to be highly resistant with a disease severity of 0.77 under greenhouse screening. Under in vitro screening LBG-645 exhibited high resistance with 3.33×10^3 conidia colony, 0.94 percent of leaf area coverage, and 1.00 colony per leaflet. These findings validated LBG-645 as a potential source for powdery mildew resistance. In another study by Khandappagol and Rangaiah (2019), genotype LBG-17 was identified with lowest per cent disease severity values of 0.82 and 0.78 under field and greenhouse, respectively. Under in vitro conditions, LBG-17 showed high resistance with 0.60×10^3 conidia colony, 1.72% of leaf area coverage, and 0.67 colonies per leaflet.

Owing to limited natural genetic variability in black gram germplasm for powdery mildew resistance, mutation breeding provides promising alternatives for isolating resistant sources. In a study by Kanimozhi et al. (2021), 59 M_3 mutant families were screened for morphological and biochemical characteristics associated with powdery mildew resistance. This study identified six moderately resistant mutants PM 27, PM 29, PM 48, PM 49, PM 42, and PM 34 based on field and glass house screening of visual scores. Further higher trichome density, leaf thickness, total phenol content, peroxidase and polyphenol oxidase activity, lower total sugar content, and stomatal density were reported in resistant mutants. Recently, Tamilzharasi et al. (2023) developed mutant populations in the background of CO6 with gamma rays and ethyl methanesulphonate (EMS). Out of the total 60 powdery mildew-resistant mutants identified, 25 high-yielding mutants were further evaluated to confirm powdery mildew resistance and yield. Nine resistant mutants (PM 13, PM 20, PM 21, PM 42, PM 53, PM 54, PM 56, PM 57, and PM 60) with significantly higher values for leaf density, trichome density, SOD, CAT, POX, PPO, total phenols, phytic acid, and silica content were reported. These elevated levels of biochemicals and enzymes can be attributed to resistance reactions exhibited by the mutants. These studies revealed a diverse spectrum of mutation-derived resistant sources, which can explored as valuable germplasm against powdery mildew.

Srivastava et al. (2013) developed an F_2 mapping population by crossing powdery mildew susceptible lineBL-20 with highly resistant line Vs. By using bulked segregant analysis, they identified two RAPD markers OPA-13_{420} and OPC-13_{480} linked to the powdery mildew resistant gene at a distance of 5.83 cM. Molecular characterization and validation of five microsatellite markers (MB-SSR238, VrCsSTS1, CEDG191, VrCsSSR1, CEDG166) linked to powdery mildew disease resistance were carried out in the mini-core set of 29 germplasm by Samal et al. (2022). Their results showed the association ofVrCsSTS1 marker with the powdery mildew resistance gene in different genetic backgrounds. In another study, a diverse set of SSR markers viz. CEDG154, CEDG290, CEDG139, CEDG259, CEDG191, CEDG024, CEDG 282, CEDG 166, CEDG 232, and CEDG 088 polymorphic between resistant mutants and the parent CO6 have been identified (Tamilzharasi et al., 2023). These studies reiterated the utilities of molecular marker resources for marker-assisted disease resistance breeding.

8 Bruchid Resistance

Bruchids or seed weevils are one of the most important storage pests of urdbean belonging to the genus *Callosobruchus*. Among the species, azuki bean weevil (*C. chinensis L.*), graham bean weevil (*C. analis F.*), and cowpea weevil (*C. maculatus F.*) are the three major menace in Asian region (Souframanien and Dhanasekar 2014). These three bruchid pests have different distribution ranges, C. *analis* occurs in Africa and Asia, C. *chinensis* occurs in Asia, while wide distribution of C. *maculatus* in Asia,

Australia, and Africa has been reported (Daglish et al. 1993). Devastation by bruchid is realized in terms of reduced seed weight, impaired germination, degrading nutritional quality, and market value, thereby rendering it unfit for human consumption, agricultural, and commercial uses (Duan et al. 2014). The infestation of this pest occurs at pod maturity and adult emergence overlaps with harvest, which leads to secondary infestations and destruction of a seed lot within 3–4 months (Banto and Sanchez 1972). Hence, host plant resistance is suggested as a suitable alternative for preventing bruchid infestation over chemical methods (Khush and Brar 1991).

Understanding of trait genetics and its inheritance pattern is crucial for the implementation of genetic measures. Dongre et al. (1996) evaluated different *Vigna* species for their resistance to infestation by *C. maculatus* and identified a wild relative *V. mungo* var. *silvestris* as a potential source of bruchid resistance. The resistance was found to be under the control of two dominant duplicate genes: *Cmr1* and *Cmr2*. Inheritance of bruchid resistance has also been studied in Trombay wild urdbean by Souframanien and Gopalakrishna (2007). They evaluated F_2, F_3, and RIL developed from a cross of TU 94-2 (mungbean yellow mosaic virus resistance) and *V. mungo* var. *silvestris* (bruchid resistance), a good-to-fit segregation ratio of 15:1 indicating the presence of two dominant duplicate genes for bruchid resistance, which is in agreement with earlier reports. Furthermore, this resistance was effective for its deployment in black gram breeding program due to cross-compatibility of this species with black gram (Dongre et al. 1996; Souframanien and Gopalakrishna 2007). Another study conducted by Subramaniyan et al. (2023) reported digenic complementary gene action in F_2 generation of MDU1 × TU68 as well as digenic duplicate gene action in F_2 generations of VBN6 × TU68 and VBN8 × TU68.

For a detailed overview of bruchid resistance screening methodology references, Souframanien and Gopalakrishna (2007) and Souframanien et al. (2010) can be referred. Predominantly decreased percentage of adult emergence, longer developmental period, and decrease in adult size are the vital indicators of bruchid resistance. Also, reduction in adult weight, longevity, reproductive capacity, and adult survival are the additional important traits that need to be considered during screening. Apart from seed morphological characteristics, such as hard seed coat, roughness on seed surface and small seed size, antibiosis mediated through various biochemical compounds including trypsin inhibitors, alpha amylase inhibitors, lectins, and arcelins also impart resistance to bruchids as reported by previous studies (Souframanien and Gopalakrishna 2007). Among the different bioassay tests conducted for screening of bruchid resistance, Duraimurugan et al. (2014) concluded "no choice" assay to be the most reliable and potentially useful test in green gram and black gram.

However, insect-based screening is difficult to implement since it requires mass rearing and artificial infestation under lab conditions. Also, it needs to be carried out after harvest; hence, it is time consuming and destructive method (Souframanien et al. 2011). In this context, identification of molecular markers linked with traits can facilitate efficient marker-assisted selection any time of year and free from seasonal variations. Some pioneering attempts of QTL mapping for bruchid resistance underpinning marker-assisted selection are available. A QTL mapping study conducted by Souframanien et al. (2010) using an inter-subspecific RIL mapping population (population size-104 individuals) generated by crossing bruchid susceptible *V. mungo* var. *mungo* (cv. TU 94-2) and bruchid-resistant *V. mungo* var. *silvestris* using 428 markers, discovered total 8 QTLs. Two QTLs, *Cmrae* 1.1 and *Cmrae* 1.2, were identified for percentage adult emergence, on linkage groups LG 3 and 4, respectively. Total of six QTLs for the developmental period were identified, of which two QTLs resided (*Cmrdp* 1.1 and *Cmrdp* 1.2) on LG 1, three QTLs (*Cmrdp* 1.3, *Cmrdp* 1.4, and *Cmrdp* 1.5) on LG 2, and one QTL (*Cmrdp* 1.6) on LG 10. Their study also revealed that about 37.8% of bruchids completed their lifecycle on seeds of *V. mungo* var. *silvestris* compared with 100% on the susceptible variety TU 94-2. Further, the total developmental period of *C. maculatus* on *Vigna mungo* var. *silvestris* was found to be considerably extended (88 days as compared to34 days on TU 94-2).

Recently, Somta et al. (2019) identified quantitative trait loci (QTLs) for *C. maculatus* resistance. They constructed a high-density linkage map (1588.7 cM) of 3675 SNP markers from specific-locus amplified fragment and sequencing using a RIL population of 150 lines from a cross between BC48 (bruchid-susceptible) and wild black gram (var. *silvestris*) TC2210 (bruchid-resistant). Two highly robust, stable QTLs *qVmunBr6.1* and q*VmunBr6.2* were jointly mapped for percentage of damaged seeds (PDS) and infestation severity progress (AUDPS) on linkage group 6.

Another study performed by Subramaniyan et al. (2021) in a biparental mapping population derived from a cross between MDU 1 (susceptible) and (TU68) resistant genotypes revealed three major QTLs responsible for the total number of adult emergence, percentage of seed damage, and developmental time. Validation of identified QTLs was also carried out through single marker analysis in other two mapping populations (VBN 6 × TU 68 and VBN 8 × TU 68). They identified and recommended two markers CEDG020 and CEDG302 to be used in bruchid resistance breeding programmes. Such QTLs and molecular markers identified in different studies can serve as useful genetic resources for the molecular breeding of black gram for bruchid resistance.

Das et al. (2021) generated a transcript profile of immature black gram pods of resistant cultivar IC-8219 against bruchid oviposition using a susceptible cultivar T-9 as a control. They studied transcript dynamics using RNAseq approach and unfolded underlying defense response mechanisms induced by salicylic acid signaling pathways and defense genes, such as defensin, genes for secondary metabolites, and enzyme inhibitors, which can serve as potential candidates for resistance to bruchids. Various promising genotypes and accessions reported by different studies for their bruchid resistance include TU 68 (Indhu et al. 2020); TU 80 and TU 68 (Swamy et al. 2016); IC424616 (Panigrahi et al. 2022); SKNU-10-03, SKNU-03-08, SKNU-07-03, SKNU-05-06, IPU-7-3, SKNU-05-05, and SKNU-08-04 (Suthar and Bharpoda 2018).

9 Superior Battering Quality Trait

Legume flours are becoming more significant as food ingredients because of the functional and sensory qualities they contribute to the end product (Vanniarajan et al. 2022). The presence of the unique mucilaginous material has made black gram as an exclusively popular pulse crop in South India as fermented food in the form of idli, dosa, and vada (Vanniarajan et al. 2022). The mucilaginous substance is a carbohydrate complex consisting of galactose, arabinose, and a trace quantity of galacturonic acid and rhamnose (Umadevi et al. 2008). The presence of the mucilaginous polysaccharide arabinogalactan and surface active globulin protein gives the leavened steamed idli a soft and spongy texture (Vanniarajan et al. 2022). The soft, spongy, honeycomb-like texture of idli is a result of this mucilaginous principle, which aids in the retention of carbon dioxide during the fermentation of the thick batter made from rice and urdbean in a 3:1 or 4:1 ratio (Susheelamma and Rao 1978). Desired physical and chemical changes during the fermentation process of idli batter increase the batter's volume and non-protein nitrogen and acidity (Vanniarajan et al. 2022).

Increased batter volume is one of the most desirable properties for idli preparation with a soft texture. However, deterioration in the battering quality within 2–3 months after the harvest is a major problem faced by farmers, traders, and consumers (Vanniarajan et al. 2022). This compels the inclusion of battering quality as one of the breeding objectives in black gram improvement. In addition to processing, battering quality is also influenced by genotype; hence, assessment of diverse genotypes for their inherent quality profile is essential. Total soluble protein content, arabinose (%), albumin (%), globulin (%), methionine (mg/g), and batter volume are the major targets of improving battering quality in black gram. A genetic divergence study consisting of 39 genotypes was carried out by Veni et al. (2016). Genotypes were evaluated for different biochemical and yield attributing traits, and cluster analysis was carried out using Mahalnobis D^2 statistics. Based on the mean performance and genetic divergence, the genotypes UH04-4, CO5, KKB05011, ACM07002 and VBG66 were identified for their use in crossing program to obtain desirable segregants for developing batter quality black gram varieties. Stability analysis of seed yield and batter quality traits was undertaken by Rajalakshmi et al. (2021). They evaluated the genotype x environment interaction of 32 black gram genotypes in three different seasons using AMMI and GGE Biplot analysis. They inferred kharif season as the most favorable environment for idli batter quality and genotype ACMBG 17-001 was identified as the most stable genotype for idly batter volume (ml). Mutation breeding has also been used to broaden the genetic variation for battering quality, which resulted in few significant achievements. Vanniarajan et al. (2019) developed gamma rays induced urdbean mutants (ACM16-30, ACM-16-29, ACM-6-17, and ACM-16-14) with good batter quality. In addition to this, genetic enhancement of the nutritional and battering qualities of varieties MDU 1 and VBN (Bg) has been carried out through electron beam and gamma ray mutants. The mutant lines, viz., ACM-014-021, ACM-015-015, ACM-15-023, ACM-015-013, ACM-015-003, ACM-015-030, ACM-014-006, and ACM-014-007, were identified to have superior albumin content, globulin content, total soluble protein, arabinose content, 100-seed weight, and seed yield per plant (Vanniarajan et al. 2021). Among the various black gram varieties available, MDU1 variety is superior in globulin content as well as batter volume. But a breakthrough was achieved by employing mutation breeding in the background of MDU 1, which led to the isolation of four mutants (ACM-16-011, ACM-16-015, ACM-16-017, and ACM-16-018) with increased arabinose, globulin, and batter volume over the parent MDU1 (Vanniarajan et al. 2022). Other recently developed black gram varieties with good battering qualities include KKM1 and ADT 6 (Pillai et al. 2017; Shanthi et al. 2019).

10 Breeding for Reducing Anti-nutritional Factors in Black Gram

Black gram is a rich source of digestible proteins, phosphoric acid, vitamins, and sulfur- containing amino acids. The nutritional profile of black gram is unique due to its high protein content (25.21 g/100 g), richness in various isoflavones and vitamin B complex, and high proportion of dietary fiber and soluble mucilaginous polysaccharides (Venkidasamy et al. 2019). However, presence of anti-nutritional factors limits its nutritional and food quality. The major anti-nutritional factors in black gram consist of flatulence causing raffinose family oligosaccharides (RFOs), saponins, phenolics, phytic acid, protease inhibitors, tannins, and lectins. RFO constitutes 31–76% of total sugar, and raffinose, stachyose, and verbascose are the most common RFOs in legumes. Additionally, higher homologs like ajugose are also present in trace amounts in black gram (Souframanien and Gopalakrishna 2014). Owing to absence of α-galactosidase in humans, anaerobic fermentation of oligosaccharides by microorganisms results in the production of carbon dioxide, hydrogen, and methane (Suneja et al. 2011).

Protease inhibitors decrease the absorption of amino acids and affect de novo protein synthesis by forming stable complexes with proteolytic enzymes; saponins interact and disrupt fluidity of bio-membranes, resulting in holes, pores, and leaky cells; phytic acid strongly chelates mineral cations (calcium, magnesium, zinc, and iron), prevents their intestinal absorption, forms insoluble complexes with proteins and starch, and reduces their bioavailability; phenols decrease digestibility of proteins, amino acids, multivalent cations, or minerals such as iron, zinc, and calcium in foods by binding with them; tannins interact with enzyme and non-enzyme proteins, resulting in tannin-protein complexes that inactivate digestive enzymes and reduce digestibility of proteins (Suneja et al. 2011).

Although a few studies are available for the assessment of genetic diversity in black gram genotypes for their nutritional and anti-nutritional profiles, their findings are helpful in the selection of genotypes to breed varieties with reduced content of anti-nutritional factors.

Suneja et al. (2011) conducted a study to characterize levels of nutritional constituents and anti-nutritional factors in 32 black gram genotypes consisting of eleven recommended cultivars and twenty-one advanced breeding lines. The reported range of different antinutritional factors in their study varied from 4.7 to 10.9 mg/g for bound fructose, 1.1–3.2 mg/g for total phenol content, 5.3–9.2 mg/g for tannin content, 71.6–106.5 units/g trypsin inhibitor activity, 9.2–21.4 mg/g saponins, and 1.7–9.0 mg/g phytic acid. This analysis lead to the identification of genotypes for low antinutritional contents, which include KUG-525 for bound fructose-sucrose and raffinose series oligosaccharides and saponin; Mash-1008 for total phenols and o-dihydroxy phenols; KUG-365, KUG-391, KUG-401, and KUG-508 for lower tannins content; KUG-511 for low trypsin inhibitor activity and KUG-365 for low phytic acid content. They also noted that genotypes with higher activity of α-galactosidase (KUG-401) and acid- and alkaline phosphatase (KUG-365,KUG-521 and KUG-525) showed lower content for bound fructose of raffinose series oligosaccharides and phytic acid contents, respectively.

Oligosaccharides are reported to play a role in plant defense and stress tolerance, and hence, any manipulation in their levels could have an adverse effect on plant growth and yield. Further, RFOs are reported to be heat stable, and cooking leads to an increase in their content. Hence, isolation of low RFO-type genotypes through breeding strategies becomes imperative for improving the nutritional quality of black gram (Souframanien et al. 2014).

A dedicated study for analysis of genetic variation for total as well as individual oligosaccharides and sucrose content was carried out by Souframanien et al. (2014). They reported predominance of verbascose among all RFOs and wide variation for individual RFOs, such as raffinose, stachyose, verbascose, and ajugose, in the range of 0.18–8.08 mg/g, 8.90–37.27 mg/g, 13.95–31.02 mg/g, and 0.30–4.48 mg/g, respectively. Low content of total RFOs was identified in genotypes TU43-1 and LBG-402. Genotype TU43-1 was also recorded to contain the lowest values for verbascose, sucrose, and ajugose concentrations, and can serve as a potential target for low RFO trait introgressions. A low concentration of hexasaccharide ajugose was reported in genotypes TU-72 (0.30 mg/g), TU 94-2 (0.33 mg/g), and TU 43-1 (0.35 mg/g), and the lowest stachyose was found in LBG-402 (8.90 mg/g).

In nutritional quality evaluation, four improved varieties of black gram (Pant U-8, Pant U-9, Pant U-31 and Pant U-35). Kanth et al. (2021) reported trypsin inhibitor content ranging from 90.33 TIU/g to 94.53 TIU/g and tannin content ranging from 6.80 to 7.24 mg per 100 gm. Variety Pant U-35 was identified with the lowest value of trypsin inhibitor (90.33 TIU/g) and tannin content (6.81 mg/g). Souframanien et al. (2021) reported genetic variation in phytic acid that ranged from 6.97 to 7.50 mg/g. Low phytic acid content was reported in genotypes KUG-365 (1.7 mg/g), Shekhar-2 (3.7 mg/g), and KUG-230 (4.0 mg/g).

Additional variability for antinutritional contents using induced mutations has been explored by Jegadeesan and Reddy (2021). They obtained black gram mutant TU43-1, which recorded low total RFOs (26.64 ± 7.28 mg/g), and also lower contents of verbascose (13.95 ± 0.25 mg/g), sucrose (3.28 ± 0.87 mg/g), and ajugose (0.35 ± 0.11 mg/g). Low stachyose content was observed in TU1-820-1-5 (11.13 mg/g). Low verbascose content was observed in TU43-1 (13.95 ± 0.25 mg/g). TU51 recorded a low stachyose content (15.11 ± 0.47 mg/g) compared with its parent TU94-2 (37.27 ± 1.07 mg/g). Low stachyose and verbascose content of 15.85 ± 1.40 mg/g and 17.02 ± 2.30 mg/g, respectively, was observed in the TU55-1 mutant. Another mutant TU1-820-1-5 recorded high verbascose (31.02 ± 0.74 mg/g) and lowest raffinose (0.18 ± 0.05 mg/g) contents. These all-black gram mutants identified with reduced RFOs, low stachyose, and verbascose will serve as valuable and unique germplasm for breeding for the reduction of anti-nutritional contents.

Genomic regions and the genetic basis for antinutritional factors in black gram remain unclear. However, several reports have characterized potential candidate genes and enzymes involved in the biosynthesis pathways of various anti-nutritional factors in other legume crops, which can be deployed in black gram breeding. In addition to this genetic, genomic and marker information developed in other *Vigna* species may also serve as valuable resources for black gram improvement.

11 Breeding for Seed Trait Improvement

On account of its contribution to yield and commercial value, large seed size remains the key target trait in modern breeding (Naito et al. 2017). Bold seeded black gram genotypes are mainly preferred by the processing industries for high dhal recovery and for minimizing processing losses (Jayamani et al. 2022). Although seed weight is a key determinant of post-harvest utility of black gram, a few studies characterizing its genetic basis are available, albeit with significant findings. Naito et al. (2017) obtained a recessive mutant designated as multiple-organ-gigantism (mog), which expressed extra-large sized leaves, stems, and seeds. Based on map-based cloning, they identified loss of function caused by an 8-bp deletion in the coding sequence of the *VmPPD* gene that was responsible for the mog phenotype. QTL mapping using mog mutant as a large-seeded parent and gene transformation studies revealed that mog gene accounted for approximately 66% of seed weight variation in the mapping population and additive gene effect of the mog increased 2.0 g of seed weight.

In another QTL mapping study, MOG phenotype was reported to contribute approximately 30% PVE for seed weight (Somta et al. 2020). Therefore, the gene *VmPPD* can serve as good target for genome editing applications for alteration of seed size. Genome-wide association mapping studies using a panel of 100 diverse black gram genotypes evaluated at two locations revealed novel alleles for seed weight (Singh et al. 2022b). They detected three QTLs viz *Q.HSW.6*, *Q.HSW.7*, and *Q.HSW.10* for 100 seed weight. Three candidate genes, putative pentatricopeptide repeat-containing protein At1g12700, peroxidase 4, and romodomain-containing protein 4B, were obtained in putative QTL regions. In addition to this, genotype IC370938 carrying superior alleles for seed weight QTLs was also identified.

Lomlek et al. (2023) detected 10 QTLs for seed weight on seven linkage groups using two F_2 populations derived from crossing CN80 × PI213017 and CN80 × TVNu 1076. These QTLs explained between 5.07% (*qSd100wt7.1+*) to 34.20% (*qSd100wt10.2+*) of the weight variation. Potential candidates underlying QTLs region consist of pentatricopeptide repeat (RPP)-containing protein, WRKY46, glutathione S-transferase U9, NAC domain-containing protein 100, calcineurin B-like protein (CBL)-interacting protein kinase, cyclin-D6-1, kinesin, histidine phosphotransfer protein, AUXIN RESPONSE FACTOR 2A, and WAVE-DAMPENED2 (WVD2)-like 4 protein.

Various breeding efforts are underway for the improvement of seed weight traits. Seed polymorphism in inter subspecific cross of black gram (*Vigna mungo* (L.) Hepper) was studied by Devi et al. (2022). They observed significant genetic variability for seed traits in 189 pre-breeding lines from the cross VBN (Bg) 5 and *Vigna mungo* var. *silvestris* 22/10. The reported range for various seed traits varies from 2.90 to 5.98 g for hundred seed weight, 4.25–5.20 mm for seed length, 3.30–4.15 mm for seed width, 1.07–1.52 for seed length-width ratio, 2.84–3.87 mm for seed thickness, and 0.75–3.00 ml for seed volume. Pillai et al. (2017) developed high-yielding variety KKM1, which had bold seeds with a mean 100 seed weight of 4.47 grams. Vanniarajan et al. (2019) isolated mutant-line ACM 16-14 from parent MDU 1 using gamma irradiation. This mutant had bold seed type and a hundred seed weight value of 6.08 gm. Recently, a new high yielding and bold seeded variety CO 7 (COBG 10-05) has been derived from an inter sub-specific cross between VBN (Bg) 5 x V. *mungo* var. *silvestris* (22/10) (Jayamani et al. 2022).

The presence or absence of the texture layer, its pigmentation, and color of testa are the key determinant of seed appearance (Souframanien and Dhanasekar 2014). Contrasting reports about the genetics of seed color are available. The brown seed coat color was reported as recessive over green seed coat color with qualitative inheritance (Sen and Jana 1964), whereas dominance of brown seed coat color over green seed color has been reported by Arshad et al. (2005). Shiny seed surface is preferred traits, and it was found to be dominant over dull seed surface (Sen and Jana 1964). A high-yielding variety ADT 6 developed by Shanthi et al. (2019) had bold, black oval shape seeds with dull luster and a mean 100 seed weight of 4.7gm. The other variety CO 7 also possessed black, globose-shaped seed with dull seed luster and test weight values ranging from 5.5 to 6.0gm (Jayamani et al. 2022).

Substantial variability for seed color and seed shape was observed in the morphological characterization of 40 black gram genotypes by Kumar et al. (2022). They observed eight different classes based on seed type and shape combinations consisting of mottled globose (PDU 1, TJU-1-30, TJU 328, TJU 103, PU 35, TJU 111, TJU 4, etc); mottled oval (TJU 18, TJU 84, TJU 55, TU-98-14, PU 19, TJU 134); black globose (TJU 339, TJU-41-1, TJU-41-2, TJU-24-10); black oval (T9, TJU 48, TJU 24, JU 30); mottled drum shaped (TJU 62, TJU 213); greenish brown globose (JU 3, TJU 262); greenish brown oval (LBG 20); greenish globose (Uradi). In addition to this, genotypes with shiny seed luster (TJU 84, TJU 24, TJU 42, LBG 20, TJI-48-3, TJU 134, TJU-41-1, TJU 213, Uradi) were also observed. This information is useful in the selection of genotypes for black gram breeding based on consumer preferences.

12 Breeding for Suitable Mechanical Harvesting in Black Gram

The conventional harvesting method in black gram is through manual cutting by labors. However, this method is lengthy and time consuming; it also adds to the cost of cultivation through expenses on wages. Timeliness in harvesting is also important in the context of frequent aberrant weather events overlapping with crop maturity and resulting in considerable damage to standing crops. Timely harvest at proper crop maturity eases land preparation and supports appropriate planting of the next season crop. Nowadays, improved harvesting tools and equipment and new inventions are coming up in machinery-based harvesting (Venkatakrishnan et al. 2019), which demands specific plant-type features to facilitate its mechanized harvesting. Certain modifications in plant architecture and pod morphology related traits are needed to make it suitable for harvesting. Plant stature is one of the important traits that needs prime consideration for mechanical harvesting. Medium to tall plant height, erect to semi-erect and determinate growth habit, and synchronous maturity are highly desirable. Stem growth habit is an essential morphological and agronomic feature, which determines height of plant, production of node, morphology of leaves, maturity, root architecture, water use efficiency, and ultimately yield (Vanniarajan et al. 2019).

It was reported that gene Terminal Flowering 1 (*TFL* 1) is known to be involved in the control of flowering time, floral architecture, and determining flowering habit in some model legumes (*Arabidopsis*, soybean, peas, and *Medicago*), the wild-type TFL 1 allele leads to axillary flowering habit resulting in indeterminate vegetative growth, while the mutant allele tfl 1 results in terminal flowering thereby promoting determinate habit (Foucher et al. 2003; Tian et al. 2010). Foucher et al. (2003) identified two homologs of (*TFL1*) in pea viz *PsTFL1a* and *PsTFL1c* each corresponding to genes *DETERMINATE* (*DET*) and *LATE FLOWERING* (LF), respectively. *DET* maintains the indeterminacy of the apical meristem during flowering, while *LF* delays the induction of flowering by lengthening the vegetative phase. This genetic basis for natural variation in flowering habit and flowering time can be explored by studying its gene homologs in black gram. Interestingly, it has been reported that the erect plant type is not completely dominant over the spreading type (Sen and Jana 1964); hence, intentional breeding and selection efforts are required for the development of erect genotypes. Among the two types of pod orientations, i.e., main stem bearing and sympodial bearing types, monogenic incomplete dominant gene action for main stem pod bearing has been reported (Nair et al. 2024).

Pillai et al. (2017) developed a high-yielding variety KKM1 (KKB05011), which is a cross derivative of COBG 653 x VBN3. It was reported to have determinate plant type and synchronized maturity. Variety VBN 8 with determinate plant type and synchronized maturity has been developed from a cross of VBN 3 x VBG 04-008 by Pandiyan et al. (2018). Another semi-erect and determinate plant type variety ADT 6 (ADBG 13004) from a cross between VBN 1 x VBG 04-006 has been developed (Shanthi et al. 2019). Vanniarajan et al. (2019) obtained promising mutant lines in the background of MDU 1 using gamma rays. Mutant line ACM-16-17 showed terminal pod bearing, while ACM-16-30 was a determinate plant-type mutant with high peduncle length (9.86 cm) compared to MDU 1. This line also possessed pods protruding up above the canopy and synchronized maturity. These types of terminal flowering habits are highly advantageous for mechanical harvesting. Geetha et al. (2020) derived variety VBN 11 (VBG 12-062) from PU 31 × CO 6. This variety has determinate plant type, medium duration, and synchronized maturity. Kumar et al. (2022) characterized 40 black gram genotypes for morphological traits using 18 DUS descriptors and identified two genotypes (PDU 1, T9) with erect and determinate growth habit, and one genotype (TJU 339) with semierect determinate plant habit. Jayamani et al. (2022) developed a new high-yielding variety CO 7 having determinate plant type, long peduncle length (10.5–11.1 cm), and synchronized maturity, which makes it suitable for single/mechanical harvest. These black gram genotypes with erect and determinate growth habit coupled with optimized vegetative growth and synchronized maturity ease the way for mechanical harvesting. This stature also ensures more responsiveness to fertilizer uptake, high crop uniformity, and increased yield and hence is highly desirable in restructuring the plant type of black gram.

13 Pre-breeding in Black Gram

Native cultivars and landraces are known to house genes against drought and heat stress offers huge scope for identification and sourcing of these untapped adaptive traits through pre-breeding efforts (Souframanien and Dhanasekar 2014). Distant hybridization has enabled us to broaden the genetic base of existing black gram cultivars. It has been evident that wild relatives of *Vigna* carry useful genes for vigorous and erect growth, sturdy stems, broad leaves, and long and profuse pods with biotic and abiotic stress tolerance, including MYMD (mungbean yellow mosaic disease) (viz., *V. sublobata*), heat and salt stress (viz., *V. angularis*, *V. luteola* (Jacq.) Benth., *V. marina* (Burm.) Merr., and *V. vexillata*) (Singh et al. 2022a). Another cultivated *Vigna* species, ricebean (*Vigna umbellata*) is closely related to black gram with the same somatic chromosome number ($2n = 22$). A successful attempt to introgress genes for desirable traits such as high pod number, seed weight, productivity and resistance to diseases from rice bean to black gram was carried out at Punjab Agricultural University, Ludhiana by Singh et al. (2013). This resulted in the development of a superior yielding black gram variety (Mash-114) with resistance to MYMV, *Cercospora* leaf spot, and bacterial leaf spot. Drought tolerance of *Vigna sublobata* with higher values of relative water content, membrane stability index, protein, proline content, leaf area, plant height, and yield traits has been reported (Bangar et al. 2019).

14 Genomic Resources in Black Gram

In recent years, the availability of draft genome sequence of black gram (Jegadeesan et al. 2021; Pootakham et al. 2021; Ambreen et al. 2022) and transcriptome sequencing, including molecular markers like SSRs and SNPs (Souframanien and Reddy 2015; Raizada and Souframanien 2019), has enriched reserve of genomic resources in black gram, which can be exploited for identifying candidate genes and developing diagnostic molecular markers for marker-assisted selection of different climate-resilient and ideal plant-type traits.

15 Conclusion

In comparison with other legumes, black gram breeding has witnessed limited efforts to address present-day challenges. We attempted to give an overview on recent progress in all major pressing issues in black gram breeding. We hope that the compilation done here will help decide the direction of future breeding efforts in black gram improvement. We believe that the conventional breeding process will surely benefit from the availability of next-generation sequencing facilities, draft genome assemblies, transcriptome data sets, and molecular marker resources to accelerate the pace of black gram breeding. Additionally, related genetic resources, pre-breeding, and cross-species learning from other legumes will be equally effective in dealing with these challenges.

References

Ahiladevi P, Ramakrishnan P, Gunasekaran M (2020) Evaluation of blackgram genotypes for resistance against major diseases. Adv Appl Res 12:108–112

Ahmad P, Umar S (2011) Oxidative stress: role of antioxidants in plants. Stadium Press, New Delhi

Ambreen H, Oraon PK, Wahlang DR, Satyawada RR, Katiyar-Agarwal S, Agarwal M, Jagannath A, Kumar A, Budhwar R, Shukla RN, Goel S (2022) Long-read-based draft genome sequence of Indian black gram IPU-94-1 'Uttara': insights into disease resistance and seed storage protein genes. Plant Genome 15:e20234

Anitha Y, Vanaja M, Kumar GV (2016) Identification of attributes contributing to high temperature tolerance in blackgram (*Vigna mungo* L. Hepper) genotypes. Int J Sci Res 5:1021–1025

Arshad M, Ghafoor A, Qureshi AS (2005) Inheritance of qualitative traits and their linkage in blackgram (*Vigna mungo* (L.) Hepper). Pak J Bot 37:41–46

Ashraf M (1989) The effect of NaCl on water relations, chlorophyll, and protein and proline contents of two cultivars of blackgram (*Vigna mungo* L.). Plant Soil 119:205–210

Bangar P, Chaudhury A, Tiwari B, Kumar S, Kumari R, Bhat KV (2019) Morphophysiological and biochemical response of mungbean [*Vigna radiata* (L.) Wilczek] varieties at different developmental stages under drought stress. Turk J Biol 43:58–69

Banto SM, Sanchez FF (1972) The biology and chemical control of *Callosobruchus chinensis* (Linn.) (Coleoptera: Bruchidae). Philipp Ent 2:167–182

Baroowa B, Gogoi N (2012) Effect of induced drought on different growth and biochemical attributes of black gram (*Vigna mungo* L.) and green gram (*Vigna radiata* L.). J Environ Res Dev 6:584–593

Baroowa B, Gogoi N (2014) Biochemical changes in black gram and green gram genotypes after imposition of drought stress. J Food Legumes 27:350–353

Baroowa B, Gogoi N (2015) Changes in plant water status, biochemical attributes and seed quality of black gram and green gram genotypes under drought. Int Lett Nat Sci 42:1–12

Basak J, Kundagrami S, Ghose T, Pal A (2005) Development of Yellow Mosaic Virus (YMV) resistance linked DNA marker in *Vigna mungo* from populations segregating for YMV-reaction. Mol Breed 14:375–383

Bhomkar P, Upadhyay CP, Saxena M, Muthusamy A, Shiva Prakash N, Pooggin M, Hohn T, Sarin NB (2008) Salt stress alleviation in transgenic *Vigna mungo* L. Hepper (blackgram) by overexpression of the glyoxalase I gene using a novel *Cestrum* yellow leaf curling virus (CmYLCV) promoter. Mol Breed 22:169–181

Biswas S, Biswas AK, De B (2020) Influence of sodium chloride on growth and metabolic reprogramming in nonprimed and haloprimed seedlings of blackgram (*Vigna mungo* L.). Protoplasma 257:1559–1583

Chaudhary S, Jha UC, Paul PJ, Prasad PV, Sharma KD, Kumar S, Gupta DS, Sharma P, Singh S, Siddique KHM, Nayyar H (2022) Assessing the heat sensitivity of Urdbean (*Vigna mungo* L. Hepper) genotypes involving physiological, reproductive and yield traits under field and controlled environment. Front. Plant Sci 13:1042999

Daglish GJ, Erbacher JM, Eelkema M (1993) Efficacy of protectants against *Callosobruchus phaseoli* (Gyll.) and *C. maculatus* (F.) (Coleoptera: Bruchidae) in mungbean. J Stored Prod Res 29:345–349

Das D, Baruah IK, Panda D, Paswan RR, Acharjee S, Sarmah BK (2021) Bruchid beetle ovipositioning mediated defense responses in black gram pods. BMC Plant Biol 21:1–22

Deshmukh S, Jha D, Patel DP (2022) Screening of genotypes and economic assessment of yield loss caused by powdery mildew in blackgram. Pharm Innov J 11:1110–1114

Devi HC, Kumari VP, Devi PS, Sinha B (2017) Screening of blackgram genotypes for molecular variability in reaction to Mungbean yellow mosaic disease. Indian Phytopathol 70:353–358

Devi M, Jayamani P, Kumar M (2022) Genetic analysis and seed polymorphism in inter sub specific cross of Blackgram (*Vigna mungo* (L.) Hepper). Madras Agric J 108:1–5

Dhaliwal SK, Gill RK, Sharma A, Kaur A, Bhatia D, Kaur S (2022) A large-effect QTL introgressed from ricebean imparts resistance to

Mungbean yellow mosaic India virus in blackgram (*Vigna mungo* (L.) Hepper). Theor Appl Genet 135:4495–4506

Dongre TK, Pawar SE, Thakare RG, Harwalkar MR (1996) Identification of resistant sources to cowpea weevil (*Callosobruchus maculatus* (F.)) in *Vigna* sp. and inheritance of their resistance in black gram (*Vigna mungo* var. mungo). J Stored Prod Res 32:201–204

Duan CX, Zhu ZD, Ren GX, Wang XM, Li DD (2014) Resistance of faba bean and pea germplasm to *Callosobruchus chinensis* (Coleoptera: Bruchidae) and its relationship with quality components. J Econ Entomol 107:1992–1999

Duraimurugan P, Pratap A, Singh SK, Gupta S (2014) Evaluation of screening methods for bruchid beetle (*Callosobruchus chinensis*) resistance in greengram (*Vigna radiata*) and blackgram (*Vigna mungo*) genotypes and influence of seed physical characteristics on its infestation. Vegetos 27:60–67

Foucher F, Morin J, Courtiade J, Cadioux S, Ellis N, Banfield MJ, Rameau C (2003) DETERMINATE and LATE FLOWERING are two TERMINAL FLOWER1/CENTRORADIALIS homologs that control two distinct phases of flowering initiation and development in pea. Plant Cell 15:2742–2754

Gandhi MK, Kumar A, Marviya GV, Paul P (2021) Evaluation of blackgram [*Vigna mungo* (L)] genotypes for saline tolerance at seedling stage using sea water. Int J Environ Clim Chang 11:136–145

Ganguli S, Dey A, Banik R, Kundu A, Pal A (2016) Analyses of MYMIV-induced transcriptome in *Vigna mungo* as revealed by next generation sequencing. Genom Data 7:226–228

Geetha S, Gnanamalar RP, Manivannan N et al (2020) VBN 11-A new high yielding and MYMV disease resistant blackgram variety. Electron J Plant Breed 11:987–992

Gupta S, Gupta DS, Anjum KT, Pratap A, Kumar J (2013) Transferability of simple sequence repeat markers in blackgram (*Vigna mungo* L. Hepper). Aust J Crop Sci 7:345–353

Gurumurthy S, Sarkar B, Vanaja M, Lakshmi J, Yadav SK, Maheswari M (2019) Morpho-physiological and biochemical changes in black gram (*Vigna mungo* L. Hepper) genotypes under drought stress at flowering stage. Acta Physiol Plant 41:1–14

Hadimani A, Konda CR, Kulkarni S, Patil R, Guru PN, Burud M (2015) Screening of blackgram genotypes under natural epiphytotic condition for powdery mildew disease resistance *Erysiphae polygoni* DC. Int J Prosthodont Endod 33:2477–2479

Hanumantha Rao B, Nair RM, Nayyar H (2016) Salinity and high temperature tolerance in mungbean [*Vigna radiata* (L.) Wilczek] from a physiological perspective. Front. Plant Sci 7:957. https://doi.org/10.3389/fpls.2016.00957

Hasan MK, Islam MS, Islam MR, Ismaan HN, El-Sabagh A (2019a) Salinity tolerance of black gram cultivars during germination and early seedling growth. Cercetări Agrono Moldova 3:51–68. https://doi.org/10.2478/cerce-2018-0025

Hasan MK, Islam MS, Islam MR, Ismaan HN, Sabagh AE, Barutçular C, Meena RS, Saneoka H (2019b) Water relations and dry matter accumulation of black gram and mungbean as affected by salinity. Thai J of Agric Sci 52:54–67

Indhu SM, Manivannan N, Mahalingam A, Kavitha Z, Souframanien J (2020) Screening for resistance against bruchid (*Callasobruchus maculatus*) damage in blackgram. Indian J Agric Sci 90:1110–1114

Jayamani P, Muthuswamy A, Kumaresan D, Latha TKS, Rajabaskar D, Senthivelu M, Anitta Fanish S, Geetha S (2022) Blackgram CO 7: a new high yielding and bold seeded variety suitable for Tamil Nadu. Electron J Plant Breed 13:49–55

Jegadeesan S, Reddy KS (2021) Radiation-induced mutations in genetic enhancement and development of new crop varieties in black gram (*Vigna mungo* (L.) Hepper). In: Mutation breeding, genetic diversity and crop adaptation to climate change. CABI, Wallingford, pp 303–311. https://doi.org/10.1079/9781789249095.0031

Jegadeesan S, Raizada A, Dhanasekar P, Suprasanna P (2021) Draft genome sequence of the pulse crop blackgram [*Vigna mungo* (L.) Hepper] reveals potential R-genes. Sci Rep 11:11247

Kanimozhi S, Kumaresan D, Thiruvengadam V, Latha TKS (2021) Morphological and biochemical characteristics associated with powdery mildew resistance in M3 population of blackgram (*Vigna mungo* (L.) Hepper). Electron J Plant Breed 12:983–989

Kanth A, Goswami K, Shukla P (2021) Nutritional quality evaluation of improved varieties of black gram (*Phaseolus mungo*). Pharm Innov J 10:201–220

Khandappagol M, Rangaiah S (2019) Evaluation of blackgram (*Vigna mungo* L. Hepper) genotypes for resistance against powdery mildew (*Erysiphe polygoni* DC) under natural and artificially inoculated conditions. Int J Curr Microbiol Appl Sci Special Issue-9:171–182

Khush GS, Brar DS (1991) Genetics of resistance to insects in crop plants. Adv Agron 45:223–274

Kumar P, Singh SK, Shukla RS, Elahi T, Prajapati SS, Rahangdale S (2022) Morphological characterization for important breeding traits in blackgram [*Vigna mungo* (L.) Hepper] genotypes. Biol Forum 14:1323–1330

Kundu A, Singh PK, Dey A, Ganguli S, Pal A (2019) Complex molecular mechanisms underlying MYMIV-resistance in *Vigna mungo* revealed by comparative transcriptome profiling. Sci Rep 9:8858

Lamaoui M, Jemo M, Datla R, Bekkaoui F (2018) Heat and drought stresses in crops and approaches for their mitigation. Front Chem 6:26

Lamichaney A, Pratap A, Katiyar PK, Singh NP (2021) Genotypic variability studies and identification of pre-harvest sprouting tolerant wild *Vigna*. Indian J Agric Sci 91:335–339

Lamichaney A, Hazra KK, Katiyar PK, Parihar AK, Gupta DS, Kumar A, Singh F (2023) Influence of seed and pod biophysical characters on pre-harvest sprouting tolerance in urdbean (*Vigna mungo* L.). Acta Physiol Plant 45:48. https://doi.org/10.1007/s11738-023-03533-8

Lomlek C, Suamuang S, Laosatit K, Tangphatsornruang S, Tanadul OU, Somta P (2023) QTL analysis of seed weight and seed dormancy in black gram (*Vigna mungo* [L.] Hepper). J Agric Nat Resour 57:397–406

Maheswari M, Sarkar B, Vanaja M, Rao SM, Prasad JVNS, Mathyam P, Ravindra Chary G, Venkateswarlu B, Ray Choudhury P, Yadava DK, Bhaskar S, Alagusundaram K (eds) (2019) Climate resilient crop varieties for sustainable food production under aberrant weather conditions. ICAR-Central Research Institute for Dryland Agriculture, Hyderabad

Maiti S, Basak J, Kundagrami S, Kundu A, Pal A (2011) Molecular marker-assisted genotyping of mungbean yellow mosaic India virus resistant germplasms of mungbean and urdbean. Mol Biotechnol 47:95–104

Naik BJ, Anuradha C, Kumar PA, Sreedhar V, Chary S (2017) Identification of simple sequence repeats (SSR) markers linked to yellow mosaic virus (YMV) resistance in blackgram [*Vigna mungo* (L). Hepper]. Agric Update 12:812–819

Nair RM, Chaudhari S, Devi N, Shivanna A, Gowda A, Boddepalli VN, Pradhan H, Schaflfleitner R, Jegadeesan S, Somta P (2024) Genetics, genomics, and breeding of black gram [*Vigna mungo* (L.) Hepper]. Front Plant Sci 14:1273363. https://doi.org/10.3389/fpls.2023.1273363

Naito K, Takahashi Y, Chaitieng B, Hirano K, Kaga A, Takagi K, Tanaka EO, Thavarasook C, Ishimoto M, Tomooka N (2017) Multiple organ gigantism caused by mutation in *VmPPD* gene in blackgram (*Vigna mungo*). Breed Sci 67:151–158

Narasimhan A, Patil BR, Datta S, Kaashyap M (2010) Genetic diversity assessment across different genotypes of mungbean and urdbean using molecular markers. Electron J Plant Breed 1:379–383

Pandiyan M, Senthil N, Balaji T, Veeramani P, Savitha BK, Senthivel V et al (2017) Studies on performance of drought tolerant genotypes under drought and normal conditions through morpho, physio and biochemical attributes of black gram (*Vigna mungo.* L.) and green gram (*Vigna radiata* L.). Int J Curr Adv Res 5:489–496

Pandiyan M, Geetha S, Gnanamalar RP, Packiaraj D, Mahalingam A, Sassikumar D, Dinakaran D, Soundararajan RP, Vijayaraghavan C, Rajarathinam P, Satya VK (2018) A new high yielding MYMV disease resistant blackgram variety VBN 8. Electron J Plant Breed 9:1272–1279

Pandiyan M, Sivaji M, Yuvaraj M, Krishnaveni A, Sivakumar C, Jamuna E (2023) Molecular and physiological approaches for effective management of Drought in black gram. In: Legumes: physiology and molecular biology of abiotic stress tolerance. Springer Nature, Singapore, pp 259–278

Pang Q, Chen S, Dai S, Chen Y, Wang Y, Yan X (2010) Comparative proteomics of salt tolerance in *Arabidopsis thaliana* and *Thellungiella halophila*. J Proteome Res 9:2584–2599

Panigrahi SK, Tripathi K, Singh R, Kumar R, Sanghamitra P, Wankhede DP, Singh N, Dubey KKD, Gupta K (2022) Evaluation of black gram (*Vigna mungo*) genepool against *Callosobruchus maculatus* and diversity analysis inter se. Indian J Agric Sci 92:915–919

Partheeban C, Vijayaraghavan H, Boominathan P (2017) The rapid and reliable method to assess the blackgram (*Vigna mungo* (l.) Hepper) genotypes for thermotolerance. Int J Bio-Resour Stress Manag 8:6–11

Pavithra N, Jayalalitha K, Sujatha T, Hari Satyanarayana N, Lakshmi NJ, Roja V (2023) Temperature induction response and multivariate Analys is approaches to screen Blackgram (*Vigna mungo* L.) genotypes for Thermotolerance. Legum Res 1:1–8. https://doi.org/10.18805/LR-5134

Pillai MA, Shunmugavalli N, Selvi B, Muthuswamy A, Anand G, Shoba RPD, Saravanan S, Binodh AK, Kannan R (2017) Blackgram KKM 1 (KKB 05011), a rice fallow variety suited for Thamirabarani tracks of Tamil Nadu. Electron J Plant Breed 8:900–906

Pootakham W, Nawae W, Naktang C, Sonthirod C, Yoocha T, Kongkachana W, Sangsrakru D, Jomchai N, U-thoomporn S, Somta P, Laosatit K, Tangphatsornruang S (2021) A chromosome scale assembly of the black gram (*Vigna mungo*) genome. Mol Ecol Resour 21:238–250

Prakash M, Elangaimannan R, Sunilkumar B, Narayanan GS (2018) Evaluation of blackgram genotypes for drought tolerance based on root dynamics and gas exchange parameters. Legum Res 41:384–391

Prasanthi L, Reddy BV, Geetha B, Jyothi R (2013) Molecular marker for screening yellow mosaic disease resistance in blackgram [*Vigna mungo* (L.) Hepper]. Electron J Plant Breed 4:1137–1141

Priyadharshini B, Vignesh M, Prakash M, Anandan R (2019) Evaluation of black gram genotypes for saline tolerance at seedling stage. Indian J Agric Res 53:83–87

Priyadharshini B, Anandan R, Manikandan S, Bharathi JK, Prakash M (2023) Evaluation of black gram (*Vigna mungo* (L.) Hepper) genotypes for coastal saline tolerance using microsatellite markers. Braz J Bot:1–23. https://doi.org/10.1007/s40415-023-00955-z

Priyanka S, Rangaiah S, Pavan R (2018) Screening black gram genotypes under artificially inoculated conditions for powdery mildew resistance. Legum Res 41:907–912

Punithavathy P, Kumaresan D, Manivannan N, Boopathi NM, Senthilraja G (2023a) Genetics of powdery mildew disease resistance in blackgram [*Vigna mungo* (L.) Hepper]. Electron J Plant Breed 14:1421–1424

Punithavathy P, Kumaresan D, Manivannan N, Boopathi NM, Senthilraja G (2023b) Variability and association analyses of powdery mildew disease scores, seed yield and component traits in blackgram [*Vigna mungo* (L.) Hepper]. Electron J Plant Breed 14:811–817

Punniyamoorthy D, Jegadeesan S (2023) Genetics and genomics of drought and heat tolerance in cowpea, mung bean and black gram. In: Legumes: physiology and molecular biology of abiotic stress tolerance. Springer Nature Singapore, Singapore, pp 203–233

Raizada A (2020) Development of genic markers for yellow mosaic virus and bruchid resistance traits in black gram *Vigna mungo* L Hepper. PhD dissertation, Homi Bhabha National Institute, Mumbai, India

Raizada A, Jegadeesan S (2023) Cross-species amplification and genetic variation among blackgram genotypes using SSR markers developed from mungbean DNA sequence scaffolds harbouring putative resistance genes. Electron J Plant Breed 14:21–30

Raizada A, Souframanien J (2019) Transcriptome sequencing, de novo assembly, characterisation of wild accession of blackgram (*Vigna mungo* var. silvestris) as a rich resource for development of molecular markers and validation of SNPs by high resolution melting (HRM) analysis. BMC Plant Biol 19:1–16

Rajalakshmi K, Manivannan N, Anand G, Vanniarajan C, Harish S (2021) AMMI and GGE biplot analysis of seed yield and batter quality traits in blackgram [*Vigna mungo* (L.) Hepper]. Legum Res 1:1–7. https://doi.org/10.18805/LR-4574

Rajasekar R, Prasad V, Djanaguiraman M, Sumathi A, Jegadeeswari D (2022) Screening Blackgram (*Vigna mungo* (L.) Hepper) genotypes for salt tolerance at germination and seedling stages. Int J Environ Clim Chang 12:383–396

Rambabu E, Anuradha C, Sridhar V, Reddy SS (2018) Genetics of mungbean yellow mosaic virus (MYMV) resistance in blackgram (*Vigna mungo* (L) Hepper). Int J Curr Microbiol App Sci 7:3869–3872

Raptan PK, Hamid A, Khaliq QA, Solaiman ARM, Ahmed JU, Karim MA (2001a) Salinity tolerance of blackgram and mungbean: I. Dry matter accumulation in different plant parts. Korean J Crop Sci 46:380–386

Raptan PK, Hamid A, Khaliq QA, Solaiman ARM, Ahmed JU, Karim MA (2001b) Salinity tolerance of blackgram and mungbean: II. Mineral ions accumulation in different plant parts. Korean J Crop Sci 46:387–394

Rather AA, Natarajan S (2021) Impact of salt stress on morphological and biochemical changes in pre flowering stage of *Vigna mungo* (L.) hepper. Int J Bot Stud 6:501–506

Roy SJ, Negrão S, Tester M (2014) Salt resistant crop plants. Curr Opin Biotechnol 26:115–124

Sai CB, Chidambaranathan P (2019) Reproductive stage drought tolerance in blackgram is associated with role of antioxidants on membrane stability. Plant Physiol Rep 24:399–409

Samal KC, Ramakrishna B, Sahoo JP, Behera L (2022) Molecular characterization and validation of micro-satellite markers linked to powdery mildew disease resistance in mini Core germplasm of Urdbean [*Vigna mungo* (L.) Hepper]. Res Biotica 4:62–68

Sen Gupta D, Basu PS, Souframanien J, Kumar J, Dhanasekar P, Gupta S, Pandiyan M, Geetha S, Shanthi P, Kumar V, Pratap Singh N (2021) Morpho-physiological traits and functional markers based molecular dissection of heat-tolerance in urdbean. Front Plant Sci 12:719381. https://doi.org/10.3389/fpls.2021.719381

Sen NK, Jana MK (1964) Genetics of black gram (*Phaseoulus mungo* L.). Genetica 34:46–57

Shanthi M, Krishna YR, Babu MR, Babu PR (2014) Screening of blackgram varieties for tolerance to saline irrigation water. Legum Res 37:439–442

Shanthi P, Iyanar K, Sassikumar D et al (2019) ADT 6-A high yielding blackgram variety suitable for rice fallow condition of cauvery Delta Zone. Electron J Plant Breed 10:1250–1254

Shanthi P, Ramesh P, Sakaravarthy KS, Vivekananthan T, Umadevi M, Sivasubramaniam K (2021a) Screening of black gram (*Vigna Mungo* L. Hepper) varieties for tolerance to salinity. Legum Res 44:911–915

Shanthi P, Ramesh P, Parameshwaran M, Sakaravarthy KS, Vivekananthan T, Umadevi M (2021b) Morphological and yield attribute of blackgram genotypes under different salinity stress conditions. Indian J Agric Res 1:1–6. https://doi.org/10.18805/IJARe.A-5697

Singh A, Khulbe RK, Panwar RK (2012) Evaluation of urdbean (*Vigna mungo*) germplasm for pre-harvest sprouting tolerance. J Food Legumes 25:183–186

Singh I, Sandhu JS, Gupta SK, Singh S (2013) Introgression of productivity and other desirable traits from ricebean (*Vigna umbellata*) into black gram (*Vigna mungo*). Plant Breed 132:401–406

Singh P, Kumar D, Sarin NB (2016) Multiple abiotic stress tolerance in *Vigna mungo* is altered by overexpression of ALDRXV4 gene via reactive carbonyl detoxification. Plant Mol Biol 91:257–273

Singh G, Gudi S, Upadhyay P, Shekhawat PK et al (2022a) Unlocking the hidden variation from wild repository for accelerating genetic gain in legumes. Front Plant Sci 13:1035878. https://doi.org/10.3389/fpls.2022.1035878

Singh L, Dhillon GS, Kaur S, Dhaliwal SK, Kaur A, Malik P, Kumar A, Gill RK, Kaur S (2022b) Genome-wide association study for yield and yield-related traits in diverse blackgram panel (*Vigna mungo* L. hepper) reveals novel putative alleles for future breeding programs. Front Genet 13:849016. https://doi.org/10.3389/fgene.2022.849016

Somta P, Chen J, Yundaeng C, Yuan X, Yimram T, Tomooka N, Chen X (2019) Development of an SNP-based high-density linkage map and QTL analysis for bruchid (*Callosobruchus maculatus* F.) resistance in black gram (*Vigna mungo* (L.) Hepper). Sci Rep 9:3930

Somta P, Chen J, Yimram T, Yundaeng C, Yuan X, Tomooka N, Chen X (2020) QTL mapping for agronomic and adaptive traits confirmed pleiotropic effect of mog gene in black gram [*Vigna mungo* (L.) Hepper]. Front Genet 11:635

Souframanien J, Dhanasekar P (2014) Advances in greengram and blackgram genomics. In: Legumes in the omic era. Springer, pp 155–184

Souframanien J, Gopalakrishna T (2006) ISSR and SCAR markers linked to the mungbean yellow mosaic virus (MYMV) resistance gene in blackgram [*Vigna mungo* (L.) Hepper]. Plant Breed 125:619–622

Souframanien J, Gopalakrishna T (2007) Source for bruchid resistance and its inheritance in Trombay wild urdbean (*Vigna mungo* var. silvestris). J Food Legumes 20:19–21

Souframanien J, Reddy KS (2015) De novo assembly, characterization of immature seed transcriptome and development of genic-SSR markers in black gram [*Vigna mungo* (L.) Hepper]. PLoS One 10:e0128748

Souframanien J, Gupta SK, Gopalakrishna T (2010) Identification of quantitative trait loci for bruchid (*Callosobruchus maculatus*) resistance in black gram [*Vigna mungo* (L.) Hepper]. Euphytica 176:349–356

Souframanien J, Gupta SK, Gopalakrishna T (2011) Breeding of black gram genotypes for resistance to bruchid beetle through molecular and conventional approaches. Indian J Genet Plant Breed 71(2):164–169

Souframanien J, Roja G, Gopalakrishna T (2014) Genetic variation in raffinose family oligosaccharides and sucrose content in blackgram [*Vigna mungo* L.(Hepper)]. J Food Legumes 27:37–41

Souframanien J, Dhole VJ, Reddy KS (2021) Breeding for low phytates and oligosaccharides in mungbean and blackgram. In: Breeding for enhanced nutrition and bio-active compounds in food legumes. Springer, pp 99–130

Spandana R, Chaurasia AK, Singh V, Pashwan VR, Dileep R (2021) Impact of salinity stress on seed quality parameters and its mitigation through seed quality enhancement in blackgram seeds (*Vigna mungo* L.). Biol Forum 13:33–37

Srivastava P, Pandey A, Singh M (2013) Identification of the RAPD marker linked to powdery mildew resistant gene (s) in black gram by using bulk segregant analysis. Res J Biotechnol 8:9–15

Subramaniyan R, Narayana M (2023) Development of bruchid pest and mungbean yellow mosaic virus disease resistance lines in blackgram [*Vigna mungo* (L.) Hepper] through marker-assisted selection. Physiol Mol. Plant Pathol 127:102105

Subramaniyan R, Narayana M, Krishnamoorthy I, Natarajan G, Gandhi K (2021) Mapping and mining of major genomic regions conferring resistance to Bruchine (*Callosobruchus maculatus*) in blackgram (*Vigna mungo*). Plant Breed 140:896–906

Subramaniyan R, Narayana M, Krishnamoorthy I, Natarajan G, Gandhi K (2022) Novel and stable QTL regions conferring resistance to MYMV disease and its inheritance in blackgram (*Vigna mungo* (L.) Hepper). J Genet 101:18

Subramaniyan R, Narayana M, Krishnamoorthy I, Natarajan G, Gandhi K (2023) Genetics of seed damage among the segregating populations of blackgram [*Vigna mungo* (L.) Hepper] against storage pest bruchine (*Callosobruchus maculatus* Fab.). Vegetos 1-6. https://doi.org/10.1007/s42535-023-00623-3

Sunanda CH, Ranganayakulu GS (2019) Effect of salt stress on plant growth and osmolyte accumulation in black gram (*Vigna mungo* L.) seedlings. Int J Life Sci 7:343–350

Suneja Y, Kaur S, Gupta AK, Kaur N (2011) Levels of nutritional constituents and antinutritional factors in black gram (*Vigna mungo* L. Hepper). Food Res Int 44:621–628

Susheelamma NS, Rao MVL (1978) Purification and characterization of the surface active proteins of black gram (*Phaseolus mungo*). Int J Pept Protein Res 12:93–102

Suthar MD, Bharpoda TM (2018) Screening of black gram varieties/genotypes against *Callosobruchus chinensis* Linnaeus. Trends Biosci 11(8):1853–1859

Swamy SG, Mahalakshmi MS, Souframanien J (2016) Evaluation of certain blackgram varieties for resistance to pulse bruchid, *Callosobruchus maculatus* (F.). J Res ANGRAU 44:8–13

Tamilzharasi M, Vanniarajan C, Karthikeyan A, Souframanien J, Meenakshisundram PMA (2020) Evaluation of urdbean (*Vigna mungo*) genotypes for mungbean yellow mosaic virus resistance through phenotypic reaction and genotypic analysis. Legum Res 43:728–734

Tamilzharasi M, Kumaresan D, Thiruvengadam V, Souframanien J, Latha TKS, Manikanda Boopathi N, Jayamani P (2023) Development and characterization of gamma ray and EMS induced mutants for powdery mildew resistance in blackgram. Int J Radiat Biol 99:1267–1284

Tian Z, Wang X, Lee R, Li Y, Specht JE, Nelson RL, McClean PE, Qiu L, Ma J (2010) Artificial selection for determinate growth habit in soybean. Proc Natl Acad Sci USA 107:8563–8568

Umadevi M, Ganesan NM, Muthiah AR (2008) Batter quality–need for traditional Indian fermented urdbean-rice food. Madras Agric J 95:440–447

Vanniarajan C, Ganeshram S, Souframanien J, Veni K, Lavanya SA, Kuralarasan V (2019) Gamma rays induced urdbbean [*Vigna mungo* (L.) Hepper] mutants with YMV resistance, good batter quality and bold seeded type. Legum Res 42:25–31

Vanniarajan C, Chandirakala R, Souframanien J (2021) Genetic and biochemical characterization of electron beam and gamma Ray induced mutants for superior battering quality trait and MYMV in Urdbean [*Vigna mungo* (L.)]. Legum Res 1:1–7. https://doi.org/10.18805/LR-4583

Vanniarajan C, Souframanien J, Lavanya SA (2022) Isolation of gamma Ray induced Urd bean [*Vigna mungo* (L.) hepper] mutants with improved batter quality. Legum Res 45:1496–1500

Veni K, Murugan E, Mini ML, Vanniarajan C, Radhamani T (2016) Genetic relationship between yield and battering quality in blackgram (*Vigna mungo* L.). Legum Res 39:355–358

Venkatakrishnan R, Raji E, Pazhaniraja J (2019) Fabrication of black gram harvesting machine. Int J Res Appl Sci Eng Technol 7:181–184

Venkidasamy B, Selvaraj D, Nile AS, Ramalingam S, Kai G, Nile SH (2019) Indian pulses: a review on nutritional, functional and biochemical properties with future perspectives. Trends Food Sci Technol 88:228–242

Verma J, Gore PG, Kumari J, Wankhede DP, Jacob SR, Thirumani Venkatesh AK, Nair RM, Tripathi K (2024) Exploring genetic diversity in black gram (*Vigna mungo* (L.) Hepper) for pre-harvest sprouting tolerance. Agron 14:197. https://doi.org/10.3390/agronomy14010197

Vishalakshi B, Umakanth B, Shanbhag AP, Ghatak A, Sathyanarayanan N, Madhav MS, Krishna GG, Yadla H (2017) RAPD assisted selection of black gram (*Vigna mungo* L. Hepper) towards the development of multiple disease resistant germplasm. 3 Biotech 7:1–6

Win KT, Oo AZ (2017) Salt-stress-induced changes in protein profiles in two blackgram (*Vigna mungo* L.) varieties differing salinity tolerance. Adv Plants Agric Res 7:206–210

Lentil (*Lens culinaris* Medikus) Breeding

Harsh Kumar Dikshit, Gyan Prakash Mishra, Muraleedhar S. Aski, Soma Gupta, Prachi S. Yadav, Suraj Mishra, Kuldeep Tripathi, and Shiv Kumar

Abstract

An essential rainfed winter season grain legume for the global diversification of cereal-based cropping systems is lentil (*Lens culinaris* Medikus). Originating in the Near East, the crop spread to various regions, establishing itself in a diverse range of agro-ecologies. More than 50 countries cultivate lentils. Lentil grains are abundant in micronutrients, prebiotic carbohydrates, protein, and vitamins. In low-income areas, lentils are a staple food that is important.

The lentil productivity is low because of low seedling vigor, high flower drop, low pod set, inadequate dry matter accumulation, and biotic and abiotic stress vulnerability tensions. Climate change-related biotic and abiotic stressors present a challenge to cultivation of lentils. Breeders now have the chance to further develop lentil varieties for increased grain yield, nutritive value, and resistance to biotic and abiotic stresses thanks to the discovery of new genes and quantitative trait loci. We go over the current obstacles and prospects for lentil development in this chapter.

Keywords

Lens culinaris · Origin · Hybridization · Breeding · Biotic and abiotic stresses · Nutritional quality

H. K. Dikshit (✉) · G. P. Mishra · M. S. Aski · S. Gupta · P. S. Yadav · S. Mishra
Division of Genetics, Indian Agricultural Research Institute, New Delhi, India

K. Tripathi
Germplasm Evaluation Division, National Bureau of Plant Genetic Resources, New Delhi, India

S. Kumar
South Asia and China Program, International Center for Agricultural Research in the Dry Areas, National Agriculture Science Complex, New Delhi, India

1 Introduction

Grown in the cool season, lentil (*Lens culinaris* subsp. *culinaris*) is a diploid crop that self-pollinates. The Old-World legume lentil was domesticated along with barley and wheat. Lentil cultivation spread from the Fertile Crescent to Central Europe, the Mediterranean region, the Nile Area, and eastward to Asia. Afterward, the lentil was brought to Australia and North America (Cubero 1984; Matny 2015). The crop covers approximately 5.01 million hectares of cultivated area worldwide, yielding 6.54 million metric tons at a productivity of 1305 kg/ha. Together, India and Canada produce almost 61.8% of the lentils produced worldwide. The average yield of lentils increased from 874.5 kg/ha to 1004.3 kg/ha in the last 20 years, indicating a significant increase in global productivity. Turkey, Nepal, India, Australia, and Canada were the top producers of lentils. Lentils are exported from Australia and Canada to South Asia.

Hidden hunger, a global crisis, especially iron and zinc malnutrition, primarily affects women, children, and newborns in low- and middle-income countries where a majority of the population depends on plant-based diets. This form of malnutrition may lead to various short- and long-term consequences, including stunted growth, heightened susceptibility to infectious diseases, and impaired development, all of which can prevent children from reaching their full potential in terms of development. The deployment of lentil varieties L4717 (Pusa Ageti) (high Zn content) and L4076 (Pusa Shivalik) (rich in Fe) and with lower phytic acid contents will address the above-mentioned health risks (Aski et al. 2023). Large ruminants and sheep have been successfully fed lentil straw because it is nutrient-rich and has a higher ruminal degradation rate than cereal straw (Mudgal et al. 2018).

Different cropping systems (rice/lentil, fallow/lentil, and pearl millet/maize/cotton/lentil) are used to grow lentils in South Asia. The Terai region, Central Highlands, Northeastern Plains, and Northwestern Plains are the main

G. Prasad Dixit et al. (eds.), *Fundamentals of Legume Breeding*, https://doi.org/10.1007/978-981-96-7838-9_7

lentil-producing regions in South Asia. The Punjab province of Pakistan, Western Uttar Pradesh, Haryana, and Punjab comprise the Northwestern Plains. Lentil varieties with small seeds, bushy growth habit, abundant branching, and a maturity period of 130–140 days are cultivated. This area makes up 10% of the entire lentil-growing area. The three main illnesses are rust (*Uromyces fabae*), *Ascochyta* blight (*Ascochyta lentis*), and vascular wilt (*Fusarium oxysporum* f.sp. *ciceri*). Bangladesh, Myanmar, Eastern Uttar Pradesh, West Bengal, and Bihar are all part of the Northeastern Plains. A total of 35–40% of the area is made up of this area. Small-seeded lentils are grown on uplands following jute, on low-lying areas (Tal area), or on rice fields that receive rain.

Rust, wilt, and *Stemphylium* blight are the main diseases. Thirty to forty percent of South Asia is made up of the Central Highlands, which include Madhya Pradesh, Uttar Pradesh, and Maharashtra. In kharif rice fallows, lentils are monocropped in this area. Preferred varieties are those with bold seeds, early maturation, long tap roots, and an upright growth habit. Rust and wilt are the main illnesses. The Himalayan foothills of India, Pakistan, and Nepal comprise the Terai region, which makes up 15% of all of South Asia. This area experiences harsh winters, a lot of humidity, and rainfall. Abiotic and biotic constraints include wilting, rusting, and frost.

2 Significance of the Crop and Achievements Made in the Last 25 Years

The variation in cultivated lentils was studied in detail by Barulina (1930), who classified lentils into two subspecies based on morphology and geographic location. The macrosperma, primarily found in the Mediterranean region and the New World, has a seed diameter ranging between 6 and 9 mm, having yellow cotyledons with little or no pigmentation. The other type, microsperma, mainly cultivated in South Asia and sub-Saharan Africa, has a seed diameter ranging from 2 to 6 mm, having red-orange or yellow cotyledons (Hawtin et al. 1980; Muehlbauer et al. 1985; Kays 1979). The indigenous South Asian microsperma type is also referred to as the pilosae type (Kumar et al. 2014). In North America, both green and red lentils are cultivated to meet international market demand. Lentils are highly nutritious grain legumes, rich in dietary protein (20.6–31.4%), vitamins, minerals, and essential amino acids. These macrosperma and microsperma were further subdivided into six narrower geographical groups. Later, based on genome-wide single-nucleotide polymorphisms, Khazaei et al. (2016) categorized lentil genotypes into three groups, viz., subtropical savannah (South Asia), Mediterranean, and northern temperate, corresponding to their geographic origin and major growing macro-environments.

Despite being grown in diverse environments, there is limited genetic diversity among South Asian and Canadian genotypes due to deliberate exclusion of unadapted lentil germplasm from different environments into the breeding programs in these regions (Khazaei et al. 2016). Further, adaptation requirements can result in founder effects, leading to genetic bottlenecks. The introduction of lentil into South Asia may have involved the introgression with wild lentil (*Lens orientalis*) containing recessive alleles for earliness, which were cyclically recombined and selected over time (Erskine et al. 2011). Enhancing genetic diversity is crucial for sustaining yield gains and is a key goal for many plant breeding programs. Therefore, understanding the adaptation constraints of diverse lentil genotypes in different environments is essential to help breeders expand genetic diversity through introducing exotic germplasm.

3 Origin, Evolution, and Distribution of *Lens* Species and Its Forms

Lentil was domesticated in the Near East "the cradle of agriculture." Lentil is among the oldest domesticated crops. Lentil (*Lens culinaris* Medikus) evolved from *Lens culinaris* ssp. *orientalis* (Boiss.). The oldest remains of lentil dated 11,000 BC were found from Greece's Franchthi caves (Helbeck 1963). Hansen and Renfrew (1978) reported that carbonized small-seeded lentil grains dating back to 8500–7500 BC were found in Tell Mureybit, Syria. The first direct evidence of domestication of lentil is from Yiftah-el. The large-seeded lentils found were dated back to 6800 BC (Garfinkel et al. 1988). Domesticated seeds larger than wild seeds found from Tepe Sabz were dated to 5500–5000 BC (Helbeck 1959). Wild species are distributed from western Turkey to Kurdistan. Lentil was initially cultivated in Egypt, Western Asia, and Southern Europe. From these regions, lentils spread Eastward to India and Northward to Europe and Southward to Ethiopia. Lentils later spread to many countries in high-altitude tropical, subtropical, and warm temperate regions. Lentil spread to the Indo-Gangetic Plain dates to around 2000 BC (Cubero 1981). The geographic distribution of variation in cultivated lentil was described by Barulina (1930). She characterized lentils based on seed size in microsperma and macrosperma types. West Asian lentils have a flat lens-like appearance, and South Asian lentils are convex in shape from both sides.

4 Cytogenetics

The *Lens* species is diploid and self-pollinated having chromosome number $2n = 2x = 14$. Several workers have studied the karyotype in cultivated lentil. The total chromosome length was reported as 39.31 μm Gupta and Singh (1981) in PL 639 and 31.77 μm by Dixit and Dubey (1986) for Type 36. Total chromatin length (53.6–121.2 μm) was reported by Jha et al. (2015a, b). Lowest chromatin length of 16.9 μm was reported by Lavania and Lavania (1983). Sindhu et al. (1983) reported three pairs of acrocentric, two metacentric, and two submetacentric chromosomes. Ladizinsky and Sakar (1982) reported that all species of genus *Lens* have similar karyotype consisting of three metacentric chromosomes, three acrocentric chromosomes, and one metacentric chromosome with a satellite. More specific details about each chromosome's karyotypes can be explored in Aski et al. (2024).

5 Plant Genetic Resources

The narrow genetic base of released cultivars results in low yield and susceptibility to biotic and abiotic stresses. Introgression of desirable genes between macrosperma and microsperma, utilization of Mediterranean and European land races, and utilization of closely related *L. orientalis* ssp. *orientalis* are suggested for broadening the genetic base. Special efforts are required to screen *Lens* germplasm in national and international gene banks to know their potential for utilization in breeding programs. Donors for yield contributing traits like number of pods developed in a plant, number of seeds produced in a pod, number of clusters formed per plant, harvest index, biological yield, early maturity, and better yield quality traits have to be identified and utilized in breeding programs. The lentil germplasm holding of different gene banks is presented in Table 1.

Large number of lentil germplasm lines (2324 cultivated lentil germplasm accessions comprising 1796 indigenous collections and 528 exotic collections) were characterized by Tripathi et al. (2022) during the winter season of 2017–2018 and 2018–2019 (Table 2).

Table 1 The lentil germplasm accession collections of different gene banks worldwide

Country	Gene bank/center	No. of accessions
Australia	Australia Temperate Field Crops Collection	5254
Bangladesh	BARI	798
Canada	PGRC	1139
Chile	Inst de Inv. Agropecuaries, Centro Regional de Investigation Carillanca	1345
China	Institute of Crop Germplasm Resources	855
Egypt	NGB	875
Ethiopia	Biodiversity Conservation and Research Institute	678
Hungary	Research Center for Agrobotany	1061
India	ICAR-NBPGR, New Delhi	2285
Iran	Seed and Plant Improvement Institute	3000
Pakistan	Plant Genetic Resources Institute	805
Morocco	ICARDA	13,958
Russian Federation	N.I. Vavilov All-Russian Scientific Research Institute of Plant Industry	2556
Spain	Centro de Recuros Fitogenetico, INIA	703
Syria	General Commission for Scientific Agricultural Research	1072
Turkey	Plant Genetic Resources Department, Aegean Agricultural Research Inst.	1095
USA	USDA	2875

Source: Dikshit et al. (2022), Sen Gupta et al. (2023)

6 Trait-Specific Germplasm

Accessions IC241532 and IC241529 exhibited very early flowering (51 days). EC267615 displayed erect plant architecture with pod set 15 cm above the ground, making it suitable for mechanical harvesting. The 100 seed weight of 7.1–7.8 g was recorded for EC499760. Stem fasciation was recorded for accession IC 241473. In addition to this, upto 16 flowers per peduncle (FPP) at multiflowering nodes have been recorded in IC 241473. Mishra et al. (2022) reported two novel multi-flowering (MF) genotypes, PMF-1 and PMF-2, forming up to five flowers per peduncle (FPP) identified from the ICARDA nursery received from Lebanon.

IC317520 was identified with extended funiculus.

EC267615 displayed erect plant architecture

Table 2 Descriptive and analyzed statistics of 2324 lentil germplasm accessions

Traits	Minimum	Maximum	Mean	Std. error	CV(%)	Skewness	Kurtosis	F statistics (treatment)	F statistics (year)
ST	1.60	2.70	2.42	0.01	6.96	1.671	2.892	1.16[a]	32.18
DFF	51.00	123.00	88.96	0.188	10.17	−1.148	1.582	5.58[a]	3928.99[a]
SD	2.73	6.80	4.18	0.02	17.41	1.734	3.276	1.08[a]	63.89
DM	93.00	140.00	120.88	0.083	3.28	−1.193	7.152	2.07[a]	3842.46[a]
PH	9.70	52.13	35.34	0.100	13.64	−0.516	0.805	2.00[a]	33.10[a]
SP	1.00	2.00	1.97	0.003	6.38	−5.094	32.724	1.13[a]	33.31
PP	20.83	571.33	206.38	1.483	34.56	0.078	0.434	1.85[a]	649.78[a]
HSW	0.75	7.6	1.81	0.011	28.50	1.811	4.521	4.90[a]	698.04[a]
SBP	6.00	51.50	28.67	0.163	27.33	−0.300	0.029	2.44[a]	3152.56[a]

ST seed thickness, *SD* seed diameter, *DFF* days to 50% flowering, *SBP* secondary branches per plant, *PH* plant height, *PP* pods per plant, *SP* seeds per pod, *DM* days to 80% maturity, *HSW* 100-seed weight

[a] Significant at 1% level

Above 50 secondary branches were recorded for IC199461 and IC201716. These accessions also exhibited above 400 pods per plant.

From the evaluation data of 2324 accessions, core set of 170 accessions was developed. This core set possessed allelic richness and diversity for major yield contributing traits. The donors identified by different researchers for different agronomic traits are presented in Table 5. The genus *Lens* is mainly distributed in the Mediterranean region. Within and between species diversity exists. Zohary et al. (2012) reported the occurrence of *L. orientalis* in Southwest Asia, Cyprus, and Central Asia. *L. ervoides* is present in Israel, Syria, the Adriatic Coast, Turkey, Spain, Southern Italy, and Algeria. *Lens ervoides* is reported from Uganda and Ethiopia. *Lens nigricans* is widely spread in the Northern Mediterranean region from Israel to Spain, Morocco, the Crimean Peninsula, Georgia, and Algeria. *L. lamottei* is reported from Morocco (Davies et al. 2007; Coyne and McGee 2013). Different wild species exist in southeastern Turkey, Palestine, Syria, Israel, and Jordan, the region for domestication of lentil (Coyne and McGee 2013). Ferguson et al. (2000) identified seven taxa and grouped them into four species:

(i) *L. culinaris* ssp. *culinaris*, *L. culinaris* ssp. *orientalis*, *L. culinaris* ssp. *tomentosus*, *L. culinaris* ssp. *odemensis*
(ii) *L. ervoides*
(iii) *L. lamottei*
(iv) *L. nigricans*

L. culinaris ssp. *orientalis* is the closest wild progenitor of *L. culinaris* ssp. *culinaris*, while L. *nigricans* is the most distant relative. As per Wong et al. (2015), four species have been classified within the genus *Lens*: *L. culinaris* (consisting of four subspecies as subsp. *culinaris*, subsp. *orientalis*, subsp. *odemensis*, and subsp. *tomentosus*), *Lens lamottei*, *Lens ervoides*, and *Lens nigricans*. *L. culinaris* ssp. *culinaris*, spp. *Odemensis*, and spp. *orientalis* constitute the primary gene pool as they are readily inter-crossed producing fertile progenies, whereas *L. nigricans* spp. *nigricans* and ssp. *ervoides* require embryo rescue for obtaining hybrids of cultigen (Ladizinsky et al. 1985) (Table 3).

Several studies have indicated compatibility among species belonging to primary and secondary gene pools, and

Table 3 Gene pool concept in lentils

Primary gene pool	Secondary gene pool	Tertiary gene pool	Quaternary gene pool	References
L. culinaris, L. orientalis	*L. odemensis*	*L. nigricans, L. ervoides*		Cubero et al. (2009)
	L. nigricans, L. ervoides Using embryo rescue			Tullu et al. (2013a, b)
	L. odemensis	*L. ervoides* and *L. nigricans*	–	Fratini and Ruiz (2006)
	L. odemensis	*L. nigricans* and *L. ervoides*	–	Ladizinsky (1993)
L. culinaris, L. orientalis, L. tomentosus	*L. lamottei, L. odemensis*	*L. ervoides*	*L. nigricans*	Wong et al. (2015)

Table 4 Key features of different species of genus *Lens*

Character	*culinaris*	*orientalis*	*odemensis*	*tomentosus*	*nigricans*	*lamottei*	*ervoides*
Toothed stipules	No	No	Slightly	No	Yes	Slightly	No
Rachis length (mm)	20–50	5–25	8–20	7–20	8–25	7–10	5–15
Leaflets per leaf	10–16	6–8	6–8	6–12	6–8	6–10	4–6
Aristate peduncle or awn rarely	Yes	Yes	Yes	Yes	Yes	Yes	Yes
Calyx teeth/corolla ratio	≠1	< 1	=	> 1	≥1	> 1	< 1
Peduncle/rachis ratio	≠1	> 1	> 1	> 1	> 1	> 1	> 1
Pod pubescent	Rarely	Rarely	No	No	No	No	Yes
Seed diameter (mm)	4–9	2.5–3.5	2.5–3.5	2.5–3.5	2.5–3.5	2.5–3.5	2–3

Cubero (1981), Ladizinsky (1987, 1993, 1997), Ferguson et al. (2000)

these genotypes can be crossed following conventional breeding methods, with or without the application of growth hormones or depending on tissue culture techniques. However, pre- and post-fertilization barriers impose difficulty in successful introgression of genes of interest, particularly from the tertiary and quaternary gene pools, into cultivated lentil.

Singh et al. (2020) evaluated 96 *Lens* accessions including *L. orientalis* (24), *L. tomentosus* (8), *L. odemensis* (16), *L. ervoides* (24), *L. nigricans* (17), *L. lamottei* (5), and *L. culinaris* (2). Principal component analysis exhibited that number of pods formed per plant, seed yield per plant, biological yield, and harvest index significantly contributed to the genetic variation *Lens* species. Wild accessions from Turkey and Syria exhibited resistance against more than one disease. The key features of different species of genus *Lens* are presented in Table 4. From two interspecific crosses, ILL8006 (*L. culinaris* ssp. *culinaris*) × ILWL62 (*L. culinaris* ssp. *orientalis*) and ILL10829 (*L. culinaris* ssp. *culinaris*) × ILWL30 (*L. ervoides*), Singh et al. (2022) reported the development of two mungbean cultivars Jammu Lentil 144 and Jammu Lentil 71. The developed cultivars exhibited a yield advantage of around 10%, earliness of 25–15 days, and resistance to root rot, fusarium wilt, aphid infestation, and pod borer resistance. The study highlights the role of interspecific hybridization for improvement of important traits (Table 5).

7 Floral Biology: Emasculation—Pollination Technique

Lentil flowers are bisexual, cleistogamous, pedicellate, and zygomorphic. Flowers exhibit an acropetal flowering pattern. Inflorescence is an axillary raceme bearing 1–16 flowers/peduncle (Tripathi et al. 2022). Small flowers (4–9 mm in size) are pink, white, purplish blue, pale blue, or purplish blue in color. A typical *Papilionaceous* flower consists of five petals (standard-outermost petal, two lateral wing petals, and keel formed by fusion of two petals enclosing stamens and carpel), five unequal sepals, androecium, and gynoecium. The calyx tube has five narrow, pointed lobes at the base, which are bent over and beyond the corolla. Less than 0.8% out crossing is reported in lentil (Wilson and Law 1972).

Hybridization involves emasculation and pollination. Emasculation followed by pollination produces highest seed set. Temperature in range of 20–25° centigrade and relative humidity of 50% along with 12–15-h light are necessary for proper seed set. For large-scale hybridization program, emasculation is carried out in the evening. Buds with 75% length of calyx lobes are identified for emasculation and pollination. One or two calyx lobes adjacent to the suture side of keel are removed using pointed steel forceps. Using forceps keel is split carefully and the stamens (9 + 1 diadelphous) are removed. Pollen is collected from young flowers after anther

Table 5 Source of resistance for biotic and abiotic stress resistance identified in wild lentil species

Specific traits	Different species of lentil	References
Biotic stress		
Rust resistance	*L. orientalis*	Gupta and Sharma (2006), Kumar et al. (2014), Singh et al. (2014, 2020), Kumari et al. (2018) and Tripathi et al. (2024)
	L. odemensis, L. ervoides, L. nigricans	Gupta and Sharma (2006), Kumar et al. (2014) and Singh et al. (2014, 2020)
	L. tomentosus	Gupta and Sharma (2006), Singh et al. (2014, 2020)
Powdery mildew	*L. orientalis, L. ervoides, L. nigricans*	Gupta and Sharma (2006), Singh et al. (2014, 2020), Coyne et al. (2020) and Tripathi et al. (2024)
	L. odemensis. L. tomentosus, L. lamottei	Singh et al. (2014, 2020), Coyne et al. (2020)
Fusarium wilt	*L. orientalis, L. ervoides*	Bayaa et al. (1995), Gupta and Sharma (2006) and Singh et al. (2014, 2020)
	L. odemensis, L. nigricans	Gupta and Sharma (2006), Singh et al. (2014, 2020)
	L. tomentosus, L. lamottei	Singh et al. (2014, 2020)
Anthracnose	*Lens ervoides, L. lamottei, L. nigricans*	Tullu et al. (2006a, b), Fiala et al. (2009), Vail et al. (2012), Singh et al. (2014), Bhadauria et al. (2017) and Coyne et al. (2020)
	L. culinaris ssp. culinaris	Buchwaldt et al. (2004), Shaikh et al. (2013)
Fusarium wilt	*L. orientalis, L. odemensis, L. ervoides, L. nigricans*	Bayaa et al. (1995), Gupta and Sharma (2006) and Singh et al. (2020)
	L. tomentosus, L. lamottei	Singh et al. (2020)
Stemphylium blight	*Lens orientalis, Lens lamottei, Lens odomensis, L. nigricans*	Podder et al. (2013), Bhadauria et al. (2017) and Rajpal et al. (2023)
	Lens ervoides	Podder et al. (2013), Bhadauria et al. (2017), Cao et al. (2019) and Rajpal et al. (2023)
Ascochyta blight	*L. orientalis*	Bayaa et al. (1994), Ye et al. (2000), Gupta and Sharma (2006), Tullu et al. (2006a, b), Tullu and Banniza (2010), Dadu et al. (2017, 2021)
	L. odemensis, L. nigricans, L. ervoides, L. lamottei	Bayaa et al. (1994), Gupta and Sharma (2006), Tullu et al. (2006a, b), Tullu and Banniza (2010) and Dadu et al. (2017)
	L. montbretti	Bayaa et al. (1994), Tullu et al. (2006a, b), Tullu and Banniza (2010) and Gupta and Sharma (2006)
Combined resistance to AB and FW	ILWL 138	Bayaa et al. (1995), Tullu et al. (2006a, b)
Combined resistance to AB and anthracnose	IG 72653, IG 72646, IG 72651	Bayaa et al. (1995), Tullu et al. (2006a, b)
Resistance to *Orobanche*	*Lens ervoides, L. odemensis, L. orientalis, Lens tomentosus*	Fernández-Aparicio et al. (2009)
Insect resistances		
Resistant to *Sitona* weevil	*L. orientalis, L. odemensis, L. ervoides, L. nigricans*	El-Bouhssini et al. (2008)
Resistant to bruchids	*L. orientalis, L. lamottei, L. nigricans*	Laserna-Ruiz et al. (2012), Gore et al. (2016)
Abiotic stresses		
Drought tolerance	*L. odemensis*	Hamdi and Erskine (1996), Gupta and Sharma (2006) and Omar et al. (2019)
	L ervoides, L. nigricans	Hamdi and Erskine (1996), Gupta and Sharma (2006) and Gorim and Vandenberg (2017)
	L. tomentosus	Gorim and Vandenberg (2017), Omar et al. (2019)
Cold tolerance and earliness	*L. orientalis, L. nigricans*	Hamdi et al. (1996)
Heat tolerance	*L. culinaris* ssp. *culinaris*	Kumar et al. (2015), Kumari et al. (2017) and Singh et al. (2019)
	L. nigricans	Hamdi and Erskine (1996), Gorim and Vandenberg (2017)
Salinity tolerance	*L. orientalis*	Singh et al. (2017)
Alkalinity tolerance	*L. culinaris* ssp. *culinaris*	Singh et al. (2022a, b)
Agronomic traits		
Seed size	*L. ervoides*	Fiala et al. (2009), Tullu et al. (2011)
Pods per plant	*L. nigricans*	Singh et al. (2014)

(continued)

Table 5 (continued)

Specific traits	Different species of lentil	References
High pod and shorter internode	*L. orientalis, L. odemensis, L. ervoides*	Singh et al. (2020)
Interspecific derivatives		
Early maturity, high yield, more pods developed per plant, substantial resistant to fusarium wilt and root rot. Jammu Lentil 71 also exhibited resistance to pest (pod borer and aphid) infestations	From two interspecific crosses ILL8006 (*L. culinaris* ssp. *culinaris*) × ILWL62 (*L. culinaris* ssp. *orientalis*) and ILL10829 (*L. culinaris* ssp. *culinaris*) × ILWL30 (*L. ervoides*) two lentil cultivars Jammu Lentil 144 and Jammu Lentil 71. These were early in maturity, high yielding, and resistant to major diseases	Singh et al. (2022a, b)

dehiscence in male parent. The emasculated buds are pollinated using golden yellow pollen in forenoon. Interspecific hybridization is being utilized for broadening the genetic base of cultivated lentil. Hybridization of cultivated lentil with secondary gene pool species causes shriveled embryos and unviable seeds. This can be overcome by embryo rescue.

8 Genetic Analysis of Qualitative and Quantitative Traits

During the last three decades, interest in lentil research increased due to international trade of the crop and focused effort of national programs and the International Centre for Agricultural Research in the dry areas. Tables 6 and 10 highlight the role of genetics in trait improvement.

9 Breeding Objectives in Lentil

9.1 Breeding for Yield Contributing Traits

9.1.1 Broadening of Genetic Base of Released Cultivars: Pre-breeding

The narrow genetic base of released lentil cultivars is the cause of yield stagnation due to susceptibility to biotic and abiotic stresses. To broaden the genetic base of lentils, ICARDA launched a pre-breeding project in India. Land races from the Mediterranean and Europe, as well as 534 lens species accessions were evaluated and used in breeding programs. In Table 7, the key breeding traits for major lentil-growing regions in the world are presented. Table 8 lists the key donors who were identified for various traits.

9.2 Breeding for Early Maturity

Earliness is required to avoid terminal moisture stress in both Central and Eastern India, the main lentil growing regions. In Central India, only three short-duration varieties, L 4717, L 4727, and L 4729, are cultivable. Eastern India requires short-duration varieties resistant to *Stemphylium* blight for cultivation in rice fallows. In Bangladesh, BARI M 4, BARI M 5, BARI M 6, BARI M 7, BARI M 8, and BARI M 9 have been released and sold. Sarker et al. (1998) reported early maturity under the control of a monogenic recessive gene. Tullu et al. (2008) reported QTLs that control lentil flowering duration. According to Shivaprasad et al. (2024a) earliness is reported to be recessive, with the non-functional LcELF3a gene playing a major role in its transmission. Two *InDel* markers, viz., I-SP-356.6 and I-SP-383.9 (chr. 3; linked to LcELF3a gene), have been reported to distinguish early and late genotypes (Shivaprasad et al. 2024b).

9.3 Breeding for Small and Bold Seed Size

Different regions have different preferences for seed size among farmers and consumers. Bold seed sizes (110 seed weight above 3 gm) are preferred in Central India, while small-seeded varieties (100 seed weight up to 2.5 gm) are preferred in Eastern and Northern India. According to reports (Abbo et al. 1991; Fedoruk et al. 2013; Verma et al. 2015), seed size is quantitatively inherited. Transcriptomic analysis, using large-seeded (L4602) and small-seeded (L830) genotypes, has been employed to investigate the pathways regulating seed size in lentils. Various differentially expressed genes have been deciphered, which were associated mainly with the hormone signaling and cell division pathways, tran-

Table 6 The genetics of agro-morphological traits in lentil

Sl. no.	Trait	Genetics/gene (s) identified	References
1	Growth habit	Incomplete dominance of bushy/erect growth habit Recessive gene *ert* was identified for erect growth habit	Ladizinsky (1979) Emami and Sharma (1999)
2	Foliage color	Green stem color *Gs* Brown leaf *Bl* Green pod *Grp* Pods with violet stripes *Pdp* Green foliage *Gl*	Ladizinsky (1979) Emami and Sharma (1996a) Vandenberg and Slinkard (1989) Havey and Muehlnauer (1989) Kumar (2002)
3	Leaflet shape, size and number	Leaf shape: Oval is dominant over acute *Ol* Leaflet size *Blf* Stipule size *Lst* Pod size *Lpd*	Kumar et al. (2004)
4	Plant pubescence	Pubescence of plant *Pub*	Kumar (2002)
5	Hypocotyl color	Hypocotyl color controlled by a single gene Green hypocotyls color is recessive to pigmentation	Verma and Nakhtore (1976)
6	Flower color	Voilet *V* and pink *P* White W Monogenic dominance of violet flowers over white	Lal and Srivastava (1975) Wilson and Hudson (1978) Kumar et al. (2005b)
7	Tendrils	Tendril formation *Tnl*	Vandenberg and Slinkard (1989)
8	Pod character	Pod size: Monogenic inheritance with incomplete dominance Pod length *Lpd*	Kumar et al. (2005a)
		Pod dehiscence*:* Dehiscent types dominate over indehiscent *Pi* Quantitative variation for pod dehiscence	Ladizinsky (1979, 1985)
9	Plant height	Plant height *Ph* Quantitative inheritance	Tahir and Muehlbauer (1994) Haddad et al. (1982)
11	Stem Fasciation	Recessive	Sharma and Sharma (1977)
12	Flowering time	Monogenic recessive inheritance *Sn*	Sarker et al. (1998, 1999)
13	Pod size	Monogenic inheritance with incomplete dominance	Kumar et al. (2005a)
14	Seed coat color	Black testa color *Blt* Gray testa color *GgC* Brown testa color *Blt* (Tgc + Ggc)	Vandenberg and Slinkard (1990)

(continued)

Table 6 (continued)

Sl. no.	Trait	Genetics/gene (s) identified	References
15	Tanin on seed coat	Tanin on seed coat monogenic dominant *Tan*	Vallincourt et al. (1986)
16	Seed coat spotting	Presence of spotting on testa is dominant over spotless *Scp*	Emami and Sharma (1996a, b)
17	Seed size	Quantitative trait	Abbo et al. (1991)
18	Seed hardness	Hard seed dominant (*L. ervoides*) and monogenically inherited *Hsc* Hard seed coat recessive (*L. orientalis)*	Ladizinsky et al. (1985)
19	Cotyledon color	Orange cotyledons double-dominant YYBB Green cotyledons double-recessive yybb	Emami and Sharma (1996a, b)

Table 7 Key breeding traits for major lentil growing regions in world

Region	Important breeding traits
South Asia and E. Africa	
1. India, Nepal, Pakistan Ethiopia 2. Bangladesh	1. Early maturity, seed yield, resistance to wilt, rust, and *Stemphylium* blight 2. Extra early maturity, seed yield, resistance to rust, and *Stemphylium* blight
Mediterranean low to medium elevation	
1. Egypt 2. Morocco 3. Below 300 mm rainfall 4. 300–400 mm rainfall	1. Early maturity, seed yield, wilt resistance, response to irrigation 2. Biomass, rust resistance, mechanical harvesting, and rust resistance 3. Earliness, high biomass 4. Biomass, wilt resistance, and mechanical harvest
Mediterranean high elevation	
1. North Africa 2. Anatolian	1. Cold tolerance and biomass 2. Moderate tolerance to cold and seed yield

scription factors, kinases, etc. (Dutta et al. 2022). Small seed size trait is very strongly governed by the locus near PBLAC449 marker, which was found to be located on chromosome 03 (Dutta et al. 2023).

9.4 Breeding for Herbicide Tolerance

Lentils grow slowly at first, and weeds pose a significant threat to their crop. It has been reported that non-parasitic weeds can lose up to 40% of their yield (Tepe et al. 2005). Using manual laborers to remove weeds is very expensive, and finding workers during prime time is a major problem. Herbicide tolerance is an alternative tactic to solve this issue. Attempts are underway in India to find herbicide-tolerant germplasm. In addition to identifying four SNP markers

Table 8 Donors for important traits identified from ICARDA germplasm

Trait	Donors identified	Trait	Donors identified
Drought tolerance	ILL 10040, ILL 10044	Pods/cluster	ILWL 15
Heat tolerance	ILL 10174, FLIP 2009-55, IG 2507, IG 4258, IG 2503, IG3101, IG 3563, IG 4079, IG4098, IG 4312, IG 4380, IG 1539, LC 279–1296, LC 270–804, LC 284–1206	Short internode	ILWL 17
Salinity resistance	ILWL 297, ILWL 368, ILWL371, ILWL 417, IG 136670	Higher primary branches	ILWL 117, ILWL 398
Fusarium Wilt resistance	ILWL 76, ILWL 79, ILWL37, ILWL 1, ILWL 138, ILL3492, ILL 4385, ILL 10956, ILL 6449, ILL 4649, ILL3796, ILL 2502, ILL 2505, ILL 5714	High biomass	IL 10732, ILL 10733, ILL 10734, ILWL 30
Rust resistance	ILWL 15, ILWL 17, ILWL19, ILWL 30, ILWL 62, ILWL81, ILWL 97	Large seed size	ILL 1005, ILL 4559, ILL 4637, ILL 4649, ILL 5488, ILL 5628, ILL 8072, ILL 8595, ILL 7946, ILL 7678, ILL 7943, ILL 9890, ILL 10734, ILL 10736, ILL 10741, IPLS-09-17, IPSL-09-23, IPLS-09-33, IPLS-09-35, ILL 45113, ILL 45101
Powdery mildew resistance	ILWL 55, ILWL 59, ILWL334, ILWL 359, ILWL 450	High grain Fe and Zn concentration	ILWL 74, ILL 8095, ILL 7981, ILL 10854, ILL 10072, ILL 88527, IG 135395, IG 135403
Root-knot nematode resistance	ILWL 18, ILWL 72	Early maturity	ILL 6002, ILL 7663, ILWL 118, ILL 2585, ILL 4605, ILL10848, LIRL-21-50-1-1-1-0, LIRL-22-46-1-1-10, LIRL-22-46-1-1-1-0, LIRL-22-46, ILL45101, ILL 45107, IPLS-09-17, IPLS-09-35

linked to traits related to imazethapyr and metribuzin tolerance, Balech et al. (2024) identified four accessions (IG323, IG5722, IG4400, IG4605) as metribuzin and imazethapyr tolerant. Canada and Australia have developed cultivars resistant to herbicides.

9.5 Breeding for Mechanized Harvesting

In order to reduce cultivation costs, mechanized harvesting is necessary. Lack of labor is a major problem during harvesting. If the crop is not harvested in a timely manner, it will shatter. Plants that grow upright are good candidates for mechanical harvesting. A minimum of 15 cm of ground clearance is needed for the combine harvester. ICARDA has worked with national programs to develop suitable varieties.

9.6 Breeding for Nutrient Use Efficiency

On marginal soils with little irrigation capacity, lentils are grown. Deficits in iron, boron, and phosphorus have been reported from Ethiopia, India, and Syria. High pH soils can cause iron deficiency. A total of 25% and 47% of the yield from India and Syria, respectively, have been reported (Ali et al. 2000; Erskine et al. 1993). Effective varieties that use nutrients efficiently will reduce cultivation costs and safeguard the environment.

10 Breeding for Biotic Stresses

The major biotic stresses of lentil include fusarium wilt, stem rot, collar rot, root rot, rust, powdery mildew, *Ascochyta* blight, *Stemphylium* blight, nematodes, black aphids, bruchids, and pea pod borers.

Wilt Fusarium wilt of lentils is caused by the soilborne fungus *Fusarium oxysporum* f. sp. lentis. Every region in the world that grows lentils has reported disease, with the exception of Australia. A 25% humidity level and a temperature range of 17–31 °C are ideal for the growth of disease. Seeds with infection show browning in the micropylar area. *Fusarium oxysporum f.* sp. *lentis* (Fol) has been recognized as a causal agent for major seed and soil-borne disease of lentils worldwide which significantly affects yield and deteriorates seed quality (Jadhav et al. 2023). The plant displays drooping of the crown, yellowing and curled leaves, and decreased plant growth. There is xylem browning. Chlamydospores, microconidia, and macroconidia are produced by the pathogen. The mycelium is septate, hyaline, and branching. The pathogen persists on the seed surface as dormant mycelium and chlamydospores. A 1995 infector technique was proposed by Bayaa et al. (1995) to screen against fusarium wilt. While Pusa 3, Pusa 4, JL 80, UPL 175, Pant 538, Pant 209, Pant 220, and Pant 234 showed significant resistance, LWS 14 and LWS 21 were found to be immune against all five strains of *Fusarium* (Kannaiyan and Nene 1975; Kannainyan and Nene 1976). By screening 2000 germplasm lines in pots, Khare et al. (1979) reported 11 resistant lines. For the inheritance of wilt resistance, a monogenic dominant gene has been reported by Kamboj et al. (1990) and Abbas (1995). Hamwieh et al. (2005) and Eujayl et al. (1998) reported molecular markers associated with wilt resistance. Linkage analysis revealed two SSR markers, PBALC203 and PBALC1409, which flanked the wilt resistance gene at 8.2 and 9.4 cM distances, respectively. PBLAC233 was found on the same linkage group at a distance of 10.2 cM from PBLAC1409. They employed BSA approach to identify linked markers in F_2 and $F_{2:3}$ populations derived from the L9–12 × ILL10965 cross. Jadhav et al. (2022) developed SCAR marker for quick detection of *Fusarium oxysporum f.* sp. *lentis* in seed during seed certification or seed health testing for long-term conservation of lentil. Fungicides can be used to control fusarium wilt, as can cultivating wilt-resistant varieties (L 4727 and L 4729) (Table 9).

Stem Rot *Sclerotinia sclerotiorum* (Lib.) de Bary is the cause of stem rot. In humid conditions, fungus cottony growth and brown sclerotia are observed on impacted plants. In terms of field symptoms, plants first turn yellow with roots and brown at the base before flowering; rotten plants then show a dry stem and eventually die. The characteristic white mycelium and the resting bodies (sclerotia) were seen on the diseased tissue near the base of the stem. As per our in vitro pathogenicity investigations, the fungus initially forms a distinctive fluffy white mycelium on the infected plant tissues. This mycelium subsequently develops into large black sclerotia, which are the most conspicuous indication of *S. sclerotiorum* infection in plants. The *S. sclerotiorum* isolates were found to be susceptible to thiophanate-methyl and triazole fungicides when it came to fungicide evaluation. The fungicides triazole and thiophanate-methyl were found to be the most successful in suppressing the disease once it broke free of the mycelium (Vagelas 2014).

Collar Rot The causal organism responsible for lentil collar rot is *Sclerotium rolfsii*. Rotting and a yellowish brown coloration are visible on the affected plants' collar regions. Disease development is aided by high levels of organic matter, sunshine, moisture, and temperature. Compared to mature plants, seedlings are more prone to collar rot. Khare et al. (1979), Kannainyan and Nene (1976), and Mohammad and Kumar (1986) have all reported on resistant sources.

Root Rot Root rot occurs due to *Rhizoctonia bataticola* and *Rhizoctonia solani*. Foliage color turns yellow. *R. bataticola* roots exhibit ashy brown mycelium, whereas *R. solani* exhibit yellowish brown mycelium. High temperature favors disease development. The RNA-seq analysis deciphered various differentially expressed genes to be associated with changes in phenolic compounds, transcription factors, antioxidants, receptor kinases, and hormone signaling. These DEGs corresponded to the cell wall modification enzymes, defense-related metabolites, and jasmonic acid (JA)/ethylene (ET) pathways (Mishra et al. 2022).

Rust Rust caused by *Uromyces viciae-fabae* (Pers.) Schroet is a foliar disease. Rust occurs globally and can cause complete crop loss. Rust is an autoecious fungus. Fungus produces aceia, uredospores, and teliospores. The initial disease symptoms are seen on lower surface of leaflets (aecial cups and yellow white pycnidia). Brown uredial pustules are later seen on all plant parts. During maturity dark brown telia are seen on branches and stem. Plants do shed leaves and appear dark brown to blackish. High humidity, cloudy weather, and temperature range of 17–22 degrees centigrade favor disease development. Screening for rust at hot spots can be carried out using infector row technique. The resistant donors for rust resistance have been reported by Nene et al. (1975), Agrawal et al. (1976), Pandey (1981), and Khare and Agrawal (1978). Single dominant gene for rust resistance is reported by Sinha and Yadav (1989) and Singh and Singh (1990). Inheritance of rust resistance in lentil was investigated in the F_1, F_2, and $F_{2:3}$ generations developed by crossing susceptible variety PL 8 with resistant variety L 4149. Genetic analysis revealed rust resistance to be have monogenic dominant inheritance. The F2 population were tagged and mapped using SSR and SRAP markers. A total of 101 SRAP and 33 SSRs were discovered to be polymorp*h*ic between parents. Of these, two SRAP and two SSR markers distinguished between resistant and susceptible bulks. SSR marker GLLC 527 is estimated to be 5.9 cM from the rust-resistant locus and can be used in marker-assisted selection for rust resistance (Dikshit et al. 2016). The lack of continuous variation in the frequency distribution of average disease severity in the RILs implies the involvement of a major gene controlling resistance. On a 1 to 9 scale, the susceptible parent, L-9–12, scored 9 at Gurdaspur, while the resistant parent, FLIP-2004-7 L, scored 3. At Langroya, the susceptible parent scored 7, whereas the resistant parent had a score of 1. The uniform spread of the inoculum in the field was indicated by the Coefficient of Variation for disease score of susceptible check at Gurdaspur and Langroya, which were 6.69% and 5.93%, respectively. Several phenotypic classes were assigned to the RILs based on the average score of both locations. RILs scoring less than 3 were placed in the resistant class, and those scoring more than 7 were

Table 9 QTLs and genes identified for agro-morpho-quality biotic and abiotic stress traits in lentil

Tait	Mapping population	Marker used	Gene/QTL/MTA	Method employed	Distance (cM)	Average marker density (cM)	PV(%) explained	References
Mn concentration	120 RILs	5385 SNP markers	6 QTL	CIM	973.1 cM	0.18 cM	15.3–24.1%	Ates et al. (2018)
Seed weight and seed size	126 F_8-RILs	Genomic SSR markers	Single QTL for seed weight and seed size (*qSW* and *qSS*) co-localized on LG4	Single marker analysis followed by composite interval mapping	–	5.48 cM	48.4% and 27.5%	Verma et al. (2015)
Agronomic traits	Diverse panel of 96 accessions	SSR markers	24 MTAs	–	–	–	7.3 to 25.8% PVE	Kumar et al. (2018a, b)
	F_2 mapping population of 173 individuals	734 DArTseq markers	9 QTLs distributed on LG6 and LG2, 7 QTLs on LG6	Interval mapping	Overall length of 1032 cM	1.41 cM, and one 10 cM	12–33.9% PVE for DTF	Rajandran et al. (2022).
Anthracnose and *Stemphylium* blight	94 F_9- RILs	2180 SNPs assembled in 543 unique haplotypes.	5 QTLs associated with resistance to *C. lentis* race 0, 6 QTLs with *C. lentis* race 1 resistance, and 3 QTLs with *S. botryosum* resistance	CIM	740.94 cM	–	–	Bhadauria et al. (2017)
Aphanomyces root rot	RIL and an association mapping population	GBS	QTL mapping identified 19 QTLs, while association mapping detected 38 QTLs (7 QTLs on 6 chromosomes, and 15 putative genes)	–	–	–	–	Ma et al. (2020)
Ascochyta blight	–	–	a major QTL with 3 QTLs on LG 5 Another QTL was identified on LG2	–	–	–	9.5–11.5%.	Dadu et al. (2021)
	3 RIL populations	SNP and SSR	2 and 3 QTLs (IH × DIG and IH × NF populations)	–	2429.61 cM	3.5 cM	52 and 69%	Sudheesh et al. (2016)
Anthracnose race 1	F_7 (a set of 102 and 139 RILs for LR-01 and LR-18)	–	Major-effect QTL (*qAnt1.Lc-3*) on chromosome 3	–	–	–	–	Gela et al. (2021)
	GWAS panel (subset of 200 genotypes)		14 SNP markers (chromosomes 3, 4, 5, and 6)	–	–	–	–	

(continued)

Table 9 (continued)

Tait	Mapping population	Marker used	Gene/QTL/MTA	Method employed	Distance (cM)	Average marker density (cM)	PV(%) explained	References
Rust	136 RILs	SSR markers	2 markers, LcSSR440 and LcSSR606 flanking the gene	–	8.3 and 8.1 cM	–	–	Singh et al. (2021)
Root and shoot traits under drought condition	RILs 132 F_{6-8} lines	SSR, SRAP, RAPD, SNP and AFLP markers	18 QTLs controlling 14 root and shoot traits	SIM and CIM	2022.8 cM	5.12 (LG V) to 9.8 cM (LG II)	4.3 ($QRSrati_{IX-77.72}$) to 28.9% ($QRSratio_{IX-2.30}$)	Idrissi et al. (2016)
Fe, Zn, phytic acid, and protein	Diverse set of 96 lentil accessions	SSR markers	31 MTAs	General Linear Model	–	–	10 to 26%	Kumar et al. (2024)
Heat	F_2, F_3, and backcross populations	SSR markers	2 major QTLs, qHt_ss, and qHt_ps	CIM	Both QTLs colocalized in the marker interval of PBA_ LC_1507 and PLC _105 covering 37.5 cM	–	12.1 and 9.23% PV explained for seedling survival and pod set, respectively	Singh et al. (2017)

Table 10 Genetic nature of different agro-morphological traits in lentil

Trait	Parents crossed	Number of gene(s) controlling	Dominant/recessive	References
Leaf pigmentation, leaf pubescence, seed cotyledon color	DPL 58 × ILWL 118	Single gene	Dominant	Kumar and Gupta (2019)
Tendril formation, seedling stem pigmentation	DPL 58 × ILWL 118	Two genes	NA	Kumar and Gupta (2019)
Tendril formation	NA	Monogenic	Dominant	Kumar (2002), Mishra (2004)
Plant height	NA		Dominant	Tahir and Muehlbauer (1994)
Leaflet shape	NA	Monogenic	Oval dominant over acute	Kumar et al. (2005a)
Broad and narrow leaflet	NA	Monogenic	Incomplete dominance	Kumar et al. (2004)
Growth habit	NA	NA	Erect phenotype is recessive	Emami and Sharma (1999), Kumar et al. (2004, 2005a)
Globe mutant	NA	NA	Recessive	Kumar (2002)

placed in the susceptible class. With a chi-square value of 0.74 and a p-value of 0.39, the segregation of resistant and susceptible RILs demonstrated a good fit to the predicted segregation ratio of 1R: 1S. This suggested that lentil rust resistance was under monogenic control (Singh et al. 2021).

Powdery Mildew Powdery mildew is caused by *Erysiphe polygoni* DC. The disease is reported from India. Ethiopia, Cyprus, Syria, Siberia, and Tanzania. Initially, symptoms are recorded on the upper surface of older leaves. A fine white powdery growth of mycelium and conidia is seen, which spreads and covers entire plant. Leaflets become dry and curled and are shredded. Moderate to high temperature and humidity favor the disease development. The disease can be controlled by foliar spray of fungicides like benomyl, aqueous sulfur, dinocarp, and ferrous bisulfide. Resistant donors for powdery mildew resistance have been reported by Mishra (1973). The best measure to control is the use of resistant varieties (Table 10).

Ascochyta Blight *Ascochyta lentis* is the casual organism of *Ascochyta* blight. *Ascochyta* blight is reported from Syria and the USA.

***Stemphylium* Blight** *Stemphylium* blight is caused by *Stemphylium botryosum* Waltr. The disease is reported from Bangladesh, Nepal, Syria, Canada, Egypt, and the USA. The disease can cause yield loss up to 62% under favorable environment. Initially, small light beige lesions are seen on the upper canopy. These lesions coalesce and enlarge. The entire foliage gets affected. Severe infection reduces the crop yield and seed size. Temperature of 18–22 degrees centigrade and relative humidity of 85% favor disease development. LL1122, ILL 10832, and P 3235 were reported as donors for *Stemphylium* resistance (Das et al. 2017).

Nematodes Nematodes are worldwide in occurrence and are important constraints in Syria, India, Turkey, and African countries. Stunting, wilting, chlorosis, and yellowing are the most common symptoms. The specific symptom of root-knot nematode is the presence of gall on roots. Cyst nematode symptom includes lemon-shaped cyst on roots. The important nematodes of lentil include root-knot nematode (*Meloidogyne incognita* Kofoid & White Chitwood and *Meloidogyne javanica* Treub) Chitwood; cyst nematode (*Heterodera cicceri* Vovlas, Greco & Di Vito. Crop rotation, soil solarization in summer, and seed treatment with nematicides can control nematodes.

Black Aphids Aphids (*Aphis craccivora* Koch) are an emerging threat worldwide. Aphid damage is common at seedling, flowering, and pod filling stage. Symptoms include deformation of leaves and shoot and black mold growth. Dry weather increases aphid infestation in lentil.

Bruchids Bruchids (*Callosobruchus chinensis, C. analis*, and *Bruchidius minutus* F.) are the storage pests of lentil. Eggs are laid on seeds/pods in the field. The pre-harvest oviposition causes infection during storage.

Pea Pod Borers Pea pod borer (*Etiella zinckenella* Treit.) causes 5–15% pod damage. Flowers and pods are attacked by young larvae. Borer consumes seeds in pod, and brown spot is visible on infested pods.

11 Breeding for Abiotic Stresses

The major abiotic stresses of lentil include drought, heat, cold, flooding and submergence, and salinity.

Drought Drought is a severe and regular constraint to lentil production. In the Mediterranean region, lentils are planted during the winter or autumn season, where the crop experiences intermittent drought during the vegetative stage and terminal drought during the reproductive stage. In South Asia, drought can occur during plant establishment, often due to delayed sowing. ICARDA characterized world lentil collections in four major groups (Erskine et al. 1989). They found that region-specific flowering response provides the basis for adaptation to climate variables of environment. Lentil can grow with 250 mm annual rainfall (Muehlbauer et al. 1985). Lentil avoids drought by forcing senescence and crop maturity due to high temperature. Drought stress affects the concentration of photosynthetic pigments, metabolism, and osmoregulation (Mishra et al. (2016); Biju et al. (2017). Drought avoidance and tolerance mechanisms are utilized by lentil to mitigate drought stress. Root-related traits are important for drought avoidance and improving production and productivity (Chen et al. 2015). To develop drought-tolerant varieties, it is necessary to study genetic variability for traits related to low moisture tolerance. Different mechanisms for escaping drought include high early vigor, early flowering, and maturity. Silim et al. (1993) reported that in dry areas of Syria, the highest yielding genotypes had high biomass, early flowering, and short grain filling stage. Early maturity is a key trait for drought escape. Early flowering and maturing genotypes were high yielding at low-rainfall and low-yielding environments in Australia (Materne 2003). Wild *Lens* species were characterized for drought tolerance by Gorim and Vandenberg (2017). Hamdi and Erskine (1996) reported that *Lens culinaris* ssp. *orientalis* (from low rainfall habitat) produced less biomass than cultivated under moisture stress; however, *L. odemensis*, L 01-827A exhibited low reduction of seed yield under drought environment. Lentil exhibits higher osmotic adjustment in comparison to other legumes (Leport et al. 1998).

Screening for drought: Osmotic adjustment is important for maintaining the turgor at low osmotic potential (Munns 2002). Relative water content, stomatal conductance, seedling vigor, and water use efficiency have been used to screen for drought by several workers (Sinclair and Ludlow 1985; Terzi et al. 2010; Dhanda et al. 2004). Hura et al. (2010) proposed drought screening based on morphological, biochemical, and physiological traits under soil condition. Screening for dehydration avoidance was proposed by Kumar et al. (2012). Singh et al. (2013) proposed rapid screening technique based on hydroponic condition. Traits like root depth, length and density, and shoot ground cover and early growth vigor are related to drought and regulate transpirational demand of the plant (Passioura 1981). Lentil genotypes PDL 1, DPL 53, JL 1 have been identified for root traits (Kumar et al. 2012). Among wild species, IG 72623 (*L. odemensis*), IG 72815 and L-01-827A (*L. ervoides*), IG 110813 (*L. lamottei*), PI 572376 and IG 72643 (*L. orientalis*) were found promising against drought stress (Gorim and Vandenberg 2017; Fang and Xiong 2015). Singh et al. (2016a, b) studied drought tolerance in lentil under the control of a single dominant gene.

11.1 Heat

The growth and development of lentil is affected by high temperature. The temperature requirement range for lentil is

18–30 degrees centigrade. Cooler weather is required at vegetative stage and warmer at maturity. Due to climate change, crop suffers from high temperature at maturity. This results in pollen sterility and flower drop. Temperature above 32 degrees centigrade at the flowering and pod filling stage adversely affects grain yield and grain quality. It is reported that heat wave temperature above 35 degrees centigrade for 6 days reduced yield by 70% across south eastern Australia (Nuttall et al. 2015). Heat stress during germination leads to poor germination, abnormal seedlings, nodule degeneration, reduction in biomass, early flowering, reduction in photosynthetic efficiency, loss of cell membrane stability, and increase in lipid peroxidation (Sehgal et al. 2017). High temperature affects development, growth productivity, and metabolism of plants. Screening for heat tolerance is carried out by late planting. Late planting ensures temperature above 35 degrees centigrade during flowering and fruiting period. Heat tolerance is a complex trait, and screening under field conditions is difficult. Basu et al. (2015) advocated for use of phenomics facility for phenotyping for heat tolerance. Phenomics ensures precise evaluation under controlled conditions. Rate of photosynthesis, membrane stability, and pollen viability are the key indicators of heat tolerance. Roy et al. (2011) screened genotypes for heat tolerance under hydroponic and field conditions. The donors for heat tolerance have been identified in lentil by Kumar et al. (2016a, b) and Kumari et al. (2017) (IG 3263, IG 3745, IG 2507, IG 4258, and FLIP 2009). Rajandran et al. (2022) identified ILL 7853 as donor for both drought and heat tolerance. Delahunty et al. (2023) identified five landraces, AGG 73838, AGG 70118, AGG 70951, AGG 70156, and AGG 70547, and one variety, Nipper, for heat tolerance. Singh et al. (2017) reported that heat tolerance in lentil is controlled by a single dominant gene.

11.2 Cold

Low temperature limits plant growth and distribution. To cope with cold stress, plants adjust their metabolism by physiological and molecular adaptations. Cold stress in lentil affects leaf development, vegetative tissues, flower initiation, seed coat development, and pod development (Gupta et al. 2019). The most sensitive stage is the flowering stage. Frost tolerance in lentil has been investigated by Erskine et al. (1981), Murray et al. (1988), Spaeth and Muehlbauer (1991), and Ali and Johnson (1999). Methodology for screening for cold tolerance was reported by Ali and Johnson (1999). Ali and Johnson (1999) identified ILL 5865 as frost tolerant, and Sarker et al. (2002) reported LL1878, ILL 875, ILL 662, ILL 1878, and ILL 975 as frost tolerant. Hamdi et al. (1996) studied winter hardiness in wild species and reported higher tolerance in *L. culinaris* spp. *orientalis* accessions. Nabati et al. (2021) MLC 409 and MLC 70 cold tolerance donors. In temperate regions, lentil is grown during the summer, when warm temperatures and long days prevail, whereas in Mediterranean regions, lentil is usually sown in winter, having cool temperatures and short days. Cooler temperature and shorter days are followed by warming and lengthening of the days after the spring equinox. In South Asia, lentil is sown in winter, benefiting from residual soil moisture, which becomes relatively warmer and short days. Under these subtropical savannah growing conditions, terminal drought is a major limitation subjecting the crop to forced maturity which ultimately reduces yield (Kumar et al. 2012, 2016a, b). Further details about trait mapping in lentil can be explored in Mishra et al. (2024).

Flooding and Submergence Tolerance Submergence and flooding cause severe loss to lentil. The damage caused depends on the severity, stage, and duration of stress. At germination, seed germination is delayed due to suppression of root growth and development. Abortion and drop of flowers and pods are due to waterlogging at flowering stage. The peculiar symptoms are leaf senescence and stunted plant growth. Waterlogging-tolerant genotypes have low biomass, early flowering, and maturity, high root porosity, and high stomatal conductance (Malik et al. 2015; Erskine et al. 2016). Adventitious roots with higher aerenchyma have been suggested for genotypes for waterlogging tolerance. Limited number of donors have been reported for waterlogging tolerance ILL 6439, ILL 6778, ILL 6793, Nugget, Bina Masur 1 (Ashraf and Chishti 1993; Malik et al. 2015; Islam et al. 2009).

12 Breeding for Nutritional Quality of Lentils

12.1 Grain Protein and Amino Acid Content in Lentils

Since lentil grains have an average crude protein content of 26%, they are an excellent source of high-quality protein. Approximately 70% of lentils' protein is made up of globulins, 16% albumins, 11% glutelins, and 3% prolamins. Numerous investigations have revealed notable variations in the protein composition of various types of lentil seeds. In one of the earliest studies on lentil protein content, Barulina (1930) discovered that different lentil cultivars ranged from 27.5% to 31.7%. When Hawtin et al. (1980) looked at the lentil germplasm worldwide, they found that there was more variation in the protein content among the 1688 accessions, ranging from 23.4% to 36.4%. Comparable variations have been found by Kumar et al. (2016a, b) and Hamdi et al. (1991) in their additional research. Important amino acids like tryptophan, threonine, phenylalanine, histidine, methionine, isoleucine, valine, and leucine are low in farmed lentils, with the exception of lysine. They do contain a lot of leucine, aspartic acid, glutamic acid, and arginine, though. They also contain small amounts of sulfur-containing amino acids, such as cysteine and methionine. Compared to animal-derived proteins, lentil proteins have a noticeably lower methionine content (0.9%) (Van Vliet et al. 2015). Since legumes have a lower methionine content but a higher lysine content, and cereals have a higher methionine content but a lower lysine content, combining cereals and legumes can optimize the anabolic qualities of plant-based diets (Van Vliet et al. 2015). Non-amino acid proteins found in lentil seeds include γ-hydroxyarginine, homoarginine, 2(S), 4(R)-4-hydroxyarginine, homoarginine (Sulser and Sager 1976), and erythro-γ-hydroxyornithine (Rozan et al. 2001). According to Grusak's (2009) evaluation of lentil nutrition, the range of protein content was 15.9% to 31.4%.

12.2 Vitamin B9/Grain Folates Content

Worldwide, millions of people consume insufficient amounts of folate (De Benoist 2008). There is a clear increase in the risk of low birth weight, developing neural tube abnormalities (NTDs), preterm delivery, and fetal growth retardation when there is insufficient folic acid intake during pregnancy. Low folate intake and elevated homocysteine levels have been related to a number of neurological disorders, cardiovascular diseases, and cancers. Eating adequate folic acid and folate can reduce plasma homocysteine levels (Blancquaert et al. 2010). While staple crops like rice, potatoes, maize, and plantains are low in folate, legume crops like lentils (*Lens culinaris* L.) and common beans (*Phaseolus vulgaris* L.) are rich in folate. Lentils are particularly high in folate, a nutrient that is crucial, with an average concentration of 255 μg/100 g (Sen Gupta et al. 2013). Moreover, lentils have a higher folate content than green field peas, yellow field peas, and chickpeas. According to Rychlik et al. (2007), the folate concentration of green lentils varies, ranging from 110 to 154 μg/100 g.

Pregnant women should take about 600 mg of folate daily, while adults should only take 400 mg. A 100 g serving of lentils contains 54–73% of the recommended daily intake (RDA) for folate (Thavarajah et al. 2008). Studies have measured eight folate monoglutamates in both wild and cultivated lentil species; higher folate concentrations (195–497 μg/100 g) are found in wild species compared to cultivated genotypes (174–361 μg/100 g) (Zhang et al. 2019). Numerous studies on lentils and other pulses have found that the most common forms of folate are 5-formyltetrahydrofolate (5-FTHF), 5-methyltetrahydrofolate (5-MTHF), and tetrahydrofolate (THF) (Jha et al. 2015a, b, 2020; Fernandez-Orozco et al. 2013). Furthermore, a large amount of 5-MTHF has been found in lentils (Hefni et al. 2010; Rychlik et al. 2007).

12.3 Prebiotic Carbohydrates in Lentil

Prebiotic carbohydrates are complex carbohydrates that are poorly absorbed in the upper gastrointestinal tract. They play a crucial role in cultivating healthy bacteria. The amount of prebiotic carbohydrates found in lentils varies widely, according to multiple studies (Bhatty 1988; Johnson et al. 2013; Chung et al. 2009; Tahir et al. 2011; Johnson et al. 2013). Lentils are primarily composed of starch, with low levels of oligo-, dis-, and monosaccharides. For every 100 g of dry lentil matter, there are 3.7 g of resistant starch, according to De Almeida Costa et al. (2006). Lentils are one food crop with a low glycemic index because their starch releases glucose slowly during digestion (Jenkins et al. 2012).

According to Johnson et al. (2013), lentils normally contain 1423 mg of sugar alcohols (SAs), 7500 mg of resistant starch (RS), 4071 mg of oligosaccharides from the raffinose family (RFOs), and 62 mg of fructo-oligosaccharides (FOSs) per 100 g. Siva et al. (2019) found that sucrose was the most common simple sugar in lentils, followed by fructose, glucose, mannose, and rhamnose. The highest concentration of SA was sorbitol, followed by mannitol and xylitol; the largest concentration of RFOs was stachyose, followed by verbascose and raffinose. For FOSs, ketose levels were higher than nystose levels. Other prebiotic carbohydrates found in lentils include cellulose, xylose, and arabinose. Johnson et al. (2015) state that the lentils' growth environment can have an impact on the quantity of prebiotic carbohydrates present in the grain.

12.4 Grain Phenols in Lentil

Dietary phenolics are flavonoids, polyphenols, and phenolic acids; of these, flavonoids account for the remaining two-thirds of total dietary consumption and phenolic acids for about one-third (Scalbert and Williamson 2000). The antioxidant qualities of phenolic compounds and their potential use as natural antioxidants in processed foods have attracted a lot of attention from researchers (Gupta et al. 2021b). The primary phenolic components of lentils are flavonoids, such as procyanidins, kaempferol glycosides, and catechin/epicatechin glucosides, according to Zhang et al. (2015), who conducted an analysis of twenty different cultivars of lentils. The phenolic composition varies with the color of the seed coat (Mirali et al. 2017; Gupta et al. 2018). Dietary phenolics are flavonoids, polyphenols, and phenolic acids; of these, flavonoids account for the remaining two-thirds of total dietary consumption and phenolic acids for about one-third (Scalbert and Williamson 2000). The antioxidant qualities of phenolic compounds and their potential use as natural antioxidants in processed foods have attracted a lot of attention from researchers (Gupta et al. 2021b). The primary phenolic components of lentils are flavonoids, such as procyanidins, kaempferol glycosides, and catechin/epicatechin glucosides, according to Zhang et al. (2015), who conducted an analysis of 20 different cultivars of lentils. The phenolic composition varies with the color of the seed coat (Mirali et al. 2017; Gupta et al. 2018). Among 96 lentil genotypes, total flavonoid content (TFC) was relatively higher in advanced breeding lines (12.21 mg QE/g) and low in Indian germplasm lines (10.68 mg QE/g) (Gupta et al. 2023). The colors of lentils can range from yellow to orange, red, green, brown, or black, depending on the cultivar, cotyledon composition, and seed coat makeup. The primary color of dehulled seeds is determined by the color of the cotyledon, which can be green, red, or yellow. The seed coat determines the color of the intact seed, which can be tan, brown, green, gray, or black. Lentil seed coats contain higher levels of proanthocyanidins, flavonols, and flavan-3-ols. This implies that lentils with gray and green seed coats may be a better option for a diet that promotes health.

12.5 Grain Micronutrients in Lentil

Numerous studies have assessed the micronutrient content of a wide range of planted lentil germplasms (Gupta et al. 2021a, b; Kumar et al. 2019). However, studies on wild relatives with high vitamin and micronutrient contents are fewer in number. There was observed to be a wide range of variability in the micronutrient concentrations in *Lamottei* (iron—64–80 mg/kg, zinc—26–40 mg/kg, calcium—311–434 mg/kg, copper—2–6 mg/kg, magnesium—754–839 mg/kg), *Lamottei nigricans* (iron—60–70 mg/kg, zinc—33–39 mg/kg, calcium—508–590 mg/kg, copper—3–4 mg/kg, magnesium—445–738 mg/kg), and *Lamottei* (iron—65 mg/kg, zinc—37 mg/kg, calcium—339 mg/kg, copper—6 mg/kg, magnesium—638 mg/kg). These results were published by Sen Gupta et al. (2016). *L. orientalis*, ILWL 118, a wild accession that tested positive for 150 ppm Fe in a recent study, is now used in pre-breeding programs (Sarker et al. 2016). When 96 wild accessions were analyzed to determine their potential for biofortification, significant differences in mineral contents were discovered between species (Kumar et al. 2018a, b). The populations of *L. culinaris* ssp. *odemensis*, *L. culinaris* ssp. *orientalis*, and *L. ervoides* showed significant diversity for Fe, Zn, Mo, Cu, and Mn. The Ca and Mg contents of *L. nigricans* and *L. culinaris* ssp. *orientalis*, as well as the Mg content of *L. ervoides*, were also found to exhibit significant heterogeneity. This variation may result from the genetic diversity of these species or from a specific gene's expression in a given environment.

However, *L. culinaris* ssp. *culinaris* showed little diversity for a few minerals, as a farmed species. Zn concentration in lentils varies significantly (36.7–50.6 mg/kg), according to Ray et al. (2014). Interactions between environment and genotype have also been reported (Khazaei et al. 2017; Vandemark et al. 2018). As per Thavarajah et al. (2008), lentils cultivated in the dark-brown and brown soil of Western Canada exhibit a high concentration of selenium (425–672 µg/kg). Selenium concentrations in lentils grown in six major countries that produce lentils were found to be significantly lower in another study (Thavarajah et al. 2011). These countries included Syria (22 µg/kg), the Northwestern USA (26 µg/kg), Morocco (28 µg/kg), Turkey (47 µg/kg), Southern Australia (148 µg/kg), and Nepal (181 µg/kg). Aski et al. (2023) conducted a multi-location study in India and found that among 16 genotypes of commercially cultivated lentils, the levels of grain iron, zinc, and phytic acid content varied significantly. These levels ranged from 114.10 to 49.90 mg/kg, 74.62 to 21.90 mg/kg, and 0.76 to 2.84 g/100 g (dw), respectively. These findings suggest that the lentils may have applications in mitigating hidden hunger. A study conducted recently by Bansal et al. (2021) found that biofortified cultivars are resistant to soil stress brought on by heavy metals (such as cadmium). Two biofortified lentil cultivars, L 4717 and IPL 220, were recently released from the Indian program. L 4717, an extra-early lentil cultivar, takes 100 days to reach maturity.

12.6 Breeding Progress/Varietal Development

N.I. Vavilov and his associates at the All-Union Institute of Agricultural Sciences in Leningrad, USSR, created the first collection of lentils between 1920 and 1930. Elena Barulina, Vavilov's spouse, oversaw the collection program and described the variety of Lens and cultivated species. The genetic variability was preserved in the European collection. The first genebank, the N.I. Vavilov Research Institute of Plant Industry, still has sizable Lens collections. With the founding of ICARDA in Aleppo, Syria, in 1977, lentil breeding gained attention on a global scale. The primary goal was the collection of wild and cultivated species and the ex situ conservation of those species. This international collection is made possible by an in-perpetuity grant from the Crop Trust. In the Svalbard Global Seed Vault, above the Arctic Circle, ICARDA has set up a gene bank. With the rise in demand for lentils, this crop spread to new regions, including Canada and Australia. To overcome the difficulties posed by the unfamiliar surroundings, breeders in these nations employed both wild and domesticated lentils. *Lens ervoides* introduced anthracnose resistance to its cultivated varieties in Canada.

Several National Research Partners received ICARDA's lentil collections. These genetic resources are important sources of resistance to illnesses, the cold, and drought. For the Middle East, Central Asia, and North Africa, a large number of lentil varieties were released. Using ICARDA material, the Bangladesh Agricultural Research Institute created about ten high-yielding, short-duration varieties. In 1924, the varietal improvement program for lentils was started. Local mixed lots were used to select individual plants. The earliest developed lentil purelines are NP1, NP47 (IARI), BR25 (Bihar), L12 (Punjab), T8, T38 (Uttar Pradesh), and B77 (West Bengal). The All India Coordinated Pulses Improvement Project, which was started in the 1960s, led to an increase in lentil research involving a multidisciplinary team of researchers. Initially, pureline selection was the main focus for improving the population. Here are the lentil varieties that were released in India up until 1985. Subsequently, bulk, single-seed descent methods, and pedigree became more prominent. Tables 11 and 12 list the lentil varieties that have been developed in India by CVRC and SVRC in different states. The released varieties of lentils have low pod set and weak seedling vigor. The main obstacles to increasing productivity, aside from biotic and abiotic stresses, are high flower drop and lodging. L 4717 and IPL 220, two biofortified varieties, have been released by the Indian program. L 4717 is the nation's first extra-early lentil variety, maturing in less than 100 days (Tables 11 and 12).

13 Breeding for Yield-Related Traits

13.1 Pre-breeding for Broadening the Genetic Base

The released cultivars in the Indian subcontinent have a limited genetic base. The primary cause of yield stagnation in hybridization programs is the persistent use of a small number of genotypes. The wild species and varied Mediterranean germplasm serve as an invaluable source of important genes absent from cultivated lentils. Mediterranean germplasm lines were assessed as part of the DAC-funded ICARDA-ICAR project "Broadening the genetic base: Pre-breeding and genetic enhancement in breaking yield barriers in kabuli chickpea and lentil." Key donors for various traits were identified.

13.2 Breeding for Nutrient Use Efficiency

Micronutrient-deficient marginal soils are used to grow lentils. There have been reports of iron and boron deficits from Ethiopia, Syria, and India. A high pH soil deficiency in fertilizers leads to yield. There have been reports of 25% and 47% yield loss in susceptible genotypes from India and Syria (Ali et al. 2000; Erskine et al. 1993). There are reports of boron deficiency from eastern India, the Terai plains of Nepal, and northern Bangladesh.

13.3 Breeding for Early Maturity and Bold Seed Size

To understand the genetics of lentil flowering time, numerous studies have been carried out (Tullu et al. 2008). Tullu et al. (2008) have reported several QTLs in lentil that control the length of flowering. A number of short-duration lentil varieties that are resistant to terminal heat have been developed in India (L 4717) and Bangladesh (BARI M4, MARI M5, MARI M6, MARI M7, MARI M8, and MARI M9). The first extra-early lentil variety to mature in less than 100 days is L 4717. In Eastern India, rice fields under fallow conditions must use early maturing lentil varieties resistant to *Stemphylium* blight. India is the home of conventionally small-seeded varieties (100 seed weight—2.5 g). Small-seeded varieties have demonstrated greater adaptability to various agroclimatic circumstances. Large-seeded varieties are only found in Central India; small-seeded varieties are more common in the country's eastern, northern, and northern hills. In lentils, seed size is a quantitative characteristic regulated by multiple genes (Abbo et al. 1991). The QTLs that regulate lentil seed size have been identified by Fedoruk et al. (2013) and Verma et al. (2015). On marginal soils, lentils are grown as a rainfed crop. Rainfall during the crop season and the remaining soil moisture from the rainy season determine the potential yield. To stabilize the production, varieties resistant to climate change are needed.

14 Thrust Areas to Be Focused to Achieve Larger Genetic Gains

14.1 Pre-breeding

Different agroclimatic conditions are used to grow lentils. Larger secondary and tertiary branches, bold seed size, vigor in seedlings, and early maturity are all expected results of hybridization with Mediterranean lines. Bold small- and large-seeded varieties have been developed using ICARDA germplasm. The IARI in New Delhi, the NBPGR in New

Table 11 Lentil varieties released through AICRP for cultivation in India

S. No.	Name of variety	Year of release	Pedigree	Average yield (kg/ha)	Maturity period	Salient features
1.	K 75 (Malika)	1986	Local sel. From Bundelkhand region	1300–1400	130–135 days	Semi spreading, bold seeded (100 seed weight-2.7 g)
2.	LH 84-8 (Sapna)	1991	L 9–12 × JLS 2	1500–1600	130–135 days	Rust resistance, semi-spreading, bold seeded (100 seed weight-2.7 g)
3.	Pant lentil 4	1993	VPL 175 × (PL 184 × P 288)	1600–1700	135–140 days	Rust and wilt resistance, semi-spreading, small seeded
4.	Lens 4076	1993	PL 234 × PL 639	1400–1500	135–140 days	Rust resistance, semi-spreading, bold seeded
5.	DPL 15 (Priya)	1995	PL 406 × L 4076	1500–1600	135–140 days	Rust and wilt resistance, bold seeded
6.	Pusa Vaibhav (L 4147)	1996	(L 3875 × P 4) × PKVL	1700–1800	130–135 days	Rust resistance, small seeded
7.	DPL 62 (Sheri)	1997	JLS 1 × LG 171	1700–1800	130–135 days	Rust and wilt resistance, bold seeded
8.	JL 3	1999	Land race Sel. From Sagar	1400–1500	110–115 days	Wilt resistance, bold seeded
9.	IPL 81 (Noori)	2000	K 75 × PL 369	1200–1300	110–115 days	Rust and wilt resistance, bold seeded
10.	KLS 218	2005	KLS 133 × L 9362	1300–1400	120–125 days	Rust resistance, small seeded
11.	HUL 57	2005	Mutant of HUL −11	1400–1500	120–125 days	Rust resistance, small seeded
12.	VL 507	2006	Sel. From ILL-7978	1200–1300	160–170 days	Wilt resistance, bold seeded
13.	VL 126	2006	LL 498 × LH 84–8	1300–1400	160–170 days	Rust resistance, small seeded
14.	IPL 406	2007	DPL 35 × EC 157634/382	1700–1800	125–130 days	Rust resistance, bold seeded
15.	WBL 77	2008	ILL 7723 × BLX 88176	1400–1500	115–120 days	Rust resistance, small seeded
16.	Pant L 8 (Pant L 063)	2010	DPL 59 × IPL 105	1500–1600	130–135 days	Moderate resistance to rust and wilt, small seeded
17.	IPL 316	2013	Sehore 74–3 × DPL −58	1400–1500	110–115 days	Rust and wilt resistance, bold seeded
18.	RVL 11–6	2017	JL 3 × DPL 62	1100–1200	107–113 days	Wilt resistance, bold seeded
19.	L 4717 (Pusa Ageti Masoor)	2017	ILL 7617 × 91516	1200–1300	96–106 days	Wilt and *Ascochyta* blight resistance, extra early type, bold seeded
20.	Kota Masoor 1 (RKL 607-1)	2017	KLB 339 × SL 94-09	1000–1400	98–107 days	Drought and high temperature tolerance, bold seeded
21.	RKL 14-20 (Kota Masoor 2)	2018	LL 1049 × RKL 11	1200–1500	97–104 days	Drought and high temperature tolerance, bold seeded
22	L 4727	2018	Sehore 74–3 × Precoz	1100–1500	92–128 days	Moderate resistant to wilt, bold seeded
23.	IPL 220	2018	(DPL 44 × DPL 62) × DPL 58	1400–1800	119–122 days	Rust and wilt resistance, small seeded
24	L- 4729	2019	SKL 259 × L 4147	1700–1800	96–110 days	Moderate resistance to wilt, bold seeded
25	LL 1373	2019	IPL406 × FLIP 2004-7 L	1500–1600	125–130 days	Moderate resistance to wilt and rust, bold seeded
26	VL Masoor-148	2019	DPL-15 × L-4076	1100–1200	145–160 days	Moderate resistance to wilt and rust, small seeded
27	Kota Masoor 3 (RKL 605–03)	2020	L 4682 × SL 73-3	1800–1900	105–110 days	Moderate resistance to wilt and rust, bold seeded
28	Kota Masoor 4 (RKL 58 F 3715)	2020	Mutant of DPL 62	1800–1900	110–115 days	Rust and *Stemphylium* blight resistance, moderately resistant to wilt, bold seeded
29	LL 1613	2023	LL 1112 × PL 02-6	1600–1700	115–120 days	Moderate resistance to wilt and rust, bold seeded
30	LL 1655	2024	LL 1112 × LL 699	1400–1500	140–145 days	Moderate resistance to wilt and rust, small seeded (100 seed weight- 2 g)
31	LH 17–19	2024	LH 07–26 × PL 01	1500–1700	130–135 days	Rust resistant and moderately resistant to wilt, small seeded (100 seed weight- 2.4 g)

Table 12 State-released varieties of lentil for cultivation with unique traits

S. No.	Name of variety	Year of release	Pedigree	Average yield (kg/ha)	Maturity period	Salient features
1.	JL 1	1991	Selection from local collections	1100–1500	110–120 days	–
2.	Narendra Masoor 1	1996	Precoz × L 9–12	2000–2200	134–140 days	Rust and wilt resistance
3.	Garima	1997	Pusa 2 × No. 4	1200–1400	140–145 days	Rust resistance, bold seeded
4.	LL 699	2002	PL 639 × PL77–2	1400–1500	140–145 days	Moderate resistance to rust and blight, tolerant to pod borer
5.	KL 320 (Shekhar 3)	2004	PL 405 × P237	2000–2200	120–125 days	Wilt resistance
6.	KLB 303 (Shekhar2)	2004	LG 60 × Sagar local	1500–1800	125–130 days	Wilt resistance
7.	Shalimar Masoor-1	2005	Single plant selection from EC-2216	1000–1100	190–200 days	Resistant to leaf spot and pod blight, moderately resistant to root rot, bold seeded
8.	Markanday	2005	Selection from Exotic material	1000–1200	160–170 days	Rust, wilt, root rot, and blight resistance
9.	HM-1	2006	K 75 × L 4076	1400–1600	135–140 days	Rust resistance, small seeded
10.	LL 931	2009	LH90–103 × LL608	1200–1300	145–150 days	Rust resistance, tolerant to pod borer
11.	VL Masoor 129	2009	VL 101 × VL 1	700–1100	145–150 days	Root rot and wilt resistance, small seeded
12.	Pant L 6	2009	Pant L 4 × DPL 55	1600–1800	125–145 days	Rust resistance, small seeded
13.	Pant L 7	2009	L 4076 × DPL 15	1600–1800	125–145 days	Rust resistance, bold seeded
14.	VL Masoor 514 (VL 514)	2011	VL 501 × VL Masoor 103	800–1000	155–160 days	Moderate resistance to rust and wilt, bold seeded
15.	RVL 31	2014	Selection from local collection from Shajahpur	1800–1900	110–115 days	Wilt resistance, bold seeded
16.	Shalimar Masoor-2 (SKUAL 9)	2015	EC-3109 (Selection from ICARDA material)	1285	200–205 days	Moderate resistance to rust, wilt, and white grub, bold seeded, protein content 23.03%
17.	KLB 2008–4 (Krati)	2015	LG 362 xDPL 62	1800–2000	115–120 days	Wilt resistance, bold seeded
18.	KLS 09–3 (Krish)	2015	L9–12 × WBL 58	1800–2000	105–110 days	Rust and wilt resistance, small seed
19.	RLG 5 (Keshwanand Masoor 1)	2016	Sel. from local germplasm	1500–1600	130 days	Moderate resistance to root-knot nematode
20.	Pant lentil—9 (PL 098)	2017	Pant L 5 × IPL 105	1300–1400	113–135 days	Rust, *Ascochyta* blight and wilt resistance, small seeded
21.	KLB 345 (Shekhar 4)	2017	Precoz × KLB 231	1800–2000	111 days	Rust and wilt resistance, bold seeded
22.	KLS 122 (Shekhar 5)	2017	KLS 564 × KL 320	1600–1800	105–115 days	Rust and wilt resistance, small seeded
23.	IPL 526	2018	DPL 62 × DPL 58	1000–1200	101–110 days	Rust and wilt resistance, bold seeded
24.	IPL 315	2018	PL 4 × DPL 62	1200–1300	135–140 days	Rust and wilt resistance, bold seeded
25.	IPL 321	2018	DPL 62 × K 75	900–1000	130–135 days	Wilt resistance, bold seeded
26.	Shalimar Masoor −3	2019	KLS 221 (Sel. from IIPR germplasm)	1150–1250	204 days	Tolerant to frost and winter chilling, bold seeded, 100 seed weight- 5.5 g
27.	IPL 534	2019	KL 178 × DPL 62	1600–1800	100–107 days	Rust and wilt resistance, bold seeded
28.	Chhattisgarh Masoor −1 (RL 3–5)	2020	Selection from Germplasm	1000–1100	90–95 days	Rust and wilt resistance, bold seeded
29.	IPL 329	2020	KL 178 × DPL 62	1000–1100	120–125 days	Rust and wilt resistance, bold seeded
30.	IPL 225	2020	(DPL 44 × DPL 62) × DPL 58	1000–1100	105–120 days	Rust and wilt resistance, small seeded
31.	Pant Lentil 11 (P L 164)	2021	DPL 15 × L 4188	1100–1200	101–132 days	Rust resistance, moderately resistant to pod borer
32.	Jammu Lentil 144	2021	ILL10829 × ILWL 30	1100–1200	120–125 days	Root rot and wilt resistance
33.	Jammu Lentil 71	2021	ILL-8006 × ILWL-62	1000–1100	140–145 days	Root rot and wilt resistance
34.	Bidhan Lentil 16 (BL 16)	2022	LL56 × L4710	1500–1600	109–132 days	Stemphylium blight and collar rot resistance, small seeded
35.	IPL 230	2022	IPL 522 × DPL 62	1600–1700	111–126 days	Rust and wilt resistance, small seeded
36.	VL Masoor 150 (VL 150)	2022	VL 501 × VL 502	1400–1500	156–161 days	Moderate resistance to wilt
37.	Pant Lentil 12	2023	PL 6 × DPL 58	800–1000	155–160 days	Rust resistance and moderate resistance to pod borer, small seeded
38.	Pusa Shweta (PSL-19)	2024	PDL 1 × L 4147	1800–2000	120–125 days	Salt tolerant and resistant to wilt and *Ascochyta* blight

Delhi, and the IIPR in Kanpur have identified new sources of resistance to biotic and abiotic stresses as a result of systematic pre-breeding efforts. Efforts are underway toward utilization of available genomic resources for mapping the relevant genes, with marker-assisted selection being employed to develop climate-resilient cultivars.

14.2 Earliness

In Eastern India's rice paddies and Central India's rice fields, earlyness is necessary to avoid terminal moisture stress. Wilt is a significant biotic stress for Central India, while *Stemphylium* blight and rust are significant for Eastern India. Extremely important tolerances include those to high temperatures during the reproductive stage and to terminal soil moisture stress. The price of manual weeding will go down thanks to herbicide tolerance. New varieties of lentils should be developed specifically for rice fallows. Two loci, viz., DTF6a and DTF6b, where dominant alleles promote early flowering in the pilosae ecotype adapted to South Asian environments have been identified. The findings show that DTF6a alone is sufficient to induce early flowering even under extremely short photoperiods. Genomic synteny among putative candidate genes allowed identification of a conserved cluster of three orthologues of the florigen (FT) gene. Expression analysis in near-isogenic lines further revealed early flowering allele to be specifically associated with strong derepression of the FTa1 gene. Extensive survey of over 350 accessions from diverse origins showed that the DTF6a allele is prevalent in South Asian material. Sequence analysis revealed a 7.4 kb deletion in the FTa1–FTa2 intergenic region in the pilosae parent (Rajandran et al. 2022).

14.3 Breeding for Stable High-Yielding and Nutritional-Dense Lentil Varieties

Micronutrient-dense types are needed to address the micronutrient deficiency. Nutrient-dense lentil varieties with high trait stability and increased availability of iron and zinc require an understanding of the roles played by genotype, environment, and genotype-by-environment interactions. Plant breeding requires a thorough understanding of the genotypic and environmental factors influencing genotype × environment interactions. Genetic stability and genotype × environment interactions have been studied in 16 commercially produced lentil cultivars across six different geographical locations in India for grain micronutrients and anti-nutritional factors (phytic acid) using GGE (Genotype, Genotype × Environment interactions) and AMMI (Additive Main effects and Multiplicative Interaction) models (Aski et al. 2023). The Fe and Zn levels of various lentil genotypes showed significant genetic variability. L 4717 and IPL 220, two varieties of lentils high in fertility, have been approved for cultivation in Central and Eastern India, respectively. It will take work to lower the phytic acid. It is desired to increase the concentration of protein.

14.4 Plant Type Restructuring

To have photosynthate partitioning, plants with higher solar light interception are needed. In order to lower harvesting losses due to shattering and lower the cost of harvesting and threshing, mechanized harvesting is necessary. Mechanized harvesting requires plants that are tall, upright, and non-lodging.

14.4.1 Speed Breeding

Plant breeders accelerate research to increase crop productivity. Improved varieties take a long time to develop using conventional breeding techniques. The development of improved varieties can proceed more quickly when the generation time is shortened. Speed breeding, which uses an extended photoperiod of 22 h, is a great tool for quickly advancing generations. Glass houses use sodium-vapor lamps (SVL), while growth chambers use metal halide and light-emitting diodes (LED) lighting (Ghosh et al. 2018). In barley, chickpea, canola, pea, and wheat, the generation time was shortened by a controlled temperature and a 22-h photoperiod.

With speed breeding, up to six generations have been produced annually. There are only a few reports (Idrissi and van Damme 2018; Idrissi et al. 2019; Mobini et al. 2016) on the application of extended photoperiod in lentil. According to a recent study by Idrissi et al. (2019), an effective protocol of prolonged light period to speed breeding in lentil can shorten the wild types' flowering time and reduce the length of the crop (for synchronization of flowering in cultivated and wild for hybridization). For this crop, however, it is necessary to optimize light intensity, spectral composition (LED, red, blue, and far-red light), and red/far-red light ratio.

14.5 Climate-Resilient Varieties

By the end of the twenty-first century, the average global temperature is predicted to have increased by 2.6–4.8 °C, up from 0.74 °C during the previous century (Leisner 2020). The global security of food and nutrition may be impacted by climate change. Researchers need to pay attention to develop climate-smart varieties to understand how climate change affects biotic and abiotic stressors, nutritional quality, and productivity. Screening protocols for abiotic stresses have been developed by IARI, New Delhi. Drought, heat, salinity, and salinity donors have all been found. QTLs controlling abiotic stress tolerance have been found, and transcriptome analysis has revealed the mechanism underlying tolerance. Genomic resources can be used to create cultivars that are resistant to abiotic stress by making use of the information currently available. To create phosphorus absorption and utilization efficient varieties, phosphorus usage efficiency in lentils is being researched. It is necessary to closely monitor pathogen virulence in light of climate change.

14.6 Bio-Fortified Varieties

Rich in potassium, iron, zinc, fiber, folate, and polyphenols, lentils are low in sodium and saturated fat. Due to their resistant starch that breaks down slowly, lentils have a low glycemic index. Prebiotics can be found in abundance in lentils. Legume protein is becoming more popular as a source of nutritious food and environmental sustainability. Blood pressure, cholesterol, and blood sugar are all lowered by lentils (Ganesan and Xu 2017; Hanson et al. 2014).

14.7 Need for Automated Harvesting (Mechanical Harvesting)

Reducing the cost of lentil production requires the use of mechanical harvesting. At maturity, a shortage of laborers can occasionally result in catastrophic losses. Tall, upright, non-lodging plants with a pod set 10 cm above the ground are needed for mechanical harvesting. Good straw and seed yield are guaranteed by these plant varieties. Erskine et al. (1991) conducted a comparative study between hand harvesting, mower cutting, and angled blade cutting. Compared to hand harvesting, both mechanical harvesting techniques produced harvest loss. There was a 16.6% loss in straw yield and an 8.6% loss in grain yield. The cost of labor during harvest offsets the loss from mechanical harvesting. An experimental harvester was created by Sidahmed and Jaber (2004) to harvest lentils. According to Sidahmed and Jaber (2004), a mechanical harvester harvests lentils 5.9 cm above ground at 22% moisture with 20% straw loss and 2% seed loss. Thirty-six genotypes were assessed by Jawad et al. (2019) for mechanical harvesting. It was discovered that two genotypes, 09S 83,184–10 and 2009S 96,518–1, from ICARDA nurseries, were superior. These have the potential to enhance the current cultivars. Lentil varieties are harvested using a combination of methods in the Mediterranean region and in Canada. To create varieties that are appropriate for South Asia, efforts are being made. Recently, ICARDA created breeding material that can be harvested mechanically, and it is now supplying national programs with this material.

14.8 Herbicide Tolerance in Lentil

Canada and Australia recently adopted the lentil. One of the main obstacles to weed control is the absence of safe and registered post-emergence weedicide. Metribuzin, also known as aminotriazinone, was tried as a weed control solution. Applied to grasses and broad-leaved weeds, metribuzin is a broad-spectrum herbicide. Reports of mutant lines displaying resistance to metribuzin herbicide have been made from Canada and Australia. 9.4×10^{-8} was a very low mutation rate that was observed. The creation of lentil cultivars that are resistant to metribuzin would guarantee crop safety, lower crop phytotoxicity, and give farmers an alternative for managing broad-leafed weeds.

15 Crop Improvement Program Modernization with Modern Tools

15.1 Molecular Marker Development

RFLP/AFLP markers were used to create the first linkage maps in lentil (Eujayl et al. 1998; Havey and Muehlnauer 1989). For RFLP markers to develop, they need good technical proficiency. Subsequently, PCR-based markers were created, including sequence-characterized amplified region (SCAR), random amplified polymorphic DNA (RAPD), and simple sequence repeats (SSR). In a number of studies (Gupta et al. 2016; Singh et al. 2016a, b, 2019a; Ates et al. 2016, 2018; Tsanakas et al. 2018; Mbasani-Mansi et al. 2019; Polanco et al. 2019), PCR-based markers were widely employed in lentil breeding programs. In 2014, Verma et al. created genomic SSR markers. Since the introduction of next-generation sequencing technologies, SNP markers have been employed in lentils to characterize the genetic diversity of germplasm collections and to associate specific markers with phenotypic traits, including seed quality, disease resistance, and micronutrient content (Lombardi et al. 2014; Sudheesh et al. 2016; Khazaei et al. 2016, 2017, 2018).

For the purpose of mapping genes and QTLs, rapid and economical SNP mining has been licensed by NGS technologies. Breeders are now favoring SNP markers over PCR-based markers. There are several methods for locating and validating SNP markers.

Unigene-derived amplicons are being resequenced using Sanger sequencing technology. Gene-based SNPs were found through the use of expressed sequence tags (EST), which were then translated into PCR-based markers. Using the Illumina Genome Analyzer, Sharpe et al. (2013) reported creating 44,879 SNPs, and Temel et al. (2014) reported finding 50,960 SNPs to build a linkage map in lentils. Using

NGS platforms for transcriptome analysis has produced a large number of SNPs from the lentil coding region (Sharpe et al. 2013; Kaur et al. 2014; Singh et al. 2019a). These SNPs have been used to create linkage maps and to characterize genetic diversity.

15.2 Molecular Markers Directed Lentil Improvement

Molecular markers interpret the gene networks governing the quantitative traits, and they have been connected to multiple economically significant traits in various crops. The use of molecular markers for lentil crop improvement has increased due to the development of PCR molecular markers, next-generation sequencing, and genomic resource enrichment. There are not many molecular markers associated with economically significant traits found in lentils (Table 9). In these studies, the use of bi-parental populations with low marker density has not revealed a very close association between molecular markers and QTLs/genes (Tables 10, 11 and 12).

References

Abbas A (1995) Variation in some cultural and physiological characters and host/pathogen interaction of *Fusarium oxysporum* f. sp. *lentis*, and inheritance of resistance to lentil wilt in Syria. PhD thesis, University of Aleppo, Syria

Abbo S, Ladizinsky G, Weedan NF (1991) Genetic-analysis and linkage study of seed weight in lentil. Euphytica 58:259–266

Agrawal SC, Khare MN, Agrawal PS (1976) Field screening of lentil lines for resistance to rust. Indian Phytopathol 29:208

Ali A, Johnson DL (1999) Association of growth habit and anthocyanin pigment with winter hardiness in lentil. Pak J Biol Sci 2:1292–1295

Ali M, Dahan R, Mishra JP, Saxena NP (2000) Towards the more efficient use of wáter and nutrients in food legume cropping. In: Knight R (ed) Proc. 3rd inter. Food legumes research conference. Kluwer Academic, Dordrecht, pp 99–105

Ashraf M, Chishti SN (1993) Waterlogging tolerance of some accessions of lentil (*Lens culinaris* medic.). Trop Agric 70(1):60–67

Aski MS, Mishra GP, Tokkas JP et al (2023) Strategies for identifying stable lentil cultivars (*Lens culinaris* Medik) for combating hidden hunger, malnourishment, and climate variability. Front Plant Sci 14:1102879

Aski MS, Mishra GP, Yadav PS, Langyan S, Kumari J, Kumar A, Thudi M, Tokkas JP, Roy J, Sinha MK, Gupta S (2024) Cytological studies of lentil genome. In: Kumar J, Sen Gupta D, Kumar S (eds) The lentil genome. Academic, pp 131–158

Ates D, Sever T, Aldemir S et al (2016) Identification QTLs controlling genes for se uptake in lentil seeds. PLoS One 11:e0149210

Ates D, Aldemir S, Yagmur B et al (2018) QTL mapping of genome regions controlling manganese uptake in lentil seed G3 (Bethesda) (8):1409–1416

Balech R, Maalouf F, Kaur S et al (2024) Identifcation of novel genes associated with herbicide tolerance in Lentil (*Lens culinaris* ssp. *culinaris* Medik.). Sci Rep 14:10215. https://doi.org/10.1038/s41598-024-59695-z

Bansal R, Priya S, Dikshit HK et al (2021) Growth and antioxidant responses in iron-biofortified lentil under cadmium stress. Toxics 9:182. https://doi.org/10.3390/toxics9080182

Barulina OH (1930) Lentil of USSR and of other countries. Bull Appl Bot Genet Plant Breed 40:1–319

Basu PS, Srivastava M, Singh P et al (2015) High-precision phenotyping under controlled versus natural environments. In: Phenomics in crop plants: trends, options and limitations. Springer, New Delhi, pp 27–40

Bayaa B, Erskine W, Hamdi A (1994) Response of wild lentil to *Ascochyta fabae* f. sp. *lentis*from Syria. Genet Resour Crop Evol 41:61–65

Bayaa B, Erskine W, Hamdi A (1995) Evaluation of a wild lentil collection for resistance to vascular wilt. Genet Resour Crop Evol 42:231–235. https://doi.org/10.1007/BF02431257

Bhadauria V, Ramsay L, Bett KE et al (2017) QTL mapping reveals genetic determinants of fungal disease resistance in the wild lentil species *Lens ervoides*. Sci Rep 7:3231. https://doi.org/10.1038/s41598-017-03463-9

Bhatty RS (1988) Composition and quality of lentil (*Lens culinaris* Medik): a review. Can Inst Food Sci Technol J 21:144–160

Biju S, Fuentes S, Gupta D (2017) Silicon improves seed germination and alleviates drought stress in lentil crops by regulating osmolytes, hydrolytic enzymes and antioxidant defense system. Plant Physiol Biochem 119:250–264

Blancquaert D, Storozhenko S, Loizeau K et al (2010) Folates and folic acid: from fundamental research toward sustainable health. Crit Rev Plant Sci 29:14–35

Buchwaldt L, Anderson KL, Morrall RAA, Gossen BD, Bernier CC (2004) Identification of lentil germ plasm resistant to *Colletotrichum truncatum* and characterization of two pathogen races. Phytopathology 94(3):236–243. https://doi.org/10.1094/PHYTO.2004.94.3.236

Cao Z, Li L, Kapoor K, Banniza S (2019) Using a transcriptome sequencing approach to explore candidate resistance genes against *stemphylium* blight in the wild lentil species *Lens ervoides*. BMC Plant Biol 19:1–16

Chen YL, Djalovic I, Rengel Z (2015) Phenotyping root traits. In: Kumar J, Pratap A, Kumar S (eds) Phenomics of crop plants: trends, options and limitations. Springer, New Delhi, pp 102–128

Chung HJ, Liu Q, Hoover R (2009) Impact of annealing and heat-moisture treatment on rapidly digestible, slowly digestible and resistant starch levels in native and gelatinized corn, pea and lentil starches. Carbohydr Polym 75:436–447

Coyne C, McGee R (2013) Lentil. In: Singh M, Upadhyaya HD, Bisht IS (eds) Genetic and genomic resources of grain legume improvement. Elsevier, pp 157–180

Coyne CJ, Kumar S, von Wettberg EJ et al (2020) Potential and limits of exploitation of crop wild relatives for pea, lentil, and chickpea improvement. Legum Sci 2(2):36. https://doi.org/10.1002/leg3.36

Cubero JI (1981) Origin, domestication and evolution. In: Webb C, Hawtin GC (eds) Lentils. Commonwealth Agricultural Bureau, Slough, pp 15–38

Cubero JI (1984) Taxonomy, distribution and evolution of the lentil and its wild relatives. In: Genetic resources and their exploitation—chickpeas, faba beans and lentils. Springer, Dordrecht, pp 187–203

Cubero JI, Perez De La Vega M, Frantini R (2009) Origin, phylogeny, domestication and spread. In: Erskine W, Muehlbauer FJ, Sarker A, Sharma B (eds) The lentil—botany, production and uses. CABI, Wallingford, pp 13–33

Dadu RHR, Ford R, Sambasivam P, Gupta D (2017) A novel *Lens orientalis* resistance source to the recently evolved highly aggressive Australian *Ascochyta lentis* isolates. Front Plant Sci 8:1038. https://doi.org/10.3389/fpls.2017.01038

Dadu RHR, Bar I, Ford R et al (2021) *Lens orientalis* contributes quantitative trait loci and candidate genes associated with *Ascochyta*

blight resistance in lentil. Front Plant Sci 12:703283. https://doi.org/10.3389/fpls.2021.703283
Das R, Nath R, Dikshit HK (2017) Host resistance of lentil genotypes against Stemphylium blight caused by *Stemphylium botryosum* Wallr. in lower Gangetic alluvial zone of West Bengal, India. J Mycopathol Res 55(2):169–172
Davies PA, LÜlsdorf MM, Ahmad M (2007) Wild relatives and biotechnological approaches. In: Yadav SS, McNeil DL, Stevenson PC (eds) Lentil. Springer, Dordrecht, pp 225–240. https://doi.org/10.1007/978-1-4020-6313-8_14
De Almeida Costa GE, da Silva Q-MK, Reis SMPM et al (2006) Chemical composition, dietary fibre and resistant starch contents of raw and cooked pea, common bean, chickpea and lentil legumes. Food Chem 94(3):327–330
De Benoist B (2008) Conclusions of a WHO technical consultation on folate and vitamin B12 deficiencies. Food Nutr Bull 29(2_suppl 1):S238–S244
Delahunty AJ, Brand JD, Nuttall JG (2023) Field screening of lentil (*Lens culinaris*) for high-temperature tolerance. Agronomy 13(7):1753
Dhanda SS, Sethi GS, Behl RK (2004) Indices of drought tolerance in wheat genotypes at early stages of plant growth. J Agron Crop Sci 190(1):6–12
Dikshit HK, Singh A, Singh D, Aski M, Jain N, Hegde VS, Basandrai AK, Basandrai D, Sharma TR (2016) Tagging and mapping of SSR marker for rust resistance gene in lentil (*Lens culinaris* Medikus subsp. *culinaris*). Indian J Exp Biol 54(6):394–399
Dikshit HK, Mishra GP, Aski MS et al (2022) Lentil breeding. In: Fundamentals of field crop breeding. Springer Nature, Singapore, pp 1181–1236
Dixit P, Dubey DK (1986) Karyotype study in lentil. Lens Newsl 13(1):8–10
Dutta H, Mishra GP, Aski MS et al (2022) Comparative transcriptome analysis, unfolding the pathways regulating the seed-size trait in cultivated lentil (*Lens culinaris* Medik.). Front Genet 13:942079. https://doi.org/10.3389/fgene.2022.942079
Dutta H, Shivaprasad KM, Aski MS, Mishra GP et al (2023) Morphobiochemical characterization of a RIL population for seed parameters and identification of candidate genes regulating seed size trait in lentil (*Lens culinaris* Medik.). Front. Plant Sci 14:1091432. https://doi.org/10.3389/fpls.2023.1091432
El-Bouhssini M, Sarker A, Erskine W, Joubi A (2008) First sources of resistance to Sitona weevil (*Sitona crinitus* Herbst) in wild *Lens* species. Genet Resour Crop Evol 55:1–4
Emami MK, Sharma B (1996a) Digenic control of cotyledon colour in lentil (*Lens culinaris*). Indian J Genet 56(3):357–361
Emami MK, Sharma B (1996b) Inheritance of brown leaf pigmentation in lentil (*Lens culinaris*). Indian J Genet 56(3):362–365
Emami MK, Sharma B (1999) Linkage between three morphological markers in lentil. Plant Breed 118(6):579–581
Erskine W, Meyveci K, Izgin N (1981) Screening a world lentil collection for cold tolerance. Lens Newsl 8:5–8
Erskine W, Adham Y, Holly L (1989) Geographic distribution of variation in quantitative traits in a world lentil collection. Euphytica 43(1):97–103
Erskine W, Diekmann J, Jegatheeswaran P, Salkini A, Saxena MC, Ghanaim A, El Ashkar F (1991) Evaluation of lentil harvest systems for different sowing methods and cultivars in Syria. J Agric Sci 117(3):333–338
Erskine W, Saxena NP, Saxena MC (1993) Iron deficiency in lentil: yield loss and geographic distribution in a germplasm collection. Plant Soil 151(2):249–254
Erskine W, Sarker A, Ashraf M (2011) Reconstructing an ancient bottleneck of the movement of the lentil (*Lens culinaris* ssp. *culinaris*) into South Asia. Genet Resour Crop Evol 58:373–381
Erskine HE, Norman RE, Ferrari AJ, Chan GC, Copeland WE, Whiteford HA, Scott JG (2016) Long-term outcomes of attention-deficit/hyperactivity disorder and conduct disorder: a systematic review and meta-analysis. J Am Acad Child Adolesc Psychiatry 55(10):841–850
Eujayl I, Baum M, Powell W et al (1998) A genetic linkage map of lentil (*Lens* sp.) based on RAPD and AFLP markers using recombinant inbred lines. Theor Appl Genet 97(1–2):83–89. https://doi.org/10.1007/s001220050869
Fang Y, Xiong L (2015) General mechanisms of drought response and their application in drought resistance improvement in plants. Cell Mol Life Sci 72(4):673–689. https://doi.org/10.1007/s00018-014-1767-0
Fedoruk MJ, Vandenberg A, Bett KE (2013) Quantitative trait loci analysis of seed quality characteristics in lentil using single nucleotide polymorphism markers. Plant Genom 6(3):1–10
Ferguson ME, Maxted N, Van Slageren M, Robertson LD (2000) A re-assessment of the taxonomy of Lens Miller (*Leguminosae*, *Papilionoideae*, *Vicieae*). Bot J Linn Soc 133:41–59
Fernández-Aparicio M, Sillero JC, Rubiales D (2009) Resistance to broomrape in wild lentils (Lens spp.). Plant Breed 128(3):266–270
Fernandez-Orozco R, Gallardo-Guerrero L, Hornero-Méndez D (2013) Carotenoid profiling in tubers of different potato (Solanum sp) cultivars: accumulation of carotenoids mediated by xanthophyll esterification. Food Chem 141(3):2864–2872
Fiala JV, Tullu A, Banniza S, Séguin-Swartz G, Vandenberg A (2009) Interspecies transfer of resistance to anthracnose in lentil (*Lens culinaris* Medic.). Crop Sci 49(3):825–830. https://doi.org/10.2135/cropsci2008.05.0260
Fratini R, Ruiz ML (2006) Interspecific hybridization in the genus *Lens* applying *in vitro* embryo rescue. Euphytica 150:271–280. https://doi.org/10.1007/s10681-006-9118-3
Ganesan K, Xu B (2017) Polyphenol-rich lentils and their health promoting effects. Int J Mol Sci 18(11):2390
Garfinkel Y, Kislev ME, Zohary D (1988) Lentil in the pre-pottery Neolithic B Yiftaḥ'el: additional evidence of its early domestication. Isr J Bot 37:49–51
Gela T, Ramsay L, Haile TA, Vandenberg A, Bett K (2021) Identification of anthracnose race 1 resistance loci in lentil by integrating linkage mapping and genome-wide association study. Plant Genome 14(3):e20131
Ghosh S, Watson A, Gonzalez-Navarro OE et al (2018) Speed breeding in growth chambers and glasshouses for crop breeding and model plant research. Nat Protoc 13:2944–2963
Gore PG, Tripathi K, Chauhan SK, Singh M, Bisht IS, Bhalla S (2016) Searching for resistance in wild *Lens* species against pulse beetle, *Callosobruchus chinensis* (L.). Legum Res 39(4):630–636
Gorim LY, Vandenberg A (2017) Evaluation of wild lentil species as genetic resources to improve drought tolerance in cultivated lentil. Front Plant Sci 8:1129. https://doi.org/10.3389/fpls.2017.01129
Grusak MA (2009) Nutritional and health-beneficial quality. In: Erskine W, Muehlbauer FJ, Sarker A, Sharma B (eds) The lentil—botany, production and uses. CABI, Wallingford, pp 368–390
Gupta D, Sharma SK (2006) Evaluation of wild *Lens* taxa for agro-morphological traits, fungal diseases and moisture stress in North Western Indian Hills. Genet Resour Crop Evol 53:1233–1241
Gupta PK, Singh J (1981) Standard karyotype in lentil var. Pant L-639. Lentil Exp News Serv 8:23
Gupta DS, Cheng P, Sablok G et al (2016) Development of a panel of unigene-derived polymorphic EST–SSR markers in lentil using public database information. Crop J 4(5):425–433
Gupta S, Singh A, Dikshit HK et al (2018) Assessment of total phenol content, total flavonoid content and anti-oxidant capacity in exotic lentil germplasm. Chem Sci Rev Lett 7(26):459–463

Gupta D, Dadu RHR, Sambasivam P et al (2019) Toward climate-resilient lentils: challenges and opportunities. In: Genomic designing of climate-smart pulse crops. Springer, Cham, pp 165–234

Gupta S, Das S, Dikshit HK et al (2021a) Genotype by environment interaction effect on grain iron and zinc content of Indian and Mediterranean lentil genotypes. Agronomy 11:1761. https://doi.org/10.3390/agronomy11091761

Gupta S, Kushwah A, Singh A et al (2021b) Nutritional profiling and antioxidant assessment of Indian and exotic lentil accessions. Indian. J Genet 81:440–449. https://doi.org/10.31742/IJGPB.81.3.11

Gupta S, Dikshit HK, Mishra GP, Aski M (2023) Association mapping of total flavonoid content and antioxidant capacities in lentil (*Lens culinaris* Medik.). Indian J Genet 83(3):326–334. https://doi.org/10.31742/ISGPB.83.3.4

Haddad NI, Bogyo TP, Muehlbauer FJ (1982) Genetic variance of six quantitative character in three lentil (*Lens culinaris* Medic.) crosses. Euphytica 31:113–120

Hamdi A, Erskine W (1996) Reaction of wild species of the genus *Lens* to drought. Euphytica 91:173–179. https://doi.org/10.1007/BF00021067

Hamdi A, Erskine W, Gates P (1991) Relationships among economic characters in lentil. Euphytica 57:109–116

Hamdi A, Küsmenoğlu I, Erskine W (1996) Sources of winter hardiness in wild lentil. Genet Resour Crop Evol 43:63–67. https://doi.org/10.1007/BF00126942

Hamwieh A, Udupa SM, Choumane W et al (2005) A genetic linkage map of *Lens* sp. based on microsatellite and AFLP markers and localization of *Fusarium* vascular wilt resistance. Theor Appl Genet 110:669–677

Hansen J, Renfrew JM (1978) Palaeolithic-Neolithic seed remains at Franchthi cave. Greece Nat 271:349–352

Hanson MG, Zahradka P, Taylor CG (2014) Lentil-based diets attenuate hypertension and large artery remodelling in spontaneously hypertensive rats. Br J Nutr 111(4):690–698

Havey MJ, Muehlnauer FJ (1989) Linkages between restriction fragment length, isozyme, and morphological markers in lentil. Theor Appl Genet 77(3):395–401

Hawtin GC, Singh KB, Saxena MC (1980) Some recent developments in the understanding and improvement of *Cicer* and *Lens*. In: International legume conference. Royal Botanical Gardens, Kew

Hefni M, Öhrvik V, Tabekha MM et al (2010) Folate content in foods commonly consumed in Egypt. Food Chem 121:540–545

Helbeck H (1959) Domestication of food plants in old world. Sci 130:365–372

Helbeck H (1963) Late Cypriote vegetable diet in Apliki. Act Instit Athen Reg Sueciae Ser 4(8):171–186

https://www.icarda.org/media/news/wild-relatives-add-significant-agronomic-improvement-and-disease-resistance-lentils

Hura T, Hura K, Grzesiak S, Banaszak Z (2010) Simulation of osmotic stress during the early stages of triticale development as a promising laboratory test for screening drought resistance. Cereal Res Commun 38(3):327–334. http://www.jstor.org/stable/23790359

Idrissi O, Van Damme P (2018) Application of extended photoperiod to lentil: towards accelerated genetic gain in breeding for rapid improved variety development. In: Proceedings of the International conference on scientific innovation for a sustainable development of African agriculture, Ghent, Belgium

Idrissi O, Udupa SM, De Keyser E et al (2016) Identification of quantitative trait loci controlling root and shoot traits associated with drought tolerance in a lentil (*Lens culinaris* Medik.) recombinant inbred line population. Front Plant Sci 7:1174

Idrissi O, Sahri A, Udupa S, Kumar S (2019) Single seed descent under extended photoperiod as a simple, rapid and efficient breeding method for accelerated genetic gain in lentil. In: Proceedings of the 3rd international legume society conference, Poznan, Poland

Islam MR, Nessa B, Haque MM et al (2009) Effect of soil flooding stress on morphology and yield of five lentil (*Lens culinaris* Medik.) genotypes. IUP J Soil Water Sci 2:48–57

Jadhav S, Kumar A, Lal SK, Akhtar J, Aski M, Mishra GP, Javeria S (2022) SCAR marker development for quick detection of *Fusarium oxysporum* f. sp. *lentis*. Indian J Agril Sci 92(2):278–280. https://doi.org/10.56093/ijas.v92i2.122252

Jadhav S, Kumar A, Lal SK, Akhtar J, Aski M, Mishra GP, Singh AK, Javeria S (2023) Effect of *Fusarium oxysporum* f. sp. *lentis* on seed quality parameters in lentil. Indian J Plant Genet Resour 36(1):70–76. https://doi.org/10.61949/09761926.2023.v36i01.10

Jawad M, Malik SR, Sarwar M et al (2019) Genetic analysis of lentil (*Lens culinaris*) exotic germplasm to identify genotypes suitable for mechanical harvesting. Pak J Agric Res 32:152–158

Jenkins DJ, Kendall CW, Augustin LS et al (2012) Effect of legumes as part of a low glycemic index diet on glycemic control and cardiovascular risk factors in type 2 diabetes mellitus: a randomized controlled trial. Arch Intern Med 172:1653–1660

Jha AB, Ashokkumar K, Diapari M et al (2015a) Genetic diversity of folate profiles in seeds of common bean, lentil, chickpea and pea. J Food Compos Anal 42:134–140

Jha TB, Mahant A, Ghorai A (2015b) Karyotype analysis of Indian lentils through EMA based Giemsa staining. Caryologia 68(4):280–288. https://doi.org/10.1080/00087114.2015.1109921

Jha AB, Gali KK, Zhang H et al (2020) Folate profile diversity and associated SNPs using genome wide association study in pea. Euphytica 216:1–16

Johnson CR, Combs GF Jr, Thavarajah P (2013) Lentil (Lens culinaris L.): a prebiotic-rich whole food legume. Food Res Int 51:107–113

Johnson CR, Thavarajah P, Fenlason A et al (2015) A global survey of low-molecular weight carbohydrates in lentils. J Food Compos Anal 44:178–185

Kamboj RK, Pandey MP, Chaube HS (1990) Inheritance of resistance to *Fusarium* wilt in Indian lentil germplasm (*Lens culinaris* Medik.). Euphytica 50(2):113–117

Kannainyan J, Nene YL (1976) Reaction of lentil germplasm and cultivars against three root pathogens. Indian J Agric Sci 46:165–167

Kannaiyan J, Nene YL (1975) Note on the effect of sowing dates on the reaction of 12 lentil varieties to wilt disease. Madras Agric J 62:240

Kaur S, Cogan NO, Stephens A et al (2014) EST-SNP discovery and dense genetic mapping in lentil (Lens culinaris Medik.) enable candidate gene selection for boron tolerance. Theor Appl Genet 127:703–713. https://doi.org/10.1007/s00122-013-2252-0

Kays DE (1979) Food legumes. Tropical development and research institute. TPI Crop Product Digest 3:48–71

Khare MN, Agrawal SC (1978) Lentil rust survey in Madhya Pradesh. In: Paper presented in All India Pulses Workshop held at Baroda

Khare MN, Agrawal SC, Jain AC (1979) Lentil diseases and their control. Technical Bulletin. JNKVV, Jabalpur

Khazaei H, Caron CT, Fedoruk M et al (2016) Genetic diversity of cultivated lentil (*Lens culinaris* Medik.) and its relation to the world's agro-ecological zones. Front. Plant Sci 7:1093. https://doi.org/10.3389/fpls.2016.01093

Khazaei H, Podder R, Caron CT et al (2017) Marker–trait association analysis of iron and zinc concentration in lentil (*Lens culinaris* Medik.) seeds. Plant Genome 10(2):1

Khazaei H, Fedoruk M, Caron CT et al (2018) Single nucleotide polymorphism markers associated with seed quality characteristics of cultivated lentil. Plant Genome 11:1. https://doi.org/10.3835/plantgenome2017.06.0051

Kumar Y (2002) Inheritance and linkage of genes for morphological trait in lentil (*Lens culinaris* Medik.). PhD thesis, Chaudhary Charan Singh University, Meerut, India

Kumar J, Gupta S (2019) Inheritance of qualitative and quantitative traits in interspecific crosses of lentil. Indian J Genet 79(03):626–631

Kumar Y, Sharma B, Mishra SK, Tyagi MC, Singh SP (2004) Inheritance of leaflet size in lentil (*Lens culinaris* Medik.). J Lentil Res 1:27–29

Kumar Y, Mishra SK, Tyagi MC, Sharma B (2005a) Inheritance of pod size in lentil (*Lentil culinaris* Medik.). J Lentil Res 2:31–33

Kumar Y, Sharma B, Mishra SK, Tyagi MC (2005b) Linkage between genes for flower colour, seed coat pattern and seed coat colour in lentil (*Lens culinaris* Medik.). J Lentil Res 2:22–26

Kumar J, Basu PS, Srivastava E et al (2012) Phenotyping of traits imparting drought tolerance in lentil. Crop Pasture Sci 63:547–554

Kumar J, Srivastava E, Singh M, Mahto D, Pratap A, Kumar S (2014) Lentil. In: Pratap A, Kumar J (eds) Alien gene transfer in crop plants. Volume 2: Achievements and impacts. Springer, Dordrecht/London, pp 191–205

Kumar S, Rajendran K, Kumar J, Hamwieh A, Baum M (2015) Current knowledge in lentil genomics and its application for crop improvement. Front Plant Sci 6:78. https://doi.org/10.3389/fpls.2015.00078

Kumar J, Kant R, Kumar S et al (2016a) Heat tolerance in lentil under field conditions. Legume Genom Genet 7:1

Kumar J, Singh J, Kanaujia R et al (2016b) Protein content in wild and cultivated taxa of lentil (*Lens culinaris* ssp. *culinaris* Medikus). Indian J Genet 76:631–634

Kumar J, Basu PS, Gupta S et al (2018a) Physiological and molecular characterisation for high temperature stress in *Lens culinaris*. Funct Plant Biol 45:474–487

Kumar S, Choudhary AK, Rana KS et al (2018b) Bio-fortification potential of global wild annual lentil core collection. PLoS One 13:e0191122

Kumar H, Singh A, Dikshit HK et al (2019) Genetic dissection of grain iron and zinc concentrations in lentil (*Lens culinaris* Medik.). J Genet 98(3):1–14. https://doi.org/10.1007/s12041-019-1112-3

Kumar J, Mir RR, Dutta A et al (2024) Estimation of iron, zinc, phytic acid concentration and protein content in lentil seeds over locations and their marker-trait association analysis. J Food Compos Anal 127:105999

Kumari S, Sehgal A, Kumar J, Kumar S, Singh S, Siddique KHM, Nayyar H (2017) Identification of high-temperature tolerant lentil (*Lens culinaris* Medik.) genotypes through leaf and pollen traits. Front Plant Sci 8:744. https://doi.org/10.3389/fpls.2017.00744

Kumari M, Mittal RK, Chahota RK, Thakur K, Lata S, Gupta D (2018) Assessing genetic potential of elite interspecific and intraspecific advanced lentil lines for agronomic traits and their reaction to rust (*Uromyces viciae-fabae*). Crop Pasture Sci 69(10):999–1008

Ladizinsky G (1979) The genetics of several morphological traits in the lentil. J Hered 70:135–137

Ladizinsky G (1985) The genetics of hard seed coatin the genus Lens. Eyphytica 34:539–543

Ladizinsky G (1987) Pulse domestication before cultivation. Econ Bot 41:60–65. https://doi.org/10.1007/bf02859349

Ladizinsky G (1993) Wild lentils. Crit Rev Plant Sci 12:169–184

Ladizinsky G (1997) A new species of Lens from South-East Turkey. Bot J Linn Soc 123(3):257–260

Ladizinsky G, Sakar D (1982) Morphological and cytogenetical characteization of *Vicia montbrettii* Fisch. & Mey. (Synonym: *Lens montbrettii* (Fisch. & Mey.) Davis & Plitmann). Bot J Linn Soc 85:209–212

Ladizinsky G, Cohen D, Muehlbauer FJ (1985) Hybridization in the genus *Lens* by means of embryo culture. Theor Appl Genet 70:97–101

Lal S, Srivastava RS (1975) Inheritance of flower colour in lentils. Indian J Genet 35(1):29–30

Laserna-Ruiz I, De-Los-Mozos-Pascual M, Santana-Méridas O, Sánchez-Vioque R, Rodríguez-Conde MF (2012) Screening and selection of lentil (*Lens* Miller) germplasm resistant to seed bruchids (*Bruchus spp.*). Euphytica 188(2):153–162

Lavania UC, Lavania S (1983) Karyotype studies in Indian pulses. Genet Agrar 37:299–308

Leisner CP (2020) Climate change impacts on food security-focus on perennial cropping systems and nutritional value. Plant Sci 293:110412

Leport L, Turner NC, French RJ, Tennant D, Thomson BD, Siddique KHM (1998) Water relations, gas exchange and growth of cool-season grain legumes in a Mediterranean-type environment. Eur J Agron 9(4):295–303. https://doi.org/10.1016/S1161-0301(98)00042-2

Lombardi M, Materne M, Cogan NO et al (2014) Assessment of genetic variation within a global collection of lentil (*Lens culinaris* Medik.) cultivars and landraces using SNP markers. BMC Genet 15:150. https://doi.org/10.1186/s12863-014-0150-3

Ma Y, Marzougui A, Coyne CJ et al (2020) Dissecting the genetic architecture of aphanomyces root rot resistance in lentil by QTL mapping and genome-wide association study. Int J Mol Sci 21(6):2129

Malik AI, Ailewe TI, Erskine W (2015) Tolerance of three grain legume species to transient waterlogging. AoB Plant 7:1

Materne MA (2003) Importance of phenology and other key factors in improving the adaptation of lentil (*Lens culinaris* Medikus) in Australia. PhD thesis, University of Western Australia, Perth, Australia

Matny ON (2015) Lentil (Lens culinaris) current status and future prospect of production in Ethiopia. Adv Plants Agric Res 2:45–53. https://doi.org/10.15406/apar.2015.02.00040

Mbasani-Mansi J, Ennami M, Briache FZ et al (2019) Characterization of genetic diversity and population structure of Moroccan lentil cultivars and landraces using molecular markers. Physiol Mol Biol Plants 25:965–974. https://doi.org/10.1007/s12298-019-00673-5

Mirali M, Purves RW, Vandenberg A (2017) Profiling the phenolic compounds of the four major seed coat types and their relation to color genes in lentil. J Nat Prod 80:1310–1317

Mishra RP (1973) Studies on powdery mildew of lentil (*Erysiphe polygoni* DC) in Madhya Pradesh. Punjabrao Krishi Vidyapeeth Res J 2:72–73

Mishra SK (2004) Final technical report on development of comprehensive genetic linkage map in lentil (*Lens culinaris* Medik.). ICAR, New Delhi

Mishra BK, Srivastava JP, Lal JP et al (2016) Physiological and biochemical adaptations in lentil genotypes under drought stress. Russ J Plant Physiol 63:695–708

Mishra GP, Aski MS, Bosamia T et al (2022) Insights into the host-pathogen interaction pathways through RNA-Seq analysis of *Lens culinaris* Medik. in response to *Rhizoctonia bataticola* infection. Genes 13:90. https://doi.org/10.3390/genes13010090

Mishra GP, Dikshit HK, Aski MS, Singh A, Tripathi K, Gupta S, Kumar S (2024) Recent advancements in trait mapping in lentil (*Lens culinaris* Medik.). In: The lentil genome. Academic, pp 93–130

Mobini SH, Lulsdorf M, Warkentin TD et al (2016) Low red: far-red light ratio causes faster in vitro flowering in lentil. Can J Plant Sci 96:908–918

Mohammad A, Kumar U (1986) Screening of lentil varieties against *Ozononium texanum* var. *parasiticum* and *Sclerotium rolfsii* causing wilt and collar rot. Indian Phytopathol 39(1):93–95

Mudgal V, Mehta MK, Rane AS (2018) Lentil straw (*Lens culinaris*): an alternative and nutritious feed resource for kids. Anim Nutr 4:417–421. https://doi.org/10.1016/j.aninu.2018.04.009

Muehlbauer FJ, Cubero JI, Summerfield RJ (1985) Lentil (*Lens culinaris* Medic.). In: Summerfield RJ, Roberts EH (eds) Grain legume crops. Collins, London, pp 266–311

Munns R (2002) Comparative physiology of salt and water stress. Plant Cell Environ 25:239–250

Murray GA, Eser D, Gusta LV et al (1988) Winter hardiness in pea, lentil, faba bean and chickpea. In: World crops: cool season food legumes. Springer, Dordrecht, pp 831–843

Nabati J, Mirmiran SM, Nezami A, Ahmadi-Lahijani MJ, Boroumand Rezazadeh E (2021) Screening lentil (*Lens culinaris* Medik.) genotypes for fall sowing and low temperature tolerance. Arch Agron Soil Sci 69. https://doi.org/10.1080/03650340.2021.1984429

Nene YL, Kannaiyan J, Saxena GC (1975) Note on performance of lentil varieties and germplasm cultures against *Uromyces fabae* (Pers.) de Bary. Indian J Agric Sci 45:177–178

Nuttall J, Nicolas M, Brand J (2015) Genotypic heat tolerance in lentil. In: Proceedings of the 17th ASA conference, pp 20–24

Omar GI, Saqer MM, Adwan GM (2019) Phylogenetic relationship among some species of the Genera *Lens, Vicia, Lathyrus* and *Pisum* (Leguminosae) in Palestine. Jordan J Biol Sci 12(3):289–296

Pandey MP (1981) Rabi pulse varieties for higher production. Indian Farm Digest 14:18–22

Passioura JB (1981) The role of root system characteristics in drought resistance of crop plants. In: Special international symposium on principles and methods of crop improvement for drought resistance: with emphasis on rice, College, Laguna (Philippines)

Podder R, Banniza S, Vandenberg A (2013) Screening of wild and cultivated lentil germplasm for resistance to *stemphylium* blight. Plant Genet Resour 11(1):26–35

Polanco C, de Miera LE, González AI et al (2019) Construction of a high-density interspecific (*Lens culinaris* x *L. odemensis*) genetic map based on functional markers for mapping morphological and agronomical traits, and QTLs affecting resistance to Ascochyta in lentil. PLoS One 14(3):e0214409

Rajandran V, Schoor ORV, Butler JK et al (2022) Genetic analysis of early phenology in lentil identifies distinct loci controlling component traits. J Exp Bot 73(12):3963–3977

Rajpal VR, Singh A, Kathpalia R, Thakur RK, Khan MK, Pandey A, Hamurcu M, Raina SN (2023) The prospects of gene introgression from crop wild relatives into cultivated lentil for climate change mitigation. Front Plant Sci 14:1127239

Ray H, Bett KE, Taran B et al (2014) Mineral micronutrient content of cultivars of field pea, chickpea, common bean, and lentil grown in Saskatchewan, Canada. Crop Sci 54:1698–1708

Roy SJ, Tucker EJ, Tester M (2011) Genetic analysis of abiotic stress tolerance in crops. Curr Opin Plant Biol 14:232–239

Rozan P, Kuo YH, Lambein F (2001) Nonprotein amino acids in edible lentil and garden pea seedlings. Amino Acids 20(3):319–324

Rychlik M, Englert K, Kapfer S et al (2007) Folate contents of legumes determined by optimized enzyme treatment and stable isotope dilution assays. J Food Compos Anal 20:411–419

Sarker A, Erskine W, Sharma B, Tyagi MC (1998) Genetics of flowering in lentil: a new major gene and polygenes. In: Abstracts of the third European conference on grain legumes. Valladolid, Spain, p 465

Sarker A, Erskine W, Sharma B, Tyagi MC (1999) Inheritance and linkage relationship of days to flower and morphological loci in lentil (*Lens culinaris* Medikus subsp. *Culinaris*). J Hered 90(2):270–275

Sarker A, Ketata H, Erskine W et al (2002) Winter lentils promise improved nutrition and income in west Asian highlands. ICARDA Caravan (ICARDA) 16:14–16

Sarker A, Kumar S, Kumar J et al (2016) Breeding pulses for nutritional quality with emphasis on biofortification. In: Abstract book of international conference on pulses. Marrakesh, Morocco, pp 18–20

Scalbert A, Williamson G (2000) Dietary intake and bioavailability of polyphenols. J Nutr 130(8):2073S–2085S

Sehgal A, Sita K, Kumar J et al (2017) Effects of drought, heat and their interaction on the growth, yield and photosynthetic function of lentil (*Lens culinaris* Medikus) genotypes varying in heat and drought sensitivity. Front Plant Sci 8:1776

Sen Gupta D, Thavarajah D, Knutson P et al (2013) Lentils (*Lens culinaris* L.), a rich source of folates. J Agric Food Chem 61:7794–7799

Sen Gupta D, Kumar J, Barpete S et al (2023) Lentils (*Lens culinaris* Medik): nutritional profile and biofortification prospects. In: Kole C (ed) Compendium of crop genome designing for neutraceuticals. Springer, Singapore, pp 719–745. https://doi.org/10.1007/978-981-19-3627-2_27-1

Shaikh R, Diederichsen A, Harrington M, Adam J, Conner RL, Buchwaldt L (2013) New sources of resistance to *Colletotrichum truncatumrace Ct0* and *Ct1* in *Lens culinaris* Medikus subsp. *culinaris* obtained by single plant selection in germplasm accessions. Genet Resour Crop Evol 60:193–201. https://doi.org/10.1007/s10722-012-9825-7

Sharma SK, Sharma B (1977) Induction of 'fasciata' mutants in lentil. Lentil Exp News Serv 4:23

Sharpe AG, Ramsay L, Sanderson LA et al (2013) Ancient orphan crop joins modern era: gene based SNP discovery and mapping in lentil. BMC Genomics 14:192. https://doi.org/10.1186/1471-2164-14-192

Shivaprasad KM, Dikshit HK, Mishra GP et al (2024a) Delineation of loci governing an extra-earliness trait in lentil (*Lens culinaris* Medik.) using the QTL-Seq approach. Plant Biotechnol J 22:2932. https://doi.org/10.1111/pbi.14415. Epub ahead of print

Shivaprasad KM, Aski M, Mishra GP, Sinha SK, Gupta S, Mishra DC et al (2024b) Genome-wide discovery of InDels and validation of PCR-based InDel markers for earliness in a RIL population and genotypes of lentil (*Lens culinaris* Medik.). PLoS One 19(5):e0302870. https://doi.org/10.1371/journal.pone.0302870

Sidahmed MM, Jaber NS (2004) The design and testing of a cutter and feeder mechanism for the mechanical harvesting of lentils. Biosyst Eng 88(3):295–304

Silim SN, Saxena MC, Erskine W (1993) Adaptation of lentil to the Mediterranean environment. I. Factors affecting yield under drought conditions. Exp Agric 29(1):9–19. https://doi.org/10.1017/S0014479700020366

Sinclair TR, Ludlow MM (1985) Who taught plants thermodynamics? The unfilfilled potential of plant water potential. Aust J Plant Physiol 12:213–217

Sindhu JS, Slinkard AE, Scoles GJ (1983) Studies on variation in Lens I. Karyotype LENS 10:14

Singh JP, Singh IS (1990) Genetics of resistance to rust in lentil. Indian J Pulses Res 3:132–135

Singh M, Rana MK, Kumar K, Bisht IS, Dutta M, Gautam NK, Bansal KC (2013) Broadening the genetic base of lentil cultivars through inter-sub-specific and interspecific crosses of *Lens* taxa. Plant Breed 132(6):667–675

Singh M, Bisht IS, Kumar S, Dutta M, Bansal KC, Karale M, Sarker A, Amri A, Kumar S, Datta SK (2014) Global wild annual Lens collection: a potential resource for lentil genetic base broadening and yield enhancement. PLoS One 9(9):107781

Singh A, Dikshit HK, Singh D et al (2016a) Use of expressed sequence tag microsatellite markers for exploring genetic diversity in lentil and related wild species. J Agric Sci 154:1254–1269. https://doi.org/10.1017/S0021859615001252

Singh D, Singh CK, Tomar RSS et al (2016b) Molecular assortment of *Lens* species with different adaptations to drought conditions using SSR markers. PLoS One 11:e0147213. https://doi.org/10.1371/journal.pone.0147213

Singh D, Singh CK, Kumari S, Tomar RSS, Karwa S, Singh R, Singh RB, Sarkar SK, Pal M (2017) Discerning morpho-anatomical, physiological and molecular multiformity in cultivated and wild genotypes of lentil with reconciliation to salinity stress. PLoS One 12(5):0177465

Singh D, Singh CK, Taunk J, Jadon V, Pal M, Gaikwad K (2019) Genome wide transcriptome analysis reveals vital role of heat responsive genes in regulatory mechanisms of lentil (*Lens culinaris*Medikus). Sci Rep 9(1):12976. https://doi.org/10.1038/s41598-019-49496-0

Singh D, Singh CK, Taunk J et al (2019a) Genome wide transcriptome analysis reveals vital role of heat responsive genes in regulatory mechanisms of lentil (*Lens culinaris* Medikus). Sci Rep 9:1–19

Singh M, Kumar S, Basandrai AK, Basandrai D, Malhotra N, Saxena DR, Gupta D, Sarker A, Singh K (2020) Evaluation and identification of wild lentil accessions for enhancing genetic gains of cultivated varieties. PLoS One 15(3):0229554. https://doi.org/10.1371/journal.pone.0229554

Singh J, Sirari A, Singh H et al (2021) Identifying and validating SSR markers linked with rust resistance in lentil (*Lens culinaris*). Plant Breed 140(3):477–485

Singh D, Singh CK, Taunk J et al (2022a) Linking genome wide RNA sequencing with physio-biochemical and cytological responses to catalogue key genes and metabolic pathways for alkalinity stress tolerance in lentil (*Lens culinaris* Medikus). BMC Plant Biol 22(1):99. https://doi.org/10.1186/s12870-022-03489-w

Singh M, Kumar S, Mehra R et al (2022b) Evaluation and identification of advanced lentil interspecific derivatives resulted in the development of early maturing, high yielding, and disease-resistant cultivars under Indian agro-ecological conditions. Front Plant Sci 13:936572. https://doi.org/10.3389/fpls.2022.936572

Sinha RP, Yadav BP (1989) Inheritance of resistance to rust in lentil. Lens Newsl 16:41

Siva N, Thavarajah P, Kumar S et al (2019) Variability in prebiotic carbohydrates in different market classes of chickpea, common bean, and lentil collected from the American local market. Front Nutr 6:38

Spaeth SC, Muehlbauer FJ (1991) Registration of three germplasms of winter hardy lentil. Crop Sci 31:1395–1395

Sudheesh S, Rodda MS, Davidson J et al (2016) SNP-based linkage mapping for validation of QTLs for resistance to Ascochyta blight in lentil. Front Plant Sci 7:604. https://doi.org/10.3389/fpls.2016.01604

Sulser H, Sager F (1976) Identification of uncommon amino acids in the lentil seed (Lens culinaris Med.). Experientia 32:422–423

Tahir M, Muehlbauer FJ (1994) Gene mapping in lentil with recombinant inbred lines. J Hered 85(4):306–310

Tahir M, Lindeboom N, Båga M et al (2011) Composition and correlation between major seed constituents in selected lentil (*Lens culinaris*. Medik) genotypes. Can J Plant Sci 91:825–835

Temel HY, Go D, Kahriman A, Tanyolac MB (2014) Construction of linkage map through genotyping-by-sequencing in lentil. In: Proceedings of plant and animal genome conference, p 358

Tepe I, Erman M, Yazlýk A et al (2005) Comparison of some winter-lentil cultivars in weed crop competition. Crop Prot 24:585–595

Terzi R, Sağlam A, Kutlu N et al (2010) Impact of soil drought stress on photochemical efficiency of photosystem II and antioxidant enzyme activities of *Phaseolus vulgaris* cultivars. Turk J Bot 34:1

Thavarajah D, Ruszkowski J, Vandenberg A (2008) High potential for selenium biofortification of lentils (*Lens culinaris* L.). J Agric Food Chem 56(22):10747–10753

Thavarajah P, Sarker A, Materne M et al (2011) A global survey of effects of genotype and environment on selenium concentration in lentils (*Lens culinaris* L.): implications for nutritional fortification strategies. Food Chem 125:72–76

Tripathi K, Kumari J, Gore PG et al (2022) Agro-morphological characterization of lentil germplasm of Indian national genebank and development of a core set for efficient utilization in lentil improvement programs. Front Plant Sci 12:751429. https://doi.org/10.3389/fpls.2021.751429

Tripathi K, Gupta D, Gore PG, Gowthami R, Mishra GP, Pamarthi RK, Dikshit HK, Kumar S (2024) Exploitation of lentil wild relatives for genetic improvement. In: Kumar J, Sen Gupta D, Kumar S (eds) The lentil genome. Academic, pp 159–194

Tsanakas GF, Mylona PV, Koura K et al (2018) Genetic diversity analysis of the Greek lentil (*Lens culinaris*) landrace 'Eglouvis' using morphological and molecular markers. Plant Genet Res 16:469–477

Tullu A, Banniza S, Tar'an B, Warkentin T, Vandenberg A (2010) Sources of resistance to ascochyta blight in wild species of lentil (*Lens culinaris* Medik.). Genet Resour Crop Evol 57:1053–1063

Tullu A, Buchwaldt L, Lulsdorf M et al (2006a) Sources of resistance to anthracnose (*Colletotrichum truncatum*) in wild *Lens* species. Genet Resour Crop Evol 53:111–119

Tullu A, Tar'an B, Breitkreutz C et al (2006b) A quantitative trait locus for resistance to Ascochyta blight Ascochyta lentis maps close to a gene for resistance to anthracnose Colletotrichum truncatum in lentil. Can J Plant Pathol 28:588–595

Tullu A, Tar'an B, Warkentin T, Vandenberg A (2008) Construction of an intraspecific linkage map and QTL analysis for earliness and plant height in lentil. Crop Sci 48(6):2254–2264. https://doi.org/10.2135/cropsci2007.11.062

Tullu A, Banniza S, Bett K, Vandenberg A (2011) A walk on the wild side: exploiting wild species for improving cultivated lentil. Grain Legum 56:13–14

Tullu A, Bett K, Banniza S et al (2013a) Widening the genetic base of cultivated lentil through hybridization of Lens culinaris 'Eston' and L. ervoides accession IG 72815. Can J Plant Sci 93:1037–1047

Tullu A, Bett K, Banniza S, Vail S, Vandenberg A (2013b) Widening the genetic base of cultivated lentil through hybridization of *Lens culinaris* 'Eston' and *L. ervoides* accession IG 72815. Can J Plant Sci 93(6):1037–1047. https://doi.org/10.4141/cjps2013-072

Vagelas I (2014) Diagnosis and management of *Sclerotinia* stem rot (white mould) of lentils in Greece. Arch Phytopathol Pflanzenschutz 47(10):1209–1217

Vail S, Strelioff JV, Tullu A, Vandenberg A (2012) Field evaluation of resistance to *Colletotrichum truncatum* in *Lens culinaris*, *Lens ervoides*, and *Lens ervoides*× *Lens culinaris* derivatives. Field Crop Res 126:145–151

Vallincourt CE, Slinkard AE, Reichert RD (1986) The inheritance of condensed tannin concentration in lentil. Can J Plant Sci 66:241–246

Van Vliet S, Burd NA, Van Loon LJ (2015) The skeletal muscle anabolic response to plant-versus animal-based protein consumption. J Nutr 145:1981–1991

Vandemark GJ, Grusak MA, McGee RJ (2018) Mineral concentrations of chickpea and lentil cultivars and breeding lines grown in the US Pacific Northwest. Crop J 6:253–262

Vandenberg A, Slinkard AE (1989) Inheritance of four new qualitative genes in lentil. J Hered 80:320–322

Vandenberg A, Slinkard AE (1990) Genetics of seed coat color and pattern in lentil. J Hered 81(6):484–488

Verma SNP, Nakhtore CL (1976) Genetic studies in lentil (*Lens culinaris* Medic). JNKVV Res J 10(3):272–273

Verma P, Goyal R, Chahota RK et al (2015) Construction of a genetic linkage map and identification of qtls for seed weight and seed size traits in lentil (*Lens culinaris* Medik.). PLoS One 10:e0139666

Wilson VE, Hudson LW (1978) Inheritance of lentil flower colour. J Hered 69:129–130

Wilson VE, Law AG (1972) Natural crossing in *Lens esculenta* Moench. J Soc Hortic Sci 97:142–143

Wong MM, Gujaria-Verma N, Ramsay L et al (2015) Classification and characterization of species within the genus *Lens* using genotyping-by-sequencing (GBS). PLoS One 10(3):e0122025

Ye G, McNeil DL, Hill GD (2000) Two major genes confer Ascochyta blight resistance in *Lens orientalis*. N Z Plant Prot 53:109–113

Zhang B, Deng Z, Ramdath DD et al (2015) Phenolic profiles of 20 Canadian lentil cultivars and their contribution to antioxidant activity and inhibitory effects on α-glucosidase and pancreatic lipase. Food Chem 172:862–872

Zhang H, Jha AB, De Silva D et al (2019) Improved folate monoglutamate extraction and application to folate quantification from wild lentil seeds by ultra-performance liquid chromatography-selective reaction monitoring mass spectrometry. J Chromatogr B 1121:39–47

Zohary D, Hopf M, Weiss E (2012) Domestication of plants in the Old World: the origin and spread of domesticated plants in Southwest Asia, Europe, and the Mediterranean basin. Oxford Academic, Oxford

Field Pea (*Pisum sativum* L.) Breeding

Ashok K. Parihar, Abhishek Bohra, Shailesh Tripathi, and Girish Prasad Dixit

Abstract

Among the cool season legumes, field pea (*Pisum sativum* L.) is known as a highly productive pulse crop which is being cultivated in more than 90 countries. Being an excellent source of macro- and micronutrients, it plays an instrumental role in providing food and nutritional security to resource-poor folks of different countries. So far, limited efforts and resources have been dedicated to designing climate-resilient, nutrient-dense field pea. However, during the past two decades, concerted efforts have been made to infuse resistance to biotic stress along with better plant types to achieve higher productivity. This chapter briefly elaborates on the importance, origin, taxonomy, botany, genetic and genomic resources, major biotic and abiotic constraints, and the current scenario of global productivity. In addition, to increase the genetic gain, priority traits i.e. yield and its attributes, biotic and abiotic stresses, lodging resistance, herbicide tolerance, and nutritional enrichment have been discussed. Furthermore, modern tools like genomic-assisted breeding, speed breeding, and genome editing; current status; and future outlooks have been deliberated. To accelerate genetic gains in yield, resistance to biotic and abiotic stresses, nutritional quality, and lodging resistance, it is essential to integrate modern tools and techniques—such as genome editing, speed breeding, genomic selection, and omics-based approaches (including proteomics, metabolomics, and epigenomics)—into conventional breeding programmes.

Keywords

Origin · Taxonomy · Chromosome · Genome · Biotic and abiotic stress · GAB · Speed breeding · Genome editing

A. K. Parihar (✉) · A. Bohra · S. Tripathi · G. P. Dixit
ICAR-Indian Institute of Pulses Research, Kanpur, Uttar Pradesh, India

1 Introduction

Field pea or dry pea (*Pisum sativum* L.) is a highly productive and post-rainy, cool season food legume cultivated in more than 90 countries across the globe in different tropical and temperate regions (Holdsworth et al. 2017; Parihar et al. 2020a, b, 2022a). Pea ranks as the fourth most important legume crop globally, covering approximately 9.77 million hectares and producing around 35.11 million tonnes. Of this total, dry pea is cultivated on about 7.15 million hectares, yielding 14.16 million tonnes with an average productivity of 1978 kg per hectare (FAOSTAT 2022). Figure 1 illustrates the current global scenario of pea cultivation in terms of area and production. The region-wise global dry pea production scenario illustrates the predominance of the European region with a 40.4% share, followed by the Americas with 34.0%, while Asia shares about 19.2%. In contrast, Asia holds a dominant position in global green pea production, accounting for 87.9% of the total output, followed by Europe (5.7%) and the Americas (3.0%). The major dry pea-producing countries are Canada, the Russian Federation, and China, while for green peas, China and India are the leading producers (FAOSTAT 2022). India contributes more than 10% and 7% to global dry pea acreage and production, respectively. Amongst pulse crops, field pea is an economically important and nutritionally rich crop, acknowledged as the 'poor man's meat' owing to its high protein, macro- and micronutrient profile, as well as its accessibility to resource-poor people at an affordable price (Amarakoon et al. 2012; Parihar et al. 2021a). Being a nutrient-dense pulse crop, it is used in different forms like vegetable, whole grain, split grain, forage, and animal feed (Bastida Garcia et al. 2011; Rubiales et al. 2019; Parihar et al. 2021a). Field pea varieties are available with yellow, green, and orange cotyledons and are employed to prepare various culinary and confectionery items, for example, chat, soup, chhola, dal, stew, snacks, sprouts, vegetables, and sweets (Parihar et al. 2016; Mahajan et al. 2018; Singh et al. 2018; Grothe et al. 2020). As a leguminous crop, green pea offers significant potential within diverse cropping

G. Prasad Dixit et al. (eds.), *Fundamentals of Legume Breeding*, https://doi.org/10.1007/978-981-96-7838-9_8

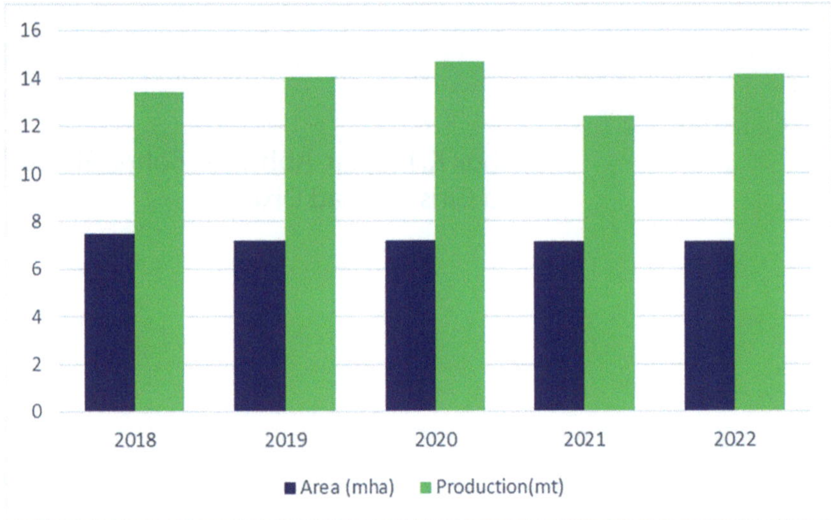

Fig. 1 Global area and production trends of dry peas during 2018–2022

systems by disrupting disease and insect pest life cycles, fixing atmospheric nitrogen, and enhancing overall soil health. It promotes the proliferation of beneficial soil microorganisms, thereby improving soil physical properties and increasing water-holding capacity. This crop has high initial vigour, which makes it most suitable for green manuring as well as a cover crop (Ingels et al. 1994; Biederbeck et al. 2005). The nutritional profile of pea holds great promise in the prevention and management of several health-related problems such as cardiovascular diseases, diabetes, breast, renal and colon cancer, hypertension, inflammation, high cholesterol, hair shedding, and appetite disorders (Sun et al. 2015; Tulbek et al. 2016; Lam et al. 2018; Liao et al. 2019; Ge et al. 2020; Grothe et al. 2020). The high yield potential, nutrient density, and wide adaptability of field pea make it a favourable crop in the prevailing context of climate change, and food and nutritional insecurity. This chapter comprehensively elaborates on the breeding of field pea.

2 Origin and Taxonomy

Pea (*Pisum sativum*) is one of the world's oldest domesticated grain legumes, and its domestication began about 11,000 years ago from its progenitor *P. humile* (syn. *Pisum sativum* subsp. *elatius*) in the Near East. The primary centre of origin for pea is the Near East and Mediterranean region (Ladizinsky and Abbo 2015; Smýkal et al. 2015, 2017, 2018). Countries like Ethiopia, Yemen, and the elevated Asiatic section of the Hindu Kush are considered secondary centres of diversity (Rubiales et al. 2019). However, Vavilov (1949) recognized Ethiopia, Mediterranean countries, and Central Asia as primary centres, while the Near East was considered a secondary centre of diversity (Chimwamurombe and Khulbe 2011).

The dicotyledonous field pea belongs to the family Fabaceae (syn. Leguminosae), sub-family Faboideae (syn. Papilionoideae), tribe Fabeae, and genus *Pisum*. The tribe Fabeae comprises five genera—*Lathyrus*, *Lens*, *Vicia*, *Pisum*, and Vavilovia formosa—and includes approximately 380 species (Schaefer et al. 2012; Mikić et al. 2013). The genus *Pisum* L. includes three species: cultivated pea (*P. sativum* subsp. *sativum*), Ethiopian pea (*P. abyssinicum*), and *P. fulvum*. Among the cultivated pea, five subspecies have been identified: *Pisum sativum* subsp. *sativum*, *arvense*, *elatius*, *humile*, and *hortense* (Maxted and Ambrose 2001; Trněný et al. 2018). Of these, *arvense* and *hortense* are considered varieties within the species *Pisum sativum* (Nasiri et al. 2009). On the basis of cross compatibility and hybridization, the *Pisum* genus has been categorized into three genepools i.e. primary, secondary, and tertiary. The primary gene pool harboured the following species *Pisum sativum* subsp. *elatius*, *Pisum sativum* subsp. *elatius* var. *brevipedunculatum*, *Pisum sativum* subsp. *elatius* var. *elatius*, and *Pisum sativum* subsp. *elatius* var. *pumilio*.

Species that produce partially fertile F_1 hybrids when crossed with members of the primary gene pool are classified as part of the secondary gene pool. While gene transfer from these species to the primary gene pool is possible, it is often challenging. In the genus *Pisum*, the species *Pisum abyssinicum* and *Pisum fulvum* are considered part of the secondary gene pool. Species that form sterile hybrids when crossed with the primary gene pool—and where gene transfer is only possible through advanced techniques such as bridge crossing, tissue culture, or genetic recombination—are classified under the tertiary gene pool. Within the genus *Pisum*, *Vavilovia formosa*, a perennial species found in hilly regions, is considered a member of the tertiary gene pool (Ben Ze'ev and Zohary 1973; Bogdanova et al. 2015; Warkentin et al. 2015).

3 Chromosome Number and Genome

Pea (*Pisum sativum*) is a self-pollinated, diploid, and economically important model plant species in genetics, with a chromosome number of $2n = 2\times = 14$ (Yarnell 1962). Its nuclear genome size is approximately 4.45 Gb and comprises seven pairs of chromosomes (Praca-Fontes et al. 2014). Chromosome numbers and their corresponding linkage groups are designated using Arabic and Roman numerals, respectively, as follows: 1 = VI, 2 = I, 3 = V, 4 = IV, 5 = III, 6 = II, and 7 = VII. Of the seven chromosomes, two (1 and 2) are sub-metacentric, while the remaining five (3, 4, 5, 6, and 7) are acrocentric (Ellis and Poyser 2002). The pea genome is predominantly composed of heterogeneous repetitive DNA sequences, accounting for approximately 75–97% of the total genomic content (Flavell et al. 1974; Novák et al. 2010). Of them, the *Ty3/gypsy* LTR-retrotransposons are the predominant followed by Ogre elements (20–33%) and *Ty1/copia* repeats (Macas et al. 2007). These repeats can be deployed as an important marker to understand chromosomal diversity within its karyotype (Smýkal et al. 2012). The recent reports also validated the occurrence of about 76% heterogenous repetitive DNA sequences of diverse families. Besides, the synteny for repetitive DNA sequences was noticed less between pea and soybean genome compared to pea and *Medicago truncatula* (Macas et al. 2015).

In recent past, pea nuclear genome was assembled into seven pseudomolecules considering NGS platform wherein initially short reads were transformed into contigs and subsequently into scaffolds. The size of 'Caméor' genome assembly measured 3.92 Gb, of which 3.16 Gb consist of contigs that represent about 82.5% of the pea genome. The different in size was owing to the occurrence of repetitive DNAs as well as transposable elements which remains as unassembled reads (Kreplak et al. 2019). However, to understand the phenotypic variation and evolution of pea genome, an enhanced genome assembly with proper annotation is needed (Roberts et al. 2013; Chaisson et al. 2015). Therefore, most recently, a de novo genome assembly of a Chinese pea variety, ZW6, was developed by using a group of sequencing platforms including Illumina NGS technologies. In the pea cultivar 'ZW6', the assembled genome size was 3.71 Gb, of which contigs accounted for 97.96%. The number of annotated genes within the pseudomolecules was higher than that in the cultivar 'Caméor', with the pseudomolecules representing approximately 1.19 Gb of the 'ZW6' genome (Yang et al. 2022). The genome of pea cv. 'ZW6' provides an improved reference genome which effectively addresses the gaps that were present in the previous reference genome. Besides, a pan-genome analysis was performed to get comprehensive understanding of pea genome diversity using individual de novo assembly of the genomes of 118 cultivated and wild pea accessions (Yang et al. 2022). The knowledge of pea genome diversity in detail will certainly enrich the scale of pea genomic resources and facilitate the genomics-assisted breeding in pea.

4 Botany of Field Pea

Field pea is an annual herbaceous plant with hollow, delicious, trailing type stems with well-developed tap root system. The plant stature ranged between 30 and 200 cm with a waxy layer (bluish-green) on all the aerial parts. In cultivated field pea, mainly two types of plant types are predominant i.e. indeterminate (tall type) or determinate (dwarf type) (MacKay 2013; Parihar and Dixit 2023; Rao et al. 2021). The leaf types are pinnately compound, wherein rachis ends in a single or branched tendril and consists of a basal pair of foliaceous stipules, one to several pairs of proximal leaflets. The stipule, leaflets, and tendrils emerge in an acropetal manner on the compound leaf primordium, which is called as marginal blastozone (Gourlay et al. 2000; Pavek 2012). The *afila* (*af*) mutant alters the leaf architecture in pea by replacing leaflets with additional tendrils. This modified leaf pattern is referred to as the tendril type or semi-leafless type. Several cultivars with semi-leafless or 'afila' type leaf pattern have been developed during recent years (Dixit and Parihar 2014; Dixit et al. 2014; Parihar and Dixit 2017, 2024; Parihar et al. 2019a).

In leafless types, the stipules transformed into rudimentary and substituted by tendrils (Mohan et al. 2013; Parihar et al. 2022a). Pea flowers are papilionaceous type developed in the leaf axils and are cleistogamous and highly self-pollinating in nature (Suso et al. 2016; Smýkal et al. 2018). The number of flowers per raceme or peduncle normally varies from one to five flowers (Sinjushin and Belyakova 2015). Individual flower comprises green compound five sepals and white, purple, or pink five petals (Majumdar 2011). The upper side petal is known as 'standard', while the two small petals in the middle which are joint together are termed as 'keel' and the bottom side two petals are called 'wings'. Field peas have reddish purple, white, pink, and blue colour flowers (Elzebroek and Wind 2008; MacKay 2013). The stamens are diadelphous with two sets of filaments (9 + 1), and gynoecium is monocarpellary with several ovules. Style connects the ovary with adhesive stigma (Ferrandiz et al. 1999). Immature pods have either green or yellow colour. Pod shapes are straight or curved with smooth and constricted surface. The pod length reached up to 5–10 cm with about 5–10 seeds. The seed harboured embryo (01), cotyledons (02), and seed coat. The mature seeds have spherical, smooth, or wrinkled shape with smooth and rough surface. The seed colour may be white, creamish, light brown, brown, green, red-orange, blue-red, dark violet, black, or dotted in colour. The cotyledon colours of the mature round seed may vary from yellow, green, and red (Rubiales et al. 2019). It has hypogeal type of seed germination (Torrey and Zobel 1977).

5 Global Scenario of Productivity

At global level the productivity of field pea varied from 300 to 5502 kg/ha in different countries. The countries with the highest productivity and substantial cultivation area are Denmark (4399 kg/ha), Sweden (3569 kg/ha), Belgium (3140 kg/ha), Germany (3017 kg/ha), Finland (2975 kg/ha), France (2841 kg/ha), Canada (2539 kg/ha), and Russian Federation (2246 kg/ha). On the other side, in China, India, Australia, Ukraine, and United states of America, productivity is quite low and oscillates between 1.0 and 2.0 t/ha (FAOSTAT 2022). Over the past five decades, the global gain in grain yield has averaged approximately 15.3 kg/ha/year, which is significantly lower compared to other major field crops. This clearly portrays the ignorance of research and developmental efforts towards dry pea improvement at global level. Canada has achieved a genetic gain of 2.0% per year for grain yield, representing the highest yield increase among major field crops globally. This achievement reflects the success of extensive research and development efforts undertaken in Canada over the years (Rubiales et al. 2019). The productivity of field pea has been fluctuated between 1585 kg/ha and 2035 kg/ha during last 10-year statistics worldwide. The highest and lowest productivity was recorded in year 2020 and 2012, respectively (Fig. 2).

6 Pea Genetic Resources Vista Worldwide

The success of any breeding programme, particularly in delivering improved varieties, primarily depends on the availability and effective utilization of genetic resources. Globally, approximately 98,000 pea accessions—including cultivars, breeding lines, landraces, wild relatives, and mutants—are conserved across various gene banks (Parihar et al. 2020a, b, 2022a). The highest number of accessions (8839) are being conserved in National Institute for Agricultural Research, Dijon, France, followed by Australian Temperate Field Crop Collection, Horsham (7432 accessions), and N. I. Vavilov Research Institute of Plant Industry (VIR), St. Petersburg, Russia (6790). The number of accessions available in different gene banks is illustrated in Fig. 3. Crop wild relatives constitute only about 2% of the total pea collection. Among them, *Pisum fulvum*, *P. sativum* subsp. *elatius*, *P. sativum* subsp. *sativum*, and *Pisum abyssinicum* have 706, 624, 1562, and 540 accessions, respectively (Smýkal et al. 2013; Rubiales et al. 2019).

In India, approximately 4608 pea accessions are maintained at the gene bank of ICAR-National Bureau of Plant Genetic Resources, New Delhi. Of these, about 3416 accessions are indigenous collections, while 1192 are exotic collections. With the advancement in technologies, numerous online platforms have been designed to share status of available germplasm as well as information of phenotypic and genotypic database. The detail list of different online platforms and web portals is given in Table 1. Through detailed

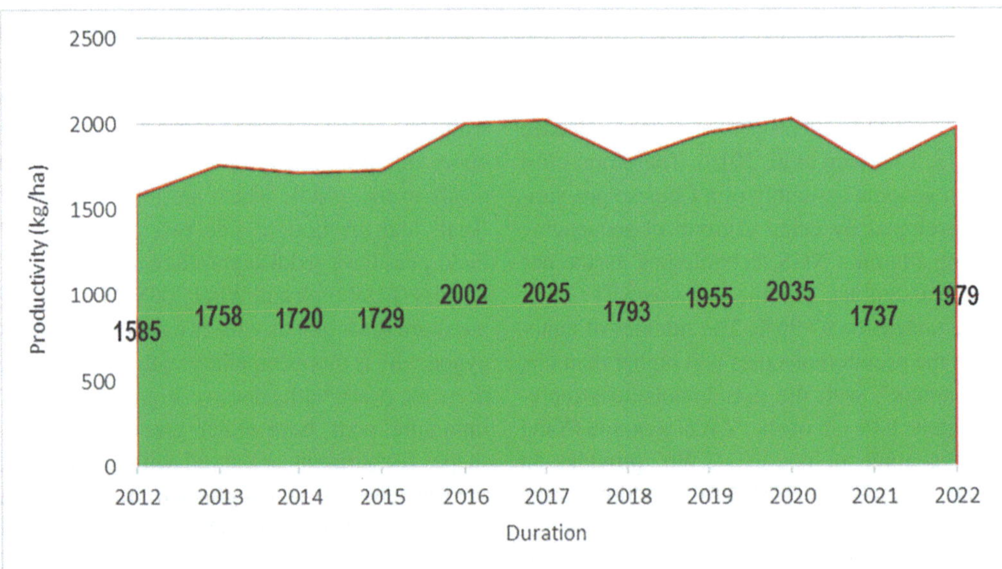

Fig. 2 Productivity scenario of field pea during past 10 years (2012–2022)

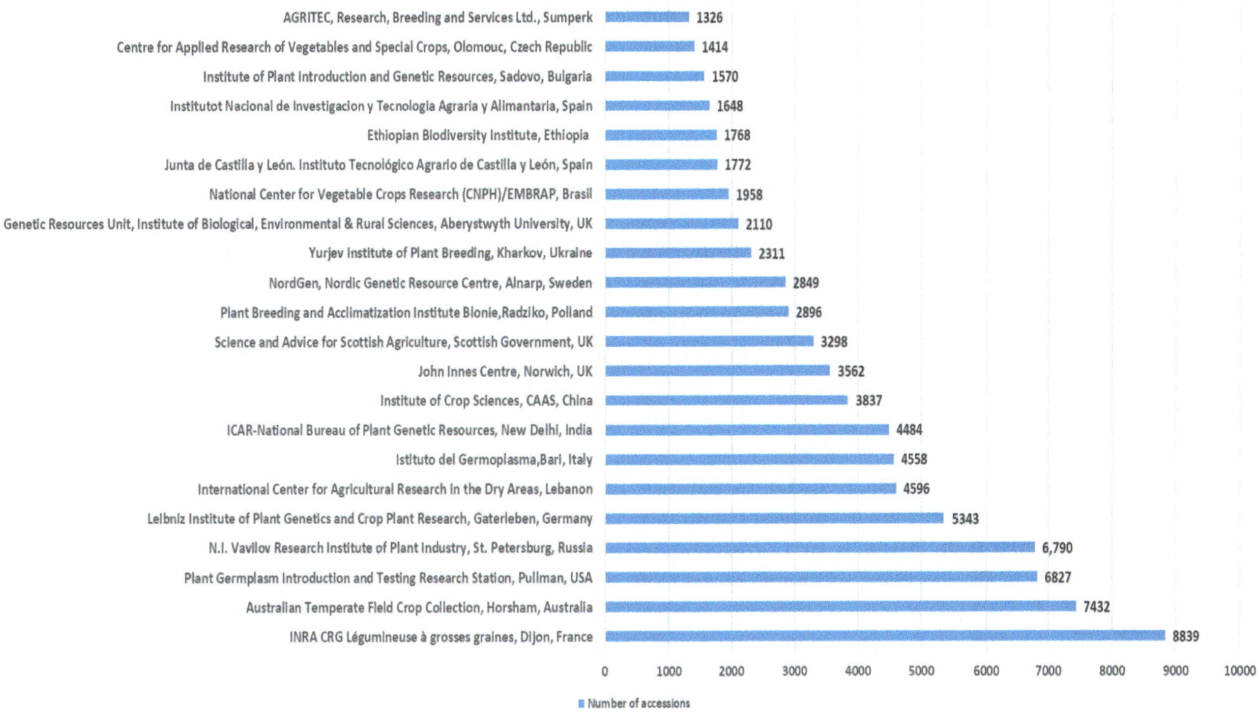

Fig. 3 Details of largest pea gene banks worldwide. (Source: Parihar et al. 2022a, b; https://www.genesys-pgr.org)

Table 1 List of different online platforms available worldwide

S. No.	Online platforms
1	Germplasm Resource Unit (GRU), John Innes Centre, Norwich, UK (https://www.jic.ac.uk/research-impact/germplasm-resource-unit/)
2	The Germplasm Resources Information Network (GRIN), USDA, (https://www.ars-grin.gov)
3	European Cooperative Program on Plant Genetic Resources (ECPGR), (https://www.ecpgr.org)
4	International Legume Database & Information Service (ILDIS), (https://www.ildis.org)
5	Legume Information System (LIS), (https://www.legumeinfo.org)
6	System-wide Information Network for Genetic Resources (SINGER) (http://singer.cgiar.org)
7	The Cool Season Food Legume Genome Database, (https://www.pulsedb.org/main)
8	KnowPulse, pulse crop breeding and genetics, (https://knowpulse.usask.ca)
9	Genesys (https://www.genesys-pgr.org/)
10	PGR Portal, NBPGR(ICAR), (http://pgrportal.nbpgr.ernet.in/)
11	ICAR-NBPGR seed genebank (http://genebank.nbpgr.ernet.in/SeedBank)

characterization of available pea germplasm accessions, based on both quantitative and qualitative traits, eight core collections have been established in Australia, France, China, Poland, the Czech Republic, Spain, the USA, and the UK to facilitate their efficient use in the regular breeding programmes of these countries (Rubiales et al. 2019). The first core set was developed based on seed quality attributes at John Innes Centre (JIC), UK (Matthews and Ambrose 1994). Subsequently, core sets were developed by USDA, Spain, France, Poland, and China, considering different traits i.e. geography, flower colour, quantitative data, isozymes, protein, and molecular diversity (Simon and Hannan 1995; Coyne et al. 2005; Sanz et al. 2011; Swiecicki et al. 2000; Baranger et al. 2004; Zong et al. 2009).

7 Major Limitations in Realization of Potential Productivity

Several biotic and abiotic stresses limit the realization of the actual yield potential of field pea under farmers' conditions (Parihar et al. 2022a). Among the biotic stresses, fungal diseases such as powdery mildew, rust, ascochyta blight, wilt, and root rots are the most devastating globally (Mahajan et al. 2018). Powdery mildew (PM), primarily caused by *Erysiphe pisi*, is the most widespread; however, in some regions, it is also caused by *Erysiphe baeumleri* and *Erysiphe trifolii* (Ghafoor and McPhee 2012; Sun et al. 2016a, b, c). Under favourable conditions, PM can cause yield losses of up to 80% (Munjal et al. 1963; Singh et al. 1978). Another major fungal disease is ascochyta blight (AB), which is caused by a complex of fungi including *Ascochyta pisi*, *Peyronellaea pinodes* (previously *Mycosphaerella pinodes*), *Phoma medicaginis* var. *pinodella*, *Phoma koolunga*, and *Phoma glomerata*. This is the most prevailing fungal disease of field peas which dispersed worldwide and in favourable circumstances causing up to 60% yield penalty (Liu et al. 2013, 2016; Tran et al. 2014). During mid 1980s, rust disease incited by *Uromyces viciae-fabae* or *U. pisi* emerged as a serious air-borne fungal disease of pea which dispersed across the countries (Barilli et al. 2010, 2018; Rubiales et al. 2015, 2019). Under warm and humid climatic conditions, *Uromyces viciae-fabae* is the predominant rust pathogen, whereas *Uromyces pisi* is more prevalent in cooler climates (Kushwaha et al. 2006; Singh et al. 2015a, b, c; Barilli et al. 2018; Rubiales et al. 2019). *U. viciae-fabae* is autoecious, completing its life cycle on a single host, while *U. pisi* is heteroecious, requiring alternate hosts such as *Euphorbia esula* and *Euphorbia cyparissias* to complete its life cycle (Rubiales et al. 2019). Yield losses due to rust infections can reach up to 30% under favourable environmental conditions (Barilli et al. 2010; Singh et al. 2015a, b, c). Corresponding to other legumes, wilt is an important soil-borne disease of field pea incited by *Fusarium oxysporum.* f. sp. *Pisi*, which has eleven different races (Gupta and Gupta 2019). Of which, two races i.e. 1 and 2 are well distributed across the field pea growing conditions, whereas races 5 and 6 are dispersed in specific regions only (Bani et al. 2018). Common root rot inflicted by *Aphanomyces euteiches* is another noteworthy soil-borne disease predominant in Canada, USA, and Europe (Chatterton et al. 2015; Wu et al. 2018).

Being nutritious crop field pea is also preferred by several insect pests viz., pod borer, leaf miners, aphids, stem fly, cut worms, and storage pests which affect the overall crop performance during different stages of pre- and post-harvest (Sharma 2000; Yadav et al. 2019, 2022). Of them, pod borer (*Helicoverpa armigera*) is a notorious pest having wide range of host plants and adaptability to diverse environmental conditions, consequently, it is prevailing in Asia, Africa and mediterranean region (EPPO 2006; Djihinto et al. 2012). Another important insect pest is pulse pod borer (*Etiella zinckenella*) which larvae feeds on the immature seed by piercing through pod wall. The overall pod damage by pod borer complex may be reached up to 40% under favourable circumstances (Dahiya and Naresh 1993). Among the seed-damaging insect pest, Bruchid (*Bruchus pisorum*) also known as pea seed beetle or weevil is an important storage insect pest which completes egg laying on pods and subsequently larvae damage the seed by piercing into pods via pod wall (Brindley et al. 1956; Clement et al. 2009). To manage this pest development of resistant varieties is the most economic approach. The resistant lines for pea weevil have been reported in *P. fulvum* and *P. sativum* ssp. *Syriacum* which can be used to develop pea weevil resistance varieties (Teshome et al. 2015a, b; Fernandez and Rubiales 2019). Leaf miner, *Phytomyza horticola* Goureau, is a polyphagous pest widely disseminated across the continent i.e. Africa, Europe, and Asia and causes up to 90% damage to crops by tender leaves mining under congenial situations (Yadav et al. 2022). Field pea crop is also damaged by the infestation of aphid (*Acyrthosiphon pisum*) through sucking of sap from tender shoots, leaves, buds, and pods which ultimately affects the overall crop performance (Ali et al. 2005). So far, only three field pea lines i.e. '061K-2P2/9/2', 'Milky', and 'Adi' have been reported as resistant reaction for aphid (Ali et al. 2005; Melesse and Singh 2012).

In the case of parasitic nematodes, cyst nematode (*Heterodera* sp.), root knot nematode (*Meloidogyne* sp.), stem nematode (*Ditylenchus* sp.), and reniform nematode (*Rotylenchulus* sp.) are the predominating which causes severe loss in field pea productivity under favourable conditions (Vovlas et al. 2011; Leach et al. 2012).

Prevailing climate change has increased the frequencies of various abiotic stresses i.e. heat stress, drought, and frost which has become serious concern in sustaining production of field pea (Guilioni et al. 2003; Sadras et al. 2012; Lamichaney et al. 2021; Parihar et al. 2022b, 2023). If the ambient temperature goes above 25 °C that negatively affects plant growth, the number of fruitful nodes, and crop duration which ultimately causes significant yield loss (Guilioni et al. 1997, 2003; Sadras et al. 2012; Bueckert et al. 2015). It has been noticed that the photosynthetic activity has detained at 40 °C (Georgieva and Lichtenthaler 1999) and at 45 °C net photosynthesis reduction reached up to 80% (Haldimann and Feller 2005). Number of studies have explained the impact of high temperature exposure on different phenological events, physiology, and overall yields of field pea (Sadras et al. 2013; Jiang et al. 2019; Mohapatra et al. 2020). The reduction in the different yield attributes like seed germination, seed setting, seed weight, and overall grain yield has been noticed under heat stress (Lamichaney et al. 2021).

In recent past, it has been reported that the exposure of field pea to elevated temperature during the flowering stage has higher negative impact on grain yield than the heat stress at the grain filling period which occurs due to the reduced pods/plant and seeds per pods. The high-temperature exposure also reduced maturity duration prominently in long-duration genotypes (Parihar et al. 2022b). Field pea has substantial variability for plant height; however, the role of plant height in grain yield stability under elevated temperature is not well understood and explained. Most recently, it has been reported that the medium-tall genotypes had longer grain-filling duration, higher pod-bearing nodes and yield, and lower yield losses as compared to semi-dwarf and dwarf genotypes under heat stress conditions that suggested higher heat tolerance capacity of tall-type plants (Parihar et al. 2023). Therefore, in future during designing of heat tolerance field pea cultivars plant stature need to be considered as important trait in regular breeding programme.

Drought or moisture stress is a crucial factor which reduces grain yield up to 25% under field conditions (Sánchez et al. 1998). Moisture stress causes changes in various morphological, physiological, and biochemical events controlling overall plant growth and productivity (Daie 1988; Iturbe-Ormaetxe et al. 1998; Alexieva et al. 2001; Guilioni et al. 2003; Sorensen et al. 2003; de Sousa-Majer et al. 2004; Karatas et al. 2012; Rubiales et al. 2019). Earlier, it was observed that flowering and pod filling stage of field pea are more sensitive to drought than the vegetative period but later on it was noticed that actual intensity of moisture stress is more important than timing (Maurer et al. 1968; Pumphrey and Schwanke 1974; Jamieson et al. 1984; Martin and Jamieson 1996). In the case of different plant types' performance under moisture stress conditions, the dwarf plant type has more resilience towards drought conditions than tall plant type (Iwaya-Inoue et al. 2003). At first, the semi-leafless plant type wherein leaflets are replaced by tendrils is considered more resilient to moisture stress as compared to conventional leafy type but later on it was observed that in semi-leafless more efficient osmotic adjustment might play crucial role in enhancing water use efficiency in moisture stress conditions (Semere and Froud Williams 2001; Gonzalez et al. 2001).

Most interestingly, the stipules of semi-leafless plants demonstrated better ability to bolster osmolarity through potassium, magnesium, and chloride accumulation than conventional leaves (Gonzalez et al. 2002). The epicuticular wax load of field pea genotype simplified markedly in moisture stress conditions along with increased residual transpiration rate (Sánchez et al. 2001). In water stress, the root lengths of field peas have been noticed more than under irrigated field conditions (Benjamin and Nielsen 2006), while PEG 6000-induced osmotic stress reduces primary root length and increases lateral root number (Kolbert et al. 2008). In the case of field pea, seed numbers are the most crucial trait and flowering and seed filling initiation is the most sensitivity period for moisture stress (Lecoeur and Guilioni 2010; Sadras et al. 2012). Overall, moisture stress in field peas reduces life cycle period, pods/plant, seeds/pod, plant height, leaflet size, and nitrogen fixation or uptake and also causes leaflet tip drying and flower bud/flower abortion. Thus, designing of drought tolerance field pea cultivars with wider adaptation ability should be the prime intent of any breeding programme in future.

Frost stress is a serious issue at vegetative and reproductive stage in pea (Shafiq et al. 2012; Liu et al. 2017). In frost, the temperature goes suddenly below 0 °C that is usually experienced by the crop before the cold acclimatisation of crop (Rubiales et al. 2019). The frost incidence permanently damaged the cell membrane system or photosynthetically active tissue (Menon et al. 2015; Liu et al. 2017). To acclimatise for cold, plants have to develop adaptation apparatus to combat frost (Xin and Browse 2000). The occurrence of frost at reproductive phase can damage flower buds, flowers, and pods and seed development which leads to drop in overall grain yield (Ridge and Pye 1985). In general, till vegetative stage field pea crop has better tolerance to frost than reproductive stage (Lejeune-Henaut et al. 1999; Meyer and Badaruddin 2001). However, genetic variability has been observed for frost tolerance at seedling, vegetative, and reproductive stages by various researcher (Bourion et al. 2003; Lejeune-Hénaut et al. 2008; Shafiq et al. 2012). The identified genotypes can be used in regular breeding as donor to infuse frost resistance in high-yielding field pea cultivars.

8 Priority Traits for Field Pea Breeding

8.1 Yield and Yield Attributing Traits

Field pea role in food as well nutritional security could be cherished by its horizontal as well as vertical expansion. The horizontal expansion is hinged on several factors including policy support. To achieve vertical gain the productivity has to be increased by breaking the yield plateau and realizing the actual potential yield at farmer's field. To increase the productivity of field pea target trait-oriented breeding efforts need to be made. The strength of genetic resources of any crop determines the extent to which genetic enhancement of it can be done through breeding. Therefore, conservation of the genetic diversity of any crop has great significance. Worldwide, a number of germplasm banks are being maintained by various institutes of different countries involved in the conservation of pea germplasm (Fig. 3).

In the case of India, NBPGR, New Delhi is the nodal agency for maintenance of pea germplasm which has a great amount of variability for various yield attributes. If the target trait is available in the primary (easily crossable) and secondary (crossing possible with some difficulty and produce partially fertile offspring) gene pool that can be easily transferred in desired good agronomic base through regular breeding, but the target trait that is available in tertiary gene pools (crossing with great difficulty and produce sterile offspring) in that case trait of interest can be introgressed using modern tools and techniques only. The range of different yield attributes and nutritional traits of various accessions evaluated by different researchers are illustrated in Table 2. This information will be useful in setting the targets while breeding for the development of nutrient-dense high yielding field pea cultivars. Breeding methods which have been routinely deployed towards genetic improvement of field pea being self-pollinated crop are pedigree selection, mass selection, pure line selection, single seed descent, backcross method and modifications of these methods.

Table 2 Glimpses of genetic variability for important yield attributes and seed quality traits in field pea

Traits	Range	References
Days to 50% flowering	41.00–168.0	Parihar et al. (2014b), Rana et al. (2017)
Days to maturity	73.00–212.0	Parihar et al. (2014b), Rana et al. (2017)
Plant height (cm)	13.33–279.8	Yadav et al. (2010), Parihar et al. (2014a, b), Umar et al. (2014) and Rana et al. (2017)
100 seeds weight	2.99–47.25	Burstin et al. (2015)
Pod length (cm)	2.90–12.10	Rana et al. (2017), Azmat et al. (2011)
Pods/plant	2.0–210	Umar et al. (2014)
Seeds/pod	2.1–10.33	Gatti et al. (2011), Ouafi et al. (2016)
Seed yield (g/plant)	0.24–213.4	Nisar et al. (2008)
Biological yield (g/plant)	9.0–452.66	Nisar et al. (2008)
Harvest index (%)	0.16–50.60	Nisar et al. (2008), Singh et al. (2018)
Protein	13.7–35.20	Tzitzikas et al. (2006), Irzykowska and Wolko (2004)
Iron (ppm)	21.90–105.2	Harmankaya et al. (2010), Kwon et al. (2012) and Grusak and Cakmak (2005)
Zinc(ppm)	11.3–106.63	Demirbas (2018), Kwon et al. (2012)
RFOs(m/g)	41.4–157.4	Gawłowska and Święcicki. (2016)
Folate (mg/100 g)	41–202	Gupta et al. (2013)

Table 3 Potential sources for powdery mildew resistance in pea

Genotypes	References
JI2302, JI2480, P660-4,NBP61, NBP72, DMR7	Bheri et al. (2016), Tiwari et al. (1997)
Highlight, AC Tamor, Tara, Mexique 4, Stratagem, JI 210, JI 1559, JI 1951, JI 2302	Tiwari et al. (1997)
Xucai 1	Sun et al. (2015)
PMR-53, Kashi Nandini, Kashi Uday, VL-7, AP-1	Gupta et al. (2023)
ILS6527, UN6651	León et al. (2020)
It-96, No. 267, JI 2302	Azmat and Khan (2013)
G0004389, G0004382, G0004400, G0004417, G0004434, G0004448, G0004450, G0002102, G0006514, G0006515, G0006516, G0006519, 20,012, Jia2, Texuan11, Hehuan66, G0001747, G0001773	Sun et al. (2019)
G0003925, Cooper,	Fu et al. (2014)
G0005576, G0003694, G0001752, G0001763, G0001764, G0001767, G0001768	Sun et al. (2016a)
G0006273	Wang et al. (2015)
PI391630	Humphry et al. (2011)
L0314, L1332, L1335	Sun et al. (2016b)
G0003895, G0003899, G0003967, G0003958	Sun et al. (2016c)
SP7, SN-1, SN-6-1, SN-7-1, SN-2, SN-5-2, SN-6-2, SN-10, SN-21, SP-281	Rana et al. (2023)
VRPMR-9, VRPMR-11, Punjab-89, PMR-62	Singh et al. (2015a, b, c)
DPP 362, JP-501-A/2, DPP-1542EP, PMR-21, KMNR-894, JI-1766	Banyal et al. (2015)

8.2 Biotic Stresses

The major bottleneck in harnessing real potential of field pea is occurrence of large number of biotic stresses i.e. powdery mildew, rust, wilt, ascochyta blight, insect-pests, and nematodes. Among these, fungal diseases are the most devastating across the different field pea growing ecologies (Reiling 1984; Bohra et al. 2014; Parihar et al. 2022a). During the last four decades, sincere efforts were made and significant progress has been made through breeding for biotic stresses in field pea. In this sequence the first powdery mildew resistance genotype i.e. Huancabamba was a land race and its inheritance was regulated by single recessive gene reported by Harland (1948). During the last seven decades, number of screenings to identify new sources of powdery mildew and their further genetic analysis were accomplished (Sun et al. 2016a, b, c; Devi et al. 2022; Parihar et al. 2022a). The inheritance studies concluded that the powdery mildew inheritance is mainly regulated by three genes i.e. *er1*, *er2*, and *Er3* (Fondevilla et al. 2013a, b), of which two are recessive (*er1*, *er2*) (Harland 1948; León et al. 2020) while one (*Er3*) is dominant in nature (Fondevilla et al. 2007a, b, c; Bobkov and Selikhova 2021). Among these three genes, the *er1* gene has been so far reported in most of the resistant pea accessions, whereas *er2* gene is only in limited lines (Tiwari et al. 1997). The dominant resistance gene (*Er3*) is identified from secondary gene pool species i.e. *P. fulvum* (Fondevilla et al. 2007a, b, c). This gene has been introgressed in the cultivated types by using as male parent in hybridization, and now it can be easily transferred to desired background in the breeding programme. Most recently, another dominant gene has been identified from the *P. fulvum* which identity yet to be confirmed (Bobkov and Selikhova 2021). However, the possibility of the quantitative mode of PM inheritance had also been advocated by several research groups (Tiwari et al. 1997; Fondevilla et al. 2006; Katoch et al. 2010; Prasanth et al. 2019). A brief list of potential sources of powdery mildew resistance is presented in Table 3. This information would be useful for designing breeding strategies for development of powdery mildew resistance high yielding field pea cultivars.

Rust is an important biotic stress which causes significant yield loss under congenial environmental conditions. Concerted efforts have been made to identify complete resistance for rust, but so far none of the genotype expressed complete resistance against rust; however, genotypic variation has been noticed for disease intensity (Singh and Srivastava 1985; Anil Kumar et al. 1994; Chand et al. 2006; Barilli et al. 2009). The inheritance of rust resistance is regulated by single gene *Ruf* which is dominant in nature; however, recent findings advocated the quantitative nature of rust resistance in pea (Singh and Ram 2001; Vijayalakshmi et al. 2005; Singh et al. 2012a, b). Partial resistance cultivars would play important role in reducing the epidemic build-up than highly susceptible cultivars. This resistance would reduce infection efficiency, pustule size, spore production, and AUDPC than the susceptible cultivars (Singh et al. 2012a, b). By deploying reported donors in regular breeding programme, breeders have developed several high yielding varieties coupled with partial rust resistance cultivars recommended for commercial cultivation. Several pea genotypes have been reported with partial or incomplete resistance for rust incidence and presented in Table 4 which can be explored in future to develop high yielding varieties with partial rust resistance.

Fusarium wilt is soil-borne fungus and considered as serious threat in pea production worldwide and causes up to 100% crop damage under congenial conditions (McClendon et al. 2002; Sharma et al. 2010; Zvirin et al. 2010; Aslam et al. 2019). This soil-borne fungus has 11 different virulent races, of which races 1 and 2 are distributed widely (Infantino et al. 2006; Bani et al. 2018; Gupta and Gupta 2019). As it is well established, the cultivation of resistant cultivars is the safest, economic, and effective approach to avoid any crop loss due to wilt incidence (Ciancio and Mukerji 2008). The inheritance mechanism of this disease is either qualitative or quantitative based on different races (McPhee et al. 2012; Rispail and Rubiales 2014; Bani et al. 2018). The oligogenic resistance has been successfully transferred in several cultivars of pea through conventional breeding (McPhee 2003), while the incorporation of polygenic resistance is complex where genomics can play important role in successful transfer of trait with minimum resources and time.

Concerted efforts have been made by researchers and subsequently various genotypes/accessions have been identified with wilt resistance. By screening of core collection for races 1 and 2, totally 39 and 62 resistance accessions were identified, respectively (McPhee et al. 1999). On similar note, several studies were conducted, and subsequently numbers of wilt resistance genotypes were identified (McPhee et al. 1999, 2012; Bani et al. 2018; Rispail and Rubiales 2014). Most recently, Deng et al. (2022) screened a total of 1311 germplasm accessions comprising 740 from China and 571 from other sources for resistance to *Fop* race 5. The results revealed that 192 accessions were resistant and rest of the accessions were susceptible. These potential sources could be used in breeding programme to develop wilt resistance pea cultivars. A brief list of wilt resistance genotypes reported by various research groups is presented in Table 5.

Table 4 Potential sources for rust resistance in pea

Genotypes/accessions	References
Pant P 11, JPBB 3, FC 1	Singh et al. (2012a, b)
PI273209, IFPI3260, BGE00471, JI199, JI198, PI347321, JI224, CGN10205, CGN10206	Osuna-Caballero et al. (2022)
IPF-2014-16, KPMR-936 and IPF-2014-13	Das et al. (2019)
Pant P 244, Pant P 42	Upadhyay et al. (2017a, b)
Bridzor, EC4294, EC9218, EC955 EC9908, NP29, Perf 3268, IC4604	Sohi et al. (1974)
PJ207508, PJ22211, EC109188	Pal et al. (1980)
Pant P11, FC1, HUDP 16, JPBB 3, HUP 14	Chand et al. (2006), Kushwaha et al. (2006)
IFPI3260, PI347321, PI347336, PI347347, PI343935, PI343965, PI347310	Barilli et al. (2014)
IFPI3260	Barilli et al. (2018)
Pant P 11,Pant P 12, Pant P14, HUDP 7 and HUDP 15	Singh et al. (2015a, b, c)
HUDP 15, Prakash, Swati, Aman, Pant P 42, IPF 5-19, IPFD 11-5, IPFD 12-2, IPFD 9-2, IPF 16-13, IPFD 13-2, IPFD 12-8	Parihar and Dixit (2017), Dixit et al. (2017) and Parihar et al. (2019a, b, 2020a, b)

Table 5 Potential sources for wilt resistance in pea

Genotypes/accessions	References
WSU23, WSU28, WSU31, A053, Zhangwan 4, Zhangwan 7, Zhangwan 11, Zhangwan 16, Zhangwan 19, Zhangwan 26, Zhangwan 178, Zhangwan 191, Zhangwan 197, 2010CQ019, L2254 etc.	Deng et al. (2022)
GP-6, GP-55, GP-942, GP-17, GP-48, GP-473, GP-941	Shubha et al. (2016)
P615, P42	Castillejo et al. (2015)
JI 1412, JI 1747, JI 2480 and New Season	Rispail and Rubiales (2015)
Garrow	Husnain et al. (2019)
Shawnee	McPhee et al. (2012)
PI 344012	McPhee et al. (1999)
JI1412, JI1559, JI1760, P23, P42, P614, P627, P633, P639, P650 and P656	Bani et al. (2012)
PI 608036	Coyne et al. (2000)
PI 614141	Okubara et al. (2005)

Table 6 Potential sources for AB resistance in pea

Accessions/genotypes	References
JI103, JI 252	Clulow et al. (1991)
JI 1006(*P. fulvum*)	Wroth (1998)
Victor, DP, Radley, P651, P658, P670(*P. fulvum*), P672 (*P.sativum ssp. elatius*	Fondevilla et al. (2005a, b)
P665 (*Pisum sativum subsp. syriacum*), P42 (*P. sativum subsp. sativum var. arvense*)	Fondevilla et al. (2007b)
PI595937, P651, W615017, PI560061(*P. fulvum*), PI344538 (*P. sativum subsp. Elatius*)	Jha et al. (2012, 2017)
P665(*P. sativum ssp. syriacum*)	Carrillo et al. (2014)
CDC Strike	Warkentin et al. (2004)

Table 7 Potential sources for CRR resistance in pea

Accessions/genotypes	References
00-2067	Conner et al. (2013)
90-2079, 90-2131, 90-2322	Kraft (1992), Pilet-Nayel et al. (2005)
PI 606694, 79-2022, 96-2058, 96-2068, 96-2198, 96-2222	Kraft and Coffman (2000a)
PI 606702, PI 606703	Kraft and Coffman (2000b)
MN313, MN314, MN 144	Davis et al. (1995), Malvick and Percich (1998)
PI180693, 552	Hamon et al. (2013)

Ascochyta blight (AB), popular as 'black spot disease', is the seed borne, most devastating, and causes more than 85% yield damage under favourable climatic conditions (Tran et al. 2014; Liu et al. 2016a, b; Sivachandra Kumar and Banniza 2017; Rubiales et al. 2019). The management of AB by chemical and cultural approaches is very difficult since its incidence under field conditions is highly influenced by micro-environment of individual plant (Banniza et al. 2005; Jha et al. 2016). Considering the above facts, the designing resistant cultivar is the most appropriate tactics to avoid the yield penalty due to AB incidence in pea (Zimmer and Sabourin 1986). Concerted efforts have been dedicated towards identification of resistance source against AB and consequently some potential genotypes identified from a pool of >3500 pea accessions (Zhang et al. 2006).

Interestingly, few accessions of *P. fulvum* with good level of resistance against AB were identified (Fondevilla et al. 2005a, b). Similarly, Jha et al. (2012) identified few promising wild accessions from *P. fulvum* and *P.sativum ssp. elatius* which demonstrated good level of resistance against AB. These accessions have been used in development of cultivars as well as different experimental populations (Sindhu et al. 2014; Jha et al. 2016). A brief detail of accession which has a good level of resistance and can be used as a potential donor in breeding programme is presented in Table 6. However, the mode of inheritance AB is quantitative which substantially hampered the varietal development programme through conventional breeding (Fondevilla et al. 2007b; Rubiales and Fondevilla 2012; Carrillo et al. 2014; Jha et al. 2017). Therefore, in such situation, marker-assisted breeding can play an instrumental role to improve precision as well as success rate of varietal development programme.

Common root rot (CRR) disease is mainly prevalent in USA and Europe; however, in recent past it has also been observed in Canada (Wicker et al. 2003; Desgroux et al. 2016). The management of CRR through conventional approaches is incapable under favourable conditions owing to long-lasting surviving ability of its oospores (Papavizas and Ayers 1974). Hence, resistant varieties development has been considered as the best option to avoid crop damage due to CRR incidence. Dedicated efforts were made by researchers to identify potential sources against CRR (Gritton 1990; Davis et al. 1995; Kraft 2000; Pilet Nayel et al. 2007; Conner et al. 2013). The identified sources have been used in breeding programme to design resistant cultivars and experimental populations (Roux-Duparque et al. 2004; Pilet-Nayel et al. 2005; Hamon et al. 2011, 2013; Lavaud et al. 2015). Due to polygenic mode of inheritance and linkage of undesirable traits i.e. long internode, anthocyanin content, and delayed flowering, breeding for common root rot resistance has become daunting task (Pilet Nayel et al. 2002). Thus, like other quantitative traits, deployment of genomics-assisted breeding is the best strategy to accelerate breeding programme accomplishment. The reported potential sources (Table 7) can be used in future in field pea breeding programme to fight against CRR incidence and reduce yield penalty.

8.3 Abiotic Stresses

As far as abiotic stresses are concerned, heat and drought stresses are the major problems in field pea. Of which, heat stress is a serious bottleneck in achieving actual yield potential of field pea, being a cool-season legume. If the ambient temperature elevated above 25 °C, especially during flowering to grain filling, that declined the overall crop performance due to reduction in above ground biomass, number of fruitful nodes, number of pods per plant, ovule to seed conversion, reproductive period, seed size and pollen functionality (Jiang et al. 2015; Bueckert et al. 2015; Jiang et al. 2019; Lamichaney et al. 2021; Parihar et al. 2022a, b, 2023). Promising heat tolerant genotypes i.e. KPF 103, DMR 15, and IPFD 4-6 have been identified on the basis of membrane stability index, biological yield, plant stature, harvest index, and grain yield (Vijaylaxmi 2013). In the recent past, it was observed that the elevated temperature at terminal crop stages decreases seed yield (24–60%) as well as seed quality in terms of germination (4–8%) also. Considering seed yield and their planting value, few promising genotypes i.e. IPFD

11-5, Pant P-72, P-1544-1, and HUDP 11 identified as the best genotypes which have the capability to develop more viable seeds under high temperature circumstances (Lamichaney et al. 2021). The identified genotypes could be utilized in designing field pea cultivars for quality seed production in heat stress-prone ecologies.

Field pea crop exposure to heat stress during flowering window has more negative impact on seed yield as compared to grain filling stage. Also noticed that the reduced sink capacity in terms of pod setting and seed setting efficiency is a major factor responsible for yield penalty in the heat stress, whereas, in normal situation yield potential is mainly contributed by the source capacity i.e. plant biomass. The elevated temperature also reduces the vegetative as well as reproductive period significantly in long-duration field pea. Interestingly, poor pod setting and high ovule abortion were noticed in heat stress environment at upper nodes. Most importantly, seed set, pods plant−1, last pod bearing node, and plant biomass have been identified as key yield-determining factors and would be used as target traits in breeding for heat stress (Parihar et al. 2022b). Earliness or short duration is a very vital character that let crop to escape or avoid prevailing terminal stresses, thereby, crop can sustain the yield potential (Parihar et al. 2021b). The significant interrelationship of plant height with grain yield, yield loss, and heat-susceptibility index under elevated temperature indicated that tall type field pea has more heat tolerance than dwarf stature. Thus, in future while designing heat tolerant field pea, the plant stature needs to be considered as important attribute (Parihar et al. 2023).

Moisture stress drought is a crucial limiting factor that affects crop performance in terms of quantity as well as quality of the seed yield (Ali et al. 1994; Sánchez et al. 1998). Initially it was reported that water stress particularly during reproductive phase caused more damage to crop than vegetative phase but later on it was observed that water stress intensity is more important than timing of stress (Pumphrey and Schwanke 1974; Martin and Jamieson 1996). In resistant cultivars, shoot-to-root ratio remains markedly lesser than the susceptible (Grzesiak et al. 1997). Most interestingly, dwarf plant types are more drought resistant to tall type (Iwaya-Inoue et al. 2003). In moisture stress, semi-leafless type field pea stipules have greater ability to boost osmolarity by gathering K, Mg, and Cl than conventional leaf architecture (Gonzalez et al. 2002). Another important trait is wax load over aerial parts which increases considerably in water stress conditions (Sánchez et al. 2001). The root length of field pea increases under water stress field condition than without stress (Benjamin and Nielsen 2006). Moisture stress reduces the number as well as the distribution of seeds, photosynthesis and transpiration efficiency, glycolate oxidase activity, chlorophyll-a, fresh & dry weight of leaf, relative leaf water content, plant growth and overall productivity (Morgan 1984; Alexieva et al. 2001; Guilioni et al. 2003; Sorensen et al. 2003; De Sousa-Majer et al. 2004), while total chlorophyll content, soluble phenols in leaves, and sucrose content in seeds increase markedly in water paucity (Alexieva et al. 2001; Sorensen et al. 2003). Conversely, the significant decrease in chlorophyll and protein contents in leaves, while an increase in the proline and malondialdehyde content in leaves has been observed. Most interestingly, superoxide dismutase, catalase and peroxidase activities increase under water stress (Karatas et al. 2012).

To sustain field pea crop production under stress prevailing environments, there are three tactics i.e. escape, avoidance, and resilience. These approaches could help in developing genotypes that can perform well under water scarcity. In the case of escape and avoidance, earliness and short duration varieties can perform well under stress conditions which is mostly preferred by breeders. But early flowering coupled with early maturity also has a disadvantage of not performing well in higher moisture conditions (Khan et al. 1996). However, a genotype can perform well if flowers early and complete pod filling when adequate moisture is available. Hence, for drought tolerance, genotype should have early vigour, flowering, and pod setting as well as greater plasticity in phenology (Khan et al. 1996; Turner et al. 2001). At the early stages of plant development turgor maintenance could be considered to identify drought-tolerant genotypes (Sanchez et al. 2004). The number of seeds can be used a selection criterion for selection drought tolerant genotypes since crop has maximum vulnerability during reproductive period (Jeuffroy et al. 2010; Sadras et al. 2012). The field pea breeders during designing of drought tolerant varieties are mainly considering earliness and extended flowering period. In addition, plant height, internode length, leaflet size, canopy colour, nitrogen uptake, flower drop, and changes in phenological events can be used for identification of drought tolerant genotypes (Rubiales et al. 2019). Being quantitative trait genomics as well as other modern biotechnological tools need to be integrated with conventional breeding to develop drought-tolerant varieties (Iglesias-Garcia et al. 2015).

8.4 Nutritional Enhancement

To increase the role of field pea in nutritional security, there is urgent need of development of nutrient-dense high-yielding varieties. The substantial variability has been reported for nutritionally important traits (Table 2). Carbohydrates (56–74%), starch (37–49%), amylose content (24.0–88.0), protein content (13.7–35.20), iron (23.16–105.2 ppm) and zinc (16.10–106.63 ppm) are prime ingredients of peas for which marked variability has been noticed for various researchers (Irzykowska and Wolko 2004; Tzitzikas et al. 2006; Dahl et al. 2012; Hoover et al. 2010; Kwon et al. 2012). Most of the nutritionally important traits are generally being quantitative in nature, highly influenced by external factors like soil profile and other climatic conditions; however, amylose content is the either least or not affected by environmental conditions (Wang et al. 2008; Hood-Niefer et al. 2012; Ray et al. 2014). Corresponding to other legumes, peas also have substantial content of raffinose family oligosaccharides which comprise stachyose (0.7.–4.1%), total a-D-galactosidase (22.6–63.4), verbascose (0.0–26.7 g/ka), raffinose (4.1–10.3 g/kg) and sucrose (11.6–25.4 g/kg) (Jones et al. 1999; Vidal-Valverde et al. 2003; Tosh and Yada 2010).

The high amount of total soluble carbohydrates and Raffinose family oligosaccharides (RFOs) has been detected in wrinkled seeds and least was in crop wild relatives i.e. *P. fulvum.* Most interestingly, positive association between stachyose and verbascose has also been noticed (Gawłowska et al. 2017). Given information clearly witnessed that significant variability exists for RFOs and that need to be explored towards the development of field pea cultivars with low RFOs which could ultimately reduce flatulence and other gastro-intestinal issues. During the last 10 years, several pea genotypes were identified for high protein (>30%) and deployed in breeding programme to design protein dense cultivars (Bing 2015; Shen et al. 2016; Demirbas 2018). Gawalko et al. (2009) noticed that yellow peas have more iron, magnesium, and manganese and less potassium to vegetable peas. The abundance of micronutrients including iron and zinc has been reported by various researchers (Gawalko et al. 2009; Amarakoon et al. 2012, 2015; Kwon et al. 2012; Ray et al. 2014; Ma et al. 2017; Demirbas 2018). In addition, pea is an excellent source of vitamins, mainly thiamine and folate (Sierra et al. 1998; Han and Tyler 2003; Gupta et al. 2013; Jha et al. 2015); however, very limited efforts have been made towards haunting the variability in field pea for these traits. Pea is containing different phytochemicals like phenolics, phytates, oxalates, and saponins, of which, most predominance are tannins, phenolics, and flavonoids. Interestingly, it has been observed that dark seed coat colour genotypes have higher concentrations of phenolics (Campos-Vega et al. 2010; Troszynska and Ciska 2002; Dueñas et al. 2004). As mentioned above, the number of genotypes has been identified with excellent amount of different macro and micro-nutrients which need to be utilized in regular breeding programme to design nutrient dense field pea cultivars as well as experimental populations to ensure nutritional security to resource poor folk of developing nations. Being quantitative in nature, breeding for nutritional traits should be location specific and seed micronutrient status may be up-scaled by increasing the bioavailability through reduction in anti-nutritional factors and increasing absorption-promoting compound levels (Hurrell and Egli 2010; Lockyer et al. 2018).

8.5 Lodging Resistance

Due to tall plant stature, weak stems, and above-ground high biomass, the pea plants lodged at the time of maturity or after pod filling. The crop lodging develops conducive micro-environment for disease development and also affects the photosynthetic activity, harvest efficiency, and overall cost of cultivation. These conditions can reduce up to 74% seed yield and also deteriorate quality of harvest (Heath and Hebblethwaite 1985; Armstrong et al. 1999; Xue and Warkentin 2001; Zhang et al. 2006). However, over the time plant height has been reduced significantly from 1 to 2 m to 0.3–0.6 m through incorporation of dwarf gene(*le-1*) (Amelin et al. 1991; Martin et al. 1997). The dwarfing gene played instrumental role in increasing productivity of pea by infusing input responsiveness and permits dense planting. But even after reducing plant height, the lodging remained serious concern. Hence, researchers have identified an alternative approach to improve standing ability through incorporation of '*afila*' leaf type in high yielding cultivars and developed 'semi-leafless' pea cultivars, which has been recognized as the greatest accomplishment of field pea breeding (Duparque 1996). This plant type has substantially increased the standing ability of cultivars which ultimately reduced yield losses (Wang et al. 2002; Singh and Srivastava 2018). Therefore, during recent years maximum cultivars developed are semi-leafless type (Mikić et al. 2006, 2013).

In Europe in 1970s first semi-leafless commercial cultivar 'Solara' was developed and till last decade in Canada, European union, and Russia, semi-leafless type cultivars shared 30–90% of varietal arsenal. The semi-leafless type cultivars played instrumental role in increase of pea cultivation in countries like Australia, Canada, China, and India (Warkentin et al. 2015). Till date, several varieties with semi-leafless plant type have been developed in India which are contributing significantly in field pea production (Parihar and Dixit 2017; Parihar et al. 2019a, b). However, there is still scope for further improvement in standing ability of field pea by developing cultivars with stiff stem. Such type of genotypes will remain erect till harvesting which will certainly helpful in mechanization of field pea. Therefore, development of erect type genotypes with stiff stem must be a key target in field pea improvement programmes (Smitchger et al. 2020).

8.6 Herbicide Tolerance

It is well known that weeds have major competition in terms of nutrient uptake, solar radiation, moisture, and space with the crops. In addition, weeds are the host plants of several insect pest and disease-causing micro-organisms. Therefore, overall productivity of the crop can reduce more than 75% by weeds incidence (Singh et al. 2016). At present, in case of field pea, a pre-emergence herbicide application and manual weeding after 30–35 days of sowing are suggested. During current scenario, manual weeding has become costly due to labour scarcity (Chaturvedi et al. 2014). In addition, being succulent plant type second manual weeding to control later flush of weed is not possible because it can cause more damage to crop than weeds. Therefore, use of post-emergence herbicides has become crucial, but, till date none of the post-emergence herbicide is recommended for controlling weeds due to its sensitivity for them. Most importantly, there is an urgent need of identification of post emergence herbicide resistant/tolerant genotypes, so that available post-emergence herbicide can be used to control weeds. The herbicide tolerance cultivars would offer more elasticity to pea growers to apply post-emergence herbicides. Till date, only one tolerant genotype i.e. P-637 and five moderately tolerant genotypes viz. P-729, P-647, P-1075, P-2016, and P-1448-2 have been identified (Parihar et al. 2019b). Thus, these genotypes may be deployed as donor in breeding programme to develop herbicide tolerant cultivars and different experimental populations. More efforts need to be diverted to identify new resistant sources against post-emergence herbicides and their further utilization to design herbicide tolerance varieties.

9 Genomics-Assisted Breeding

As it is well known that quantitatively inherited traits are difficult to handle, being highly affected by external environmental conditions, the use of molecular markers has become indispensable towards mapping genetic loci or QTLS regulate polygenic traits. During the last two decades, significant progress has been made by various national and international institutions regarding development of valuable genomic resources using cutting-edge sequencing and genotyping technologies (Loridon et al. 2005; Gong et al. 2010; Leonforte et al. 2013; Sindhu et al. 2014; Tayeh et al. 2015; Cheng et al. 2015). In the case of pea, different types of DNA-based markers like STMS: sequence-tagged microsatellite sites (Pandian et al. 2000; Xu-Xiao et al. 2008), ISSR: inter simple sequence repeat (Kuznetsova et al. 2005; Mishra et al. 2009), SRAP: sequence-related amplified polymorphism (Esposito et al. 2007; Kole et al. 2015; Guindon et al. 2016), SNP: single nucleotide polymorphisms (Deulvot et al. 2010; Duarte et al. 2014; Sindhu et al. 2014), IRAP: Inter-retrotransposon amplified polymorphism (Smýkal 2006; Smýkal et al. 2008), RBIP: retrotransposon based insertion polymorphism (Jing et al. 2010), EST-SSR: Expressed sequence tags—simple sequence repeat (Mishra et al. 2012; Teshome et al. 2015a, b; Nisar et al. 2017), and SSR: simple sequence repeat (Gong et al. 2010; Sun et al. 2014; Jain et al. 2014) are developed and employed for genetic diversity studies, mapping and tagging of genes. By using these markers, linkage maps were developed and subsequently used for mapping genes/QTLs regulating different biotic and abiotic stresses. Details of molecular mapping of different traits including biotic and abiotic stress resistance in pea using different types of markers is presented in Table 8.

Likely to other crops, SSR has been extensively utilized being more reliable, co-dominant, economic, rapid, and polymorphic (Snowdon and Friedt 2004; Gong et al. 2010). The next generation sequencing accelerated the discovery of SNPs and array development for pea genotyping (Kaur et al. 2012; Duarte et al. 2014; Boutet et al. 2016). The second-generation sequencing has its own challenges during the assemble of the entire pea genome but transcriptomics assisted scientists to perform de novo assembly at transcript level (Srivastava et al. 2012; Alves-Carvalho et al. 2015). The third-generation sequencing and cost cutting of second-generation sequencing play instrumental role in the development of the first chromosome-level reference genome of pea (Kreplak et al. 2019). Afterward, to understand the yellow pod trait reference genome of pea variety 'JI128' has been developed (Shirasawa et al. 2021). In this sequence, the combination of third-generation sequencing and Hi-C mounting has been employed to assemble the improved reference genome of Chinese pea cultivars 'ZW6' with more accuracy (Yang et al. 2022). The abundance of linked genetic loci or QTLs with the trait of interest is indispensable to embrace genomic-assisted breeding in regular conventional breeding. The engagement of genomic resources with conventional breeding would certainly speed-up the varietal development to assist in early generation selection of preferred individuals based on associated genetic loci (Tayeh et al. 2015). Various marker associated with major biotic stress are presented in Table 9. These markers could play instrumental role in adoption of genomics assisted breeding in pea to accelerate development of climate resilient cultivars to ensure nutritional and food security. In addition, the above-mentioned reference genome would certainly hold great promise towards availability of more genes with excellent breeding value and suitability for high-throughput genotyping.

10 Speed Breeding and Genome Editing

The application of conventional plant breeding methodologies to develop improved varieties is time consuming (10–12 years) and costly approach. But, the prevailing climate change owing to global warming has increased frequency of several biotic and abiotic stresses, which has

Table 8 Molecular mapping for various agronomic, biotic, and abiotic stresses in pea using different molecular markers

Population type/size	Pedigree	Marker types	Mapped traits	References
RIL/51	Slow x JI1794	RAPD, RFLP	PM	Timmerman Vaughan et al. (1994)
RIL/106	Kaspa × Yarrum	SSR, SNP	PM	Sudheesh et al. (2014)
RIL/106	Kaspa × ps1771	SSR, SNP	PM	Sudheesh et al. (2014)
F2/100	C2 × Messire	RAPD, SCAR	PM	Fondevilla et al. (2008)
F2/111	Almota' × '88 V1.11	RAPD, RFLP	PM	Timmerman Vaughan et al. (1994)
F2/111	Lincoln/JI2480	SSR	PM	Katoch et al. (2010)
$F_{2:3}$/99	Radley × highlight	RAPD/SCAR	PM	Tiwari et al. (1998)
F2/208	$PG3^{HFP\,4}$ × PG 3	RAPD, SCAR	PM	Srivastava et al. (2012)
F2/192	Majoret × 955180	SSR	PM	Ek et al. (2005)
F2/230	Solara × Frilene-derived mutant	ISSR, RAPD, AFLP	PM	Pereira et al. (2010)
F2/102	Bawan 6 × DDR-11	SCAR, SSR	PM	Sun et al. (2016a, b, c)
F2:3/120	WSU 28 × G0004389	SCAR, SSR	PM	Sun et al. (2019)
F2:3/119	Bawan 6 × G0004400	SCAR, SSR	PM	Sun et al. (2019)
F2:3/71	G0001778 × Bawan 6	SSR	PM	Sun et al. (2016a, b, c)
F2/91	Qizhen 76 × Xucai 1	SSR	PM	Sun et al. (2014)
F2/161	Xucai 1× Bawan 6	SSR	PM	Sun et al. (2014)
F3/94	IFPI3260 × IFPI3251	RAPD, STSs	Rust	Barilli et al. (2010)
RIL/136	HUVP 1 × FC 1	SSR, RAPD, STS	Rust	Rai et al. (2011)
RIL/84	IFPI3260 × IFPI3251	DArT, SNP, SSR, STS	Rust	Barilli et al. (2018)
F2/174	Erygel × 661	RFLP, RAPD	DF, PH, no. of nodes, AB	Dirlewanger et al. (1994)
RIL/133	A88 × Rovar	RFLP, RAPD, AFLP	AB, DM	Timmerman Vaughan et al. (2002)
RIL/88	Carneval × MP1401	AFLP, RAPD, STS	DM, lodging, PH, AB, SY, PC	Taran et al. (2003, 2004)
F2/148	A26 × Rovar	RAPD, RFLP, AFLP, STS	AB	Timmerman Vaughan et al. (2004)
RIL/135	JI296 × DP	RAPD, SSR, STS	AB, PH, DF	Prioul et al. (2004), Loridon et al. (2005) and Tayeh et al. (2015)
RIL/111	P665 × Messire	SSR	AB	Fondevilla et al. (2008)
RIL/111	P665 × Messire	SSR	AB	Fondevilla et al. (2012), Carrillo et al. (2014)
RIL/51	Alfetta × P651	SNP	AB	Jha et al. (2017)
RIL/134	Carerra × CDC Striker	SNP	AB	Gali et al. (2018)
RIL/139	K586 × Torsdag	RAPD	FW	Laucou et al. (1998)
RIL/393, 187	Lifter'/'Radley' Shawnee'/'Bohatyr	CAPS, SSR	FW	Jain et al. (2015)
RIL/187	Shawnee × Bohatyr	RAPD, SSR	MC, FW	McPhee et al. (2012)
RIL/80	Green Arrow × PI 179449	TRAP	FW	Kwon et al. (2013)
RIL/127	Puget × 90–2079	AFLP, RAPD, SSR, ISSR, STS	CRR	Pilet Nayel et al. (2002), Loridon et al. (2005) and Hamon et al. (2013)
RIL/127	Puget × 90–2079	AFLP, RAPD	CRR	Pilet-Nayel et al. (2005)
RIL/356	Baccara × PI180693, Baccara × 552	SSRs, RAPD, RGA	CRR	Hamon et al. (2011)
RIL/178	Baccara × PI180693	SNP	CRR, EF	Duarte et al. (2014), Tayeh et al. (2015)
RIL/111	DSP × 90–2131	RAPD, RFLP, SSR	EF, CRR, PH	Hamon et al. (2013)
Diverse panel /175	Pea- Aphanomyces collection	SNP	CRR	Desgroux et al. (2016)
Diverse panel /266	Pea accessions	SNP	CRR	Desgroux et al. (2018)
F2/ 227	Primo × OSU442-15	RAPD, RFLP, AFLP, STS,	SW, SC, SY & its attributes, nodes, flowering node	Timmerman-Vaughan et al. (2005), Gilpin et al. (1997)

(continued)

Table 8 (continued)

Population type/size	Pedigree	Marker types	Mapped traits	References
F2/92	NGB5839 × JI1794	–	Flower initiation node	Weller et al. (2012)
RIL/110	Wt11238 × Wt3557	AFLP, RAPD, STS, CAPS, ISSR, SSR	SY attributes, PC	Krajewski et al. (2012)
RIL/101	Wt10245 9 Wt11238	AFLP, RAPD, STS, CAPS, ISSR, SSR	SY attributes, PC	Krajewski et al. (2012)
RIL/104	Wt10245 × Wt11238	AFLP, ISSR, STS, CAPS, RAPD	PH	Irzykowska et al. (2002)
RIL/114	Wt10245 × Wt11238	AFLP, ISSR, STS, CAPS, RAPD	SY attributes, PC	Irzykowska and Wolko (2004)
RIL/164	Champagne × Terese	SNP, SSCP	Photoperiod, frost tolerance	Lejeune-Hénaut et al. (2008), Dumont et al. (2009) and Tayeh et al. (2015)
RIL/211	VavD265 × Cameor	SSR, SNP	PC & composition	Bourgeois et al. (2011), Tayeh et al. (2015)
RIL/207	Cameor × Ballet	SSR, SNP	DF, LA, RUE, CC, shoot length, SW, PC & quality, BNF	Bourion et al. (2010), Bourgeois et al. (2011) and Tayeh et al. (2015)
RIL/211	VavD265 × ballet	SSR, SNP	PC & its quality	Bourgeois et al. (2011), Tayeh et al. (2015)
RIL/ 129	China × Cameor	SSR, SNP	DF, BP, LA, CC, PH, SY attributes, BY, HI, PC, frost	Deulvot et al. (2010), Klein et al. (2014) and Tayeh et al. (2015)
RIL/139	Térèse × K586	RAPD, SSR	SY attributes, PC, DF, PH, no. of basal branches, BY, NC, HI	Loridon et al. (2005), Aubert et al. (2006) and Burstin et al. (2007)
RIL/90	Orb × CDC striker	AFLP, SSR, SNP	Cotyledon color bleaching	Ubayasena et al. (2010), Sindhu et al. (2014)
RIL/120	Alfetta × CDC bronco	AFLP, SSR	Seed appearance attributes	Ubayasena et al. (2011)
RIL/134	Kaspa × Parafield	SNP	Salinity tolerance	Leonforte et al. (2013))
RIL/163	1-2347-144 × CDC meadow	SNP	Phytic acid, phosphorus content	Sindhu et al. (2014), Shunmugam et al. (2015)
RIL/362	P665 × messire	RAPD, STS, EST, SSR, SNP	AB, broomrape, bacterial blight,AGB, RL, earliness, DT	Valderrama et al. (2004), Fondevilla et al. (2008, 2012) and Iglesias-Garcia et al. (2015)
RIL/47	MN313 × JI 1794	RAPD	Pod dehiscence	Weeden et al. (2002)

PM Powdery mildew, *AB* Ascochyta blight, *CRR* Common root rot, *FW* Fusarium wilt, *DF* Days to flowering, *DM* Days to maturity, *PH* Plant height; *SY* Seed yield, *PC* Protein content, *MC* Micronutrient content, *EF* early flowering, *LA* Leaf area, *RUE* radiation use efficiency, *DT* Drought tolerance, *SW* Seed weight, *SC* seed colour, *BY* Biological yield, *HI* Harvest index, *BP* Branching pattern, *BNF* Biological nitrogen fixation, *AGB* Above ground biomass, *RL* root length, *CC* Chlorophyll content, *NC* Nitrogen content

become an impediment in achieving food and nutritional security at global level. Hence, the need of the hour is the development of climate resilient nutrient dense varieties in a short span of time. However, as mentioned through traditional breeding methods which takes several years to develop cultivars is inadequate to cater the future food demand of alarmingly increasing population. During the recent years, efforts have been made to speed up the varieties and experimental population development by using single seed descent (SSD), off-season nurseries, and double haploid (DH) by rapid generation advancement; however, this could not succeed due to several disadvantages i.e. inability to perform selection, loss of segregating materials, and costly and limited recombinations (Croser et al. 2006; Brummer et al. 2011; Germanà 2011; Liu et al. 2016a, b; Atlin et al. 2017; Saxena et al. 2019). Attempt was also made with other in-vitro methodology like inducing flowering through manipulation of photoperiod and application of hormones towards rapid generation advancement (Ochatt et al. 2002; Surma et al. 2013; Ribalta et al. 2014).

The embryo rescue along with the flowering acceleration in hydroponic system with photo and thermo period manipulation and growth reduction through anti-gibberelin has also been performed in pea (Mobini and Warkentin 2016). The high cost as well as need of specialized staff are limiting factors in deployment of in vitro techniques in traditional breeding programme to accelerate generation advancement; thus, urgent need was felt for the technologies which can improve the efficiency as well as decrease the cost of breeding programme (Varshney et al. 2019; Hickey et al. 2019). Hence, most recently, a powerful technique called as Speed Breeding (SB) has been designed in different crop species by manipulation of photope-

Table 9 Molecular markers associated with resistance to major biotic stresses in field pea

Trait	Marker	References
Rust	AA446, AA505, AD146, AA416	Rai et al. (2011), Singh et al. (2015)
	AA121, AD147	Rai et al. (2016)
	A446-AA505, AD146-AA416	Rai et al. (2017), Upadhyay et al. (2017b)
	AD280, 3,567,800, 3,563,695, 3,569,323	Barilli et al. (2018)
Fusarium wilt	PSMPSAD171	McPhee et al. (2004)
	AD134_213, AA5_225, AA5 _235, AB111_166, AD73, AB30 AD85_178	Loridon et al. (2005)
	AC22_185, AD171_197, AB70_203, AD180_161, AB85-284	McPhee et al. (2012)
	Fw_Trap_480, Fw_Trap_340 and Fw_Trap_220,	Kwon et al. (2013)
	Aux1.SNP1, Hlhrep_SNP6, Hlhrep_SNP1, Cwi1_SNP3, Cwi1.SNP1, PPT2.SNP1, FVE.SNP6, PM34like.SNP2, ProteasB.SNP2, PFK_SNP1, Subt_SNP2, Sus3_SNP8, Trans_SNP1, TE002G22_SNP3	Cheng et al. (2015)
	TC112650, TC112533	Jain et al. (2015)
	A016180	Deng et al. (2024)
Powdery mildew	PSMAD51, PSMPSAA369, PSMPA5, PSMPSAA374, PSMPSAD60	Ek et al. (2005)
	A5	Pereira et al. (2010), Reginah et al. (2024)
	er1-5/HRM54	Pavan et al. (2013)
	AD60; c5DNAmet	Sun et al. (2014)
	PSMPSAD60, PSMPSA5,	Sun et al. (2016a, b, c)
	SNP1121	Sun et al. (2016a, b, c)
	KASPar-er1-1, KASPar-er1-3, KASPar-er1-4, KASPar-er1-5, KASPar-er1-6, KASPar-er1-7, KASPar-er1-10, KASPar-er1-11	Ma et al. (2017)
	AD61	Cobos et al. (2018)
	c5DNAmet, AA200, PSMPSAD51, OPX04-880,	Sun et al. (2019)
	KASPar-*er1*-8, KASPar-*er1*-9	Sun et al. (2019)
	KASPar-*er1*-13, KASPar-*er1*-14	Sun et al. (2022)
	AD237, AD14	Gupta et al. (2023)
Common root rot	AF0164458, AA176, A08_2000, X03_1000, E12_1100	Hamon et al. (2011)
	X03_1000, AB70, A19_800, AF016458, AA430942, E8M2_280, IJB174, J14_850,AB122b	Hamon et al. (2013)
	AA446-486, PA8, AB23-376, AA430942, AB145-364, AD57-300, AA175-282, AB112-402, AD83, AC75-297, PD21-226	Lavaud et al. (2015, 2016)
	AA122, AA387, AB101	Desgroux et al. (2016)
	Ps115429	Desgroux et al. (2018)
	PsCam037549_22628_1642, PsCam026054_14999_2864	Wu et al. (2021, 2022)
Ascochyta Blight	PSm PSAA175, PSMPSAA 163.2, PSMPSAA399,	Prioul et al. (2004)
	OPM6598/OPW5387, OPAI141353/OPW21157, OPAI141273/OPAI141353, OPRS4782, OPK6818, OPB111477	Fondevilla et al. (2008)
	OPAI14_1353/AA175, OPAI14_1273/OPAI14_1353	Fondevilla et al. (2011)
	OPM4_490/OPK6_887, agpl1_SNP2/MSU515_SNP3, OPZ10_576/Sugtrans_SNP3, sut1_SNP1/ OPRS4_699	Carrillo et al. (2014)
	PsDof1p308, RGA-G3Ap103	Jha et al. (2015)
	PsC8780p118, PsC22609p103, PsC8031p219, PsC20818p367,PsC7497p542,PsC13000p248,PsC4701p407	Jha et al. (2016)
	Sc33287_25420, Sc34405_60551, Sc33468_44352, Sc12023_67096	Jha et al. (2017)
	PsC1846p336 - Sc5317_256613, Sc3030_71736 - PsC7000p195, Sc8865_149928 - Sc7388_112888	Gali et al. (2018)
	sC8780p118	Jha et al. (2019)

riod, temperature, and humidity under controlled conditions (Collard et al. 2017; Nagatoshi and Fujita 2018; Idrissi 2020). In the case of pea speed breeding, protocols have been developed to reduce the time required to complete breeding cycle. By deploying SB tool in pea up to five generation per year can be achieved (Ghosh et al. 2018; Watson et al. 2018; Cazzola et al. 2020). These different SB protocols along with modern biotechnological tools available in pea would accelerate the designing of climate resilient nutrient dense cultivars.

During past five decades, genetic enhancement of pulse crops has been done through conventional breeding approaches. However, in recent years modern genomics and phenomic techniques have been used to speed up genetic gain in legumes crops which resulted delivery of few cultivars with important agronomic traits (Varshney et al. 2018). The remarkable increase of genomic resources has been noticed during last decade including whole genome sequencing in different pulses (Schmutz et al. 2014, 2016; Kang et al. 2014; Bett et al. 2014; Kreplak et al. 2019; Jegadeesan et al. 2021), which has prepared an excellent platform for deployment of different cutting-edge technologies, including genome editing for genetic enhancement. Genome editing has been used effectively in different pulses crops like cowpea, chickpea, and pea (Ji et al. 2019; Badhan et al. 2021; Baloglu et al. 2022; Bhowmik et al. 2023). The preference of field pea in the food industries is limited owing to the undesirable flavour causes by Lipoxygenase enzymes activities which led to formation of specific volatile organic compounds from certain polyunsaturated fatty acids (Jelen and Wasowicz 2011; Ma et al. 2015). To rectify off-flavour issue in field pea, genome editing has been performed through Clustered Regularly Interspaced Short Palindromic Repeats (CRISPR) approach to knockout *PsLOX2* which is responsible for the formation of off-flavour in field pea (Bhowmik et al. 2023). With the help of genome editing, multiple SNPs can be edited without changing the non-targeted traits of a cultivar. Genome-editing technologies like TALEN (transcription activator-like effector nucleases), ZFN (zinc finger nuclease), and CRISPR/Cas9 [the clustered regularly interspaced short palindromic repeat (CRISPR)/CRISPR-associated protein 9 nuclease (Cas9)] holds great promise towards the genetic improvement of agronomically important traits in pea breeding (Baloglu et al. 2022).

11 Conclusion

Field pea crops have to be climate resilient and nutrient dense to ensure they contribute to food as well as nutritional security of resource-poor folk of developed and developing countries. However, concerted efforts have been dedicated towards its genetic improvement through conventional breeding. In conventional breeding major intent was to incorporate a few traits like semi-leafless plant type, dwarf plant stature and resistance to powdery mildew which ultimately rumpled the genetic base; thus, now time has come to engage multiple diverse parents in hybridization to substantially expand the existing genetic base as well as to develop multiple disease-resistant cultivars with high nutritional profile. Unfortunately, the productivity of field pea in countries like Australia, China, and India is still quite low than global productivity which needs more attention in coming years to achieve significant gain in productivity. During the recent year, significant progress has been made to generate genomic resources and subsequently accurate reference genome has been assembled that would certainly hold great promise towards availability of more genes with excellent breeding value and suitability for high-throughput genotyping. Besides, the existing plenty of variation for protein, minerals, vitamins, and anti-nutritional factors needs to be explored to design nutrient dense cultivars. The abundance of biotic, abiotic, and yield attributes specific genetic loci or QTLs need to be embraced aggressively to accelerate genomic-assisted breeding that would certainly speed-up the delivery of climate resilient and nutrient dense varieties. The regular breeding programme should be judiciously aligned with available high-throughput SNP genotyping tools, reference genome, speed breeding, genome editing and other omics approaches i.e. transcriptomics, proteomics, metabolomics, and epigenomics. Notably, the integration of modern and traditional approaches should be done to speed up the genetic enhancement of field pea.

References

Alexieva V, Sergiev I, Mapelli S, Karanov E (2001) The effect of drought and ultraviolet radiation on growth and stress markers in pea and wheat. Plant Cell Environ 24:1337–1344

Ali K, Louw SM, Swart WJ (2005) Components and mechanisms of resistance in selected field pea *Pisum sativum* lines to the pea aphid *Acyrthosiphonpisum* (Homoptera: Aphididae). Int J Trop Insect Sci 25(2):114–121

Ali SMB, Sharma B, Ambrose MJ (1994) Current status and future strategy in breeding pea to improve resistance to biotic and abiotic stresses. Euphytica 73: 115–126.

Alves-Carvalho S, Aubert G, Carrere S, Cruaud C, Brochot AL, Jacquin F, Klein A, Martin C, Boucherot K, Kreplak J, Da Silva C (2015) Full-length de novo assembly of RNA-seq data in pea (*P isum sativum* L.) provides a gene expression atlas and gives insights into root nodulation in this species. Plant J 84(1):1–19

Amarakoon D, McPhee K, Thavarajah P (2012) Iron-, zinc-, and magnesium-rich field peas (Pisum sativum L.) with naturally low phytic acid: a potential food-based solution to global micronutrient malnutrition. J Food Compos Anal 27(1):8–13

Amarakoon D, Gupta DS, McPhee K, DeSutter T, Thavarajah P (2015) Genetic and environmental variation of seed iron and food matrix factors of North-Dakota-grown field peas (Pisum sativum L.). J Food Compos Anal 37:67–74

Amelin AV, Obraztsov AS, Lakhanov AP, Uvarov VN (1991) Possibility of increasing the resistance of pea to lodging. Selektsiya-i-Semenovodstvo-Moskva 2:21–23. Abstract from CAB Abstract, AN: 911698886

Anil Kumar TB, Rangaswamy KT, Ravi K (1994) Assessment of tall field pea genotypes for slow rusting resistance. Legum Res 17:79–82

Armstrong E, Matthews P, Carpenter D, Cassells J (1999) Desiccation and harvest monitor moisture and maturity for quality seed. Farming Ahead No: 93

Aslam S, Ghazanfar MU, Munir N, Hamid MI (2019) Managing fusarium wilt of pea by utilizing different application methods of fungicides. Pak J Phytopathol 31:81–88

Atlin GN, Cairns JE, Das B (2017) Rapid breeding and varietal replacement are critical to adaptation of cropping systems in the developing world to climate change. Glob Food Sec 12:31–37

Aubert G, Morin J, Jacquin F, Loridon K, Quillet MC, Petit A et al (2006) Functional mapping in pea, as an aid to the candidate gene selection and for investigating synteny with the model legume *Medicago truncatula*. Theor Appl Genet 112:1024–1041. https://doi.org/10.1007/s00122-005-0205-y

Azmat MA, Khan AA (2013) Inheritance of er1-based broad-spectrum powdery mildew resistance in pea (Pisum sativum L.). Notulae Botanicae Horti Agrobotanici Cluj-Napoca 41(2):485–490

Azmat MA, Nawab NN, Khan AA, Ashraf M, Niaz S, Mahmood K (2011) Characterization of pea germplasm. Int J Veg Sci 17(3):246–258

Badhan S, Ball AS, Mantri N (2021) First report of CRISPR/Cas9 mediated DNA-free editing of 4CL and RVE7 genes in chickpea protoplasts. Int J Mol Sci 22:396. https://doi.org/10.3390/ijms22010396

Baloglu MC, Celik Altunoglu Y, Baloglu P, Yildiz AB, Türkölmez N, Özden Çiftçi Y (2022) Gene-editing technologies and applications in legumes: Progress, evolution, and future prospects. Front Genet 13:859437

Bani M, Rubiales D, Rispail N (2012) A detailed evaluation method to identify sources of quantitative resistance to fusarium oxysporum f. sp. Pisi race 2 within a Pisum spp. germplasm collection. Plant Pathol 61:532–542. https://doi.org/10.1111/j.1365-3059.2011.02537.x

Bani M, Pérez De Luque A, Rubiales D, Rispail N (2018) Physical and chemical barriers in root tissues contribute to quantitative resistance to fusarium oxysporum f. sp. Pisi in Pea. Front. Plant Sci 9:199. https://doi.org/10.3389/fpls.2018.00199

Banniza S, Hashemi P, Warkentin TD, Vandenberg A, Davis AR (2005) The relationship among lodging, stem anatomy, degree of lignification and susceptibility to mycosphaerella blight in field pea (Pisum sativum L.). Can J Bot 83:954–967. https://doi.org/10.1139/b05-044

Banyal DK, Chaudhary JAYA, Singh A (2015) Evaluation of pea (Pisum sativum) germplasm for inheritance of resistance to powdery mildew (Erysiphe pisi). Indian Phytopathol 68(2):166–171

Baranger A, Aubert G, Arnau G, Laine AL (2004) Genetic diversity within Pisum sativum using protein and PCR-based markers. Theor Appl Genet 108:1309–1321

Barilli E, Sillero JC, Serrano A, Rubiales D (2009) Differential response of pea (Pisum sativum) to rusts incited by Uromyces viciae-fabae and U. Pisi. Crop Prot 28:980–986

Barilli E, Satovic Z, Rubiales D et al (2010) Mapping of quantitative trait loci controlling partial resistance against rust incited by Uromyces pisi (Pers.) Wint. In a *Pisum fulvum* L. Intraspecific Cross Euphytica 175:151–159

Barilli E, Sillero JC, Prats E, Rubiales D (2014) Resistance to rusts (Uromyces pisi and U. Viciae-fabae) in pea. Czech J Genet Plant Breed 50(2):135–143

Barilli E, Cobos MJ, Carrillo E, Kilian A, Carlin J, Rubiales D (2018) A high-density integrated DArTseq SNP-based genetic map of *Pisum fulvum* and identification of QTLs controlling rust resistance. Front Plant Sci 9:167. https://doi.org/10.3389/fpls.2018.00167

Bastida Garcia JL, González-Ronquillo M, Dominguez Vara IA, Romero-Bernal J, Castelan Ortega O (2011) Effect of field pea (Pisum sativum L.) level on intake, digestion, ruminal fermentation and in vitro gas production in sheep fed maintenance diets. Anim Sci J 82(5):654–662

Ben Ze'ev N, Zohary D (1973) Species relationships in the genus Pisum L. Israel J Bot 22:73–91

Benjamin JG, Nielsen DC (2006) Water deficit effects on root distribution of soybean, field pea and chickpea. Field Crop Res 97:248–253

Bett K, Ramsay L, Sharpe A, Cook D, Penmetsa RV, Verma N (2014) Lentil genome sequencing: establishing a comprehensive platform for molecular breeding. In: proceedings of international food legumes research conference (IFLRC-VI) and ICCLG-VII. Crop Development Center, Saskatoon, p 19

Bett K, Ramsay L, Chan C, Sharpe AG, Cook DR, Penmetsa RV et al (2016) Lentil 1.0 and beyond. In: Plant & animal genome conference XXIV. TSNN, San Diego, p W222

Bheri M, Fareeda G, Makandar R (2016) Assessing host specialization of *Erysiphe pisi* on garden pea germplasm through genotypic and phenotypic characterization. Euphytica 212:1–14. https://doi.org/10.1007/s10681-015-1511-3

Bhowmik P, Yan W, Hodgins C, Polley B, Warkentin T, Nickerson M, Ro DK, Marsolais F, Domoney C, Shariati-Ievari S, Aliani M (2023) CRISPR/Cas9-mediated lipoxygenase gene-editing in yellow pea leads to major changes in fatty acid and flavor profiles. Front Plant Sci 14:1246905

Biederbeck VO, Zenter RP, Campbell CA (2005) Soil microbial populations and activities as influenced by legume green fallow in a semiarid climate. Soil Biol Biochem 37:1775–1784

Bing DJ (2015) Breeding field pea cultivars with improved protein content. In: Eucarpia international symposium on protein crops. Pontevedra, Spain, April 5–7, Abstracts, pp 31–32

Bobkov SV, Selikhova TN (2021) Introgession of powdery mildew resistance into cultural pea from wild accession of P. Fulvum. IOP Conf Ser Earth Environ Sci 650:012091. https://doi.org/10.1088/1755-1315/650/1/012091

Bogdanova VS, Zaytseva OO, Mglinets AV, Shatskaya NV, Kosterin OE, Vasiliev GV (2015) Nuclear-cytoplasmic conflict in pea (Pisum sativum L.) is associated with nuclear and plastidic candidate genes encoding acetyl-CoA carboxylase subunits. PLoS One 10(3):e0119835

Bohra A, Pandey MK, Jha UC, Singh B, Singh IP, Datta D, Chaturvedi SK, Nadarajan N, Varshney RK (2014) Genomics-assisted breeding in four major pulse crops of developing countries: present status and prospects. Theor Appl Genet 127:1263–1291. https://doi.org/10.1007/s00122-014-2301-3

Bourgeois M, Jacquin F, Casseculle F et al (2011) A PQL (protein quantity loci) analysis of mature pea seed proteins identifies loci determining seed protein composition. Proteomics 11:1581–1594

Bourion V, Lejeune-He'naut I, Munier-Jolain N, Salon C (2003) Cold acclimation of winter and spring peas: carbon partitioning as affected by light intensity. Eur J Agron 19:535–548

Bourion V, Rizvi SMH, Fournier S, de Larambergue H, Galmiche F, Marget P et al (2010) Genetic dissection of nitrogen nutrition in pea through a QTL approach of root, nodule, and shoot variability. Theor Appl Genet 121:71–86. https://doi.org/10.1007/s00122-010-1292-y

Boutet G, Alves Carvalho S, Falque M, Peterlongo P, Lhuillier E, Bouchez O, Lavaud C, Pilet-Nayel ML, Rivière N, Baranger A (2016) SNP discovery and genetic mapping using genotyping by sequencing of whole genome genomic DNA from a pea RIL population. BMC Genomics 7:121

Brindley TA, Chamberlin JC, Schopp R (1956) The Pea weevil and methods for its control, farmers' bulletin no. 1971. U.S. Department of Agriculture, Washington, DC

Brummer FC, Barber WT, Collier SM, Cox TS, Johnson R, Murray SC, Olsen RC (2011) Plant breeding for harmony between agriculture and the environment. Front Ecol Environ 9:561–568

Bueckert RA, Wagenhoffer S, Hnatowich G, Warkentin TD (2015) Effect of heat and precipitation on pea yield and reproductive performance in the field. Can J Plant Sci 95:629–639

Burstin J, Marget P, Huart M et al (2007) Developmental genes have pleiotropic effects on plant morphology and source capacity, eventually impacting on seed protein content and productivity in pea. Plant Physiol 144:768–781

Burstin J, Salloignon P, Chabert-Martinello M, Magnin-Robert JB, Siol M, Jacquin F, Chauveau A, Pont C, Aubert G, Delaitre C, Truntzer C (2015) Genetic diversity and trait genomic prediction in a pea diversity panel. BMC Genomics 16(1):1–17

Campos-Vega R, Loarca-Piña G, Oomah BD (2010) Minor components of pulses and their potential impact on human health. Food Res Int 43(2):461–482
Carrillo E, Satovic Z, Aubert G, Boucherot K, Rubiales D, Fondevilla S (2014) Identification of quantitative trait loci and candidate genes for specific cellular resistance responses against Didymellapinodes in pea. Plant Cell Rep 33(7):1133–1145
Castillejo MÁ, Bani M, Rubiales D (2015) Understanding pea resistance mechanisms in response to fusarium oxysporum through proteomic analysis. Phytochemistry 115:44–58
Cazzola F, Bermejo CJ, Guindon MF, Cointry E (2020) Speed breeding in pea (Pisum sativum L.), an efficient and simple system to accelerate breeding programs. Euphytica 216(11):178
Chaisson MJP et al (2015) 2015. Resolving the complexity of the human genome using single-molecule sequencing. Nature 517:608–611
Chand R, Srivastava CP, Singh BD, Sarode SB (2006) Identification and characterization of slow rusting components in pea (*Pisum sativum* L.). Genet Resour Crop Evol 53:219–224
Chatterton S, Bowness R, Harding MW (2015) First report of root rot of field pea caused by Aphanomyces euteiches in Alberta, Canada. Plant Dis 99:288
Chaturvedi SK, Muraleedhar A, Gaur PM, Mishra N, Singh K, Nadarajan N (2014) Genetic variations for herbicide tolerance (imazethapyr) in chickpea (*Cicer arietinum*). Indian J Agric Sci 84(8):968–970
Cheng P, Holdsworth W, Ma Y, Coyne CJ, Mazourek M, Grusak MA et al (2015) Association mapping of agronomic and quality traits in USDA pea single-plant collection. Mol Breed 35:1–13
Chimwamurombe PM, Khulbe RK (2011) Domestication. In: Pratap A, Kumar J (eds) Biology and breeding of food legumes. CABI, Massachusetts, pp 19–34
Ciancio A, Mukerji KG (2008) Integrated management of diseases caused by fungi, phytoplasma and bacteria. Springer, Dordretch. https://doi.org/10.1007/978-1-4020-8571-0
Clement SL, McPhee KE, Elberson LR, Evans MA (2009) Pea weevil, *Bruchuspisorum* L. (Coleoptera: Bruchidae), resistance in *Pisum sativum* x *Pisum fulvum* interspecific crosses. Plant Breed 128:478–485
Clulow SA, Lewis BG, Matthews P (1991) A pathotype classification for *Mycosphaerellapinodes*. J Phytopathol 131(4):322–332
Cobos MJ, Satovic Z, Rubiales D, Fondevilla S (2018) Er3 gene, conferring resistance to powdery mildew in pea, is located in pea LGIV. Euphytica 214:203. https://doi.org/10.1007/s10681-018-2292-2
Collard BCY, Beredo JC, Lenaerts B et al (2017) Revisiting rice breeding methods—evaluating the use of rapid generation advance (RGA) for routine rice breeding. Plant Prod Sci 20:337–352. https://doi.org/10.1080/1343943X.2017.1391705
Conner RL, Chang KF, Hwang SF, Warkentin TD, McRae KB (2013) Assessment of tolerance for reducing yield losses in field pea caused by Aphanomyces root rot. Can J Plant Sci 93:473–482
Coyne CJ, Inglis DA, Whitehead SJ, McClendon MT, Muehlbauer FJ (2000) Chromosomal location of Fwf, the fusarium wilt race 5 resistance gene in Pisum sativum. Pisum Genet 32:20–22
Coyne CJ, Razai L, Baik B-K et al (2005) Variation for pea seed protein concentration in USDA Pisum core collection. Pisum Genet 37:7–11
Croser JS, Ulsdorf M, Davies PA et al (2006) Toward doubled haploid production in the Fabaceae: progress, constraints, and opportunities. Crit Rev Plant Sci 25:139–157
Dahiya B, Naresh JS (1993) Bio-efficacy of some insecticides against pea pod borer in field pea. In: National conference on ecofriendly approchesin the management of pests, diseases and industrial effluents. University of Agriculture and Technology, Kanpur, pp 20–22
Dahl WJ, Foster LM, Tyler RT (2012) Review of the health benefits of peas (Pisum sativum L.). Br J Nutr 108(S1):S3–S10
Daie J (1988) Mechanism of drought induced alteration in assimilate partitioning and transport in crops. Crit Rev Plant Sci 7:117–137
Das A, Parihar AK, Saxena D, Singh D, Singha KD, Kushwaha KPS et al (2019) Deciphering genotype-by-environment interaction for targeting test environments and rust resistant genotypes in field pea (Pisum sativum L.). Front. Plant Sci 10:825. https://doi.org/10.3389/fpls.2019.00825
Davis DW, Fritz VA, Pfleger FL, Percich JA, Malvick DK (1995) MN 144, MN 313 and MN 314: garden pea lines resistant to root rot caused by Aphanomyces euteiches Drechs. HortSci 30(3):639–640
De Sousa-Majer MJ, Turner NC, Hardie DC, Morton RL, Lamont B, Higgins TJV (2004) Response to water deficit and high temperature of transgenic peas (*Pisum sativum* L.) containing a seed-specific alpha-amylase inhibitor and the subsequent effects on pea weevil (*Bruchuspisorum* L.) survival. J Exp Bot 55:497–505
Demirbas A (2018) Micro and macronutrients diversity in Turkish pea (Pisum sativum) germplasm. Int J Agric Biol 20(4):701–710
Deng D, Sun S, Wu W, Zong X, Yang X, Zhang X, He Y, Duan C, Zhu Z (2022) Screening for Pea germplasms resistant to fusarium wilt race 5. Agronomy 2022(12):1354
Deng D, Sun S, Wu W, Duan C, Wu X, Zhu Z (2024) Fine mapping and identification of a fusarium wilt resistance gene FwS1 in Pea. Theor Appl Genet. https://doi.org/10.1007/s00122-024-04682-1
Desgroux A, L'anthoëne V, Roux-Duparque M, Rivière JP, Aubert G, Tayeh N, Moussart A, Mangin P, Vetel P, Piriou C, McGee RJ (2016) Genome-wide association mapping of partial resistance to Aphanomyces euteiches in pea. BMC Genomics 17(1):1–2
Desgroux A, Baudais VN, Aubert V, Le Roy G, de Larambergue H, Miteul H et al (2018) Comparative genome-wide-association mapping identifies common loci controlling root system architecture and resistance to Aphanomyces euteiches in pea. Front Plant Sci 8:2195. https://doi.org/10.3389/fpls.2017.02195
Deulvot C, Charrel H, Marty A, Jacquin F, Donnadieu C, Lejeune-Hénaut I et al (2010) Highly-multiplexed SNP genotyping for genetic mapping and germplasm diversity studies in pea. BMC Genomics 11:468–478. https://doi.org/10.1186/1471-2164-11-468
Devi J, Mishra GP, Sagar V, Kaswan V, Dubey RK, Singh PM, Sharma SK, Behera TK (2022) Gene-based resistance to Erysiphe species causing powdery mildew disease in peas (Pisum sativum L.). Genes 13(2):316
Dirlewanger E, Isaac P, Ranade S, Belajouza M, Cousin R, Devienne D (1994) Restriction fragment length polymorphism analysis of loci associated with disease resistance genes and developmental traits in Pisum sativum L. Theor Appl Genet 88:17–27
Dixit GP, Parihar AK (2014) Aman-A high yielding fieldpea variety with better standing ability. E-Publication on http://krishisewa.com/cms/varieties/pulse-varieties/368-fieldpea-variety.html
Dixit GP, Parihar AK, Gupta S (2014) Perspective for increasing field pea production in India. In: Handbook on minor and imported pulses of India, pp 37–39. Published by commodityindia.com
Dixit GP, Parihar AK, Singh NP (2017) "Variety central fieldpea IPFD 11-5" notification of crop varieties and registration of germplasm. Indian J Genet 77(3):438–439
Djihinto AC, Katary A, Djaboutou MC, Prudent P, Menozzi P, Atachi P (2012) Variation in biological parameters of cypermethrin resistant and susceptible strains of Helicoverpaarmigera from Benin Republic, West Africa. Int J Biol Chem Sci 6:931–940
Duarte J, Rivière N, Baranger A, Aubert G, Burstin J, Cornet L, Lavaud C, Lejeune-Hénaut I, Martinant JP, Pichon JP, Pilet-Nayel ML (2014) Transcriptome sequencing for high throughput SNP development and genetic mapping in Pea. BMC Genomics 15(1):1–15
Dueñas M, Estrella I, Hernandez T (2004) Occurrence of phenolic compounds in the seed coat and the cotyledon of peas (Pisum sativum L.). Eur Food Res Tech 219:116–123
Dumont E, Fontaine V, Vuylsteker C, Sellier H, Bodele S, Voedts N, Devaux R, Frise M, Avia K, Hilbert JL, Bahrman N, Hanocq E,

Lejeune-Henaut I, Delbreil B (2009) Association of sugar content QTL and PQL with physiological traits relevant to frost damage resistance in pea under field and controlled conditions. Theor Appl Genet 118:1561–1571

Duparque M (1996) Main history steps of the pea improvement. Grain Legum 12:18

Ek M, Eklund M, Von Post R, Dayteg C, Henriksson T, Weibull P, Ceplitis A, Isaac P, Tuvesson S (2005) Microsatellite markers for powdery mildew resistance in pea (Pisum sativum L.). Hereditas 142:86–91

Ellis THN, Poyser SJ (2002) An integrated and comparative view of pea genetic and cytogenetic maps. New Phytol 153:17–25

Elzebroek T, Wind K (2008) Guide to cultivated plants. CAB International, Oxfordshire

EPPO (2006) Distribution maps of quarantine pests. *Helicoverpaarmigera*. On-line available at www.eppo.org/QUARANTINE/insects/Helicoverpa_armigera/HELIAR_map.htm

Esposito MA, Martin EA, Cravero VP, Cointry E (2007) Characterization of pea accessions by SRAP'S markers. Sci Hortic 113(4):329–335

FAO (2022) Food and agriculture organization statistics. https://www.fao.org/faostat/en/#data/QCL

Fernandez TA, Rubiales D (2019) Flower and pod source influence on Pea weevil (Bruchuspisorum) oviposition capacity and preference. Front Plant Sci. https://doi.org/10.3389/fpls.2019.00491

Ferrandiz C, Navarro C, Gomez MD, Canas LA, Beltran JP (1999) Flower development in Pisum sativum: from the war of the whorls to the battle of the common primordia. Dev Genet 25(3):280–290

Flavell RB, Bennett MD, Smith JB, Smith DB (1974) Genome size and the proportion of repeated nucleotide sequence DNA in plants. Biochem Genet 12:257–269

Fondevilla S, Avila CM, Cubero JI, Rubiales D (2005a) Response to Ascochyta pinodes in a germplasm collection of Pisum spp. Plant Breed 124:313–315. https://doi.org/10.1111/j.1439-0523.2005.01104.x

Fondevilla S, ÅVila CM, Cubero JI, Rubiales D (2005b) Response to Mycosphaerella pinodes in a germplasm collection of Pisum spp. Plant Breed 124(3):313–315

Fondevilla S, Carver TLW, Moreno MT, Rubiales D (2006) Macroscopical and histological characterization of genes er1 and er2 for powdery mildew resistance in pea. Eur J Plant Pathol 115:309–321

Fondevilla S, Carver TLW, Moreno MT, Rubiales D (2007a) Identification and characterization of sources of resistance to *Erysiphe pisi* Syd. In Pisum spp. Plant Breed 126:113–119

Fondevilla S, Cubero JI, Rubiales D (2007b) Inheritance of resistance to Mycosphaerellapinodes in two wild accessions of Pisum. Eur J Plant Pathol 119:53–58

Fondevilla S, Torres AM, Moreno MT et al (2007c) Identification of a new gene for resistance to powdery mildew in Pisum fulvum, a wild relative of pea. Breed Sci 57:181–184

Fondevilla S, Rubiales D, Moreno MT, Torres AM (2008) Identification and validation of RAPD and SCAR markers linked to the gene Er3 conferring resistance to Erysiphe pisi DC in pea. Mol Breed 22:193–200

Fondevilla S, Krajinski F, Kuster H, Cubero JI, Rubiales D (2011) Identification of genes differentially expressed in a resistance reaction to Mycospherellapinodes in pea using micro-array technology. BMC Genomics 12:28

Fondevilla S, Martin-Sanz A, Satovic Z, Dolores Fernandez-Romero M, Rubiales D, Caminero C (2012) Identification of quantitative trait loci involved in resistance to pseudomonas syringaepv. Syringae in pea (Pisum sativum L.). Euphytica 186:805–812. https://doi.org/10.1007/s10681-0110592-x

Fondevilla S, Rotter B, Krezdorn N, Jüngling R, Winter P, Rubiales D (2013a) Identification of genes involved in resistance to Didymellapinodes in pea by deep super SAGE genome-wide transcriptome profiling. In: Book of abstracts of first legume society conference, p 148

Fondevilla S, Chattopadhyay C, Khare N, Rubiales D (2013b) Erysiphe trifolii is able to overcome er1 and Er3, but not er2, resistance genes in pea. Eur J Plant Pathol 136:557–563. https://doi.org/10.1007/s10658-013-0187-6

Fu H, Sun S, Zhu Z, Duan C, Yang X (2014) Phenotypic and genotypic identification of powdery mildew resistance in pea cultivars or lines from Canada. J Plant Genet Resour 15:1028–1033. (In Chinese with English abstract). [Google Scholar]

Gali KK, Liu Y, Sindhu A, Diapari M, Shunmugam AS, Arganosa G et al (2018) Construction of high-density linkage maps for mapping quantitative trait loci for multiple traits in field pea (Pisum sativum L.). BMC Plant Biol 18:172. https://doi.org/10.1186/s12870-018-1368-4

Gatti I, Esposito MA, Almiron P, Cravero VP, Cointry EL (2011) Diversity of pea (Pisum sativum) accessions based on morphological data for sustainable field pea breeding in Argentina. Genet Mol Res 10(4):3403–3410

Gawalko E, Garrett RG, Warkentin T, Wang N, Richter A (2009) Trace elements in Canadian fi eld peas: a grain safety assurance perspective. Food Addit Contam A 26:1002–1012

Gawłowska M, Święcicki W, Lahuta L, Kaczmarek Z (2017) Raffinose family oligosaccharides in seeds of Pisum wild taxa, type lines for seed genes, domesticated and advanced breeding materials. Genet Resour Crop Evol 64(3):569–578

Gawłowska M, Święcicki W (2016) The fa2 gene and molecular markers mapping in the gp segment of the Pisum linkage group V. J Appl Genet 57:317–322. https://doi.org/10.1007/ s13353-015-0335-0

Ge J, Sun CX, Corke H, Gul K, Gan RY, Fang Y (2020) The health benefits, functional properties, modifications, and applications of pea (Pisum sativum L.) protein: current status, challenges, and perspectives. Compr Rev Food Sci Food Saf 19(4):1835–1876

Georgieva K, Lichtenthaler HK (1999) Photosynthetic activity and acclimation ability of pea plants to low and high temperature treatment as studied by means of chlorophyll fluorescence. J Plant Physiol 155(3):416–423

Germanà MA (2011) Anther culture for haploid and doubled haploid production. Plant Cell Tissue Organ Cult 104:283–300. https://doi.org/10.1007/s11240-010-9852-z

Ghafoor A, McPhee K (2012) Marker assisted selection (MAS) for developing powdery mildew resistant pea cultivars. Euphytica 186:593–560

Ghosh S, Watson A, Gonzalez-Navarro OE et al (2018) Speed breeding in growth chambers and glasshouses for crop breeding and model plant research. Nat Protoc 13:2944–2963

Gilpin BJ, McCallum JA, Frew TJ, Timmerman-Vaughan GM (1997) A linkage map of the pea (Pisum sativum L.) genome containing cloned sequences of known function and expressed sequence tags (ESTs). Theor Appl Genet 95:1289–1299

Gong YM, Xu SC, Mao WH, Hu QZ, Zhang GW, Ding J, Li YD (2010) Developing new SSR markers from ESTs of pea (Pisum sativum L.). J Zhejiang Univ Sci B 11:702–707

Gonzalez EM., Arrese Igor C, Aparicio Tejo PM, Royuela M, Koyro HW (2001) Osmotic adjustment in different leaf structures of semileafless pea (Pisum sativum L.) subjected to water stress. In: Plant nutrition. Food security and sustainability of agro-ecosystems through basic and applied research. Developments in plant and soil sciences, Vol. 92. WJ Horst, MK Schenk, AM Bürkert, N Claassen, H Flessa, WB Frommer, H Goldbach, HV Olfs, V Romheld, B Sattelmacher, U Schmidhalter, S Schubert, van Wiren N, L Wittenmayer, eds. (Springer, Berlin), pp. 374–375

Gonzalez EM, Arrese Igor C, Aparicio Tejo PM, Royuela M, Koyro HW (2002) Solute heterogeneity and osmotic adjustment in different leaf structures of semi-leafless pea (*Pisum sativum* L.) subjected to water stress. Plant Biol 4:558–566

Gourlay CW, Hofer JM, Ellis TN (2000) Pea compound leaf architecture is regulated by interactions among the genes UNIFOLIATA, COCHLEATA, AFILA, and TENDRIL-LESS. Plant Cell 12(8):1279–1294

Gritton ET (1990) Registration of five pea root rot resistant germplasm lines of processing pea. Crop Sci 30:1166–1167

Grothe T, Wandrey F, Schuerch C (2020) Clinical evaluation of pea sprout extract in the treatment of hair loss. Phytother Res 34(2):428–431

Grusak MA, Cakmak I (2005) Methods to improve the crop-delivery of minerals to humans and livestock. In: Broadley MR, White PJ (eds) Plant nutritional genomics. Blackwell Publishing, Oxford, pp 265–286

Grzesiak S, Iijima M, Kono Y, Yamauchi A (1997) Differences in drought tolerance between cultivars of field bean and field pea. A comparison of drought-resistant and drought-sensitive cultivars. Acta Physiol Plant 19(3):349–357

Guilioni L, Wery J, Lecoeur J (2003) High temperature and water deficit may reduce seed number in field pea purely by decreasing plant growth rate. Funct Plant Biol 30:1151–1164. https://doi.org/10.1071/FP03105

Guilioni L, Wery J, Tardieu F (1997) Heat stress-induced abortion of buds and flowers in pea: is sensitivity linked to organ age or to relations between reproductive organs? Ann Bot 80: 159–168. doi: 10.1006/anbo.1997.0425

Guindon MF, Martin E, Zayas A, Cointry E, Cravero V (2016) Evaluation of SRAP markers for mapping of Pisum sativum L. Crop Breed Appl Biotechnol 16:182–188

Gupta SK, Gupta M (2019) Fusarium wilt of pea-A mini review. Plant Dis Res 34(1):1–9

Gupta DS, Thavarajah D, Knutson P, Thavarajah P, McGee RJ, Coyne CJ, Kumar S (2013) Lentils (Lens culinaris L.) a rich source of folates. J Agric Food Chem 61:7794–7799

Gupta A, Singh B, Kumar M, Sirohi U, Yadav SK, Sharma VR (2023) Identification of SSR markers linked to powdery mildew resistance in table Pea (*Pisum sativum* var. Hortense L.). Legume Res Int J 1:7

Haldimann P, Feller URS (2005) Growth at moderately elevated temperature alters the physiological response of the photosynthetic apparatus to heat stress in pea (Pisum sativum L.) leaves. Plant Cell Environ 28(3):302–317

Hamon C, Baranger A, Coyne CJ, McGee RJ, LeGoff I, L'Anthoëne V et al (2011) New consistent QTL in pea associated with partial resistance to Aphanomyces euteiches in multiple French and American environments. Theor Appl Genet 123:261–281. https://doi.org/10.1007/s00122-011-1582-z

Hamon C, Coyne CJ, McGee RJ, Lesné A, Esnault R, Mangin P et al (2013) QTL meta-analysis provides a comprehensive view of loci controlling partial resistance to Aphanomyces euteiches in four sources of resistance in pea. BMC Plant Biol 13:45. https://doi.org/10.1186/1471-2229-13-45

Han JY, Tyler RT (2003) Determination of folate concentrations in pulses by a microbiological method employing trienzyme extraction. J Agric Food Chem 51:5315–5318

Harland SC (1948) Inheritance of immunity to mildew in Peruvian forms of Pisum sativum. Heredity 2:263–269

Harmankaya M, Özcan MM, Karadaş S, Ceyhan E (2010) Protein and mineral contents of pea (Pisum sativum L.) genotypes grown in central Anatolian region of Turkey. S West J Hortic Biol Environ 1(2):159–165

Heath MC, Hebblethwaite PD (1985) Agronomic problems associated with the pea crop. In: Hebblethwaite PD, Heath MC, Dawkins TCK (eds) The Pea crop: a basis for improvement. Butterworth, London, pp 19–30

Hickey LT, Hafeez NA, Robinson H et al (2019) Breeding crops to feed 10 billion. Nat Biotechnol 37(7):744–754. https://doi.org/10.1038/s41587-019-0152-9

Holdsworth WL, Gazave E, Cheng P, Myers JR, Gore MA, Coyne CJ, McGee RJ, Mazourek M (2017) A community resource for exploring and utilizing genetic diversity in the USDA pea single plant plus collection. Hortic Res 4:17017. https://doi.org/10.1038/hortres.2017.17

Hood-Niefer SD, Warkentin TD, Chibbar RN, Vandenberg A, Tyler RT (2012) Effect of genotype and environment on the concentrations of starch in, and protein and the physicochemical properties of starch from, field pea and fababean. J Sci Food Agric 92(1):141–150. https://doi.org/10.1002/jsfa.4552

Hoover R, Hughes T, Chung HJ et al (2010) Composition, molecular structure, properties, and modification of pulse starches: a review. Food Res Int 43:399–413

Humphry M, Reinstädler A, Ivanov S, Bisseling T, Panstruga R (2011) Durable broad-spectrum powdery mildew resistance in pea er1 plants is conferred by natural loss-of-function mutations in PsMLO1. Mol Plant Pathol 12:866–867

Hurrell R, Egli I (2010) Iron bioavailability and dietary reference values. Am J Clin Nutr 91:1461S–1467S

Husnain SK, Hussain SH, Atiq M, Rajput NA, Abbas W, Mohsin M (2019) Screening of peas (Pisum sativum) varieties/lines against fusarium wilt (fusarium oxysporum f. sp. Pisi) and in vitro evaluation of fungicides against mycelial growth of pathogen. Pakistan. J Phytopathol 31(1):89–96

Idrissi O (2020) Application of extended photoperiod in lentil: towards accelerated genetic gain in breeding for rapid improved variety development. Moroccan J Agric Sci 1(1):14–19

Iglesias-Garcia R, Prats E, Fondevilla S, Satovic Z, Rubiales D (2015) Quantitative trait loci associated to drought adaptation in pea (Pisum sativum L.). Plant Mol Biol Report 33:1768

Infantino A, Kharrat M, Riccioni L, Coyne CJ, McPhee KE, Grunwald NJ (2006) Screening techniques and sources of resistance to root diseases in legumes. Euphytica 147:201–221. https://doi.org/10.1007/s10681-006-6963-z

Ingels C, VanHorn M, Bugg RL, Miller PR (1994) Selecting the right cover crop gives multiple benefits. Calif Agric 48(5):43–48

Irzykowska L, Wolko B (2004) Interval mapping of QTLs controlling yield-related traits and seed protein content in Pisum sativum. J Appl Genet 45:297–306

Irzykowska L, Wolko B, Swiecicki W (2002) Interval mapping of QTLs controlling some morphological traits in pea. Cell Mol Biol Lett 7:417–422

Iturbe-Ormaetxe I, Escuredo PR, Arrese-Igor C, Becana M (1998) Oxidative damage in pea plants exposed to water deficit or paraquat. Plant Physiol 116:173–181

Iwaya-Inoue M, Motooka K, Ishibashi Y, Fukuyama M (2003) Characteristic water status in dwarf pea in relation to drought resistance. J Fac Agric Kyushu Univ 48:29–38

Jain S, Kumar A, Mamidi S, McPhee K (2014) Genetic diversity and population structure among pea (Pisum sativum L.) cultivars as revealed by simple sequence repeat and novel genic markers. Mol Biotechnol 56:925–938

Jain S, Weeden NF, Kumar A, Chittem K, McPhee K (2015) Functional codominant marker for selecting the Fw gene conferring resistance to fusarium wilt race 1 in Pea. Crop Sci 55:2639–2646. https://doi.org/10.2135/cropsci2015.02.0102

Jamieson PD, Wilson DR, Hanson R (1984) Analysis of responses of field peas to irrigation and sowing date. Models of growth and water use. Proc Agron Soc N Z 14:75–81

Jegadeesan S, Raizada A, Dhanasekar P, Suprasanna P (2021) Draft genome sequence of the pulse crop blackgram [Vigna mungo (L.) Hepper] reveals potential R-genes. Sci Rep 11(1):11247

Jelen H, Wasowicz E (2011) Lipid-derived flavor. In: Jelen H (ed) Food flavors: chemical, sensory, and technological properties. CRC Press, Boca Raton, pp 65–93

Jeuffroy MH, Lecoeur J, Roche R (2010) The seed number. In: Munier-Jolain N, Biarnès V, Chaillet I, Lecoeur J, Jeuffroy MH (eds) Physiology of the pea crop. Enfield, CRC Press, pp 104–131

Jha AB, Warkentin TD, Gurusamy V, Taran B, Banniza S (2012) Identification of ascochyta blight resistance in wild Pisum species for use in pea breeding. Crop Sci 52:2462–2468. https://doi.org/10.2135/cropsci2012.04.0242

Jha AB, Ashokkumar K, Diapari M et al (2015) Genetic diversity of folate profiles in seeds of common bean, lentil, chickpea and pea. J Food Compos Anal 42:134–140

Jha AB, Taran B, Stonehouse R, Warkentin TD (2016) Identification of QTLs associated with improved resistance to ascochyta blight in an interspecific pea recombinant inbred line population. Crop Sci 56:2926–2939. https://doi.org/10.2135/cropsci2016.01.0001

Jha AB, Gali KK, Taran B, Warkentin TD (2017) Fine mapping of QTLs for Ascochyta blight resistance in pea using heterogeneous inbred families. Front Plant Sci 8:765

Jha AB, Gali KK, Banniza S, Warkentin TD (2019) Validation of SNP markers associated with ascochyta blight resistance in pea. Can J Plant Sci 99:243–249. https://doi.org/10.1139/cjps-2018-0211

Ji J, Zhang C, Sun Z, Wang L, Duanmu D, Fan Q (2019) Genome editing in cowpea Vigna unguiculata using CRISPR-Cas9. Int J Mol Sci 20(10):2471

Jiang Y, Lahlali R, Karunakaran C, Kumar S, Davis AR, Bueckert RA (2015) Seed set, pollen morphology and pollen surface composition response to heat stress in field pea. Plant Cell Environ 38(11):2387–2397

Jiang Y, Davis AR, Vujanovic V, Bueckert RA (2019) Reproductive development response to high daytime temperature in field pea. J Agro Crop Sci:1–10. https://doi.org/10.1111/jac.12328

Jing R, Vershinin A, Grzebyta J, Shaw P, Smykal P, Marshall D, Ambrose MJ, Ellis THN, Flavell AJ (2010) The genetic diversity and evolution of field pea (Pisum) studied by high throughput retrotransposon based insertion poly-morphism (RBIP) marker analysis. BMC Evol Biol 10:44

Jones DA, Dupont MS, Ambrose MJ, Frias J, Hedley CL (1999) The discovery of compositional variation for the raffinose family of oligosaccharides in pea seeds. Seed Sci Res 9(04):305–310

Kang YJ, Kim SK, Kim MY, Lestari P, Kim KH, Ha BK, Jun TH, Hwang WJ, Lee T, Lee J, Shim S (2014) Genome sequence of mungbean and insights into evolution within Vigna species. Nat Commun 5(1):5443

Karatas I, Ozturk L, Demir Y, Unlukara A, Kurunc A, Duzdemir O (2012) Alterations in antioxidant enzyme activities and proline content in pea leaves under long-term drought stress. Toxicol Ind Health:1–8. https://doi.org/10.1177/0748233712462471

Katoch V, Sharma S, Pathania S, Banayal DK, Sharma SK, Rathour R (2010) Molecular mapping of pea powdery mildew resistance gene er2 to pea linkage group III. Mol Breed 25:229–237. https://doi.org/10.1007/s11032-009-9322-7

Kaur S, Pembleton LW, Cogan NO, Savin KW, Leonoforte T, Paull J, Materne M, Forste JW (2012) Transcriptome sequencing of field pea and faba bean for discovery and validation of SSR genetic markers. BMC Genomics 13:104

Khan TN, French RJ, Hardie DC (1996) Breeding field peas for Western Australia: progress and problems. Pisum Genet 28:5–12

Klein A, Houtin H, Rond C, Marget P, Jacquin F, Boucherot K et al (2014) QTL analysis of frost damage in pea suggests different mechanisms involved in frost tolerance. Theor Appl Genet 127:1319–1330. https://doi.org/10.1007/s00122-014-2299-6

Kolbert Z, Bartha B, Erdei L (2008) Osmotic stress-and indole-3-butyric acid-induced NO generation are partially distinct processes in root growth and development in Pisum sativum. Physiol Plant 133:406–416

Kole PR, Sharma MK, Kumar S, Kumar A, Singh S, Sengar RS (2015) Assessment of genetic diversity in Indian cultivated pea (Pisum sativum L.) by using SRAP markers. Int J Plant Res 28(3):84

Kraft JM (1992) Registration of 90-2079, 90-2131 and 90-2322 pea germplasms. Crop Sci 32:1076

Kraft JM (2000) AEP workshop on screening for disease resistance in grain legumes. In: Monreal AR, Redondo RL (eds) AEP workshop on disease evaluation. Standardisation diseases resistance screening in grain legumes germplasm banks. Junta de Castilla y Leon, Servicio de Investigatcion y Tecnologia Agraria, Mata Digital, Valladolid, pp 61–65

Kraft JM, Coffman VA (2000a) Registration of 96-2052, 96-2058, 96-2068, 96-2198 and 96-2222 pea germplasms. Crop Sci 40:301–302

Kraft JM, Coffman VA (2000b) Registration of 97-261 and 97-1254 pea germplasms. Crop Sci 40:302–303

Krajewski P, Bocianowski J, Gawłowska M, Kaczmarek Z, Pniewski T, Święcicki W, Wolko B (2012) QTL for yield components and protein content: a multienvironment study of two pea (Pisum sativum L.) populations. Euphytica 183:323–336

Kreplak J, Madoui MA, Cápal P, Novák P, Labadie K, Aubert G, Bayer PE, Gali KK, Syme RA, Main D, Klein A (2019) A reference genome for pea provides insight into legume genome evolution. Nat Genet 51(9):1411–1422

Kushwaha C, Chand R, Srivastava C (2006) Role of aeciospores in outbreaks of pea (Pisum sativum) rust (Uromyces fabae). Eur J Plant Pathol 115:323–330

Kuznetsova OI, Ash OA, Hartina GA, Gostimskij SA (2005) RAPD and ISSR analyses of regenerated pea Pisum sativum L. plants. Russ J Genet 41:60–65

Kwon S-J, Brown A, Hu J, McGee R, Watt C, Kisha T, Timmerman-Vaughan G, Grusak M, McPhee K, Coyne C (2012) 2012. Genetic diversity, population structure and genome-wide marker-trait association analysis emphasizing seed nutrients of the USDA pea (Pisum sativum L.) core collection. Genes Genom 34(3):305–320

Kwon SJ, Smýkal P, Hu J, Wang M, Kim SJ, McGee RJ et al (2013) User-friendly markers linked to fusarium wilt race 1 resistance Fw gene for marker-assisted selection in pea. Plant Breed 132:642–648. https://doi.org/10.1111/pbr.12085

Ladizinsky G, Abbo S (2015) The Pisum genus. In: Ladizinsky G, Abbo S (eds) The search for wild relatives of cool season legumes. Springer, Cham, pp 55–69

Lam ACY, Karaca AC, Tyler RT, Nickerson MT (2018) Pea protein isolates: structure, extraction and functionality. Food Rev Intl 34(2):126–147. https://doi.org/10.1080/87559129.2016.1242135

Lamichaney A, Parihar AK, Hazra KK, Dixit GP, Katiyar PK, Singh D, Singh AK, Kumar N, Singh NP (2021) Untangling the influence of heat stress on crop phenology, seed set, seed weight, and germination in field Pea (*Pisum sativum* L.). Frontiers. Plant Sci 12:437

Laucou V, Karine H, Ellis N, Rameau C. (1998) Genetic mapping in pea. 1. RAPD-based genetic linkage map of Pisum sativum. Theoretical and Applied Genetics 97: 905–915

Lavaud C, Lesné A, Piriou C, Le Roy G, Boutet G, Moussart A, Poncet C, Delourme R, Baranger A, Pilet-Nayel ML (2015) Validation of QTL for resistance to Aphanomyces euteiches in different pea genetic backgrounds using near isogenic lines. Theor Appl Genet 128:2273–2288

Lavaud C, Baviere M, Le Roy G, Hervé MR, Moussart A, Delourme R et al (2016) Single and multiple resistance QTL delay symptom appearance and slow down root colonization by Aphanomyces euteiches in pea near isogenic lines. BMC Plant Biol 16:166. https://doi.org/10.1186/s12870-016-0822-4

Leach M, Agudelo P, Lawton-Rauh A (2012) Genetic variability of Rotylenchulusreniformis. Plant Dis 96:30–36

Lecoeur J, Guilioni L (2010) Abiotic stresses. In: Munier-Jolain N, Biarnès V, Chaillet I, Lecoeur J, Jeuffroy MH (eds) Physiology of the pea crop. Enfield, CRC Press, pp 136–181

Lejeune-Henaut I, Bourion V, Eteve G, Cunot E, Delhaye K, Desmyter C (1999) Floral initiation in field-grown forage peas is delayed to a greater extent by short photoperiods, than in other types of European varieties. Euphytica 109:201–211

Lejeune-Hénaut I, Hanocq E, Bethencourt L, Fontaine V, Delbreil B, Morin J et al (2008) The flowering locus Hr colocalizes with a major QTL affecting winter frost tolerance in Pisum sativum L. Theor Appl Genet 116:1105–1116. https://doi.org/10.1007/s00122-008-0739-x

León DP, Checa ÓE, Obando PA (2020) Inheritance of resistance of two pea lines to powdery mildew. Agron J 112:2466–2471. https://doi.org/10.1002/agj2.20253

Leonforte A, Sudheesh S, Cogan NOI, Salisbury PA, Nicolas ME, Materne M et al (2013) SNP marker discovery, linkage map construction and identification of QTLs for enhanced salinity tolerance in field pea (Pisum sativum L.). BMC Plant Biol 13:161. https://doi.org/10.1186/1471-2229-13-161

Liao W, Fan H, Liu P, Wu J (2019) Identification of angiotensin converting enzyme 2 (ACE2) up-regulating peptides from pea protein hydrolysate. J Funct Foods 60:103395

Liu JF, Cao TS, Feng J, Chang KF, Hwang SF, Strelkov SE (2013) Characterization of the fungi associated with ascochyta blight of field pea in Alberta, Canada. Crop Prot 54:55–64

Liu H, Zwer P, Wang H, Liu C, Lu Z, Wang Y, Yan G (2016a) A fast generation cycling system for oat and triticalebreeding. Plant Breed 135(5):574–579

Liu N, Xu S, Yao X, Zhang G, Mao W, Hu Q, Feng Z, Gong Y (2016b) Studies on the control of Ascochyta blight in field peas (Pisum sativum L.) caused by Ascochyta pinodes in Zhejiang Province, China. Front Microbiol 7:481. https://doi.org/10.3389/fmicb.2016.00481

Liu R, Fang L, Yang T, Zhang X, Hu J, Zhang H, Han W, Hua Z, Hao J, Zong X (2017) Marker-trait association analysis of frost tolerance of 672 worldwide pea (Pisum sativum L.) collections. Sci Rep 7:5919

Lockyer S, White A, Buttriss JL (2018) Biofortified crops for tackling micronutrient deficiencies—what impact are these having in developing countries and could they be of relevance within Europe? Nutr Bull 43:319–357

Loridon K, McPhee KE, Morin J, Dubreuil P, Pilet-Nayel ML, Aubert G, Rameau C, Baranger A, Coyne CJ, Lejeune-Hénault I, Burstin J (2005) Microsatellite marker polymorphism and mapping in pea (Pisum sativum L.). Theor Appl Genet 111:1022–1031

Ma X, Zhang Q, Zhu Q, Liu W, Chen Y, Qiu R et al (2015) A robust CRISPR/Cas9 system for convenient, high-efficiency multiplex genome editing in monocot and dicot plants. Mol Plant 8:1274–1284. https://doi.org/10.1016/j.molp.2015.04.007

Ma Y, Coyne CJ, Main D, Pavan S, Sun S, Zhu Z et al (2017) Development and validation of breeder-friendly KASPar markers for er1, a powdery mildew resistance gene in pea (Pisum sativum L.). Mol Breed 37:151. https://doi.org/10.1007/s11032-017-0740-7

Macas J, Neumann P, Navrátilová A (2007) Repetitive DNA in the pea (Pisum sativum L.) genome: comprehensive characterization using 454 sequencing and comparison to soybean and Medicago truncatula. BMC Genomics 8. https://doi.org/10.1186/1471-2164-8-427

Macas J, Novák P, Pellicer J, Čížková J, Koblížková A, Neumann P, Fukova I, Doležel J, Kelly LJ, Leitch IJ (2015) In depth characterization of repetitive DNA in 23 plant genomes reveals sources of genome size variation in the legume tribe Fabeae. PLoS One 10(11):e0143424

MacKay P (2013) Mojave Desert wildflowers: a field guide to wildflowers, trees, and shrubs of the Mojave Desert, including the Mojave National Preserve, Death Valley National Park, and Joshua tree national Park. Rowman & Littlefield

Mahajan R, Dar AA, Mukthar S, Zargar SM, Sharma S (2018) Pisum improvement against biotic stress: current status and future prospects. In: Pulse Improvement. Springer, Cham, pp 109–136

Majumdar DK (2011) Pulse crop production: principles and technologies. PHI Learning Pvt. Ltd

Malvick DK, Percich JA (1998) Genotypic and pathogenic diversity among pea-infecting strains of Aphanomyces euteiches from the central and western united sates. Phytopathology 88:915–921

Martin RJ, Jamieson PD (1996) Effect of timing and intensity of drought on the growth and yield of field peas (Pisum sativum L.). N Z J Crop Hortic Sci 24(2):167–174. https://doi.org/10.1080/01140671.1996.9513949

Martin DN, Proebsting WM, Hedden P (1997) Mendel's dwarfing gene: cDNAs from the Le alleles and the function of the expressed proteins. Proc Natl Acad Sci USA 94:8907–8911

Matthews P, Ambrose MJ (1994) Development and use of a 'core' collection for the John Innes Pisum collection. In: Balfourier F, Perretant MR (eds) Evaluation and exploitation of genetic resources, Proceedings of the genetic resources section meeting of EUCARPIA. Clermont Ferrand, France, pp 99–107

Maurer AR, Ormrod DP, Fletcher HF (1968) Response of peas to environment IV. Effect of five soil water regimes on growth and development of peas. Can J Plant Sci 48:129–137

Maxted N, Ambrose M (2001) Peas (Pisum sativum L.). In: Maxted N, Bennett SJ (eds) Plant Genetic Resources of Legumes in the Mediterranean. Kluwer Academic Publishers, Dordrecht, pp 181–190

McClendon MT, Inglis DA, McPhee KE, Coyne CJ (2002) DNA markers for fusarium wilt race 1 resistance gene in pea. J Am Soc Hortic Sci 127:602–607

McPhee K (2003) Dry pea production and breeding – a mini-review. Food, Agric Environ 1:64–69

McPhee KE, Tullu A, Kraft JM, Muehlbauer FJ (1999) Resistance to fusarium wilt race 2 in the Pisum core collection. J Am Soc Hortic Sci 124:28–31

McPhee KE, Inglis DA, Coyne CJ (2004) Linkage map location of Fusarium wilt race 2 (Fnw) in pea. In: Proceedings of the 5th European Conference on Grain Legumes, June 7–11, 2004, Dijon, p 342

McPhee KE, Inglis DA, Gundersen B, Coyne CJ (2012) Mapping QTL for fusarium wilt race 2 partial resistances in pea (Pisum sativum). Plant Breed 131:300–306. https://doi.org/10.1111/j.1439-0523.2011.01938.x

Melesse T, Singh SK (2012) Effect of climatic factors on pea aphid, Acyrthosiphonpisum Harris (Homoptera: Aphididae) population and its management through planting dates and biopesticides in field pea (Pisumsativum L.). J Agric Technol 8(1):125–132

Menon M, Barnes WJ, Olson MS (2015) Population genetics of freeze tolerance among natural populations of Populus balsamifera across the growing season. New Phytol 207:710–722

Meyer DW, Badaruddin M (2001) Frost tolerance of ten seedling legume species at four growth stages. Crop Sci 41:1838–1842

Mikić A, Mihailović V, Milić D, Vasiljević S, Katić S, Ćupina B (2006) The role of af, det and le genes in increasing grain yield of feed pea (Pisum sativum L.). In: Serbia and Montenegro. Abstracts, 3rd international conference on legume genomics and genetics, Brisbane, Australia, vol 109

Mikić A, Smýkal P, Kenicer G, Vishnyakova M, Sarukhanyan N, Akopian J, Vanyan A, Gabrielyan I, Smýkalová I, Sherbakova E, Zorić L (2013) The bicentenary of the research on 'beautiful'vavilovia (Vaviloviaformosa), a legume crop wild relative with taxonomic and agronomic potential. Bot J Linn Soc 172(4):524–531

Mishra RK, Kumar A, Chaudhary S, Kumar S (2009) Mapping of the multifoliate pinna (mfp) leaf-blade morphology mutation in grain pea Pisum sativum. J Genet 88:227–232

Mishra RK, Gangadhar BH, Nookaraju A, Kumar S, Park SW (2012) Development of EST-derived SSR markers in pea (Pisum sativum) and their potential utility for genetic mapping and transferability. Plant Breed 131:118–124

Mobini SH, Warkentin TD (2016) A simple and efficient method of in vivo rapid generation technology in pea (Pisum sativum L.) in vitro. Cell Dev-Pl 52:530–536

Mohan N, Aghora TS, Wani MA, Divya B (2013) Garden pea improvement in India. J Hortic Sci 8(2):125–164
Mohapatra C, Chand R, Tiwari JK, Singh AK (2020) Effect of heat stress during flowering and pod formation in pea (Pisum sativum L.). Physiol Mol Biol Plants 26(6):1119–1125
Morgan JM (1984) Osmoregulation and water stress in higher plants. Annu Rev Plant Physiol 35(1):299–319
Munjal RL, Chenulu VV, Hora TS (1963) Assessment of losses due to powdery mildew (Erysiphe polygoni) on pea. Indian Phytopathol 19:260–267
Nagatoshi Y, Fujita Y (2018) Accelerating soybean breeding in aCO2-supplemented growth chamber. Plant Cell Physiol 60:77–84. https://doi.org/10.1093/pcp/pcy189
Nasiri J, Haghnazari A, Saba J (2009) Genetic diversity among varieties and wild species accessions of pea (Pisum sativum L.). Afr J Biotechnol 8:3405–3417
Nisar M, Ghafoor A, Ahmad H, Khan M, Qureshi A, Ali H (2008) Evaluation of genetic diversity of pea germplasm through phenotypic trait analysis. Pak J Bot 40:2081–2086
Nisar M, Khan A, Wadood SF, Shah AA, Hanci F (2017) Molecular characterization of edible pea through EST-SSR markers. Turk J Bot 41(4):338–346
Novák P, Neumann P, Macas J (2010) Graph-based clustering and characterization of repetitive sequences in next-generation sequencing data. BMC Bioinformatics 11(1):1–12
Ochatt S, Sangwan RS, Marget P et al (2002) New approaches towards the shortening of generation cycles for faster breeding of protein legumes. Plant Breed 121:436–440
Okubara PA, Keller KE, McClendon MT, Inglis DA, McPhee KE, Coyne CJ (2005) Y15 999Fw, a dominant SCAR marker linked to the fusarium wilt race 1 (Fw) resistance gene in pea. Pisum Genet 37:30–33
Osuna-Caballero S, Rispail N, Barilli E, Rubiales D (2022) Identification and characterization of novel sources of resistance to rust caused by Uromyces pisi in Pisum spp. Plan Theory 11(17):2268
Ouafi L, Alane F, Rahal-Bouziane H, Abdelguerfi A (2016) Agro-morphological diversity within field pea (Pisum sativum L.) genotypes. Afr J Agric Res 11(40):4039–4047
Pal AB, Brahmappa HS, Rawal RD, Ullasa BA (1980) Field resistance of pea germplasm to powdery mildew (Erysiphe polygoni) and rust (Uromyces fabae). Plant Dis 64:1085–1086
Pandian A, Ford R, Taylor PWJ (2000) Transferability of sequence tagged microsatellite site (STMS) primers across four major pulses. Plant Mol Biol Report 18:395. https://doi.org/10.1007/BF02825069
Papavizas GC, Ayers WA (1974) Aphanomyces species and their root diseases in pea and sugarbeet: a review, Report No.:1485. Agricultural Research Service, US Department of Agriculture, Washington, DC
Parihar AK, Dixit GP (2017) 'Variety central fieldpea IPFD 12-2' notification of crop varieties and registration of germplasm. Indian J Genet 77(4):584–585
Parihar AK, Dixit GP (2023) Potential varieties of field Pea: current scenario and future outlook in India. The Agriculture Magazine, 2(3):241-247
Parihar AK, Dixit GP (2024) High-yielding and multiple stress resistance field pea varieties suitable for cultivation in north-eastern plains of India. Indian Farming 74(2):09–11
Parihar AK, Dixit GP, Pathak V, Singh D (2014a) Assessment of the genetic components and trait associations in diverse set of fieldpea (Pisum sativum L.) genotypes. Bangladesh J Bot 43(3):323–330
Parihar AK, Dixit GP, Pathak V, Singh D (2014b) Genetic diversity and trait inter-relationship studies in a diverse set of fieldpea (Pisum sativum L.) genotypes. J Food Legumes 27(4):297–301
Parihar AK, Bohra A, Dixit GP (2016) Nutritional benefits of winter pulses with special emphasis on peas and rajmash. In: Biofortification of food crops. Springer, New Delhi, pp 61–71
Parihar AK, Dixit GP, Singh NP (2019a) 'Variety Fieldpea IPFD 2014-2' notification of crop varieties and registration of germplasm. Indian J Genet 79(3):639
Parihar AK, Kumar N, Nath CP, Singh AK, Chaturvedi SK, Dixit GP, Singh NP (2019b) Genetic variations for post emergence herbicide tolerance in field pea (Pisum sativum). Indian J Agric Sci 89(4):634–639
Parihar AK, Dixit GP, Singh NP (2020a) 'Variety IPF16-13 (Harit)' notification of crop varieties and registration of germplasm. Indian J Genet 80(1):124
Parihar AK, Dixit GP, Bohra A, Gupta DS, Singh AK, Kumar N, Singh D, Singh NP (2020b) Genetic advancement in dry pea (Pisum sativum L.): retrospect and prospect. In: Accelerated plant breeding, vol 3. Springer, Cham, pp 283–341
Parihar AK, Dixit GP, Singh U, Singh AK, Kumar N, Gupta S (2021a) Potential of field Pea as a nutritionally rich food legume crop. In: Breeding for enhanced nutrition and bio-active compounds in food legumes. Springer, Cham, pp 47–82
Parihar AK, Dixit GP, Singh AK, Kumar N, Singh NP (2021b) Development of extra early field pea genotypes using inter-specific hybridization. Indian J Agric Sci 91(2)
Parihar AK, Yadav R, Lamichaney A, Mishra RK, Chandra A, Gupta DS, Tripathi K, Hazra KK, Dixit GP (2022a) Field pea breeding. In: Fundamentals of field crop breeding. Springer Nature Singapore, Singapore, pp 1237–1321
Parihar AK, Hazra KK, Lamichaney A, Dixit GP, Singh D, Singh AK, Singh NP (2022b) Characterizing plant trait (s) for improved heat tolerance in field pea (Pisum sativum L.) under subtropical climate. Int J Biometeorol 66(6):1267–1281
Parihar AK, Hazra KK, Lamichaney A, Singh AK, Dixit GP (2023) Delineating the role of plant stature towards heat stress tolerance in field pea (Pisum sativum L.). Heliyon 9(3):e14539
Pavan S, Schiavulli A, Appiano M, Miacola C, Visser RG, Bai Y, Lotti C, Ricciardi L (2013) Identification of a complete set of functional markers for the selection of er1 powdery mildew resistance in Pisum sativum L. Mol Breed 31:247–253
Pavek PLS (2012) Plant guide for pea (Pisum sativum L.). USDA-Natural Resources Conservation Service, Pullman
Pereira G, Marques C, Ribeiro R, Formiga S, Dâmaso M, Sousa MT et al (2010) Identification of DNA markers linked to an induced mutated gene conferring resistance to powdery mildew in pea (Pisum sativum L.). Euphytica 171:327–335
Pilet Nayel ML, Muehlbauer FJ, McGee RJ, Kraft JM, Baranger A, Coyne CJ (2002) Quantitative trait loci for partial resistance to Aphanomyces root rot in pea. Theor Appl Genet 106(1):28–39
Pilet Nayel ML, Coyne C, Hamon C, Lesne A, Le Goff I, Esnault R, Lecointe R, Roux-Duparque M, McGee R, Mangin P, McPhee K, Moussart A, Baranger A (2007) Understanding genetics of partial resistance to Aphanomyces root rot in pea for new breeding prospects. In: Third international Aphanomyces workshop on legumes, 07–09 November (2007), Rennes, France, p 36
Pilet-Nayel ML, Muehlbauer FJ, McGee RJ, Kraft JM, Baranger A, Coyne CJ (2005) Consistent quantitative trait loci in pea for partial resistance to Aphanomyces euteiches isolates from the United States and France. Phytopathology 95(11):1287–1293
Praca-Fontes MM, Carvalho CR, Clarindo WR (2014) Karyotype revised of Pisum sativum using chromosomal DNA amount. Plant Syst Evol 300:1621–1626. https://doi.org/10.1007/s00606-014-0987-y
Prasanth K, Varalakshmi B, Venugopalan R, Sriram S (2019) Screening of bitter gourd germplasm and advanced breeding lines against powdery mildew. Indian Phytopathol 72:15–22. https://doi.org/10.1007/s42360-018-0070-7
Prioul S, Frankewitz A, Deniot G, Morin G, Baranger A (2004) Mapping of quantitative trait loci for partial resistance to *Mycosphaerellapinodes* in pea (*Pisum sativum* L.) at the seedling and adult plant stages. Theor Appl Genet 108:1322–1334

Pumphrey FV, Schwanke RK (1974) Effects of irrigation on growth, yield, and quality of peas for processing. J Am Soc Hortic Sci 100:507–509

Rai R, Singh AK, Singh BD, Joshi AK, Chand R, Srivastava CP (2011) Molecular mapping for resistance to pea rust caused by Uromyces fabae (Pers.) de-Bary. Theor Appl Genet 123:803–813. https://doi.org/10.1007/s00122-0111628-2

Rai R, Singh AK, Chand R, Srivastava CP, Joshi AK, Singh BD (2016) Genomic regions controlling components of resistance for pea rust caused by Uromyces fabae (Pers.) de-Bary. J Plant Biochem Biotechnol 25:133–141. https://doi.org/10.1007/s13562-015-0318-6

Rai R, Singh AK, Chand R, Srivastava CP, Joshi AK, Singh BD (2017) Genomic regions controlling components of resistance for pea rust caused by Uromyces fabae (Pers.) de-Bary. J Plant Biochem Biotechnol 25:133–141

Rana JC, Rana M, Sharma V, Nag A, Chahota RK, Sharma TR (2017) Genetic diversity and structure of pea (Pisum sativum L.) germplasm based on morphological and SSR markers. Plant Mol Biol Report 35:118–129

Rana C, Sharma A, Rathour R et al (2023) In vivo and in vitro validation of powdery mildew resistance in garden pea genotypes. Sci Rep 13:2243. https://doi.org/10.1038/s41598-023-28184-0

Rao S, Armstrong R, Silva-Perez V, Tefera AT, Rosewarne GM (2021) Pulse root Ideotype for water stress in temperate cropping system. Plan Theory 10(4):692

Ray H, Bett K, Tar'an B, Vandenberg A, Thavarajah D, Warkentin T (2014) Mineral micronutrient content of cultivars of field pea, chickpea, common bean, and lentil grown in Saskatchewan, Canada. Crop Sci 54(4):1698–1708

Reginah P, Singh KN, Rajesh T, Mayurakshee M, Samuthirapandi S, Lyngdoh AA, Chamroy T (2024) Genetic variability and association study of SSR markers for yield and powdery mildew disease in pea (Pisum sativum). Indian J Agric Sci 94(3):258–262

Reiling TP (1984) Powdery mildew. In: Hagedorn DJ (ed) Compendium of pea diseases. APS, St. Paul, pp 21–22

Ribalta FM, Croser JS, Erskine W, Finnegan PM, Lulsdorf MM, Ochatt SJ (2014) Antigibberellin-induced reduction of internode length favors in vitro flowering and seed-set in diferent pea genotypes. Biol Plant 58:39–46. https://doi.org/10.1007/s10535-013-0379-0

Ridge PE, Pye DL (1985) The effects of temperature and frost at flowering on the yield of peas grown in a Mediterranean environment. Field Crop Res 12:339–346

Rispail N, Rubiales D (2014) Identification of sources of quantitative resistance to fusarium oxysporum f. sp. medicaginis in Medicago truncatula. Plant Dis 98:667–673. https://doi.org/10.1094/pdis-03-13-0217-re

Rispail N, Rubiales D (2015) Rapid and efficient estimation of pea resistance to the soil-borne pathogen fusarium oxysporum by infrared imaging. Sensors 15(2):3988–4000

Roberts RJ, Carneiro MO, Schatz MC (2013) The advantages of SMRT sequencing. Genome Biol 14:405

Roux-Duparque M, Boitel C, Decaux B, Moussart A, Alamie J, Pilet-Nayel ML, Muel F (2004) Breeding peas for resitance to Aphanomyces root rot: current main outputs of three breeding programmes. In: Proceedings of the 5th European conference on grain legumes, 7–11 June 2004, Dijon, France, p 133

Rubiales D, Fondevilla S (2012) Future prospects for ascochyta blight resistance breeding in cool season food legumes. Front Plant Sci 3:27. https://doi.org/10.3389/fpls.2012.00027

Rubiales D, Fondevilla S, Chen W, Gentzbittel L, Higgins TJV, Castillejo MA, Singh KB, Rispail N (2015). Achievements and challenges in legume breeding for pest and disease resistance. Crit Rev Plant Sci 34(1–3): 195–236. doi:10.1080/07352689.2014.898445.

Rubiales D, González Bernal MJ, Warkentin T, Bueckert T, Vaz Patto MC, McPhee K, McGee R, Smýkal P (2019) Advances in pea breeding. In: Hochmuth G (ed) Achieving sustainable cultivation of vegetables. Burleigh Dodds Science Publishing, Cambridge. https://doi.org/10.19103/AS.2019.0045.28. ISBN: 978 1 78676 236 8; www.bdspublishing.com

Sadras VO, Lake L, Chenu K, McMurray LS, Leonforte A (2012) Water and thermal regimes for field pea in Australia and their implications for breeding. Crop Pasture Sci 63(1):33–44. https://doi.org/10.1071/CP11321

Sadras VO, Lake L, Leonforte A, McMurray LS, Paull JG (2013) Screening field pea for adaptation to water and heat stress: associations between yield, crop growth rate and seed abortion. Field Crop Res 150:63–73. https://doi.org/10.1016/j.fcr.2013.05.023

Sánchez FJ, Manzanares M, de Andres EF, Tenorio JL, Ayerbe L (1998) Turgor maintenance, osmotic adjustment and soluble sugar and proline accumulation in 49 pea cultivars in response to water stress. Field Crop Res 59(3):225–235

Sánchez FJ, Manzanares M, De Andrés EF, Tenorio JL, Ayerbe L (2001) Residual transpiration rate, epicuticular wax load and leaf colour of pea plants in drought conditions: influence on harvest index and canopy temperature. Eur J Agron 15(1):57–70. https://doi.org/10.1016/S1161-0301(01)00094-6

Sanchez FJ, de Andres EF, Tenorio JL, Ayerbe L (2004) Growth of epicotyls, turgor maintenance and osmotic adjustment in pea plants (Pisum sativum L.) subjected to water stress. Field Crop Res 86:81–90

Sanz AM, Saldaña CC, Jing R et al (2011) Genetic diversity among Spanish pea (Pisum sativum L.) landraces, pea cultivars and the world Pisum sp. core collection assessed by retrotransposon based insertion polymorphisms (RBIPs). Span J Agric Res 1:166–178

Saxena KB, Saxena RK, Hickey LT, Varshney RK (2019) Can a speed breeding approach accelerate genetic gain in pigeonpea? Euphytica 215:202. https://doi.org/10.1007/s10681-019-2520-4

Schaefer H, Hechenleitner P, Santos-Guerra A, Menezes de Sequeira M, Pennington RT, Kenicer G et al (2012) Systematics, biogeography, and character evolution of the legume tribe Fabeae with special focus on the middle-Atlantic Island lineages. BMC Evol Biol 12:250

Schmutz J, McClean PE, Mamidi S, Wu GA, Cannon SB, Grimwood J, Jenkins J, Shu S, Song Q, Chavarro C, Torres-Torres M (2014) A reference genome for common bean and genome-wide analysis of dual domestications. Nat Genet 46(7):707–713

Semere T, Froud Williams RJ (2001) The effect of pea cultivar and water stress on root and shoot competition between vegetative plants of maize and pea. J Appl Ecol 38:137–145

Shafiq S, Mather DE, Ahmad M, Paull JG (2012) Variation in tolerance to radiant frost at reproductive stages in field pea germplasm. Euphytica 186(3):831–845. https://doi.org/10.1007/s10681-012-0625-0

Sharma SD (2000) Insect pests of pea (Pisumsativum L.) in Himachal Pradesh. Insect Environ 6(3):113

Sharma A, Rathour R, Plaha P, Katoch V, Khalsa GS, Patial V, Singh Y, Pathania NK (2010) Induction of fusarium wilt (fusarium oxysporum f. sp. pisi) resistance in garden pea using induced mutagenesis and in vitro selection techniques. Euphytica 173(3):345–356

Shen S, Hou HW, Ding CB, Bing DJ, Lu ZX (2016) Protein content correlates with starch morphology, composition and physicochemical properties in field peas. Can J Plant Sci 96(3):404–412

Shirasawa K, Sasaki K, Hirakawa H, Isobe S (2021) Genomic region associated with pod color variation in pea (Pisum sativum). G3 11(5):jkab 081

Shubha K, Dhar S, Choudhary H, Dubey SC, Sharma RK (2016) Identification of resistant sources and inheritance of fusarium wilt resistance in garden pea (Pisum sativum ssp. hortense). Indian J Hortic 73(3):356–361

Shunmugam ASK, Liu X, Stonehouse R, Bett KE, Taran B, Warkentin TD (2015) Mapping of seed phytate concentra¬tion and iron bioavailability in pea recombinant inbred line population. Crop Sci 55:828–836. https://doi.org/10.2135/cropsci2014.08.0544

Sierra I, Vidal-Valverde C, Kozlowska H (1998) Effect of ripening stage on thiamin and riboflavin levels in lupin, pea and faba bean seeds. Zeitschriftf€ur Lebensmitteluntersuchung und -Forschung A 206:126–129

Simon CJ, Hannan RM (1995) Development and use of core subsets of cool-season food legume germplasm collections. HortScience 30(4):907C

Sindhu A, Ramsay L, Sanderson LA, Stonehouse R, Li R, Condie J et al (2014) Gene-based SNP discovery and genetic mapping in pea. Theor Appl Genet 127:2225–2241. https://doi.org/10.1007/s00122-014-2375-y

Singh R, Ram H (2001) Inheritance of days to flowering and rust resistance in peas. Res Crops 2:414–418

Singh RM, Srivastava CP (1985) Evaluation classification and usefulness of pea germplasm for quantitative characters. Legum Res 8:68–73

Singh AK, Srivastava CP (2018) Effect of plant types on grain yield and lodging resistance in pea (Pisum sativum L.). Indian J Genet 75(1):69–74. https://doi.org/10.5958/0975-6906.2015.00008.5

Singh L, Narsinghani VG, Kotasthane SR, Tiwari AS (1978) Yield losses caused by powdery mildew in different varieties of peas. Indian J Agric Sci 48:86–88

Singh AK, Rai R, Srivastava CP, Singh BD, Kushwaha C, Chand R (2012a) A quantitative analysis of rust (Uromyces fabae) resistance in pea (Pisum sativum) using RILs. Indian J Agric Sci 82:190–192

Singh VK, Srivastava CP, Chand R, Kushwaha C (2012b) Slow rusting resistance in pea (Pisum sativum). Indian J Agric Sci 82(10):887–889

Singh AK, Rai R, Singh BD, Chand R, Srivastava CP (2015a) Validation of SSR markers associated with rust (Uromyces fabae) resistance in pea (Pisum sativum L.). Physiol Mol Biol Plants 21(2):243–247. https://doi.org/10.1007/s12298-015-0280-8

Singh AK, Singh VK, Chand R, Kushwaha C, Srivastava CP (2015b) Evaluation of slow rusting components in pea. J Plant Pathol:87–92

Singh J, Dhall RK, Aujla IS (2015c) Characterization of resistance response of garden pea (Pisum sativum L.) against powdery mildew (Erysiphe pisi DC) in sub-tropical plains of India. SABRAO J Breed Genet 47(4):384–393

Singh M, Rakesh K, Kumar S, Kumar V (2016) Critical period for weed control in field pea. Legum Res 39:86–89

Singh R, Babu S, Avasthe RK, Singh A, Yadav GS, Pashte V, Singh JK (2018) Screening of field pea varieties for rice-fallow areas under organic management conditions in NE Himalayas. Ann Agric Res 39(3):246–250

Sinjushin AA, Belyakova AS (2015) Ontogeny, variation and evolution of inflorescence in tribe Fabeae (Fabaceae) with special reference to genera Lathyrus, Pisum and Vavilovia. Flora 211:11–17

Sivachandra Kumar NT, Banniza S (2017) Assessment of the effect of seed infection with Ascochyta pisi on Pea in Western Canada. Front Plant Sci 8:933. https://doi.org/10.3389/fpls.2017.00933

Smitchger JA, Weeden N, Akin I, Warkentin T (2020) Stress equation for a cantilever beam: A model of lodging resistance in field pea. Int Agrophys 34:213–222. https://doi.org/10.31545/intagr/118318

Smýkal P (2006) Development of an efficient retrotransposon-based fingerprinting method for rapid pea variety identification. J Appl Genet 47(3):221–230

Smýkal P, Horáček J, Dostálová R, Hýbl M (2008) Variety discrimination in pea (Pisum sativum L.) by molecular, biochemical and morphological markers. J Appl Genet 49(2):155–166

Smýkal P, Aubert G, Burstin J, Coyne CJ, Ellis NTH et al (2012) Pea (Pisum sativum L.) in the genomic era. Agronomy 2:74–115. https://doi.org/10.3390/agronomy2020074

Smýkal P, Coyne C, Redden R, Maxted N (2013) Peas. In: Genetic and genomic resources of grain legume improvement. Elsevier, pp 41–80

Smýkal P, Coyne CJ, Ambrose MJ, Maxted N, Schaefer H, Blair MW, Berger J, Greene SL, Nelson MN, Besharat N, Vymyslický T (2015) Legume crops phylogeny and genetic diversity for science and breeding. Crit Rev Plant Sci 34(1–3):43–104

Smýkal P, Hradilová I, Trněný O, Brus J, Rathore A, Bariotakis M, Das RR, Bhattacharyya D, Richards C, Coyne CJ, Pirintsos S (2017) Genomic diversity and macroecology of the crop wild relatives of domesticated pea. Sci Rep 7(1):1–12

Smýkal P, Nelson MN, Berger JD, Wettberg EJB (2018) The impact of genetic changes during crop domestication. Agron 8(7):119. https://doi.org/10.3390/agronomy8070119

Smýkal P, Trněný O, Brus J, Hanáček P, Rathore A, Roma RD, Pechanec V, Duchoslav M, Bhattacharyya D, Bariotakis M, Pirintsos S (2018) Genetic structure of wild pea (Pisum sativum subsp. elatius) populations in the northern part of the Fertile Crescent reflects moderate cross-pollination and strong effect of geographic but not environmental distance. PLoS One 13(3):e0194056

Snowdon RJ, Friedt W (2004) Molecular markers in brassica oilseed breeding: current status and future possibilities. Plant Breed 123:1–8. https://doi.org/10.1111/j.1439-0523.2003.00968.x

Sohi HS, Sokhi SS, Rawal RD (1974) Varietal reaction of peas to powdery mildew (Erysiphe polygoni) and rust (Uromyces fabae). Mysore J Agric Sci 8:529–532

Sorensen JN, Edelenbos M, Wienberg L (2003) Drought effects on green pea texture and related physical-chemical properties at comparable maturity. J Am Soc Hortic Sci 128:128–135

Srivastava RK, Mishra SK, Singh AK, Mohapatra T (2012) Development of a coupling-phase SCAR marker linked to the powdery mildew resistance gene 'er1'in pea (Pisum sativum L.). Euphytica 186:855–866. https://doi.org/10.1007/s10681-012-0650-z

Sudheesh S, Lombardi M, Leonforte A, Cogan NOI, Materne M, Forster JW et al (2014) Consensus genetic map construction for field pea (Pisum sativum L.), trait dissection of biotic and abiotic stress tolerance and development of a diagnostic marker for the er1 powdery mildew resistance gene. Plant Mol Biol Report 33:1391–1403. https://doi.org/10.1007/s11105-014-0837-7

Sun X, Yang T, Hao J, Zhang X, Ford R, Jiang J, Wang F, Guan J, Zong X (2014) SSR genetic linkage map construction of pea (Pisum sativum L.) based on Chinese native varieties. Crop J 2(2–3):170–174

Sun S, Wang Z, Fu H, Duan C, Wang X, Zhu Z (2015) Resistance to powdery mildew in the pea cultivar Xucai 1 is conferred by the gene er1. Crop J 3(6):489–499

Sun S, Deng D, Wang Z, Duan C, Wu X, Wang X, Zong X, Zhu Z (2016a) A novel er1 allele and the development and validation of its functional marker for breeding pea (Pisum sativum L.) resistance to powdery mildew. Appl Genet 129:909–919. [Google Scholar]

Sun S, He Y, Dai C, Duan C, Zhu Z (2016b) Two major er1 alleles confer powdery mildew resistance in three pea cultivars bred in Yunnan Province, China. Crop J 4:353–359

Sun S, Fu H, Wang Z, Duan C, Zong X, Zhu Z (2016c) Discovery of a novel er1 allele conferring powdery mildew resistance in Chinese pea (Pisum sativum L.) landraces. PLoS One 11:e0147624

Sun S, Deng D, Duan C, Zong X, Xu D, He Y, Zhu Z (2019) Two novel er1 alleles conferring powdery mildew (Erysiphe pisi) resistance identified in a worldwide collection of Pea (Pisum sativum L.) germplasms. Int J Mol Sci 20:5071. https://doi.org/10.3390/ijms20205071

Sun S, Deng D, Wu W, He Y, Luo G, Du C, Duan C, Zhu Z (2022) Molecular characterizations of the er1 alleles conferring resistance to *Erysiphe pisi* in three chinese pea (Pisum sativum L.) landraces. Int J Mol Sci 23(19):12016

Surma M, Adamski T, Swiecicki S et al (2013) Preliminary results of in vitro culture of pea and lupin embryos for the reduction of generation cycles in single seed descent technique. Acta Soc Bot Pol 82(3):231–236. https://doi.org/10.5586/asbp.2013.021

Suso MJ, Bebeli PJ, Christmann S, Mateus C, Negri V, Pinheiro de Carvalho MA, Torricelli R, Veloso MM (2016) Enhancing legume ecosystem services through an understanding of plant–pollinator interplay. Front Plant Sci 7:333

Swiecicki WK, Wolko B, Apisitwanich S et al (2000) An analysis of isozymic loci polymorphism in the core collection of the polish Pisum genebank. Genet Resour Crop Evol 47:583–590

Taran B, Warkentin T, Somers D, Miranda D, Vandenberg A, Blade S, Woods S, Bing D, Xue A, DeKoeyer D, Penner G (2003) Identification of quantitative trait loci for plant height, lodging resistance and reaction to mycosphaerella blight in pea (Pisum sativum L.) using an AFLP-based linkage map. Theor Appl Genet 107:1482–1491

Taran B, Warkentin T, Somers DJ, Miranda D, Vandenberg A, Blade S et al (2004) Identification of quantitative trait loci for grain yield, seed protein concentration and maturity in field pea (Pisum sativum L.). Euphytica 136:297–306. https://doi.org/10.1023/B:EUPH.0000032721.03075.a0

Tayeh N, Aluome C, Falque M, Jacquin F, Klein A, Chauveau A et al (2015) Development of two major resources for pea genomics: the GenoPea 13.2K SNP Array and a high density, high resolution consensus genetic map. Plant J. https://doi.org/10.1111/tpj.13070

Teshome A, Mendesil E, Geleta M, Andargie D, Anderson P, Rämert B, Seyoum E, Hillbur Y, Dagne K, Bryngelsson T (2015a) Screening the primary gene pool of field pea (Pisum sativum L. subsp. sativum) in Ethiopia for resistance against pea weevil (Bruchuspisorum L.). Genet Resour Crop Evol 62(4):525–538

Teshome A, Bryngelsson T, Dagne K, Geleta M (2015b) Assessment of genetic diversity in Ethiopian field pea (Pisum sativum L.) accessions with newly developed EST-SSR markers. BMC Genet 16(1):1–12

Timmerman Vaughan GM, Frew TJ, Weeden NF (1994) Linkage analysis of er1, a recessive Pisum sativum gene for resistance to powdery mildew fungus (Erysiphe pisi D.C). Theor Appl Genet 88:1050–1055

Timmerman Vaughan GM, Frew TJ, Russell AC, Khan T, Butler R, Gilpin M, Murray S, Falloon K (2002) QTL mapping of partial resistance to field epidemics of Ascochyta blight of peas. Crop Sci 42:2100–2111

Timmerman Vaughan GM, Frew TJ, Butler R, Murray S, Gilpin M, Falloon K, Johnston P, Lakeman MB, Russell A, Khan T (2004) Validation of quantitative trait loci for Ascochyta blight resistance in pea (Pisum sativum L.), using populations from two crosses. Theor Appl Genet 109:1620–1631

Timmerman-Vaughan GM, Mills A, Whitfield C, Frew T, Butler R, Murray S, Lakeman M, McCallum J, Russell A, Wilson D (2005) Linkage mapping of QTL for seed yield, yield components, and developmental traits in pea. Crop Sci 45(4):1336–1344

Tiwari KR, Penner GA, Warkentin TD (1997) Inheritance of powdery mildew resistance in pea. Can J Plant Sci 77:307–310

Tiwari KR, Penner GA, Warkentin TD (1998) Identification of coupling and repulsion phase RAPD markers for powdery mildew resistance gene er1 in pea. Genome 41:440–444

Torrey JG, Zobel R (1977) Root growth and morphogenesis. In: Sutcliffe JF, Pate JS (eds) The physiology of the garden Pea. Academic, New York, pp 119–152. ISBN: 0-12-677550-8

Tosh SM, Yada S (2010) Dietary fibres in pulse seeds and fractions: characterization, functional attributes, and applications. Food Res Int 43:450–460

Tran HS, You MP, Lanoiselet V, Khan TN, Barbetti MJ (2014) First report of Phoma glomerata associated with the ascochyta blight complex on field pea (Pisum sativum) in Australia. Plant Dis 98:427. https://doi.org/10.1094/PDIS-08-13-0809-PDN

Trněný O, Brus J, Hradilová I, Rathore A, Das RR, Kopecký P, Coyne CJ, Reeves P, Richards C, Smýkal P (2018) Molecular evidence for two domestication events in the pea crop. Genes 9(11):535. https://doi.org/10.3390/genes9110535

Troszynska A, Ciska E (2002) Phenolic compounds of seed coats of white and coloured varieties of pea (Pisum sativum L.) and their total antioxidant activity. Czech J Food Sci 20:15–22

Tulbek MC, Lam RSH, Wang YC, Asavajaru P, Lam A (2016) Pea: a sustainable vegetable protein crop. In: Nadathur SR, Wanasundara JPD, Scanlin L (eds) Sustainable protein sources. Academic, San Diego, pp 145–164

Turner NC, Wright GC, Siddique KHM (2001) Adaptation of grain legumes (pulses) to water-limited environments. Adv Agron 71:193–231. https://doi.org/10.1016/S0065-2113(01)71015-2

Tzitzikas EN, Vincken JP, de Groot J, Gruppen H, Visser RGF (2006) Genetic variation in pea seed globulin composition. J Agric Food Chem 54(2):425–433. https://doi.org/10.1021/jf0519008

Ubayasena L, Bett K, Tar'an B et al (2010) Genetic control and QTL analysis of cotyledon bleaching resistance in green field pea (Pisum sativum L.). Genome 53(5):346–359

Ubayasena L, Bett K, Taran B, Warkentin T (2011) Genetic control and identification of QTLs associated with visual quality traits of field pea (Pisum sativum L.). Genome 54(4):261–272

Umar HMI, Ur-rehman S, Bilal M, Atif S, Naqvi H, Manzoor SA, Ghafoor A, Khalid M, Iqbal MT, Qayyoum A, Ahmad F, Irshad MA (2014) Evaluation of genetic diversity in pea (Pisum sativum) based on morpho-agronomic characteristics for yield and yield associated traits. J Biodivers Environ Sci 4(5):323–328

Upadhyay V, Kushwaha KPS, Pandey P (2017a) Evaluation of pea germplasms for rust disease resistance. J Pure Appl Microbiol 11(1):307–314

Upadhyay V, Kushwaha KPS, Pandey P (2017b) Molecular screening of pea germplasm for rust disease resistance using SSR marker. J Pure Appl Microbiol 11:343–348

Valderrama MR, Román B, Satovic Z, Rubiales D, Cubero JI, Torres AM (2004) Locating quantitative trait loci associated with Orobanche crenata resistance in pea. Weed Res 44(4):323–328

Varshney RK, Thudi M, Pandey MK, Tardieu F, Ojiewo C, Vadez V et al (2018) Accelerating genetic gains in legumes for the development of prosperous smallholder agriculture: integrating genomics, phenotyping, systems modelling and agronomy. J Exp Bot 69(13):3293–3312. https://doi.org/10.1093/jxb/ery088

Varshney RK, Pandey MK, Bohra A et al (2019) Toward the sequence-based breeding in legumes in the post-genome sequencing era. Theor Appl Genet 132:797–816. https://doi.org/10.1007/s00122-018-3252-x

Vavilov NI (1949) The phytogeographic basis of plant breeding. In: Chester KS (ed) The origin, variation, immunity, and breeding of cultivated plants. Chronica Botanica, Waltham, pp 13–54

Vidal-Valverde C, Frias J, Hernández AM, Martín-Alvarez PJ, Sierra I, Rodríguez C, Blazquez I, Vicente G (2003) Assessment of nutritional compounds and Antinutritional factors in Pea (Pisum Sativum) seeds. J Sci Food Agric 83:298–306

Vijayalakshmi S, Yadav K, Kushwaha C, Sarode SB, Srivastava CP, Chand R, Singh BD (2005) Identification of RAPD markers linked to the rust (Uromyces fabae) resistance gene in pea (Pisum sativum). Euphytica 144:265–274

Vijaylaxmi (2013) Effect of high temperature on growth, biomass and yield of field pea genotypes. Legum Res 36:250–254

Vovlas N, Troccoli A, Palomares-Rius J et al (2011) Ditylenchus gigas spp. parasitizing broad bean: a new stem nematode singled out from the Ditylenchus dipsaci species complex using apolyphasic approach with molecular phylogeny. Plant Pathol 60:762–777

Wang FB, Fu JF, Dong LF (2002) Study on tendril inheritance of semi-leafless pea and its application to pea breeding. J Hebei Vocation Tech Teach Coll 16:6–8

Wang N, Hatcher DW, Gawalko EJ (2008) Effect of variety and processing of nutrients and certain anti-nutrients in field peas (Pisum

sativum). Food Chem 111:132–138. https://doi.org/10.1016/j.foodchem.2008.03.047
Wang Z, Fu H, Sun S, Duan C, Wu X, Yang X, Zhu Z (2015) Identification of powdery mildew resistance gene in pea line X9002. Acta Agron Sin 41(4):515–523
Warkentin TD, Vandenberg A, Banniza S, Slinkard AE (2004) CDC Striker field pea. Can J Plant Sci 84:230–240
Warkentin TD, Smykal P, Coyne CJ, Weeden N, Domoney C, Bing D, Leonforte A, Xuxiao Z, Dixit G, Boros L et al (2015) Pea (Pisum sativum L.). In: De Ron AM (ed) Grain legumes, Series handbook of plant breeding. Springer Science Business Media, New York, pp 37–83
Watson A, Ghosh S, Williams M et al (2018) Speed breeding is a powerful tool to accelerate crop research and breeding. Nat Plants 4:23–29
Weeden NF, Brauner S, Przyborowski JA (2002) Genetic analysis of pod dehiscence in pea (Pisum sativum L.). Cell Mol Biol Lett 7:657–663
Weller JL, Liew LC, Hecht VF, Rajandran V, Laurie RE, Ridge S, Wenden B, Vander Schoor JK, Jaminon O, Blassiau C, Dalmais M (2012) A conserved molecular basis for photoperiod adaptation in two temperate legumes. Proc Natl Acad Sci 109(51):21158–21163
Wicker E, Moussart A, Duparque M, Rouxel F (2003) Further contributions to the development of a differential set of pea cultivars (Pisum sativum) to investigate the virulence of isolates of Aphanomyces euteiches. Eur J Plant Pathol 109:47–60
Wroth JM (1998) Possible role for wild genotypes of Pisum spp. to enhance ascochyta blight resistance in pea. Aust J Exp Agric 38:469–479. https://doi.org/10.1071/ea98024
Wu L, Chang KF, Conner RL, Strelkov S, Fredua Agyeman R, Hwang SF, Feindel D (2018) Aphanomyces euteiches: A threat to Canadian field Pea production. Engineering 4:542–551. https://doi.org/10.1016/j.eng.2018.07.006
Wu L, Fredua-Agyeman R, Hwang SF, Chang KF, Conner RL, McLaren DL, Strelkov SE (2021) Mapping QTL associated with partial resistance to Aphanomyces root rot in pea (Pisum sativum L.) using a 13.2 K SNP array and SSR markers. Theor Appl Genet 134(9):2965–2990
Wu L, Fredua-Agyeman R, Strelkov SE, Chang KF, Hwang SF (2022) Identification of novel genes associated with partial resistance to Aphanomyces root rot in field pea by BSR-Seq analysis. Int J Mol Sci 23(17):9744
Xin Z, Browse J (2000) Cold comfort farm: the acclimation of plants to freezing temperatures. Plant Cell Environ 23:893–902
Xue AG, Warkentin TD (2001) Partial resistance to Mycosphaerellapinodes in field pea. Can J Plant Sci 81:535–540
Xu-Xiao ZONG, Jian-Ping GUAN, Shu-Min WANG, Qing-Chang LIU (2008) Genetic diversity among Chinese pea (Pisum sativum L.) landraces as revealed by SSR markers. Acta Agron Sin 34(8):1330–1338
Yadav P, Singh AK, Srivastava CP (2010) Genetic variability and character association in diverse collection of Indian and exotic germplasm lines of pea (Pisum sativum L.). Vegetable. Science 37(1):75–77
Yadav A, Singh V, Yadav A, Singh H (2019) Studies on succession of insect pest complex associated with pea at Bikaner. J Entomol Zool Stud 7(3):1606–1608
Yadav RK, Chandra A, Parihar AK, Sujayanand GK, Kumar V, Bandi SM, Arya M (2022) Escape resistance in fieldpea against Phytomyzahorticola Goureau, as influenced by sowing time and its correlation with plant metabolites. Pest Manag Hortic Ecosyst 28(2):119–128
Yang T, Liu R, Luo Y, Hu S, Wang D, Wang C, Pandey MK, Ge S, Xu Q, Li N, Li G (2022) Improved pea reference genome and pangenome highlight genomic features and evolutionary characteristics. Nat Genet 54(10):1553–1563
Yarnell SH (1962) Cytogenetics of the vegetable crops. III. Legumes. A. Garden peas, Pisum sativum L. Bot Rev 28:465–537
Zhang R, Hwang SF, Chang KF, Gossen BD, Strelkov SE, Turnbull GD et al (2006) Genetic resistance to Ascochyta pinodes in 558 field pea accessions. Crop Sci 46:2409–2414. https://doi.org/10.2135/cropsci2006.02.0089
Zimmer MC, Sabourin D (1986) Determining resistance reaction of field pea cultivars at the seedling stage to *Mycosphaerellapinodes*. Phytopathology 76:878–881. https://doi.org/10.1094/Phyto-76-878
Zong X, Redden R, Liu Q et al (2009) Analysis of a diverse Pisum sp. collection and development of a Chinese P. sativum core collection based on microsatellite markers. Theor Appl Genet 118:193–204
Zvirin T, Herman R, Brotman Y, Denisov Y, Belausov E, Freeman S et al (2010) Differential colonization and defence responses of resistant and susceptible melon lines infected by Fusarium oxysporum race 1.2. Plant Pathol 59:576–585. https://doi.org/10.1111/j.1365-3059.2009.02225.x

Lathyrus (*Lathyrus sativus* L.) Breeding

K. R. Ramya, Rinky Resma Panda, Padmavati G. Gore, Arpita Das, Surendra Barpete, A. K. Parihar, and Kuldeep Tripathi

Abstract

Grasspea (*Lathyrus sativus* L.) is a versatile cool-season legume that is commonly used as food and fodder. It is a hardy legume that is adaptable to varied fragile agroecosystems as it can tolerate drought, salinity, and flood and has multiple uses in the pharmaceutical and industrial sectors apart from its use as food. It is a rich source of protein, folic acid, and minerals and the only dietary source of L-homoarginine. The cultivation of this crop was undermined because of the presence of β-ODAP in its seed, which may cause neurotoxic effects when it is consumed as a sole diet for longer periods. The development of improved varieties with earliness, high biomass, and low ODAP content are the pipeline breeding objectives in this underutilized legume. This chapter deals with the importance of this crop, its wild relatives, cytogenetics, breeding programs, and future perspectives.

Keywords

Grasspea · Improved varieties · Breeding objectives · ODAP content · Wild relatives

K. R. Ramya · P. G. Gore · K. Tripathi (✉)
ICAR-National Bureau of Plant Genetic Resources, New Delhi, India

R. R. Panda
The Graduate School, ICAR- Indian Agricultural Research Institute, New Delhi, India

A. Das
Bidhan Chandra Krishi Viswavidyalaya, Mohanpur, West Bengal, India

S. Barpete
International Centre for Agricultural Research in the Dry Areas, Amlaha, Madhya Pradesh, India

A. K. Parihar
ICAR-Indian Institute of Pulses Research, Kanpur, India

1 Introduction

The rate of human population growth far exceeds the rate of increase in food production per hectare of land, and the population is expected to be around 8.5–12 billion in 2100 (Arora 2019; Singh et al. 2019). The complex interplay between population growth and climate change emphasizes the significant impact of these factors on global food security, with climate change being a key driver of recent rises in global hunger (Molotoks et al. 2021). Looking ahead, a major challenge will be producing adequate food to meet rising demands while minimizing environmental harm. One potential solution is to promote agricultural practices and climate-smart crops that effectively integrate technological advances and traditional knowledge to boost yields in ecologically sustainable and socially responsible ways. The adoption of science-based, eco-friendly farming approaches and underutilized/neglected crops with huge nutritional potential can take account of the socioeconomic considerations that could serve as a foundation for sustainably fulfilling future global nutritional needs and also be able to meet sustainable development goals (Singh et al. 2019; Ramya et al. 2022).

Grasspea (*Lathyrus sativus* L.) is one such nutritious underutilized legume crop that can thrive in challenging agro-ecosystems and withstand extreme conditions like drought, flooding, heat waves, infertile soils, and other stresses (Pratap and Gupta 2022; Tripathi et al. 2021). This annual cool-weather legume is native to South and Southeast Asia but plays an economic and ecological role primarily in South Asia and sub-Saharan Africa, with smaller production in West and Central Asia, North Africa, Southern Europe, and South America. It is cultivated mostly for human consumption in India, Bangladesh, Nepal, Pakistan, and Ethiopia and for animal feed and fodder elsewhere (Sarwar et al. 1996; Campbell 1997; Kumar et al. 2011; Kumar et al. 2013a, b). Approximately 1.5 million hectares are planted with grasspea worldwide, generating around 1.2 million tons annually, and in India out of 11.6 million hectares left fallow

G. Prasad Dixit et al. (eds.), *Fundamentals of Legume Breeding*, https://doi.org/10.1007/978-981-96-7838-9_9

annually, grasspea could easily be cultivated on approximately 0.5 million hectares to improve land productivity and boost farmer income as a rotational pulse crop with rice in eastern India (Kumar et al. 2013a, b; Ramya et al. 2022). Grasspea is one of the most affordable and protein-rich crops after soybean and can fix nitrogen and withstand both drought and flooding benefiting subsequent crops (Schulz et al. 1999; Kumar et al. 2020). The protein content in the seed ranges from 8.6% to 34.6% (Barpete et al. 2012; Tamburino et al. 2012; Gupta et al. 2021; Ramya et al. 2022). Apart from protein, grasspea has high pharmaceutical values as it is the only dietary source of L-homoarginine and contains high folic acid content among all other legume crops (Tamburino et al. 2012; Arslan et al. 2017), and it is used in the treatment of Alzheimer's disease, diabetes, thrombopoiesis, and hypoxia and as a wound healing hemostatic agent (Kuo et al. 2003; Lan et al. 2013; Ding et al. 2018; Sharma et al. 2018; Lambein et al. 2019). The morphological adaptations and minimal input with better ecological values such as high-water use efficiency, high nitrogen fixation, and controlling soil erosion make this crop an amicable choice as climate-smart crop (Brunet et al. 2008; Di Fusco et al. 2011; Ebrahimi et al. 2016; Ramya et al. 2022). The multifaceted uses of the grasspea crop are given in Fig. 1.

Despite its value as both human food and animal feed, most Indian farmers hesitate to grow grasspea on a large scale for commercial sale due to the presence of the neurotoxin (β-N-oxalyl-l-α, β diaminopropionic acid, β-ODAP) which causes paralysis when grasspea was taken as a sole diet for continuous 3 months (Mathur et al. 1999). Moreover, some Indian states have prohibited commercial grasspea production, citing food safety concerns. As a result, India's total land area devoted to grasspea farming has declined to a few lakh hectares over the past decade. Moreover, knowledge of a potential neurological disorder (lathyrism) in humans with excessive intake has further discouraged adaptive research on this overlooked crop. Overcoming negative perceptions and reassuring consumers of grasspea's safety will be vital to realizing its potential contribution to nutrition and food security in India. However, the crop has seen limited advancement due to scarce research on its genetic and genomic resources in global genebanks. Grasspea's robustness under challenging conditions makes it a useful "insurance crop" for resource-poor farmers in developing nations facing a changing climate.

Its ability to produce reliable yields with minimal inputs on marginal lands can help ensure economic, social, and nutritional security for vulnerable rural populations (Vaz Patto and Rubiales 2014). Given these traits, grasspea is well-suited to provide a measure of food security even as environmental stresses intensify. Expanding grasspea cultivation thus represents an adaptation strategy to maintain agricultural outputs and farmer livelihoods in marginal areas despite climate shifts or weather extremes. Thus, conservation and sustainable utilization of grasspea's genetic resources are extremely important for continued improvement efforts. This chapter deals with its origin, genetic resources, floral biology, breeding activities, and emerging challenges with its future perspectives discussed to bring back this golden legume into food platter.

2 Origin, Evolution, and Distribution of Species and Forms: Wild Relatives

Grasspea belongs to the Fabaceae (Leguminosae) family within the Faboideae (Papilionoideae) subfamily and the Fabeae (Vicieae) tribe, in the genus *Lathyrus*. The name "*Lathyrus*" comes from the ancient Greek word "lathuros," meaning "exciting," a reference to the alleged aphrodisiac qualities of grasspea (Loudon 1855). This hardy legume goes by numerous common names across different countries, including chickling vetch, chickling pea, and dog toothed pea (Britain and America), *khesari* (Bangladesh), *san lee do* (China), *fovetta* (Cyprus), *sabberi* (Ethiopia), and *gilban* (Sudan) among others (Campbell 1997). Currently, there are 184 taxonomically accepted *Lathyrus* species, encompassing over 160 annual and perennial types and subspecies that fall into 15 divisions based on morphological traits (Allkin 1986; Smartt et al. 1994; Plitmann et al. 1995; Chtourou-Ghorbel et al. 2001) but further reclassification may be needed as more genetic analysis is done across the numerous species (Asmussen and Liston 1998; Badr et al. 2002).

Lathyrus species have a predominantly Northern Hemisphere distribution, with a high diversity in temperate zones of Europe, Asia, and North America (Kupicha 1983). Grasspea is cultivated extensively across a wide altitudinal range of 1200–3000 meters in the Indian subcontinent, where it is found throughout the country (Gautam et al. 1999). The different species of this genus are given in Table 1. The origins of key grain crops like wheat, pea, grasspea, and lentil trace back to the Fertile Crescent of West Asia during the Neolithic period around 4000 BCE. Domestication began in the vicinity of the Irano-Turanian regions. Over subsequent millennia, cultivation diffused outward to temperate zones bordering the Mediterranean before eventually reaching tropical and sub-tropical areas across northern latitudes, including East Africa and South Asia. Gradually, these annual crops spread across hemispheres, taking hold as naturalized species in parts of the New World (Kupicha 1983). The precise center of origin of grasspea is unclear, and it is thought to have first been domesticated in the Mediterranean area before spreading to other regions (Smartt 1984). While Central Asia and Abyssinia were previously proposed as centres of origin, archaeological proof points to cultivation initiated in the Balkan Peninsula amid the early Neolithic age.

Fig. 1 Multifaceted use of grasspea (*L. sativus* L.)

Table 1 Classification of *Lathyrus* genus and its distribution

Section	Species	Important species	Geographical distribution
Orobus	54		Europe, West and EastAsia, Northwest Africa, North and Central America
Lathyrostylis	20		Central and Southern Europe, West Asia, Northwest Africa
Orobon	1		Anatolia, Caucasia, Crimea, Iran
Lathyrus	33	*L. annuus*, *L. cicera*, *L. sativus*, *L. sylvestris*, *L. tingitanus*, *L. tuberosus*, *L. gorgoni*, *L. hirsutus*, *L. latifolius*, *L. odoratus*, *L. rotundifolius*, *L. blepharicarpus*	Europe, Canaries, West and Central Asia, North Africa
Pratensis	6	*L. pratensis*	Europe, West and Central Asia, North Africa
Aphaca	2	*L. aphaca*	Europe, West and Central Asia, North Africa
Clymenum	3	*L. clymenum*, *L. ochrus*	Mediterranean
Orobastrum	1		Mediterranean, Crimea, Caucasia
Viciopsis	1		Southern Europe, Eastern Anatolia, North Africa
Linearicarpus	7		Europe, West and Central Asia, North and East Africa
Nissolia	1		Europe, West and Central Asia, Northwest Africa
Neurolobus	1		West Crete
Notolathyrus	23		Temperate South America, Southeast USA

Source: Adapted from Kupicha (1983)

Grasspea likely spread from Southwest/Central Asia to the Eastern Mediterranean via trade and immigration during the bronze era (Mahler-Slasky and Kislev 2010). The earliest grasspea traces come from eighth millennium BCE Neolithic sites in Northern Iraq near Turkey/Iran (Jackson and Yunus 1984; Lambein et al. 2019), with relics also turning up in adjacent Iranian areas and the Gangetic plains around 2000–1500 BCE, hypothetically signifying Middle Eastern introgression (Renfrew 1969; Saraswat 1980). Several archaeobotanical and phytogeographical pieces of evidence revealed that grasspea was first domesticated on the Balkan Peninsula during the early Neolithic era (Kislev 1989).

3 Plant Genetic Resources

Plant genetic resources are the basic raw materials that form the foundation for agricultural advancement and contain a reservoir of variability that enables adaptation to environmental changes. Their conservation is thus critical for global food security in the face of future uncertainties (Sharma et al. 2013). The need to conserve and strategically harness the potential of underutilized grains like grasspea is reinforced by the manifold roles their diverse traits can play in sustainably advancing food security and human development outcomes. The most economically important cultivated *Lathyrus* species fall within section *Lathyrus*: *L. sativus*, *L. cicera*, and the ornamental sweet pea, *L. odoratus*. However, very limited genetic improvement initiatives exist globally for these species compared to other major crops. The success of any breeding program depends on access to adequate genetic resources and diversity. Recently, conserving *Lathyrus* germplasm has received greater attention given the climatic resilience demonstrated by this genus and its potential value under climate change (Kumar et al. 2013a, b, 2020).

International collaborative efforts to assemble and preserve *Lathyrus* diversity started in the 1990s with the creation of a "*Lathyrus* Genetic Resources Network" (Mathur et al. 1998). Grasspea was also included among the food legumes prioritized for conservation and equitable access and benefit sharing of genetic resources under the International Treaty on Plant Genetic Resources for Food and Agriculture (ITPGRFA) framework. Similarly, the Global Crop Diversity Trust (GCDT) in collaboration with ICARDA initiated a program specifically targeting the long-term preservation of *L. sativus*, *L. cicera*, and *L. ochrus*. The Himalayan region of India contains a diversity of *Lathyrus* species, harboring nine taxa including cultivated grasspea (*L. sativus*) and wild types like *L. aphaca* L., *L. pratensis* L., *L. sphaericus* Retz., *L. inconspicuus* L., *L. odoratus* L., *L. altaicus* Ledeb., *L. luteus* Baker., *L. imphalensis*, and *L. sativus*. Across India, nine *Lathyrus* species and subspecies have been documented as follows: *L. aphaca* is widespread, *L. hirsutus* and *L. cicera* occur in Jammu and Kashmir, *L. laevigatus* through *L. erectus* appear in northern mountain areas, *L. odoratus* grows in central and northern plains, while *L. sphaericus* inhabits most of India apart from the south and northeast. The different uses of a few species are given in Fig. 2.

Ex situ collections of grasspea germplasm exist globally at several sites, with major genebanks located in India and other South Asian countries, the United States of America, the United Kingdom, Russia, and Bangladesh. The different components of PGR activities like collection, conservation, characterization, evaluation, documentation, regeneration, maintenance and seed propagation protocols, duplication efforts, and bringing samples to international conservation standards are priorities for effective long-term preservation. As part of a global backup plan, the Svalbard Global Seed Vault holds 4935 *Lathyrus* accessions of 45 species from 19 depositors (accessed https://seedvault.nordgen.org/ on Feb 28, 2024).

Networking initiatives like the "*Lathyrus* Genetic Resources Network" have facilitated grasspea conservation and pre-breeding work internationally. Targeted multistakeholder support for grasspea in India and South Asia could alleviate protein deficiencies sustainably if approached systematically. While genomic resources have advanced in some food legumes, relatively little progress has occurred for grasspea, likely due to its large complex genome and poorly characterized germplasm. The list of major genebanks holding *Lathyrus* germplasm is listed in Table 2.

4 Floral Biology: Emasculation—Pollination Techniques

Grasspea is an annual herb with a well-branched taproot system with modified stems and leaves which has a ratooning capacity (Ramya et al. 2022). Grasspea is primarily a self-pollinating species, although some degree of outcrossing does occur depending on genetic or environmental factors. Reported outcrossing rates range from 2.2% to 27.8%, with significant variation among genotypes (Rahman et al. 1995; Chowdhury and Slinkard 1997). Blue-flowered types showed higher outcrossing around 27.8% versus 19.4% for pink flowers and 9.8% for white (Rahman et al. 1995). Bees are thought to be the main outcrossing vectors in grasspea (Rahman et al. 1995; Chowdhury and Slinkard 1997). Somewhat split, twisted keels provide an opening for the entry of bees. Erect, clawed standard petals, surmount ovate, clawed and obtuse wing petals along with a lighter-colored boat-shaped keel that splits near its base to enable pollinator access. The diadelphous stamens (9 + 1) have filiform vexillary stamens and bright yellow, elliptical anthers, while the enlarged, upturned stigmas and glandular-papillate spatulate

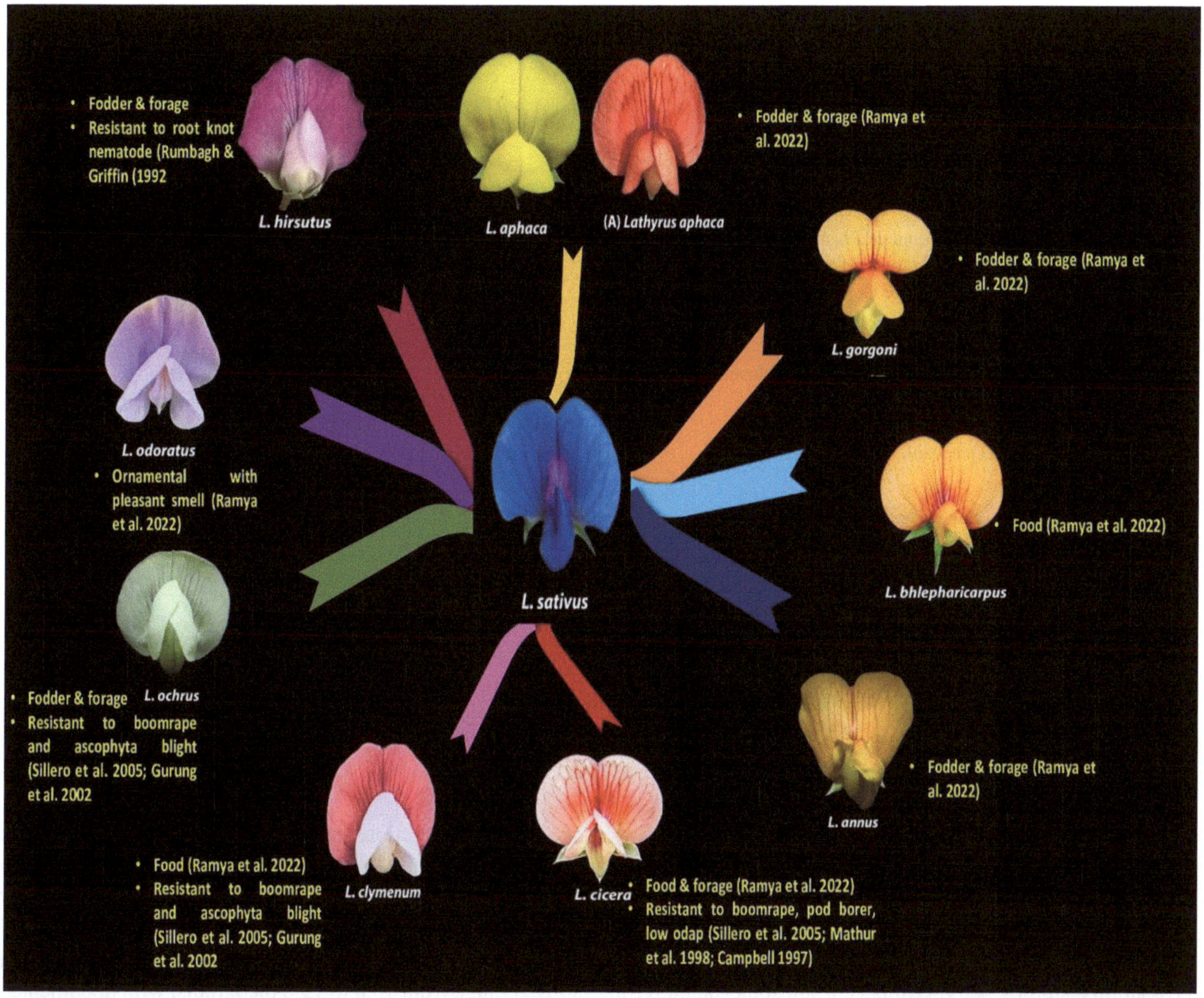

Fig. 2 Different species of *Lathyrus* genus and its uses

terminals sit atop sessile ovaries with 5–8 ovules. The oblong, flattened pods measure 2.5–4.5 cm long and 0.6–1 cm wide, with slightly curled tips and small bulges over the seeds. The flower variations of different species of *Lathyrus* are given in Fig. 3, and in Fig. 4a, b, the variation in young and mature pods across different species are given.

Regarding appearance, attractive blue, pink, red, and white colored flowers occur in grasspea, sometimes in combinations. Blue-flowered plants predominate in South and Southeast Asia and show greater outcrossing tendencies (Polignano et al. 2005; Kumar et al. 2013a, b). White flowers with lower outcrossing rates are more common around the Mediterranean basin (Smartt 1984). Flowers arise from leaf axils, either singularly or in pairs, with colors including blue, dark pink, light yellow, and white, occasionally with purple stripes. Blue dominates, with variation attributable to four pigmentation genes. Controlled crossing in greenhouses or nets prevents unwanted outcrossing by insects during grasspea breeding. Flowers are emasculated by removing anthers in the late bud stage before blooming. The next morning after anther dehiscence on desired parent plants, their pollen is used to manually fertilize previously emasculated flowers. As grasspea sets multiple seeds per pollination, this labor-intensive but reliable method yields sufficient crossed seeds. Early research identified male sterility in some plants with nonfunctional stamens and pollen (Srivastava and Somayajulu 1981). Both cytoplasmic and genetic male sterility occur in grasspea, controlled by single and double restorer genes (Quader 1987). The floral dissection of *L. sativus* is given in Fig. 5.

Table 2 *Lathyrus* germplasm in major genebank holdings

S. no.	Institute	Total accessions	References
1	CBNPMP (Conservatoire Botanique National des Pyrénées et de Midi-Pyrénées, France)	4477	Patto and Rubiales (2014), Ramya et al. (2022)
2	ICARDA (International Center for Agricultural Research in Dry Areas, Morocco)	4473	LBN002 https://www.genesys-pgr.org/a/v26DkEDkXgD
3	ICAR-NBPGR (Indian Council of Agricultural Research–National Bureau of Plant Genetic Resources)	2661	http://genebank.nbpgr.ernet.in/SeedBank/CropgroupCrops.aspx
4	BARI (Plant Genetic Resource Centre (PGRC), Bangladesh)	2661	Mathur et al. (2005), Ramya et al. (2022)
5	INIA (Instituto Nacional de Investigacióղ Agraria, Chile)	1824	Patto and Rubiales (2014), Ramya et al. (2022)
6	N.I. Vavilov All-Russian Scientific Research Institute of Plant Industry, Saint Petersburg, Russia	1208	RUS001 https://www.genesys-pgr.org/a/v26DkEDkXgD
7	Australian Grains Genebank, Australia	1477	AUS165 https://www.genesys-pgr.org/a/v26DkEDkXgD
8	MSB (Royal Botanic Gardens Kew, United Kingdom)	1439	GBR004 https://www.genesys-pgr.org/a/v26DkEDkXgD
9	Ustymivka Experimental Station of Plant Production, Ukraine	1215	UKR008 1302 https://www.genesys-pgr.org/a/v26DkEDkXgD
10	USDA (United States Department of Agriculture) National Plant Germplasm System	871	USA022 877 https://www.genesys-pgr.org/a/v26DkEDkXgD

5 Molecular Cytogenetics and Breeding

Several approaches have examined phylogenetic relationships within *Lathyrus*, including morphological analysis, crossability studies, karyotyping, chromosome banding, in situ hybridization, and molecular markers. Cross-breeding experiments by Yunus delineated interspecies affinities (Yunus 1990). Karyotypes mirrored taxonomic sections proposed by Kupicha. Isozyme patterns unexpectedly clustered *L. pubescens* with *L. sativus*. Storage protein genes grouped *L. sativus*, *L. annuus*, *L. cicera*, and *L. tingitanus* in a monophyletic clade, while *L. latifolius* aligned with *L. clymenum* and *L. ochrus* (de Miera et al. 2008). Cytological research confirms that like most *Lathyrus* species, grasspea possesses a diploid chromosome number of $2n = 14$ (Hao et al. 2017). Some perennial *Lathyrus* exhibit higher ploidy levels, including tetraploid and hexaploid types (*L. palustris* $2n = 42$; *L. venosus* $2n = 28$) (Narayan and Durrant 1983; Talukdar and Biswas 2008). Documented autopolyploids and aneuploids in *L. sativus*, *L. odoratus*, *L. pratensis*, and *L. venosus* retained the base chromosome number, indicating polyploidy is an uncommon anomaly in *Lathyrus*, which largely maintains consistent ploidy (Kalmt and Wittmann 2000; Seijo and Fernandez 2001).

Despite identical chromosome counts, variations in size, centromere position and number, and size and location of secondary constrictions manifest between and within species (Broich 1989; Battistin and Fernández 1994). Differences in chromosome size frequently stem from amplification or removal of chromatin segments over evolutionary trajectories (Rees 1997). Reported 2C nuclear DNA contents in *L. sativus* range from 13.8 to 15.6 picograms, with documented variation in euchromatin versus heterochromatin and repetitive versus nonrepetitive sequences. While chromosome number holds steady, sizable chromosome size deviations have driven genomic evolution among *Lathyrus* species, correlating to a fourfold 2C DNA quantity span (Narayan and Rees 1976).

6 Genetic Studies of Qualitative and Quantitative Traits

Genetic studies of qualitative and quantitative traits in legumes have significantly advanced our understanding of the genetic architecture of key agricultural characteristics in these important crops. Researchers have elucidated the inheritance patterns and identified specific genes controlling qualitative traits such as seed color, flower color, and pod shape, providing insights into the mechanisms governing these phenotypic variations. Furthermore, investigations into quantitative traits like yield have leveraged approaches such as QTL mapping and GWAS to uncover the genetic basis of

Fig. 3 Flower color variation in different species of *Lathyrus* genus

complex traits and pinpoint genomic regions associated with desired agronomic traits. The availability of genomic resources and advanced molecular tools has accelerated these genetic studies, facilitating marker-assisted breeding efforts aimed at developing legume varieties with improved traits, ultimately contributing to enhanced crop productivity, nutritional value, and resilience to environmental stresses in agriculture.

Very few works have been done on studies of these traits of *Lathyrus* so far. In this crop genetic diversity and phenotypic characterization are crucial initial steps for evaluating and categorizing germplasm collections, enabling their effective utilization in crop breeding programs (Bisht et al. 2005; Tripathi et al. 2018). Principal component analysis (PCA) and clustering are widely used multivariate methods for estimating genetic diversity, with applications in assessing diversity across various *Lathyrus* species (Grela et al. 2010a, b).

For qualitative traits, Mendelian principles play a significant role, with genes typically located at a single locus and exhibiting dominant-recessive patterns (Vizmanos et al. 2020). Understanding these traits involves techniques like Punnett squares, monohybrid and dihybrid crosses, and test

Fig. 4 (**a**) Immature pod variation in different species of *Lathyrus* genus (**b**) Mature pod variation in different species of *Lathyrus* genus

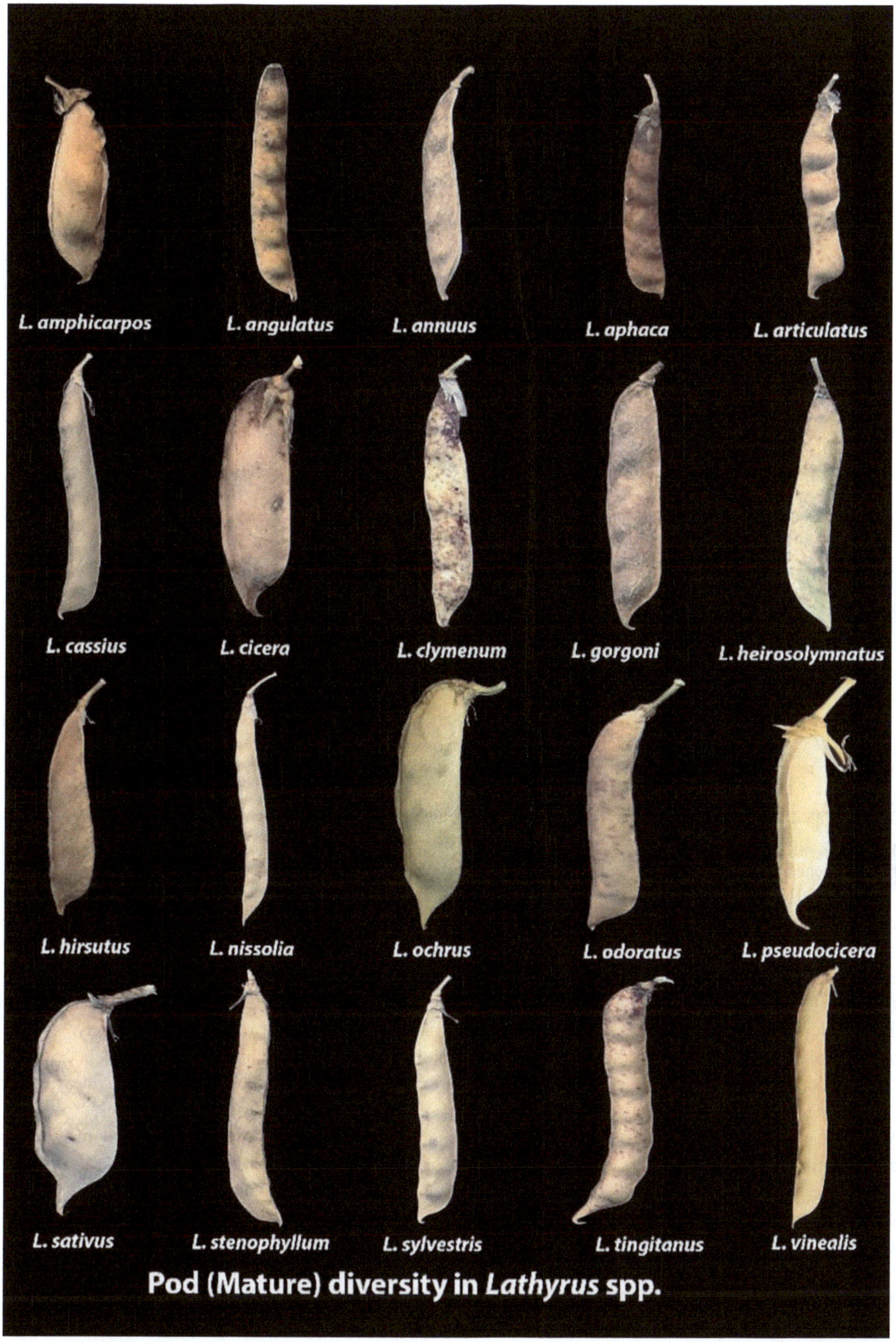

Fig. 4 (continued)

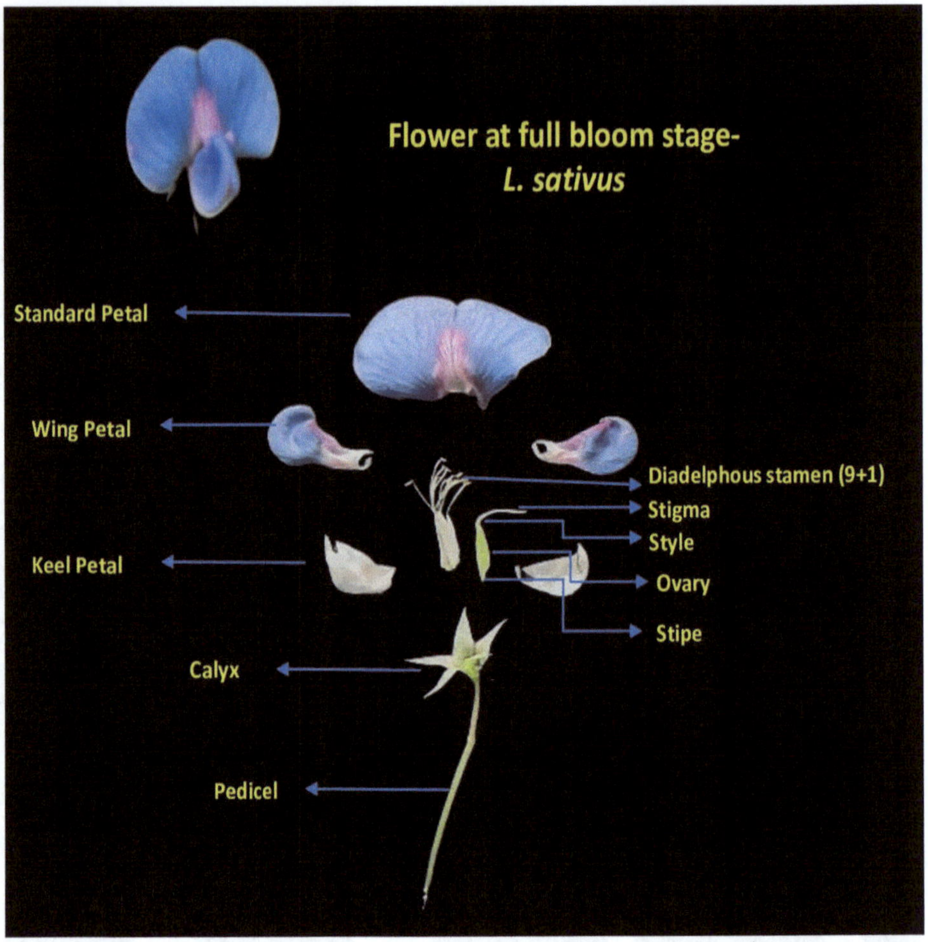

Fig. 5 Floral dissection of *L. sativus* flower

crosses. The study revealed that the F2 progenies followed a 13:3 segregation ratio, suggesting the involvement of two genes with inhibiting gene interactions. Additionally, it was observed that the blue flower exhibited dominance over white (Tiwari and Campbell 1996). The study indicated that seed coat color segregated in a 3:1 phenotypic ratio, suggesting a single gene controlling the trait. It was found that the brown color of the seed coat exhibited complete dominance over other seed coat colors (Talukder and Biswas 2005). For breeders, the precise dominant state of seed color may not hold practical importance because it is a monogenic trait, meaning controlled by a single gene. Such traits can be fixed relatively easily through minimal inbreeding. Thus, the focus would be more on identifying and selecting the desired trait (brown seed color) rather than on understanding the intricacies of dominance.

Breeding for qualitative traits involves selecting individuals with desired characteristics and propagating those traits through controlled crosses. In contrast, quantitative traits are influenced by multiple genes and environmental factors, leading to continuous variation within populations (Farrall 2004). Many economically important traits in grasspea are quantitatively inherited. These include seed size (Mondal and Puteh 2014), protein content, seed weight (Bisignano 2002), primary branch number, seed number per pod, and seed yield (Parihar et al. 2015). Polygenic inheritance is characteristic of these traits, and they often require advanced techniques like QTL mapping and genomic selection for analysis and improvement (Würschum 2012). Breeding for quantitative traits involves selecting individuals with superior phenotypic performance and utilizing genomic information to predict their breeding value more accurately.

In both cases, advancements in molecular genetics have revolutionized the study of genetics in grasspea, allowing for more efficient selection and breeding strategies. By understanding the genetic basis of both qualitative and quantitative traits, breeders can develop improved crop varieties with desirable traits, contributing to enhanced agricultural productivity and sustainability.

7 Breeding Objectives

Breeding objectives for grasspea include optimizing yield, enhancing stress resistance, improving quality traits, exploiting heterosis, and advancing hybrid development. This involves developing varieties with increased productivity, resilience to environmental stresses, superior nutritional and agronomic characteristics, and leveraging hybrid vigor for enhanced performance and adaptability.

7.1 Yield

Lathyrus breeding employs genetic selection to enhance yield by optimizing traits like biomass accumulation and stress tolerance. This approach aims to develop high-yielding varieties with improved agronomic performance through scientific methodologies. To obtain varieties with increased yield, genetic enhancement efforts in the region predominantly relied on traditional breeding techniques. This involved evaluating germplasm across multiple sites and selecting candidates primarily based on their agronomic performance (Dejene Girma and Lijalem Korbu 2012). Mutation breeding in grasspea produced viable diploid mutants, including the non-winged internode mutant (NWIM), derived from BioR-231. NWIM exhibits erect, determinate, semidwarf traits with increased primary and secondary branches, higher grain yield, and reduced seed β-ODAP content (Dixit et al. 2016). Molecular-assisted selection (MAS) offers a promising avenue for enhancing grasspea by cultivating high-yielding varieties tailored to local conditions, fortified with robust disease resistance, and featuring reduced toxin levels (Hillocks and Maruthi 2012). Six significant marker-trait associations were identified using SSR markers and mixed linear model analysis for plant phenology and yield-related traits. Marker regions were linked to genes encoding proteins like homeobox-leucine zipper protein ATHB-6-like, rubredoxin family protein, and cationic peroxidase, suggesting potential targets for grasspea crop improvement toward higher yield (Soren et al. 2020a, b).

7.2 Quality Characters Including Biofortification

Biofortification refers to the process of enriching nutrients in food crops that offers a sustainable and cost-effective means of boosting the concentration of micronutrients in major staple crops (Singh et al. 2015). This strategy is poised to address the prevalence of malnutrition on a global scale. While biofortified foods resulting from crop enhancement may not provide the same quantity of minerals/vitamins per day as typically found in supplements or fortified foods, they contribute to enhancing the daily intake of micronutrients, especially among economically poor individuals (Bouis et al. 2013). Meanwhile, pulses contain high levels of antinutrient compounds that hinder the absorption of micronutrients. These findings underscore the importance of fortifying pulses to ensure balanced diets and combat malnutrition. Biofortification of crops can be achieved through traditional methods or novel breeding techniques (NBTs).

Traditional approaches include mineral fertilization by foliar or soil application, microbe-mediated enhancement of nutrient uptake, and conventional breeding to combine genes for improved nutrient uptake and availability. On the other hand, NBTs utilize gene silencing, gene editing, overexpression, and gene transfer from other species to develop mutant plants with enhanced nutritional profiles (Shahzad et al. 2021). Recently, modern molecular tools like DNA markers and marker-assisted selection (MAS) schemes have become available to accelerate the development of nutrient-rich genotypes. A 0.5% $FeSO_4$ treatment on *Lathyrus* resulted in a 21% reduction in phytic acid (PA) concentration, a key inhibitor of mineral absorption in humans. This decrease led to significant reductions in PA:Zn and PA:Fe molar ratios. Additionally, a maximum yield increase of 13% was also observed (Ghosh et al. 2023). Where antinutrients in grasspea hinder the absorption of essential micronutrients like Ca, Fe, Zn, and folates. These include tannins, phytates, polyphenols, protease inhibitors, saponins, α-amylase inhibitors, lectins, and lathyrogens. Processing methods like soaking, cooking, germination, autoclaving, and milling are effective in reducing these antinutrients to enhance nutrient bioavailability (Srivastava 2016).

7.3 Biotic Stresses

Due to climate change, grasspea faces shifting disease patterns, prompting the need for resistant cultivars. Grasspea exhibits resistance to several diseases like powdery mildew and rust (Vaz Patto and Rubiales 2009). Resistance mechanisms have been elucidated through transcriptome analyses and genome-wide association studies (GWAS), aiding in the identification of potential resistance genes. Additionally, grasspea displays resilience to pests such as thrips and aphids (Vaz Patto et al. 2006). As climate change intensifies, the development of resilient grasspea varieties becomes increasingly crucial, with pre-breeding efforts essential for identifying adaptive traits and genetic targets for precision breeding. Resistance to powdery mildew (*Erysiphe polygoni*) has been documented in *L. sativus*, *L. ochrus*, *L. clymenum*, *L. aphaca*, and hybrid sweet peas (*L. odoratus* × *L. belinensis*) (Poulter et al. 2003). Downy mildew resistance has been reported in *L. aphaca* (Campbell 1997) and *L. sativus* (Asthana and Dixit 1997). Ascochyta blight resistance has been recorded in *L. sativus* (Pang et al. 2000; Weimer 1947a, b; Skiba et al. 2004a, b; Gurung et al. 2002a, b), *L. clymenum*, and *L. ochrus* (Gurung et al. 2002a, b).

High levels of resistance to broomrape have been found in *L. clymenum* and *L. ochrus* (Linke et al. 1993). Molecular markers facilitate the identification and analysis of these resistance loci with quantitative inheritance, enabling

Table 3 Biotic stresses causing damage to *Lathyrus sativus*

Stress	Source of resistance	References
Powdery mildew (*Erysiphe* sp.)	*L. sativus* and *L. cicera*	Vaz Patto et al. (2006)
Downy mildew (*Peronospora* sp.)	*L. sativus* and *L. aphaca*	Asthana and Dixit (1997), Campbell (1997)
Broomrape (*Orobanche* sp.)	*L. ochrus* and *L. clymenum*	Linke et al. (1993)
Gray mold (*Botrytis* sp.)		
Rust (*Uromyces* sp.)	*L. sativus* and *L. cicera*	
Ascochyta blight (*Ascochyta pisi* and *Mychospaerella pinodes*)	*L. sativus*, *L. ochrus*, and *L. clymenum*	Weimer (1947a, b), Pang et al. (2000), Gurung et al. (2002a, b), Skiba et al. (2004a, b) and Gurung et al. (2002a, b)
Thrips (*Caliothrips indicus*)	Indian accessions JRL6 and JLR41	Asthana (1995)

genome mapping and quantitative trait loci (QTL) analysis. This accelerates the grasspea breeding process by allowing the selection of resistant offspring containing desired genetic traits, ultimately linking phenotypic characteristics to specific genomic loci. The various sources of biotic stresses causing damage to *L. sativus* are given in Table 3.

7.4 Abiotic Stress Resistance Including Climate Change

Breeding efforts in grasspea focus on developing varieties resilient to abiotic stresses such as drought, salinity, and extreme temperatures (Gonçalves et al. 2022). Traits associated with stress tolerance, such as deep root systems and efficient water and nutrient uptake, are targeted for genetic improvement. Molecular techniques aid in identifying stress-resilient genes and markers, enabling precise breeding strategies. Ultimately, the goal is to enhance the adaptability of *Lathyrus* cultivars to adverse environmental conditions, ensuring sustainable production in challenging environments. A study with 194 grasspea accessions showed diverse responses to water treatments, emphasizing leaf water content and gas exchange as pivotal traits. Light, large-seeded accessions had higher dry biomass, while dark, small-seeded ones fared better under water stress (Sanches et al. 2024). The study found that the performance index on absorption basis (PIABS) is the most effective parameter for identifying genotypes with enhanced performance and grain production under water deficit (Silvestre et al. 2014). A cytogenetic study investigated cytomictic variations induced by abiotic stresses (temperature, heavy metal Pb, water stress) on meiotic cells of *L. sativus* L. (Kumar and Tripathi 2008).

Table 4 Different parental combinations of crossing in *Lathyrus* between different species

S. no.	Parental combinations	References
1	*L. sativus* * *L. pseudocicera*	Addis and Narayan (2000)
2	*L. articulatus* * *L. clymenum* *L. articulatus* * *L. ochrus* *L. hirsutus* * *L. odoratus* *L. sylvestris* * *L. latifolius*	
3	*L. sativus* * *L. amphicarpus* *L. sativus* * *L. cicera*	Yunus and Jackson (1991)
4	*L. annuus* * *L. hierosolymilanus* *L. odoratus* * *L. belinenesis*	Hammett et al. (1996)
5	*L. articulates* * *L. clymenus* *L. sylvestris* * *L. latifolius*	Davies (1958)
6	*L. cicero* * *L. blepharicarpus* *L. marmoratus* * *L. pseudocicera* *L. gorgoni* * *L. pseudociceraL. marmoratus* * *L. blepharicarpus*	Kearney (1993), Yamamoto et al. (1989)
7	*L. hirsutus* * *L. odoratus*	Davies (1958), Trankovskij (1963), Khawaja (1988) and Yamamoto et al. (1989)

7.5 Exploitation of Heterosis and Hybrid Development

Grasspea breeding aims to exploit heterosis through hybrid development strategies, leveraging genetic diversity to enhance traits such as yield, disease resistance, and stress tolerance. In *Lathyrus* breeding, parent Prateek displayed superior general combining ability for both green forage yield and low β-ODAP content. Cross Nirmal × Prateek showed the highest specific combining ability for green forage yield, while Prateek × JCL-3 was optimal for low β-ODAP content. Significant heterosis was observed in favorable directions for most traits. The cross JCL2 × JCL-3 exhibited the best heterosis for green forage yield, whereas Prateek × JCL3 displayed the highest heterosis for β-ODAP content (Lyngdoh et al. 2022). Successful hybrid development happened in the following interspecific crosses. The different parental combinations of crossing in *Lathyrus* between different species are given in Table 4.

8 Breeding Approaches: Conventional and Non-conventional Including the Use of Genomic Tools

The common conventional approaches include hybridization through controlled crosses between selected parents, induction of mutations using chemicals or radiation to generate genetic variation, and exploitation of existing genetic diversity from cultivated types or wild crop relatives. Advanced techniques now augment conventional breeding methods. Marker-assisted selection using molecular markers linked to valued genes guides crossing decisions. Genomic selection models whole-genome marker profiles to predict phenotypes for selection. Ultimately, both traditional and modern approaches jointly enable pulse breeders to craft superior-performing, locally adapted varieties that meet current and emerging agricultural challenges. A toxin-free gene identified from *L. tingitanus* provides a means to remove neurotoxins and breed safer grasspea varieties (Zhou and Arora 1996). Useful traits absent in cultivated germplasm but present within wild relatives include broomrape resistance from *L. ochrus* and *L. clymenum* (Sillero et al. 2005) as well as early maturity and cold tolerance from *L. cicera*. Introgression of such valuable characteristics from these wild species can potentially improve grasspea's agronomic performance and commercial viability. Comprehensively characterizing grasspea's close and distant wild relatives along with centers of geographical origin provides key insights for accelerating breeding efforts, as outlined previously.

Grasspea has seen limited genetic advancement from conventional breeding in the past, with very few molecular biology efforts undertaken due to a lack of genomic resources and reliable molecular markers representing the entire genome (Yang et al. 2014). Indeed, *Lathyrus* trails behind other major pulse crops regarding the availability of genomic tools and resources. To date, just three linkage maps using molecular markers exist for any *Lathyrus* species (Chowdhury and Slinkard 1999; Skiba et al. 2004a, b; Santos et al. 2018). Approaches like genetic mapping, QTL analysis, marker-assisted selection (MAS), and association mapping (AM) in bi-parental or diverse germplasm sets are accelerating the discovery of trait genetic control. This enables trait-marker identification for downstream applications like MAS or genomic selection (GS) models for grasspea improvement (Kulwal et al. 2011). The number of molecular markers available in *Lathyrus*, especially *L. cicera* and *L. sativus*, has increased substantially in the past decade (Lioi et al. 2011; Shiferaw et al. 2012; Soren et al. 2020a, b).

Modernizing grasspea breeding by embracing advanced genomics tools and techniques is urgently needed to accelerate genetic gains. The emergence of "highly scalable" methods in recent years has also sparked the creation of next-generation sequencing (NGS) machines with higher precision and throughput where resequencing, sequence alignment, and comparison of their results with reference genomes are carried out. 454 Life Sciences commercialized the first of its kind, which relied on pyrosequencing and did not require cloning (Margulies et al. 2005). Sequencing-based allele mining may eventually lead to rapid allelic data at a lower cost and longer read length. The published draft genome estimates of the grasspea range from 6.52 to 8.90 gigabase pairs (Gbp) based on different studies (Nandini et al. 1997; Ghasem et al. 2011; Ochatt et al. 2013; Macas et al. 2015; Rajarammohan et al. 2023), highlighting the large and complex nature of its genome.

The availability of a genome sequence, albeit still in draft form, opens up new possibilities for identifying genes and regulatory pathways controlling traits of interest. This includes the capacity to pinpoint genes and allelic variants modulating β-ODAP toxin biosynthesis or other agronomic and nutritional characteristics. Comparative genomic analyses with sequences from related legume species will further enable the construction of improved genetic and physical maps for grasspea. Such genomic resources can then facilitate the deployment of advanced approaches like marker-assisted selection, genomic selection via high-density marker platforms, or targeted genome editing to efficiently introgress beneficial alleles and accelerate grasspea improvement (Emmrich et al. 2020).

Continued elucidation of genomic architecture and clarifying genotype-phenotype linkages through integrated structural, functional, and molecular analyses will lay the foundation for knowledge-based enhancement of this underutilized crop. The mining of the target genes in the selected accessions by sequencing would significantly speed up precision breeding tactics for agricultural crop plants. Faster sequencing platforms have made the sequencing procedure easier and less expensive. Genotyping-by-sequencing (GBS) approaches can help discover novel genes and allelic variants for priority traits (Kole et al. 2015). Genome-wide association studies (GWAS) also offer potential, enabling the identification of genomic regions governing valuable characteristics by screening diverse germplasm collections using high-density marker panels. By integrating genomic tools, breeders can more accurately select desired traits, predict breeding values, and accelerate the development of improved varieties.

9 Emerging Challenges at the National and International Levels

Lathyrus genetic diversity has undergone substantial erosion due to agricultural intensification, overgrazing, declining permanent pastures, and loss of Mediterranean sclerophyllous trees, maquis, and garrigue habitats. Many weedy species depend on vanishing traditional farming approaches. Drylands across the Central and West Asia and North Africa region face intensifying climate change pressures exacerbating biodiversity declines (Maxted and Bisby 1987). In South Asia, the genetic diversity of *Lathyrus* has suffered greatly as a result of the government's restriction on its sale, causing major erosion of landraces in the region. Germplasm curators are increasingly interested in the in situ and ex situ conservation of plant genetic resources (Ramya et al. 2022). Past sporadic attempts occurred to conserve *Lathyrus* genetic diversity via ex situ and in situ approaches, but rising interest reflects the genus' potential value under climate change. The Global Crop Diversity Trust, collaborating with ICARDA, developed inclusive legume conservation strategies assigning *Lathyrus* as priority status. Nevertheless, previous explorations enabled extensive *Lathyrus* germplasm accruals in various national and international banks. India maintains over 2650 grasspea accessions, and the list of major genebanks holding *Lathyrus* germplasm were already presented in Table 2.

Limited germplasm use arises from insufficient characterization and duplication. Core collections help overcome this by capturing the most variation from the entire set, which can be further distilled into mini-core subsets focused on specific traits. However, grasspea's modest global holdings may not currently warrant mini-cores. Moreover, core sets often inadequately represent adaptive traits. An alternative called FIGS instead identifies subsets based on targeted parameters like heat/drought tolerance using climatic data, enabling a higher probability of capturing desired traits for direct utilization while reducing regeneration needs. Ultimately, conventional breeding has thus far modestly capitalized on conserved allelic diversity. However, modern biotechnological advances offer tremendous potential to unlock these resources, with *Lathyrus* well positioned to confront climate challenges given available genetic wealth. Streamlined evaluation can elucidate variability for deployment alongside bolstered conservation underpinning crop improvement activities.

10 Breeding Progress/Varietal Development

Grasspea, an ancient crop, was domesticated long ago and has primarily been cultivated as a pulse crop since the discovery of its seeds during Balkan excavations (Lambein et al. 2019). Its adaptation and utilization as a forage crop have hindered significant evolutionary advancements when cultivated for grains. However, grasspea has received less attention in genetic research compared to other pulse crops (Rahman 1991; Almeida et al. 2015a, b). Systematic grasspea breeding efforts started during 1900, and numerous countries of South Asia, viz., India (Lal et al. 1986; Pandey et al. 1996), Bangladesh (Sarwar et al. 1996), Pakistan (Haqqani and Arshad 1996), Nepal (Yadav 1996; Upadhyay et al. 2017), China (Yang and Zhang 2005; Yan et al. 2006), as well as East Africa (Tadesse and Bekele 2003; Girma and Korbu 2012; Fikre 2017), Syria (Abd et al. 2000; Agha Alikhani et al. 2014), Australia (Hanbury et al. 1995; Siddique et al. 2006), Chile (Mera et al. 2003), Canada (Campbell et al. 1994), Europe (Grela et al. 2010a, b) have established conventional grasspea breeding programs.

Recently grasspea has gained attention owing to its climate resiliency, and still, efforts toward grasspea improvement program are sparse across the globe, and a limited number of scientific personnel are working on this crop. The breeding and evaluation program in grasspea encompasses minimizing β-ODAP levels, enriching nutritional quality, enhancing yield, adjusting phenology, and bolstering resistance to both biotic and abiotic stressors (Das et al. 2021; Banerjee et al. 2022; Sharma et al. 2022). Since 1991, numerous breeding programs involving introduction, characterization, selection, pre-breeding, hybridization, and mutation have been attempted to reduce neurotoxin, β-ODAP, while maximizing yield. This section will delve into the breeding initiatives and advancements aimed at transforming grasspea from orphan to mainstream leguminous crop.

Despite the significant potential that grasspea holds, relatively limited efforts have been directed toward its improvement, largely due to the stigma of β-ODAP (Patto et al. 2006; Mahapatra et al. 2020). During the early phase, grasspea breeding program was directed toward the development of high-yielding cultivars with low β-ODAP content. Initially, the majority of grasspea varieties were developed by opting for selection from available germplasm and local landraces (Tripathi et al. 2022). The breeding program was started in the 1940s while the first phase was committed toward yield improvement through the selection of superior single plants from local landraces followed by their release in the traditional grasspea growing areas of India (Dixit et al. 2016). As grasspea is widely accepted for both food and fodder, the primary emphasis lies on genotypes with high potential for yield and substantial biological yield (Patto et al. 2006). As grain yield is highly influenced by environmental factors with low heritability, thus, breeders prefer to concentrate on indirect selection relying on yield-attributing traits. Additionally, focus was given to the selection of characters like double podding, number of seeds per pod, growth habit, earliness, etc. (Campbell et al. 1994; Abd et al. 2000; Patto et al. 2011).

The second phase, spanning from 1974–1990s, was dedicated to creating high-yielding varieties with β-ODAP content of less than 0.2% (Das et al. 2021). The first Indian landmark variety with low β-ODAP content (0.2%) was released in 1974, which was a direct selection from the landrace of Bihar. The variety was suitable for upland conditions, garnering widespread acceptance (Lal et al. 1986; Pandey et al. 1996). This variety served as base material in conventional breeding programs for developing several varieties in India and abroad. Following subsequent research, a series of varieties well-suited for upland conditions (LSD1, LSD2) and for rice fallow niche (LSD3, LSD6, Pusa-305, and

Selection 1276) were developed, maintaining low β-ODAP contents (Sethi et al. 1981; Pandey et al. 1996; Gautam et al. 1998). In Chile, the grasspea cv. Quila-blanco was released in 1983, featuring uniform maturity, large white seeds, and a commendable protein concentration, selected from the diverse local population (Campbell et al. 1994). Likewise, the variety 'Luanco-INIA,' with a white seed coat, was developed through selection from local germplasm accessions (Mera et al. 2003).

Through hybridization followed by selection, several breeding lines such as BioR 202, BioL 203, BioL 208, Mahateora, and Prateek have been developed in India (Gautam et al. 1998). In Canada, LS82046 was derived from Pusa 24, exhibiting a low β-ODAP (0.03%) level in its seeds (Campbell and Briggs 1987). The International Centre for Agricultural Research on Dry Areas (ICARDA) launched an expansive grasspea breeding initiative, collaborating with national partners, yielding 30 high-yielding varieties that have been released and cultivated worldwide. In Ethiopia, low β-ODAP cv. Wasie; in Kazakhstan, cv. AliBar; and in Turkey, cv. Gurbuz 2001 have been released (Kumar et al. 2021). Through hybridization followed by selection, two high-yielding varieties with low β-ODAP content (BARI Khesari-1, BARI Khesari-2) have been developed by Bangladesh Agricultural Research Institute (BARI) (Malek et al. 1996). Among the popular cultivars with low β-ODAP concentration are CLIMA pink, 19A' and 20B in Nepal; LS 8246 and AC-Greenfix in Canada; Derek and Krab 9 in Poland; Strandja in Bulgaria; and Gurbuz-1 in Turkey (Hillocks and Maruthi 2012; Sarkar et al. 2019; Parihar et al. 2022).

Later, *Lathyrus* crop wild relatives (CWR) like *L. cicera* and *L. ochrus* with very low β-ODAP content have been utilized to develop many low β-ODAP cultivars across the globe like IFLLO 185 in Jordan, Chalus and Ceora in Australia, and Jaboulah in Lebanon (Campbell 1997; Gixhari et al. 2016). Later, mutation breeding was deployed for the detection of mutant lines with low β-ODAP content. Mutant lines of high yield potential with low β-ODAP, namely Hoogly local and Poltavskaya, were generated by subjecting them to EMS (ethyl methane sulfonate) treatment at concentrations of 0.15% and 0.1%, respectively. Additionally, BioR 231 in India and BINA Khesari-1 in Bangladesh were developed through exposure to gamma radiation (Kumar et al. 2011; Rizvi et al. 2016). Moreover, the induction of somaclonal variation in in vitro culture led to the development of low β-ODAP varieties from Pusa-24 named and released as BioL 212 (Ratan) (ICAR 2019). Simultaneous research efforts identified NGOG 5, a bold seeded mutant with high yield potential and reduced β-ODAP content (Tripathy et al. 2016).

Over the last five decades (1940–1990), attention was given to developing improved grasspea cultivars with low β-ODAP content, while insufficient attention was given to developing cultivars having resistance to key biotic and abiotic stresses. Consequently, in the third phase (from the 1990s onward), the primary objectives shifted to achieving optimal yield, low β-ODAP levels, and introgressing tolerance against key biotic vagaries like powdery mildew, rust, Ascochyta blight thrips, and pea aphids. Many *Lathyrus* CWRs act as sources of resistance against these diseases. Notably, *L. odoratus*, *L. belinensis*, *L. clymenum*, *L. ochrus*, and *L. aphaca* were detected as resistant against powdery mildew (Pandey et al. 1995; Gurung et al. 2002a, b; Poulter et al. 2003). Distant hybridization followed by selection has been attempted in India and abroad, and many resistant cultivars (RLS 1, RLS 2, JRS 115, and JRL 16) have been detected through artificial selection against this disease. Ascochyta blight is another serious menace in grasspea cultivation, causing huge crop loss (Barilli et al. 2016). The accessions of cultivated species *L. sativus* along with *L. ochrus* and *L. clymenum* are detected as valuable sources of resistance against this disease (Weimer 1947a, b; Gurung et al. 2002a, b; Skiba et al. 2004a, b; Almeida et al. 2015a, b).

In Ethiopia, several grasspea accessions, viz., 46034, 46047, 46048, 46095, 201556, 214802, 46075/l/Ad87, 46072/1 lAd87, 46076/l/Ad87, 46075, 201545, 226007, and 46098, exhibited resistances to pea aphids (Barpete et al. 2023). In India, JRL 6 and JLR 41 were identified as resistant against thrips (Pandey et al. 1995; Asthana 1995). Through a mutation breeding program, one mutagenic line IVAT 15-290-40KR was identified in Ethiopia showing resistance to blue butterflies (Damte and Dawd 2003). In recent decades, emphasis has been given to developing grasspea cultivars enriched with micronutrients (iron, zinc, and methionine) and amenable to mechanical harvesting through integrating traditional breeding approaches with modern biotechnological tools. A list of grasspea varieties released worldwide is given in Table 5.

11 Coordinated System of Testing

The Crop Improvement program, applicable to any crop, demands a coordinated, interdisciplinary, and multi-sectoral approach that harnesses recent groundbreaking scientific and technological advancements. Under the aegis of the Indian Council of Agricultural Research, notable progress has been achieved by launching All-India Coordinated Crop Improvement Projects (AICCIPs) in 1957, with a targeted approach for individual crops. The AICCIPs are characterized by a multidisciplinary strategy for varietal development and addressing crop production issues. They involve collective planning and testing, the exchange of germplasm and breeding lines, operational flexibility that transcends administrative boundaries, and connections with the Consortium of International Agricultural Research Centers (CGIAR).

Table 5 Grasspea varieties released across the globe

Varieties	Specific features	Breeding methods	Country of release
Ceora	Yield potential is 1.5 t/ha, β-ODAP content is 0.06%, white flowered	K̄33 × 8604	Australia
Chalus	High yielding with low B-ODAP content (0.09%), early to medium duration	Selection from IFLA 1279	
BARI Khesari-1	Yield potential is 1.7 t/ha, β-ODAP content 0.06%, 115 days maturity	P-24 × local landrace	Bangladesh
BINA Khesari-1	Yield potential is 2.4 t/ha low β-ODAP content (<0.2%), blue flowered	Mutation breeding (gamma radiation)	
BARI Khesari-2	Yield potential 1.7 t/ha, β-ODAP content 0.14%, 115 days maturity	P-24 × local	
BARI Khesari-3	Yield potential 1.8–2.0 t/ha, duration: 120–125 days, β-ODAP content 0.04%, blue flowered variety	Sel 190	
BARI Khesari-4	Yield potential 0.7–1.08 t/ha, low β-ODAP content, 118 days maturity, white flowered, powdery mildew resistant	Sel 1337	
BARI Khesari-5	Yield potential 1.4–1.7 t/ha, β-ODAP content 0.08%, 121–125 days maturity, large seeded, blue colored flower, downy mildew tolerant	Germplasm selection	
BINA Khesari-2	Released during 2021, high yielding with low β-ODAP content	Mutation breeding	
Strandja	High yielding (2.5 t/ha), low β-ODAP content, early maturing, white seeded	Selection from landrace (VIL)	Bulgaria
LS 8246	High yielding (2.0 t/ha), low β-ODAP content (0.03%), 110–130 days maturity	Selection from Pusa 24	Canada
Luanco-INIA	Yield potential is 4 t/ha, low β-ODAP content (0.18%), white and bold seeded, number of seeds per pod is 1–3	Pure line selection from LS 0027	Chile
Quila-blanco	High yielding, bold and white seeded, synchronous maturity, with good protein content	Selection from germplasm	
Wasie (ILAT-LS-LS-B2)	Yield potential is 1.7 t/ha, low β-ODAP content (0.08%), blue flowered, 110 days maturity, resistant to powdery mildew	SC5 × PGRC 46071	Ethiopia
Pusa 24	High yielding, with low β-ODAP (0.2%), suitable for upland situation	Selection from Bihar landrace	India
Nirmal	High yielding (1.1 t/ha), β-ODAP content (0.3%)	Germplasm selection	
BioL-212 (Ratan)	Yield potential is 1.5–1.6 t/ha, low β-ODAP (0.1%), bold seeded, blue flower, powdery mildew resistant	Somaclone of Pusa 24	
Prateek	Yield potential is 1.5–1.6 t/ha, low β-ODAP (0.10%), blue flower, tolerant to powdery mildew	LS 82046 × A 60	
Mahateora	Yield potential is 1.5–1.6 t/ha, low β-ODAP (0.07%), pink flower, creamy white seeded, early maturing	Ratan × JRL 2	
Bidhan Khesari-1	Yield potential is 1.5–1.6 t/ha, short duration (108–116 days), low β-ODAP (0.10%), white flowers with white seed coat	Prateek × Ratan	
Ali bar	Yield potential: 1.2 t/ha, β-ODAP content 0.01%, 156 days duration, high biomass, white seeded	Selection from germplasm accession IFLLS 554	Kazakhstan
CLIMA 2 pink	High yielding (1.5 t/ha), low β-ODAP content (< 0.04%), 130 days maturity, pink flowered	Introduction	Nepal
19A	High yielding (1.08 t/ha), low β-ODAP content (< 0.04%), 130 days maturity, blue flowered	Selection	
20B	High yielding (750 kg/ha), low β-ODAP content (< 0.04%), blue flowered	Selection	
Derek	High yielding (1.9 t/ha), low β-ODAP content, 110–112 days maturity, white seeded	Selection from Der	Poland
Krab	High yielding (2.3 t/ha), low β-ODAP content, 110 days maturity, white seeded	Selection from Kra	
Studenica	High yielding (3.1 t/ha), high biomass, high crude protein content, suitable for dual purpose	Polish cultivar × Serbian land race followed by pedigree selection	Serbia
Gurbuz 1	High yielding (1.2 t/ha) with low β-ODAP content (0.01%), 156 days maturity, white seeded	Selection from IFLLS 554	Turkey

Source: Adapted from Kumar et al. (2011)

Initially, research on pulses was characterized by independent endeavors focused on enhancing locally adapted but genetically diverse populations. Primarily employing breeding methods like pure line and mass selections, the emphasis was placed on traits other than yield. Following this, the initiation of the All-India Co-ordinated Pulses Improvement Project (AICPIP) in 1967 aimed to facilitate access to germplasm accession and breeding materials for pulse breeders and to evaluate improved breeding lines across multiple locations nationwide.

Later, grasspea was included under the purview of the distinct All India Co-ordinated Crop Improvement Project (AICRP) on MULLaRP (Mungbean, Urdbean, Lentil, *Lathyrus*, Rajmash, and Pea), which commenced operations in November 1995 administered by the ICAR–Indian Institute of Pulses Research (IIPR), Kanpur. This program encompasses a wide network of 28 AICRP centers spread across 21 major pulse-growing states of the country (Anonymous 2021). These centers actively engage in strategic research focused on crop improvement, production, and protection. Additionally, AICRP on MULLaRP coordinates the production of nucleus and breeder seeds to fulfill the nation's demand for high-quality seeds. This testing system enables multilocational evaluation of advanced breeding lines of grasspea to elucidate complex G × E interaction and identify stable breeding materials to be released as a variety for the specific agro-climatic zone.

Additionally, AICRP on MULLaRP facilitates the formulation of an improved package of practices of grasspea and other pulses tailored for diverse agro-ecological conditions. Recently, the AICRP on Rabi Pulses was formed by combining the AICRP on Chickpea with Rabi-MULLaRP crops, namely Lentil, Fieldpea, and *Lathyrus*. In 2002, a standardized set of guidelines for testing, identifying, releasing, and notifying crop varieties across the nation was established and published as the Guidelines for Crop Variety Testing under AICCIPs which was revised by Tandon et al. (2015). As per these guidelines, the newly submitted entries undergo sequential testing for three consecutive years across three trials: Initial Varietal Trials (IVTs), Advanced Varietal Trials-I (AVT-I), and Advanced Varietal Trials-II (AVT-II) conducted across multiple locations in different zones.

During the first year of trial under AICRP, the IVTs consist of newly submitted entries from collaborating institutions, along with check varieties to follow an appropriate experimental design without any compromise. During the second year, AVT-I will be formulated with the promoted entries from IVT following the specific criteria of promotion along with check varieties for specific agro-ecological zones. During the third year of testing, AVT-II will be constituted with the entries promoted from AVT-I based on superiority considering yield performance over the checks along with other specified characteristics. All procedures outlined for the IVT and AVT-I must be adhered to during the AVT-II trials as well. The plot size and number of replications must remain consistent across all testing locations. During these 3 years of testing, observations on major diseases and insect pests need to be recorded to elucidate the performance of the entries against biotic stresses.

During AVT-II testing, an Agronomy trial is generally formulated to judge the performance of entries on differential dates of sowing, spacing, fertilizer doses, etc. In the case of grasspea, data on various quality parameters like β-ODAP (<0.1%) and protein content is essential. Finally, selected entries based on their superiority at AVT-II will be identified at the meeting of the Varietal Identification Committee (VIC) for their release and notification to the Central Sub-Committee on Crop Standard Notification and Release of Varieties which is under the aegis of Central Seed Committee (CSC). Across the globe, the coordinated research program on grasspea is conducted by the National Agricultural Research System (NARS) in multiple countries of South Asia, including India, Bangladesh, Pakistan, Nepal, China, the Middle East, Ethiopia, East Africa, Australia, Europe, Canada, and Russia. The International Centre on Agricultural Research on Dry Areas in collaboration with NARS partners is taking a frontier role in sharing germplasm, breeding materials, and release of high-yielding low β-ODAP cultivars in grasspea across the growing zones.

12 Maintenance Breeding

Elevating the quality of seed production is considered a crucial pillar in reshaping Indian agriculture. This can be achieved by introducing recently developed, high-yielding cultivars to farmers to enable cultivation within their adopted regions with minimum resources. To maintain seed quality, the Government of India enacted the "Seed Act" in 1966, and subsequently, a Central Seed Committee was established under the aegis of the Ministry of Agriculture. Upon notification of a newly released variety in the Gazette by the Central Variety Release Committee (CVRC), the variety becomes part of the seed production chain, progressing through breeder seed, foundation seed, and certified seed stages (Chand et al. 2021). Maintenance breeding is the maintenance of variety in its pure form. This program delves into production of nucleus seed which is the source of breeder seed. Therefore, the highest genetic purity should be maintained during nucleus seed production.

As a member of Papiolionaceae, the floral biology of grasspea strictly favors self-pollination. However, depending on the flower color the chances of outcrossing are varied in grasspea and may extend up to 30%. Previous reports stated that the outcrossing was highest in blue (30%) followed by red (27.8%), pink (19.40%), and white (9.8%) (Rahman et al. 1995; Almeida et al. 2015a, b). Thus, proper isolation distance or growing of buffer crops or by caging the unopened buds to prevent entomophily is crucial to maintain the genetic purity of this kind of often cross-pollinated crop (Chowdhury and Slinkard 1999). The land for grasspea seed production should be well-drained, fertile, and free from volunteer plants to prevent natural outcrossing. The land should also be maintained in weed-free conditions. Minimum isolation distance must be maintained to separate the seed production plot from potential sources of contamination.

In grasspea for maintaining genetic purity, the minimum isolation distance of 10 m and 5 m should be maintained for foundation and certified seed production, respectively. Potential sources of contamination can include volunteer plants within the field, different varieties of grasspea in neighboring fields, the same variety that does not meet genetic purity standards, and cross-compatible species such as crops or weed plants. In grasspea, reserve row method is more effective than the plant row method. A minimum of two inspections, one at pre-flowering and another at flowering to pod formation stage are crucial to detect off-type plants and their roguing will be done to maintain genetic purity (Trivedi and Gunasekaran 2013). At the final inspection, the permissible percentage of off-types is capped at 0.10% for foundation seed production plots and 0.20% for certified seed production plots. Seed and pieces of seed with the presence of seed coat are considered pure seed, and physical purity should be 98% for both foundation and certified seed. However, in most of the grasspea-growing countries, informal seed systems are still operating where resource-poor farmers are dependent on landraces and saved seeds from their farms.

13 Future Thrust Area

Breeding grasspea cultivars with no or minimal β-ODAP toxin levels is a major priority. Broader awareness and technical capacity for detoxification and processing could expand usage beyond niche subsistence scenarios. A multi-stakeholder collaboration involving government, researchers, producers, and industry around generating improved lines and demonstrating preparation techniques may help to overcome negative perceptions. Positioning grasspea as a functional food or source of pharmaceutical compounds in addition to their viability as animal feed and industrial crops highlights their multifaceted utility. Continued applied research and policy support rather than neglecting this climate-resilient crop will be prudent in given uncertain future conditions. Promoting viable processing methods alongside mainstreaming adapted low/no-toxin varieties can enable safe, expanded grasspea utilization at household and commercial levels. An integrated, proactive approach focused simultaneously on product refinement and market growth will be key to unlocking benefits from this under-researched legacy crop. The possible ways to promote grasspea cultivation are given in Fig. 6.

Studies on consumption patterns conducted 50 years ago in Chhattisgarh districts like Bilaspur, Durg, and Raipur reported β-ODAP levels of 0.5–2.5 g/100 g in most grasspea

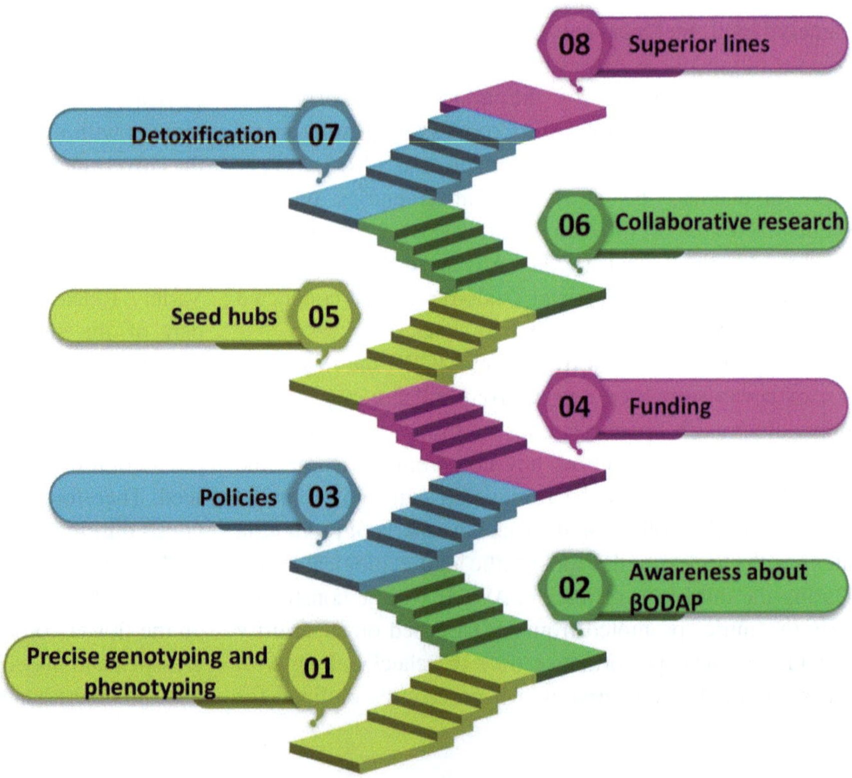

Fig. 6 Possible ways to promote grasspea cultivation

lines and local varieties (Nagarajan and Gopalan 1968). Revisiting the same regions in 2018 showed significant declines in β-ODAP content (0.63–0.65 g/100 g) along with negligible incidence of neurolathyrism (Khandare et al. 2018). A similar minimal prevalence was seen in Maharashtra villages, with only three late-onset paralysis cases weakly correlated to grasspea reported in Uttar Pradesh (Chaurasia et al. 2018; Khandare et al. 2018). The reduced neurolathyrism likely stems from the greater adoption of low-toxin varieties distributed by state agencies. With the first draft of the grasspea genome now available, opportunities exist to deploy advanced genomic tools and multi-omics technologies like transcriptomics, proteomics, and metabolomics to elucidate complex trait control mechanisms, including physiological and molecular regulation of β-ODAP accumulation, which remains unclear. Collaborative research across institutions and disciplines is imperative to bring grasspea into mainstream pulse crop status and solve key bottlenecks like neurotoxin biosynthesis through genomics-enabled breeding approaches.

Realizing the full potential of grasspea germplasm conserved globally requires systematically analyzing available diversity, removing duplicates, and enhancing documentation to facilitate access and utilization in crop improvement programs. Molecular characterization serves to enrich utility and refine the management of genebank holdings. Expanding collection efforts for wild relatives as well as cultivated forms will further widen the trait repertoire to drive continual genetic gain. The capacity to address food security and adaptation challenges posed by climate change will ultimately hinge on coordinated action to assemble and pre-breed with grasspea biodiversity via both conventional breeding and emerging biotechnological interventions.

References

Addis G, Narayan RKJ (2000) Interspecific hybridisation of *Lathyrus sativus* (Guaya) with wild Lathyrus species and embryo rescue. Afr Crop Sci J 8(2):129–136

Agha Alikhani M, Behnamfar F, Rezaie EE, Bannayan M (2014) Assessment of sowing date impacts on productivity of different Grass pea ecotypes. Int J Farming Allied Sci 3:798–801

Allkin R (1986) Names and synonyms of species and subspecies in the 'Vicieae': issue 3. Vicieae Database Project, University of Southampton, Biology Department, Southhampton

Almeida NF, Krezdorn N, Rotter B, Winter P, Rubiales D, Vaz Patto MC (2015a) *Lathyrus sativus* transcriptome resistance response to *Ascochyta lathyri* investigated by deepSuper SAGE analysis. Front Plant Sci 6:178

Almeida NF, Rubiales D, Vaz Patto MC (2015b) Grass pea. Grain legumes. In: De Ron AM (ed) Grain legumes, handbook of plant breeding, vol 10, pp 251–265. https://doi.org/10.1007/978-1-4939-2797-5_8

Anonymous (2021) Project coordinator report, All India Coordinated Research Project on MULLaRP (Mungbean, Urdbean, Lentil, Lathyrus, Rajmash& Pea). ICAR-Indian Institute of Pulses Research, Kanpur

Arora NK (2019) Impact of climate change on agriculture production and its sustainable solutions. Environ Sustain 2:95–96. https://doi.org/10.1007/s42398-019-00078-w

Arslan M, Oten M, Erkaymaz T, Tongur T, Kilic M, Elmasulu S, Cinar A (2017) β-N-oxalyl-L-2, 3-diaminopropionic acid, L-homoarginine, and asparagine contents in the seeds of different genotypes *Lathyrus sativus* L. as determined by UHPLC-MS/MS. Int J Food Prop 20(sup1):S108–S118

Asmussen CB, Liston A (1998) Chloroplast DNA characters, phylogeny, and classification of Lathyrus (Fabaceae). Am J Bot 85:387–401

Asthana AN (1995) Grasspea cultivation in problem areas: present approaches. In: Arora RK, Mathur PN, Riley KW, Adham Y (eds) Lathyrus genetic resources in Asia. Proceedings of a regional workshop. December, Raipur, India, pp 27–29

Asthana AN, Dixit GP (1997) Utilization of genetic resources in *Lathyrus*. In: Mathur PN, Rao VR, Arora RK (eds) *Lathyrus* genetic resources network. Proceedings of a IPGRI-ICARDA ICAR regional working group meeting, 8–10 December, New Delhi, India, pp 64–70

Badr A, Shazly HE, Rabey HE, Watson LE (2002) Systematic relationships in Lathyrus sect. Lathyrus (Fabaceae) based on amplified fragment length polymorphism (AFLP) data. Can J Bot 80(9):962–969

Banerjee J, Das A, Parihar AK, Sharma R, Pramanik K, Barpete S (2022) Genomic designing towards development of abiotic stress tolerant grass pea for food and nutritional security. In: Genomic designing for abiotic stress resistant pulse crops. Springer International Publishing, Cham, pp 345–381

Barilli E, Cobos MJ, Rubiales D (2016) Clarification on host range of *Didymellapinodes* the causal agent of pea Ascochyta blight. Front Plant Sci 7:592

Barpete S, Dhingra M, Parmar D, Sairkar P, Sharma NC (2012) Intraspecific genetic variation in eleven accessions of Grass Pea using seed protein profile. Sci Secure J Biotechnol 1:21–27

Barpete S, Das A, Kahriz PP, Kahriz MP, Khawar KM, Qunale X, Tripathi K, Parihar AK, Kumar S (2023) Disease resistance breeding in *Lathyrus sativus* L. In: Diseases in legume crops: next generation breeding approaches for resistant legume crops, pp 233–256

Battistin A, Fernández A (1994) Karyotypes of four species of South America natives and one cultivated species of *Lathyrus* L. Caryologia 47(3–4):325–330

Bisht IS, Bhat KV, Lakhanpaul S, Latha M, Jayan PK, Biswas BK, Singh AK (2005) Diversity and genetic resources of wild Vigna species in India. Genet Resour Crop Evol 52:53–68

Bisignano V (2002) Variation for protein content and seed weight in grass pea (*Lathyrus* spp.) germplasm. Bulletin de Resso Phytogénétiques

Bouis H, Low J, McEwan M, Tanumihardjo S (2013) Biofortification: evidence and lessons learned linking agriculture and nutrition. The Food and Agriculture Organization of the United Nations (FAO) and the World Health Organization (WHO), p 18

Broich SL (1989) Chromosome number of North American Lathyrus (Fabaceae). Madrona 36:41–48

Brunet J, Repellin A, Varrault G, Terryn N, Fodil YZ (2008) Lead accumulation in the roots of grass pea (*Lathyrus sativus* L.): a novel plant for phytoremediation systems. Comptes Rendus Biol 331:859–864. https://doi.org/10.1016/j.crvi.2008.07.002

Campbell CG (1997) Grass pea (*Lathyrus sativus* L.) promoting the conservation and use of underutilized and neglected crops, vol 18. Institute of Plant Genetics and Crop Plant Research. Gatersleben/ International Plant Genetic Resources Institute, Rome

Campbell CG, Briggs CJ (1987) Registration of low neurotoxin content Lathyrus germplasm LS-8246. Crop Sci 27:821

Campbell CG, Mehra RB, Agrawal SK, Chen YZ, Abdel Moneim AM, Khawaja HI, Yadov CR, Tay JU, Araya WA (1994) Current status and future strategy in breeding grasspea (*Lathyrus sativus*). In: Expanding the production and use of cool season food legumes: a global perspective of persistent constraints and of opportunities and strategies for further increasing the productivity and use of pea, lentil, faba bean, chickpea and grasspea in different farming systems, pp 617–630

Chand S, Chandra K, Champa I, Khatik L (2021) Varietal release, notification and denotification system in India. Intech Open, London, p 73656. https://doi.org/10.5772/intechopen.94212

Chaurasia RN, Pathak A, Singh S, Joshi D, Mishra VN (2018) Study of knowledge, attitude, and practice in participants with regular intake of Lathyrus, but no spastic paraparesis. J Neurosci Rural Pract 9:011–013. https://doi.org/10.4103/jnrp.jnrp_305_17

Chowdhury MA, Slinkard AE (1997) Natural outcrossing in grasspea. J Hered 88:154–156

Chowdhury MA, Slinkard AE (1999) Linkage of random amplified polymorphic DNA, isozyme and morphological markers in grasspea (*Lathyrus sativus*). Agric Sci 133(4):389–395

Chtourou-Ghorbel N, Lauga B, Combes D, Marrakchi M (2001) Comparative genetic diversity studies in the genus *Lathyrus* using RFLP and RAPD markers. Lathyrus Lathyrism Newsl 2:62–68

Damte M, Dawd T (2003) Chickpea, lentil and grass pea insect-pest research in Ethiopia: a review. In: Food and Forage Legumes of Ethiopia: Progress and Prospects, 260. Proceedings of the Workshop on food and forage legume. 22–26 September, 2003, Addis Ababa, Ethiopia. Published by ICARDA

Das A, Parihar AK, Barpete S, Kumar S, Gupta S (2021) Current perspectives on reducing the β-ODAP content and improving potential agronomic traits in grass pea (*Lathyrus sativus* L.). Front Plant Sci 12:703275

Davies AJS (1958) A cytogenetic study in the genus*Lathyrus* (Doctoral dissertation, The University of Manchester (United Kingdom))

de Miera LES, Ramos J, Pe'rez de la Vega M (2008) A comparative study of convicilin storage protein gene sequences in species of the tribe *Vicieae*. Genome 51:511–523

Dejene Girma DG, Lijalem Korbu LK (2012) Genetic improvement of grass pea (*Lathyrus sativus*) in Ethiopia: an unfulfilled promise. Plant Breed. https://doi.org/10.1111/j.1439-0523.2011.01935.x

Di Fusco M, Federico R, Boffi A, Macone A, Favero G, Mazzei F (2011) Characterization and application of a diamine oxidase from *Lathyrus sativus* as component of an electrochemical biosensor for the determination of biogenic amines in wine and beer. Anal Bioanal Chem 401:707–716. https://doi.org/10.1007/s00216-011-5131-z

Ding S, Wang M, Fang S, Xu H, Fan H, Tian Y, Zhai Y, Lu S, Qi X, Wei F, Sun G (2018) D-dencichine regulates thrombopoiesis by promoting megakaryocyte adhesion, migration and proplatelet formation. Front Pharmacol 9:297

Dixit GP, Parihar AK, Bohra A, Singh NP (2016) Achievements and prospects of grass pea (*Lathyrus sativus* L.) improvement for sustainable food production. Crop J 4(5):407–416

Ebrahimi SE, Koocheki A, Milani E, Mohebbi M (2016) Interactions between *Lepidium perfoliatum* seed gum–Grass pea (*Lathyrus sativus*) protein isolate in composite biodegradable film. Food Hydrocoll 54:302–314. https://doi.org/10.1016/j.foodhyd.2015.10.020

Abd El-Moneim AM, Van Dorrestein B, Baum M, Mulugeta W (2000) Improving the nutritional quality and yield potential of grasspea (*Lathyrus sativus* L.). Food Nutr Bull 21(4):493–496

Emmrich PM, Sarkar A, Njaci I, Kaithakottil GG, Ellis N, Moore C, Edwards A, Heavens D, Waite D, Cheema J, Trick M (2020) A draft genome of grass pea (*Lathyrus sativus*), a resilient diploid legume. BioRxiv 2020-04

Farrall M (2004) Quantitative genetic variation: a post-modern view. Hum Mol Genet 13(suppl_1):R1–R7

Fikre A (2017) Research and development with grasspea (*Lathyrus sativus* L) in Ethiopia. LAP LAMBERT Academic Publishing

Gautam PL, Singh IP, Karihaloo JL (1998) Need for a crop network on Lathyrus genetic resources for conservation and use. In: Mathur PN, Ramanatha Rao V, Arora RK (eds) Lathyrus genetic resources network. (Proceedings of the IPGRI-ICARDA-ICAR regional working group meeting, 8–10 December 1997, New Delhi, India). International Plant Genetic Resources Institute-Office for South Asia, New Delhi, pp 15–21

Gautam PL, Singh IP, Karihaloo JL (1999) Need for a crop network on Lathyrus genetic resources for conservation and use. Lathyrus Genet Resour Netw 8:15

Ghasem K, Danesh-Gilevaei M, Aghaalikhani M (2011) Karyotypic and nuclear DNA variations in *Lathyrus sativus* (Fabaceae). Caryologia 64:42–54. https://doi.org/10.1080/00087114.2011.10589763

Ghosh A, Nalia A, Reja MH, Mukherjee S, Nath R (2023) Zinc and iron foliar feeding on different lathyrus varieties influencing the biofortification, bio-availability and productivity. J Plant Nutr:1–14

Girma D, Korbu L (2012) Genetic improvement of grass pea (*Lathyrus sativus*) in Ethiopia: an unfulfilled promise. Plant Breed 131(2):231–236

Gixhari B, Doko A, Hobdari V, Vrapi H (2016) Diversity of grass pea (*L. sativum*) landraces for sustainable field grass pea breeding in Albania. Int J Ecosyst Ecol Sci 6(1):81–88

Gonçalves L, Rubiales D, Bronze MR, Vaz Patto MC (2022) Grass pea (*Lathyrus sativus* L.) – a sustainable and resilient answer to climate challenges. Agronomy 12(6):1324

Grela ER, Rybinski W, Klebaniuk R, Mantras J (2010a) Morphological characteristics of some accessions of grass pea (*Lathyrus sativus* L.) grown in Europe and nutritional traits of their seeds. Genet Resour Crop Evol 57:693–701

Grela ER, Rybiński W, Klebaniuk R, Matras J (2010b) Morphological characteristics of some accessions of grass pea (*Lathyrus sativus* L.) grown in Europe and nutritional traits of their seeds. Genet Resour Crop Evol 57:693–701. https://doi.org/10.1007/s10722-009-9505-4

Gupta DS, Barpete S, Kumar J, Kumar S (2021) Breeding for better grain quality in Lathyrus. In: Gupta DS, Gupta S, Kumar J (eds) Breeding for enhanced nutrition and bio-active compounds in food legumes. Springer, Cham, pp 131–156. https://doi.org/10.1007/978-3-030-59215-8_6

Gurung AM, Pang ECK, Taylor PWJ (2002a) Examination of Pisum and *Lathyrus* species as sources of ascochyta blight resistance for field pea (*Pisum sativum*). Aus Plant Pathol 31:41–45

Gurung AM, Pang ECK, Taylor PWJ (2002b) Examination of Pisum and Lathyrus species as sources of Ascochyta blight resistance for field pea (Pisum sativum). Aust Plant Pathol 31:41–45

Hammett KR, Murray BG, Markham KR, Hallett IC, Osterloh I (1996) New interspecific hybrids in *Lathyrus* (Leguminosae): *Lathyrus annuus× L. hierosolymitanus. Bot J Linn Soc* 122(2):89–101

Hanbury CD, Sarker A, Siddique KHM (1995) Evaluation of Lathyrus germplasm in a Mediterranean type environment in South-Western Australia. In: CLIMA occasional publication (cooperative research Centre for Legumes in Mediterranean agriculture) no 8, Perth, p 23

Hao X, Yang T, Liu R et al (2017) An RNA sequencing transcriptome analysis of grass pea (*Lathyrus sativus* L.) and development of SSR and KASP markers. Front Plant Sci 8:1873. https://doi.org/10.3389/fpls.2017.01873

Haqqani AM, Arshad M (1996) Crop status and genetic diversity of grasspea in Pakistan. In: Lathyrus genetic resources in Asia: proceedings of a regional workshop, 27–29 December 1995. Indira Gandhi Agricultural University, Raipur, India. IPGRI Office for South Asia, New Delhi, p 59

Hillocks RJ, Maruthi MN (2012) Grass pea (*Lathyrus sativus*): is there a case for further crop improvement? Euphytica 186(3):647–654

ICAR (2019) Project Coordinator's report of all India coordinated research project on mungbean, urdbean, lentil, lathyrus, rajmash and pea. Indian Council of Agricultural Research (ICAR), New Delhi

Jackson MT, Yunus AG (1984) Variation in the grass pea (*Lathyrus sativus* L.) and wild species. Euphytica 33:549–559. https://doi.org/10.1007/BF00021156

Kalmt A, Wittmann MT (2000) Karyotype morphology and evolution in some Lathyrus (Fabaceae) species of southern Brazil. J Genet Mol Biol 23:463–467

Kearney JP (1993) Wild *Lathyrus* species as genetic resources for improvement of grasspea (L. Sativus). PhD thesis, University of Southampton, UK

Khandare AL, Kumar RH, Meshram II, Arlappa N, Laxmaiah A, Venkaiah K, Rao PA, Validandi V, Toteja GS (2018) Current scenario of consumption of *Lathyrus sativus* and lathyrism in three districts of Chhattisgarh State, India. Toxicon 150:228–234. https://doi.org/10.1016/j.toxicon.2018.06.069

Khawaja HIT (1988) A new interspecific *Lathyrus* hybrid to introduce the yellow flower character into sweet pea. Euphytica 37:69–75

Kislev ME (1989) Origins of the cultivation of *Lathyrus sativus* and *L. cicera* (Fabaceae). Econ Bot 43:262–270. https://doi.org/10.1007/BF02859868

Kole C, Muthamilarasan M, Henry R, Edwards D, Sharma R, Abberton M, Batley J, Bentley A, Blakeney M, Bryant J, Cai H (2015) Application of genomics-assisted breeding for generation of climate resilient crops: progress and prospects. Front Plant Sci 6:563

Kulwal P, Thudi M, Varshney RK (2011) Genomics interventions in crop breeding for sustainable agriculture. Springer, New York

Kumar G, Tripathi R (2008) Induced cytomictic variations through abiotic stresses in grasspea (*Lathyrus sativus* L.). Indian J Genet Plant Breed 68(01):58–64

Kumar S, Bejiga G, Ahmed S, Nakkoul H, Sarker A (2011) Genetic improvement of grass pea for low neurotoxin (β-ODAP) content. Food Chem Toxicol 49(3):589–600

Kumar S, Gupta P, Barpete S, Choukri H, Maalouf F, Sarkar A (2013a) Grass pea. In: Singh M, Upadhyaya HD, Bisht IS (eds) Genetic and genomic resources of grain legume improvement. Elsevier, Amsterdam, pp 269–292. https://doi.org/10.1016/B978-0-12-397935-3.00011-6

Kumar S, Gupta P, Barpete S, Sarker A, Amri A, Mathur PN, Baum M (2013b) Genetic and genomic resources of grain legume improvement: 11. Grass Pea

Kumar S, Gupta P, Barpete S, Choukri H, Maalouf F, Sarkar A (2020) "Grass pea": the beans and the peas, from orphan to mainstream crops. In: Pratap A, Gupta S (eds) p. The Beans and the Peas, Woodhead Publishing, Sawston, pp 273–287

Kumar S, Gupta P, Barpete S, Choukri H, Maalouf F, Sarkar A (2021) Grass pea. In: The beans and the peas. Woodhead Publishing, pp 273–287

Kuo YH, Ikegami F, Lambein F (2003) Neuroactive and other free amino acids in seed and young plants of Panax ginseng. Phytochemistry 62:1087–1091. https://doi.org/10.1016/S0031-9422(02)00658-1

Kupicha FK (1983) The infrageneric structure of Lathyrus. Notes RBG Edinburgh 41:209–244

Lal MS, Agrawal I, Chitale MW (1986) Genetic improvement of chickling vetch (*Lathyrus sativus* L.) in Madhya Pradesh, India. In: Kaul AK, Combes D (eds) Lathyrus and Lathyrism. Third World Medical Research Foundation, New York, pp 146–160

Lambein F, Travella S, Kuo YH, Montagu MV, Heijde M (2019) Grass pea (*Lathyrus sativus* L.): orphan crop, nutraceutical or just plain food. Planta 250:821–838. https://doi.org/10.1007/s00425-018-03084-0

Lan G, Chen P, Sun Q, Fang S (2013) Methods for treating hemorrhagic conditions. U.S. Patent. 8,362,081. Washington, DC: Patent US. and Trademark Office

Linke KH, Abd El-Moneim AM, Saxena MC (1993) Variation in resistance of some forage legumes species to *Orobanche crenata*Forsk. Field Crop Res 32:277–285

Lioi L, Sparvoli F, Sonnante G, Laghetti G, Lupo F, Zaccardelli M (2011) Characterization of Italian grasspea (*Lathyrus sativus* L.) germplasm using agronomic traits, biochemical and molecular markers. Genet Resour Crop Evol 58:425–437

Loudon JC (1855) Encyclopaedia of plants, new edition edited by Jane Loudon. Longman, Brown, Green & Longmans, London

Lyngdoh AA, Neog SB, Akashi S, Khousik D, Priyanka D (2022) Heterosis and Combining ability studies in a diallel cross of grass pea (*Lathyrus sativus* L.) Genotypes for green forage yield and its component traits, pp 183–188

Macas J, Novák P, Pellicer J, Čížková J, Koblížková A, Neumann P, Fukova I, Doležel J, Kelly LJ, Leitch IJ (2015) In depth characterization of repetitive DNA in 23 plant genomes reveals sources of genome size variation in the legume tribe Fabeae. PLoS One 10:e0143424. https://doi.org/10.1371/journal.pone.0143424

Mahapatra NS, Das A, Bhattacharyya P, Bhattacharya S, Pal S, Barpete S (2020) Studies on genetic variability, divergence and association of characters in grass pea. J Crop Weed 16(1):155–161

Mahler-Slasky Y, Kislev ME (2010) Lathyrus consumption in late bronze and iron age sites in Israel: an Aegean affinity. J Archaeol Sci 37:2477–2485

Malek MA, Sarwar CD, Sarker A, Hassan MS (1996) Status of grass pea research and future strategy in Bangladesh. In: Lathyrus genetic resources in Asia: proceedings of a regional workshop, 27–29 December 1995, Indira Gandhi Agricultural University, Raipur, India. IPGRI Office for South Asia, New Delhi, p 7

Margulies M, Egholm M, Altman WE, Attiya S, Bader JS, Bemben LA, Berka J, Braverman MS, Chen YJ, Chen Z, Dewell SB (2005) Genome sequencing in microfabricated high-density picolitre reactors. Nature 437(7057):376–380

Mathur PN, Ramanatha Rao V, Arora RK (1998) Lathyrus genetic resources network: proceeding of a IPGRI-ICARDA-ICAR regional working group meeting, 8–10 Dec 1997. NBPGR, IPGRI Office for South Asia, New Delhi

Mathur PN, Rao RV, Arora RK (1999) Lathyrus genetic resources network. Bioversity International, Rome

Mathur PN, Alercia A, Jain C (2005) Lathyrus germplasm collections directory. Bioversity International, Rome

Maxted N, Bisby FA (1987) IBPGR final report-wild forage legume collection in Turkey/phase I-South-West Turkey. IBPGR, Rome

Mera M, Tay J, France A, Montenegro A, Espinoza N, Gaete N, Barrientos L (2003) Luanco-INIA, a large-seeded cultivar of *Lathyrus sativus* released in Chile. Lathyrus Lathyrism Newsl 3:26

Molotoks A, Smith P, Dawson TP (2021) Impacts of land use, population, and climate change on global food security. Food Energy Secur 10(1):e261

Mondal MMA, Puteh AB (2014) Spectrum of variability in seed size and content of protein and β-ODAP in grass pea *(Lathyrus sativus* L.) germplasm. Legume Res Int J 37(5):479–482

Nagarajan V, Gopalan C (1968) Variation in the neurotoxin β-(N)-oxalylamino-alanine content in *Lathyrus sativus* samples from Madhya Pradesh. Indian J Med Res 56:95–99

Nandini AV, Murray BG, Obrien IEW, Hammett KRW (1997) Intra-and interspecific variation in genome size in Lathyrus (Leguminosae). Bot J Linnean Soc 125:359–366. https://doi.org/10.1111/j.1095-8339.1997.tb02265.x

Narayan RKJ, Durrant A (1983) DNA distribution in chromosomes of Lathyrus species. Genetica 61:47–53

Narayan RK, Rees H (1976) Nuclear DNA variation in Lathyrus. Chromosoma 54:141–154

Ochatt SJ, Conreux C, Jacas L (2013) Flow cytometry distinction between species and between landraces within Lathyrus species and assessment of true-to-typeness of in vitro regenerants. Plant Syst Evol 299:75–85. https://doi.org/10.1007/s00606-012-0704-7

Pandey RL, Chitale MW, Sharma RN, Rastogi N (1995) Status of Lathyrus research in India. In: Arora RK, Mathur PN, Riley KW, Adham Y (eds) Lathyrus genetic resources in Asia. Proceedings of a Regional Workshop. December, Raipur, India, pp 27–29

Pandey RL, Chitale MW, Sharma RN, Rastogi N (1996) Status of Lathyrus research in India. In: Lathyrus genetic resources in Asia: proceedings of a regional workshop, 27–29 December 1995, Indira Gandhi Agricultural University, Raipur, India. IPGRI Office for South Asia, New Delhi, p 45

Pang ECK, Croser JS, Imberger KT, McCutchan JS, Taylor PWJ (2000) Tissue culture and protoplast fusion of cool-season pulses: pea (*Pisum sativum* L.) and chickpea (*Cicer arietinum* L.). Curr Plant Sci Biotechnol Agric 34:429–427

Parihar AK, Dixit GP, Singh D (2015) Genetic variability analysis for quantitative traits in a germplasm set of grasspea (*Lathyrus* spp.). Legume Res Int J 38(4):461–464

Parihar AK, Barpete S, Das A, Lamichaney A, Gupta S (2022) Lathyrus breeding. In: Fundamentals of field crop breeding. Singapore, Springer Nature Singapore, pp 1323–1389

Patto CMV, Rubiales D (2014) Lathyrus diversity: available resources with relevance to crop improvement–sativus L. and L. cicera as case studies. Ann Bot 113:895–908. https://doi.org/10.1093/aob/mcu024

Patto MV, Skiba B, Pang EC, Ochatt SJ, Lambein F, Rubiales D (2006) Lathyrus improvement for resistance against biotic and abiotic stresses: from classical breeding to marker assisted selection. Euphytica 147:133–147

Patto MC, Hanbury CD, Van Moorhem M, Lambein F, Ochatt S, Rubiales D (2011) Grass pea. Vega de la MP., Torres, AM, Cubero JI. and Kole, C.(eds.) Genetics, genomics and breeding of cool season grain legumes. Science, Lebanon, 151-204

Plitmann U, Gabay R, Cohen O (1995) Innovations in the tribe Vicieae (Fabaceae) from Israel. Israel J Plant Sci 43:249–258. https://doi.org/10.1080/07929978.1995.10676609

Polignano GB, Uggenti P, Alba V, Bisignano V, Della Gatta C (2005) Morpho-agronomic diversity in grasspea (*Lathyrus sativus* L.). Plant genetic. Resources 3(1):29–34

Poulter R, Harvey L, Burritt DJ (2003) Qualitative resistance to powdery mildew in hybrid sweet peas. Euphytica 133:349–358

Pratap A, Gupta S (eds) (2022) The beans and the peas: From orphan to mainstream crops. Woodhead Publishing

Quader M (1987) Male sterility in *Lathyrus sativus* L. Bangladesh J Bot 16(1):9–13

Rahman MM (1991) Progress and prospects of minor pulses in Bangladesh. In: Advances in pulses research in Bangladesh: proceedings of the second national workshop on pulses, 6–8 June 1989, Joydebpur, Bangladesh. Patancheru, pp 43–50

Rahman MM, Kumar J, Rahaman MA, Afsal MA (1995) Natural outcrossing in *Lathyrus sativus* L. Indian J Genet 55:204–207

Rajarammohan S, Kaur L, Verma A, Singh D, Mantri S, Roy JK, Sharma TR, Pareek A, Kandoth PK (2023) Genome sequencing and assembly of *Lathyrus sativus*-a nutrient-rich hardy legume crop. Sci Data 10(1):32

Ramya KR, Tripathi K, Pandey A, Barpete S, Gore PG, Raina AP, Khawar KM, Swain N, Sarker A (2022) Rediscovering the potential of multifaceted orphan legume Grasspea-a sustainable resource with high nutritional values. Front Nutr 8:826208

Rees H (1997) Evolutionary DNA variation in Lathyrus. Chromosomes Today 6:131–139

Renfrew JM (1969) The archaeological evidence for the domestication of plants: methods and problems. In: Ucko PJ, Dimbleby GW (eds) The domestication and exploitation of plants and animals. Transaction Publisher, London, pp 149–172

Rizvi AH, Sarker A, Dogra A (2016) Enhancing grasspea (*Lathyrus sativus* L.) production in problematic soils of South Asia for nutritional security. Indian J Genet Plant Breed 76(04):583–592

Sanches M, Sampaio AM, Araújo S, van Eeuwijk F, Van Breusegem F, Patto MCV (2024) Grass pea (*Lathyrus sativus*) interesting panoply of mechanisms to cope with contrasting water stress conditions- a controlled study of sub populational differences in a worldwide collection of accessions. Agric Water Manag 292:108664

Santos C, Almeida NF, Alves ML, Horres R, Krezdorn N, Leitão ST, Aznar-Fernández T, Rotter B, Winter P, Rubiales D, Vaz Patto MC (2018) First genetic linkage map of Lathyrus cicera based on RNA sequencing-derived markers: key tool for genetic mapping of disease resistance. Hortic Res. https://doi.org/10.1038/s41438-018-0047-9

Saraswat KS (1980) The ancient remains of the crop plants at Atranjikhera (c. 2000-1500 B.C.). J Indian Bot Soc 59:306–319

Sarkar A, Emmrich PM, Sarker A, Zong X, Martin C, Wang TL (2019) Grass pea: remodeling an ancient insurance crop for climate resilience. In: Genomic designing of climate-smart pulse crops, pp 425–469

Sarwar CDM, Malek MA, Sarker A, Hassan MS (1996) Genetic resources of grasspea (*Lathyrus sativus* L.) in Bangladesh. In: Lathyrus Genetic Resources in Asia: proceedings of a regional workshop, 27–29 December, 1995. Indira Gandhi Agricultural University/IPGRI Office for South Asia, Raipur/New Delhi, p 13

Schulz S, Keatinge JDH, Wells GJ (1999) Productivity and residual effects of legumes in rice-based cropping systems in a warm-temperate environment: I. Legume biomass production and N fixation. Field Crop Res 61:23–35. https://doi.org/10.1016/S0378-4290(98)00146-4

Seijo G, Fernandez A (2001) Cytogenetic analysis of Lathyrus japonicus willd. (Leguminosae). Caryologia 54(2):173–179

Sethi KL, Mehra RB, Lal BM (1981) Studies on variability and correlation between toxin content and seed weight in *Lathyrus sativus* L. [legume crop]. Sci Cult 47:263

Shahzad R, Jamil S, Ahmad S, Nisar A, Khan S, Amina Z, Zhou W (2021) Biofortification of cereals and pulses using new breeding techniques: current and future perspectives. Front Nutr 8:721728

Sharma S, Upadhyaya HD, Varshney RK, Gowda CL (2013) Pre-breeding for diversification of primary gene pool and genetic enhancement of grain legumes. Front Plant Sci 4:309

Sharma D, Singh P, Singh SS (2018) B-N-oxalyl-l-α β-Diaminopropionic acid induces wound healing by stabilizing HIF-1α and modulating associated protein expression. Phytomedicine 44:9–19. https://doi.org/10.1016/j.phymed.2018.04.024

Sharma R, Paramanik K, Banerjee J, Das A, Bhan K (2022) Genomic designing for biotic stress resistance in Grasspea. In: Genomic designing for biotic stress resistant pulse crops, pp 415–441

Shiferaw E, Pè ME, Porceddu E, Ponnaiah M (2012) Exploring the genetic diversity of Ethiopian grass pea (*Lathyrus sativus* L.) using EST-SSR markers. Mol Breed 30:789–797

Siddique KH, Hanbury CL, Sarker A (2006) Registration of 'Ceora' grass pea. Crop Sci 46(2):986–987

Sillero JC, Cubero JI, Fernández-Aparicio M, Rubiales D (2005) Search for resistance to crenate broomrape (*Orobanche crenata*) in Lathyrus. Lathyrus Lathyrism Newsl 4:7–9

Silvestre S, de Sousa Araújo S, Vaz Patto MC, Marques da Silva J (2014) Performance index: an expeditious tool to screen for improved drought resistance in the *Lathyrus* genus. J Integr Plant Biol 56(7):610–621

Singh U, Praharaj CS, Singh SS, Bohra A, Shivay YS (2015) Biofortification of pulses: strategies and challenges. Int J Econ Plants 2(3):140–144

Singh R, Srivastava P, Singh P, Upadhyay S, Raghubanshi AS (2019) Human overpopulation and food security: challenges for the agriculture sustainability. In: Urban agriculture and food systems: breakthroughs in research and practice. IGI Global, pp 439–467

Skiba B, Ford R, Pang EC (2004a) Construction of a linkage map based on a *Lathyrus sativus* backcross population and preliminary investigation of QTLs associated with resistance to ascochyta blight. Theor Appl Genet 109:1726–1735

Skiba B, Ford R, Pang ECK (2004b) Genetics of resistance to *Mycosphaerellapinodes* in *Lathyrus sativus*. Aust J Agric Res 55(9):953–960

Smartt J (1984) Evolution of grain legumes. I. Mediterranean pulses. Exp Agric 20:275–296. https://doi.org/10.1017/S0014479700017968

Smartt J, Kaul A, Araya WA, Rahman MM, Kearney J (1994) Grasspea (*Lathyrus sativus* L.) as a potentially safe legume food crop. In: Muehlbauer FJ, Kaiser WJ (eds) Expanding the production and use of cool season food legumes. Curr Plant Sci Biotechnol Agric, vol 19. Springer, Dordrecht, pp 144–155. https://doi.org/10.1007/978-94-011-0798-3_7

Soren KR, Konda AK, Gangwar P, Tiwari VA, Shanmugavadivel PS, Parihar AK, Dixit GP, Singh NP (2020a) Development of SSR markers and association studies of markers with phenology and yield-related traits in grass pea (*Lathyrus sativus*). Crop Pasture Sci 71(8):768–775

Soren KR, Konda AK, Gangwar P, Tiwari VA, Shanmugavadivel PS, Parihar AK, Singh NP (2020b) Development of SSR markers and association studies of markers with phenology and yield-related traits in grass pea (*Lathyrus sativus*). Crop Pasture Sci 71(8):768–775

Srivastava RP (2016) Antinutrients restraining biofortification. In: Biofortification of food crops, pp 333–348

Srivastava YC, Somayajulu PLN (1981) Male sterility in Lathyrus. Indian J Genet 41(1):1964–1966

Tadesse W, Bekele E (2003) Phenotypic diversity of Ethiopian grass pea (*Lathyrus sativus* L.) in relation to geographical regions and altitudinal range. Genet Resour Crop Evol 50:497–505

Talukdar D, Biswas AK (2008) Seven different primary trisomics in grass pea (*Lathyrus sativus* L.). II Pattern of transmission. Cytologia 73:129–136

Talukder D, Biswas AK (2005) Induced seed coat colour mutations and their inheritance in grasspea (*Lathyrus sativus* L.). Indian J Genet Plant Breed 65(02):135–136

Tamburino R, Guida V, Pacifico S, Rocco M, Zarelli A, Parente A, Di Maro A (2012) Nutritional values and radical scavenging capacities of grass pea ('*Lathyrus sativus*' L.) seeds in Valle Agricola district, Italy. Aust J Crop Sci 6(1):149–156

Tandon JP, Sharma SP, Sandhu JS, Yadava DK, Prabhu KV, Yadav OP (2015) Guidelines for testing crop varieties under the all-India coordinated crop improvement projects. Indian Council of Agricultural Research, New Delhi

Tiwari KR, Campbell CG (1996) Inheritance of neurotoxin (β-ODAP) content, flower and seed coat colour in grass pea (*Lathyrus sativus* L.). Euphytica 91(2):195–203

Trankovskij DA (1963) Interspecific hybridization in the genus *Lathyrus*. Agronomie. https://doi.org/10.1051/agro:19990609

Tripathi K, Bhardwaj R, Bhalla S, Kaur V, Bansal R, Yadav R, Gangopadhyay KK, Kumar A, Chaudhury R (2018) Plant genetic resources evaluation: principles and procedures. ICAR-National Bureau of Plant Genetic Resources, New Delhi, p 50

Tripathi K, Pamarthi RK, Gowthami R, Gore PG, Gayacharan C, Barpete S, Singh N, Sarker A, Kumar A (2021) Deciphering morpho-taxonomic variability in Lathyrus species. Indian J Plant Genet Resour 34(02):279–289

Tripathi K, Sadhukhan R, Das A, Jana K, Semwal DP, Kumar A, Sarker A, Ahlawat SP (2022) Khesari (*Lathyrus sativus* L.), an ancient legume for future gain: an expedition collection from parts of West Bengal state of eastern India. Indian J Tradit Knowl 21(2):395–403

Tripathy SK, Panda A, Nayak PK, Dash S, Lenka D, Mishra DR, Kar RK, Senapati N, Dash GB (2016) Somaclonal variation for genetic improvement in grasspea (*Lathyrus sativus* L.). Legum Res 39:329–335. https://doi.org/10.18805/lr.v39i2.9530

Trivedi RK, Gunasekaran M (2013) Indian minimum seed certification standards. The Central Seed Certification Board, Department of Agriculture and Cooperation, Ministry of Agriculture, Government of India, New Delhi, pp 401–402

Upadhyay N, Timsina KP, Gairhe S, Sapkota S, Acharya Y, Khadka S, Plaza S (2017) Pulse in Nepal: a retrospective analysis. In: Proceedings of 30th national winter crops workshop, vol 15, p 484

Vaz Patto MC, Rubiales D (2009) Identification and characterization of partial resistance to rust in a germplasm collection of *Lathyrus sativus* L. Plant Breed 128:495–500

Vaz Patto MC, Skiba B, Pang ECK, Ochatt SJ, Lambein F, Rubiales D (2006) Lathyrus improvement for resistance against biotic and abiotic stresses: from classical breeding to marker assisted selection. Euphytica 147:133–147

Vizmanos JL, Cross NC, Novo FJ (2020) The nature of traits, genes and variation. In: Principles of nutrigenetics and nutrigenomics. Academic, pp 3–9

Weimer JL (1947a) Resistance of *Lathyrus* spp. and *Pisum* spp. to *Ascochyta pinodella* and *Mycosphaerellapinodes*. J Agric Res 75:181–190

Weimer JL (1947b) Resistance of Lathyrus spp. and Pisum spp. to Mycosphaerella*Ascochyta pinodes*. J Agric Res 75:181–190

Würschum T (2012) Mapping QTL for agronomic traits in breeding populations. Theor Appl Genet 125:201–210

Yadav CR (1996) Genetic evaluation and varietal improvement of grasspea in Nepal. In: Lathyrus genetic resources in Asia: proceedings of a regional workshop, 27–29 December, 1995. Indira Gandhi Agricultural University, Raipur, India. IPGRI Office for South Asia, New Delhi, p 21

Yamamoto K, Fujiware T, Blumenreich L (1989) Isozymic variation and interspecific crossability in annual species of the genus Lathyrus L. In: Kaul AK, Combes D (eds) Lathyrus and Lathyrism. Third World Medical Research Foundation, New York, pp 118–121

Yan ZY, Spencer PS, Li ZX, Liang YM, Wang YF, Wang CY, Li FM (2006) *Lathyrus sativus* (grass pea) and its neurotoxin β-ODAP. Phytochemistry 67(2):107–121

Yang HM, Zhang XY (2005) Considerations on the reintroduction of grass pea in China. Lathyrus Lathyrism Newsl 4:22–26

Yang T, Jiang J, Burlyaeva M, Hu J, Coyne CJ, Kumar S, Redden R, Sun X, Wang F, Chang J, Hao X (2014) Large-scale microsatellite development in grasspea (*Lathyrus sativus* L.), an orphan legume of the arid areas. BMC Plant Biol 14(1):1–2

Yunus AG (1990) Biosystematics of Lathyrus section Lathyrus with special reference to the grasspea, *L. sativus* L. Unpublished PhD thesis, University of Birmingham, UK

Yunus AG, Jackson MT (1991) The gene pools of the grasspea (*Lathyrus sativus* L.). Plant Breed 106(4):319–328

Zhou M, Arora RK (1996) Conservation and use of underutilized crops in Asia. In: Arora RK, Mathur PN, Riley KW, Adham Y (eds) Lathyrus genetic resources in Asia. International Plant Genetic Resources Institute, Rome, pp 91–96

Cowpea [*Vigna unguiculata* (L.) Walp.] Breeding

Dhanasekar Punniyamoorthy
and Souframanien Jegadeesan

Abstract

Cowpea [*Vigna unguiculata* (L.) Walp.] is a multifaceted proteinaceous legume crop laden with resilience to changing climatic conditions. Despite its hardiness and high plasticity, cowpeas have largely been neglected as an orphan crop, not receiving adequate research attention commensurate with its potential. The expanding genetic and genomic resources in the recent past have rekindled the researcher's interest, and this momentum has to be sustained and channelized for broadening the genetic base and developing smart cowpea varieties. In doing so, the contribution of cowpeas in alleviating food and nutritional security and the outlook towards this crop could be drastically moulded for human betterment. This chapter deliberates on the various constraints in cowpea breeding and the means and resources for overturning the breeding shortcomings. Various aspects related to germplasm resources, genetics, genomics, mutation breeding, speed breeding, pre-breeding, ideotype breeding, biotechnological and modern tools for cowpea breeding are discussed. The perspectives on research priorities encompassing yield and yield attributing traits, resistances to fungal, bacterial and viral diseases, pests, parasitic weeds, nematodes, and tolerance to drought, high temperature, waterlogging, salinity including nutritional and antinutritional traits have been revisited. The chapter aims at enlightening comprehensively the breeding aspects of cowpea.

Keywords

Biotechnology · Genomics · Ideotype · Modern breeding tools · Pre-breeding · Speed breeding

D. Punniyamoorthy (✉) · S. Jegadeesan
Nuclear Agriculture and Biotechnology Division, Bhabha Atomic Research Centre, Trombay, Mumbai, Maharashtra, India
e-mail: sekar@barc.gov.in

1 Introduction

Cowpea (*Vigna unguiculata*) is a C3 short-day all-purpose food legume crop widely cultivated in the tropical and subtropical regions of the world and primarily concentrated in the tropics between 35N to 30S of the equator, spanning Africa, Asia, Oceania, the Middle East, Southern Europe, Central and South America and the Southern United States (Boukar et al. 2018). It is a prerogative crop especially in hotter regions of arid and semi-arid agro-climates as it is highly plastic and capable of thriving in extremes of water, heat and soil conditions. This herbaceous, warm-season annual legume requires temperatures not less than 18 °C all through its developmental period with an optimal requirement of about 28 °C (Craufurd et al. 2010). Being rich in proteins and predominantly cultivated by resource-poor farmers, it is befittingly referred to as the 'poor man's meat' (Boukar et al. 2018). It also immensely contributes to soil fertility by virtue of its biological nitrogen fixation capability and therefore, plays a pivotal role in environmental sustainability (Omomowo and Babalola 2021). Cowpea has been under domestication for ages (Chivenge et al. 2017) and its popularity as cow feed in yesteryears corroborates with the etymology. It is widely known by various vernacular names such as 'lobia'/'chowli' in India; 'niébé' in francophone Africa; 'seub' or 'niao' in Senegal; 'wake' or 'bean' in Nigeria; 'lubahilu' in Sudan; 'southern pea', 'crowder pea' or 'black eye pea' in the United States; 'feijãocaupe' in Brazil, 'long beans' or 'asparagus bean' in China; and also by a host of other regional names across the globe.

The nutritional richness of cowpea could be appreciated from the fact that the leaves of cowpea have been divulged to be more nutrient-dense than many of the green-leafy vegetables (Gonçalves et al. 2016; Gayacharan et al. 2023). The grains are not only rich sources of proteins, calories, vitamins, minerals and resistant starch but are also low in unhealthy fats (Dhanasekar et al. 2021). Unlike the sulfurous amino acid-rich cereals, cowpeas, akin to other pulses, are lysin- and tryptophan-rich and, with more than two-fold pro-

G. Prasad Dixit et al. (eds.), *Fundamentals of Legume Breeding*, https://doi.org/10.1007/978-981-96-7838-9_10

tein content, play a crucial role in complementing the cereal carbs in imparting a balanced diet.

Current estimates indicate that it is cultivated globally over an area of about 15.2 million ha, with an annual production of over 9.8 million tons and a productivity of 643 kg/ha (FAOSTAT 2022). In the last five decades, the 18.9% annual grain cowpea production growth rate has been accounted for by annual growth in area of 5.3% and a mere 3.8% annual growth in yield (FAOSTAT 1972–2022). In India, though grain cowpea production statistics are not available per se, it is one of the largest producers in Asia, and together with Bangladesh, Indonesia, Myanmar, Nepal, Sri Lanka, Pakistan, the Philippines, Thailand and other far eastern countries, more than 1.5 million ha are under grain cowpea cultivation (Steele and Mehra 2009).

The versatile utilities, nutritional richness, and climate-smartness of cowpeas make them a propitious crop in the present context of climate change, food and nutritional insecurity. This chapter comprehensively elaborates on the breeding of cowpeas.

2 Origin and Taxonomy

The dicotyledon cowpea, botanically known as *Vigna unguiculata*, belongs to the order Fabales of Clade Rosids, family Fabaceae (Syn. Leguminosae), subfamily Faboideae (Syn. Papillionoideae), tribe Phaseoleae, subtribe Phaseolinae, genus *Vigna* Savi, subgenus *Vigna* and section Catiang (Baudoin and Maréchal 1985). The genus *Vigna* is subdivided into seven subgenera, of which the subgenus *Vigna* is constituted by six sections (Maréchal et al. 1978). The two non-intercrossable species *unguiculate* and *nervosa* belong to the section Catiang of subgenus *Vigna* and genus *Vigna*. All the annual cowpeas are housed in the subspecies *unguiculata*, one of the five subspecies under the species *unguiculata*. This subspecies is further composed of two biovars: the cultivated *Vigna unguiculata* subsp. *unguiculata* var. *unguiculata* and the wild inter-crossable form *V. u.* subsp. *u.* var. *spontanea*.

The biovar *unguiculata* is in turn composed of five cultivar groups: *unguiculata* (grain and forage types with more than 16 ovules per pod), *melanophthalmus* (American black-eyed peas with less than 17 ovules per pod), *biflora* (Indian catiang type with smooth seeds in short, erect pods), *sesquipedalis* (vegetable type with very long pods) and *textilis* (rare African type with long peduncles once cultivated for fibre) based mainly on pod, seed and ovule characteristics (Pasquet 1999). However, presently the *Vigna unguiculata* species complex is divided into 11 subspecies (Padulosi 1993; Padulosi and Ng 1997), ten of which are perennial and one, cowpea, is annual. The wild cowpea, *Vigna unguiculata* ssp. *unguiculata* var. *spontanea*, is supposedly the likely progenitor of cultivated cowpea (Singh 2005).

Africa, with the highest genetic diversity, is considered to be the primary centre of origin of cowpea, although the exact place of domestication is widely contested. Ethiopia, Central Africa, South Africa and West Africa are considered probable centres of domestication (Ng and Marechal 1985). Padulosi and Ng 1997) postulated that the South African Transvaal region prevalent with most primitive subspecies may have been the centre of speciation. Of late, parallel double centres of cowpea domestication as supported by molecular evidence: Southeast Africa or West and Central Africa are suggested as the primary centres of the species origin (Panzeri et al. 2022). Probably, a second event of domestication for fresh pod consumption ensuing its introduction in Asia gave rise to the cultigroup *sesquipedalis*, and consequently, India is considered to be the secondary centre of origin for the species (Huynh et al. 2013).

3 Chromosome Number and Genome

Cowpea is a self-pollinated diploid species with a chromosome number of $2n = 2x = 22$. The nuclear genome size of cowpea has been previously estimated to be 613 Mb (Arumuganathan and Earle 1991) and 841 Mb (Parida et al. 1990). The cowpea genome exhibits a high degree of synteny with other tropical legumes like *Phaseolus vulgaris* (common bean) and *Glycine max* (soybean) (Vasconcelos et al. 2015; Muñoz-Amatriaín et al. 2017). In recent times, Lonardi et al. (2019) have re-estimated the genome size of cowpea and have produced genome assembly of cowpea using single-molecule real-time next-generation sequencing. This study involving cytometry reported the 2C nuclear DNA content in cowpea to be 1.310 ± 0.026 pg DNA (mean ± sd) corresponding to a haploid nuclear genome size of 640.6 Mbp. The final assembly consisted of 519 Mb, primarily of repetitive elements enriched within recombination-poor pericentromeric regions.

The genes were abundantly found in the distal regions of chromosomes associated with high recombination events. The chromosome numbering (Vu01 to Vu11) was also revised by Lonardi et al. (2019) based on the synteny with common bean. Chromosome Vu03 was the longest with an estimated size of about 64 Mb while chromosome Vu02 was the shortest with an estimated size of about 34 Mb (based on reference genome assembly GCF_004118075.2 in NCBI). A total of 29,773 genes or protein-coding loci were annotated, along with 12,514 alternatively spliced transcripts. The average gene length was reported to be 3881 bp with an average exon length of 313 bp and an average of 6.29 exons per gene. The GC content in coding exons (40.82%) was higher than in introns (24.27%). The published genome consisted of 49.5% repetitive elements, 39.2% transposable elements, 4% simple sequence repeats, and 5.7% unidentified low-complexity sequences (Lonardi et al. 2019).

4 Botany of Cowpea

Cultivated cowpeas are herbaceous annuals with well-developed tap root systems, while the wild types are over-wintering perennials with fleshy roots. Cowpeas exhibit erect (upright stems), semi-erect, prostrate (trailing on the ground) or climbing (twining onto supports) growth habits with largely glabrous stems and heights reaching up to 2 m. Most of the varietal cultivars present indeterminate growth habits, though determinate and semi-determinate types are also common. The germination being epigeal, the cotyledonary leaves are seen which are simple and opposite. The subsequent true leaves are compound, trifoliate and alternate with varying degrees of globose, hastate, ovate or lanceolate shapes and entire margins. Unbranched axillary racemose inflorescence bears flowers at the terminal end of peduncles in alternating pairs. Peduncles varying in length from 5 cm to 60 cm are slightly twisted and ribbed.

Calyx is of gamosepalous type, tubular, ribbed and lobbed. The papilionaceous corolla has a large, erect vexillum or standard petal which spreads at anthesis and is white, yellow, pale blue or violet coloured. The two wing or alae petals adhere to the two keels which in turn are stuck together to form a boat-shaped structure that encloses the androecium and gynoecium. The stamens are diadelphous with two sets of filaments (9 + 1, nine fused and one free) and the gynoecium has a sessile ovary with many ovules. The style is bearded along the inside and ends in an oblique stigma. Pods are smooth and cylindrical, 8–30 cm long, pendulous or erect, with 8–20 seeds. The immature pods are green or purple that tan or brown upon maturing. The seeds are oval, rhomboid or kidney shaped, depending on the restriction within the pods, and vary greatly in colour (white, yellow, brown, pink, green or black) with solid, eyed or Holstein seed coat patterns.

5 Reasons for Low Productivity

The average global productivity of cowpea is little above 600 kg/ha (FAOSTAT 2022), which is dismally low in comparison to its estimated optimum potential yield of 1500–2500 kg/ha (Kamara et al. 2018). Cowpea is largely cultivated by subsistence farmers (Omomowo and Babalola 2021) in rainfed and marginal lands with little to no inputs and following poor agronomic practices (Horn et al. 2022). The second major issue responsible for low productivity is the non-availability of quality seeds despite the development of elite improved varieties, and even when available, the high cost associated with quality seeds forbids the adoption of these varieties by subsistence farmers (Rabé et al. 2023). Furthermore, as opposed to being cultivated as sole crops, the majority of cowpeas are planted as intercrops resulting in possible yield penalties of up to 58% (Sibhatu et al. 2015). High temperatures are a common phenomenon in tropical arid and semi-arid regions during reproductive stages leading to flower or ovule abortion, pollen and ovule sterility, impaired fertilization and, consequently, reduced yield (Sita et al. 2017).

Besides, the yield of cowpea is seriously affected by a number of biotic stresses like fungal (anthracnose by *Colletotrichum* spp., *Fusarium* wilt, *Rhizoctonia* root-rot, rust by *Uromyces* spp., powdery mildew by *Erysiphe polygoni*, *Macrophomina* blight, *Sclerotium* blight, *Cercospora* leaf spot), bacterial (bacterial blight by *Xanthomonas campestris* pv. *vignicola* and bacterial pustule by *Xanthomonas campestris* pv. *Vigna eunguiculatae*) and viral diseases (cowpea aphid borne mosaic (CABMV), cucumber mosaic virus (CMV), cowpea mild mottle virus (CPMMV), cowpea severe mosaic virus (CPSMV), cowpea golden mosaic virus, cowpea chlorotic mottle virus, black-eye cowpea mosaic virus, cowpea mottle virus, cowpea yellow mosaic virus, cowpea stunt virus); insect pests both in field (*Maruca* pod borer, gram pod borer (*Helicoverpa armigera*), aphids (*Aphis craccivora*), leaf hopper (*Empoasca kerri*), blue butterfly (*Lampides boeticus*), pod bugs (*Riptortus pedestris*), thrips (*Megalurothrips sjostedti*) and at storage (cowpea weevil, *Callosobruchus maculatus*); nematodes (root knot nematodes: *Meloidogyne incognita* and *M. javanica*); and parasitic weeds (*Striga gesnerioides* and *Alectra vogelii*) (Singh and Allen 1979). The abiotic stresses like drought, salinity, high temperature, heavy metals, water logging, nutrient deficiencies, low pH and mineral toxicity do have adverse impacts on cowpea productivity (Timko et al. 2007).

6 Constraints in Cowpea Breeding

Genetic diversity forms the basis of any crop improvement programme. Genetic improvement of crops leverages the extent of variability existing in the gene pool of a crop for hybridization and selection to maximize the likelihood of obtaining desirable genotypes with an appropriate combination of alleles and traits. The narrow cowpea gene pool and reduced genetic variability remain a bottleneck and greatly hamper the improvement of cowpea. The reproductive behaviour favouring self-pollination in cowpea, single domestication event, its evolution from narrow wild germplasm, and limited exchange of genetic resources between the wild and cultivated types largely impact the genetic variability of the crop (Boukar et al. 2020). The breeding methodology of repeatedly using the best breeding lines or varieties in hybridization programmes and repeated backcrossing involving limited donors have further limited the genetic base of the crop varieties. The narrow genetic base not only jeopardized the improved varieties from genetic vulnerabilities but has also resulted in the plateauing of grain yield and minuscule genetic gains (Boukar et al. 2020). Robust cross-compatibility barriers between valuable wild species such as *V. vexillata* outside of the Catiang subsection and the cultivated cowpea (Fatokun 2002) have forbidden the introgression of many beneficial genes in varietal development. Cowpea improvement has predominantly been through classical breeding methods, and the non-availability of appropriate donors or resistance genes has significantly impacted the outcome.

7 Manual Hybridization in Cowpea

Among the legumes, cowpeas have large flowers, exhibit lower rate of abortion and are relatively easier to cross. The success of crossing is relatively higher under conditions of moderate temperature and high humidity. Anthesis occurs between 0630 and 0900 h and may continue up to 1200 h depending upon the environmental conditions. Anthesis is completed earlier under hot and dry conditions in comparison to cold and humid conditions. Dehiscence of anthers occurs much before anthesis mostly between 2200 and 0045 h. The pollen remains viable for 12–15 h after anthesis, while stigma remains receptive from 12 h pre-anthesis to 6 h post-anthesis. A rapid and effective method of crossing has been established by the International Institute of Tropical Agriculture (IITA) (Myers 1996), with a minimum of 50% success. Mature floral buds (full size with slight paling compared to the deep green immature buds) destined to open the following morning are appropriate for emasculation. Though emasculation and pollination could be carried out any time of the day, it is advisable to do late afternoon as cooler nights are favourable for fertilization compared to hotter days. For emasculation, a partial cut is given in the centre of the bud on the straight edge side such that it runs two-thirds the width of the unopened bud (Fig. 1a).

The upper portion of the buds from the cut region is removed without injuring the style and the stigma with pointed forceps. This exposes the upper portion of style, stigma and stamens. The 9 + 1 pollen sacs of the stamen are removed with the forceps carefully without touching the receptive stigma (Fig. 1b, c). The forceps can be decontaminated with 75–90% ethanol in between emasculations. Pollination can be done immediately or the following morning. For pollination, the mature flowers from the male parent are removed, anther sacs exposed and the mass of pollen on the hairy style could be used as a brush to deposit pollen grains on the receptive stigma (Fig. 1d). The crossed bud is thereafter tagged appropriately. Care has to be taken to prevent self-pollination by larvae and adult insects during and immediately following pollination. Unfertilized flowers drop off within 24 h while unfertilized ovary remains attached up to 48 h post-anthesis.

8 Priority Traits for Cowpea Breeding

8.1 Yield and Yield-Attributing Traits

Cowpeas' contribution to food security could be appreciated only by enhancing the yield and productivity of the cowpea by breaking the yield plateau and realizing the potential yield. Towards this end, breeding for yield and yield-attributing traits gains top priority to fulfil the growing demands of the exploding population. The genetic diversity of any crop determines the extent to which it can be genetically improved through breeding. Therefore, germplasm conservation, preserving the genetic diversity of any crop, assumes significance. Across the globe, a number of gene banks are being maintained by various institutes of different countries involved in the conservation of cowpea germplasm (Table 1). Most of the collections overlap to varying extents, and as per the Genesys database (https://www.genesys-pgr.org/), the top five countries contributing to the germplasm include: Nigeria (11,804 accessions), India (3214), Niger (1521), the USA (1473) and Botswana (1366). Cowpea exhibits a great amount of variability in plant and seed morphology.

The primary (easily crossable producing fertile offspring), secondary (crossable with some difficulty and produce partially fertile offspring) and tertiary gene pools (crossable with great difficulty and produce sterile offspring) providing gene sources that can be utilized in cowpea breeding are given in Table 2. The range and mean of different yield and yield attributing traits of various accessions held at NBPGR, New Delhi, is enumerated in Table 3 along with those accessions which have been registered for their novel traits. These data shall be useful in setting the targets while breeding for genetic improvement of cowpea. A number of researchers have reported lines which have been found promising for high yield characteristics: 'Fahari', 'IT93K129-4', 'Glenda' and vegetable cowpea 'Dakama Cream' (Gerrano et al. 2015); 'Bubebe', 'CP411', 'CP421', 'CP645', 'Chimponogo' and 'MS1–8–1-4' (Nkhoma et al. 2020), 'IT96D-604', '93K-619-1', 'IT97K-569-9' and 'IT99K-1060' (Goa et al. 2022). Breeding methods that have been routinely implemented in the genetic improvement of cowpea as any other self-pollinated crops include: pedigree selection, mass selec-

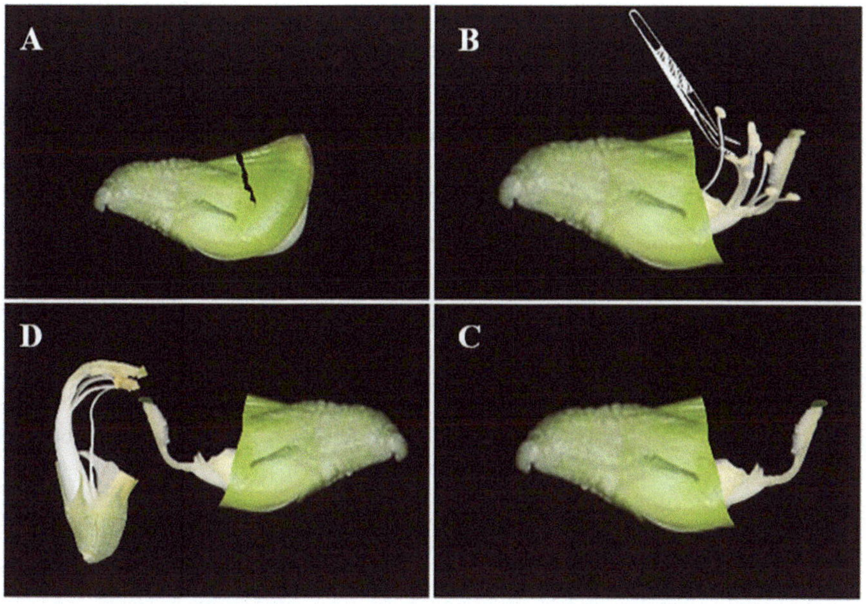

Fig. 1 Panel representing the manual emasculation technique in cowpea (Developed by IITA, Myers 1996). (**A**) Mature cowpea bud is given a two-third cut from the straight edge side as indicated by a black line, (**B**) the upper portion of buds from the cut region is removed exposing stamens and stigma, (C) the stamens are removed with fine forceps, (**D**) pollination of emasculated bud is done using the hairy style of donor male flower

Table 1 Cowpea germplasm conserved at various genebanks worldwide

S. No.	Genebank	Country	No. of accessions	Remarks
1.	International Institute of Tropical Agriculture (IITA)	Nigeria	18,684	From about 90 countries and also has about 590 wild species, includes core collection of 2120 and mini core of 376 accessions (majorly of landraces: 15383, breeding lines: 730, wild types: 2061, of unknown biological status: 434, wild and weedy: 38, research material: 38), subset for drought tolerance: 200; includes 2314 from India
2.	United States Department of Agriculture-Genetic Resources Information Network (USDA-GRIN)	USA	8244	Including 192 wild types and 791 from India
3.	University of California, Riverside	USA	~6000	Substantially overlaps with USDA collection; mini core collection of 368 (242 landraces, 98 breeding lines, 3 weedy lines and 25 uncategorized) overlaps with IITA mini core
4.	National Bureau of Plant Genetic Resources	India	4045	Includes 2671 indigenous, 1093 exotic and *V. unguiculata* ssp. *sesquipedalis* accessions
5.	World Vegetable Center (previously known as Asian Vegetable Research and Development Center (AVRDC)	Taiwan	1637	Includes 591 accessions of Yard Long Bean (*Vigna unguiculata* ssp. *sesquipedalis*)
6.	N.I. Vavilov Institute of Plant Genetic Resources	Russia	1356	
7.	Australian Grains Genebank	Australia	889	
8.	National Centre for Genetic Resources and Biotechnology (NACGRAB)	Nigeria	908	
9.	International Livestock Research Institute (ILRI)	Ethiopia	691	
10.	SADC Regional Gene Bank (SRGB)	Zambia	640	
11.	Plant Genetic Resources Research Institute (PGRRI)	Ghana	853	
12.	Istituto di GeneticaVegetale (IGV)	Italy	600	Largely Mediterranean and African landraces
13.	Genetic Resources Research Institute (GeRRI)	Kenya	565	

Table 2 Primary, secondary and tertiary gene pools of cowpea

Primary gene pool	Secondary gene pool	Tertiary gene pool	Reference
V. unguiculata var. *unguiculata* (L.) Walp (cv. *unguiculata*, cv. *biflora*, cv. *sesquipedalis*, cv. *melanophthalmus*, cv. *textilis*), *V. unguiculata* var. *spontanea* (Schweinf.), *V. unguiculata* subsp. *alba*, *V. unguiculata* subsp. *dekinditiana* (Harms.), *V. unguiculata* subsp. *pubescence*, *V. unguiculata* subsp. *stenophylla*, *V. unguiculata* subsp. *tenuis*	*V. unguiculata* subsp. *aduensis* *V. unguiculata* subsp. *baoulensis* *V. unguiculata* subsp. *burundiensis* *V. unguiculata* subsp. *letouzeyi* *V. unguiculata* subsp. *pawekiae*	*V. angularis* *V. dalzelliana* *V. glabrescens* *V. grandis* *V. umbellata* *V. vexillata*	Kumar et al. (2011)

Derived from Tripathi et al. (2020a)

tion, pure line selection, single seed descent, backcross method and modifications thereof.

8.2 Disease Resistance

Cowpea is extensively infected by a number of fungal, bacterial and viral diseases culminating in severe yield losses depending on the time and duration of incidence. One of the challenges in developing biotic stress-resistant cowpea is the interaction between different stresses, such as drought and insect pests, which can affect the plant's defence mechanisms. Therefore, understanding the molecular and physiological responses of cowpea to combined stresses is important for improving its adaptation and productivity. Even though a number of studies have gone into exploring the resistance sources against various diseases, no concrete donors have been identified for use in introgression breeding. The genetic resources are poor in cowpea, and they lack a sufficient level of resistance against pathogens. This has resulted in poor resistance against various races of the pathogens in released cultivars resulting in differences in the potential and actual yields. The sources of resistance to various diseases and the genes or QTLs identified for major diseases of cowpea are listed in Table 4.

8.3 Fungal Diseases

About 40 fungal species are known to infect cowpea leading to yield losses ranging from 20–100% (Horn and Shimelis 2020). Among the soilborne fungus, southern blight or stem

Table 3 Mean and range of yield and other yield attributing traits of cowpea accessions at NBPGR, New Delhi, India (http://pgrportal.nbpgr.ernet.in)

S. no.	Trait	Range		Mean	Important accessions
		Min	Max		
1.	Plant height (cm)	6.8	396	103.71	IC402184, EC767710, IC249591, IC325996, IC326749 (~30 cm)
2.	No. of primary branches	1	43	5.41	EC528395, IC243501, IC249585, IC383486 (15-43)
3.	Length of main branch (cm)	13.63	204.32	81.46	IC347378, IC373450, IC402135 (35-37)
4.	Clusters/plant	2.30	35.60	13.17	IC402137, IC402157, IC402167, IC402169, IC402201 (33-35.6)
5.	Peduncle length (cm)	1.60	192	23.89	EC528384, EC528403, EC528407, IC398455 (44-50)
6.	Pod length (cm)	2.80	115.66	16.06	EC528411, IC536572, IC536636, IC536572 (>45)
7.	Pods/plant	1	556	23.85	IC202762, IC390242, IC390257, IC390261(>300)
8.	80% Maturity (days)	10	163	92.43	EC394736, EC472274, IC402148 (45-50)
9.	Seeds/pod	1	25	12.20	EC572715, IC402172, IC536572 (24-25)
10.	Yield/plant (g)	0.60	99	15.05	EC458411, EC472266, IC391564, EC528385, IC397847 (70-99)
11.	100 seed weight (g)	1.57	32.19	14.38	IC202790, EC528428, EC528393 (31-33)
12.	50% flowering (days)	30	105.33	54.24	EC394731, EC472287, EC458451, EC517129 (30-32)
13.	Leaf length (cm)	2.70	18.50	10.95	IC262051, EC458487, IC337287, IC106838, IC415590 (>16.5)

(continued)

Table 3 (continued)

14.	Leaf width (cm)	3.24	15.60	7.53	EC390220, IC106838, IC108442, IC328967, IC202790 (>12)
15.	Green biomass per plant (g)	1	787.90	60.51	IC5969, IC15667, IC259063, EC18734 (>300)

Cowpea germplasm registered with NBPGR, New Delhi for novel traits (http://genebank.nbpgr.ernet.in)

No.	Novel trait	Donor institute	Accession no.
1.	Black testacolour	CAZRI, Jodhpur	IC299972
2.	Bold seed	NBPGR, New Delhi	IC202803
3.	Blackeye cowpea mosaicvirus resistance	NBPGR, New Delhi	IC519745
4.	Root-knot nematode resistance	NBPGR, New Delhi	EC723686, EC725122 (from IITA core set)
5.	Dense pubescence	NBPGR, New Delhi	EC762384 (exotic germplasm)

Source: http://pgrinformatics.nbpgr.ernet.in/

rot (*Sclerotium rolfsii*), damping-off (*Pythium* sp.), seedling or web blight (*Rhizoctonia solani*), charcoal rot or dry root rot (*Macrophomina phaseolina*) and *Fusarium* wilt (*F. oxysporum* f.sp. *tracheiphilum*) are important. Other fungal diseases include rust (*Uromyces phaseoli* var. *vignae*), *Cercospora* leaf spot (*Pseudocercospora cruenta*), powdery mildew (*Erysiphe polygony*), brown blotch (*Colletotrichum capsica*). Even though a number of management approaches are available for containment of the diseases, utilization of host plant resistance is the most effective, economical and environmentally safe approach. *Macrophomina phaseolina* causing root-rot is assuming the status of a major disease in cowpea attributed to climate change. This soil- and seed-borne polypathogenic fungal infection clogs the vascular bundles depriving the plants of water and nutrients ultimately leading to death and severe yield losses. The infection being aggravated by high temperature and drought conditions presumes importance under changing climate. Use of host plant resistance is advocated for effective control of the pathogen. There are a number of reports wherein resistance to *Macrophomina* has been reported.

8.4 Bacterial Diseases

Among the bacterial diseases, bacterial blight and bacterial pustule are of wide occurrence globally in cowpea. Bacterial blight caused by *Xanthomonas axonopodis* pv. *vignicola* is of major concern, while bacterial pustule caused by *Xanthomonas axonopodis* pv. *Vigna eunguiculatae* is less prevalent and less destructive (Boukar et al. 2018). Under severe infection, the bacterial blight eventually results in stem canker leading to mortality of the infected plant. Both the pathogens are seed transmitted, while rain and insects are also responsible for secondary spread. The diseases are characterized by water-soaked lesions (with yellow halos in case of blight) culminating in defoliation. Bacterial blight is controlled by additive gene action and the genotypes 'WC26' and 'NE31' have been identified as good general combiners or donors for resistance against this disease in Uganda (Nantale et al. 2023a). Dan Ila, a cultivar from IITA, has been widely used in breeding programmes as a resistance source. The bacterial blight resistance was associated with four SNP markers (Shi et al. 2016).

8.5 Viral Diseases

Among the three types of pathogens, viruses either individually or in combination are the major pathogens causing widespread yield losses in cowpea. The overlapping type of symptoms and the ability of the viruses to mutate and evolve rapidly have led to the breakdown of resistances in short periods of time. Moreover, the involvement of vectors in their dissemination and infection has aggravated the problems in understanding resistance mechanisms. Viruses cause severe yield losses to the tune of 10–100% in cowpea (Kareem and Taiwo 2007). Twenty of the almost 140 viruses known to infect cowpea are widespread and more of economic concern (Thottappilly and Rossel 1992). Cowpea aphid-borne mosaic virus (CABMV), cowpea chlorotic mottle virus (CPCMV), cowpea severe mosaic virus (CPSMV), cowpea mosaic virus (CPMV), cowpea mild mottle virus (CPMMV), cucumber mosaic virus (CMV), cowpea chlorotic mosaic virus (CCMV), cowpea golden mosaic virus (CGMV), blackeye cowpea mosaic virus (BCMV) and even some strains of mungbean yellow mosaic virus are known to infect cowpea widely in India and elsewhere. Aphids, whiteflies and beetles are the major vectors transmitting cowpea viruses, while most of them are also transmitted variably through seeds.

The resistance to viruses can also depend on the environmental factors interacting with the host. For instance, it was found that when cowpeas were exposed to salt stress either concurrently or a day prior to exposure to CPSMV, the plants became susceptible to the virus as the salt stress favoured the infection and proliferation of the virus (Varela et al. 2019). Likewise, drought also enhanced the susceptibility of cowpea to CPSMV at the early stages of viral infection (Silva et al. 2016). The resistance to BCMV was controlled by duplicate recessive genes while CMV and southern bean mosaic virus resistances were governed by duplicate domi-

Table 4 Recent sources of genetic resistance and genes/QTLs reported in cowpea against various diseases

S. No.	Disease	Screening method	Source of resistance	Country	References	Genes/QTLs	References
1	Cowpea aphid borne mosaic virus	Mechanical inoculation, PTA-ELISA	Tvu 379, Tvu 382, TE94-256-2E and TE97-309G-9 IT85F-2687, Patativa, TE97-309G-22, TE97-309G-9, Tvu 379, Tvu 382, Tvu 410, and Tvu 966	Paraguay Brazil	Guillermo et al. (2023), de Oliveira et al. (2012), Barros et al. (2013)	Single to many recessive genes; polygenic with additive effect	Orawu et al. (2013), Silva et al. (2021a)
2	Cowpea severe mosaic virus	Mechanical inoculation, PTA-ELISA	CNCX-698-128F, Tvu 379, Tvu 382, Tvu-3961, TE97-309G-9 and TE97-309G-3	Paraguay	Guillermo et al. (2023)	–	–
3	Cowpea mosaic virus	Field conditions	IC202786, IC202809, Bellary Local	India	Deshpande et al. (2010)	Six SNP markers (3 for immune and 3 for hypersensitive response)	Bhattarai et al. (2017)
4	YMV	Hotspots under field conditions	C38, C44, C159, C180, C182, C222, C249, C355, C385, C422, C463, C480, C483, C492, C533, C566, C568, C685, C687, V130	India	Mishra et al. (2008)	–	–
5	Mixed virus infection (bean common mosaic virus-blackeye cowpea mosaic strain (BCMV-BlCM), southern bean mosaic virus (SBMV) and cucumber mosaic virus (CMV)	Artificial inoculation	IT-98 K-1092-1	Nigeria	Ogunsola et al. (2021)	–	–
6	Root-rot (*Macrophomina phaseolina*)	Sick pot	Suvita 2, Abagbaala, IT97K573-1-1, IT93K-503-1-1	Ghana	Lamini et al. (2022)	3 QTLs: CoBB-1, CoBB-2, CoBB-3 in LG3, LG5 & LG9, respectively	Agbicodo et al. (2010)
		Greenhouse conditions – artificial inoculation	B05-5a, B27 07a, CB27, SP369 A Profil-39B and SP88 Profil-13A	Burkina Faso	Ouédraogo et al. (2021)	Eight QTLs (Mac-1 to Mac-8), Mac-2	Muchero et al. (2011)
		Field conditions over multiple years	T98K-499-39, Suvita 2, IT93K-503-1 and Mouride	USA	Muchero et al. (2011)	Single dominant, Quantitative	Fery and Dukes (2011), Lima et al. (2017)

(continued)

Table 4 (continued)

S. No.	Disease	Screening method	Source of resistance	Country	References	Genes/QTLs	References
7	Phytophthora root rot (*Phytophthora vignae*)		Caloona	Australia	Ralton et al. (1988)	Single dominant gene	Ralton et al. (1988)
8	Bacterial blight (*Xanthomonas axonopodis* pv. *vignicola*)	Artificial inoculation	Dark Green-28, Dark green 1028, and Felon local Danila, IT00K-1263, IT03K-324-9 and 11 others	Bangladesh Nigeria	Saha et al. (2022) Boukar et al. (2018)	–	–
		Screenhouse under artificial inoculation	Tvu 58, Tvu 64, and Tvu 102 (immune)	Nigeria	Durojaye et al. (2019)	–	–
		Field conditions	NE 32, WC 32A, WC 26 and NE 44	Uganda	Nantale et al. (2023b)	–	–
9	Multiple disease resistance (bean common mosaic virus, rust and anthracnose diseases)	Field and artificial inoculation	VBN-3	India	Ganesh et al. (2021)	–	–
10	Powdery mildew (*Erysiphe polygony*)	Hotspots under field conditions	C7, C200, C265, C347, C402 IT90K-59, Nsukka-BA and IT89KD-374-57	India Nigeria	Mishra et al. (2008) Amazue and Adewale (2016)	–	–
		Field conditions in hotspots	IC202280 and EC528412	India	Reddy et al. (2023)	–	–
11	Web blight (*Rhizoctonica solani*)	Hotspots under field conditions Sick pots	C380, C453, C505, C568, C578, C614, C851 DCS 47-1, TC 161, VCP 12-007 DC7-15, CPD229	India India	Mishra et al. (2008) Priyanka et al. (2018)	–	–
12	Cercospora leaf spot (*Pseudocercospora cruenta*)	Hotspots under field conditions	C380, C405, C406, C443, C453, C476, C569, C614, C620	India	Mishra et al. (2008)	–	–
		Greenhouse and field conditions	CP-383, CP-219	Pakistan	Hussain et al. (2019)	–	–
		Field conditions	IC257420, IC 27502, IC91556, IC198330, IC202797, IC219574, Bellary local (white), Guntur local, IC 202791 VRB-10	India West Indies	Deshpande et al. (2010) Booker and Umaharan (2007)	Single dominant or oligogenic (3 genes)	Booker and Umaharan (2008)

(continued)

Table 4 (continued)

S. No.	Disease	Screening method	Source of resistance	Country	References	Genes/QTLs	References
13	Rust (*Uromyces phaseoli* var. *vignae*)	Hotspots under field conditions	C240, C12, C28, C33, C46, C52, C104, C181, C238, C252, C277, C347, C402	India	Mishra et al. (2008)	–	–
		Field conditions	IC206240, IC214834, IC214835, IC219871, Guntur Local, Bellary Local	–	Deshpande et al. (2010)	–	–
		Field conditions	C152×V16, 59, K4-2, PKB4-3, Kashiunnati, IC206240, PKB6-2, V-16, EC458483, PKB6-4, EC458480, KBC-2, IC402180, 21-2, IC58905, IC249593	India	Bamabalwad et al. (2019)	–	–
14	Brown blotch (*Colletotrichum capsica*)	Greenhouse under artificial inoculation	KN-1	Burkina Faso	Thio et al. (2017)	Dominant (*Rcc2*), semi dominant (*Rcc1*) or recessive gene (*rcc2*), single dominant QTL qBBDR2 tightly linked with 102 SNP markers	Thio et al. (2021)
15	*Fusarium* wilt (*Fusarium oxysporum* f. sp. *tracheiphilum*)	Greenhouse using modified root-dip method of inoculation (LFD and VD indicators)	G136, X480, X486, X487, X503, X436, X450	China	Dong et al. (2022)	3 SNPs for leaf damage and 7 for vascular discolouration (on chromosomes 3,4,5,6 and 9)	Dong et al. (2022)
		Screenhouse	TVu134, TVu410, TVu901-1 and MNCO1-649F-2-1	–	Omoigui et al. (2018), Silva et al. (2021b)	Monogenic dominant	Silva et al. (2021b)
		Field and greenhouse conditions	CB46 and CCB88 (against race 3) CB27 and CB50 (race 3 and race 4)	USA	Ehlers et al. (2000), Ehlers et al. (2009)	11 SNPs for leaf damage and 7 SNPs for vascular discolouration distributed on 9 linkage groups	Wu et al. (2015)
16	Aphids (*Aphis craccivora*)	Artificial infestation	SARC 1-57-2 and SARC 1-91-1	Ghana	Kusi et al. (2010, 2020)	–	–
		Artificial screening cages	TVu-6464, TVu-1583, and TVu-15445	Nigeria	Togola et al. (2020)	–	–
17	Parasitic weed (*Striga gesnerioides*)	Natural infestation	IT93K-693-2, IT99K-573-1-1 and IT98K-205-8	Niger	Salifou et al. (2019)	–	–
		Field and pot screening under artificial inoculation	27 RILS of IT97K 499-35 x Sanzi	Ghana	Leandre et al. (2018)	–	–
		Screenhouse pot screening with artificial inoculation	TVu-1272 and Tvu-16514	Nigeria	Omoigui et al. (2017a)	–	–

nant genes (Ogunsola et al. 2023). The different sources of virus resistance and genetics of resistance have been reviewed by Hampton et al. (1997).

8.6 Pest Resistance

Cowpea is highly infested by insect pests throughout the crop growth stages and therefore, is commonly used as a trap crop. Of the 85-insect species known to attack cowpea, legume pod borer (*Maruca vitrata*), aphids (*Aphis craccivora*), pod bug (*Clavigralla tomentosicollis*) and flower thrips (*Megalurothripss jostedti*) are the most important and widespread across the world (OECD 2015). Apart from resistance genes, various physical, biochemical and physiological defence factors also contribute to insect resistance. Physical factors comprising of morphological and anatomical features like trichomes or hairs (long and dense trichomes are associated with *Maruca* and pod bug resistance), pod wall thickness (strength and hardness determine the susceptibility to pod borers), thick and compact collenchyma cells (floral pests), pod surface texture (aphids), pod sclereids (pod bugs) and plant pigmentation (*Maruca* borers) contribute to non-preference for certain insects. The biochemical factors are mostly secondary metabolites that act as repellents, toxins or deterrents.

The high levels of antibiosis in 'TVNu1158' make them aphid resistant (Souleymane et al. 2013). Polyphenolic compounds, terpenoids and flavonoids contribute to thrips and pod borer resistance (Alabi et al. 2011). *Maruca* also exhibited averseness to genotypes with high ethyl-acetate content in stems (Soundararajan et al. 2013). High concentration of volatile compounds also desists seed weevils attack during storage (Rymal and Chambliss 1981). In the case of *Clavigralla*, the genotype 'IT86D-716' exhibited high concentrations of cyanogenic heterosides, flavonoids, tannins and trypsin inhibitors (Dabire-Binso et al. 2010). Serine and cysteine protease inhibitors also inhibit the growth and development of lepidopteran and coleopteran pests (Machuka 2002). Cowpea vicilins, trypsin inhibitors, chymotrypsin, isolectin and α-amylase inhibitors, E-64 cysteine protease inhibitors, phenolics such as hydroxycinnamic acids and para-aminophenylalanine compounds through their action on chitin, inhibition of digestive proteases, growth and development or through their antimetabolite activities adversely affect development of *Callosobruchus maculatus*, an important storage pest of cowpea (Machuka 2002; Huesing et al. 1990; Birch et al. 1986; Gatehouse et al. 1986; Murdock 1992; Togola et al. 2017).

Among the phenolic compounds, p-coumaric acid, ferulic acid and quercetin are especially effective against aphids (Lattanzio et al. 1990). The physiological parameters such as single pods above canopy (Usua and Singh 1979) and open canopies (MacFoy et al. 1983) disfavour *Maruca* pod borers, while those with profuse flowering act as buffer to compensate for the losses occurring through thrips (Alabi et al. 2003). Cowpea aphids also happen to be the vector for 14 viral diseases (Thottappilly et al. 1990). Anusha et al. (2016) have reviewed the genotypes available in the Indian subcontinent that have tolerance against major insect pests like aphids, *Maruca* pod borer, pod bugs and thrips. Consequent to the breakdown of *Rac*-dominant aphid resistance gene, explorations for new sources of resistance have resulted in identification of new cultivated ('Tvu-6464', 'Tvu-1583', and 'Tvu-15445'; Togola et al. 2020) and wild accessions ('TVNu-1158'; Souleymane et al. 2013) with potent resistance genes. The resistance in the cultivated lines has been attributed to their low sucrose content and high phenolic aglycones like kaempferol and quercetin.

Accessions with high flavonoid glycosides were also found to be aphid resistant (Lattanzio et al. 1997). In addition, IITA has developed cultivars such as 'TVu-801', 'TVu-3000', 'IT81D-1020', 'IT82E-1-108' (Togola et al. 2020), 'TVu2897', 'IT84S-2246' (Omoigui et al. 2017b) with high levels of resistance to aphids. Resistance genes in 'TVNu2876' and in 'SARC1-57-2' and 'TVNu1158' were non-allelic (Mofokeng and Gerrano 2021). Aphid resistance seems to be controlled by a single dominant gene or QTLs housed in linkage groups 1 and 7 (Mofokeng and Gerrano 2021). Tripathi et al. (2020b) identified 'EC528425' and 'EC528387' as sources of bruchid resistance, while seven cowpea genotypes, 'IT85 F-2687', 'MN05-841 B-49', 'MNC99-508-1', 'MNC99-510-8', 'TVu 1593', 'Canapuzinho-1-2' and 'SanziSambili' were found to deter bruchids from ovipositing and feeding (de Castro et al. 2013).

The IITA genotype 'IT84S-2246-4' was reported to be one of the promising lines exhibiting multiple resistance for bruchids, aphids, thrips and 10 other diseases (Singh and Singh 1990). Kpoviessi et al. (2019) in their review have enlisted the genotypes with potential tolerance to bruchids. The inheritance for bruchid resistance hints at the involvement of a few major genes (Tripathy 2016), while studies in related *Vigna* sp. point towards QTLs (Somta et al. 2007; Souframanien et al. 2010). Dhanasekar et al. (2023) identified three cowpea mutants with possible resistance against *Maruca* pod borer and through differential gene expression analysis postulated the involvement of isoprene synthase and phenylalanine ammonia-lyase in imparting *Maruca* resistance. In recent times, a transgenic developed pod borer-resistant cowpea variety ('SAMPEA 20-T') has been released for commercial cultivation for the first time in Nigeria using the Bt*Cry1AB* gene (Barrero et al. 2021). Research is also in the advanced stage for developing transgenic variety against the notorious storage cowpea weevil (bruchids).

8.7 Nematode Resistance

Among the nematodes, root-knot nematode (RKN), particularly *Meloidogyne incognita* and *M. javanica*, cause considerable yield losses in cowpea by impeding water and nutrient uptake and also interfering with the translocation of photo-assimilates. Till recently, the mono-dominant (*Rk*) narrow-based resistance has been conferring protection against RKN nematodes in cowpea globally, especially against the avirulent forms (Ehlers et al. 2009). The discovery of virulent forms led to the identification of new resistance genes *Rk2* (with dominant effect) and *rk3* with recessive and additive effect (Roberts et al. 1996) which contributed to broad-based resistance against RKN. The *Rk* locus was mapped onto a QTL localized on chromosome Vu04 (Huynh et al. 2016). A broad-based resistance against RKN was identified in a cowpea accession 'FN-2-9-04' from Mozambique (Ndeve et al. 2018) conferring higher resistance against both *M. incognita* and *M. javanica* than *Rk* gene alone.

The 'FN-2-9-04' accession possessed two major QTLs with partial dominance associated with resistance to root-galling and egg-mass production which were found localized on chromosomes Vu01 and Vu04. It was observed that the resistance locus on Vu04 corresponded to the *Rk* locus and was allelic. The novel locus on Vu01 conferred resistance to virulent *M. javanica*, while both the resistance loci contributed resistance against avirulent *M. incognita*, with Vu04 playing a major role (Ndeve et al. 2019). IITA has also identified cowpea varieties like 'CE-31', 'Frade Preto', 'CE-28', 'CE-01', 'CE-315' and 'CE-237' with nematode resistance (Oliveira et al. 2012). Two cowpea accessions 'US-1136' and 'Zipper Pea' were very effective against both *M. incognita* and *M. enterolobii* (Dareus et al. 2021). R-genes involved in the resistance against these nematodes in cowpea were found to be activated by the interaction of NBS-LRR and WRKY genes. Besides, protease inhibitors and actin cytoskeleton were also essential for RKN resistance (Ribeiro et al. 2022). In India, the cultivars 'Arka seeds', 'TVX-944' and 'IT-38956-1' were found highly resistant to *M. incognita* (Mahantheshwara and Nayak 2018).

8.8 Parasitic Weed Resistance

Parasitic weed infection on cowpea is more common in the African continent and South East Asia including India especially where cowpeas are grown under marginal environments in poor soil (sandy) and low rainfall conditions. The two most important root parasites are *Striga gesnerioides* (Willd.) Vatke and *Alectravogelii* (Benth.), which greatly affect cowpea production (Horn et al. 2015). The combined infection of both parasitic weeds could result in complete yield loss (Emechebe et al. 1988). The seeds of these parasitic weeds are dispersed through wind, water, animals and along with grain harvest. The seeds are so persistent that they can even survive for 15–20 years. Resistance to both the weeds is controlled by non-allelic and independent genes (Atokple et al. 1993) in a race-specific manner. Although many resistance and breeding lines have been identified, 'B301' ('TVu-13950') and the breeding line 'IT81D-994' have been well characterized for *Alectra* resistance. Both genotypes were also found to have resistance against *Striga* (Atokple et al. 1995). The *Alectra* resistance in 'B301' is reported to be under the control of duplicate dominant genes *Rav1* and *Rav2* (Singh et al. 1993), while that in 'IT81D-994' is controlled by a third dominant non-allelic gene *Rav3* (Atokple et al. 1995).

Contrary to the proposed two dominant genes *Rav1* and *Rav2*, it was established that the QTL associated with the former dominant gene is localized on chromosome Vu04 while that with the latter is recessive and lies on Vu11, both of which can be exploited by breeders for their independent contribution to *Alectra* immunity (Ohlson and Timko 2022). It was reported that the roots of some resistant cowpea lines exuded alectrol or strigolatone chemicals which are prerequisites for germination of these parasitic weeds (Phiri et al. 2018). The IITA line 'IT99K-494-6' with high yield and resistance to *Alectra* was released in Malawi (Kabambe et al. 2013). Of the initial five biotypes in *Striga*, 'B301' and 'IT82D-849' exhibited resistance to four excepting Mali among the mini-core collection as additional sources of resistance to *Striga* race 3 (SG3) (as of now seven races are known; SG1 from Burkina Faso, SG2 from Mali, SG3 from Nigeria and Niger, SG4 from Benin, SG5 from Cameroon, SG6 from Senegal and SG4Z from Zakpota in Benin), non-allelic to the gene in 'B301'. 'B301' carries two resistance genes *Rsg-1* and *Rsg-2* while it lacks the third known resistance gene *Rsg-3*. Some of the wild *Vigna* accessions of the species *ambacensis*, *parkeri*, *oblongifolia* and *reticulata* showed high immunity to *Striga* (Oyatomi et al. 2016). Ohlson and Timko (2020) reported that none of the cowpea lines were resistant to all *Striga* races and that no *Striga* race could breach the resistance in all the cowpea lines. They advocated stacking of multiple resistance genes for developing durable broad-spectrum resistance in cowpea.

8.9 Drought Resistance

Drought or water stress, arising due to inequality in water availability to the plants and the evapotranspiration losses occurring from plants, is highly detrimental to crop production across the globe. Drought stress is a highly complex situation affecting the morpho-physiology, molecular-biochemistry and edaphic symbiotic microbial interactions of the plants. Depending on the stage of plant development and the extent and duration of the water scarcity, yield losses in the range of 34–68% in cowpea have been reported (Farooq et al. 2017). Reproductive phase is the most critical stage which is severely impacted by water stress. Two types of drought tolerance, Type 1 and Type II, exist in cowpea as described by Mai-Kodomi et al. (1999b). In Type 1, the wilting symptoms were delayed both in the primary and true leaves, whereas in Type II the cotyledonary leaves alone wilted and the true leaves were comparatively more tolerant. One of the most important parameters for assessing drought tolerance is the stress tolerance index deciphered through biplot analysis (Batieno et al. 2016).

Genetics of water stress tolerance is highly intricate and is majorly a quantitative trait under environmental influence (Boukar et al. 2016; Muchero et al. 2009b, 2013). The genetics of Type 1 and Type 2 drought tolerance as determined through the extensively used wooden box technique (Mai-Kodomi et al. 1999a, b) were found to be under the governance of single dominant genes, *Rds1* and *Rds2*, the former being dominant to the latter. Yield and drought adaptive traits exhibited additive and non-additive gene actions in cowpea, although dominance and/or dominance × dominance effects primarily controlled the traits (Olajide and Ilori 2018). Boukar et al. (2018) identified 10 lines from 1200 germplasm lines which were highly tolerant to water stress. Breeding for water stress tolerance could be through alteration of root morphology or by improving the water use efficiency or its tolerance capabilities (Sofi et al. 2019).

Multi-parental approaches such as MAGIC intercrosses (Ravelombola et al. 2021a) should be adopted rather than relying on bi-parental crosses. Besides, interspecific, intergenic, pre-breeding and mutation breeding approaches should be exploited for effective breeding against this complex trait. A range of morphological, biochemical, physiological and molecular parameters have been reported for drought tolerance screening (Dhanasekar and Souframanien 2023), albeit none of them could be useful singly. A combination of screening parameters has to be adopted to identify useful genotypes tolerant for water stress. A number of QTLs associated with water stress have been identified: 10 for seedling drought tolerance (Muchero et al. 2009b), five for stay-green trait (Muchero et al. 2013) and three for delayed senescence, biomass and grain yield under water stress (Muchero et al. 2013). Ravelombola et al. (2021a) performed GWAS in a cowpea MAGIC population and identified 14, 18, 5, 5 and 35 SNP markers linked with drought tolerance indexes for plant growth habit, maturity, blooming time, seed size and grain yield, respectively. The list of candidate genes attributing drought tolerance in cowpea has been reviewed by Dhanasekar and Souframanien (2023). Metabolites like proline, galactinol and quercetin have been found to be associated with the acclimatization of cowpea and contributed to yield under water stress conditions (Goufo et al. 2017). Metabolites like γ-aminobutyric acid (GABA) have also been found to contribute to drought and heat stress tolerance (Shelp et al. 2021).

8.10 High Temperature Resistance

High temperature or heat stress is increasingly becoming one of the most important abiotic stresses in the present context of climate change. Heat stress causes irreversible harm to plant developmental physiology when the soil and air temperatures remain above a threshold level for a minimum time period (Lamaoui et al. 2018). Plants in particular the male reproductive organs are more vulnerable to heat stress, especially during the pre-fertilization reproductive phase. The pollen viability, stigma receptivity and pollen tube growth are highly impacted resulting in elevated ovule abortion, irregular pod filling and shrivelled seeds ultimately causing severe yield losses (Dhanasekar and Souframanien 2023). In cowpea, Hall (2004) observed that with every one-degree increment in night temperatures beyond 16 °C, pod filling and seed yield were severely impacted to an extent of 4%–14%.

Elevated temperatures coinciding with critical stages of blooming and pod filling acutely affected cowpea grain yields (Ntare 1991). At temperatures beyond 40 °C, cowpeas exhibited inferior pod setting, diminished harvest index, increased floral drop, male sterility, poor pollen shedding, tanned testa and decreased biological nitrogen fixation (Jha et al. 2017). Cowpea exhibits significant genetic variability for heat stress tolerance at podding and blooming stages under short- and long-day conditions. 'Prima', 'B 89-200', 'TN 88-63', 'California Blackeye No. 27' ('CB27'), 'TVu 4552' (Ehlers and Hall 1998), 'IT82D889-1', 'IT98K-1111' and 'IT98K-205-8' in the early maturing group, and 'IAR-48', 'IT98K-277-2' and 'IT98K-1092-1' in the medium maturing group are some of the genotypes reported to possess heat stress tolerance. Five quantitative trait loci (QTLs)—Cht-1, Cht-2, Cht-3, Cht-4 and Cht-5—explaining 11.5–18.1% of the phenotypic variation were tagged with 48 SNP markers, four of which were contributed by 'CB27' and a lone QTL was derived from the other donor 'IT82E-18' (Lucas et al. 2013). Three QTLs—Hbs-1, Hbs-2 and Hbs-3—responsible for heat-induced tanning of cowpea testa that explained up to 77.3%, 12.3% and 6.8% of the total phenotypic variation, respectively, were discerned (Pottorff et al. 2014).

The transcription factors NAC1/NAC2 were found to impart heat tolerance in transgenic cowpea (Srivastava et al. 2023). Initial studies on the inheritance of heat stress tolerance traits suggested the involvement of a single dominant, a single recessive, or double dominant genes; however, subsequent findings demonstrated that QTLs control the traits (Lucas et al. 2013). Selections for heat tolerance based on abundance of blooming and pod set under high nocturnal temperatures and long-day growing conditions (Marfo and Hall 1992) or based on depressed leaf-electrolyte-leakage (Hall 2004) were promising. Discriminations of genotypes based on stomatal function, proline and anthocyanin content (Carvalho et al. 2019), photosynthetic rate, pollen viability, and membrane stability, leaf water potential and relative water content for high temperature tolerance have been reported (Janni et al. 2020; Kumar et al. 2020; Siddiqui et al. 2015).

8.11 Waterlogging and Salinity Stress Tolerance

Under the present context of climate change, the vagaries in the timing and duration of rainfall often result in waterlogging conditions in the field, and this factor also needs to be given priority while breeding cowpea. Cowpeas, in general, are tolerant to water deficit and high temperature stresses, while they are highly sensitive to waterlogging conditions. No extensive studies have been carried out towards this aspect and only scanty reports are available. Olorunwa et al. (2022a, b, 2023) in their study on cowpea genotypes for 10 days of waterlogging observed plant stature, leaf area, fresh and dry weight to be affected drastically consequent to anaerobic conditions. Cowpeas adapted to waterlogging by enhancing the formation of adventitious roots. Photosynthetic parameters, especially ΦPSII, were affected the most and waterlogging tolerance was more explainable in terms of leaf chlorophyll and gas exchange parameters. They identified the genotypes 'Dagupan Pangasinan', 'UCR 369' and 'Negro' to possess better waterlogging tolerance at the seedling stage. The reproductive stage was the most sensitive stage to waterlogging.

Salinity stress is another important abiotic stress constraining cowpea productivity as cowpeas are moderately tolerant to salinity stress. Cowpeas are most sensitive to salinity stress during the vegetative stage seriously impacting seed yield up to the extent of 50%. The tolerance level of cowpeas increases gradually in the post-vegetative stage (Maas and Poss 1989). The main manifestation of salinity on cowpea is through the reduction in pod number and number of seeds per pod (Praxedes et al. 2022). A number of accessions tolerant to salinity at varying salt concentrations of 150 mM NaCl at germination stage (Ravelombola et al. 2017) and 100 mM to 200 mM NaCl at seedling stages (Ravelombola et al. 2018, 2021b; Dong et al. 2019; Akter et al. 2021) have been identified. Three and seven SNP markers highly associated with salt tolerance at germination and seedling stages, respectively, were identified (Ravelombola et al. 2018). In addition, a number of SNPs were identified for various salt tolerance-related traits based on a MAGIC population along with candidate genes encoding for Na^+/Ca^{2+} K^+ independent exchanger and H^+/oligopeptide symporter ion transport proteins that are associated with salt tolerance (Ravelombola et al. 2022).

8.12 Nutritional Quality

Cowpeas are highly nutritious and are rich sources of proteins, carbohydrates, minerals and vitamins with a low lipid profile. Cowpeas with good amount of resistant starch, dietary fibres, polyphenols and peptides endorse it as a candidate nutraceutical option for global food and nutritional security. The key health benefits of cowpea include their ability to regulate blood cholesterol levels, reduce inflammation, combat diabetes and prevent cancer. Dhanasekar et al. (2021) and Raizada et al. (2023) have reviewed in detail the nutritional and nutraceutical properties of cowpea. Padhi et al. (2022) in their quest to identify nutri-dense cowpea accessions evaluated over 120 germplasm lines and reported large variability for protein content (19.4–27.9%), starch content (27.5–42.7 g 100 g^{-1}), amylose content (9.65–21.7 g 100 g^{-1}), total dietary fibre-TDF (13.7–21.1 g 100 g^{-1}) and total soluble sugar content-TSS (1.30–8.73 g100 g^{-1}). The high protein-containing germplasm lines 'EC390248' (27.9%), 'EC169879' (27.7%) and 'EC240667' (26.6%) identified in the study could be used for genetic improvement of protein content.

The bulk of storage proteins in the cowpea seeds consists of globulins, albumins and glutelins with very little prolamins. Moreover, the lysine and tryptophan-rich amino acid profile of cowpea is lacking in sulfur-containing amino acids and therefore, concerted breeding efforts should be aimed at not only increasing the protein content but also at enhancing the proportions of methionine and cysteine amino acids. About nine African accessions with elevated protein contents in the range of 30–40% have also been reported in cowpea (Dakora and Belane 2019), the highest being in 'Bengpla' landrace. Cowpeas with starch contents as high as 60 g 100 g^{-1} have been testified from Nigeria (Gonçalves et al. 2016). The legume starch in contrast to cereal starch is hard to digest and hence has low glycaemic index (GI). Accessions with high amylose (>20%, 'EC472284', 'IC128727', 'EC244077') (resistant starch and low GI) are suitable for diabetic meals, while those with low amylose (~10%, 'IC249141', 'EC240900A' and 'IC214752') have good cooking quality owing to their softening consequent to increased swelling upon soaking. Accessions like 'EC724033' and 'EC244138'with high TDF (>20%) are ideal sources of dietary fibres which could prove beneficial in increasing bowel movements, reducing the risks associated with heart ailments and also could assist in weight loss (Padhiet al. 2022).

The sugar content in terms of TSS contributes to tiding over abiotic stresses and enhances the shelf life of seeds by improving their desiccation tolerance apart from improving the palatability of food. Apart from this, cowpea also exhibited broad variability for most of the mineral nutrients: iron (51–109 ppm), zinc (33–51 ppm), calcium (581–1252 ppm), potassium (12084–15,133 ppm), magnesium (1611–2052 ppm), phosphorus (3867–4922 ppm) and sulfur (1880–2354 ppm) (Singh 2016). Cowpea genotypes from IITA with high protein and rich in Fe, Zn, Ca and K ('IT97K-1042-3' and 'IT98K-205-8') and those with high antioxidant activity ('IT 97K-1042-3') could be used in breeding programmes for genetic improvement of nutritional quality of cowpea (Singh 2016).

8.13 Anti-nutritional Factors

Cowpeas like other legumes also contain certain constituents which are considered to be anti-nutritional and undesirable. These are the by-products of secondary metabolites produced by the plants per se and when consumed, their gradual build-up could have detrimental repercussions. The major anti-nutritional factors in cowpea include oligosaccharides, phytic acid, polyphenols, flavonoids, tannins, protease inhibitors and lectins (Avanza et al. 2013) among others.

8.13.1 Phytic Acid

Cowpeas contain phosphorus mainly in the form of phytic acid (PA), which potentially chelates essential nutrients like Ca, Fe, Mg and Zn, making them unavailable for uptake. They also reduce the digestibility of starch and proteins by their inhibitory action on proteases and amylases (Gonçalves et al. 2016). The PA content in cowpeas has been observed to vary from 0.69 to 1.88% with a mean of about 1.10%. Some of the exotic collections ('EC723824', 'EC240697', 'EC004218', 'EC240808', 'EC390215') in the study exhibited low PA in the range of 0.69% to 0.85% (Padhi et al. 2022). PA could be reduced by adopting various physical processing methods such as sprouting, drying, storing, dehulling, soaking, milling, cooking, fermenting or roasting (Gonçalves et al. 2016; Devi et al. 2015).

8.13.2 Oligosaccharides

The flatulence-causing raffinose family of oligosaccharides (RFOs) including raffinose, stachyose and verbascose are thermostable and recalcitrant to physical processing methods. These sugars remain undigested in the small intestines of humans and other monogastric animals lacking α-galactosidase enzyme but are metabolized by the microflora present in the large intestine producing gases causing stomach discomfort. The RFO content in cowpea is reported to vary between 4.27 and 9.7 m moles/100 gm, with a mean of 7.77 m moles/100 gm. Some of the genotypes like 'TC-501' and 'IT86F2K14' identified with low RFOs (4.27–6.46 m moles/100gm) could be used in the breeding programmes (Dhanasekar and Reddy 2015a). On average, the cowpea genotypes contained 3.4, 15.4 and 31.5 mg/g of raffinose, sucrose and stachyose, respectively. Improved breeding lines ('IT93K-452-1', 'IT90K-277-2' and 'IT98K-205-8') have been identified with low RFOs and favourable agronomic and nutrient contents (Muranaka et al. 2016).

8.13.3 Polyphenols

Cowpeas are rich in polyphenolic compounds, tannins and flavonoids and, therefore, exhibit considerable antioxidant and radical scavenging activities. These phenolic compounds are more pronounced in coloured seeds rather than white-seeded genotypes as the phenolic compounds are largely restricted to the seed coat. They are known to interfere with protein and starch digestibility and also chelate metals. The primary polyphenol found in cowpeas is the protocatechuic acid followed by ferulic acid and *p*-coumaric acid. The total phenolic content was reported to be in the range of 0.026–0.883% with a mean value of 0.272% (Padhi et al. 2022). Some of the exotic germplasm ('EC724674', 'EC724681', 'EC240908', 'EC472284', 'EC724556', 'EC108722') in the study possessed low polyphenols (<0.1%) which could be exploited in breeding for reduced antinutritional polyphenolic content. Brown- and cream-coloured genotypes (0.192–0.196%) contained comparatively more phenolic compounds than white-coloured genotypes (0.099%) (Gonçalves et al. 2016). Cooking and dehulling have been found to reduce total phenolic content to a certain extent. Variation for tannin content (measured as gallic acid equivalents 'GAE') was observed to vary from 55.12 mg to 233.92 mg GAE/100 g dry weight, and in the case of flavonoids (measured as rutin equivalents 'RE'), it varied from 363.64 mg to 453.93 mg RE/100 g dry weight (Maina et al. 2020). These antinutrients exhibited high broad-sense heritability (0.86–0.99) and low narrow-sense heritability (0.06–0.50). They were reported to be under the control of additive and non-additive gene action with duplicate epistatic interactions suggestive of the fact that breeding for increased tannins, flavonoids and phytate contents in cowpea seeds would be quite efficient through recurrent selections in advanced generations.

8.13.4 Protease Inhibitors (PI) and Lectins

Protease inhibitors block proteases, which are crucial enzymes involved in the digestion of proteins. On the other hand, lectins are glycoproteins involved in selective and reversible binding with carbohydrates. Once considered as serious anti-nutritional factors, modern-day breeding does not explicitly stress breeding for these factors, as they can easily be destroyed while cooking owing to their heat lability (Boukar et al. 2015). However, consideration of PI contents becomes necessary in situations where cooking is not involved (like fodder cowpeas). The important PI found in cowpeas are trypsin and chymotrypsin inhibitors. Trypsin inhibitor (TI) levels varied considerably in the range of 2.2–4.2 units/mg of protein across cowpea varieties, while chymotrypsin inhibitor (CI) content also varied in the range of 2.3–3.8 units/mg of protein (Carvalho et al. 2012). Sprouting considerably influenced TI levels, whereas it had no effect on CI levels. Lectins bind to the gut wall leading to reduced nutrient absorption and are usually measured in terms of haemagglutination activity per gram of flour. Their activity in cowpea was reported to vary between 40 and 640 with a mean of 220 (Carvalho et al. 2012).

8.14 Mechanical Harvesting Compliant Ideotype

Non-availability of timely labour is becoming a common occurrence across the globe resulting in great economic losses for farmers. The existing cowpea varieties are largely indeterminate with viny growth habits and non-synchronous maturity, making them non-amenable to machine harvest. The best way forward is to develop varieties with ideotypes that are compliant with mechanical harvesting. The important traits contributing to compliance include a determinate and erect plant type with synchronous maturity, long peduncles with pods above canopy, pod bearing branches emerging at least 5–7 cm from the ground level. In recent times, allelic variants of terminal flowering gene (TFL) housing SNPs responsible for determinate plant types have been identified (Dhanasekar and Reddy 2015b; Eswaramoorthy et al. 2023). These could be suitably exploited through genome editing for developing determinate plant types with synchronous maturity in elite genetic backgrounds. The cowpea variety 'V 38' has long peduncles and the pods are placed above the crop canopy that can be utilized in breeding programmes to incorporate this trait (Singh et al. 1997). Ribeiro et al. (2014) observed that additive and epistatic gene interactions governed the plant architectural traits influencing mechanical harvesting in cowpea.

Not much research efforts have been directed towards developing mechanical harvest suitable cowpea varieties, yet this is one of the important research priorities for cowpea breeding in the present context of labour shortage. Some of the other traits that could be targeted for machine harvest suitability based on studies in other legumes like chickpea are tall plant height, branch angle of 70–80°, first pod height above 30 cm, stem thickness, stem bio-density and seeds per plant (Shridevi and Naik 2017; Arya et al. 2023). The height to first pod (HFP) has been found to be a key trait for machine harvest suitability in legumes that varies with crop: 12 cm in soybean (Ramteke et al. 2012), 15 cm in common bean (Bisognin et al. 2019) and 25–29 cm in chickpea (Petrova 2021). The HFP is influenced by the node number of the first flower and the internodal length (Kuzbakova et al. 2022). QTLs associated with HFP include genes associated with auxin transport and signal transduction in *Glycine max*, MADS box *SOC1* gene in *Medicago* and *WD40* genes in *Phaseolus vulgaris*. These parameters need to be standardized in cowpea also for breeding ideotypes suitable for combined harvesting.

8.15 Market Ideotype

Market ideotype breeding encompasses those traits such as growth habits, seed size and shape, seed coat colour, etc., that determine farmer's and consumer's preferences (Nkhata et al. 2010). To augment varietal adoption and marketability of the varieties, it is pertinent to include these traits while breeding and perform selections cognizant of the preferred traits. White and brown seeded varieties with rough seed coats are preferred in West and Central Africa due to the ease of removal of seed coats for local food preparation, while red, black, white or brown seeded varieties with smooth seed coats are preferred in East and Southern Africa and parts of Central and South America where cowpeas are taken in boiled form without removing the seed coats. In Central African countries like Cameroon, people disdain cowpeas with coloured (especially black) eyes. In Cuba and other Central American countries, black-seeded varieties as a substitute for black beans are used to prepare delicacies. Green seeded varieties are preferred for frozen food in the United States (Timko and Singh 2008). Countries like Mexico, Guatemala, Nicaragua, Costa Rica and Cuba prefer black-seeded cultivars, while red is the choice in El Salvador, Venezuela, Honduras and Jamaica. In Asia, white- and cream-seeded cultivars are most preferred and black the least (Singh 2016). The preferences even vary within the same country. For instance, northern Nigerians prefer white-seeded varieties while brown-seeded varieties are the choice in Southern Nigeria (Boukar et al. 2018). Likewise, in India, people from the northern regions prefer white testa varieties, while in the southern region, they prefer tan-coloured varieties with small seed size. Large seeds are preferred in the Konkan region (south-west) where cowpea is extensively used in all sorts of delicacies. The people of Jammu and Uttarakhand in the Himalayas are not colour particular and even devour black-seeded varieties. Considering the ease of harvesting, African women especially in western regions like Ghana prefer varieties exhibiting pods above canopy with long peduncles (Jinbaani et al. 2023). The use of mutants in

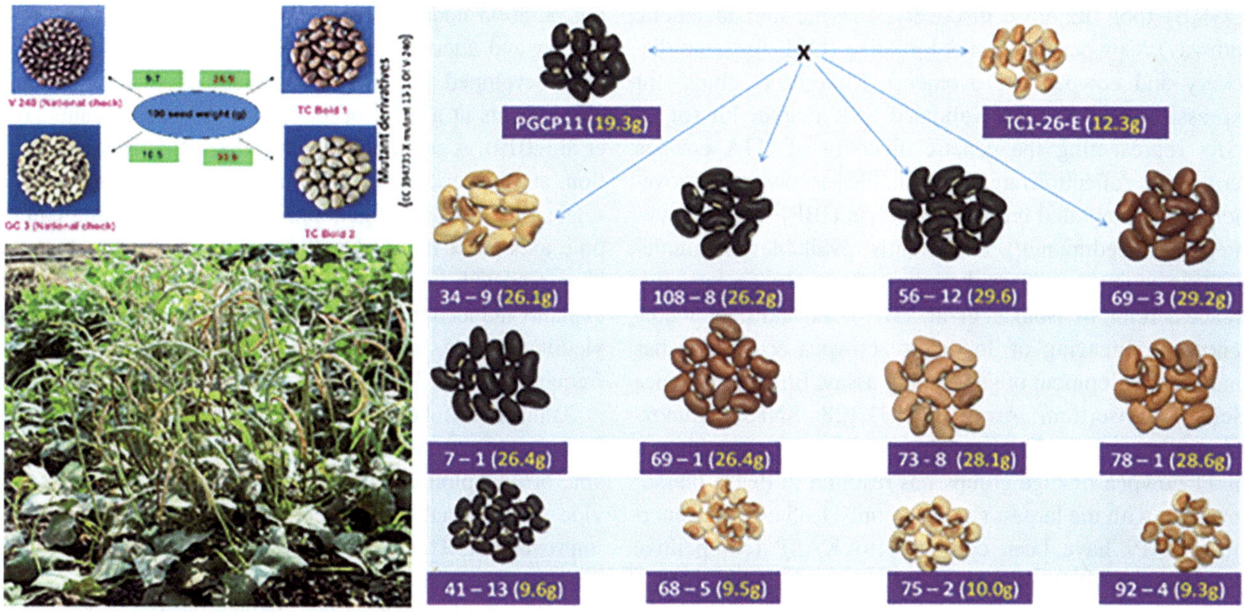

Fig. 2 Panel exemplifying the use of induced mutants in developing market ideotypes for seed size, seed colour and pods above canopy in cowpea

developing market ideotypes in cowpea is shown in Fig. 2. The inheritance of seed coat and pod colour shows the dominance of black over red or brown and white with digenic supplementary gene interactions (Patel et al. 2016).

9 Genomics in Cowpea Breeding

Cowpea, until recently, has largely been an orphan crop and the pace of genetic improvement has been on the downtrend owing to limited genetic and genomic resources. With the advent of next-generation sequencing and the availability of cowpea genome (Lonardi et al. 2019), the utilization of genomics has gained momentum in cowpea breeding. Prior efforts have largely been concentrated on unravelling molecular diversity and establishing genetic linkage maps. A number of molecular markers like allozymes, seed proteins, chloroplast DNA markers, RFLPs (Restriction Fragment Length polymorphism), AFLPs (Amplified Fragment Length Polymorphism), RAPDs (Random Amplification of Polymorphic DNA), DAFs (DNA Amplification Fingerprinting), SSRs (Simple Sequence Repeats), ISSRs (Inter Simple Sequence Repeats), STMS (Sequence-Tagged Microsatellite Sites) and SNPs (Single Nucleotide Polymorphism) have been employed in establishing the genetic diversity in cowpea (reviewed in Boukar et al. 2016). The deployment of marker systems for genetic improvement is gaining pace off late. Many linkage maps including consensus maps have been established in cowpea (Menendez et al. 1997, Ouedraogo et al. 2002, Ubi et al. 2000, Muchero et al. 2009a, Lucas et al. 2011) since the first in 1993 (Fatokun et al. 1993).

Further updates of linkage maps with higher marker density and physical maps are now available via HarvEST:Cowpea (http://harvest.ucr.edu/). The maps have been used to identify QTLs for desirable traits like thrips resistance, cowpea yellow mosaic virus resistance, *Striga* resistance, bacterial blight resistance, several agronomic traits such as days to flower, maturity, pod length, seeds per pod, leaf length, leaf width, primary leaf length, seed size, etc. The resistance genes for BICMV, SBMV, CPSMV, CPMV, RKN, chilling tolerance, *Fusarium* resistance have also been mapped. SNP-based linkage maps have also been used to map QTLs associated with seedling stage drought, stem greenness and recovery dry weight post water stress, *Macrophomina* resistance, stay-green trait, heat stress, aphid resistance, etc. Many of these QTLs would be applied to marker-assisted recurrent selection (MARS) and marker-assisted backcrossing (MABC) populations. MARS and MABC have been implemented in cowpea for improving drought, staygreen trait, heat tolerance, large seed size, and resistance to *Striga*, *Macrophomina* and nematodes. Using MABC, foreground and background selections have been employed for combining traits such as *Striga* resistance with seed size, grain quality and nematode resistance.

Apart from this a number of genomic resources like 'BreedITSNP Selector' (http://breedit.org), 'ParentChecker' (for parental inference for QTL mapping), 'KBioConverter' and 'Backcross selector' are available for managing genotyping data. Cowpea Genespace/Genomics Knowledge Base

(CGKB) tool (for gene discovery, enzyme and metabolic pathway), Cowpea Genomics Initiative (CGI, for gene discovery and cowpea improvement), Microarray chips (for expression analysis) and validated SSR marker kit (of 20 SSRs representing the genetic diversity of IITA cowpea germplasm collection) are also available for cowpea improvement. The integrated breeding platform (IBP) (https://www.integratedbreeding.net/) is currently available for implementing cowpea molecular breeding efficiently (for back references, refer to Boukar et al. 2016). In addition, whole genome sequencing of 36 diverse cowpea accessions has enabled development of genotyping assay, Illumina Cowpea iSelect Consortium Array for 51,128 SNPs (Munoz-Amatriain et al. 2017). Using this platform, fine mapping of all 11 cowpea linkage groups has resulted in dense marker coverage with the largest gap being only 1.85 cM. The enormous SNPs have been converted to KASP (competitive allele-specific PCR) markers (Boukar et al. 2018).

10 Biotechnological Tools in Cowpea Breeding

Akin to most of the legumes, cowpea is not readily amenable for regeneration owing to its recalcitrant nature. Lack of proper regeneration protocols in cowpea makes applications of genetic engineering in its improvement, still challenging. A number of reports involving the selection of appropriate ex-plants like mature seeds (Raveendar et al. 2009), cotyledonary nodal cuttings (Chaudhury et al. 2007), mature embryos (Penza et al. 1991), plumules (Aasim et al. 2010) and shoot apices or tips (Mao et al. 2006) for in vitro propagation though available are not promising. Many attempts to regenerate cowpea through somatic embryogenesis and multiple shoot organogenesis have not been able to make a breakthrough (Chaudhury et al. 2007; Raveendar et al. 2009; Behura et al. 2015). Popelka et al. (2006) gave the impetus to genetic transformation of cowpea by developing a successful transformation protocol and demonstrated the stable inheritance of the *Bt* transgene against *Maruca* pod borer. Consequent to this there were several successful attempts to transfer genes into cowpea including those for bruchids (Solleti et al. 2008; Higgins et al. 2012), *Maruca vitrata* pod borer (Higgins et al. 2012; Bett et al. 2017), imazapyr herbicide tolerance (Citadinet al. 2013), CPSMV and CABMV resistance (Curz and Aragao 2014).

Transformation protocols based on *Agrobacterium*-mediated and biolistic guns have satisfactorily introduced transgenes related to pest, virus and herbicide resistance into cowpea, frequently with neomycin phosphotransferase II (NPT II) as the selectable marker. Recently, an improved *Agrobacterium*-mediated transformation protocol involving imbibed mature cowpea seeds and sonication followed by regeneration under a selection regime with selectable nptII marker and alternating kanamycin-geneticin antibiotics has been developed resulting in successful production of transgenic plants at a level of 1% of the starting explants (Bett et al. 2019). A rapid and highly efficient cowpea regeneration, agrobacterium-mediated transformation using spcN as selectable marker and genome editing system using embryonic axis from imbibed mature seeds as explants has also been reported. Excision of shoot apical meristem from the explants induced multiple shoot organogenesis from the cotyledonary node and transgenic plantlets were recovered at frequencies of 4–37% (Che et al. 2021).

Transgenics, keeping aside the regulatory restrictions, have the potential to introduce novel genes which are unavailable or unexploitable in the natural germplasms. They provide a sustainable and eco-friendly approach to crop improvement. Transgenics have been mostly exploited for development of pest and disease resistance in cowpea. Like in other crops, the *Cry* (crystalline) proteins of *Bacillus thuringiensis* (Bt), especially the *Cry1Ab* followed by *CryAa*, *Cry2Aa* and *Cry1Ac* have been found promising against the lepidopteran pests (Srinivasan 2008). Recently, a transgenically developed *Maruca*-resistant cowpea variety ('SAMPEA 20-T') has been released for commercial cultivation for the first time in Nigeria involving the Bt*Cry1AB* gene (Barrero et al. 2021). Similarly, *Cry2AB* and *Cry1AC* transgenes in cowpea also provided complete protection against *Maruca*. Vegetative insecticidal proteins (*Vips*) produced by vegetative cells of *Bacillus thuringiensis* exhibit exceptional insecticidal activity in comparison to other insecticidal proteins and unlike Cry proteins are effective at minuscule doses. The *BtVip3Ba* transgene expressed in cowpea conferred complete protection against *Maruca* larva (Bett et al. 2017). Likewise, cowpea transgenes expressing common bean alpha-amylase inhibitors (*αAI-1* and/or *αAI-2*) using a seed-specific promoter phytohaemagglutinin (dlec2) exhibited complete resistance against *Callosobruchus maculatus* storage pest in cowpea (Solleti et al. 2008; Lüthi et al. 2013).

11 Mutation Breeding

Mutation breeding is one of the important tools for generating genetic variability for breeding better crop varieties. More like conventional breeding, mutation breeding creates variation within the existing genome of the organism and does not involve 'foreign genes' as is the case of genetically modified organisms (GMOs), and therefore, no regulatory restrictions are in place for mutation breeding (Parry et al. 2009). Mutation breeding involves induction, screening and confirmation of mutation prior to release of potent mutants as varieties or use as potential donors in classical breeding. Mutation breeding comes to the rescue of the breeder, especially when there is no or little variability for the trait of interest in the gene pool or there is a need to correct one or two shortcomings in otherwise promising genotypes. Prior to the start of any mutation breeding programme, it is imperative to select proper mutagen(s) and the optimal dose to maximize mutant recovery and minimize unsolicited damages such as mortality.

Optimal dose chosen based on LD_{50} values (50% lethality), though common, should be avoided as beneficial mutations could either be lost or overlooked owing to plant death or poor agronomic presentation (Oldach 2011). Instead, doses with sub 50% lethality (LD_{20} or LD_{30}) or showing 50% growth reduction (GR_{50}) should be adopted. Globally, a total of 3433 varieties have been released through mutation breeding (IAEA Mutant Variety Database 2024). In cowpea, 22 mutant varieties have been released worldwide of which 16 are from India (Souframanien and Dhanasekar 2023). Gamma rays in the dose range of 100–250 Gy were found successful in varietal development (Table 5). Most of the mutant varieties released in India have been developed through gamma rays, although a few of the early varieties involved use of the chemical mutagen, DMS. Bhabha Atomic Research Centre, Mumbai, has made significant contribution in the release of two direct cowpea mutant varieties in India. The variety 'TRC77-4' ('Kalleshwari') is a dual-purpose, semi-dwarf mutant suitable for rice fallows, while 'TC-901' released in 2018 is India's first central summer-suitable cowpea variety with synchronous maturity and field resistance to cowpea mosaic and root-rot diseases (Dhanasekar and Reddy 2018). The other significant Indian stakeholders include Indian Agricultural Research Institute, New Delhi, and Tamil Nadu Agricultural University, Coimbatore. Further, Costa Rica, Kenya, Zimbabwe and Zambia are the main global contributors of cowpea mutant varieties (Table 5). Souframanien and Dhanasekar (2023) and Raina et al. (2023) may be referred for a detailed review of mutation breeding in cowpea.

12 Speed Breeding

Speed breeding is a powerful tool for accelerating crop research and breeding by enabling a significant reduction in crop generation time through manipulation of physiological parameters, especially photoperiod, light intensity, light quality and temperature, so as to hasten floral induction and seed production (Xu et al. 2023; Watson et al. 2018). It also entails the technique of using dried immature seeds for propagation in a controlled environment to curtail the generation time (Edet and Ishii 2022). The speed breeding technique is primarily aimed at expediting attainment of homozygosity and stability, saving breeding time and resources through rapid advancement of generations. Speed breeding results in ~3 to 9 generations as against 1 to 2 generations per year through classical breeding. Speed breeding has been successfully established in many long-day crops (wheat, oats, barley, lentils, faba bean, chickpea), while its exploitation has been limited so far in the case of day-neutral (peanuts) and short-day plants (rice, soybean, pigeonpea). Synergistic integration of this technique with cutting-edge techniques involving genomic tools, molecular markers, genome editing, high-throughput phenomics and other rapid generation advancement practices like doubled haploidy would further shorten the breeding cycle resulting in rapid genetic gains and speedy varietal development.

In cowpea, a speed breeding protocol capable of accommodating eight generations per year has been established primarily by using oven-dried (39 °C for 2 days in presence of silica gels) seeds from 11-day-old immature pods in regulated growth chambers (Edet and Ishii 2022). Crossing success was enhanced under optimal growth conditions (230–420 μmol m^{-2} s^{-1} light intensity, 70/80% (day/night) relative humidity, 10 h photoperiod, and 22–23/25–29 °C (night/day) temperature. Metal halide chamber-grown plants grew faster, larger and flowered 2–3 days earlier than those in LED chambers, while LED favoured manual hybridization and development of pods. Therefore, it has been recommended to grow cowpea plants in metal halide chambers until floral bud initiation and switch over to LED chambers for successful pollination and pod development. However, CO_2 supplementation in this study failed to elicit a selective advantage in cowpea. Cowpea breeding cycles can also be shortened by elevated temperatures (Cavalcante et al. 2016), but only at the cost of flower abortion and a remarkable reduction in pod setting (Angelotti et al. 2020). The universal centromere-specific histone 3 (CENH3) protein can be beneficially manipulated to induce centromere-mediated genome elimination for production of haploid plants. In cowpea, two such CENH3 variants with unequal contributions to centromeric function have been identified (Ishii et al. 2020) and could be potentially used for rapid production of doubled haploids for accelerated breeding.

13 Pre-breeding

The use of wild relatives in cowpea improvement has largely remained unexploited even though many of these cross-compatible wild species are known to source important resistance genes. This could be attributed to the number of unpleasant agronomic attributes present in the wild species that could be dragged along with the useful genes when used in introgression breeding. Contrarily, of the two QTLs contributing to number of seeds per pod, the one present on chromosome Vu09 has been donated by a wild parent that is associated with ovule development (Wei et al. 2015). Thus,

Table 5 List of cowpea mutant varieties released for commercial cultivation in India and other countries

Mutant variety	Country	Year of release	Mutagen	Salient features
India				
V16 (Amba)	India	1981	DMS	High yield, resistance to fungal and bacterial diseases
V37 (Shreshta)	India	1981	DMS	High yield, luxuriant vegetative growth
V38 (Swarna)	India	1981	DMS	High yield, early maturity, synchronous flowering, better quality pods and grains, resistance to bacterial blight, *Macrophomina* and PM
V240	India	1984	DMS	High yield, resistance to fungal, viral and bacterial diseases, tolerant to moisture stress
GC-1	India	1984	–	Yield, early maturity, root knot resistance
Co-5	India	1986	Gamma rays	High yield, suitable for forage
Cowpea-88	India	1992	Gamma rays	Large seed size, resistant to YMV and anthracnose
KBC-2	India	1999	–	Resistant to rust
PUSA SAMPADA (V-585)	India	1999	–	Erect, bushy, multiple disease resistant
COCP 702 [CoVu 702 and CO(CP) 7]	India	2002	Gamma rays (200 Gy)	High yield and good quality
Co (FC) 8	India	2004	Gamma rays	High green fodder, indeterminate growth, resistant to cowpea YMV and root rot
V578 (Pusa-578)	India	2004	Gamma rays	Resistant to YMV, determinate
TRC77-4 (Kalleshwari)	India	2007	Gamma rays (200 Gy)	Semi-dwarf, high yield, suitable for rice fallows, dual purpose
Co-9	India	2016	Gamma rays	Higher protein content, reduced fibre confers increased digestibility, palatability and intake rate; MR to YMV and resistant to major pests
TC-901	India	2018	Gamma rays (250 Gy)	Semi-determinate mutant, high yielding, increased seed size, pod size and pod number, resistance to cowpea mosaic and root-rot diseases
DCS-6	India	–	–	–
Other countries				
ICV 11	Kenya	1985	–	Semi-erect, large leaves, maturity in 65 days, yield 1100 kg/ha, resistance to cowpea aphids
ICV 12	Kenya	1985	–	High yield and resistance to cowpea aphids

(continued)

Table 5 (continued)

Mutant variety	Country	Year of release	Mutagen	Salient features
Uneca-Gama	Costa Rica	1986	Gamma rays (100 Gy)	High yield
CBC5	Zimbabwe	2017	Gamma rays (150 Gy)	High grain and fodder yield (at least 18% yield advantage), good dynamic stability across environments (sites and seasons), Increased seed size
Lukusuzi	Zambia	2018	Gamma rays (150 Gy)	Good for mechanization, tolerant to *Ascochyta* and *Cercospora*; and medium tolerant to cowpea aphid borne mosaic virus, early maturity, yield
Lunkhwakwa	Zambia	2018	Gamma rays (150 Gy)	Aluminium toxicity, tolerance to aphids and bruchids, yield

regardless of the small seed size of the wild cowpeas, they are potential contributors to increased yields. Trichomes present in some of the wild species were found to contribute to insect resistance, especially *Maruca*. Therefore, with the increasing availability of modern tools, pre-breeding activities need to be rejuvenated for introgressing the resistance genes into the cultivated species. Some of the wild species and the promising resistance genes possessed by them are: *Vigna vexillata* (*Maruca* pod borer, *Callosobruchus maculatus* weevils, *Clavigralla* pod bug and parasitic weed *Striga gesnerioides*); *V. unguiculata* ssp. *dekindtiana* (aphids, drought and heat); *V. oblongifolia*, *V. macrosperma*, *V. reticulata* (*Maruca* pod borer); *V. luteola*, *V. macrosperma*, *V. angustifolia* (*Clavigralla tomentosicollis*), *V. luteola* (*Callosobruchus maculatus*); *V. ambacensis*, *V. parkeri*, *V. oblongifolia* and *V. reticulata* (*Striga* parasitic weed); *V. vexillata* (high protein, >29%, low protein digestibility), *V. reticulata*, *V. ambacensis*, *V. dekindtiana* (low trypsin inhibitors, tannins and lectins); *V. vexillata*, *V. macrosperma*, *V. luteola*, *V. oblongifolia*, *V. unguiculata dekindtiana*, *V. racemosa*, *V. reticulata* and *V. ambacensis* (high sulphur containing AA); ssp. *tenuis* and *stenophylla* (drought) (Boukar et al. 2020).

14 Modern Tools in Cowpea Breeding

Modern tools in plant breeding are techniques that use advanced science and technology to improve crop traits and accelerate breeding processes. Some modern tools that have been successfully employed in cowpea improvement include post-transcriptional gene silencing (PTGS) like RNA interference (RNAi) and MicroRNAs (miRNAs), nuclease-induced genome editing, gene stacking, etc. RNAi technology has been successfully implemented in developing cowpea resistant to viruses MYMIV (Mungbean yellow mosaic India virus), CPSMV and CABMV by downregulation of important viral genes through transformation. The gene constructs harbouring sense, anti-sense or self-complementary hairpin RNA (hpRNA) containing sequences analogous to the target gene were used, resulting in the formation and degradation of double-stranded RNA (dsRNA) and, thereby, the silencing of the concerned gene. In cowpea, the insect metabolism-specific genes TPS (trehalose phosphate synthase), AChE (required for nerve transmission) and CHS (chitin biosynthesis) were downregulated for controlling *Helicoverpa armigera* and *M. vitrata* (Kumar et al. 2009) in host plants. RNAi has also been successfully employed for controlling MYMIV, a Begomovirus, causing golden mosaic and leaf curl diseases in cowpea. The RNAi constructs targeting *AC2*, *AC4* and a combination of both (*AC2*+*AC4*) genes have been used for RNAi-derived resistance (Kumar et al. 2017). Likewise, the CPSMV proteinase cofactor and CABMV coat protein genes were silenced using RNAi technology to develop cowpea transgenics resistant to these viruses (Cruz and Aragão 2014).

Site-directed nucleases-induced genome editing tools are increasingly becoming powerful techniques for elucidating gene functions and for generating allelic variants of important traits including tolerance to biotic and abiotic stresses. The nucleases involved in genome editing precisely bind and cleave a sequence-specific nucleic acid region by inducing double-stranded breaks (DSBs) at or near the target site. These nucleases can be either meganucleases, zinc finger nucleases (ZFNs), transcription activator-like effector nucleases (TALENs), or clustered regularly inter-spaced palindromic repeats/CRISPR-associated 9 (CRISPR/Cas9) (Stella and Montoya 2016). The CRISPR/Cas9 system is increasingly being used to target viral genomes or host factors that aid in viral replication, transcription or translation, thereby curbing virus infecting the host. The eukaryotic initiation factor 4E (eIF4E) of the host on which the viruses depend for protein translation has been knocked off using CRISPR/Cas9 in *Arabidopsis* against Turnip mosaic virus (Pyott et al. 2016). In cowpea, however, CRISPR applicability has only been demonstrated for disruption of symbiotic nitrogen fixation genes (Ji et al. 2019). In days to come, this technology has the potential to change the way genomics could be made use of to alleviate the lacunas in genetic crop improvement through classical breeding.

15 Conclusions

Cowpea until recently has been considered an orphan crop receiving comparatively less research attention. However, owing to its hardiness and inherent tolerance to abiotic stresses such as high temperature and drought, it is considered a potential crop in the present scenario of climate change. The modest progress made in the genetic improvement of cowpea needs to be accelerated through various "omics" approaches and modern biotechnology tools currently available. A number of genomic resources have been generated in the recent past, which have to be fully exploited to the core for its genetic improvement. The narrow genetic base of the crop needs to be widened up by adopting methodologies such as mutation breeding, pre-breeding, genome editing, TILLING and transgenics. Varieties with various genes combined for different agronomical traits and biotic and abiotic stress resistances could be generated through gene stacking. Introgressive breeding involving multiple parents acting as donors for different traits could be used to develop powerful MAGIC populations converging different permutations and combinations of the various traits in elite genetic backgrounds. Considerable research progress made in the recent past needs to be channelized and synchronized with modern breeding technologies for developing sustainable, efficient and smart cowpea varieties and, thereby, enables cowpeas to effectively contribute to food and nutritional security.

References

Aasim M, Khawar KM, Ozcan S (2010) Efficient in vitro propagation from preconditioned embryonic axes of Turkish cowpea (*Vigna unguiculata* L.) cultivar Akkiz. Arch Biol Sci 62:1047–1052. https://doi.org/10.2298/ABS1004047A

Agbicodo EM, Fatokun CA, Bandyopadhyay R et al (2010) Identification of markers associated with bacterial blight resistance loci in cowpea [*Vigna unguiculata* (L.) Walp.]. Euphytica 175:215–226. https://doi.org/10.1007/s10681-010-0164-5

Akter M, Mahmud A, Akter M et al (2021) Dissecting the salt tolerance potential of cowpea genotypes based on morpho-physiology and yield-related attributes. Ann Appl Biol 180:428–437

Alabi OY, Odebiyi JA, Jackai LEN (2003) Field evaluation of cowpea cultivars (*Vignaunguiculata* [L.] Walp.) for resistance to flower bud thrips (*Megalurothripssjostedti*Trybom) (Thysanoptera: Thripidae). Int J Pest Manag 49(4):287–291

Alabi OY, Odebiyi JA, Tamo M et al (2011) The roles of plant secondary metabolites from cowpea floral structures in resistance to the flower bud thrips. J Agril Sci Tech A 1:262–269

Amazue U, Adewale D (2016) A search for candidate gene for cowpea powdery mildew resistance in the Southern Guinea ecology of Nigeria. Ekin J Crop Breed Genetic 2(2):87–94

Angelotti F, Barbosa LG, Barros JRA et al (2020) Cowpea development under different temperatures and carbon dioxide concentrations. Pesqui Agropecu Trop. https://doi.org/10.1590/1983-40632020v5059377

Anusha C, Natikar PK, Balikai RA (2016) Insect pests of cowpea and their management - a review. J Exp Zool (India) 19:635–642

Arumuganathan K, Earle ED (1991) Nuclear DNA content of some important plant species. Plant Mol Biol Report 9:208–218. https://doi.org/10.1007/BF02672069

Arya M, Prasanth C, Maring KM et al (2023) Selection parameters for mechanical harvesting chickpea and its advantages over manual harvesting. Pharm Innov SP-12:1994–1998

Atokple IDK, Singh BB, Emechebe AM (1993) Independent inheritance of Striga and *Alectra* resistance in cowpea genotype B301. Crop Sci 33:714–715

Atokple IDK, Singh BB, Emechebe AM (1995) Genetics of resistance to *Striga* and *Alectra* in cowpea. J Hered 86:45–49

Avanza M, Acevedo B, Chaves M et al (2013) Nutritional and anti-nutritional components of four cowpea varieties under thermal treatments: principal component analysis. LWT- Food Sci Technol 51:148–157. https://doi.org/10.1016/j.lwt.2012.09.010

Bamabalwad J, Pankaja NS, Sugeetha G et al (2019) Investigations on the rust disease prevalence on cowpea in Mandya district and evaluation of cowpea genotypes for its resistance. Int J Curr Microbiol Appl Sci 8(11):1126–1133. https://doi.org/10.20546/ijcmas.2019.811.132

Barrero JM, MacKenzie DJ, Goodman RE et al (2021) A biotech shield for cowpea against the major insect pests. Agriculture News https://www.openaccessgovernment.org/major-insect-pests/111590/. Accessed on 06.06.2023.

Barros GB, Nogueira M, do S da R, Oliveira CRR et al (2013) Obtenção de plantas de feijão-caupiresistentesao Cowpea severe mosaic virus e ao Cowpea aphid-borne mosaic virus. Summa Phytopathol 39:130–136. https://doi.org/10.1590/S0100-54052013000200009

Batieno BJ, Danquah E, Tignegre J-B et al (2016) Application of marker-assisted backcrossing to improve cowpea (*Vignaunguiculata* L. Walp) for drought tolerance. J Plant Breed Crop Sci 8:273–286. https://doi.org/10.5897/JPBCS2016.0607

Baudoin JP, Maréchal R (1985) Genetic diversity in Vigna. In: Singh SR, Rachie KO (eds) Cowpea research, production and utilization. Wiley, Chichester, pp 3–11

Behura R, Kumar S, Saha B et al (2015) Cowpea [*Vignaunguiculata* (L.) Walp]. Methods Mol Biol 223:255–264. https://doi.org/10.1007/978-1-4939-1695-5_20

Bett B, Gollasch S, Moore A et al (2017) Transgenic cowpeas (*Vigna unguiculata* L. Walp) expressing *Bacillus thuringiensis* Vip3Ba protein are protected against the *Maruca* pod borer (*Marucavitrata*). Plant Cell Tissue Organ Cult 131:335–345. https://doi.org/10.1007/s11240-017-1287-3

Bett B, Gollasch S, Moore A et al (2019) An improved transformation system for cowpea (*Vignaunguiculata* L. Walp) via sonication and a kanamycin-geneticin selection regime. Front. Plant Sci 10:219. https://doi.org/10.3389/fpls.2019.00219

Bhattarai G, Shi A, Qin J et al (2017) Association analysis of cowpea mosaic virus (CPMV) resistance in the USDA cowpea germplasm collection. Euphytica 213:230. https://doi.org/10.1007/s10681-017-2015-0

Birch ANE, Fellows LE, Evans SV et al (1986) Para-aminophenylalanine in *Vigna*: possible taxonomic and ecological significance as a seed defence against bruchids. Phytochemistry 25(12):2745–2749

Bisognin MB, Pias OHC, Vian AL et al (2019) Seed spacing variability reduces common bean yield. Pesqui Agropecu Trop 49:e55134. https://doi.org/10.1590/1983-40632019v4955134

Booker HM, Umaharan P (2007) Identification of resistance to *Cercospora* leaf spot of cowpea. Eur J Plant Pathol 118:401–410. https://doi.org/10.1007/s10658-007-9152-6

Booker HM, Umaharan P (2008) Quantitative resistance to *Cercospora* leaf spot disease caused by *Pseudocercosporacruenta* in cowpea. Euphytica 162:167–177. https://doi.org/10.1007/s10681-007-9490-7

Boukar O, Fatokun CA, Roberts PA et al (2015) Cowpea. In: De Ron AM (ed) Grain Legumes. Springer, New York, pp 219–250. https://link.springer.com/content/pdf/10.1007%2F978-1-4939-2797-5_7.pdf

Boukar O, Fatokun CA, Huynh BL et al (2016) Genomic tools in cowpea breeding programs: status and perspectives. Front Plant Sci 7:757

Boukar O, Belko N, Chamarthi S et al (2018) Cowpea (*Vignaunguiculata*): genetics, genomics and breeding. Plant Breed 138:415–424. https://doi.org/10.1111/pbr.12589

Boukar O, Abberton M, Oyatomi O et al (2020) Introgression breeding in cowpea [*Vignaunguiculata* (L.) Walp.]. Front Plant Sci 11:567425. https://doi.org/10.3389/fpls.2020.567425

Carvalho AFU, de Sousa NM, Farias DF et al (2012) Nutritional ranking of 30 Brazilian genotypes of cowpeas including determination of antioxidant capacity and vitamins. J Food Compos Anal 26:81–88

Carvalho M, Castro I, Moutinho-Pereira J et al (2019) Evaluating stress responses in cowpea under drought stress. J Plant Physiol 241:153001. https://doi.org/10.1016/j.jplph.2019.153001

Cavalcante EG, de Medeiros JF, Sobrinho JE (2016) Development and water requirements of cowpea under climate change conditions in the Brazilian semi-arid region. Rev Bras Eng Agr Amb 20(9):783–788

Chaudhury D, Madanpotra S, Jaiwal R et al (2007) *Agrobacterium tumefaciens*-mediated high frequency genetic transformation of an Indian cowpea (*Vignaunguiculata* L. Walp) cultivar and transmission of transgenes into progeny. Plant Sci 172:692–700. https://doi.org/10.1016/j.plantsci.2006.11.009

Che P, Chang S, Simon MK et al (2021) Developing a rapid and highly efficient cowpea regeneration, transformation and genome editing system using embryonic axis explants. Plant J 106(3):817–830. https://doi.org/10.1111/tpj.15202

Chivenge P, Mabhaudhi T, Modi AT et al (2017) The potential role of neglected and underutilised crop species as future crops under water scarce conditions in sub-Saharan Africa. Int J Environ Res Public Health 12(6):5685–5711. https://doi.org/10.3390/ijerph120605685

Citadin CT, Cruz AR, Aragão FJ (2013) Development of transgenic imazapyr-tolerant cowpea (*Vignaunguiculata*). Plant Cell Rep 32(4):537–543. https://doi.org/10.1007/s00299-013-1385-6

Craufurd PQ, Summerfield RJ, Ell RH et al (2010) Photoperiod, temperature and the growth and development of cowpea (*Vigna unguiculata*). In: Singh BB et al (eds) Advances in cowpea research. Intl Inst Tropical Agric (IITA) and Japan Intl Res Center Agric Sci (JIRCAS), Sayce, Devon, UK

Cruz ARR, Aragão FJL (2014) RNAi-based enhanced resistance to cowpea severe mosaic virusand cowpea aphid-borne mosaic virus in transgenic cowpea. Plant Pathol 63(4):831–837. https://doi.org/10.1111/ppa.12178

Dabire-Binso CL, Ba NM, Sanon A et al (2010) Resistance mechanism to the pod-sucking bug *Clavigrallatomentosicollis* (Hemiptera: Coreidae) in the cowpea IT86D-716 variety. Int J Trop Insect Sci 30:192–199

Dakora FD, Belane AK (2019) Evaluation of protein and micronutrient levels in edible cowpea (*Vignaunguiculata* L. Walp.) leaves and seeds. Front Sustain Food Syst 3:70. https://doi.org/10.3389/fsufs.2019.00070

Dareus R, Porto ACM, Bogale M et al (2021) Resistance to *Meloidogyne enterolobii* and *Meloidogyne incognita* in cultivated and wild cowpea. Hortic Science 56(4):460–468

de Castro MJ-P, Baldin ELL, Cruz PL (2013) Characterization of cowpea genotype resistance to *Callosobruchus maculatus*. Pesq Agrop Brasileira 48(9):1201–1209

de Oliveira CRR, Freire Filho FR, Nogueira M, do S da R et al (2012) Reação de genótipos de feijão-caupirevelaresistênciaàscoinfecçõe spelo Cucumber mosaic virus, Cowpea aphid-borne mosaic virus e Cowpea severe mosaic virus. Bragantia 71:59–66. https://doi.org/10.1590/S0006-87052012005000007

Deshpande KS, Patil BR, Salimath PM et al (2010) Evaluation of native and collected germplasm for earliness seed traits and resistance to rust, CMV and leaf spot in cowpea [*Vignaunguiculata* (L.) Walp]. Electron J Plant Breed 1:384–392

Devi CB, Kushwaha A, Kumar A (2015) Sprouting characteristics and associated changes in nutritional composition of cowpea (*Vignaunguiculata*). J Food Sci Technol 52:6821–6827. https://link.springer.com/content/pdf/10.1007%2Fs13197-015-1832-1.pdf

Dhanasekar P, Reddy KS (2015a) Genotypic variation for raffinose family oligosaccharides in cowpea. In: Proceedings of the national conference on pulses: challenges and opportunities under changing climatic scenario, ISPRD, JNKVV, Jabalpur, 20 September–October 1, 2014, pp 316–320

Dhanasekar P, Reddy KS (2015b) A novel mutation in TFL1 homolog affecting determinacy in cowpea (*Vignaunguiculata*). Mol Gen Genomics 290(1):55–65. https://doi.org/10.1007/s00438-014-0899-0

Dhanasekar P, Reddy KS (2018) Development of Trombay cowpea variety 'TC-901' amenable for summer cultivation by induced mutagenesis. In: abstracts of FAO/IAEA international symposium on plant mutation breeding and biotechnology, Vienna, Austria, 27–31 August 2018. IAEA-CN-263-170

Dhanasekar P, Souframanien J (2023) Genetics and genomics of drought and heat tolerance in cowpea, mung bean and black gram. In: Samy MA et al (eds) Legumes: physiology and molecular biology of abiotic stress tolerance. Springer, Singapore, pp 203–233. https://doi.org/10.1007/978-981-19-5817-5_8

Dhanasekar P, Souframanien J, Suprasanna P (2021) Breeding cowpea for quality traits: a genetic biofortification perspective. In: Gupta DS et al (eds) Breeding for enhanced nutrition and bio-active compounds in food legumes. Springer, Cham, pp 157–180. https://doi.org/10.1007/978-3-030-59215-8_7

Dhanasekar P, Souframanien J, Dhole VJ et al (2023) Physical mutagenesis for induction of resistance against legume pod borer (*Marucavitrata* F.) in cowpea [*Vignaunguiculata* (L.) Walp]. In: Souvenir of international conference on pulses: smart crops for agricultural sustainability and nutritional security, New Delhi, India, 10–12 February 2023, S6.46, p 519

Dong L, Ravelombola W, Weng Y et al (2019) Seedling salt tolerance for above ground–related traits in cowpea (*Vignaunguiculata* (L.) Walp). Euphytica 215(3):22–53. https://doi.org/10.1007/s10681-019-2379-4

Dong J, Song Y, Wang B (2022) Identification of genomic regions associated with *Fusarium* Wilt resistance in cowpea. Appl Sci 12(14):6889. https://doi.org/10.3390/app12146889

Durojaye HA, Moukoumbi YD, Dania VO et al (2019) Evaluation of cowpea (*Vignaunguiculata* (L.) Walp.) landraces to bacterial blight caused by *Xanthomonas axonopodis*pv. *vignicola*. Crop Prot 116:77–81

Edet OU, Ishii T (2022) Cowpea speed breeding using regulated growth chamber conditions and seeds of oven-dried immature pods potentially accommodates eight generations per year. Plant Methods 18:106. https://doi.org/10.1186/s13007-022-00938-3

Ehlers J, Hall A (1998) Heat tolerance of contrasting cowpea lines in short and long days. Field Crop Res 55:11–21

Ehlers JD, Hall AE, Roberts P (2000) Registration of 'California Blackeye 27' cowpea. Crop Sci 40:854–855

Ehlers JD, Sanden BL, Frate C et al (2009) Registration of 'California Blackeye 50' cowpea. J Plant Regist 3:236–240. https://doi.org/10.3198/jpr2009.01.0039crc

Emechebe A, Singh BB, Leleji O (1988) Cowpea-striga problems and research in Nigeria. In: Proceedings of the International Workshop on Combating Striga in Africa, Ibadan, Nigeria, 22–24 August 1988, pp 18–28

Eswaramoorthy V, Kandasamy T, Thiyagarajan K (2023) Characterization of terminal flowering cowpea (*Vignaunguiculata* (L.) Walp.) mutants obtained by induced mutagenesis digs out the loss-of-function of phosphatidylethanolamine-binding protein. PLoS One 18(12):e0295509. https://doi.org/10.1371/journal.pone.0295509

FAOSTAT, (2022) *Food and Agriculture Organization Corporate Statistical Database*, https://www.fao.org/faostat/en/#data/QCL

Farooq M, Gogoi N, Barthakur S et al (2017) Drought stress in grain legumes during reproduction and grain filling. J Agron Crop Sci 203:81–102

Fatokun CA (2002) Breeding cowpea for resistance to insect pests: attempted crosses between cowpea and *Vigna*vexillata. In: Fatokun CA et al (eds) Challenges and opportunities for enhancing cowpea production. Proceedings of the world cowpea conference III, International Institute of Tropical Agriculture (IITA), Ibadan, Nigeria, 4–8 September 2000, pp 52–61

Fatokun CA, Dariush D, Menancio-Hautea DI et al (1993) A linkage map for cowpea [*Vigna unguiculata* (L.) Walp.] based on DNA markers. In: O'Brien SJ (ed) Genetic maps: locus maps of complex genomes, 6th edn. Cold Spring Harbour Laboratory Press, New York, pp 256–258

Fery RL, Dukes PD (2011) Southern blight (*Sclerotium rolfsii*Sacc.) of cowpea: genetic characterization of two sources of resistance. Int J Agron 2011:1–6. https://doi.org/10.1155/2011/6524042011

Ganesh SK, Packiaraj D, Geetha S et al (2021) VBN 3: a new high yielding multiple disease resistant cowpea variety. Electron J Plant Breed 12:1375–1379

Gatehouse AM, Fenton KA, Jepson I et al (1986) The effects of a-amylase inhibitors on insect storage pests: inhibition of a-amylase in vitro effects on development in vivo. J Sci Food Agric 37:727–734

Gayacharan P, Mondal SK, Yadav N et al (2023) Mining legume germplasm for genetic gains: an Indian perspective. Front Genet 14:996828. https://doi.org/10.3389/fgene.2023.996828

Gerrano AS, Adebola PO, Rensburg WSJ et al (2015) Genetic variability in cowpea (*Vigna unguiculata* (L.) Walp.) genotypes. S Afr J Plant Soil 32(3):165–174. https://doi.org/10.1080/02571862.2015.1014435

Goa Y, Mohammed H, Worku W et al (2022) Genotype by environment interaction and yield stability of cowpea (*Vigna unguiculata* (L.) Walp.) genotypes in moisture limited areas of Southern Ethiopia. Heliyon 8(3):e09013. https://doi.org/10.1016/j.heliyon.2022.e09013

Gonçalves A, Goufo P, Barros A et al (2016) Cowpea (*Vignaunguiculata* L. Walp), a renewed multipurpose crop for a more sustainable agri-food system: nutritional advantages and constraints. J Sci Food Agric 96(9):2941–2951. https://doi.org/10.1002/jsfa.7644

Goufo P, Moutinho-Pereira JM, Jorge TF et al (2017) Cowpea (*Vigna unguiculata* L. Walp.) metabolomics: Osmoprotection as a physiological strategy for drought stress resistance and improved yield. Front Plant Sci 8:586. https://doi.org/10.3389/fpls.2017.00586

Guillermo A, Pablo D, Camelo-García M et al (2023) Screening cowpea genotypes for resistance to cowpea aphid borne mosaic virus (CABMV) and cowpea severe mosaic virus (CPSMV) in Paraguay. Agric Sci Dig 43(5):593–597. https://doi.org/10.18805/ag.D-362

Hall AE (2004) Breeding for adaptation to drought and heat in cowpea. Eur J Agron 21(4):447–454. https://doi.org/10.1016/j.eja.2004.07.005

Hampton R, Thottappilly G, Rossel H (1997) Viral diseases of cowpea and their control by resistance-conferring genes. In: Singh BB et al (eds) Advances in cowpea research. IITA, Ibadan, pp 159–175

Higgins TJV, Gollasch S, Molvig L et al (2012) Insect-protected cowpeas using gene technology. In: Boukar O et al (eds) Innovative research along the cowpea value chain. Proceedings of the fifth world cowpea conference on improving livelihoods in the cowpea value chain through advancement in science. International Institute of Tropical Agriculture, Ibadan, pp 131–137

Horn LN, Shimelis H (2020) Production constraints and breeding approaches for cowpea improvement for drought prone agro-ecologies in Sub-Saharan Africa. Ann Agric Sci 65:83–91. https://doi.org/10.1016/j.aoas.2020.03.002

Horn L, Shimelis H, Laing M (2015) Participatory appraisal of production constraints, preferred traits and farming system of cowpea in the northern Namibia: implications for breeding. Legume Res J 38:691–700

Horn LN, Nghi-tuwamhata SN, Isabella U (2022) Cowpea production challenges and contribution to livelihood in Sub-Saharan region. Agric Sci 13:25–32. https://doi.org/10.4236/as.2022.131003

Huesing JE, Murdock LL, Shade RE (1990) Effects of wheat germ isolectins on development of cowpea weevil. Phytochemistry 30:785–788

Hussain A, Gondal MR, Hayat S et al (2019) Evaluation of cowpea (*Vignaunguiculata*) genotypes for fodder yield and resistance against leaf spot (*Pseudocercosporacruenta*). J Plant Breed Genet 7(1):01–07. https://doi.org/10.33687/pbg.007.01.2816

Huynh B-L, Close TJ, Roberts PA et al (2013) Gene pools and the genetic architecture of domesticated cowpea. Plant Genome 6:1–8. https://doi.org/10.3835/plantgenome2013.03.0005

Huynh BL, Matthews WC, Ehlers JD et al (2016) A major QTL corresponding to the Rk locus for resistance to root-knot nematodes in cowpea (*Vignaunguiculata* L. Walp.). Theor Appl Genet 129(1):87–95. https://doi.org/10.1007/s00122-015-2611-0

Ishii T, Juranic M, Maheshwari S et al (2020) Unequal contribution of two paralogous CENH3 variants in cowpea centromere function. Commun Biol 3(1):775

Janni M, Gullì M, Maestri E et al (2020) Molecular and genetic bases of heat stress responses in crop plants and breeding for increased resilience and productivity. J Exp Bot 71:3780–3802. https://doi.org/10.1093/jxb/eraa034

Jha UC, Bohra A, Parida SK et al (2017) Integrated "omics" approaches to sustain global productivity of major grain legumes under heat stress. Plant Breed 36:437–459. https://doi.org/10.1111/pbr.12489

Ji J, Zhang C, Sun Z et al (2019) Genome editing in cowpea *Vigna unguiculata* using CRISPR-Cas9. Int J Mol Sci 20(10):2471. https://doi.org/10.3390/ijms20102471

Jinbaani AN, Owusu EY, Mohammed A-R et al (2023) Gender trait preferences among smallholder cowpea farmers in northern Ghana: lessons from a case study. Front Sociol 8:1260407. https://doi.org/10.3389/fsoc.2023.1260407

Kabambe VH, Tembo YLB, Kazira E (2013) Awareness of the parasitic weed *Alectravogelii* (Benth.) amongst extension officers in three districts in Malawi. Am J Exp Agric 3:432–442

Kamara A, Omoigui L, Kamai N et al (2018) Improving cultivation of cowpea in West Africa. In: Achieving sustainable cultivation of grain legumes, vol 2. Burleigh Dodds Science Publishing Limited, pp 235–252

Kareem KT, Taiwo MA (2007) Interactions of viruses in cowpea: effects on growth and yield parameters. Virol J 4:15

Kpoviessi D, Agbahoungba S, Agoyi EE et al (2019) Resistance of cowpea to cowpea bruchid (*Callosobruchus maculatus* Fab.): knowledge level on the genetic advances. J Plant Breed Crop Sci 11(8):185–195

Kumar A, Kumar A, Satpathy S et al (2009) Legume pod borer (*Maruca testulalis* Geyer) and their relative yield losses in cowpea cultivars. Progress Hortic 41(2):233–235

Kumar S, Imtiaz M, Gupta S et al (2011) Distant hybridization and alien gene introgression. In: Pratap A, Kumar J (eds) Biology and breeding of food legumes. CABI, Oxfordshire, pp 81–110

Kumar S, Tanti B, Patil BL et al (2017) RNAi-derived transgenic resistance to Mungbean yellow mosaic India virus in cowpea. PLoS One 12(10):e0186786. https://doi.org/10.1371/journal.pone.0186786

Kumar J, Sen Gupta D, Djalovic I (2020) Breeding, genetics, and genomics for tolerance against terminal heat in lentil: current status and future directions. Legum Sci 2(3):e38. https://doi.org/10.1002/leg3.38

Kusi F, Obeng-Ofori D, Asante SK et al (2010) New sources of resistance in cowpea to the cowpea aphid (*Aphis craccivora* KOCH) (Homoptera: Aphididae). J Ghana Sci Assoc 12:95–104

Kusi F, Nboyine JA, Attamah P et al (2020) Stability of sources of resistance to cowpea aphid (*Aphis craccivora* Koch, Hemiptera: Aphididae) across major cowpea production zones in Ghana. Int J Agron 869334:1–8. https://doi.org/10.1155/2020/8869334

Kuzbakova M, Khassanova G, Oshergina I et al (2022) Height to first pod: a review of genetic and breeding approaches to improve combine harvesting in legume crops. Front Plant Sci 13:948099. https://doi.org/10.3389/fpls.2022.948099

Lamaoui M, Jemo M, Datla R et al (2018) Heat and drought stresses in crops and approaches for their mitigation. Front Chem 6:26. https://doi.org/10.3389/fchem.2018.00026

Lamini S, Kusi F, Cornelius EW et al (2022) Identification of sources of resistance in cowpea lines to *Macrophomina* root rot disease in Northern Ghana. Heliyon 8(12):e12217. https://doi.org/10.1016/J.HELIYON.2022.E12217

Lattanzio V, Linsalata V, Maruca G et al (1990) Pre-existing phenols and resistance to aphid in cowpea (*Vignaunguiculata* (L.) Walp.). In: Ng QN, Monti LM (eds) Cowpea genetic resources. International Institute of Tropical Agriculture, Ibadan, pp 128–136

Lattanzio V, Cardinali A, Linsalata V et al (1997) Flavonoid HPLC fingerprints of wild *Vigna* species. In: Singh BB et al (eds) Advances in cowpea research. International Institute of Tropical Agriculture (IITA) and Japan International Research Center for Agricultural Sciences (JIRCAS), IITA, Ibadan, pp 66–74

Leandre SP, Franci K, Richard A et al (2018) Screening for resistance to *Striga gesnerioides* and estimation of yield loss among cowpea (*Vignaunguiculata* (L.) Walp.) progenies in the Upper East region of Ghana. Afr J Agric Res 13(28):1430–1442

Lima LRL, Damasceno-Silva KJ, Noronha MA et al (2017) Diallel crosses for resistance to *Macrophominaphaseolina* and *Thanatephoruscucumeris* on cowpea. Genet Mol Res 16:gmr16039804

Lonardi S, Muñoz-Amatriaín M, Liang Q et al (2019) The genome of cowpea (*Vigna unguiculata* [L.] Walp.). Plant J 98(5):767–782. https://doi.org/10.1111/tpj.14349

Lucas MR, Diop NN, Wanamaker S et al (2011) Cowpea–soybean synteny clarified through an improved genetic map. Plant Genome 4:218–225. https://doi.org/10.3835/plantgenome2011.06.0019

Lucas MR, Ehlers JD, Huynh BL et al (2013) Markers for breeding heat-tolerant cowpea. Mol Breed 31:529–536. https://doi.org/10.1007/s11032-012-9810-z

Lüthi C, Alvarez-Alfageme F, Ehlers J et al (2013) Resistance of αAI-1 trans-genic chickpea (*Cicer arietinum*) and cowpea (*Vigna unguiculata*) dry grains to bruchid beetles (Coleoptera: Chrysomelidae). Bull Entomol Res 103:373–381

Maas EV, Poss JA (1989) Salt sensitivity of cowpea at various growth stages. Irrig Sci 100:313–320

MacFoy GA, Dabrowski ZT, Okech S (1983) Studies on the legume pod-borer, Marucatestulalis (Geyer)—VI. Cowpea resistance to oviposition and larval feeding. Int J Trop Insect Sci 4(1–2):147–152

Machuka J (2002) Potential role of transgenic approaches in the control of cowpea insect pests. In: Fatokun CA et al (eds) Challenges and opportunities for enhancing sustainable cowpea production. Intl Inst Trop Agric, Ibadan, pp 213–232

Mahantheshwara B, Nayak DK (2018) Screening and evaluation of cowpea varieties against root nematode, *Meloidogyne incognita*. Int J Pure App Biosci 6(1):136–140

Mai-Kodomi Y, Singh BB, Terao T et al (1999a) Inheritance of drought tolerance in cowpea. Indian J Genet Plant Breed 59:317–323

Mai-Kodomi Y, Singh BB, Myers O et al (1999b) Two mechanisms of drought tolerance in cowpea. Indian J Genet Plant Breed 59:309–316

Maina AN, Souina D, Mathias JH et al (2020) Generation means analysis of three seeds antinutrients in cowpea (*Vignaunguiculata* (L.) Walp.). Int J Environ Agric Biotech 5(5):1389–1396. https://doi.org/10.22161/ijeab.55.28

Mao JQ, Zaidi MA, Arnason JT et al (2006) In vitro regeneration of *Vignaunguiculata* (L.) Walp. Cv. Blackeye cowpea via shoot organogenesis. Plant Cell Tissue Organ Cult 87:121–125. https://doi.org/10.1007/s11240-006-9145-8

Maréchal R, Mascherpa JM, Stainier F (1978) Etude taxonomique d'un grouped'espèces des genres Phaseolus et *Vigna* (*V. unguiculata*) sur la base de donnéesmorphologiques et polliniques, traitées pour l'analyseinformatique. Boissiera 28:1–273

Marfo KO, Hall AE (1992) Inheritance of heat tolerance during pod set in cowpea. Crop Sci 32(4):912–918. https://doi.org/10.2135/cropsci1992.0011183X003200040015x

Menendez CM, Hall AE, Gepts P (1997) A genetic linkage map of cowpea (*Vigna unguiculata*) developed from a cross between two inbred, domesticated lines. Theor Appl Genet 95:1210–1217

Mishra SK, Sharma B, Tyagi MC et al (2008) Screening of cowpea germplasm for field tolerance against biotic and abiotic stresses. Indian J Genet Plant Breed 68(04):446–448

Mofokeng MA, Gerrano AS (2021) Efforts in breeding cowpea for aphid resistance: a review. Acta Agric Scand B-Soil Plant Sci 71:489–497. https://doi.org/10.1080/09064710.2021.1923797

Muchero W, Diop NN, Bhat PR et al (2009a) A consensus genetic map of cowpea [*Vigna unguiculata* (L.) Walp.] and synteny based on EST-derived SNPs. Proc Natl Acad Sci USA 106:18159–18164. https://doi.org/10.1073/pnas.0905886106

Muchero W, Ehlers JD, Close TJ et al (2009b) Mapping QTL for drought stress-induced premature senescence and maturity in cowpea [*Vignaunguiculata* (L.) Walp.]. Theor Appl Genet 118:849–863. https://doi.org/10.1007/s00122-008-0944-7

Muchero W, Ehlers JD, Close TJ (2011) Genic SNP markers and legume synteny reveal candidate genes underlying QTL for *Macrophominaphaseolina* resistance and maturity in cowpea [*Vigna unguiculata* (L) Walp.]. BMC Genomics 12:8

Muchero W, Roberts PA, Diop NN et al (2013) Genetic architecture of delayed senescence, biomass, and grain yield under drought stress in cowpea. PLoS One 8:e70041. https://doi.org/10.1371/journal.pone.0070041

Munoz-Amatriain M, Mirebahim H, Xu P (2017) Genome resources for climate-resilient cowpea, an essential crop for food security. Plant J 89(5):1042–1054. https://doi.org/10.1111/tpj.13404

Muranaka S, Shono M, Myoda T (2016) Genetic diversity of physical, nutritional and functional properties of cowpea grain and relationships among the traits. Plant Genet Res 14(1):67–76. https://doi.org/10.1017/S147926211500009X

Murdock LL (1992) Improving insect resistance in cowpea through biotechnonology: initiatives at Purdue University, USA. In: Thottappilly G et al (eds) Biotechnology: enhancing research on tropical crops in Africa. CTA/IITA co-publication, Ibadan, pp 313–320

Myers GO (1996) Hand crossing of cowpeas. IITA Research Guide, 42. Training Program, Third Edition. International Institute of Tropical Agriculture (IITA), Ibadan, Nigeria

Nantale G, Wasswa P, Edgar M, Richard T et al (2023a) Combining ability of cowpea (*Vignaunguiculata* (L) Walp) genotypes for resistance to cowpea bacterial blight in Uganda. Am J Agric Sci Eng Technol 7(1):21–28. https://doi.org/10.54536/ajaset.v7i1.1174

Nantale G, Wasswa P, Tusiime R et al (2023b) Response of cowpea germplasm to bacterial blight in Uganda. J Sci Agric 7:17–27. https://doi.org/10.25081/jsa.2023.v7.8061

Ndeve AD, Matthews WC, Santos JRP et al (2018) Broad-based root-knot nematode resistance identified in cowpea gene-pool two. J Nematol 50(4):545–558. https://doi.org/10.21307/jofnem-2018-046

Ndeve AD, Santos JR, Matthews WC et al (2019) A novel root-knot nematode resistance QTL on chromosome Vu01 in cowpea. G3 Genes Genom Genet 9(4):1199–1209. https://doi.org/10.1534/g3.118.200881

Ng NQ, Maréchal R (1985) Cowpea taxonomy, origin and germ plasm. In: Singh SR, Rachie KO (eds) Cowpea research, production and utilization. Wiley, Chichester, pp 11–21

Nkhata W, Shimelis H, Melis R et al (2010) Population structure and genetic diversity analyses of common beans germplasm collections of East and Southern Africa using morphological traits and high-density SNP markers. PLoS One 12:e0243238. https://doi.org/10.1371/journal.pone.0243238

Nkhoma N, Shimelis H, Laing MD et al (2020) Assessing the genetic diversity of cowpea [*Vigna unguiculata* (L.) Walp.] germplasm collections using phenotypic traits and SNP markers. BMC Genet 21(1):110–116. https://doi.org/10.1186/s12863-020-00914-7

Ntare BR (1991) Variation in reproductive efficiency and yield of cowpea under high temperature conditions in a Sahelian environment. Euphytica 59(1):27–32. https://doi.org/10.1007/BF00025358

OECD (2015) Consensus document on the biology of cowpea (*Vignaunguiculata* (L.) Walp.). Series on harmonisation of regulatory oversight in biotechnology No. 60

Ogunsola KE, Ilori C, Fatokun CA et al (2021) Disease incidence and severity in cowpea lines evaluated for resistance to single and multiple infections of endemic viruses in Nigeria. J Crop Improv 35:427–452. https://doi.org/10.1080/15427528.2020.1824952

Ogunsola KE, Fatokun CA, Boukar O et al (2023) Inheritance of resistance to three endemic viral diseases of cowpea in Nigeria. J Crop Improv 37:291–308. https://doi.org/10.1080/15427528.2022.2090476

Ohlson EW, Timko MP (2020) Race structure of cowpea witchweed (*Striga gesnerioides*) in West Africa and its implications for *Striga* resistance breeding of cowpea. Weed Sci 68:125–133

Ohlson EW, Timko MP (2022) Mapping and validation of *Alectravogelii* resistance in the cowpea landrace B301. Agronomy 12(11):2654. https://doi.org/10.3390/agronomy12112654

Olajide AA, Ilori CO (2018) Genetic analysis of seedlings characters associated with drought tolerance in cowpea under a controlled environment. Plant Genet Resour Char Util 16:218–227. https://doi.org/10.1017/S1479262117000235

Oldach KH (2011) Mutagenesis. In: Pratap A, Kumar J (eds) Biology and breeding of food legumes. CABI, Wallingford, pp 208–219

Oliveira JTA, Andrade NC, Martins-Miranda AS et al (2012) Differential expression of antioxidant enzymes and PR-proteins in compatible and incompatible interactions of cowpea (*Vignaunguiculata*) and the root-knot nematode *Meloidogyne incognita*. Plant Physiol Biochem 51:145–152

Olorunwa OJ, Adhikari B, Shi A et al (2022a) Screening of cowpea (*Vignaunguiculata* (L.) Walp.) genotypes for waterlogging tolerance using morpho-physiological traits at early growth stage. Plant Sci 315:111136. https://doi.org/10.1016/j.plantsci.2021.111136

Olorunwa OJ, Adhikari B, Brazel S et al (2022b) Growth and photosynthetic responses of cowpea genotypes under waterlogging at the reproductive stage. Plants (Basel) 11(17):2315. https://doi.org/10.3390/plants11172315

Olorunwa OJ, Adhikari B, Brazel S et al (2023) Waterlogging stress reduces cowpea (*Vignaunguiculata* L.) genotypes growth, seed yield, and quality at different growth stages: implications for developing tolerant cultivars under field conditions. Agric Water Manag 284:108336. https://doi.org/10.1016/j.agwat.2023.108336

Omoigui LO, Kamara AY, Alunyo GI et al (2017a) Identification of new sources of resistance to *Striga gesnerioides* in cowpea *Vignaunguiculata* accessions. Genet Resour Crop Evol 64:901–911. https://doi.org/10.1007/s10722-016-0410-3

Omoigui LO, Ekeuro GC, Kamara AY et al (2017b) New sources of aphids [*Aphis craccivora* (Koch)] resistance in cowpea germplasm using phenotypic and molecular marker approaches. Euphytica 213(8):178

Omoigui LO, Danmaigona CC, Kamara AY et al (2018) Genetic analysis of *Fusarium* wilt resistance in cowpea (*Vigna unguiculata* (L.) Walp.). Plant Breed 137:773–781. https://doi.org/10.1111/pbr.12628

Omomowo OI, Babalola OO (2021) Constraints and prospects of improving cowpea productivity to ensure food, nutritional security and environmental sustainability. Front Plant Sci 12:751731. https://doi.org/10.3389/fpls.2021.751731

Orawu M, Melis R, Laing M et al (2013) Genetic inheritance of resistance to cowpea aphid-borne mosaic virus in cowpea. Euphytica 189:191–201. https://doi.org/10.1007/s10681-012-0756-3

Ouédraogo JT, Gowda BS, Jean M et al (2002) An improved genetic linkage map for cowpea (*Vigna unguiculata* L.) combining AFLP, RFLP, RAPD, biochemical markers, and biological resistance traits. Genome 45:175–188. https://doi.org/10.1139/g01-102

Ouédraogo N, Zida EP, Oué HM et al (2021) Identification of resistant cowpea (*Vignaunguiculata* (L.) Walp.) genotypes against charcoal under artificial inoculation. Asian J Plant Sci 20(2):271–280. https://doi.org/10.3923/ajps.2021.271.280

Oyatomi O, Fatokun C, Boukar O et al (2016) Screening wild *Vigna* species and cowpea (*Vignaunguiculata*) landraces for sources of resistance to *Striga gesnerioides*. In: Maxted N et al (eds) Enhancing crop genepool use: capturing wild relatives and landrace diversity for crop improvement. CAB International, Wallingford, pp 27–31. https://doi.org/10.1079/9781780646138.0027

Padhi SR, Bartwal A, John R et al (2022) Evaluation and multivariate analysis of cowpea [*Vignaunguiculata* (L.) Walp] germplasm for selected nutrients—mining for nutri-dense accessions. Front Sustain Food Syst 6:888041. https://doi.org/10.3389/fsufs.2022.888041

Padulosi S (1993) Genetic diversity, taxonomy and ecogeographic survey of the wild relatives of cowpea (*Vignaunguiculata* L. Walpers), PhD thesis, Univ. Catho. de Louvain, Belgium

Padulosi S, Ng NQ (1997) Origin, taxonomy, and morphology of *Vignaunguiculata* (L.) Walp. In: Singh BB et al (eds) Advances in cowpea research. Japan International Research Center for Agricultural Sciences and IITA, Ibadan, pp 1–12

Panzeri D, Guidi Nissim W, Labra M et al (2022) Revisiting the domestication process of African *Vigna* species (Fabaceae): background, perspectives and challenges. Plan Theory 11:532. https://doi.org/10.3390/plants11040532

Parida A, Raina S, Narayan R (1990) Quantitative DNA variation between and within chromosome complements of *Vigna* species (Fabaceae). Genetica 82:125–133

Parry MAJ, Madgwick PJ, Bayon C et al (2009) Mutation discovery for crop improvement. J Exp Bot 60(10):2817–2825

Pasquet RS (1999) Genetic relationships among subspecies of *Vignaunguiculata* (L.) Walp. based on allozyme variation. Theor Appl Genet 98:1104–1119

Patel JB, Patel H, Sharma SC et al (2016) Genetics of seed related attributes in cowpea [*Vignaunguiculata* (L.) Walp.]. Legum Res 39(1):1–6. https://doi.org/10.18805/lr.v39i1.8855

Penza R, Lurquin PF, Filippone E (1991) Gene transfer by co-cultivation of mature embryos with *Agrobacterium tumefaciens*: application to cowpea (*Vignaunguiculata* L. Walp). J Plant Physiol 138:39–43. https://doi.org/10.1016/S0176-1617(11)80727-3

Petrova S (2021) Chickpea plant model for the climatic conditions of the Sadovo region according to the yield components. Bulg J Agric Sci 27:531–535

Phiri C, Kabambe V, Bokosi J (2018) Screening for resistance mechanisms in cowpea genotypes on *Alectravogelii*. Am J Plant Sci 9:1362–1379. https://doi.org/10.4236/ajps.2018.96099

Popelka JC, Gollasch S, Moore A et al (2006) Genetic transformation of cowpea (*Vignaunguiculata* L.) and stable transmission of the transgenes to progeny. Plant Cell Rep 25(4):304–312. https://doi.org/10.1007/s00299-005-0053-x

Pottorff M, Roberts PA, Close TJ et al (2014) Identification of candidate genes and molecular markers for heat-induced brown discoloration of seed coats in cowpea [*Vignaunguiculata* (L.) Walp]. BMC Genomics 15(1):328. https://doi.org/10.1186/1471-2164-15-328

Praxedes SSC, Ferreira Neto M, Loiola AT et al (2022) Photosynthetic responses, growth, production and tolerance of traditional varieties of cowpea under Ssalt stress. Plan Theory 11(14):1863. https://doi.org/10.3390/plants11141863

Priyanka GSL, Meena AK et al (2018) Field reaction of cowpea genotypes against *Rhizoctonia solani* Kuhn inciting web blight disease. Bull Env Pharmacol Life Sci 8:133–136

Pyott DE, Sheehan E, Molnar A (2016) Engineering of CRISPR/Cas9-mediated potyvirus resistance in transgene free *Arabidopsis* plants. Mol Plant Pathol 17:1276–1288

Rabé MM, Baoua IB, Baributsa D (2023) Enhancing cowpea productivity in the Sahel: exploring seed access among smallholder farmers in South-Central Niger. Horticulturae 9(12):1287. https://doi.org/10.3390/horticulturae9121287

Raina A, Laskar RA, Wani MR et al (2023) Role of mutation and molecular breeding in the improvement of cowpea, a primary pulse crop. In: Bhat TA, Hakeem KR (eds) Biotechnologies and genetics in plant mutation breeding, vol 3. Taylor & Francis, pp 53–84. https://doi.org/10.1201/9781003305101

Raizada A, Punniyamoorthy D, Jegadeesan S et al (2023) Breeding cowpea: a nutraceutical option for future global food and nutritional security. In: Kole C (ed) Compendium of crop genome designing for nutraceuticals. Springer, Singapore. https://doi.org/10.1007/978-981-19-3627-2_26-1

Ralton JE, Howlett BJ, Clarke AE et al (1988) Interaction of cowpea with *Phytophthoravignae*: inheritance of resistance and production of phenylalanine ammonia-lyase as a resistance response. Physiol Mol Plant Pathol 32:89–103. https://doi.org/10.1016/S0885-5765(88)80008-0

Ramteke R, Singh D, Murlidharan P (2012) Selecting soybean (*Glycine max*) genotypes for insertion height of the lowest pod, the useful trait for combine harvester. Ind J Agric Sci 82:511–515

Raveendar S, Premkuma A, Sasikumar S et al (2009) Development of a rapid, highly efficient system of organogenesis in cowpea *Vignaunguiculata* (L.) Walp. S Afr J Bot 75:17–21. https://doi.org/10.1016/j.sajb.2008.05.009

Ravelombola WS, Shi A, Weng Y et al (2017) Evaluation of salt tolerance at germination stage in cowpea [*Vignaunguiculata* (L.) Walp]. Hortic Sci 52(9):1168–1176. https://doi.org/10.21273/hortsci12195-17

Ravelombola W, Shi A, Weng Y et al (2018) Association analysis of salt tolerance in cowpea (*Vignaunguiculata* (L.) Walp) at germination and seedling stages. Theor Appl Genet 131(1):79–91. https://doi.org/10.1007/s00122-017-2987-0

Ravelombola W, Shi A, Huynh BL (2021a) Loci discovery, network-guided approach, and genomic prediction for drought tolerance index in a multi-parent advanced generation intercross (MAGIC) cowpea population. Hortic Res 8. https://doi.org/10.1038/s41438-021-00462-w

Ravelombola W, Dong L, Barickman TC et al (2021b) Evaluation of salt tolerance in cowpea at seedling stage. Euphytica 217:116. https://doi.org/10.1007/s10681-021-02832-w

Ravelombola W, Shi A, Huynh BL et al (2022) Genetic architecture of salt tolerance in a Multi-Parent Advanced Generation Inter-Cross (MAGIC) cowpea population. BMC Genomics 23:100. https://doi.org/10.1186/s12864-022-08332-y

Reddy BR, Nagendran K, Pandey M et al (2023) Field screening for powdery mildew resistance (*Erysiphe polygo*ni DC) in cowpea. Int J Plant Soil Sci 35(19):963–967. https://doi.org/10.9734/ijpss/2023/v35i193631

Ribeiro HLC, Boiteux LS, Santos CAF (2014) Genetic parameters of earliness and plant architecture traits suitable for mechanical harvesting of cowpea ('*Vignaunguiculata*'). Aust J Crop Sci 8(8):1232–1238

Ribeiro DG, Mota APZ, Santos IR et al (2022) NBS-LRR-WRKY genes and protease inhibitors (PIs) seem essential for cowpea resistance to root-knot nematode. J Proteome 261:104575. https://doi.org/10.1016/j.jprot.2022.104575

Roberts PA, Matthews WC, Ehlers JD (1996) New resistance to virulent root-knot nematodes linked to the Rklcus of cowpea. Crop Sci 36(4):889–894. https://doi.org/10.2135/cropsci1996.0011183X0036000400012x

Rymal KS, Chambliss OL (1981) Influence of cultivar and maturity on pod wall strength in the southernpea [Cowpeas]. HortScience 16(2):186–187

Saha S, Romi IJ, Khatun F et al (2022) Assessing high-yielding cowpea varieties for bacterial blight resistance in Bangladesh: a step towards an environment-friendly and sustainable solution. Saudi J Biol Sci 29(8):103365. https://doi.org/10.1016/j.sjbs.2022.103365

Salifou M, Souleymane O, Hamidou M et al (2019) Screening of cowpea germplasm for resistance to *Striga gesnerioides* in Niger. Afr Crop Sci J 27(4):641. https://doi.org/10.4314/acsj.v27i4.7

Shelp BJ, Aghdam MS, Flaherty EJ (2021) γ-Aminobutyrate (GABA) regulated plant defense: mechanisms and opportunities. Plants (Basel) 10(9):1939. https://doi.org/10.3390/plants10091939

Shi A, Buckley B, Mou B et al (2016) Association analysis of cowpea bacterial blight resistance in USDA cowpea germplasm. Euphytica 208(1):143–155

Shridevi BM, Naik VR (2017) Dissection of genetic variability for traits related to mechanical harvesting and yield components in chickpea (*Cicer arietinum* L.). Res Environ Life Sci 10:111–114

Sibhatu B, Belete K, Tessema T (2015) Effect of cowpea density and nitrogen fertilizer on a sorghum-cowpea intercropping system in Kobo, Northern Ethiopia. Int J Agri For 5(6):305–317. https://doi.org/10.5923/j.ijaf.20150506.02

Siddiqui MH, Al-Khaishany MY, Al-Qutami MA (2015) Morphological and physiological characterization of different genotypes of faba bean under heat stress. Saudi J Biol Sci 22(5):656–663. https://doi.org/10.1016/j.sjbs.2015.06.002

Silva RG, Vasconcelos IM, Martins TF (2016) Drought increases cowpea (*Vigna unguiculata* [L.] Walp.) susceptibility to cowpea severe mosaic virus (CPSMV) at early stage of infection. Plant Physiol Biochem 109:91–102. https://doi.org/10.1016/j.plaphy.2016.09.010

Silva JA, Costa AF da, Benko-Iseppon AM (2021a) Inheritance of cowpea resistance to Cowpea aphid-borne mosaic virus. Pesq Agrop Brasileira 56:e02109. https://doi.org/10.1590/S1678-3921.pab2021.v56.02109

Silva RCZ, Lôbo RA, Nicoli A et al (2021b) Inheritance of genetic resistance to *Fusarium* wilt in cowpea. Trop Plant Pathol 46:481–486. https://doi.org/10.1007/s40858-021-00423-7

Singh BB (2005) Cowpea (*Vignaunguiculata* (L.) Walp.). In: Singh RJ (ed) Genetic resources, chromosome engineering and crop improvement. CRC Press, Boca Raton, pp 117–162

Singh BB (2016) Genetic enhancement for yield and nutritional quality in cowpea [*Vigna unguiculata* (L.) Walp.]Indian. J Genet 76(4):568–582

Singh SR, Allen DJ (1979) Cowpea pest and diseases. Manual series No. 2. International Institute of Tropical Agriculture, Ibadan

Singh BB, Singh SR (1990) Breeding for bruchid resistance in cowpea. In: Fujii K et al (eds) Bruchids and legumes: economics, ecology and coevolution. Kluwer Academic Publishers. International Institute ofTropical Agriculture (IITA), Ibadan, pp 219–228

Singh BB, Emechebe AM, Atokple IDK (1993) Inheritance of *Alectra* resistance in cowpea genotype B301. Crop Sci 33:70–72

Singh B, Chambliss O, Sharma B (1997) Recent advances in cowpea breeding. In: Singh BB et al (eds) Advances in cowpea research. IITA, Ibadan, pp 30–49

Sita K, Sehgal A, Hanumantha Rao B et al (2017) Food legumes and rising temperatures: effects, adaptive functional mechanisms specific to reproductive growth stage and strategies to improve heat tolerance. Front Plant Sci 8:1658. https://doi.org/10.3389/fpls.2017.01658

Sofi PA, Ara A, Gull M et al (2019) Canopy temperature depression as an effective physiological trait for drought screening. In: Drought-detection and solutions. IntechOpen

Solleti SK, Bakshi S, Purkayastha J et al (2008) Transgenic cowpea (*Vignaunguiculata*) seeds expressing a bean alpha-amylase inhibitor 1 confer resistance to storage pests, bruchid beetles. Plant Cell Rep 27(12):1841–1850. https://doi.org/10.1007/s00299-008-0606-x

Somta P, Ammaranan C, Ooi PAC et al (2007) Inheritance of seed resistance to bruchids in cultivated mungbean (*Vigna* radiata (L.) Wilczek), vol 155. Kluwer Academic Publishers, Netherlands, pp 47–55

Souframanien J, Dhanasekar P (2023) Potential of mutation breeding in genetic improvement of pulse crops. In: Penna S, Jain SM (eds) Mutation breeding for sustainable food production and climate resilience. Springer, Singapore, pp 445–485. https://doi.org/10.1007/978-981-16-9720-3_15

Souframanien J, Gupta SK, Gopalakrishna T (2010) Identification of quantitative trait loci for bruchid (*Callosobruchus maculatus*) resistance in black gram [*Vigna* mungo (L.) Hepper]. Euphytica 176:349–356

Souleymane A, Aken'Ova ME, Fatokun CA et al (2013) Screening for resistance to cowpea aphid (*Aphis craccivora* Koch) in wild and cultivated cowpea (*Vignaunguiculata* L. Walp.) accessions. Int J Sci Environ Tech 2:611–621

Soundararajan RP, Chitra N, Geetha S (2013) Host plant resistance to insect pests of grain legumes–a review. Agric Rev 34(3):176–187

Srinivasan R (2008) Susceptibility of legume pod borer (LPB), *Marucavitrata* to δ-endotoxins of *Bacillus thuringiensis* (Bt) in Taiwan. J Invertebr Pathol 97:79–81

Srivastava R, Kobayashi Y, Koyama H et al (2023) Cowpea NAC1/NAC2 transcription factors improve growth and tolerance to drought and heat in transgenic cowpea through combined activation of photosynthetic and antioxidant mechanisms. J Integr Plant Biol 65(1):25–44. https://doi.org/10.1111/jipb.13365

Steele WM, Mehra KL (2009) Structure, evolution and adaptation to farming systems and environments in *Vigna*. In: Summerfield RJ, Bunting AH (eds) Advances in legume science. Royal Botanic Gardens, Kew, pp 393–404

Stella S, Montoya G (2016) The genome editing revolution: a CRISPR-Cas TALE off-target story. BioEssays 38:S4–S13. https://doi.org/10.1002/bies.201670903

Thio GI, Elisabeth PZ, Fidele BN (2017) Differential reaction of cowpea genotypes to brown blotch disease (*Colletotrichum capsici*) in Burkina Faso. Afr J Agric Res 12(36):2773–2782. https://doi.org/10.5897/AJAR2017.12554

Thio IG, Jean BT, Inoussa D et al (2021) Inheritance and detection of QTL in cowpea resistance to brown blotch disease. J Plant Breed Crop Sci 13(3):123–135. https://doi.org/10.5897/JPBCS2021.0959

Thottappilly G, Rossel HW (1992) Virus diseases of cowpea in tropical Africa. Int J Pest Manag 38:337–348

Thottappilly G, Rossel HW, Reddy DVR et al (1990) Vectors of virus and mycoplasma diseases: an overview. In: Singh SR (ed) Insect pests of tropical food legumes. Wiley, Chichester, pp 323–342

Timko MP, Singh B (2008) Cowpea, a multifunctional legume. In: Moore PH, Ming R (eds) Genomics of tropical crop plants. Plant genetics and genomics: crops and models, vol 1. Springer, New York. https://doi.org/10.1007/978-0-387-71219-2_10

Timko MP, Ehlers JD, Roberts PA (2007) Cowpea. In: Kole C (ed) Pulses, sugar and tuber crops. Genome mapping and molecular breeding in plants, vol 3. Springer, pp 49–67. https://doi.org/10.1007/978-3-540-34516-9_3

Togola A, Bouka O, Belko N et al (2017) Host plant resistance to insect pests of cowpea (*Vignaunguiculata* L. Walp): achievements and future prospects. Euphytica 213:239

Togola A, Boukar O, Servent A et al (2020) Identification of sources of resistance in cowpea mini core accessions to *Aphis craccivora* Koch (Homoptera: Aphididae) and their biochemical characterization. Euphytica 216(6):88. https://doi.org/10.1007/s10681-020-02619-5

Tripathi K, Gore PG, Singh M et al (2020a) Legume genetic resources: status and opportunities for sustainability. In: Mirza H (ed) Legume crops - prospects, production and uses. IntechOpen. https://doi.org/10.5772/intechopen.91777

Tripathi K, Prasad TV, Bhardwaj R et al (2020b) Evaluation of diverse germplasm of cowpea [*Vignaunguiculata* (L.) Walp.] against bruchid [*Callosobruchus maculatus* (Fab.)] and correlation with physical and biochemical parameters of seed. Plant Genet Resour 18(3):120–129. https://doi.org/10.1017/s1479262120000180

Tripathy SK (2016) Bruchid resistance in food legumes-an overview. Res J Biotechnol 11(7):97–105

Ubi BE, Mignouna H, Thottappilly G (2000) Construction of a genetic linkage map and QTL analysis using a recombinant inbred population derived from an inter-subspecific cross of a cowpea [*Vigna unguiculata* (L.) Walp.]. Breed Sci 50:161–173. https://doi.org/10.1270/jsbbs.50.161

Usua EJ, Singh SR (1979) Behaviour of the cowpea pod borer *Marucatestulalis* Geyer. Niger J Entomol 3:231–239. https://doi.org/10.36108/NJE/9791/30.0211

Varela ALN, Oliveira JTA, Komatsu S et al (2019) A resistant cowpea (*Vigna unguiculata* [L.] Walp.) genotype became susceptible to cowpea severe mosaic virus (CPSMV) after exposure to salt stress. J Proteome 194:200–217. https://doi.org/10.1016/j.jprot.2018.11.015

Vasconcelos EV, Fonsêca AFA, Pedrosa-Harand A et al (2015) Intra- and interchromosomal rearrangements between cowpea [*Vigna unguiculata* (L.) Walp.] and common bean (*Phaseolus vulgaris* L.) revealed by BAC-FISH. Chromosom Res 23:253–266. https://doi.org/10.1007/s10577-014-9464-2

Watson A, Ghosh S, Williams MJ et al (2018) Speed breeding is a powerful tool to accelerate crop research and breeding. Nat Plants 4(1):23–29. https://doi.org/10.1038/s41477-017-0083-8

Wei B, Zhang J, Pang C et al (2015) The molecular mechanism of sporocyteless/nozzle in controlling *Arabidopsis* ovule development. Cell Res 25:121. https://doi.org/10.1038/cr.2014.145

Wu X, Wu X, Xu P et al (2015) Association mapping for *Fusarium* wilt resistance in Chinese asparagus bean germplasm. Plant Genome 8(2):eplantgenome2014.11.0082. https://doi.org/10.3835/plantgenome2014.11.0082

Xu Y, Luo H, Zhang H (2023) Feeding the world using speed breeding technology. Trends Plant Sci 28:372–373

Arid Legumes Breeding

D. Kumar, Manish Sharma, H. R. Mahala, Kuldeep Kumar, Sudhir Kumar Jha, and Meenal Rathore

Abstract

This chapter focuses on a group of four crops classified as arid legumes, viz. guar/cluster bean [*Cyamopsis tetragonoloba* L. (Taub)], moth bean [*Vigna aconitifolia* L. (Jacq) Marechal], cowpea [*Vigna unguiculata* (L.) Walpers], and horsegram [*Macrotyloma uniflorum* L.], have been shown to mitigate droughts, maintain soil health, diversify agriculture, and produce a range of high-quality organic feed and industrial by-products like guar gum, guar meal, and guar *churi* having export value. These legumes are thereby known as natural climate-resilient organic resource crops. These crops suffer from some inherent bottlenecks like longer maturity, slow growth, poor source-sink relationship, rudimentary plant types, and sometimes being susceptible to diseases and insect pests under uncongenial situations with some quality inhibitory traits. Efforts have been made to alter their plant types, shorten growth periods, and improve grain yield returns. For instance, targeted breeding efforts have successfully reduced the maturity period of several arid legumes: from 120 to 85 days in guar, 95 to 58 days in moth bean, 130 to 60 days in cowpea, and 125 to 85 days in horse gram. Additionally, the plant architecture of cowpea and moth bean has been improved to include erect, semi-erect, and non-viny types, which are physiologically more efficient and better suited for modern cultivation practices. Floral biology and structure of arid legumes are not convenient for emasculation and pollination, therefore restricting hybridization procedures. Selection and mutation breeding techniques have made a big dent in these crops. For instance, all notified cultivars of horse gram and moth bean except CAZRI Moth-2 have been developed using these breeding techniques. A large number of economic genes appear to be accumulated in their genomes, which easily respond to selection and mutation techniques. Deliberate and targeted breeding efforts have resulted in the development of 39 guar, 32 cowpea, 20 moth bean, and 21 horsegram cultivars up to the year 2022, all aimed at improving yield potential and agronomic performance. Significant research has been conducted on these crops involving molecular marker development, genetic diversity analysis, and quantitative trait loci (QTL) mapping. For instance, minor QTLs for gum content (qGcI.1) mapped on a single linkage group 102.41 were identified. These crops are embedded with sizeable opportunities: adaptation to low inputs, source of organic produces, soil enrichment virtues, export-worthy significance, etc. These legumes suffer from specific challenges, which need to be addressed. For instance, there is no cultivar of guar maturing in less than 80 days. The increasing use of chemical substitutes for guar gum in export driven markets, coupled with the presence of anti-nutritional factors in guar meal that limit its use as animal feed, poses significant challenges to the sustainability and competitiveness of the guar industry. The area of moth bean is gradually decreasing, and cowpea is cultivated as a companion crop mainly and not as a sole crop. There is a need to breed guar for 70–75 days growth period, developing cultivars of guar with gum content <30% and representing food- and industrial-grade gum viscosity. Enhancing the economic viability of moth bean cultivation requires not only improving yields and market access but also exploring alternative uses such as value-added food products, nutraceuticals, and climate-resilient cropping systems.

Keywords

Guar gum · Guar gum viscosity · Climate-resilient crops · Poor source-sink relationship · Nutrition

D. Kumar (✉) · H. R. Mahala
Central Arid Zone Research Institute, Jodhpur, Rajasthan, India

M. Sharma
Sardarkrushinagar Dantiwada Agricultural University, Dantiwada, Gujarat, India

K. Kumar · S. K. Jha · M. Rathore
Indian Institute of Pulses Research, Kanpur, Uttar Pradesh, India

G. Prasad Dixit et al. (eds.), *Fundamentals of Legume Breeding*, https://doi.org/10.1007/978-981-96-7838-9_11

1 Introduction

Arid legumes, viz. guar/cluster bean moth bean, cowpea, and horsegram, have pivotal and unparalleled roles in harsher growing environments. These legumes look naturally adapted and tamed in such adverse situations. Due to their need-based adaptation and adjustment to harsh weather and soil conditions, so-called arid legumes have been identified as a low-cost source of income for farmers who are surviving with limited resources. Of the many legumes that are grown in India in a variety of agroclimatic, soil, and input conditions to satisfy the diverse needs of people and animals, arid legumes are the most immediate and critical farming resource.

These crops are cultivated across a wide geographic range—from the arid western states like Gujarat to the humid eastern states such as Odisha and West Bengal, and from the northern regions of Jammu and Kashmir to the southern states of Karnataka and Kerala. Their widespread distribution reflects their adaptability to diverse agro-climatic conditions. In fragile arid and semi-arid ecosystems, these crops play an essential role, providing multiple products and by-products for daily use. Grown mostly under subsistence conditions with minimal use of chemical fertilizers or plant protection measures, their produce is often considered organic by default. In livestock-based farming systems, particularly in arid and semi-arid regions, these crops contribute significantly to soil regeneration and nutrient enrichment, while also enhancing overall system resilience and diversification.

These hardy legumes are key components of diverse farming systems. They are commonly cultivated as sole crops, mixed crops, or intercrops and are also integrated into agroforestry, agri-pastoral, and silvopastoral systems, serving a wide range of purposes. These crops may also be grown on undulating lands, sand dunes, rocky sub-soils, sub-plains, and sub-mountain zones (like horsegram) in almost the whole of country in arid to sub-humid zones (like cowpea). Moth bean is primarily cultivated for its use in various traditional and industrial food products, including confectionery items such as *bhujia*, *papad*, *dalmoth*, and as a soaked and boiled morning snack. Similarly, guar is valued for its high endosperm galactomannan gum content, which makes it a key raw material for numerous industrial applications. With substantial export potential for both its primary and secondary products, guar is widely grown in arid and semi-arid regions. These stress-loving legumes, known for minimum agronomic inputs and restricted aftercare, are loaded with many socioeconomic virtues and climate-resilient attributes. This group of four legumes is universally known to tilt toward survival values, not for vertical production returns. These crops face several inherent challenges, including long maturity periods, slow growth rates, poor source-sink relationships, and rudimentary plant architecture. They often exhibit low yield potential and can be susceptible to diseases and insect pests, especially under unfavorable conditions. Additionally, they require specific habitat conditions, possess certain quality-inhibiting traits, and frequently show limited responsiveness to agricultural inputs and varying agro-environments. As a result, the actual yields achieved in the field are often significantly lower than their theoretical potential. Thus, these crops appear to suit more to fodder behavior than grain yield potential behavior.

Deliberate breeding efforts of conventional nature have resulted in developing new genotypes released through the Central Variety Release Committee and certain ones through the State Variety Release Committee, considerably enhancing the yield potential of these crops. For example, in the case of guar, newly developed cultivars exhibit a yield potential of 1000–1400 kg ha^{-1}, nearly double that of older cultivars released between 1970 and 1998, which yielded only 500–700 kg ha^{-1}. In the same view, new cultivars of cowpea could yield 800–1000 kg ha^{-1} over old cultivation (1981–1996), which could yield only 300–500 kg ha^{-1}. New promising early maturing moth bean varieties have a yield potential of 600–1000 kg ha^{-1} over old traditional varieties (200–400 kg^{-1}). The situation also improved in horsegram, wherein new varieties (1997–2015) had a yield potential to the tune of 900–1200 kg ha^{-1} in comparison to old, late, slow-growing cultivars that used to yield hardly 500–600 kg^{-1}. Thus, yield potential increased almost two times with the same management and input application.

Thus, these crops responded much toward traditional breeding tools. The genomes of these crops appear to contain conserved genes related to yield, and significant genetic variability can be rapidly uncovered and utilized through modern breeding techniques. The germplasm selection from natural stress–hit regions proved beneficial in expressing positive yield alleles, which had been accumulated for decades under multi-stress situations. Similarly, mutation breeding has proven to be an effective tool for inducing beneficial genes related to growth regulation and yield components. For example, reducing the growth period of moth bean by 25–30 days through mutation breeding did not result in any yield penalty, likely due to enhancements in photosynthetic efficiency and improved source-sink relationships.

2 Origin, Evolution, and Distribution of Arid Legumes

Two of the arid legumes, cowpea and moth bean, belong to the family Fabaceae, which includes more than 200 wild and cultivated species, including the genus *Vigna* (Pratap et al. 2014). Tropical and subtropical areas of Africa, Asia, Australia, and the Pacific region are home to the cultivation of *Vigna* species

(Verma et al. 2022). For cowpeas, Ethiopia was first suggested by Vavilov (1926) as the primary center of its domestication, with China and India serving as secondary sites. Later, the wild ancestors of cowpeas were only identified in Africa, ruling out India and China as potential domestication sites (Herniter et al. 2020). Faris (1965) suggested that domestication of cowpeas may have taken place in Central or West Africa. Subsequent research led to the conclusion that either Northeastern or West Africa was the location where cowpea domestication occurred (Xiong et al. 2018).

As isoenzymatic and ethnobotanical investigations failed to identify diversity in West African wild cowpea forms, the notion supporting cowpea domestication in that site was eventually disproved (Feleke et al. 2006). However, the theory of cowpea domestication in the Northeastern African region was based on two main points (Baudoin and Marechal 1985). First, there is no true ecotype of the wild cowpea in the West African region. Second, there is a great deal of variability in the wild cowpea forms in the region, which range from Ethiopia to Southern Africa. Additionally, studies on ethnobotany and linguistic groups have provided support for the Northeastern African theory of domestication. According to these findings, cowpea is related to the Chadic and Nilo-Saharan language-speaking populations, two linguistic groups that are prevalent in Northeastern Africa. This notion was further supported by molecular marker-based genetic characterization investigations of a global collection of cowpea landraces and wild genotypes (Coulibaly et al. 2002).

The main variable characteristics that established the taxonomy of cowpeas were the morphology of the seeds and pods, the length of the oral stalk, the shape of the leaves, the anthocyanin coloration of the internodes, and photosensitivity. In cultivated forms, the traits of seeds and pods are highly varied and are frequently employed to characterize and identify cultivars and cultigroups. According to Baudoin and Marechal (1989), the cowpea is classified as a herbaceous legume that is a member of the dicotyledon group, the order Fabales, family Fabaceae, subfamily Faboideae, tribe Phaseoleae, subtribe Phaseolinae, genus *Vigna*, species *unguiculata*, and subspecies *unguiculata*. In *V. unguiculata* ssp. *unguiculata*, four cultigroups were identified: unguiculata, bifurcata (or cylindrica), sesquipedalis, and textilis (Ng and Marechal 1985). Cultivated cowpea was long disputed as to its taxonomic status; however, years of study have led to the classification of cultivated cowpea as *V. unguiculata* ssp. *unguiculata* (Table 1). With a variable number of subspecies, it is considered a complex of species. There are presently 11 known wild subspecies of *V. unguiculata* included in the species complex: *aduensis*, *alba*, *baoulensis*, *burundiensis*, *letouzeyi*, *pawekiae*, *kindtiana*, *protracta*, *pubescens*, *stenophylla*, and *tenuis*. It is believed that among these, *V. unguiculata* ssp. *dekindtiana* var. *spontanea* is the direct progenitor of cultivated cowpea (Padulosi and Ng 1997).

Table 1 Taxonomy of cowpea

Vigna unguiculata (L.) Walpers
ssp. *unguiculata* cv. gr. *unguiculata* cv. gr. *biflora* cv. gr. *sesquipedalis* cv. gr. *textiles*
ssp. *dekindtiana* (Harms) Verdcourt var. *dekindtiana* (Harms) Verdcourt
var. *mensensis* (Schweinf.) Marechal, Mascherpa & Stainer
var. *pubescens* (Wilczec) Marechal, Mascherpa & Stainer
var. *protracta* (E. Mey.) Verdcourt
ssp. *tenuis* (E. Mey.) Marechal, Mascherpa & Stainer
ssp. *stenophylla* (Harvey) Marechal, Mascherpa & Stainer

It is widely distributed on the edges of cultivated lands throughout Africa (Feleke et al. 2006). According to gene pool studies conducted on *V. unguiculata* using traditional breeding techniques, the species' primary gene pool is made up of wild allo-autogamous forms as well as cultivated and autogamous spontaneous forms, while the secondary gene pool only includes allogamous forms. Kouadio et al. (2007) supplied additional evidence in favor of this conclusion regarding the *V. unguiculata* gene pool.

The spread of the domesticated cowpea has been reconstructed with the use of genetic, textual, and archaeological resources by Herniter et al. (2020). Before 2500 BCE, the cowpea was likely domesticated in both West and East Africa. By 400 BCE, it had become well-established throughout the modern Old World's major production regions, including sub-Saharan Africa, the Mediterranean Basin, India, and Southeast Asia. The Columbian Exchange transported African germplasm to the Caribbean and the southeastern United States and South American and Mediterranean germplasm to Cuba, the southwestern United States, and Northwest Mexico.

Moth bean may be the most primitive of all crops in the genus *Vigna* (Yundaeng et al. 2019). It is a diploid crop with $2n = 2x = 22$. Arid and semi-arid areas of India, Pakistan, Afghanistan, Nepal, Sri Lanka, Myanmar, and several African nations are home to moth beans. However, due to their high tolerance to drought, they may also be grown in other water-limited areas of Africa, particularly given the anticipated future climate (Baath et al. 2018). Moth bean is a member of genus *Vigna*, subgenus *Ceratotropis*, and section Aconitifoliae, Tomooka and Maxted (Bisht et al. 2005). In addition to the moth bean, the following five *Vigna* species are also categorized under section Aconitifoliae: *V. khandalensis* (Santapau) Raghvan et Wadhwa; *V. trilobata* (L.) Verdc.; and *V. stipulacea* (Lam.) Kuntze. These species are *V. aridicola* N. Tomooka & Maxted; *V. indica* T.M. Dixit, K.V. Bhat, and S.R. Yadav. The moth bean is thought to be native to India; studies by Tomooka et al. (2009) and Takahashi et al. (2016) have documented the distribution of wild moth bean forms in Southeastern India, lending credence to this theory.

There was uncertainty about the moth bean domestication and its wild progenitor until recently. Actually, *Vigna trilobata* (L.) Verdc. was once thought to be its wild form. Eventually, though, morphological and molecular analyses made it possible for scientists to discriminate between domesticated and wild moth beans with precision. Consequently, it was discovered that *V. trilobata* belonged to a completely different species (Takahashi et al. 2016). A molecular analysis of the moth bean accessions gathered from Tamil Nadu, India, showed that a small number of the accessions were essentially similar to cultivated moth beans but had some differences, meaning they could be thought of as the moth bean's wild progenitors. For instance, the leaflet shapes of domesticated and wild accessions are similar, but the wild accessions' seeds had a semi-transparent seed coat, while the seeds of domesticated accessions have a larger seed size, a water-permeable seed coat, and non-shattering pods (Takahashi et al. 2016). Furthermore, the fact that wild moth bean accessions may be found in southern India suggests that this plant was likely domesticated there before spreading to other regions of India and other countries.

The most economically significant plant in the *Cyamopsis* genus, the guar, belongs to the Fabaceae family and has a diploid genome ($2n = 2x = 14$). A working theory regarding the trans-domestication of guar was put forth by Hymowitz (1972). The idea of trans-domestication describes how a wild species is brought by humans from its native habitat to a new location where it is then domesticated. The cultivated guar's ancestor may have been the drought-tolerant African wild species *C. senegalensis* Guill. et Perr., which was brought to the Indo-Pakistan subcontinent by Arab-Indian traders between the ninth and thirteenth centuries A.D. and was distributed in the semi-arid savannah zone of the Sahara from Senegal to Saudi Arabia. The guar was domesticated approximately a millennium ago in the arid regions of the northwest of the subcontinent of India and Pakistan. According to Gopala Krishnan et al. (2011), the wild ancestors of the crop, *Cyamopsis serrata* Schinz and *Cyamopsis senegalensis* Guill. & Perr., most likely originated in Africa. The crop was first domesticated in the northwest of the Indo-Pakistan region. The theory that *C. senegalensis* is the wild ancestor of guars has, however, come under scrutiny because no wild form of the species is found in the subcontinent, and no reports of weedy forms have been made from the northwest region of India to date. Furthermore, little is known about the process by which the wild species was domesticated (Ravelombola et al. 2021).

Since ancient times, rural and tribal populations in Asian and African nations have grown horsegram, a member of the Fabaceae family, for food, medicine, and fodder. This priceless ancient legume has strong adaptation to risky traditional farming systems in marginal environments of semi-arid and arid regions, in addition to providing diversity and resilience to agro-ecosystems (Aditya et al. 2019). The three economic species in this *Macrotyloma* genus where horsegram is currently found are: *M. uniflorum*, *M. axillare* (Meyer) Verdcourt, a fodder crop, and *M. geocarpum* (Harms) Mare'chal and Baudet, the Bambara, or African groundbean or bambara groundnut (Isely 1983). Based on a limited evaluation of herbarium specimens from the Botanical Survey of India, Pune, and the distribution of archaeobotanical evidence, Fuller and Harvey (2006) deduced a likely South Indian origin and possibly a distinct northwest Indian origin. The regions of South Deccan, Gujarat, and Haryana are all likely sites of early domestication or cultivation. According to a recent study by Fuller and Murphy (2018), horsegram was domesticated in ancient India. The evidence of surviving wild populations today points to the distribution of the wild progenitors in the semi-arid savannah or scrub woodland zones of western and peninsular India, as well as in some areas of the lower slopes of the western Himalayas. These zones include the margins of tropical dry deciduous woodlands. This suggests that there were two main geographically separated wild distribution areas: one in the western Himalayas and another in the savannah corridor of western and peninsular.

3 Floral-Biology, Emasculation, and Pollination

3.1 Cowpea

It is a self-pollinating species, which is encouraged by the arrangements of the floral parts. The inflorescence of cowpea is an unbranched auxiliary raceme bearing several flowers at the terminal end of peduncles. The flower is bisexual, white, purple, or pale violet. With five petals and five sepals, it is a complete flower. The standard petal ranges in color from white to violet, while the keel petal is straight. The calyx is made up of five gamosepalous sepals. The arrangement of the sepals is (2 + 3) given that the two are large and the three are small. The corolla is papilionaceous with an erect standard spreading at the time of flower opening. The androecium is enclosed by the wings, which are attached to the boat-shaped keel. Diadelphous stamens (9 + 1) are present. The ovary has numerous ovules and is unilocular and monocarpellary.

The floral biology and pollination ecology of three cowpea varieties were studied by Ige et al. (2011). The result showed that cowpea flowers open between 6:00 and 6:30 am and close between 11:30 am and 12:00 pm. Similarities in the morphology of the pollen between the cowpea varieties were revealed by pollen analysis. Recently a study was carried out by Araméndiz-Tatis et al. (2022) in order to evaluate the percentage of viable crosses in cowpea as well as stigmatic receptivity. In the morning, the greatest level of stigmatic receptivity was observed. Between 7:00 and 9:00 am,

Fig. 1 Hybridization steps in cowpea

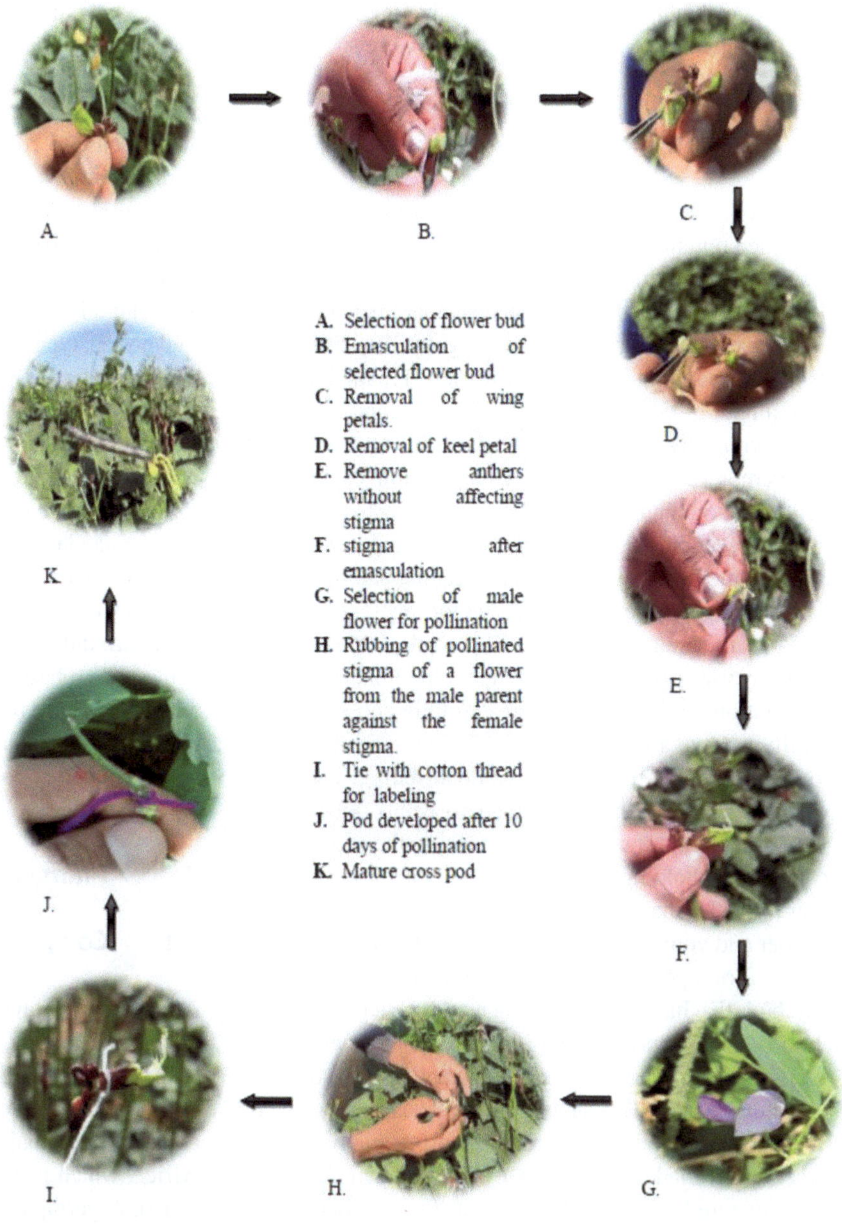

controlled pollinations yielded a higher percentage of viable crosses. Because of the increased stigmatic receptivity, artificial hybridizations should be carried out in the morning when the environment is more conducive to pollen grain germination. The process of artificial hybridization is detailed in Fig. 1.

3.2 Moth Bean (*Vigna aconitifolia*)

Cleistogamy enforces rigorous self-pollination in this crop. Moth bean inflorescence is axillary. It has standard light-yellow flowers, bright yellow wings, twisted keel, diadelphous stamen (9 + 1), basal ovary, and flat stigma flat. Moth bean species have limited recombination breeding support due to their small flower size and sensitivity to manipulation. Handling moth bean flowers can be laborious due to their small size and delicate peduncles, which can be 2–5 cm or 10–15 cm long. Mahla and Sharma (2022) attempted direct pollination without emasculation. Forceps were inserted between normal petals, moved down through wing petals, and tilted to penetrate the keel, forcing the stigma out of the keel. Sessile flowers detach from the peduncle with minimal force since they are delicately connected. The feathery stigma was easily pollinated because it was exposed from the bud by the upward forceps swapping. Dehisced anthers were gently rubbed against the feathery stigma in order to facilitate pollination. About 2–5% of hybrid seeds were successfully produced and identified using contrasting traits. About 8% of attempts in the hybrid pod scenario were successful.

3.3 Guar (*Cyamopsis tetragonoloba*)

In this crop, ten stamens are divided into two groups, with one stamen remaining free (diadelphous), and the other nine filaments are fused together to form a tube around the style. A keel that is encircled by wing and standard petals encloses all of the stamens, stigma, and style tightly. The entire complex is a closed, minuscule system designed to facilitate easy self-pollination. For artificial crossing, a recently emerging raceme with fully developed lower buds is preferred. Mahla and Sharma (2022) attempted emasculation of immature green buds from buds that were just beginning to turn yellow or light green. Simultaneous emasculation and pollination were also attempted after sunrise. Buds that had been emasculated during the evening and early morning (5–6 am) were pollinated after sunrise (7–9 am). This work yielded less than 1% of hybrid pods. They attempted pollination without emasculation, in which buds were carefully chosen for ideal maturity to have a receptive stigma but not yet shed pollen, in order to increase the frequency of pod setting. These buds opened easily by pushing the standard petals outward with forceps, tilting out the wings, and pushing the stigma out of the keel while holding the buds between the thumb and forefinger of the left hand.

Older and younger buds were extracted from the inflorescence to form open flowers. When pollination occurred right away, buds that had anthesis during maneuvering also yielded hybrid seeds. For pollination, it is sufficient to gently press forceps against the petals of relatively mature buds, exposing the stigma head. After successful anthesis has been confirmed, dehisced anthers are rubbed on the stigma head by extracting the diadelphous stamens and gynoecium complex from a mature bud of the donor plant (Fig. 1). With further refinement, it is possible to achieve a 10–30% pod setting from cross-pollinated buds. After 3 years of application, this method yielded about 28% success in pod setting and 60% hybrid seeds, as indicated by the emergence of a dominant male trait in F1 and segregation in F2.

3.4 Horsegram (*Macrotyloma uniflorum*)

The flowers are zygomorphic, papilionaceous, bisexual, and complete. The calyx was gamosepalous, with five sepals that joined to form an involucre at the base. The corolla displayed five petals in a vexillary estivation, making it polypetalous. Ten stamens, all of which were in diadelphous condition, made up the androecium. The gynoecium was monocarpellary and basic. The standard petal, filament length, carpel length, and number of ovules per flower were all highly variable. Recombination breeding is necessary to increase a crop's genetic potential over time. In order to enable the plant breeder to create appropriate crossing techniques, it makes sense to collect data on floral attributes such as flowering phenology, pollen biology, and stigma receptiveness of the crop in order to address the challenges of hybridization. Isiaka et al. (2023) studied horsegram accessions from six different countries to study flowering phenology, pollen biology, and stigma receptiveness. There were two main floral and pollen types present in those accessions. One with a white standard, a white petal, and five similarly shaped and sized ovules in a carpel. In the polar view, the pollen is triangular or circular, and in the equatorial view elliptical and tricolporate in nature. And another with green-yellow petals, a carpel containing six to seven ovules of varying sizes, and a yellow standard with a purple dot. Triporate pores and an echinulate structure with tuberous spines make up the pollen. Pollens were generally very viable from 7:00 am to 3:00 pm, but at around 7:00 pm, they became less viable. At the bud dehiscence phase and during anthesis, stigma was found receptive. These information on the reproductive biology of horsegram can be utilized to inform future hybridization initiatives aimed at optimizing crop productivity.

4 Plant Genetic Resources

4.1 Cowpea

At the International Institute of Tropical Agriculture (IITA), in Ibadan, Nigeria, the Genetic Resources Center (GRC) is home to the largest collection of cowpea germplasm. Although the IITA collection includes germplasm from 90 countries, the majority of the collection comes from West Africa, which is the crop's center of diversity (Boukar et al. 2015). A germplasm set's core collection is usually created to include 5–10% of the total collection or a significant portion of the diversity found in a number of accessions that are easier to phenotype. Molecular markers and agromorphological traits are frequently combined to create this. Using agronomic, botanical, and geographic descriptors, IITA has created a "core collection" of 2062 cultivated cowpea accessions. A "mini-core" collection of cowpeas with 374 accessions was created as an additional refinement of this core collection concept, supported by the CGIAR's Generation Challenge Program (GCP) (Boukar book). Huynh et al. (2013) used 422 cowpea landraces that were gathered from 56 different countries.

In this study, over 1200 single nucleotide polymorphism markers were used to genotype a global collection of cowpea landraces and African ancestral wild cowpea. Two main gene pools were found in African cowpeas that were cultivated, according to Bayesian inference. While the bulk of gene pool 2's landraces are found in eastern Africa, most of gene pool 1's landraces are found in western Africa. The closest relationship between each gene pool and the wild cowpea occurs within the same geographic area, suggesting that different domestication processes resulted in the formation of the two gene pools. Recently, Muñoz-Amatriaín et al. (2021) studied molecular diversity in a minicore set of cultivated cowpea accessions. Drawing from the UCR cowpea collection, a minicore known as the "University of California, Riverside (UCR) Minicore" has been assembled, comprising 368 global accessions of cultivated cowpea. High-density genotyping with 51,128 SNPs, principal component, and genetic assignment studies identified six subpopulations in the UCR Minicore. These subpopulations were primarily distinguished by cultivar group and place of origin. The presence of all six subpopulations to varying degrees in West African material suggests that the region is a center of diversity for cowpea cultivation. Identification of trait-specific germplasm and their registration as genetic stock will enhance the utilization of available germplasm for crop improvement.

In order to assess 1288 lines for drought tolerance in Ibadan, Fatokun et al. (2012) randomly selected them from cowpea germplasm collections kept at the International Institute of Tropical Agriculture. Five weeks after sowing, irrigation was stopped, causing a drought. Drought shortened the days to flowering by an average of 12 days and decreased the mean grain yield per plant by 67.28%. Even though some of the cowpea lines produced no pods at the end of the study, some of them continued to be green for up to 6 weeks after irrigation was stopped. Six lines that might serve as prospective parents for creating breeding lines with increased drought tolerance were found through additional assessment of the 142 drought-tolerant lines that were chosen for the screenhouse. Recently, Nkhoma et al. (2020) used a 10 × 10 alpha lattice design with two replications to assess 100 cowpea genotypes for agronomic traits in two chosen sites in Zambia. A total of 14,116 SNP markers were used to profile 94 test genotypes. The finest parental lines for breeding were determined to be the most genetically diverse and high-yielding genotypes, specifically Bubebe, CP411, CP421, CP645, Chimponogo, and MS1-8-1-4. In addition to identifying prospective cowpea genetic resources for systematic conservation and effective breeding, this study established baseline data. Effective screening led to isolation of insect pests–resistant lines at different sources (Table 2).

Table 2 Sources of resistance to different insect pests in cowpea

Insects pests	References
Pod border	
TVu946, TVu4557 (VITA-5) TVu946	Singh (1977) Jackai (1982)
Pod sucking bug	
TVu 1977 (VITA-4), TVu7274 TVu3342–01 Cowpea curculio Fla. 453–01	Singh (1977) Ehlers (1994) Todd and Canerdy (1968)
Cowpea storage weevil	
TVu 2027 RS 9 JC 5, 25/8/2/2, RS-9, CG-28, Sel 1476 EC 1697 16–17, EC 169720–22	Singh (1977) Dabi et al. (1979) Vir 1983 NBPGR 1987
Root knot nematode	
Bruchus sp.	
T2	Chandola et al. (1969)

4.2 Moth Bean

The largest collection of moth bean germplasm is held by ICAR-NBPGR, New Delhi, mostly from Indian states like Rajasthan, Gujarat, Punjab, and Maharashtra, as well as the union territory of Jammu and Kashmir (Bisht et al. 2005). A small number of moth bean germplasm studies have recently been the subject of genetic diversity studies, some of which have used agro-morphological traits as well as molecular markers. Vir and Singh (2015a, b) evaluated 44 accessions in order to assess genetic variability, identify an appropriate accession for summer cultivation, compare genotype performance, and determine inter-character associations, from accessions that were collected in the hot, dry climate of western Rajasthan, Gujarat, and M.P. Under field conditions, the accession IC 39786 demonstrated complete genetic resistance to the crinkle virus disease. In terms of seed yield, accessions IC-39675, IC-36607, IC-251908, IC 36245, and IC-36563 outperformed best check during the kharif season, but accessions IC-36245, IC-36555, IC 36667, IC 36577, and IC 36604 showed a yield advantage over best check during the summer.

Molecular markers are important tools to identify and characterize the plant genetic resources of any crop. In moth bean too, inter simple sequence repeat (ISSR) markers were used by Bhadkaria et al. (2020) to evaluate the genetic diversity among 25 accessions. Geographical origin and molecular data were shown to be logically similar. The accessions grown on India's western coast were more closely related to each other than those grown in the country's arid northwest. They recommended to use accessions such as Maru, Jadiya, and Jawala as parents in future breeding programs to obtain more variation complexes in moth bean genetic stocks.

4.3 Guar

Numerous studies have evaluated guar germplasm for yield and its components, biotic and abiotic stresses as well as biochemical parameters. Morris (2010) assessed the morphological diversity of 73 guar accessions that came from 1298 accessions that were provided by the Plant Genetic Resources Conservation Unit of the Agricultural Research Service of the United States Department of Agriculture. Guar accessions were categorized into three groups (clusters) using cluster analysis according to their low, intermediate, or high seed numbers. Indian guar genotypes that can withstand heat were found by Vir and Singh (2015a, b) through field assessments. For heat stress, a total of 44 guar genotypes have been examined. Among the genotypes that were evaluated, a significant variation in agronomic traits under warm environments was found. In order to gain a better understanding of the response mechanisms to heat stress, An RNA-Seq study on guar leaves was carried out by Alshameri et al. (2020) to determine the genes involved in heat resistance and to quantify gene abundance. Two hundred and seventy-two of the 1551 up- and 1466 down-regulated genes identified in the results may be new guar-specific genes with unknown functions.

In order to characterize the salt tolerance of 25 different guar genotypes, Suthar et al. (2018) assessed the growth and seed yield of the genotypes under field conditions while irrigating them with waters of varying salinity. The growth and yield of the guar crop were significantly impacted by salinity levels, according to the results. Only four of the 25 guar genotypes (BR-99, S-5608, 2/1, S-5615, and S-8885) were deemed tolerant based on the geometric mean productivity (GMP) and salinity tolerance index (STI). Six elite guar genotypes were assessed by Purkayastha et al. (2006) for resistance to four isolates of *M. phaseolina*, which causes the disease known as charcoal rot. Both field and greenhouse conditions were used for the evaluation. The findings indicated that charcoal rot had an impact on the guar plant during every stage of growth. In the susceptible genotype, plant death was observed, and in certain genotypes, the disease was not as severe. These results imply that the germplasm of guars contains sources of innate disease resistance. Manivannan et al. (2017) estimated the differences in the contents of 42 Indian-origin guar genotypes with regard to total protein, fat, fiber, fat, carbohydrate, and gum. Reddy et al. (2019) examined 51 guar genotypes that were gathered from various parts of India, looking at protein and gum content in addition to a number of agronomic features. Considerable genetic influences were found in the traits that were assessed. The average content of gum, fiber, and protein was 4.93%, 56.99%, and 17.88%, respectively. Fiber content had the highest broad-sense heritability (94.40%) of all the evaluated traits. They concluded that it would be difficult to select both protein and guar gum at the same time, though.

Evaluation of cluster bean genetic resource led to identification of sources showing tolerance to Powdery Mildew (GAUG 63), Bacterial Leaf Blight (GCI, RGC 1027, GAUG 9406, GAG 75, G8, G102, and G225), Alternaria Blight (GC1, GAUG 9406, GAUG 9003, and GAUG 9005), and root rot such as GAUG 9406, GG1, and HGS 844 (Kumar 2005). Among these, GAUG 9406 was shown to be resistant to or tolerant of a number of diseases. Using RAPD markers, Punia et al. (2009) and Pathak et al. (2010) examined the genetic diversity of cluster bean varieties and cultivars.

4.4 Horsegram

As a focal institution for the collecting, protection, and documentation of horsegram germplasm, the National Bureau of Plant Genetic Resources (NBPGR), located in New Delhi, India, has the largest horsegram accessions preserved in its gene bank. Several studies on germplasm resources, their characterization, and evaluation are available. In various Indian research institutes, 1161 accessions were methodically characterized and assessed between 1999 and 2004 (Chahota et al. 2013). Recently 12 quantitative traits on 252 genotypes in horsegram to conduct genetic relatedness found a wide range of variability for yield components. Based on estimates of the D^2 statistic, 25 clusters of genotypes were identified. There was greater genetic diversity among accessions, as indicated by the magnitude variation of the intra- and inter-cluster distances. A broad range of genetic diversity was found in the horsegram accessions used for the study, and numerous distinct clusters containing promising accessions were formed that can be used to improve particular yield components.

Gautam and Chahota (2022) evaluated the metabolic diversity of 96 different horsegram lines that were obtained from 700 germplasm accessions across Pan India. In the chosen panel, a variable range of protein content was observed, ranging from 13% to 40%. With a protein content of 40%, the related wild species *Macrotyloma sar-gharwalensis* had the highest content. The lines IC-280031 and IC-139356 exhibited the greatest quantity of metabolites (44). Among the chosen panel of horsegram germplasm, lines IC-280031 and IC-139356 were determined to be the most nutritive due to their highest concentration of quantifiable metabolites. The nutritional profile of 139 horsegram germplasm lines kept at the National Gene Bank of India and sourced from 16 different Indian states was evaluated by Kumari et al. in 2023. Protein content (ranging from 21.8 to 26.7 g/100 g), total soluble sugars (ranging from 0.86 to 12.1 g/100 g), phenolics (ranging from 3.38 to 11.3 mg), gallic acid equivalents (GAEs)/g), and phytic acid content (ranging from 1.07 to 21.2 mg/g) were among the essential nutrients that showed significant variability in the study.

A large number of germplasm lines (>1500) collected from NBPGR, New Delhi, have been screened and evaluated at Akola, Trissur, Coimbatore, Piyaur, and Bangalore. Out of these lines, some promising varieties at the national or state level have been developed directly through selection or thereafter. Some of them are Bihar (BR-5, BR-10, and Madhu), Himachal Pradesh (HPK-2, HPK-4, HPK-5, and HPK-6), Andhra Pradesh (PDM-1) and VZM-1), Jharkhand (K-82 and Birsa Kulth), Orissa (S-27, S-39, and S-1264), Rajasthan (Maru Kulthi, KS-2, AK-1, and AK-42), Uttaranchal (VL Ghat-1), Karnataka (Hebbal Hurali-2, PHG-9, and KBH-1), Tamil Nadu (Co-1, 35–5-122, and 35–5-123), and still more (Kumar and Rodge 2012). The varietal improvement has led to curtail life cycle and with more productivity potential of the newly bred varieties. For instance, promising cultivars maturing in 85–90 days, 90–100 days, and 105–112 days have been developed. The curtailment of the growth period from 125–135 days (VLG-1) to 80–85 (AK-42) while maintaining the same level of grain yield (850–110 kg ha^{-1}) has been notable progress in the genetic improvement of horsegram. Dehusking of seeds resulting in an improvement of grain protein efficiency ratio, digestibility, and cooking time has been considerably reduced from almost 2 h to 30–45 min (Kumar 2007).

5 Genetic Studies and Quantitative Inheritance

The success of any crop improvement program depends on the amount of genetic variety that is retained in the germplasm. According to Kumar et al. (2017), cluster beans have little genetic diversity based on DNA markers. Furthermore, the small flower size and extremely poor seed setting in manually hybridized buds make it difficult to harness the abundant diversity through hybridization (Mahla et al. 2010). In order to create diversity and isolate desired mutants for a particular purpose, mutation breeding may be the best option in these circumstances (Mahla et al. 2022). Numerous studies have shown that cluster beans are mutable (Bhosale and Kothekar 2010; Mahla et al. 2010; Velu et al. 2012; Patil and Rane 2015), but systematic research is still lacking. The ability to produce desired alterations with the fewest negative effects is essential for the successful application of induced mutation. Therefore, basic information must be created in order to properly apply mutagenesis in cluster bean enhancement.

Gamma radiation was therefore used in a study to produce beneficial genetic variation in cluster beans. A systematic detailed study through graded doses (100 Gy intervals) of gamma rays (100–1000 Gy) at ICAR-Central Arid Zone Research Institute, Jodhpur, revealed peculiar behavior of three distinct cluster bean genotypes to gamma radiation. No significant differences were observed in germination, but after 15 days of sowing, yellowing of leaf tips started and severe mortality was observed beyond 700 Gy. Only a few plants survived in 1000 Gy treatment that too with stunted growth. A gradual reduction in seedling parameters (root length, shoot length, root fresh weight, shoot fresh weight, root dry weight, and shoot dry weight) was recorded in the M_1 generation; however, the differential behavior of varieties is conspicuous. Besides seedling parameters, a clear-cut reduction in the number of pods, pod length, seeds per pod, and seed yield per plant was recorded beyond 500 Gy. A few chlorophyll-deficient and unifoliate leaves as well as morphological mutants also observed in the M_2 generation. There has never been a record of 800 Gy and greater levels causing severe harm without compromising germination. This might be a result of the endosperm's galactomannan, a gummy polysaccharide that offers some protection and nourishment during seed germination. Any dose between 100 and 700 Gy may be taken into consideration for boosting the diversity for yield and related attributes in cluster beans when taking into account mutation rate, efficiency, and effectiveness (Mahla et al. 2018).

5.1 Genetic Study

While there is unanimity on numerous characteristics, there are differing findings about branching inheritance. According to Chaudhary and Lodhi (1981), branching outnumbered nonbranching. In a cross between two nonbranching parents, Chaudhary and Lodhi (1981) reported an F2 ratio of 9:7 for branching to nonbranching, further complicating matters and suggesting the involvement of complementary gene effects. Bhagwan et al. (1983) discovered that environmental variables, not genetic ones, influenced the gum proportion in the seed. For gum content, Saini et al. (1982) reported a genotype × environmental interaction.

Systematic attempts to improve crops after realizing their economic significance led to the introduction of roughly 40 cultivars in India (Bhatt et al. 2016). Surprisingly, all improved cluster bean varieties cultivated in India for seed are violet-flowered and pubescent (hairy leaves and stems), with the exception of a handful (RGC-936), which has white flowers. Breeding programs that aim to increase yield must comprehend the inheritance pattern of both quantitative (such as yield) and qualitative (monogenic) features.

Simple Mendelian monogenic inheritance is seen in cluster beans for both flower color and pubescence. The co-inheritance of monogenic features, such as flower color and pubescence/hairiness, with polygenic quantitative traits, however, is poorly understood. In order to obtain comprehensive data on the genetics and co-inheritance of flower color and leaf hairiness, Sharma et al. (2021) used an F2 population to perform a linkage analysis of both of these qualitative traits with yield-associated parameters.

Flower color was found to be significantly correlated with pods per cluster, cluster on the main stem, pod length, and seed per pod. Similarly, there was a correlation between hairiness and days to flowering, plant height, and leaf area. The findings indicated that the released varieties' selection has been favored by the mild effects of hairiness and violet flower on yield and yield-attributing characteristics. The foundation of recombination breeding is a robust hybridization program that enables new combinations of desired traits to increase yield. In order to find genes, biochemical pathways, and precision breeding, systematic segregating populations are also crucial for genetic and molecular investigations of characteristics. Like most leguminous crops, particularly cluster beans, the small blossom size and sensitivity to manipulation have hindered a robust recombination breeding.

Successful use in programs aimed at improving the genetics of legumes requires a high percentage of pod settings during artificial hybridization. An essential genetic strategy for taking advantage of the largely nonadditive gene action is heterosis breeding. Utilizing hybrid vigor in self-pollinated crops is dependent on a technique that generates F1 seeds profitably. Similar to other crops, cowpea hybrid vigor is dependent on the particular parents employed in the hybrid combination. Several studies on cowpea have shown considerable heterosis for many characters. For instance, heterosis was demonstrated for leaves per plant, dry and fresh stem weight and dry and fresh fodder yield (Singh et al. 1986), days to flowers, vine length, stem girth, leaf length, leaf breadth, green fodder, yield dry matter yield (Lodhi et al. 1990), branches per plant and plant height, number of cluster per plant (Viswanatha et al. 1998), number of pods (Viswanatha et al. 1998), number of seeds per pod (Viswanatha et al. 1998), seed weight (Viswanatha et al. 1998), grain yield (Bhor et al. 1997; Viswanatha et al. 1998), protein content, and in vitro dry matter digestibility of fodder (Lodhi et al. 1990).

5.2 Genomics Studies in Arid Legumes

5.2.1 Development of Molecular Markers in Arid Legumes

Development of molecular markers in arid legumes: By utilizing molecular markers, breeders can speed up the process of developing improved varieties of arid legumes that are better adapted to challenging environmental conditions, thereby contributing to sustainable agriculture in dry regions. In essence, molecular markers play a pivotal role in accelerating the breeding process, enhancing the precision of selection, and understanding the genetic basis of traits in arid legumes, ultimately contributing to the development of improved varieties adapted to challenging environmental conditions. Several studies have been conducted to develop molecular markers in arid legumes, a few of which are listed in Table 3.

5.3 Genetic Diversity Studies in Arid Legumes

A diverse gene pool in crops helps protect against crop failures due to pests, diseases, or adverse weather conditions. Further, preserving genetic variation within species is crucial for maintaining healthy populations and ensuring their long-term survival, especially for endangered species or those at

Table 3 Some selected studies conducted to develop molecular markers in four arid legumes

S. no	Crop	Type of marker	Development strategy	References
1.	*Cyamopsis tetragonoloba*	18,792 SSRs, 5999 SNPs, and 249 InDels	Transcriptome assembly	Thakur and Randhawa (2018)
		1859 genomic SSRs	Genome assembly	Tribhuvan et al. (2019)
		4907 common SNPs and 327 InDels	RADseq genotyping	Grigoreva et al. (2021)
		27,066 SSRs	Transcriptome assembly	Fahad et al. (2019)
2.	*Vigna aconitifolia*	12,096 SSRs	Transcriptome assembly	Suranjika et al. (2022)
		1287 SSRs	Transcriptome assembly	Tiwari et al. (2018)
3.	*Vigna unguiculata*	1071 SSRs	Unigene dataset	Gupta and Gopalakrishna (2010)
		1536 SNPs	GoldenGate genotyping assay	Huynh et al. (2013)
		5828 SNPs	Genotyping by sequencing	Xiong et al. (2016)
		4600 SNPs	Dart	Gbedevi et al. (2021)
4.	*Macrotyloma uniflorum*	SSR and intron length polymorphism	EST	Sharma et al. (2015)
		SSR	Sequencing	Chahota et al. (2017)

risk of extinction. Assessing and understanding genetic diversity involves various techniques such as DNA sequencing, genotyping, population genetics analysis, and phylogenetic studies. These tools help researchers and scientists quantify, measure, and comprehend the patterns of genetic variation within and between populations.

Several genetic diversity studies have been conducted on arid legumes. Sultan et al. (2013) used 12 RAPD markers, which clustered the 30 guar genotypes into 2 groups. In their study of genetic diversity and cross-transferability, Trihuvan et al. (2019) used 82 genomic SSRs to divide 54 *Cyamopsis* accessions into four primary clusters, two of which had two subclusters, while the wild accessions constituted a distinct cluster. Boghara et al. (2016) found that SSR amplification with 17 primers revealed five distinct clusters at a 0.68 cut-off value. The cluster analysis showed low genetic variation among the genotypes studied. Similarly, in moth bean, only a handful of studies have been conducted to access its genetic diversity and mixed results have been obtained, where Yadav et al. (2023) concluded that Indian genotypes have higher diversity while other groups revealed the presence of substantial diversity (Chaudhari et al. 2021; Sahoo et al. 2022).

Among the arid legumes, cowpea is the most explored in terms of diversity analysis. Nkhoma et al. (2020) used 14116 SNP markers to study 90 genotypes and infer that the accessions were classified into four groups, irrespective of the source of collection. Similar results were obtained by Gomes et al. (2020) on African cultivars using SSR markers, where they concluded central Mozambique as hotspot of genetic variation. Gbedevi et al. (2021) studied the genetic diversity and population structure analysis of 255 accessions using Dart comprising 4600 SNPs and inferred the presence of low diversity based on the heterozygosity parameters and concluded that most of the lines are inbred based on the average inbreeding coefficient. In the context of Indian cultivars, mixed results have been obtained in terms of variability by several workers (Devi and Jayamani 2020; Kumar et al. 2023).

In order to evaluate the genetic diversity and cross-transferability of the molecular markers created in the related species, numerous investigations have been carried out in horsegram. In one study, Chahota et al. (2017) used 30 different SSR primers amplifying 170 polymorphic alleles and 24 different morphological traits on 360 horsegram accessions to find genetic diversity and population structure. They observed polymorphism information content (PIC) in the range of 0.15 to 0.76 with an average of 0.50 and classified the germplasm into two groups based on the dendrogram suggesting that the SSR markers used were suitable for the characterization of horsegram germplasm. Similarly, Kaldate et al. used next-generation sequencing (NGS) technology for the development of 2458 novel SSR markers, which were characterized in 48 diverse lines of horsegram, revealing high diversity based on measures, such as mean values of expected heterozygosity (He; 0.54) and observed heterozygosity (Ho; 0.64), further clustering the genotypes into two categories.

5.4 QTL Mapping Studies in Arid Legumes

Arid environments pose significant challenges to crop production, especially for legumes that are crucial sources of protein and nutrients. To address the complexities of these arid conditions, scientists employ advanced genetic techniques such as QTL mapping, which serves as a powerful tool to identify regions in the genome associated with specific traits, like drought resistance, yield, etc. QTLs are genomic regions that play a key role in determining quantitative traits, such as those related to stress tolerance, yield, and other agronomically important characteristics. QTL mapping aims to pinpoint these locations on the genome, allowing researchers to associate them with positive or negative responses to various conditions.

Arid legumes must cope with water scarcity, making drought resistance a critical trait. QTL mapping enables the identification of genomic regions linked to enhanced drought tolerance, paving the way for breeding programs focused on developing resilient varieties. Similarly, the challenges of arid environments often result in reduced crop yields. QTL mapping helps unravel the genomic regions associated with improved yield under water-limited conditions, facilitating the development of high-yielding legume varieties suited for arid climates.

QTL mapping involves analyzing the genetic variation within populations of arid legumes. This is typically achieved through the creation of mapping populations, which are then subjected to controlled environmental conditions, including drought stress. The phenotypic variations observed within these populations are correlated with genetic markers, revealing the locations of QTLs associated with the targeted traits. QTL mapping provides a foundation for marker-assisted selection (MAS) in breeding programs. Breeders can use this information to selectively breed legume varieties with improved drought resistance and higher yields. Moreover, QTL mapping contributes to our understanding of the genetic basis of complex traits, guiding molecular breeding strategies for sustainable legume production in arid regions.

Four minor QTLs for plant height (qPH1.1), 100-seed weight (qHSW1.1), disease resistance (qPDI1.1), and gum content (qGC1.1) were found in an interspecific cross in cluster beans. These QTLs were mapped on a single linkage

Table 4 Some selected QTL mapping studies in arid legumes

Crop	Population	Trait	References
Cyamopsis tetragonoloba	F3:4	Plant height, 100 seed weight, gum content, disease resistance	Priti et al. (2023)
Vigna aconitifolia	F2	Domestication-related traits	Yundaeng et al. (2019)
	F2	Bruchid resistance	Somta et al. (2018)
	Germplasm collection	Flowering trait	Yadav et al. (2023)
Cowpea	RILs	Root knot nematodes	Santos et al. (2018)
	RILs	Domestication-related traits	Lo et al. (2018)
	MAGIC and RILs	Seed coat pattern	
Horsegram	RILs	Drought- and yield-related traits	
	RILS	Days to flowering and maturity	Katoch et al. (2022)
	Germplasm Collection	Root volume, root length, and relative water content	Choudhary et al. (2022)

group at 39.41, 123.41, 112.41, and 102.41 cM, respectively (Priti et al. 2023). In moth beans, three genes and 50 QTLs linked to 20 features related to domestication were found. The majority of QTLs were associated with five LGs: 1, 2, 4, 7, and 10. Large-effect QTLs (PVE >20%) dominated one or two minor QTLs, controlling important domesticated features, including seed dormancy and pod cracking (Yundaeng et al. 2019). Using information on egg production, a significant QTL in cowpea, identified as QRk-vu9.1, linked to resistance to *Meloidogyne javanica* reproduction, was found and mapped to linkage group LG9 at position 13.37 cM (Santos et al. 2018).

In the face of increasing environmental challenges, QTL mapping emerges as a valuable tool for enhancing the resilience and productivity of arid legumes. The insights gained from this genetic approach pave the way for developing climate-smart and high-yielding legume varieties, ensuring food security in regions prone to water scarcity. Information on some important QTL mapping is given below (Table 4).

6 Opportunities

Guar crop is recognized as a source of industrial guar gum with great export potential. There is, therefore, a great need to upgrade guar for industrial purposes, having higher gum content (> 30%) and a wide viscosity profile (cP) for food-grade and industrial-grade grain products. Guar is characterized by high capacity to recover once stress is relieved. Water stress situation, however, restricts N-fixation capacity. However, nodules are reduced less in guar than in other legumes due probably to meristematic types of nodules present in guar. Appreciably N_2-ase activities may be reversible even if water content of guar nodules decreases to 60–70%, the same ability may not be applicable to other legumes. The threshold limit for N_2-ase activities was reported to the extent of −2.4, −2.2, and −1.8 MPa in guar, mung bean, and moth bean, respectively. Guar is, therefore, a physiologically efficient legume in reference to water stress.

Cowpea is characterized by initial fast growth, with drooping leaf behavior, therefore, can suppress the weeds at the initial stage. It has good regeneration potential and is grown in *kharif*, *rabi*, and summer seasons. Cowpea is known for a strong storehouse of wide variability of all sorts: growth period, plant types, grain shape size and color, pod length and shape, types of leaves, and erect, bushy, and winy plant types. With the shortest reproductive phase, it is meant for various farming system situations with a maturity period of 60–140 days. It is known for specific physiological capabilities:

- Cowpea stomata are more sensitive than leaf growth in SMS.
- Cowpea stomata generally close down at 45% soil moisture depletion.
- Cowpea osmotic potential of leaves drops at 55–60% soil moisture depletion.
- Leaf growth of cowpea stops at 80–85% soil moisture depletion.
- Phospholipase-D (PLD) enzyme responsible for lipid degradation may be influenced even at moderate water stress. Tolerant genotypes show remarkable stability in PLD gene expression.
- Dry matter distribution is more important in cowpea at maturity. Large dry matter is partitioned toward canopy and roots but less toward sinks.

While dry fodder is superior to cowpea and guar, moth bean green fodder is nearly on par with alfalfa. Hay from moth beans can yield 7–8 t ha^{-1}. The crop is known for rudimentary plant type having more biomass with poor conversion (Fig. 2). Old varieties used to spread like met on the ground, covering inter-row spacing, even if planting is done 60 cm apart. Basically, this crop is fodder type with extreme indeterminate growth habit.

- Compared to mung bean and guar, moth bean is more sensitive to N-fixation in rising water stress situations.
- It requires more thermal times or heat degree days to reach from flowering to maturity.
- The crop is known to have adequate dry matter, but its distribution toward reproductive plant is slow, and there is also a low rate of NAR, therefore making it poor yielder.

Fig. 2 Typical plant type of guar with profused podding, new deemed plant type of moth bean with 60 days maturity

- It requires a strict and specific arid habitat.

Horsegram is probably the toughest legume, used as preparatory crops to improve soil fertility of newly reclaimed soils and cleared forests, adding organic carbon through leaf shed. It is also grown in peninsular India, and even up to 5000 feet in Himachal, Uttaranchal, and Nepal. Almost 93% area of this crop is confirmed in six states, viz., Karnataka, Orissa, Maharashtra, Andhra Pradesh, Madhya Pradesh, and Tamil Nadu. It is a component of *panchdhani*: a mixture of five crops (horsegram, Indian bean, cowpea, niger, and castor) in Karnataka to combat drought. It is known for very high content of crude protein (29–30%). It is a gradient of dietary and medicinal uses loaded with anti-calciferous properties. Horsegram is known for lush green soft attractive foliage even in drought-prone situations.

7 Challenges

Some of the challenges are manageable at short-term and long-term basis; some are researchable issues, and others are planning-related issues at the government levels. Social issues are also associated with these crops. Some of the critical challenges have been outlined below:

Technical Issues

These crops have been technically investigated at different levels by government and nongovernment agencies, but some issues have not been resolved till date. The same has become a chief bottleneck and hence deserves immediate documentation as below:

Unstable and Still Productivity

This group of crops is known for adaptability, sustainability, and survival values, not for reasonable productive values. Under extreme or even uncongenial situations, productivity behavior does not change considerably. These crops hardly touch 8–10 quintals of yield level per ha in any situation. Thus, they appear to carry inherent poor productivity genes. These crops need special and specific emphasis on scaling up their yield levels in reference to input supply under varied geographic situations by inviting and taming yield genes. Technical interventions of a deliberate nature therefore required to make these crops more popular and a source of livelihood security for arid inhabitants.

Unproductive Plant Types

These crops bear spreading, mating, and bushy plant types with more foliage but less fruits and pods. These crops are also characterized by longer vegetative but shorter reproductive phases ultimately finding less time for expression of yield traits. Such branching behavior helps sustain soil and water conservation for the whole life cycle not allowing propagation of weeds but such behavior is for survival significance only. These crops need erect-upright branching behavior, allowing more plants per unit area and capturing adequate solar radiation, reducing canopy competition, and discouraging the spreading of insect pests by reduced canopy by contact.

Longer Life Cycles

Longer life cycle period of these crops is not suitable for these crops for assured production under rainfed conditions. Efforts are, therefore, required to curtail their life spans to the regional requirements. It is suggested that breeding for earliness and altered plant types must go hand in hand; either effort would not be accepted in the regional ecosystem. The big challenge is to curtail guar life span from 80–85 days to 70–75 days, horsegram 80–85 days to again 60–65 days, and cowpea genotype of 65–70 days are required to fit them in sole cropping during summer after harvest of mustard and other intercropping sequences in main *kharif* season. Thus, deliberate, region, and season-specific efforts are essentially required to curtail their life spans accordingly. Some early maturing sources could become good donors. In guar, for instance,

Cyamopsis senegalensis and *Cyamopsis serrata* wild types are good sources of 45–50 days of maturity (Singh and Arora 2002).

Disease Resistance

These crops do not suffer drastically from diseases of different kinds because they are grown in dry and warm climates, which prove uncongenial for their expression and severe spreads. However, under extremely dry situations, dry root rot, and during continued rains, unclear sky with 30–32 °C, bacterial disease Bacterial leaf blight and angular leaf spot (ALS) diseases may prove havoc, and Yellow Mosaic Virus and Anthracnose in cowpea deserve emphasis for their genetic resistance. In moth beans, Golden Yellow and Stunted virus are the main problem; no resistant variety against these biotic stresses is available. Horsegram crops may suffer from powdery mildew and yellow mosaic virus in humid situations. Genetic efforts for breeding against these pathogen cities are required.

8 Nutritional and Quality Considerations

Guar plant, its seeds, and *korma* are the source of abundant availability of crude protein from 18% to 45%. However, guar protein is also known for its low contents of sulphur containing amino acids (methionine and cysteine). The combined proportion of these amino acids in guar is found at 1.11% compared to 4.0% found in soybean. The alternative sources of galactomannan guar like *Sesbania bispinosa*, *Sesbania sesban*, *Cassia accidentalis*, *Cassia sophera*, and *Cassia tora* with abundant availability of gum content with similar quality are posing serious threats to *Cyamopsis tetragonoloba* (guar). The proportion of endosperm (the location of guar powder) is higher in these legumes than in guar. For instance, *C. tora* has 33.0, and *S. bispinosa* has 37 compared to 32% found in the endosperm of guar (Table 5). Viscosity is also found more in these legumes, for instance, 3160 cP in *S. sesban* and 3800 cP in *S. bispinosa* compared to 2800 cP viscosity values in guar. Chemical source of guar gum (CMC) is an additional threat to local domesticated guar. Thus, per se increase in gum content to 35% and gum yield through increased productivity seem alternative measures to thwart challenges against guar crop.

Legumes contain some anti-nutritional factors (ANFs); some of them are protease inhibitors, anti-vitamins, phytase, saponins, flatulence factors, high trypsin activity, lecithins, etc. The same is nontoxic but can develop adverse physiological disorders. Some legumes like cowpea and horsegram contain trypsin inhibitors, which hamper protein digestibility. In case of cowpeas, anti-nutritional problems can be resolved by appropriate dehulling and heat treatments. In legumes, polyphenolic compounds are found abundantly. Till recently some of them (tannins) are considered anti-nutrients due to their adverse effects on protein digestibility. Horsegram seeds contain abundant seed protein to the extent of 20–30% (Mahadevappa and Raina 1978). However, crop suffers from anti-nutritional factors. The slow digestion of horsegram seed protein is attributed to the presence of trypsin inhibitors (Sudha et al. 1995). The formation of calcium oxalate, crystals, and change in the fine structure of crystal cells in the seed testa of horsegram were observed during the ripening stage (Solbrig 1983). Moth bean grains also contain some amounts of anti-nutrients such as trypsin inhibitors phytohemagglutinins, phytic acid, saponins, etc. Moth beans may display up to 3560 units/ml of hemagglutinating activity. Seeds of this crop may also exhibit a flatulence ratio of 11.4 (ratio of flatus from a meal of 100 g dry weight for 3 h period). Various preparations are made from moth bean grains like *Dal*, *Kheech*, *Rabri*, *Papad*, *Bhujia*, *Mangori (Nugget)*, *Vada*, *Chapatti*, and Sprouts.

9 Progress Made in the Past 25 Years

Basically, the arid legumes are considered the crops of regional niches adapted to local situations, catering to regional requirements and supporting the main crops. The

Table 5 Comparative carbohydrate composition (%) and rheological attributes of various galactomannan gum-containing legume seeds

Gum containing legumes	Endosperm (%)	Galactomannan (%)	Galactose: Mannose ratio (G/M)	Viscosity[a] $(n)^2$ cps	Rheological attributes		
					Consistency index value "K"	Flow behavior index, value "n"	Rheological behavior
Guar (*Cyamopsis tetragonoloba*)	32.5 ± 2.0	31.2 ± 2.0	1:20	2800.0	4.26	0.47	Pseudoplastic
Sesbania bispinosa	37.5 ± 1.5	34.5 ± 1.5	1:2.20	3800.0	3.40	0.46	Pseudoplastic
Sesbania sesban	34.5 ± 1.5	31.8 ± 0.5	1:1.80	3160.0	2.98	0.45	Pseudoplastic
Cassia tora	33.0 ± 1.5	30.5 ± 0.5	1:1.50	1880.0	1.72	0.78	Pseudoplastic
Cassia sophera	36.0 ± 2.5	32.5 ± 1.25	1:2.30	1900.0	1.60	0.68	Pseudoplastic
Cassia accidentalis	35.2 ± 2.5	31.5 ± 1.0	1:2.60	1650.0	1.21	0.66	Pseudoplastic

Source: Kumar and Singh (2002)

[a]The values represent at 1.0 w/v gum solution, at shear rate 4.51 s^{-1}, and temperature 30 ± 1.5 °C

same are still behaving for survival value but less for desired productivity. Broadly, the initial efforts toward the genetic improvement of these crops have focused on collecting germplasm from diverse natural habitats, followed by extensive evaluation and screening of large germplasm collections maintained by crop-specific breeders, the National Bureau of Plant Genetic Resources (NBPGR), and its regional research stations (Mishra et al. 2004) and have proved quite successful and played great dividends. Conventional hybridization approaches boosted improvement in the growth of guar. Mutation breeding—employing physical mutagens, chemical agents, or their combinations—has proven highly effective in developing improved cultivars. These studies and other attempts led to induce new, early, and altered plant types in moth bean and guar, which provided the base for future breeding (Fig. 2).

Guar has assumed status of international export-oriented cash crop due to great consumption pattern of this crop by the export market, particularly petroleum industry, during the past two decades (APEDA 2023). India is largest producer of guar. This crop is grown in Northwest part of country. Being a rainfed crop, its status has been fluctuating over the years. Rajasthan is a major guar-growing state with a contribution of 70.0–91.5% in the area and 61.4–87.7% in production during the past 12 years. Notably, guar output and yield have grown with CAGRs of 4.5 and 5.1%, respectively, despite negative CAGR for acreage during the past 20 years (Chauhan et al. 2023).

Considerable work on genetic improvement of guar has been done under NARS, AICRP, and National Network Research Project on guar (later on arid legumes) at NBPRG, RAU; Bikaner, CAZRI, and IIPR; and Kanpur during various stages of shifting of this project since 1993 till date. Cluster bean is largely a self-pollinated crop with only 2% outcrossing; hence, breeding methodologies employed are as for self-pollinated crops like selection, pure line selections, pedigree method of selections, hybridization, and mutagenesis, but interspecific hybridization method has totally been restricted. In view of transfer earliness in *C. tetragonoloba* cultivated species from wild types (*S. senegalensis* and *S. serrata*), Sandhu made an effort in 1988.

Nevertheless, he noted that traditional methods had failed. Additionally, it has been reported that other strategies such as bud pollination, amputation of stigma and style, and the use of organic solutes were unsuccessful in overcoming stigmatic incompatibility obstacles (Kumar and Rodge 2012). Efforts made at Haryana Agricultural University, Hissar (forage section) also proved futile. Scope for interspecific hybridization, therefore, appears limited in guar. This prompts the use of nonconventional tools of hybridization including ovary rescue, protoplasm fusion, etc. for genetic improvement.

According to Bhatt et al. (2015), the methods being used to generate high-yielding guar varieties include germplasm collection from arid regions, summer screening to find tolerant genotypes, hybridizing high yielders with tolerant donors, and inter-varietal hybridization.

For recombination breeding, which provides a variety of segregating populations to influence selection and aid in gene identification, intensive hybridization is crucial. However, in desert legumes, cluster beans, and moth beans, cleistogamy, small flowers, and sensitivity to treatments that function as deterrents to a robust recombination breeding support make hybridization difficult and time-consuming. In comparison to previous hybridization approaches, Mahla and Sharma (2022) observed that direct pollination without emasculation produced about 28% seed setting and 60% hybrid seeds in cluster beans. They also found that this method was relatively simple and effective. This method would also make powerful recombination breeding and genetic research easier.

The development of vegetable-type varieties with high green pod production, early maturity, and enhanced pod quality—such as long, green, succulent, tender, and low fiber content—has been the main goal of the breeding effort. The development of a single stem is likewise emphasized, and it may be grown in both summer and irrigated conditions.

Since the first variety (FS 277) was made public in 1976, 39 forage and grain-type varieties of cluster beans have been released and notified in India as a result of NARS's coordinated research efforts. The most recent notification, issued on March 6, 2023, included Karan Guar 14 at the 90th meeting of the Central Subcommittee on Crop Standards, Notification and Release of Varieties for agricultural crops, which was held on May 2, 2023 (Seedneet 2023). Efforts have been made to increase gum content to more than 32%, develop varieties with varied viscosity profiles, and remove anti-nutritional factors from guar meal and meal korma so as to make it perfect fodder and feed. Efforts are also being made to increase and isolate guar protein and to be used as a secondary and organic source for humankind (Tiwari et al. 2020).

9.1 Moth Bean

In India, it used to occupy an area of about 1.8 m ha with a total production of 0.4 mt (Kharbet et al. 1987). However, the area of moth bean is currently showing a decreasing trend to 0.9 m ha. The same is of great concern. A lot of studies on variability, correlation, and path analysis have been conducted (Tikka et al. 1977, 1980; Sharma and Mathur 1982; Saini et al. 1983), including high co-heritability/plant to seed yield, high heritability estimates, and genetic gains 1000-seed weight, days to flowering maturity, and grain yield. In spite of the cultivation of moth bean in 1.5 million hectares in India and its vast potential both as food and fodder crops, inadequate attention has been paid to its improvement. Most of the available commercial varieties being secondary selections from local types provide marginal superiority for economic attributes. Saini and Malik were the first to identify T-3 as a fodder variety for Punjab State. Govade (1950) identified Baleshwar-12, Medhi-33, and MG-1 for Gujarat state, while Ramanujam (1972) found T-9 as the most adaptable variety. Release of Jadia in 1980 provided a variety adapted to major moth bean–growing states (Sharma and Mathur 1983). It was followed by a yellow mosaic virus-resistant variety Jwala released in 1985 (Sharma et al. 1986).

In 1988, another variety IPCMO-880 was identified for the west zone. However, all these varieties take 80–100 days to mature, and this behavior does not fit the short rainy season available in the moth growth area of the state. There was a demand for short-duration, high-yielding variety with resistance to disease. The development of the variety RMO-40, which was identified in 1973, provided a partial solution to the problem. This variety matured in 55–60 days and was resistant to yellow mosaic virus. Attempts to develop new varieties followed. Notably Maru Moth-1, Jadia representing a late group (85–95 days) was also released. Thereafter, many varieties (RMO-225, RMO-435, RMO-257, RMO-423, RMB-25, RMO-2251) from Bikaner and CAZRI Moth-1, CAZRI Moth-2 CAZRI Moth-3, CAZRI Moth-4, and CAZRI Moth-5 from ICAR-CAZRI were also developed; FMO-96 from Fatehpur was developed; and two registrations CZM-32 and RMM-12 were also developed from ICAR-CAZRI, Jodhpur (Figs. 2 and 3).

Influx of Nutrients

Productivity of crops not only depends on morpho-physiological architecture, plant type, and required earliness in dry areas but also nutritional uptake and also the capabilities of the specialized root sphere. The nutritional uptake from root sphere to root is a genetically governed trait; hence, the varietal difference if found would prove a great help in the isolation of high-yielding genotypes. Such variations had been observed in moth bean (Kumar et al. 1998; Tarafdar and Kumar 2003). For instance, N inflow was lowest (11.54 mol cm^{-1} root $second^{-1} \times 10^{-15}$) in the genotype GMO-9120, but it was maximum (25.89×10^{-15}) in Exp. 109 being 2.2 times more yielder than the former. The genotypes Exp. 213 and RMM-12 also behaved so. These results precisely revealed that high-yielding genotypes (Exp.109 and Exp. 203) were, in general, characterized by the highest rates of N, P, and K inflow from soil to the roots, whereas poor yielders (GMO-9105 and GMO-9120) were marked with the lowest inflow. Hence, the possibility for screening genotypes of moth bean for high-yielding capabilities existed in moth bean following nutrient inflow.

9.2 Cowpea

It is the oldest source of food for human beings and a fine source of fodder for livestock. Cowpea is called by many synonyms like *black-eyed pea, crowder pea, southern pea, lobia, niebe, coupe, frijole, chawla*, etc. It is the crop of tropics and subtropic regions: Asia, Africa, Central and Southern America, parts of South Europe, and the USA. The Republic of South Africa could be the center of specifications of *V. unguiculata* and moved toward Mozambique and Tanzania. Globally cowpea is grown in 9.8 m ha of which 9.0 m ha is grown in western Africa itself. India grows 50% of Asia. World cowpea productivity is 337 kg ha^{-1} whereas productivity in Nigeria is 417 kg ha^{-1}. Out of total production of 3.3 mt world over, Nigeria alone produces 2.1 mt, niger 0.65 mt, and 0.11 mt by Mali. In India, cowpea is grown as intercrops and mixed crop and less as sole crop.

Uses Almost 80% cowpea seeds are consumed as dal/fried grains (whole, split/milled). The processes used in cowpea preparations are soaking, cooking, roasting, puffing, extraction, cooking, germination, fermentation, canning, etc. The utilization pattern of cowpea can be grouped as below:

- Whole/green cowpea (fresh, boiled, and roasted)
- Sprouted and germinated (boiled and fried)
- Puffed and roasted (special/salted)
- Milled and cooked (steamed, soiled, and fried)
- Fermented products (idly, dhokla, etc.)

Use and marketability of cowpea depend on their type, shape, size, color, etc. as below. Breeding crops are directed to specific classes of seeds so as to be used for different purposes as per market demands.

Black and Purple Eye The shelled peas are visually appealing, have a moderate flavor, and can be processed. The black, pink, or pale red items that surround the white hilum are utilized as food items.

Brown Eye Pods may vary in color from green to lavender and have a wide range of lengths. The immature seeds on cooking turn medium to dark brown in color with a delicate flavor.

Crowder Seeds are closely crowded in the pods and tend to be globular shaped.

Cream Seeds of these types are generally cream colored and have no noticeable eye (the helium is inconspicuous).

Clay These are older types with medium-to-dark brown color and kidney-shaped seeds.

Fig. 3 Root growth of moth bean during extreme unbranched 60 days maturity in a drought situation in 45 days registered genotype of moth bean

White Acre The kidney-shaped peas have a blunt tip. This kind of crowder is rather small and typically tan in color. Pods are quite stiff.

Interspecific hybridization in cowpea, particularly hybridization of cultivated species with wild species, is required for transfer of resistance against pod-borers and sucking bugs from wild to cultivated species. Fertile hybrids could be obtained by crossing cultivated cowpea [*Vigna unguiculata* ssp. *unguiculata* (L.) Walp.] with wild relatives belonging to *V. unguiculata* ssp. *dekindtiana* (Harms) Verdc., *Vigna unguiculata* ssp. *stenophylla* (Harvey) M. M. & S., and *Vigna unguiculata* ssp. *tenuis* (E. Mey.) M.M. & S. Related cowpea species *Vigna vexillata* (L.) possesses substantial resistance against pod-sucking bugs, pod-borer, and some other insects. However, successful crosses with cultivated species with *Vigna vexillata* have not been reported. To transfer resistance against pod-borer and pod-sucking bugs, IITA has bred cultivated cowpea with wild species (*Vigna unguiculata* ssp. *dekindtiana*). The characteristics of recurrent parents were fixed as early as in IBC2 due to introgressive backcrossing (IBS). With every IBC generation, fertility rose.

Congruity backcrossing (CBC), a novel technique that involves crossings with each parent species in alternate generations, has been shown to boost fertility in following generations of the species used as the female in the original crosses. Mahudeswaran et al. (1973) derived a 'Co2' variety from three-way crosses involving *V. catjang* (Burm. F.) Walp., *V. unguiculata* and *V. sequipedalis* (L.) Fruw. The same was high-yielding and quite fertile. Similarly, seeds were produced if *V. unguiculata* (L.) Walp. ssp. *dekindtiana* var-*mensensis* Schweinf was the female parent in crosses with subsp. *dekindtiana*, but seeds were shriveled and reduced in weight. No viable interspecific hybrids have been observed among other *Vigna* species, viz. *V. umbellate* (Thumb.) Ohwi & Ohasi, *V. vexillata*, *V. mungo* (L.) Hepper, *V. radiata* (L.) R. Wilczek, *V. aconitifolia*, and *V. angularis* (Wild.) Ohwi and Ohosi Wild cowpea sp. carry gene for pod-borer and pod-sucking pests resistance. Fortune is that hybrids between cultivated (*V. unguiculata* (L.) Walp.) vs. wild sp. like *V. dekindtiana* (Harms), *V. stenophylla* (Harvey), and *V. tenuis* (E. Mey.) are possible. However, hybridization between cultivated spp. and *V. vexillata* have not been reported.

Enormous breeding strategies have been undertaken at various institutions, particularly at IITA, Nigeria; IARI, New Delhi; CCS HAU, Hissar; TNAU, Coimbatore; BARC, Mumbai; and S. K. Nagar, Dantiwada, Gujarat. Various but usual methods like selection, hybridization-induced mutations, and many promising cultivars satisfying the desired traits including disease and insect pest resistance and grain quality referring to marketability of cowpea seeds have been employed. The following list indicates some of the promising varieties.

9.3 Horsegram

It is a traditional and indigenous tropical grain legume, recognized for drought hardiness and adaptability toward structurally and nutritionally poor soils and much adverse climatic situations, which may prove uncongenial to other crops for cultivation (Kumar 2006). It is extensively grown in peninsular India and is cultivated up to 5000 feet elevation in Himachal and Uttaranchal states and in Nepal. At present, it is grown in almost 1.7 m ha with a total production of 0.74 mt with a productivity of 494 kg ha^{-1} against 539 kg ha^{-1} for grain legumes as a whole. Almost 93% of the total area of horsegram in India is consumed by six states, viz., Karnataka, Maharashtra, Orissa, Andhra Pradesh, Tamil Nadu, and Madhya Pradesh. Karnataka is the mega state for area (5.90 lakh ha) and production (2.15 lakh t), with low productivity (370 kg ha^{-1}). Productivity of horsegram is maximum in Kerala (740 kg ha^{-1}).

Approximately 0.81 million tons of horsegram are produced in India each year, accounting for 7% of the nation's total grain legume production. Nonetheless, horsegram accounts for over 28% of the total production of grain legumes in Karnataka, 18% in Tamil Nadu, and roughly 10% in each of Maharashtra, Orissa, Andhra Pradesh, and Madhya Pradesh.

Uses

Although horsegram is mostly used as animal feed, a sizable portion of the rural population also eats it as a human diet, either as whole seeds, sprouts, or entire meal combined with cereal flour. The use of dry seeds of horsegram is very sporadic due to its poor cooking quality. This legume is said to have several medicinal properties for treating cough and bronchitis, kidney trouble, and irregular periodicity of menstrual cycle in ladies. Although not valued so much as the other pulses, it is equally nutritive. The impoverished take it in big quantities since it is a less expensive pulse. This bean is used to make a variety of delicious dishes in South India, including curry, pappad, and more. Farmers in Karnataka plant punchdhani, an amalgam of five crops (horsegram, Indian bean, cowpea, niger, and castor), as a typical drought-fighting strategy.

Anti-nutritional Factors

Oligosaccharides (raffinose, stachyose, and verbascose) present in horsegram seeds are involved in flatulence production in human beings and animals. However, there is a reduction in these oligosaccharides on soaking of seeds. Anti-nutritional factors present in horsegram are trypsin inhibitors, hemagglutinins, phytic acid, and α-amylase inhibitors. However, on soaking and cooking, the contents of these anti-nutritional factors are significantly reduced. Dehusking results in marked improvement in protein.

Breeding Objectives

Arid legumes are known as climate-resilient source crops with limited inputs, grown on poor fertile soils as major and minor crops in varied ecological zones. These crops are also recognized as the cheapest source of food, fodder, feed, and organic produce, loaded with high-quality nutrients. Livelihood, nutritional, environmental, social, livestock, and economic securities are therefore expected from these magic crops. Being food for tribals, a source of export-oriented products and sub-products, commerce utility, and associated with nonbanking benefits are embodied with typical immediate and long-term potentialities. Breeding objectives of these groups of crops are thus crucial and need to be framed to justify each and every requirement as below:

- Breaking stagnated yield barriers of these crops in the national and regional context under existing poor and organic input situations with achievable yield levels of 8000–10,000 kg ha^{-1} and 12,000–15,000 kg ha^{-1} in improved and assured situation.
- Breeding for sustained production to the tune of 500–6000 kg ha^{-1} under worst situations to provide continued livelihood security for the inhabitants in arid and hyper-arid situations. Providing more attention toward sustenance vis-à-vis production in famine situations, which are of common occurrence here.
- Curtailing growth cycle in reference to the rainfall pattern and intensities in rainfed habitat of arid legumes. Rainy days last 50–65 days in arid Rajasthan, Haryana, Gujarat, and Maharashtra, but longer in other regions. Growth periods of guar and horsegram can be reduced to 65–70 days; no cultivar is available so far. The required growth periods of moth bean and cowpea is 55–60 days for extreme hyper-arid situations for both sole and intercropping situations. Irrigated guar requires 100–105 days of maturity. Thus, growth cycle of the arid legumes is required to be adjusted accordingly without the same blanket recommendations.
- Breeding for physiologically efficient plant types suited for rainfed and irrigated situations. For rainfed situations, semi-erect to semi-bushy plant types characterized with space compensatory tendency are required. Upright erect canopies with more fertile pods are required so as to harvest solar radiation nutritious and space efficiently. For irrigated and assured input situations, these crops need erect upright behavior only. Space compensatory behavior is not required here. Vegetable-type guar varieties require unbranched behavior with complete podding tendency of the main shoot.
- Simultaneously breeding for curtailed growth period and altered plant types with effective source-sink relationship assuring harvest index to the extent of 40–45%.

- Quality breeding for enhancing guar gum content per se from 29–30% currently to 34–35% and developing varieties for distinct food grade (1000–3000 cP) viscosity and industrial grade (5000–8000 cP) viscosity to enhance the utility of guar galactomannan gum for various industries. Enhancing the size of endosperm (lactation of guar gum) of guar grains.
- Breeding for making guar meal, *korma* and *churi* as perfect poultry feed and concentrate (40–45% cP) for livestock by overcoming anti-nutritional factors (saponins, trypsin inhibitors, and odor smell) of these guar subproducts. Efforts for making horsegram grain crude protein more palatable, digestible, and acceptable by the poor tribal people by lowering anti-nutritional factors, in addition to enhancing S-containing amino acids in other arid.
- Breeding vegetable guar for enhancing grain yield from the current level of 200–300 q ha^{-1} and making minerals profile more nutritive by enhancing Fe content from the present value of 4–5 mg g^{-1} to 10–15 mg g^{-1} at least.
- Chemical inputs are not served to the arid legumes, which is why the same are organic crops by default; hence, breeding against biotic agents is quite crucial. In case of guar, Dry Root Rot (*Macrophomina phaseolina*) is very disastrous; besides, Bacterial Leaf Blight (*Xanthomonas campestris* pv. *cyamopsidis*) is also a serious disease, and *Alternaria* Leaf Spot (*Alternaria brassicae* (Bark) and Powdery Mildew (*Erysiphe polygoni*) are also common. Breeding for resistance against these diseases is required to have exportable guar gum products of organic nature.
- In cowpea *anthracnose* (*Colletotrichum lindemuthianum*) fungal disease, and among viral diseases, Cowpea Yellow Mosaic Virus (CYMV) and Southern Bean Mosaic Virus (SBMV) are important ones; hence, breeding against these diseases for resistance potential is required. Anthracnose and powdery mildew diseases are also of great concern; hence, breeding for transferring resistance genes is required in horsegram. Breeding for disease resistance is also crucial against Yellow Mosaic Virus (YMV) and Leaf Curl Virus (LCV). Management of *Helicoverpa armigera* causing severe damage as pod borer in cowpea is difficult; thus, nonconventional intervention is required to develop BT cowpea.
- Efforts on hybridization in general and interspecific crosses particularly have been quite limited due to unfavorable and poor manageability of tiny flowers in arid legumes. Interspecific crosses need to be attempted on priority for transferring disease resistance, dwarfing, and early maturing genes, for developing new plant types.
- In case of guar wild spices *C. serrata* and *C. senegalensis* are good sources of early maturity (55–60 days) with a more drought-hardy nature. Thus, interspecific hybridization between *C. tetragonoloba* × *C. serrata* and *C. titragonoloba* × *C. senegalensis* using nonconventional and biotechnology tools is required to break the barrier of longer maturity. A substantial amount of genetic potential exists in wild cowpea species, which remains largely underutilized. The cowpea gene pool includes three cultivated subspecies—*unguiculata*, *cylindrica*, and *sesquipedalis*—and two wild relatives, *dekindtiana* and *mensensis*, offering valuable traits for crop improvement. The crosses between cultivated × cultivated, wild × wild, and cultivated × wild will prove quite useful, yielding new genotypes of awaited uses and wide adaptations. In moth bean, two species *V. aconitifolia* and *V. trilobata* are available. Efforts through nonconventional methods need to be attempted to give more lucrative plant types. Crosses must be attempted between related species of horsegram. The same can be *Macrotyloma uniflorum* (grain type), *M. axillare* (source of YMV), *M. africanum*, and *M. daltonii* (forage types).
- Breeding for facing climate change through arid manipulation of legume genomes by conventional and nonconventional approaches. Guar and moth bean can face drought period of 45–50 days and atmospheric temperature >45 °C by standing on hungry and thirsty arid sandy soil. The need is to elevate tolerance potential against these abiotic factors by incorporating specific peptides and heat shock genes by shuttle screening in natural habitats. Efforts could be made screening and evaluating genetic materials during summer season by lowering SM to 4–5%. Aridity and salinity hand in hand and efforts are also required to breed for salt tolerance also. Breeding for higher uptake and nutrient-efficient arid legumes must also be a priority.

Emerging Challenges at National and International Levels

Challenges in agriculture have also been there in view of crop competition of crop produce in the market, fluctuating demands and supply, and sudden emerging demand of products and sub-products (as in the case of guar gum in 2011–2014) at the international level. Also, there is every livelihood of intervention of climate change, particularly in rainfed situations. Moth bean, the lifeline of hyperarid areas of northwestern districts of the *Thar* Desert, is losing ground, and the area has been decreasing for the last half decade (9.26 lakh ha). It is competing with lucrative, high-priced early maturing varieties of mung bean. Seeds of early maturing varieties of moth bean are not available to the farmers, and no MSP is provided for their crops. Its utility is daily snacks, and it is not finding additional national and international demands. This crop does not have a recognized international center of origin, but its primary cultivation and genetic diversity are concentrated in the dry tracts of western India. Guar is more stable and less influenced by biotic and abiotic agents and is becoming the choice of moth bean farmers.

Guar is competing with alternate sources, viz., *Sesbania bispinosa*, *S. sesban*, *Cassia tora*, *C. sophera*, and *C. accidentalis* (Kumar and Singh 2002), and chemical source (CMC) of guar gum is available, and international demands of guar gum powder are not picking up. Domestic uses are not being explored due to the nonavailability of various derivatives used in different industrial products. Guar gum with different viscosity (cP) values is not available; gum content (29–30%) has been stable for the last five to seven decades; there is no MSP on guar production, and cultivation of guar has spread to Maharashtra, A.P., Karnataka, and T.N. rainfed situation as supplementary crop. It is now cultivated in the rainy season, rabi-season, and summer season in different parts of India. Vegetable guar is getting more acceptance and popularity due to being more stable and consumed in months of summer and rainy seasons. Cowpea does not occupy the status of major pulse and is not cultivated as a sole crop but in inter-crop, mixed cropping systems. It grows very fast as fodder but quickly bows down to water stress situations. Compared to urad bean, mung bean and pigeon pea digestibility is also a problem. The number of early maturing varieties of 60–65 days and the availability of seeds are great problems in this legume. Unlike in some African regions, this crop is not widely utilized in India as a weed-suppressing mulch or as a means to save one or two irrigations, despite its potential for such uses. Cowpea faces the main challenge of very poor sink and a lot of foliage but photosynthates not converting to sink.

Horsegram crop of sub-hills and foothills of Uttaranchal and Himachal Pradesh but largely grown in Karnataka and Orissa lacks early to medium early varieties maturing in 90–105 days. In view of the large scattered area, the number of promising cultivars is limited. It has medicinal homeopathic uses; such virtues have not been technically and commercially exploited. The presence of certain anti-nutrients (trypsin inhibitors), calcium oxalate, crystals, oligosaccharides, phytic acid, and amylase inhibitors (Kumar 2006) and less quantity of S-containing amino acids like cysteine and methionine provide this pulse a low score. Poor yield potential due to obvious reasons of poor and nonresponsive plant types are major challenges to this crop. It is also suffering from photo- and thermosensitivity, which is also a main barrier.

Conventional Breeding Approaches Leading to Varietal Development

Of four arid legumes, guar is the most investigated legume. Sincere efforts were undertaken at ICAR-CAZRI, RAU Durgapura; CCS HAU, Hisar; ICAR-IGFRI, Jhansi; and S. K. Nagar, Dantiwada. Breeding had been done for grain, fodder, and vegetable purposes. Work initiated in 1974 (CV FS-277) till date 2023 (Karan Guar-14), 39 improved varieties have been developed in respect of improved plant types, curtailing life cycle, better adaption, resistance against major disease, and finally improving productivity per se. Through targeted breeding efforts, modern guar varieties have been developed that mature within 85–105 days. These varieties feature synchronized growth, compact plant architecture, and a higher number of pods, all while maintaining a reduced canopy size for improved agronomic efficiency (RGC-1002, RGC-1017, RGM-12, RGC-1038, HG-870, Karan Guar-14) and have yield potential to the extent of 1200–1400 kg ha^{-1} against late maturing (105–120 days) ones with loose canopy and unsystematically growth habits that used to yield 500–700 kg ha^{-1}. The cultivars of arid legumes have been bred in reference to the rainfall intensity of the regions and have been tabulated in Tables 6, 7, 8 and 9 (Anonymous 2023).

The Grain Type (Used for Industrial Purpose) Varieties Ageti Guar 112 (1983), early maturating RGC-1031 (2005) late sown conditions, Guar Kungal-1033 (2011), RGC-1066 (2007) single stem; RGC-1055 (2007), GG-3 (2021), HG-365 (1998) early maturity, HG-2-20 medium late, HG-563 (2004), early maturity, HG-870 (2010) high viscosity, HG-884 (2010), Karan Guar-1 (2018), shattering resistant, lodging resistant, Karan Guar-14 (2023), high cP values, Maru Guar (1986) for late planting RGC-1038 (2009), RGC-197 (1970), unbranched, RGC-936 (1999) most popular in Rajasthan, RGC-1002 (1999), RGC-1003 (1999), RGC-1017 (2002), RGM-112 (2005) early, X-10 (2021).

Fodder Type Agaita Guar (1978), Bundel Guar-1 (1993), Bunder Guar-2 (1995), Bunder Guar-3 (1999), Guara-80 (1983), HFG-11 (1982), HFG-258 (1988).

Vegetable Type Pusa Nav Bahar (1967), Pusa Mausani, Pusa Sadabhar, Durgapura Safed, FS-277, Sardarbahar, HG-75. There are 25 varieties that fall in the vegetable category.

Dual Purpose Durgajay (1980), Durgapur safed (1980), HFG-119 (1982), HG-75 (1982), HG-182 (1984), RGC-471 (1995).

Irrigated Condition RGC-986 (1999), RGC-1066 (2007), unbranched late maturity behavior.

9.4 Cowpea

It has been established that cowpeas are native to Africa. According to Marechal et al. (1978a, b), the majority of the

Table 6 Varieties of guar released and notified from 2000 to 2022

S. no.	Variety	Year	Guar growing area	Days to maturity	Grain yield (kg/ha)
1.	HG-563	2001	Haryana and Rajasthan	85–90	1400–1600
2.	RGC-1017	2002	Rajasthan, M.P., Punjab, U.P.	92–95	1200–1400
3.	RGC-1038	2006	Rajasthan	95–100	1200–1500
4.	RGC-1031	2006	Rajasthan	110–114	1500–2000
5.	RGC-1055	2006	Rajasthan	96–106	1500–2000
6.	RGC-1066	2006	Rajasthan	97–95	1000–1400
7.	HG-864	2010	Haryana, Gujarat, Rajasthan	90–95	1400–1500
8.	HG-2-20	2010	Haryana, Gujarat, U.P.	90–100	2000–2200
9.	HG-870	2010	Haryana	85–95	2000–2200
10.	Guar Kunjal (RGC-1033)	2010	Rajasthan	95–100	1500–2000
11.	Karan Guar-I (RGr 12-1)	2018	Rajasthan	90–100	1500–2000
12.	Gujarat Guar-3	2021	Gujarat State	98–100	1300–1400
13.	Karan Guar-I (RGr-12-1)	2018	Rajasthan	90–100	1500–2000
14.	Gujarat Guar-3	2021	Gujarat State	98–100	1300–1400
15.	Karan Guar 14 (RGR 18-1)	2022	Rajasthan, Gujarat, Haryana, and Maharashtra	95–96	1200–1300

variability among wild species is limited to South Africa, although the range of wild forms spans much of tropical Africa. It has been highlighted that the subspecies *unguiculata* is common in all cowpea-growing regions of the world, whereas the cultivated forms, cylindrical and sesquipedalis, have evolved in Asia. Extensive research has been conducted on developing improved cowpea varieties using conventional breeding approaches. These efforts have led to enhancements in plant architecture, modifications in growth duration, and better environmental adaptation, ultimately resulting in increased yield potential. A summary of these improved varieties is presented in Table 6. A complete listing of major cowpea varieties released and notified for different regions is provided below in Table 7.

Uses Almost 80% of cowpea seeds are consumed as dal/fried grains (whole, split/milled). The processes used in cowpea preparations are soaking, cooking, roasting, puffing, extraction, cooking, germination, fermentation, canning, etc. The utilization pattern of cowpea can be grouped as below:

- Whole/green cowpea (fresh, boiled, and roasted)
- Sprouted and germinated (boiled and fried)
- Puffed and roasted (special/salted)
- Milled and cooked (steamed, soiled, and fried)
- Fermented products (idly, dhokla, etc.)

Use and marketability of cowpea depend on their type, shape, size, color, etc. as below. Breeding crops are directed to specific classes of seeds so as to be used for different purposes as per market demands.

Black and Purple Eye The shelled peas are attractive, mild flavored, and suitable for processing. The white hilum is surrounded by black, pink, or light red and used as food products.

Brown Eye Pods may vary in color from green to lavender and have a wide range of lengths. The immature seeds on cooking turn medium to dark brown in color with a delicate flavor.

Crowder Seeds are closely crowded in the pods and tend to be globular-shaped.

Cream Seeds of these types are generally cream-colored and have no noticeable eye (the helium is inconspicuous).

Clay These are older types with medium to dark brown color and kidney-shaped seeds.

White Acre The peas are kidney-shaped with a blunt end. This type is a semi-crowder, generally tan in color and somewhat small. Pods are quite stiff.

9.5 Moth Bean

The most important and crucial crop grown very well in hyper-arid situations is moth bean, which conserves soil and soil ecology. It has been intensely investigated for needful genetic improvement at ICAR-CAZRI, Jodhpur, S.K. RAU, Bikaner, S.K. RAU, Fatehpur, and S. K. Nagar, Dantiwada. Until the 1990s, traditional moth bean plant types were characterized by a spreading, mat-forming growth habit, covering up to a one-meter radius. These types produced abundant biomass with long peduncles bearing relatively few pods and exhibited poor reproductive efficiency. While rich in fodder value, they were highly susceptible to yellow

Table 7 Varieties of cowpea released and notified from 2000 to date

S. no.	Variety	Year	Cowpea growing area	Days to maturity	Grain yield (kg/ha)
1.	RC-101	2001	Rajasthan	60–65	750–850
2.	CoVu-702, Co (CP-7)	2002	Tamil Nadu, Karnataka, Andhra Pradesh, Kerala	67–73	950–1250
3.	GC-5	2005	Gujarat	94–100	1100–1200
4.	Pant Lobia-1	2009	Uttarakhand	65–70	1500–2000
5.	Pant Lobia-2	2010	Uttarakhand	75–80	1400–1800
6.	Hissar cowpea 46 (HC-98-46)	2010	Haryana	65–70	1000–1200
7.	DCS 47-1	2014	Southern region	80	1200–1400
8.	Pant Lobia-4	2015	All cowpea-growing regions	60–65	1400–1800
9.	Pant Lobia-3	2016	All cowpea-growing regions	65–70	1400–1800
10.	(Phule CP-05040)	2016	Maharashtra	70–80	1000–1200
11.	Pant Lobia-5	2017	Uttarakhand	65–70	1600–2000
12.	Phule Rakhumai PCP 0306-1	2017	Southern zone	70–80	1000–1102
13.	TPIC 29 (Tirupati Cowpea-1)	2017	Andhra Pradesh, Karnataka, Tamil Nadu, Orissa	80–90	1000–1100
14.	D-15	2017	Karnataka, Andhra Pradesh, Tamil Nadu and Kerala	75–80	1000–1300
15.	KBC-9	2018	Karnataka, Andhra Pradesh, Tamil Nadu and Kerala	80–85	1100–1200
16.	TC-901	2018	Northern part of India	69–75	900–1000
17.	CPD-119	2018	Rajasthan	65–70	800–1000
18.	Pant Lobia-7 (PGCP-24)	2021	Northern part of the country for spring summer season	70–75	1100–1200
19.	Jammu Lobia Super 60	2021	Jammu province of union territory of Jammu & Kashmir	60–65	1200–1300
20.	PGCP-6	2021	Karnataka state	70–75	900–1000
21.	KBC-11	2021	Karnataka province of union territory of Jammu & Kashmir	95–100	1100–1200
22.	Jammu Lobia Supper 60 (J&K)	2021	Karnataka state	60–65	1200–1300
23.	PGCP-6 (UAS, Bangalore)	2021	Karnataka state	70–75	900–1000
24.	KBC-11 (UAS, Bangalore)	2021	Gujarat	95–100	1100–1200
25.	GC-7 (Banas Tejas) GC-1501	2022	Maharashtra	69–70	1000–1100
26.	Phule Sonoi	2022	J&K	72–75	1300–1400
27.	Shaimar Cowpea-2 (SKUA WCP-149)	2022	J&K	110–115	1200–1300

mosaic virus (YMV) and terminal drought stress. They also had a long maturity period of 100–110 days and yielded only 200–350 kg ha^{-1} under typical conditions. The effective selection procedures and mutation breeding efforts were put in by Saini and Malik, Govade (1950), Ramanujam (1972), Sharma and Mathur (1983), Sharma et al. (1986), and Tikka et al. (1980). The efforts led to the development of 23 new varieties between the period of 1950 and 2021. Maturity has been curtailed from 90–95 days to 75–80 days (CAZRI Moth-1) in 1999, 62–65 days (CAZRI Moth-2) in 2003, and 60 days (CAZRI Moth-3) in 2004. RMO-40 (1992) was induced and matured in 58–60 days. It was the earliest variety released in 1992 from Jawala. Besides, several varieties of RMO-225 (1999), RMO-435, Maru Moth-1 (1989), RMO-2576 (1997), and FMM-96 (1999) were developed. These varieties have erected upright, branching habit, earliness (58–70 days), synchronized maturity, and yield potential of up to 8000–10,000 kg ha^{-1}. These varieties are less affected by terminal drought stress and possess an inherent ability to escape adverse climatic conditions, making them more resilient to the impacts of climate change (Table 8).

9.6 Horsegram

It is a traditional topical grain legume, known for hardiness and adaptability to unfavorable climatic conditions on poor soils, which may prove unsuitable to many crops. It thrives well in almost all types of soils; however, light-textured sandy soils are most suited for this legume. Poor productivity (494 kg ha^{-1}) of this crop compared to 540 kg ha^{-1} for the pulses as a whole warrants its genetic improvement. The same has been practiced mainly through single-plant selection. Germplasm collected from the natural habitats has been effectively utilized in the development of promising cultivars as below:

Bihar:	BR-5, BR-10, Madhu
Himachal Pradesh:	HPK-2, HPK-4, HPK-5, HPK-6
Andhra Pradesh:	POM-1, POM-2, VZMI
Rajasthan:	Maru Kulthi, AK-21, AK-42
Orissa:	DS-2-1-1, DS-2-2, OGH-11, OGH=60-1, S-27, S-28, S-1264
Maharashtra:	Dapali local, KBR-1, Local Red, SINA, Man
Tamil Nadu:	Paiyar-1, Co-1, PDP-1236, Kurunkollu 35-5-122, 35-5-123

Table 8 Varieties of moth bean released and notified during last two decades

S. no.	Variety	Year	Moth growing area	Maturity (days)	Grain yield (kg/ha)
1.	CAZRI Moth-1	1999	Whole country	72–75	550–650
2.	RMO-435	2002	Rajasthan	62–67	550–600
3.	CAZRI Moth-2	2003	Rajasthan, Haryana, Gujarat, Maharashtra	65–67	1000–1200
4.	RMB-25	2004	Rajasthan	67–70	600–700
5.	RMO-423	2004	Rajasthan	67–70	550–600
6.	RMO-257	2005	Rajasthan	62–67	500–550
7.	CAZRI Moth-3	2005	Rajasthan, Haryana, Gujarat, Maharashtra	65–75	550–750
8.	RMO 2251	2016	Rajasthan, Gujarat	60–67	500–600
9.	CAZRI-4	2022	Rajasthan, Gujarat	65–72	500–800
10.	CAZRI-5	2022	Rajasthan, Gujarat	65–72	600–700

Karnataka:	KBH-1, PHG-9, Herbal Hurali-1, and Hebbal Hurali-2
Uttaranchal:	VLG-1
Jharkhand:	K-82

The cultivars developed from the hills are long duration (120–130 days), but those selected from arid Rajasthan, particularly Bheelara, Rajasthan (AK-24, AK-42), and Jodhpur (Maru Kulthi), are early to mature in 80–90 days. Hybridization programming has not been taken up in horsegram. Newly developed cultivars (PHG-9, AK-21, AK-42, CRIDA-1-18R) are less winey, have short in stature, compact nature, and have a good yield potential of almost 800–12,000 kg ha^{-1} (Table 9).

10 Maintenance Breeding

Maintenance breeding is a sort of genetic preparedness for providing genetic relief to the cultivars subjected to various environment lethal shocks. It is a nut and bolt type repair mechanism of the old reckoned otherwise ruling genetic material (like variety, hybrid, synthetic, composite, and farmers material) of the region. It is readiness and providing F_2 and M_2 material of the ruling variety to compensate for and overcome segmental defects arising due to management and environmental adversities. It is continued and parallel breeding to provide short-term genetic relief to the serving ruling cultivar(s) of the region. It would not allow to proliferate genetic vacuum for productivity. Besides, maintenance breeding being an alternative to the existing genetic resource, it amounts to faster breeding speed, curtailing the breeding period. The aim is to keep ready genetic material, population to be immediately used in the wake of an emergency arising due to lost, damaged spoiled genetic material by natural or unnatural unexpected reasons and causes.

This approach of ever-preparedness and continued breeding to face any eventually is too valid in arid legumes. Reason is their rainfed nature of cultivation, location-specific genetic buffering, few varieties of large areas of coverage and stressful situations of environment and inputs management, poor land support, few institutions working with these crops, and non-willingness of big seed companies. Slow release of arid legume varieties justifies genetic parallel preparedness. For instance, till date no variety of moth bean after 2004 (CAZRI Moth-2), a variety like RGC-936, or guar is most popular in arid areas, and no substitute in adaptation and earliness seems to exist in the whole of Rajasthan. The only early maturing cowpea variety is RC-101 (60 days) for the last 20 years. In horsegram, there is no suitable early maturing high-yielding alternative to AK-21 and AK-41. These varieties and other ones in other states like HG-365 and HG-563 guar varieties in Haryana are 25 years old. The same has lost genetic purity, and their potentials are creating a vacuum. Such a genetic gap could have been filled up had we initiated maintenance breeding of these popular varieties or kept genetic material ready to compensate for or replace these varieties. Parallel efforts in the mother institutes of these cultivars should have been continued to create genetic purity or improve these cultivars, besides breeding new material as new notified varieties.

Thus, maintenance breeding, genetic repairing of existing ruling varieties, and breeding for new varieties (for facing new challenges) must go hand in hand with the mandatory arid legume crops so as to keep productivity alive in the arid dry lands. Maintenance breeding is to be done for disease susceptibility, which decreases grain yield and reduces the age of a cultivar. Multiline concept can be operated in these crops, proportionally adding or decreasing on a grain weight basis in newly synthesized maintained variety. All attempts may be made against Bacterial Leaf Blight in guar, Anthracnose, and Powdery Mildew in cowpea and horsegram and yellow mosaic virus in cowpea and moth bean.

Table 9 Varieties of horsegram released and notified from 2000 to date

S. no.	Variety	Year	Horsegram growing area	Maturity period (days)	Grain yield (kg/ha)
1.	VL Gahat-8	2005	North India	115–135	900–1000
2.	VL Gahat-10	2006	Northern India	113–117	700–800
3.	CRIDA 1-18R	2007	South India	90–100	800–850
4.	VL Gahat-15	2009	North & Central India	90–100	600–700
5.	VL Gahat-19	2010	North India	100–105	850–900
6.	Criddatha (CRHG-4)	2010	South India	110–112	780–800
7.	India Kulthi 1 (IKGH-05-01)	2011	Chhattisgarh	95–100	800–850
8.	Gujarat Dantiwada Horsegram-1 (GRHG-5)	2012	Gujarat, Rajasthan, U.P. and U.K., JH	94	550–575
9.	Cridehasha (CRHG-19)	2014	AP, KR, KL & TN	100	700–900
10.	Pratap Kulth-2 (AK 53)	2016	RJ, GJ, UK, CG & MH	80	500–650
11.	Cridavardhan (CRHG-22)	2016	South India	94–100	800–900
12.	Phule Sakos (SHG-0628-4)	2016	Maharashtra	110–120	1000–1100
13.	Chhattisgarh Kulth-3 (BHG-03)	2017	Chhattisgarh	85–90	500–600
14.	Chhattisgarh Kulthi-2 (BHW-1)	2017	Chhattisgarh & neighboring states	104–112	600–750
15.	Blasa Kulth-1 (BSP-15-1)	2018	Chhattisgarh & neighboring states	100–107	900–1000
16.	Alokh Kulthi (BSP 17-1)	2021	Northern parts of the country	105–106	900–1000
17.	Sabri Kulthi (BSP 17-3)	2021	Northern part of the country	100–103	900–1000
18.	Anantha Vulava (ATPHG 11)	2021	Northern part of the country	105–110	800–1100

Registration of genetic material is given in Table 10

Table 10 Registration of arid legumes

Crop	National identity	Donor identity	INGR No	Year	Pedigree	Institute	Novel feature
Guar	IC-299971	GDM-1 (Guar determinate mutant)	O4037	2004	Mutant from FS-277	CAZRI, Jodhpur	Determinate type
Cowpea	IC-299972	CAZC-B	04038	2004	Charodi-1	CAZRI, Jodhpur	Black seed coat color
Horsegram	IC-212722	IC-212722/GN-1882	02007	2002	Natural	Garhwal Univ. Srinagar & NBPGR Bhowali	High grain protein (35.0%)
Moth bean	IC-296803	CZM-32	O1924	2001	Mutant from Jadia	CAZRI, Jodhpur	Drought hardy, erect, early maturity
Moth bean	IC-432859	CZM-12	04095	2004	Mutant from RMO-40	CAZRI, Jodhpur	Single stem, 60 days maturity, high influx of Na

11 Future Thrust

These crops need to be addressed, managed, and improved as natural genetic climate-resistant resource crops. They have considerable drought and heat tolerance potentials under poor nourishment. Applying repeated shuttle screening in natural habitats and mining more heat shock genes (HSGs) through biotechnological means, their existing status tolerance can be raised to 45 °C and 50 days drought in open fields. Being the cheapest organic source of food, fodder, feed, and industrial products, these crops must be provided organic tags for specific pocket zones, ensuring financial gains to the financially ridden arid and semi-arid farmers. Along with removing hurdles in the export of guar gum powder and its sub-products (earning US$ 617.14 M annually), stress must also be given to the manufacturing of chemical derivatives (high-value, low-volume, chemical products) used as raw material in various industries. It would encourage financial gains by three- to fourfold. Promising, semi-erect to erect, physiologically efficient plant types with a life span of 65–70 days are required for these crops, particularly for guar (no variety available). The rainy days in their habitats are hardly 60–65 days; thus, matching the two will allow productive filling of arid legumes. Arid regions are not suited to cultivars with more than 70 days.

The area of moth bean is shrinking, coming down to 9.0 lakh ha vis-à-vis 15.0 lakh ha two decades before. Mung bean, due to early maturity (55–60 days), attractive grains, and better marketing avenues, is competing better with moth bean. Thus, lucrative moth bean grains with alternative uses are required to restore the popularity of moth beans. Availability of high-quality seeds of moth bean must be ensured in hyper-arid districts like Bikaner, Barmer, Jaisalmer, Phalodi, and Churu in Rajasthan; Bhiwani and Rewari districts of Haryana; and Banaskantha and Bhuj districts of Gujarat. Being an organic source, helping soil fertility, and other virtues, the arid legumes must be genetically loaded with resistance against important and devastating diseases like Bacterial Leaf blight (BLB) and Alternaria Leaf Spot (ALS) in guar. Anthracnose in cowpea and horsegram and YMV in moth bean and cowpea. The same would ensure long-term stability.

Arid legumes are important components of the farming system, having versatile utility; thus, breeding for multi-purpose varieties (grain, fodder, feed, vegetable, and export) is done to explore multi-market and multi-diversity in fragile agri-ecosystem. Attention must be given to the presence of anti-nutritional factors in horsegram—such as saponins, phytic acid, trypsin inhibitors, and α-amylase inhibitors—which can limit its nutritional utilization. Mitigating these compounds is essential to ensure that the crop's abundant protein content (29–30%) can be more widely and effectively utilized in human diets. There is a need to improve the cooking quality of grain. Authentication and validation of multi-medicinal uses of horsegram in kidney stones, better sleep, irregular monthly periods in ladies, cough, acidity, and constipation are needed. But cowpea is required for effective control of *Helicoverpa armigera* (pod borer). Guar varieties representing specific guar viscosity levels of 1000–3000 cP values and 3000–8000 cP values are used in food-grade industries and other non-food-grade industries, respectively, and need to be developed; in other words, breeding for industrial uses is required.

References

Aditya JP, Bhartiya A, Chahota RK, Joshi D, Chandra N, Kant L, Pattanayak A (2019) Ancient orphan legume horse gram: a potential food and forage crop of future. Planta 250:891–909

Alshameri A, Al-Qurainy F, Gaafar AR, Khan S, Nadeem M, Alansi S (2020) Identification of heat-responsive genes in guar [*Cyamopsis tetragonoloba* (L.) Taub]. Int J Genomics 2020:3126592

Anonymous (2023) Project coordinator S report 2022-23 AICRP on Kharif Pulses, IIPR, Kanpur

APEDA (2023). https://agriexchange.apeda.gov.in. Accessed 30 July 2023

Araméndiz-Tatis H, Cardona-Ayala C, Espitia-Camacho M, Hernández-Murillo JR (2022) Stigmatic receptivity and hybridization in cowpea beans (*Vigna unguiculata* L.(Walp.)). Revista Colombiana de Ciencias Hortícolas 16(2)

Baath GS, Northup BK, Gowda PH, Turner KE, Rocateli AC (2018) Mothbean: a potential summer crop for the Southern Great Plains. Am J Plant Sci 9(07):1391

Baudoin JP, Marechal R (1985) Genetic diversity. In: Singh SR, Rachie KO (eds) Vigna-cowpea research, production and utilization. Wiley

Baudoin JP, Marechal R (1989) Taxonomy and evolution of the genus Vigna. Mungbean, p 1

Bhadkaria A, Gupta N, Narvekar DT, Bhadkariya R, Saral A, Srivastava N, Koul KK, Bhagyawant SS (2020) ISSR-PCR approach as a means of studying genetic variation in moth bean (*Vigna aconitifolia* (Jacq.) Maréchal). *Biocatalysis and Agricultural*. Biotechnology 30:101827

Bhagwan D, Solanki KR, Hooda MS (1983) Studies on genotype environment interactions for gum content in clusterbean. Agric Sci Digest 3(2):81–84

Bhatt RK, Jukanti AK, Roy MM (2015) Clusterbean [*Cyamopsis tetragonoloba* (L.). Taub.] an industrial arid legume: A review. Legumes Res:1–8, Online ISSN:0976-0571/print ISSN:0250-5371

Bhatt RK, Sharma R, Mahla HR, Kalia RK (2016) Guar production, utilization and marketing: current scenario and future prospects. Central Arid Zone Research Institute, Jodhpur, p 40

Bhor TJ, Kute NS, Dumbre AD, Sarode ND (1997) Heterosis and inbreeding depression in cowpea. Indian J Agric Res 31:122–126

Bhosale SS, Kothekar VS (2010) Mutagenic efficiency and effectiveness in clusterbean (Cyamopsistetragonoloba (L.) Taub). J Phytology 2:21–27

Bisht IS, Bhat KV, Lakhanpaul S, Latha M, Jayan PK, Biswas BK, Singh AK (2005) Diversity and genetic resources of wild Vigna species in India. Genet Resour Crop Evol 52:53–68

Boghara MC, Dhaduk HL, Kumar S, Parekh MJ, Patel NJ, Sharma R (2016) Genetic divergence, path analysis and molecular diversity analysis in cluster bean (*Cyamopsis tetragonoloba* L. Taub.). Ind Crop Prod 89:468–477

Boukar O, Fatokun CA, Roberts PA, Abberton M, Huynh BL, Close TJ, Kyai-Boahen S, Higgins TJV, Ehlers JD (2015) Cowpea. In: Grain legumes. Springer, pp 219–250

Chahota RK, Sharma TR, Sharma SK, Kumar N, Rana JC (2013) Horsegram. In: Genetic and genomic resources of grain legume improvement. Elsevier, pp 293–305

Chahota RK, Shikha D, Rana M, Sharma V, Nag A, Sharma TR, Isobe S (2017) Development and characterization of SSR markers to study genetic diversity and population structure of horsegram germplasm (Macrotyloma uniflorum). Plant Mol Biol Report 35:550–561

Chandola RP, Trehan KB, Bagrecha LR (1969) Varietal resistance of *Bruchus* sp in cowpea (*Vigna sinensis*) under storage conditions. Curr Sci 38:370–371

Chaudhary BS, Lodhi GP (1981) Studies on the inheritance of five qualitative characters in clusterbean [Cyamopsistetragonoloba (L.) Taub.]. Euphytica 30:161–165

Chaudhari SR, Dahat DV, Rajput HJ, Chavan BH (2021) Genetic diversity and variability studies in moth bean [Vigna aconitifolia (Jacq.)]. J Pharmacogn Phytother 10(1):1338–1341

Chauhan JS, Singh KH, Choudhury PR, Singh RK, Mishra JP (2023) Growth, varietal development and cultivar diversity in seed production chain of clusterbean (*Cyamopsis tetragonoloba*) in India. Indian J Agric Sci 93(8):811–820

Choudhary MK, Sharma R, Mahla HR, Singh D, Sarolia DK (2021) Study the inheritance of branching behaviour and bearing pattern in clusterbean (*Cyamopsis tetragonoloba* (L.) Taub.). J Agric Ecol 12:94–97

Choudhary S, Isobe S, Chahota RK (2022) Elucidation of drought tolerance potential of horsegram (*Macrotyloma uniflorum* Var.) germplasm using genome wide association studies. Gene 819:146241. https://doi.org/10.1016/j.gene.2022.146241

Coulibaly S, Pasquet RS, Papa R, Gepts PAFLP (2002) AFLP analysis of the phenetic organization and genetic diversity of Vigna unguiculata L. Walp. reveals extensive gene flow between wild and domesticated types. Theor Appl Genet 104:358–366

Dabi RK, Gupta HC, Sharma SK (1979) Relative susceptibility of some cowpea varieties to pulse beetle, *Callosobruchus macuhulatus* (*Fabriecius*). Indian J Agric Sci 49:48–50

Devi SM, Jayamani P (2020) Molecular genetic diversity in cowpea {Vigna unguiculata (L.) Walp.} genotypes. J Food Legumes 33(1):6–9

Ehlers JD (1994) Correlation of performance of sole crop and intercrop cowpea with and without protection from insect-pests. Field Crop Res 36:133–143

Fahad A-Q, Alshameri A, Gaafar A-R, Khan S, Nadeem M, Alameri AA, Tarroum M, Ashraf M (2019) Comprehensive stress-based De Novo transcriptome assembly and annotation of Guar (*Cyamopsis tetragonoloba* (L.) Taub.): an important industrial and forage

crop. Int J Genomics 2019:Article ID 7295859, 14. https://doi.org/10.1155/2019/7295859
Faris DG (1965) The origin and evolution of the cultivated forms of Vigna sinensis. Can J Genet Cytol 7(3):433–452
Fatokun CA, Boukar O, Muranaka S (2012) Evaluation of cowpea (Vigna unguiculata (L.) Walp.) germplasm lines for tolerance to drought. *Plant Genetic*. Resources 10(3):171–176
Feleke Y, Pasquet RS, Gepts P (2006) Development of PCR-based chloroplast DNA markers that characterize domesticated cowpea (Vigna unguiculata ssp. unguiculata var. unguiculata) and highlight its crop-weed complex. Plant Syst Evol 262:75–87
Fuller DQ, Harvey EL (2006) The archaeobotany of Indian pulses: identification, processing and evidence for cultivation. Environ Archaeol 11(2):219–246
Fuller DQ, Murphy C (2018) The origins and early dispersal of horsegram (Macrotyloma uniflorum), a major crop of ancient India. Genet Resour Crop Evol 65:285–305
Gautam M, Chahota RK (2022) Metabolite profiling and protein quantification to a large library of 96 horsegram (Macrotyloma uniflorum) germplasm. Sci Rep 12(1):7865
Gbedevi KM, Boukar O, Ishikawa H, Abe A, Ongom PO, Unachukwu N et al (2021) Genetic diversity and population structure of cowpea [*Vigna unguiculata* (L.) Walp.] germplasm collected from Togo based on DArT markers. Genes 12(9):1451
Gomes AMF, Draper D, Talhinhas P, Santos PB, Simões F, Nhantumbo N et al (2020) Genetic diversity among cowpea (Vigna unguiculata (L.) Walp.) landraces suggests central Mozambique as an important hotspot of variation. Agronomy 10(12):1893
Gopala Krishnan S, Dwivedi NK, Singh JP (2011) Primitive weedy forms of guar, adak guar: possible missing link in the domestication of guar [Cyamopsis tetragonoloba (L.) Taub.]. Genet Resour Crop Evol 58:961–966
Govade GK (1950) Improved varieties of pulses and lesser known millets for Gujarat. Indian FMG 11:194–196
Grigoreva E, Barbitoff Y, Changalidi A, Karzhaev D, Volkov V, Shadrina V, Safronycheva E, Ben C, Gentzbittel L, Potokina E (2021) Development of SNP Set for the Marker-Assisted Selection of Guar (*Cyamopsis tetragonoloba* (L.) Taub.) Based on a custom reference genome assembly. Plan Theory 10:2063. https://doi.org/10.3390/plants10102063
Gupta SK, Gopalakrishna T (2010) Development of unigene-derived SSR markers in cowpea (*Vigna unguiculata*) and their transferability to other Vigna species. Genome 53(7):508–523
Herniter IA, Muñoz-Amatriaín M, Close TJ (2020) Genetic, textual, and archeological evidence of the historical global spread of cowpea (Vigna unguiculata [L.] Walp.). *Legume*. Science 2(4):e57
Huynh BL, Close TJ, Roberts PA, Hu Z, Wanamaker S, Lucas MR et al (2013) Gene pools and the genetic architecture of domesticated cowpea. Plant Genome 6:1–8
Hymowitz T (1972) The trans-domestication concept as applied to guar. Econ Bot 26(1):49–60
Ige OE, Olotuah OF, Akerele V (2011) Floral biology and pollination ecology of cowpea (Vigna unguiculata L. Walp). Mod Appl Sci 5(4):74
Isely D (1983) A revision of Macrotyloma (Leguminosae) B. Verdcourt. 138 pp. illus. Hooker's Icones Plantarum 38 (4). Spottiswoode Ballantyne, London, 1982
Isiaka AI, Agossou CO, Agbolade JO, Adje CA, Hotegni NVF, Tossou MG, Oselebe H, Achigan-Dako EG (2023) Phenology, floral organs ultra-structure, traits correlation, stigma receptiveness, pollen viability and germinability in horsegram (*Macrotyloma uniflorum* Lam.) Verdc. S Afr J Bot 161:444–453
Jackai LE (1982) Field screening technique for resistance of cowpea (*V. unguiculata*) to the pod borer, *Maruca testulalis* (Geyer) (*Liepidopotera*): Pyralidae. Bull Entomol Res 72:145–156
Katoch M, Mane RS, Chahota RK (2022) Identification of QTLs Linked to Phenological and Morphological Traits in RILs Population of Horsegram (*Macrotyloma uniflorum*). Front Genet 12:762604
Kharbet RPS, Singh VP, Tomer YS (1987) Moth bean (*Vigna aconitifolia* (Jacq) Marechal) – a review. Forage Res 13(2):113–132
Kouadio D, Echikh N, Toussaint A, Pasquet RS, Baudoin JP (2007) Organisation du pool génique de Vigna unguiculata (L.) Walp.: croisements entre les formes sauvages et cultivées du niébé
Kumar D (2005) Status and direction of arid legumes research in India. Indian J Agric Sci 77:375–391
Kumar D (2006) Horsegram research – an introduction. In: Kumar D (ed) Horsegram in India. Indian Arid Legumes Society, Jodhpur, pp 1–10, p 144
Kumar D (2007) Production technology for horsegram. Central Arid Zone Research Institute, p 20
Kumar D, Rodge AB (2012) Status, scope and strategies of arid legumes research in India – a review. J Food Legumes 25(4):255–272
Kumar D, Singh NB (2002) Guar-an introduction. In: Kumar D, Singh NB (eds) Guar in India. Indian Arid Legumes Society, Jodhpur, pp 1–10, 223
Kumar D, Tarafar JC, Kumar P (1998) Variability of growth, yield and nutrient concentration in moth bean. Ann Arid Zone 37:395–400
Kumar S, Modi AR, Parekh MJ, Mahla HR, Sharma R, Fougat RS, Yadav D, Yadav NR, Patil GB (2017) Role of conventional and biotechnological approaches for genetic improvement of cluster bean. Indus Crops Prod 97:639–648
Kumar S, Sagar P, Pandey J, Kumar K, Pratap A, Singh NP, Rathore M (2023) Genetic diversity and population structure analysis in popular cowpea (Vigna unguiculata (L.) Walp) cultivars. Indian J Genet Plant Breed 83(01):143–145
Kumari M, Padhi SR, Chourey SK, Kondal V, Thakare SS, Negi A, Gupta V, Arya M, Yasin JK, Singh R, Bharadwaj C (2023) Unveiling Diversity for Quality Traits in the Indian Landraces of Horsegram [Macrotyloma uniflorum (Lam.) Verdc.]. Plan Theory 12(22):3803
Lo S, Muñoz-Amatriaín M, Boukar O, Herniter I, Cisse N, Guo YN et al (2018) Identification of QTL controlling domestication-related traits in cowpea (*Vigna unguiculata* L. Walp). Sci Rep 8(1):6261
Lodhi GP, Boora KS, Jain S, Balchand (1990) Heterosis for fodder yield and quality characters in cowpea (*V. unguiculata* (L.) Walp). Crop Res (Hissar) 3:66–73
Mahadevappa VG, Raina PL (1978) Nature of some Indian legumes lipids. J Agric Food Chem 26:1241–1243
Mahla HR, Sharma R (2022) An easy method of artificial hybridization in two arid legumes, guar (*C. tetragonoloba* Taub.) and moth bean (*V. aconilifolia* (Jacq) Marechal). Indian J Genetic Plant Breed 82(1):109–112
Mahla HR, Kumar D, Shekhawat A (2010) Effectiveness and efficiency of mutagens and induced variability in cluster bean (Cyamopsistetragonoloba). Indian J Agric Sci 80:1033–1037
Mahla HR, Sharma R, Bhatt RK (2018) Effect of gamma irradiations on seed germination, seedling growth and mutation induction on cluster bean [*Cyamopsis tetragonoloba* (L.) Taub.]. Indian J Genet Plant Breed 78(2):261–269
Mahla HR, Sharma R, Rathore VS, Meena SC, Swami S (2022) Trait specific genetic stocks for Clusterbean [Cyamopsis tetragonoloba (L.) Taub.]. Improv Ann Arid Zone 61(3&4):187–192
Mahudeswaran K, Veeraswamy R, Palaniswamy GA, Rathnaswamy R, Rajasekaran VPA, Marappan PV (1973) Co-2 a vegetable-cum-grain cowpea. Madras Agric J 60:1844–1845
Manivannan A, Anandakumar CR, Yuvaraja A, Sritharan N (2017) Genetic variability and interrelationship among biochemical traits of Indian clusterbean (*Cyamopsis tetragonoloba* (L.) Taub.) germplasm. Electron J Plant Breed 8(3):842–848
Marechal R, Mascherpa JM, Stainer (1978a) Combinations and new names in the genera *Phaseolus, Minkelassia, Macroptilium Ramirezella* and *Vigna*. Taxon 28:199–202

Marechal R, Mascherpa JM, Stainier F (1978b) Etude taxonomique d'um grope d'especes des generes *phaseolus* et *Vigna* (*Papilionaceae*) sur la base donnees morphologique et polliniques traitees pour analyse Informatique. Boissiera 28:1–273

Mishra SK, Choudhary AB, Chand D (2004) Cowpea genetic improvement. In: Kumar D, Singh NB (eds) Cowpea in India. Indian Arid Legumes Society, Jodhpur, pp 37–48, p 281

Morris JB (2010) Morphological and reproductive characterization of guar (Cyamopsis tetragonoloba) genetic resources regenerated in Georgia, USA. Genet Resour Crop Evol 57:985–993

Muñoz-Amatriaín M, Lo S, Herniter IA, Boukar O, Fatokun C, Carvalho M, Castro I, Guo YN, Huynh BL, Roberts PA, Carnide V (2021) The UCR Minicore: a resource for cowpea research and breeding. Legume Sci 3(3):e95

Ng NQ, Marechal R (1985) Cowpea taxonomy, origin and germplasm. In: Cowpea research, production and utilization, pp 11–21

Nkhoma N, Shimelis H, Laing MD, Shayanowako A, Mathew I (2020) Assessing the genetic diversity of cowpea [Vigna unguiculata (L.) Walp.] germplasm collections using phenotypic traits and SNP markers. BMC Genet 21(1):1–16

Padulosi S, Ng NQ (1997) Origin, taxonomy, and morphology of Vigna unguiculata (L.) Walp. In: Advances in cowpea research, p 1

Pathak R, Singh SK, Singh M, Henry A (2010) Molecular assessment of genetic diversity of clusterbean (*C. tetragonoloba* L.). J Genet 89:243–246

Patil BM, Rane GM (2015) Gamma radiation induced chlorophyll mutations in clusterbean (Cyamopsis tetragonoloba (L.) Taub) Var. NCB-12. Int J Allied Pract Res Rev 2:75–85

Pratap A, Basu PS, Gupta S, Malviya N, Rajan N, Tomar R, Madhavan L, Nadarajan N, Singh NP (2014) Identification and characterization of sources for photo-and thermo-insensitivity in Vigna species. Plant Breed 133(6):756–764

Priti, Panchta R, Yadav RC, Pahuja SK, Yadav NR (2023) Linkage mapping and identification of quantitative trait loci associated with bacterial leaf blight resistance and gum content in cluster bean using the interspecific population (HG563× *Cyamopsis serrata*). Genet Resour Crop Evol 70(5):1501–1516

Punia A, Yadav R, Arora P, Choudhary A (2009) Molecular and morphological characterization of superior clusterbean (*Cyamopsis tetragonoloba*) varieties. J Crop Sci Biotechnol 12:143–148

Purkayastha S, Kaur B, Dilbaghi N, Chaudhury A (2006) Evaluation of cluster bean genotypes for resistance to charcoal rot caused by Macrophomina phaseolina using different host inoculation methods. J Crop Improv 15(1):67–79

Ramanujam S (1972) Some salient results of pulse research. Indian FMG 21(2):12–14

Ravelombola W, Manley A, Adams C, Trostle C, Ale S, Shi A, Cason J (2021) Genetic and genomic resources in guar: a review. Euphytica 217:1–19

Reddy DR, Saidaiah P, Reddy KR, Pandravada SR, Geetha A (2019) Genetic variability for growth, pod and quality attributes in germplasm of cluster bean [Cyamopsis tetragonoloba (L.) Taub.]. Legume Res Int J 42(5):620–624

Sahoo S, Sanadya SK, Sharma AK, Shekhawat SS (2022) Assessment of genetic diversity of mothbean (Vigna aconitifolia (Jacq.) Marechal) germplasm. J Food Legumes 35(2):95–99

Saini ML, Paroda RS, Arora SK (1982) Genotype environment interaction for gum content in clusterbean (Cyamopsis tetragonoloba). Genetics Agriculture. 36 (1/2): 57–62. Saini, H.R. and Malik, H.C. 1947. Pulses in Punjab. Indian J Agric Sci 17:281–288

Saini ML, Malik JS, Solanki KR, Singh JV (1983) G×E interactions for green fodder yield in moth bean. Indian J Agric Sci 53:83–84

Santos JRP, Ndeve AD, Huynh BL, Matthews WC, Roberts PA (2018) QTL mapping and transcriptome analysis of cowpea reveals candidate genes for root-knot nematode resistance. PLoS One 13(1):e0189185

Seedneet (2023) https://seednet.gov.in. Accessed 6–8 July 2023

Sharma RC, Mathur JR (1982) Studies of flower shedding, shedding behavior in moth bean. Sci Cult 48:64–65

Sharma RC, Mathur JR (1983) Jadia a promising cultivar of moth bean. Indian FMG 32(ii):13–15 & 16

Sharma RC, Bhati SD, Mathur JR (1986) Jwala – a new improved variety of moth bean. Kheti 39(2):27–28

Sharma V, Rana M, Katoch M, Sharma PK, Ghani M, Rana JC et al (2015) Development of SSR and ILP markers in horsegram (Macrotyloma uniflorum), their characterization, cross-transferability and relevance for mapping. Mol Breed 35:1–22

Sharma R, Mahla HR, Kumar S, Gaikwad K (2021) Study of correlation, path coefficient and linkage of flower colour and hairiness with yield controlling quantitative traits in segregating population of cluster bean. Curr Plant Biol 26:100202

Singh SR (1977) Cowpea cultivars resistant to insect-pests in world germplasm collection. Trop Grain Legumes Bull 9:3–7

Singh JV, Arora RN (2002) Genetics and breeding. In: Kumar D, Singh NB (eds) Guar in India. Indian Arid Legumes Society, Jodhpur, pp 11–34, 223

Singh R, Singh S, Paroda RS (1986) A study of heterosis, heritability and genetic advances in cowpea. Forage Res 12:47–47

Solbrig H (1983) Anatomical studies on *Dolichos bioflorus* L. with special reference to the ontogeny of calcium oxalate crystals in the seeds. Dissertation Fachbereich Biologie Universitat, Hamburn, p 119

Somta P, Jomsangawong A, Yundaeng C, Yuan X, Chen J, Tomooka N, Chen X (2018) Genetic dissection of Azuki bean weevil (Callosobruchus chinensis L.) resistance in moth bean (Vigna aconitifolia [Jaqc.] Maréchal). Genes 9(11):555

Sudha N, Begam JM, Shambulingappa KP, Babu CK (1995) Nutrients and some anti-nutrients in horsegram. Indian J Exp Biol 26:618–620

Sultan M, Zakir N, Rabbani MA, Shinwari ZK, Masood MS (2013) Genetic diversity of guar (*Cyamopsis tetragonoloba* L.) landraces from Pakistan based on RAPD markers. Pak J Bot 45(3):865–870

Suranjika S, Pradhan S, Nayak SS et al (2022) *De novo* transcriptome assembly and analysis of gene expression in different tissues of moth bean (*Vigna aconitifolia*) (Jacq.). Marechal BMC Plant Biol 22:198. https://doi.org/10.1186/s12870-022-03583-z

Suthar JD, Rajpar I, Ganjegunte GK (2018) Evaluation of Guar (Cyamopsis tetragonoloba L.) genotypes performance under different irrigation water salinity levels: Growth parameters and seed yield. Ind Crop Prod 123:247–253

Takahashi Y, Somta P, Muto C, Iseki K, Naito K, Pandiyan M, Natesan S, Tomooka N (2016) Novel genetic resources in the genus Vigna unveiled from gene bank accessions. PLoS One 11(1):e0147568

Tarafdar JC, Kumar D (2003) Moth bean (*V. aconitifolia*) cultivars compared for root nutrient uptake efficiency in arid N.W. India. J Arid Environ 55:215–222

Thakur O, Randhawa GS (2018) Identification and characterization of SSR, SNP and InDel molecular markers from RNA-Seq data of guar (*Cyamopsis tetragonoloba*, L. Taub.) roots. BMC Genomics 19(1):1–14. https://doi.org/10.1186/s12864-018-5205-9

Tikka SBS, Asawa BM, Kumar S (1977) Correlated response to selection in mothbean. Gujarat Agric Univ Res J 3:1–4

Tikka SBS, Asawa BM, Sharma RK, Kumar S (1980) Selection indices in moth bean (*V. aconitifolia* Jacq Marechal). Tropical Grain Legume Bull 17(18):46–48

Tiwari B, Kalim S, Tyagi N, Kumari R, Bangar P, Barman P et al (2018) Identification of genes associated with stress tolerance in moth bean [*Vigna aconitifolia* (Jacq.) Marechal], a stress hardy crop. Physiol Mol Biol Plants 24:551–561

Tiwari AK, Shivhare AK, Tikle A (2020) Clusterbean. Directorate of Pulses Development, Government of India, Bhopal (www.dpd.gov.in), p 2

Todd JW, Canerdy TD (1968) Reactin of Southern peas to the cowpea curculio. J Econ Entomol 61:1327–1329

Tomooka N, Pandiyan M, Senthil N, Ramamoorthi N, Kaga A (2009) Collection and conservation of crops and their wild relatives in Tamil Nadu, India. Ann Rep Explor Introd Plant Genet Resour 25:83–109

Tribhuvan KU, SV AM, Sharma P, Das A, Kumar K, Tyagi A et al (2019) Identification of genomic SSRs in cluster bean (*Cyamopsis tetragonoloba*) and demonstration of their utility in genetic diversity analysis. Ind Crop Prod 133:221–231

Vavilov NI (1926) Studies on the origin of cultivated plants… Institut de Botanique Appliquée et d'Amélioration des Plantes

Velu S, Mullainathan L, Arulbalachandran D (2012) Induced morphological variations in cluster bean (Cyamopsistetragonoloba (L.) Taub). Int J Curr Trends Res 1:48–55

Verma SK, Singh CK, Taunk J, Chandra Joshi D, Kalia S, Dey N, Singh AK (2022) Vignette of Vigna domestication: from archives to genomics. Front Genet 13:960200

Vir O, Singh AK (2015a) Moth bean [Vigna aconitifolia (Jacq.) Marechal] germplasm: evaluation for genetic variability and inter characters relationship in hot arid climate of western Rajasthan, India. Legume Res Int J 38(6):748–752

Vir O, Singh AK (2015b) Variability and correlation analysis in the germplasm of cluster bean [Cyamopsis tetragonoloba (L.) Taub.] in hyper hot arid climate of Western India. Legume Res Int J 38(1):37–42

Viswanatha KP, Balaraju, Chakravarthy KK (1998) Heterosis and inbreeding depression in cowpea. [*V. unguiculata* (L.)] Walp. Mysore J Agric Sci 32:181–185

Xiong H, Shi A, Mou B, Qin J, Motes D, Lu W et al (2016) Genetic diversity and population structure of Cowpea (*Vigna unguiculata* L. Walp). PLoS One 11(8):e0160941. https://doi.org/10.1371/journal.pone.0160941

Xiong H, Qin J, Shi A, Mou B, Wu D, Sun J, Shu X, Wang Z, Lu W, Ma J, Weng Y (2018) Genetic differentiation and diversity upon genotype and phenotype in cowpea (Vigna unguiculata L. Walp.). Euphytica 214:1–15

Yadav AK, Singh CK, Kalia RK, Mittal S, Wankhede DP, Kakani RK, Singh AK (2023) Genetic diversity, population structure, and genome-wide association study for the flowering trait in a diverse panel of 428 moth bean (*Vigna aconitifolia*) accessions using genotyping by sequencing. BMC Plant Biol 23(1):228

Yundaeng C, Somta P, Amkul K, Kongjaimun A, Kaga A, Tomooka N (2019) Construction of genetic linkage map and genome dissection of domestication-related traits of moth bean (Vigna aconitifolia), a legume crop of arid areas. Mol Gen Genomics 294:621–635

Underutilized Grain Legumes Breeding

Kuldeep Tripathi, U. S. Thendral, Rinky Resma Panda, Padmavati G. Gore, Ravi Pamarthi, V. P. Sahi, and G. P. Singh

Abstract

Grain legumes play a crucial role in global food security as rich sources of protein, carbohydrates, minerals, and vitamins. Despite their nutritional value, grain legumes remain underutilized compared to major cereals. This chapter explores the challenges and opportunities in breeding underutilized grain legumes to enhance their productivity and nutritional quality. This chapter highlights the nutritional importance of grain legumes and their potential to contribute to global food and nutrition security. It discusses the factors limiting their widespread adoption, including low yield, susceptibility to pests and diseases, and limited genetic diversity. The recent advances in grain legume breeding, with a focus on innovative approaches to improve yield, disease resistance, and nutritional content, is reviewed. It examines genetic studies on qualitative and quantitative traits, breeding objectives, and the progress made in varietal development through conventional and modern approaches, including marker-assisted breeding, genomic tools, CRISPR-Cas technology, and de novo domestication. It discusses the application of genomic tools, participatory breeding, and molecular marker-assisted selection in accelerating the development of improved varieties. Finally, the research priorities and future directions for underutilized grain legume breeding are identified, and it emphasizes the need for collaborative efforts among researchers, farmers, and policymakers to promote the wider adoption of these crops.

Keywords
Underutilized grain legumes · Breeding · Yield · Disease resistance · Nutritional quality · Food security

1 Introduction

Globally, there are approximately 4,00,000 plant species worldwide (Vasconcelos et al. 2020), among them only 12 crops account for most of the present global food production. (Singh et al. 2019). Additionally, a mere dozen animal species supply 90% of the global animal protein, and three crop species (rice, wheat, and maize) furnish 50% of the plant-based calories in the average person's diet, indicating a significant decline in both the gene pool and the genetic resources of both plants and animals (FAO 2009). It is noteworthy that despite the existence of over a thousand wild plant species, including legumes, a large number remain untapped and undomesticated as potential food crops. These staple crops catered to a large proportion of the global population's food and nutritional needs. Meanwhile, other consumable plant species are often overlooked or considered insignificant. Despite their significant nutritional value and potential contribution to nutrition security, these neglected and underutilized crop species are not adequately recognized in national diet and nutrition programs. The significance of "underutilized" crop species is drawing increasing attention from the international research community and organizations focused on food security policy. Neglected and underutilized crop species (NUS) have been prioritized as "Future Smart Foods" (FSFs). NUS are nutritionally dense, climate-resilient, economically viable, and locally available or adaptable food crops (FAO 2018). These crops, referred to by multiple terms like orphan crops, minor crops, traditional, indigenous crops, potential crops, or heirloom crops, among others, offer promising opportunities to address food scarcity

K. Tripathi · P. G. Gore · R. Pamarthi · G. P. Singh (✉)
ICAR-National Bureau of Plant Genetic Resources, New Delhi, India
e-mail: gp.singh@icar.gov.in

U. S. Thendral · R. R. Panda
The Graduate School, ICAR- Indian Agricultural Research Institute, New Delhi, India

V. P. Sahi
Department of Genetics and Plant Breeding, Sam Higginbottom University of Agriculture Technology and Sciences, Prayagraj, India

G. Prasad Dixit et al. (eds.), *Fundamentals of Legume Breeding*, https://doi.org/10.1007/978-981-96-7838-9_12

and nutritional deficiencies. The challenges faced by minor underutilized crops are given in Fig. 1.

Grain legumes, part of the Fabaceae family, are recognized for their rich protein content, significantly impacting nutrition, diets, and global health. Legumes are classified into major and minor species based on their economic importance and utility. Major legumes are well known and widely cultivated and have established practices for domestication, cultivation, utilization, and conservation. Examples include chickpea (*Cicer arietinum* L.), soybean (*Glycine max* L.), cowpea (*Vigna unguiculata* L.), groundnut (*Arachis hypogaea* L.), common beans (*Phaseolus vulgaris* L.), and pea (*Pisum sativum* L.), among others. On the other hand, minor legumes are lesser known, underexploited, neglected, and often considered underutilized. Several species in this category include (i) grasspea (*Lathyrus sativus* L.); (ii) Lupin (*Lupinus* L.); (iii) horsegram (*Macrotyloma uniflorum* (Lam.) Verdc.); (iv) limabean (*Phaseolus lunatus* L.); (v) fababean (*Vicia faba* L.); (vi) winged bean (*Psophocarpus tetragonolobus* (L.) DC.), (vii) ricebean (*Vigna umbellata* (Thunb.) Ohwi and Ohashi); (viii) Bambara groundnut (*Vigna subterranea* (L.) Verdc); (ix) Adzuki bean (*Vigna angularis* (Willd.) Ohwi & H.Ohashi); (x) lablab bean (*Lablab purpureus* (L.) Sweet); (xi) mothbean (*Vigna aconitifolia* (Jacq.) Maréchal); (xii) tepary bean (*Phaseolus acutifolius* A. Gray); (xiii) cluster bean (*Cyamopsis tetragonoloba* (L.) Taub.); (xiv) Minni payaru (*Vigna stipulacea* (Lam.) Kuntz.); and (xv) Jackbean (*Canavalia ensiformis*).

Breeding programs for minor grain legumes aim to develop improved varieties that are more resilient to climate change, have higher yields, and possess desirable nutritional qualities such as increased protein content and improved amino acid profiles. These breeding programs also focus on enhancing the agronomic traits of minor grain legumes, such as early maturity, drought tolerance, and resistance to diseases and pests. These legumes offer an affordable and protein-rich alternative, especially during periods when the consumption of animal products is restricted due to potential health concerns. This has led to a rise in advocating for and supporting plant-based protein as a viable solution to meet people's dietary needs (Cheng et al. 2019). Diversifying cultivation practices to include these neglected crops is crucial for promoting food security and harnessing a wider range of nutritional resources. Grasspea is a popular food and feed crop in Asian and African countries. In challenging climatic conditions where other crops may falter, grasspea emerges as a resilient option, often becoming the sole accessible food source for impoverished communities. It proves to be versatile, serving multiple roles including animal feed and fodder. Its seeds boast a notable protein content of 18–34%, while mature leaves have an impressive 17% protein. This dual nutritional benefit makes it a valuable resource not only for sustaining livestock but also as a protein-rich component in human diets (Rizvi et al. 2016). These values are higher than those of field pea (*Pisum sativum* subsp. *arvense* (L.) Asch) (23%), (Petterson et al. 1997).

Plants of *Lupinus* spp. are generally grown for pasture, as green manure, to fix atmospheric nitrogen in the soil and for controlling soil erosion, as well as for ornamentation in some instances (Cowling et al. 1998). Lupin seeds have high protein (Lampart-Szczapa 1996), high dietary fiber (Petterson

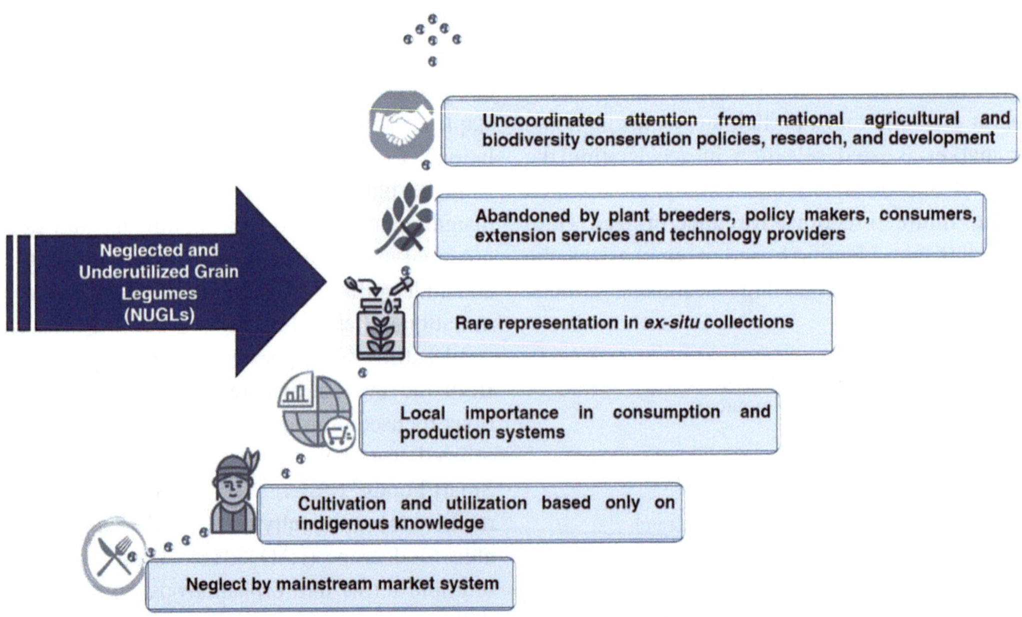

Fig. 1 Challenges associated with underutilized crops

1998; El-Adawy et al. 2001), and higher amounts of oleic, linolenic, and arachidic acids (Uzun et al. 2007). Horsegram has been recognized as a potential arid legume with a good source of protein and nutrients (National Academy of Sciences 1979). There is a wide variety of fatty acids in the seeds of winged bean (Gu et al. 2022). Limabeans are a nutritious food source, rich in protein, fiber, and carbohydrates. They provide essential minerals, balanced amino acids, and energy, making them beneficial for food formulation (Obiakor 2015) However, limabeans also contain anti-nutrients, including phytohaemagglutinin, trypsin inhibitor, tannins, oxalate, and phytic acid, which can have both positive and negative effects on the body (Ezeagu and Ibegbu 2010; Adebo 2023). Fababeans, similar to other pulses, can undergo processing to yield protein, starch, and fiber. Notably, they boast a high protein content ranging from 28% to 32%, surpassing field peas with their 24% protein content (Merga et al. 2019). Despite having a longer shelf life in comparison, this legume faces limited utilization due to the abundance of anti-nutritional factors (ANFs) such as amylase inhibitors, protease inhibitors, convicine, and vicine (Rahate et al. 2021).

Winged bean is a versatile legume used for its fresh leaves, flowers, green pods, seeds, and tuberous roots. Mature seeds contain 15–18% oil content. The plant also provides substantial amounts of phosphorus, iron, and various vitamins (Mohanty et al. 2020). According to Bamshaiye et al. (2011), Bambara nut seeds have high essential amino acids like isoleucine, lysine, methionine, phenylalanine, threonine, and valine (Massawe et al. 2002). The high protein content of Bambara nut provides a nutritional advantage, addressing disorders in both humans and animals (Tripathi et al. 2022a, b). Ricebean nutritional quality is higher as compared to many other *Vigna* family legumes (Katoch 2013). It is known for its nutritional benefits, and it serves as a valuable source of high-quality feed for livestock (Katoch 2013). Moreover, ricebean has the potential to not only address food and nutritional deficiencies but also play a key role in promoting environmental sustainability (Katoch 2013). Adzuki beans are a vital nutritional source which are often dubbed as "weight loss bean" due to low calorie and fat content; they boast easily digestible protein and abundant bioactive compounds (Kitano-Okada et al. 2012).

Notably, adzuki bean extract exhibits the highest total antioxidant activity among seven beans, including fababean, broad bean, pea, and lentil (Agarwal and Chauhan 2019). Mothbean is a source of nutrition both by way of macro- and micronutrients as seeds of mothbean contain high protein and carbohydrates associated with an adequate amount of minerals (Mg, Ca, Fe, Mn, and Zn), vitamins (ascorbic acid and niacin), unsaturated fatty acids and essential amino acids (Bhadkaria et al. 2021; Tresina et al. 2017). Minni payaru is mainly used for food, livestock feed, and green manure. It is used as an alternative source of blackgram in Indian cuisine (Tripathi et al. 2021). This indigenous crop stands out for its resilience in the face of global climate change (Gore et al. 2021). Notably, it demonstrates the ability to thrive in harsh environments and showcases resistance to a range of biotic and abiotic stresses (Nagaraj et al. 1981; Gore et al. 2022).

Lablab is a valuable protein source for human diets, particularly its seeds, containing balanced amino acids like lysine (6–7%) (Omondi 2011). The leaves are protein-rich (28%) and stand out for iron content among legume leaves (National Research Council 2006). However, red-colored Lablab seeds have been noted for high anti-nutritional factors, specifically 12.56 mg/g of tannins, potentially contributing to a bitter taste that could impact consumer acceptance (Omondi 2011; Minde et al. 2021). Tepary bean is a highly nutritious crop with a range of composition and other characteristics similar to those of common bean (Porch et al. 2017). Cluster bean is used in traditional medicine for the treatment of various diseases and metabolic disorders (Pathak and Roy 2015). Cluster bean is being utilized as arid legume and also a notable source of soluble dietary fiber, primarily in the form of guar gum extracted from the ground endosperm. The galactomannan-rich guar gum, with its hydrophilic gel-forming attributes, serves various purposes in the food industry (Srinivasan 2020).

Like other minor legumes, Jackbean has also anti-nutritional and toxic factors, including trypsin inhibitors, hemagglutinins, and cyanogen glucosides poses considerations for its consumption (Akpapunam and Sefa-Dedeh 1997). The nutritional composition of certain legume seeds may be influenced by specific molecules unique to each species. For instance, fababean seeds contain vicine and convicine, two pyrimidine glycosides (Khazaei et al. 2019). The aglycone forms of these compounds, divicine and isouramil, are responsible for causing favism—a hemolytic anemia affecting males with a particular genetic defect in the gene encoding erythrocyte-located glucose-6-phosphate dehydrogenase. Vicine and convicine also exhibit anti-nutritional effects in the diets of monogastric animals. Various initiatives, including breeding and processing treatments, have been implemented to mitigate their presence in seeds (Crépon et al. 2010). Fababeans, along with *Mucuna pruriens* (velvet bean), are noteworthy for containing L-3,4-dihydroxyphenylalanine (L-dopa), a naturally occurring compound. While excessive L-dopa intake can be toxic, leading to hallucinations and gastrointestinal issues, it is a crucial component in medicines treating Parkinson's disease (Han et al. 2022) In fababeans, L-dopa is concentrated in the germinating seed's embryo axis, reaching approximately 75 mg/g dry weight after 9 days. Conversely, *Mucuna* species seeds contain lower L-dopa levels (3.1–6.7% dry weight), with potential up to 9%. These findings underscore the significance of these legumes in both nutrition and pharmaceutical applications. In grasspea seeds, a nonprotein amino acid called β-N-oxalyl-I-α,β-diaminopropionic acid (ODAP) is present, exhibiting neuroexcitatory properties (Ghosh et al. 2015).

Prolonged and excessive consumption of this protein-rich seed in a repetitive diet leads to a disorder known as neurolathyrism. This condition results in the degeneration of upper motor neurons, causing irreversible paralysis of the legs in up to 6% of affected individuals. ODAP content varies among grasspea germplasm, with South Asian varieties containing higher amounts (0.7–2.4% dry weight) compared to those from North Africa, Syria, Turkey, and Cyprus, which exhibit significantly lower quantities (0.02–1.2%). In legumes, the amount of phenolics ranges from 0.325 to 6.378 mgGAE/g (Ramya et al. 2021). However, these compounds are considered anti-nutritional because they interact with proteins, leading to the inactivation or reduced solubility of important enzymes (trypsin, lipase, and amylase). This interference affects the digestion of proteins and carbohydrates, as well as the availability of vitamins and minerals (Tripathi et al. 2022a, b). Notably, horsegram seeds have higher levels of tannins and phenolics compared to other legumes, intensifying these anti-nutritional effects (Aditya et al. 2019).

Over the past 25 years, efforts have been made to integrate these crops into mainstream research. Although genomics progress on these crops lags behind other legumes, efforts are underway to enhance genetic research, breeding, and genomic studies. Molecular markers like SSRs and SNPs aid in creating detailed linkage maps to pinpoint genes or QTLs controlling essential traits. Current emphasis is on refining these maps using accurate breeding tools like NGS and RNA-seq. Translational genomics, compared with model species, may identify key genes. Collaboration between academia and breeding programs aims to accelerate the discovery of valuable genes, potentially leading to the release of improved varieties of orphan grain legume crops in the near future. Steps have been taken to prioritize the preservation of genetic diversity in underutilized crops. The importance of the underutilized grain legumes is given in Fig. 2. Genebank and conservation initiatives are actively working to protect the distinct genetic resources of these crops, ensuring their resilience and availability for future generations. Efforts to develop markets for underutilized crops have been underway. This includes creating awareness among consumers, promoting these crops in urban markets, and integrating them into food value chains. Breeding programs have been initiated to improve the agronomic traits of underutilized legumes. This includes efforts to enhance yield, pest resistance, and adaptation to different environmental conditions.

Over time, earnest initiatives have been directed toward genetically improving the grasspea, aiming to mitigate ODAP content and boost its production capabilities (Parihar et al. 2022). Significant advancements have recently been made in the realm of quality traits for fababeans. The reduction of vicine-convicine and seed coat tannins, which are the primary anti-nutritional factors hindering the widespread utilization of fababean seeds, has been successfully accomplished through the discovery of key genes (Khazaei et al. 2019).

2 Origin, Evolution, and Distribution of Species and Forms: Wild Relatives

The origin, evolution, and distribution of crops and their wild relatives play a crucial role in understanding the genetic diversity of underutilized grain legumes. These wild relatives serve as the foundation for the development of cultivated varieties, providing a rich source of genetic material that can contribute to enhanced traits such as resistance to pests, diseases, and environmental stresses. The evolutionary history of these legumes reveals the adaptive mechanisms that have allowed them to thrive in diverse ecological niches. Additionally, studying the distribution patterns of wild relatives helps in identifying potential areas for germplasm collection and conservation. Grasspea, also known as *Lathyrus sativus*, is believed to have originated in the Mediterranean region and parts of Asia. The genome size of grasspea is 6620 Mbp (Rajarammohan et al. 2023). Its closest wild relative and likely progenitor is *Lathyrus cicera* (Red Pea or Red Vetchling), found in various regions including Europe, Asia, and the Mediterranean. Other notable wild relatives include *L. hirsutus* (Hairy Vetchling), *L. aphaca* (Yellow Pea or Yellow Vetchling), *L. sylvestris* (Flat Pea), and *L. nissolia* (Grass Vetchling) which are distributed across different habitats and contribute to maintaining biodiversity.

Lupins, belonging to the genus *Lupinus*, have several significant wild relatives that are crucial for breeding programs. Some examples are *Lupinus angustifolius* (narrow-leafed lupin), *L. albus* (white lupin), *L. polyphyllus* (bigleaf lupin), *L. perennis* (sundial lupin), and *L. mutabilis* (pearl or Andean lupin). These wild species are native to various regions, including the Mediterranean, North America, and the Andes, and some have been domesticated as major cultivated lupin crops. Cluster bean, commonly known as guar, is believed to have originated in West Africa and India. Its wild relatives, such as *Cyamopsis serretais* from South Africa and *C. senegalensis* from dry tropical regions of Africa, have adapted to diverse environments like deserts and shrublands (Tyagi et al. 2019). These wild species contribute to the overall genetic diversity of cluster beans. The origin, evolution, and distribution of species and wild relatives of other underutilized legume crop species have been provided in Table 1.

Fig. 2 Importance of underutilized grain legumes

3 Plant Genetic Resources

Plant genetic resources (PGR) serve as the foundation for advancements in agriculture and function as a reservoir of genetic variability, providing a crucial defense against environmental changes (Gautam et al. 2023). The depletion of these resources poses a significant risk to global food security. Recognizing the imperative to protect and harness plant genetic resources as a precautionary measure for an uncertain future is widely acknowledged. PGR of underutilized legumes are highly important for biodiversity conservation, adaptation to changing environments, enhanced stress resilience, nutritional security, crop improvement and breeding programs, sustainable agriculture (Pathirana and Carimi 2022), and market diversification. The seed morphological diversity of underutilized grain legumes is given in Fig. 3. The need for ex situ conservation of underutilized grain legume germplasm is underscored by the essential role played by these genetic resources in attaining Sustainable Development Goals (SDGs). In Table 2, the list of conserved plant genetic resources of underutilized grain legumes from various genebanks is given.

4 Floral Biology: Emasculation—Pollination Techniques

Grain legumes exhibit a characteristic papilionate flower structure (Fig. 4), featuring a pentamerous structure (Suso et al. 2015). The flowers display bilateral symmetry, with the corolla comprising five petals. Notably, a prominent upper petal, referred to as the standard petal, dominates the composition. This standard petal possesses a semi-cylindrical base that envelops and compresses two smaller lateral wings. These wings, in turn, enclose a pair of keel petals, situated slightly lower but interior to them, exhibiting concave sides mirroring the shape of the wings (Reddy 2007). The keel petals are typically fused at their bases, forming a boat-shaped structure that encases the essential flower organs: the androecium and gynoecium (Gupta 1900). These flowers commonly exhibit a vexillary aestivation, characterized by overlapping petals. The stamens often display a diadelphous morphology, uniting in two sets with distinct filament shapes. The inferior set forms a membranous sheath at its base, enveloping the single superior pistil within a tube that subsequently divides into nine filaments tracing the base of the

Table 1 List of origin, distribution, and wild relatives of underutilized grain legumes

Crop	Botanical name	Genome size	Origin	Distribution	Wild relatives
Horsegram ($2n = 20, 22$)	*Macrotyloma uniflorum* (Lam.) Verdc.	400 Mbp (Bhardwaj et al. 2013)	Southern Asia	India, Sri Lanka, Malaysia, West Indies	*M. axillare* *M. uniflorum* var. *uniflorum* *M. uniflorum* var. *stenocarpum* *M. uniflorum* var. *verrucosum* *M. uniflorum* var. *benadirianum* (Fuller and Murphy 2018)
Limabean ($2n = 22$)	*Phaseolus lunatus* L.	546 Mbp (https://www.pulsedb.org/organism/988181)	Meso- and South America	Peru, Colombia, Brazil, Mexico	*P. augusti* *P. pachyrrhizoides* (Maquet et al. 1999)
Fababean ($2n = 12$)	*Vicia faba* L.	13,000 Mbp (Ellwood et al. 2008)	Middle-East	Northern Africa and Southwest Asia	*V. narbonensis* *V. palaestina* (Caracuta et al. 2016)
Winged bean ($2n = 18$)	*Psophocarpus tetragonolobus* (L.) DC.	2400 Mbp (Vatanparast et al. 2016)	Southern Asia	Tropical southern Asia—Papua New Guinea, Mauritius, Madagascar, and India	*P. lukafaensis* *P. lancifolius* *P. grandifloras* *P. scandens* *P. palustris* *P. obovalis* *P. lecomtei* *P. monophyllus* (Yang et al. 2018)
Ricebean ($2n = 22$)	*Vigna umbellata* (Thunb.) Ohwi and Ohashi	414 Mbp (Kaul et al. 2019)	Indochina	Nepal, Bhutan, and Northeast India, Myanmar, Southern China, Northern Thailand, Laos, Vietnam, Indonesia	*V. hirtella* *V. exilis* *V. tenuicaulis* *V. napalensis* *V. minima* *V. nakashimae* *V. ruikiuensis* (Doi et al. 2002)
Bambara groundnut ($2n = 22$)	*Vigna subterranea* (L.) Verdc	520 Mbp (https://www.pulsedb.org/organism/983374)	West Central Africa	Sub-Saharan Africa, Malaysia, South, and Central America, some parts of Northern Australia, Sri Lanka, and Indonesia	*V. subterranea* var. *spontanea* (Basu et al. 2007)
Adzuki bean ($2n = 22$)	*Vigna angularis* (Willd.) Ohwi & H.Ohashi	538 Mbp (Kang et al. 2015)	China	Northern to southern East Asia including Taiwan, India, Korean peninsula	*V. angularis* var. *nipponensis* (Siriwardhane et al. 1991)
Lablab bean ($2n = 20, 22, 24$)	*Lablab purpureus* (L.) Sweet	367 Mbp (Missanga et al. 2021)	India	Southeast Asia, Africa, tropical and subtropical countries	*L. purpureus* subsp. *uncinatus* (Maass et al. 2005)
Mothbean ($2n = 22$)	*Vigna aconitifolia* (Jacq.) Maréchal	409 Mbp (Suranjika et al. 2023)	India	India, Pakistan, Myanmar, and Sri Lanka	*V. trilobus* (Pathak 2023)
Tepary bean ($2n = 22$)	*Phaseolus acutifolius* A. Gray	682 Mbp (https://www.pulsedb.org/organism/983374)	Southwestern United States and Mexico	United States, Mexico	*P. angustissimus*, *P. carteri* *P. filiformis* (Muñoz et al. 2006)
Minni payaru	*Vigna stipulacea* (Lam.) Kuntze	445.1 Mbp (Takahashi et al. 2019)	Southern Asia	India, Indonesia, Bhutan, Myanmar, Pakistan, Sri Lanka, Taiwan, Vietnam	

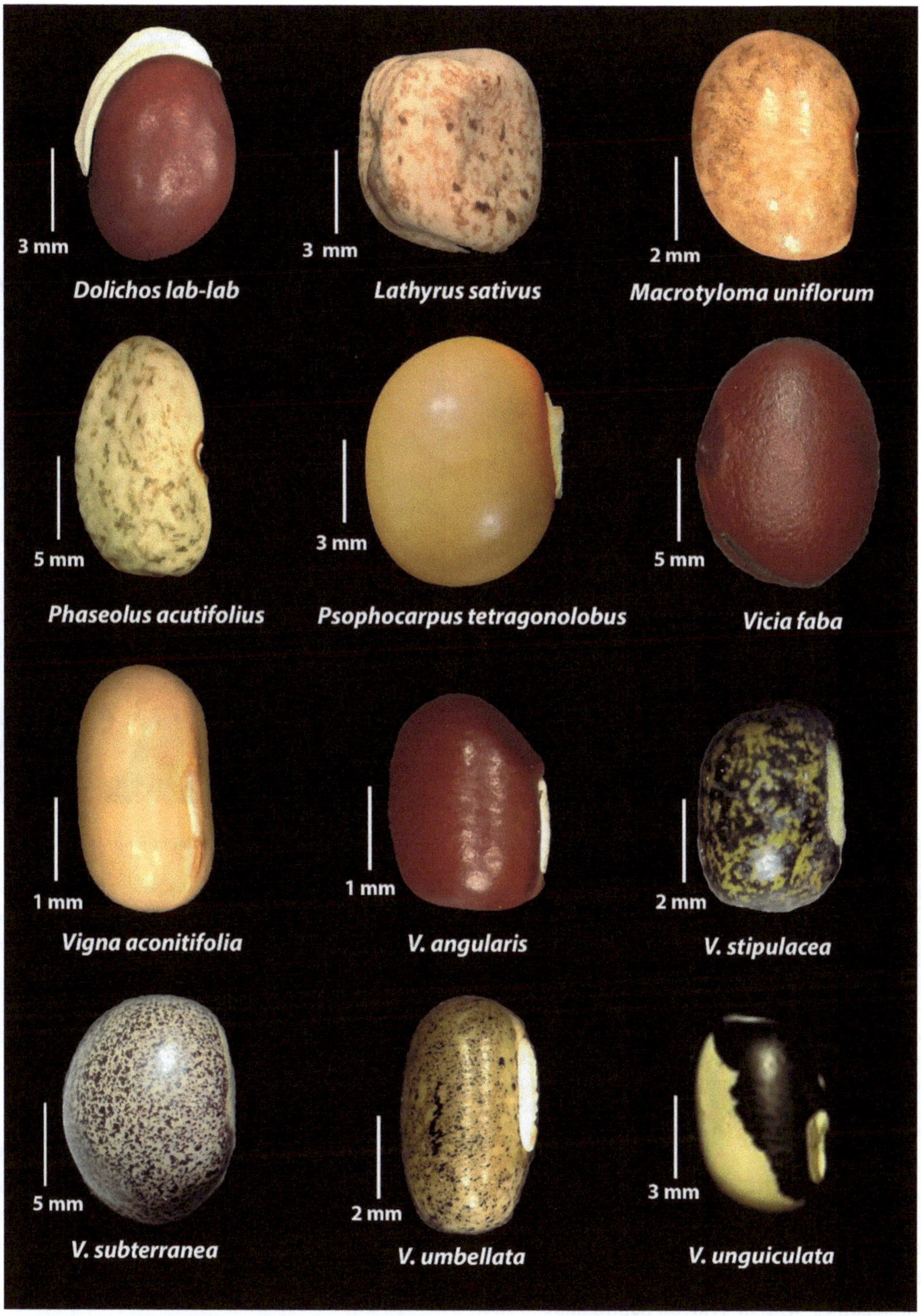

Fig. 3 Morphological diversity of seeds of underutilized grain legumes

enclosing keel petals. A detailed dissection image of Fababean is given in Fig. 4. Notably, a solitary free filament lies above them, completing the intricate floral arrangement. A comprehensive exploration of floral biology and associated mechanisms is imperative for elucidating the pollination modes in diverse crop species. The floral diversity of different underutilized grain legumes is given in Fig. 5.

Emasculation and pollination techniques are essential steps in the controlled breeding of grain legumes to achieve desired traits (Salimath et al. 2007). Emasculation involves

Table 2 List of conserved plant genetic resources of underutilized grain legumes in top three genebanks

Crop	Top three genebanks		
Grasspea	2650 ICAR-NBPGR	4853 ICARDA	3191 SGSV
Lupin	1821 SGSV	69 CGN	24 BRGV
Horsegram	3167 ICAR-NBPGR	57 SGSV	4 NARO
Limabean	37861 CIAT	3775 SGSV	921 WVC
Fababean	13090 ICARDA	5557 SGSV	911 ICAR-NBPGR
Winged bean	299 WVC	254 SGSV	192 ICAR-NBPGR
Ricebean	2239 ICAR-NBPGR	543 NARO	489 SGSV
Bambara groundnut	21045 IITA	16401 WVC	977 CIAT
Adzuki bean	3277 SGSV	2300 NARO	206 ICAR-NBPGR
Lablab bean	1681 ICAR-NBPGR	809 SGSV	594 WVC
Mothbean	21045 IITA	16401 WVC	1516 ICAR-NBPGR
Tepary bean	37884 CIAT	921 WVC	467 SGSV
Cluster bean	4017 ICAR-NBPGR	887 NCGRP	110 SGSV

BRGV Bank of Plant Genetic Resources, Romania; *CGN* Centre for Genetic Resources, Netherlands; *CIAT* International Center for Tropical Agriculture, Colombia; *ICAR-NBPGR* Indian Council of Agricultural Research–National Bureau of Plant Genetic Resources, New Delhi; *IITA* International Institute of Tropical Agriculture (IITA), Nigeria; *NARO* National Agriculture and Food Research Organization, Japan; *NCGRP* National Center for Genetic Resources Preservation, USA; *SGSV* Svalbard Global Seed Vault, Norway; and *WVC* World Vegetable Centre, Taiwan. (https://www.nbpgr.ernetin/PGR_Databases.aspx; https://www.ecpgr.cgiar.org/aegis/european-collection/european-accessions; https://www.genesys-pgr.org/; https://www.gene.affrc.go.jp/databases-plant_search_en.php; https://seedvault.nordgen.org/)

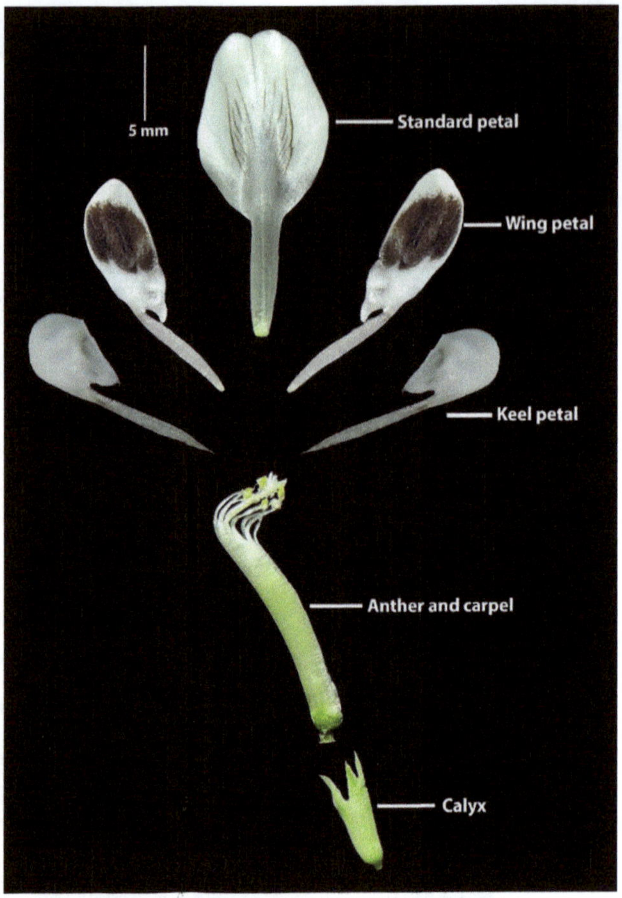

Fig. 4 Dissected floral parts of *Vicia faba*

the removal of the male reproductive organs (anthers) from a flower, preventing self-pollination and ensuring that only the desired pollen will be used in the pollination process. In the case of grain legumes with a typical papilionade flower structure, emasculation is performed by carefully excising the anthers, often before they release pollen, to avoid accidental self-pollination. This delicate process requires precision to maintain the integrity of the flower and facilitate successful controlled pollination. Pollination in grain legumes can be achieved through various techniques, including artificial pollination, where pollen from a selected donor plant is manually transferred to the stigma of the emasculated flower. This controlled pollination ensures that specific genetic traits are passed on to the next generation. In some cases, natural pollination by insects or wind may be undesirable due to the risk of unintended cross-pollination. Therefore, isolation measures may be implemented to prevent unwanted pollen from reaching the emasculated flowers, preserving the genetic purity of the selected plant. The species outcrossing rate is important for inheritance and linkage experimental design studies and the selection of appropriate breeding methods for crop improvement. Normally underutilized grain legumes are self-pollinating in nature except grasspea (Chowdhury and Sllnkard 1997) and fababean (Drayner 1959) which show often cross-pollination.

5 Molecular Cytogenetics and Breeding

The utilization of comparative cytogenetic mapping emerges as a potent strategy for unraveling the genomic organization of orphan crops, which lack a fully sequenced genome (Oliveira et al. 2020). The integration of molecular cytogenetics into the breeding processes of underutilized grain legumes represents a cutting-edge approach that holds significant promise for enhancing crop improvement strategies.

Fig. 5 Floral diversity of underutilized grain legumes

The utilization of advanced multicolor karyotyping methods enables the thorough examination of intricate chromosome rearrangements, offering a potential avenue for discovering novel recurrent structural alterations. These techniques facilitate the comprehensive dissection of complex chromosomal changes, enhancing the identification of previously unknown patterns in genetic rearrangements (Kearney 2001). By employing advanced molecular techniques, such as fluorescence in situ hybridization (FISH) and molecular markers, researchers can gain detailed insights into the chromosomal organization, structural variations, and genetic composition of these legumes (Jiang and Gill 2006). This molecular cytogenetic information becomes instrumental in identifying key genomic regions associated with desirable traits, facilitating the development of improved varieties through targeted breeding efforts. Recent advancements in comparative genomic hybridization to DNA microarrays hold considerable promise for achieving high-resolution and automated screening to detect chromosomal imbalances. This scientific inquiry contributes significantly to the exploration of the potential within minor grain legume crops, positioning them as active sources for target genes and facilitating targeted breeding efforts.

By the use of molecular cytogenetic techniques, analysis of the cowpea genome reveals two major retrotransposons in pericentromeric regions and a 455-bp tandem repeat at seven centromere pairs that suggest unique chromosomal structures evolved post-cowpea divergence from the common bean. This distinctive repetitive fraction contributes to the genomic specificity of cowpea (Iwata-Otsubo et al. 2016). The cytogenomic evolution of significant *Vigna* and *Phaseolus* beans has been explored through the construction of a BAC-FISH map (fluorescent in situ hybridization of the bacterial artificial chromosome) for mothbean. This map was then compared with the maps of *V. unguiculata* (Vu, subgenus *Vigna*) and *Phaseolus vulgaris* (Pv), providing a comparative analysis of the cytogenomic features across these bean species (Oliveira et al. 2020). *Vigna* plants distributed in Korea were analyzed by molecular cytogenetic fluorescence in situ hybridization (FISH), genomic in situ hybridization (GISH), and rDNA ITS/NTS sequences. FISH analysis showed variable 45S rRNA gene loci (ranging from one to four) at chromosome terminals, while all examined species, except ricebean, displayed two conserved 5S rRNA gene loci, with ricebean having a single locus (Choi et al. 2013). Fuchs et al. (1998) carried out a comprehensive survey of the molecular cytogenetic features of the *Vicia faba* chromosome. Molecular cytogenetic characterization and karyotypic studies marked *M. axillare* as the closest parent of the cultivated horsegram (Mishra et al. 2023).

Distinct fluorochrome binding patterns with CMA3 and DAPI highlighted the prevalence of higher GC content in *V. aconitifolia* compared to *V. umbellata*, indicating potential variations in species interrelationships, possibly influenced by the comparatively elevated AT content in *V. umbellata* (Shamurailatpam et al. 2015). SNPs derived from transcript analysis in six common bean and two tepary bean accessions were employed to create a 768-SNP Illumina GoldenGate assay for each species. Utilizing the tepary bean assay enabled the evaluation of diversity in both wild and cultivated tepary beans, leading to the construction of the first gene-based map of the tepary bean genome (Gujaria-Verma et al. 2016). The comparative cytogenetic map between common bean (*Phaseolus vulgaris*) and limabean (*P. lunatus*) revealed significant synteny conservation, with identified chromosomes and mapped single-copy clones (Bonifácio et al. 2012). Karyotype analysis of three improved grasspea varieties (Nirmal, BioL-203, and BioR-231) revealed diploidy with $2n = 14$, characterized by metacentric and submetacentric chromosomes. Distinct variations in GC-specific CMA and AT-specific DAPI banding patterns highlighted differences among the varieties, with Nirmal exhibiting the highest percentage of GC-rich sequences, indicating karyomorphological polymorphism in the studied materials (Talukdar 2011). The study highlighted genotypic-level structural variations in the karyotype of grasspea, with variations in total haploid chromosome length, arm ratio, and centromeric index among the accessions (Barpete et al. 2012). Still limited research has been conducted on numerous underutilized grain legumes. There is a need to prioritize and conduct detailed studies on their diversity patterns to enhance our understanding of these crops.

6 Genetic Studies: Qualitative and Quantitative Traits

Several important morphological and phenological properties show substantial natural variation in legume germplasm, which can be leveraged in breeding efforts. These properties encompass characteristics such as growth habit, plant height, pod cross-section, number of pods in the plant, pod curvature, hypocotyl color, flower color, days to flowering, node numbers, seeds per pod, number of flower buds, and 100-seed weight (Okii et al. 2014). It is worth noting that morpho-physiological and reproductive traits exhibit remarkable consistency across different legume species (Talukdar 2013). When it comes to monogenic traits, such as color, shape, texture, and the presence or absence of specific characteristics, conventional breeding approaches have been successfully employed to control and manipulate these traits. However, the situation becomes more complex when dealing with multigenic traits. These traits, including plant yield and resistance to abiotic stresses, are significantly influenced by environmental factors and G × E interactions. Breeding techniques aimed at improving these traits require more time and precision due to the intricate nature of their genetic control (Tahir et al. 2019).

In plant breeding, the measurement of heritability is commonly used by plant breeders to quantify the proportion of phenotypic variance present in a population. Heritability is estimated based on the additive genetic effects in the narrow sense (NSH), which includes only the additive component of genetic variation. However, to account for a more comprehensive understanding of genetic effects, heritability can also be defined in a broad sense (BSH), which considers the sum of additive, dominance, and epistasis effects (Piepho and Möhring 2007).

In Fababean, various agronomic and yield-related traits have been studied across different environments. It has been found that certain traits, such as seed weight and days to flowering, are minimally affected by environmental conditions. Additionally, the number of pods per plant also remains relatively stable across different environments. These findings suggest that these traits exhibit a strong degree of genetic stability. On the other hand, traits such as seed yields and the number of stems per plant are strongly influenced by environmental factors. These traits show significant variation depending on the specific conditions in which the fababean plants are grown (Toker 2004). Another important aspect of conventional fababean breeding is the development of frost tolerance. Studies have indicated that frost tolerance in fababeans exhibits a high degree of heritability after undergoing a hardening process (Link et al. 2014). This means that this trait can be effectively selected for and improved through breeding efforts.

Inheritance studies have revealed that some traits in cluster bean (*Cyamopsis tetragonoloba*) are controlled by relatively simple genetic factors. For example, foliage color, branching pattern, and pollen fertility were found to be governed by single genes (Saharan et al. 2004). However, other traits like pod length are quantitative and influenced by multiple genes. Efforts to breed cluster bean for increased seed-gum content have faced challenges due to both additive and nonadditive gene effects. Seed-gum content also shows negative associations with seed weight but a positive correlation with seed yield (Pathak et al. 2011).

Understanding the genetic basis and correlations of key traits will inform breeding approaches in the minor legumes. Simple traits controlled by one or a few genes can be improved through selection, while complex quantitative traits may require more sophisticated breeding schemes or the use of modern tools like marker-assisted selection. Overcoming unfavorable correlations will also be key to making simultaneous gains in yield, quality, and adaptation.

7 Breeding Objectives

7.1 For Yield

Maximizing yield is critical in plant breeding programs for minor grain legumes. Increasing yields can contribute significantly to meeting the growing global demand for food while making these nutritious legumes accessible to more people worldwide. To enhance yield potential in minor grain legume crops, breeders employ various strategies such as selecting genotypes with high-yielding traits (Kumar et al. 2020), optimizing agronomic practices, employing precision agriculture techniques, utilizing molecular markers associated with high-yielding traits, incorporating genetic diversity from wild relatives or landraces into breeding populations through introgression, using advanced genomic tools for marker-assisted selection (MAS) or genome-wide association studies (GWAS), and implementing genomic selection methods which enable prediction of breeding values based on genotypic information (Varshney et al. 2013).

7.2 Quality Characters Including Biofortification

The breeding objectives for nutritional quality in minor grain legumes focus on enhancing the content and bioavailability of essential nutrients such as protein, vitamins, minerals, and bioactive compounds. This can be achieved by selecting varieties with higher protein content and improved amino acid profiles, increased levels of micronutrients such as iron and zinc, and enhanced concentrations of phytochemicals with potential health benefits, reducing anti-nutritional factors, and so on. An example of biofortification work done in the Adzuki bean by Oliveira et al. (2018) has aimed to improve the micronutrients such as iron and selenium in its sprouts. There is a significant diversity among these underutilized grain legumes in terms of nutritional composition, hydration properties, and textural characteristics. Yadav et al. (2018) found substantial variation in seed weight (75–148 g per 1000 seeds), protein content (19–24%), and physicochemical and cooking properties among different adzuki bean accessions. The hardness of soaked adzuki beans ranged from 69 to 120 Newtons and was positively correlated with gumminess and chewiness.

Recent studies have revealed the rich phytochemical composition and antioxidant properties of adzuki bean seed coats, indicating their potential nutritional and health benefits. Johnson et al. (2022) quantified a wide range of total metabolites (16–945 mg/g) and phenolics (0.80–57.35 mg/g) among different accessions, demonstrating significant diversity. Analyses by Desta et al. (2022) identified the predominant anthocyanins in black seedcoats as delphinidin-3-O-glucoside and delphinidin-3-O-galactoside. Further characterization of seedcoat color types by Zhao et al. (2022) revealed anthocyanins to be the primary pigments, with no carotenoids or pelargonidin derivatives detected. They also elucidated how proanthocyanidin and anthocyanin combinations determine the diverse hues. Overall, these recent studies reveal adzuki bean seed coats as a rich source of phenolics, antioxidants, and pigments, with diversity in phytochemical profiles across accessions.

Recent studies have revealed an interesting diversity and antioxidant properties in the phenolic profiles of Bambara groundnut accessions. Analyses by Tsamo et al. (2018) of landraces showed considerable variation in total phenolics (0.75–17.71 mg/g), flavonoids (0.01–2.51 mg/g), and anthocyanins (0.03–1.31 mg/g), with caffeic acid and catechin differentiating accessions. Adedayo et al. (2021) also found variation in phenolics between whole and dehulled cultivars of different seed colors, ranging from 3.6 to 11.0 and 2.7 to 3.2 mg/g, respectively. Notably, Chinnapun and Sakorn (2022) identified a new antioxidant compound, luteolin ($C_{20}H_{18}O_9$), suggesting Bambara groundnut could provide natural antioxidants to reduce inflammation. Recent studies have characterized the nutritional composition and antinutritional factors of grasspea (*L. sativus* and *L. cicera*) seeds. Analyses of accessions have found protein contents ranging from 25% to 30% with high-quality amino acid profiles, except for low methionine (Grela et al. 2012; Rajendran et al. 2019). While seeds contain little fat (<1%), they are rich in polyunsaturated fatty acids. Darker-seeded accessions of *L. sativus* and *L. cicera* exhibit higher tannin content, with average concentrations of 3.3 g/kg and 6.4 g/kg, respectively (Grela et al. 2012). Rajendran et al. (2019) found significant variation in seed protein (28.82–30.72%) and β-ODAP (0.32–0.53%) among 702 *L. sativus* accessions.

Analyses of lablab bean (*Lablab purpureus*) grains indicate they can provide good sources of protein, minerals, and essential fatty acids such as linoleic acid and linolenic acid while containing low levels of the anti-nutrient phytate (Kumari et al. 2022). The latter suggests lablab beans may have high mineral bioavailability according to Dwivedi and Ortiz (2021). Significant variability has also been reported among mothbean (*Vigna aconitifolia*) accessions for trypsin inhibitors (0.041–0.078 TIU/mg), phytic acid (0.26–1.22 mg/g DM), antioxidant activity (5.63–38.26 μmol TE/g), and tannins (1.4–3.2 mg/g DM) (Gupta et al. 2016). Ricebean (*Vigna umbellata*) germplasm found significant variation among accessions for seed protein content, ranging from 2.88 to 31.79 mg/g dry matter, and antioxidant activity, ranging from 9.36 to 18.56% (Bhagyawant et al. 2019). Some accessions exhibited notably high in vitro protein and starch digestibility at 63.84% and 44.6%, respectively.

A study conducted by Deepika et al. (2022) in adzuki bean accessions revealed that Significant genetic variation was observed across traits like moisture, ash, protein, starch, sugars, phytic acid, phenols, and antioxidant activity. Protein content ranged from 18 to 23.9 g/100 g, while antioxidant activity was between 11.4 and 19.7 mg/100 g GAE. Cluster analysis identified nutritionally dense accessions like IC360533 with high ash (>4.2 g/100 g), protein (>23.4 g/100 g), and low phytic acid (0.652 g/100 g), promising for developing improved food/feed varieties.

Recent studies characterizing nutritional quality in tepary bean (*Phaseolus acutifolius*) have revealed some interesting attributes and diversity. Analyses of wild and domesticated accessions by Waines (1978) found higher protein content but lower 100-seed weight among wild types, though domesticated lines showed a greater range in protein values (specific values). Comparison to common bean (*P. vulgaris*) by Porch et al. (2017) indicated that tepary bean has reduced fat and ash but increased sucrose, along with shorter cooking times and higher water uptake (specific percentages). However, some 'hard-shell' tepary bean lines exhibited low water uptake and longer cooking. Overall, tepary bean appears highly nutritious with compositional and cooking properties similar to common bean. Interspecific breeding efforts have also shown promise, with some tepary x common bean congruity-backcross lines achieving seed iron levels over 100 mg/kg (Burbano-Erazo et al. 2021), though yields were poor.

Another crucial objective of quality improvement in lupin is achieving low seed-alkaloid content, with a target of less than 0.02%. Researchers have made substantial progress in identifying genes and markers associated with low alkaloid content in lupin. For instance, Li et al. (2011) identified a microsatellite-anchored fragment length polymorphism (lucLi) marker, located 0.9 cM away from the low-alkaloid locus *iucundus*. Similarly, Lin et al. (2009) developed a sequence-specific PCR marker (PauperM1) closely linked (1.4 cM) to the low-alkaloid locus pauper in white lupin (*Lupinus albus* L.).

Gore et al. (2021) conducted a study on an underutilized legume, *Vigna stipulacea*, for micronutrient and protein content over two locations in India. They found significant genetic variation in iron, zinc, calcium, protein, and phytate levels. They recorded iron content ranging from 29.35 to 130.96 mg/kg, zinc content ranging from 19.44 to 74.20 mg/kg, calcium content ranging up to 251.50 mg/kg, and protein levels ranging up to 25.73%. Accession IC622867 had the lowest phytate levels, desirable for bioavailability.

Breeding programs for underutilized grain legumes aim to improve taste, nutritional content, cooking quality, and shelf-life traits. Breeders can use conventional breeding techniques such as pedigree selection or hybridization between parental lines to achieve this. Additionally, modern genomic tools like marker-assisted backcrossing (MABC) can speed up the process by allowing breeders to efficiently introgress targeted genes associated with improved quality traits (Singh et al. 2020).

7.3 Biotic Stresses

Underutilized grain legume crops face significant challenges due to biotic stresses that can cause diseases and lead to substantial yield losses. These stresses are mainly caused by fungi, bacteria, viruses, and pests such as insects or nematodes. Breeding for improved resistance to biotic stresses such as diseases, insects, and nematodes is critical to enhancing productivity and resilience in minor grain legumes. Substantial progress has been made in recent years to identify diverse sources of genetic resistance and tolerance and incorporate these into adapted, high-yielding cultivars through breeding programs.

Biotic stresses like insect pests and diseases are the biggest yield-limiting factors for underutilized grain legumes globally. However, the increased availability of genomic resources and molecular markers has enabled the identification of genome regions and genes linked to disease resistance. For example, in cowpea, Amkul et al. (2019) identified a major QTL (*qCc_PDS6.1*) providing resistance against the bean weevil (*Callosobruchus chinensis*) and another QTL (*qCm_PDS6.1*) conferring resistance to the cowpea weevil (*Callosobruchus maculatus*). Likewise, in white lupin, Rychel-Bielska et al. (2020) discovered four major QTLs (*antr04_1*, *antr05_1*, *antr04_2*, and *antr05_2*) that control anthracnose disease resistance, explaining 14–25% of the phenotypic variation. In narrow-leafed lupin, Yang et al. (2013) used SNP markers to identify candidate R-genes for phomopsis stem blight disease resistance. Recently, Zhou et al. (2021) mapped a major QTL in narrow-leafed lupin associated with two candidate genes, *LOC109334326* and *LOC109334327*, that provide resistance against destructive gray leaf spot disease.

To address these issues breeding efforts are focused on developing resistant varieties through traditional approaches such as recurrent selection or pyramiding multiple resistance genes into a single genotype via MAS (Cobb et al. 2019). This involves screening germplasm collections or wild relatives for natural sources of resistance and crossbreeding them into elite cultivars through traditional plant breeding methods (Bohra et al. 2022). The ultimate goal is to create crop varieties that possess genetic resistance against pathogens or pests, which will help to increase crop productivity and reduce the negative effects of biotic stresses on minor grain legume crops (Mangena 2020).

7.4 Abiotic Stress Resistance/Tolerance

Climate change is causing a rise in abiotic stresses such as drought, heatwaves, and soil salinity, making it increasingly challenging to sustain agriculture systems globally. In regions where these stressors frequently occur, achieving resilience is crucial for ensuring sustainable food production under changing environmental conditions (Nedumaran et al. 2015). To achieve this, breeders use various approaches such as identifying tolerant genotypes through phenotypic selection under stress conditions, utilizing advanced genomic tools like quantitative trait loci (QTL) mapping combined with MAS, exploring transgenic approaches targeting key genes involved in abiotic stress response pathways, and using molecular markers linked to abiotic stress tolerance attributes during selection processes (Varshney et al. 2013). Breeding resilient varieties adapted to abiotic stresses like drought, heat, and soil salinity is key to stabilizing the productivity of minor legumes grown in marginal environments by smallholder farmers. Recent advances have been made in several underexploited crops.

In recent years, the impact of abiotic stresses, particularly drought, has been a significant concern for underutilized legumes, leading to significant yield losses (Liu et al. 2017; Chahota et al. 2020). Extensive research has been conducted to identify genetic factors that contribute to drought tolerance in mungbean (Liu et al. 2017) and horsegram (Chahota et al. 2020), resulting in the discovery of several quantitative trait loci (QTLs). The identification of QTLs, markers, and candidate genes provides valuable tools for breeding programs to enhance the cultivar's resilience to abiotic stresses and improve its quality characteristics. Further research in this field will undoubtedly contribute to the development of more productive and adaptable legume varieties.

7.5 Exploitation of Heterosis and Hybrid Development

Heterosis is a biological phenomenon in which offspring produced by crossbreeding two genetically diverse parents display superior characteristics such as higher yields, improved quality traits, or other desirable qualities (Priyadarshan 2019). This phenomenon is not fully utilized as a successful breeding strategy to enhance the yield potential of underutilized grain legumes. The process of hybrid development involves the crossbreeding of elite parental lines, followed by selecting superior hybrid combinations with improved performance over open-pollinated varieties. This approach allows breeders to select desired traits and enhance genetic diversity within a population, leading to increased vigor, yield stability, and adaptability (Nayak et al. 2022).

8 Breeding Approaches: Conventional and Non-conventional Methods

Conventional breeding methods have been used for generations to improve minor grain legumes. Techniques such as phenotypic selection, pedigree selection, and recurrent selection are still valuable tools for breeders. By utilizing diverse germplasm resources and genetically diverse parents, breeders can enhance genetic variation and broaden the breeding pool. However, the use of modern genomic tools has revolutionized the efficiency of breeding programs. Genomic tools such as MAS, GWAS, MABC, and genomic selection have enabled breeders to select and develop superior varieties more efficiently (Mangena 2020). MAS involves selecting individuals based on specific DNA markers associated with desirable traits. GWAS uses a large number of markers to identify genomic regions that are associated with traits of interest. MABC combines the advantages of traditional breeding with the precision of molecular markers to introduce desirable traits into elite cultivars. Genomic selection uses genomic information to predict the breeding values of individuals and select superior individuals for breeding. By integrating genomic tools, breeders can more accurately select desired traits, predict breeding values, and accelerate the development of improved varieties.

8.1 Leveraging Speed Breeding Techniques to Upgrade Orphan Legume Crops

Speed breeding presents an innovative approach to accelerate genetic gains in plant breeding efforts. By optimizing conditions like photoperiod, temperature, and day length, speed breeding aims to dramatically shorten generation times and plant growth cycles (Watson et al. 2018). This allows a more rapid turnover of breeding populations compared to traditional long-season field breeding. So far, speed breeding protocols have been successfully demonstrated in major legume crops like soybean (Fang et al. 2021), chickpea (Samineni et al. 2019), pigeon pea (Saxena et al. 2019), and pea (Mobini and Warkentin 2016). However, this approach has yet to be explored in underutilized grain legumes that have seen limited breeding improvements to date. Establishing optimized speed breeding systems for neglected legume species could significantly increase the rate of genetic gain for key traits like yield, resilience, and nutrition. This would enable more rapid development of improved varieties to benefit smallholder farmers relying on these climate-smart yet overlooked food crops. Realizing the potential of speed breeding in minor legumes, it will require collaborative efforts between breeders, growers, and researchers to adapt protocols and infrastructure for specific crops. However rapid generation cycling holds great promise to accelerate genetic progress in legumes that have lagged due to a lack of research investment.

8.2 Marker-Assisted Breeding

The progress made in DNA technology has expanded our knowledge of the enormous potential found in the genomes of numerous plant species, especially the underutilized legumes. Plant ancestries, origins, and phylogenetic relationships have been identified, traced, and confirmed using DNA-based techniques, which have proven to be credible (Takahashi et al. 2016). Utilizing DNA barcoding, certain underutilized and neglected legumes, including *Lablab purpureus*, *Tylosema esculentum*, *Vigna subterranea*, *V. vexillata* (Wojciechowski et al. 2004; Popoola et al. 2017), and *Vicia* and *Phaseolus* species (Madesis et al. 2012), have been identified and characterized. Genomics has enabled the use of DNA-based molecular markers for the establishment of MAS in plant breeding programs, which apply genotypic selection rather than phenotypic selection which is used in conventional breeding. MAS combines two primary approaches, such as major gene or candidate gene localization and QTL mapping. These techniques are based on linkage disequilibrium (LD), which studies a segregating offspring of parental lines that differ in specific traits, and analyses of association, which study the attributes in a large and diverse population (Flint 2011).

The utilization of molecular markers in marker-assisted selection (MAS) or breeding programs has held substantial significance in appraising the extent of genetic diversity or interrelatedness observed among various species belonging to underutilized legumes (Ojuederie et al. 2014). On a global scale, diverse DNA techniques such as random amplified polymorphic DNA (RAPD), amplified fragment length polymorphisms (AFLPs), inter simple sequence repeats (ISSR), single-nucleotide polymorphisms (SNP), and assessments involving the chloroplast gene RBCL, among others, have been applied to assess the genetic relationships and diversities present among neglected and underutilized species (Boukar et al. 2016).

8.3 Use of Genomic Tools

Genomic tools have revolutionized plant breeding by providing insights into complex traits underlying genetics, enabling more efficient selection processes based on molecular markers linked to these traits. Marker-assisted selection (MAS) integrates marker data associated with target genes or quantitative trait loci (QTLs) during selection processes allowing breeders to identify individuals possessing desirable combinations of alleles without having to wait for phenotypic expression. The list of QTLs and other candidate genes available in the underutilized minor grain legumes are listed in the table below. Genomics tools like GWAS enable breeders to explore the broader genetic architecture underlying complex traits by examining associations between markers and phenotypic variations across diverse germplasm collections. The advent of next-generation sequencing technologies has facilitated whole-genome sequencing and genotyping-by-sequencing, enabling genome-wide approaches in plant breeding for minor grain legumes (Majidian 2022) (Table 3).

Table 3 List of QTLs and other candidate genes available in the underutilized minor grain legumes

S. no	Crop name	Trait	QTLs/Putative candidate genes/TFs	References
1	Grasspea (*Lathyrus sativus* L.)	Salt stress	*LsbHLHD4, LsbHLHD5, LsbHLHR6, LsbHLHD8, LsbHLHR14, LsbHLHR68, LsbHLHR86*	Alsamman et al. (2023)
2	Lupin (*Lupinus* L.)	Flower initiation	*Lup019134, Lup015264,* Lup*021909, Lup021911*	Plewiński et al. (2019)
		Maturity	*Lup015264,Lup004734*	
		Plant height	*Medtr1g030750*	
		Yield traits	*Lup021835, Lup022535*	
		Anthracnose resistance	*antr04_1, antr05_1, antr04_2, antr05_2*	Rychel-Bielska et al. (2020)
		Low-seed alkaloid content	*LOC109339893, LOC109339862, LOC109339875, LOC109339876*	Wang et al. (2021a)
3	Horsegram (*Macrotyloma uniflorum* (Lam.) Verdc.)	Drought stress	*MuWRKY3, MuNAC4, MuMYB96, MuHSP70*	Kiranmai et al. (2018), Masand and Yadav (2016), Pandurangaiah et al. (2014)
		Drought and yield	*qDFW01, qDFW02, qDTM01*	Chahota et al. (2020)
4	Limabean (*Phaseolus lunatus* L.)	Flowering time	*PlTFL1*	Garcia et al. (2021)
		Cyanide content	A major QTL & two other minor QTL	
5	Fababean (*Vicia faba* L.)	*Orobanche crenata* resistance	*Oc2, Oc3, Oc4, Oc5*	Díaz-Ruiz et al. (2010)
		Ascochyta blight resistance	*Af3, Af4, Af5, Af6, Af7, Af8*	Avila et al. (2004)
		Frost tolerance	*Côte d'Or/1, BPL4628*	Arbaoui et al. (2008)
		Drought tolerance	Putative aquaporin gene *VfPIP1*	Cui et al. (2008)
6	Winged bean (*Psophocarpus tetragonolobus* (L.) DC.)	Pod/seed size and quality	31 QTL linked with pod length, pod color, pod anthocyanin content, and flower and seed color	Chankaew et al. (2022)
7	Ricebean (*Vigna umbellata* (Thunb.) Ohwi and Ohashi)	pH and aluminum tolerance	C_2H_2-type zinc finger transcription factor gene VuSTOP1	Fan et al. (2015)
8	Bambara groundnut (*Vigna subterranea* (L.) Verdc)	Internode length	One major QTL	Ho et al. (2017)
9	Adzuki bean (*Vigna angularis* (Willd.) Ohwi & H.Ohashi)	Increased anthocyanin synthesis	VaSDC1 (R2R3-MYB Transcription factor)	Chu et al. (2021)
		Seed size	12 seed size-related QTLs	Wang et al. (2021b)
10	Lablab bean (*Lablab purpureus* (L.) Sweet)	Growth phenological and fruit Traits	41 main effect QTLs (22 for growth phenological traits and 19 for fruit traits)	Yuan et al. (2009)
		Inflorescence length traits	40 QTLs (8.1–55.0% variation)	Yuan et al. (2011)
		Drought and salt tolerance	A novel R2R3-MYB factor gene LpMYB1	Yao et al. (2016)
		Photoperiod responsive flowering	*PvTFLy1* locus	Ramtekey et al. (2019)
11	Mothbean (*Vigna aconitifolia* (Jacq.) Maréchal)	Bruchid resistance	*qVacBrc2.1, qVacBrc5.1*	Somta et al. (2018)
12	Cluster bean (*Cyamopsis tetragonoloba* (L.) Taub.)	Bacterial blight resistance	*L19, D1, AB7, QLTY3*	Bajaj (2011), Dhaliwal et al. (2020)
		Drought tolerance	*OPQ20, OPD10, OPD14, OPQ12, OPAC8, OPF9*	

8.4 Potential Applications of CRISPR-Cas in Underutilized Food Legumes

As a result of negligence and lack of technology, the successful transfer of genes from high-yielding crop varieties is restricted by regulatory restrictions and environmental biosafety concerns (Zhang et al. 2018). In legumes, genome editing is characterized by its importance in improving productivity, quality, and stress resistance. The technology of genome editing has been used in various legumes, including model legumes such as alfalfa and lotus as well as crops with industrial significance like soybean and cowpea. The widespread dissemination of this technology is bound to encounter many challenges and regulatory hurdles. Genome editing to be found in legumes will allow precise editing of legume genomes to improve characteristics such as yield, resistance to stress, and nutritional quality. For example, genome editing can enhance genetic resistance to both biotic and abiotic stress, thereby increasing the resilience and productivity of the growing crop. Genome editing can potentially contribute to improved food safety, higher nutritional value of dietary protein, and reduced food waste. As a result, genome editing can contribute to improved food security, particularly in low-middle-income countries. Despite genome editing's promise, its adoption was impeded by stumbling blocks such as low transformation efficiency and recalcitrant regeneration in many legume crops.

Minor underutilized grain legumes are a rich source of dietary protein and offer tremendous potential for sustainable agriculture. However, these crops often face many challenges during the breeding process, in part because of complex genetics often being polyploid and because limited genomic resources are available for breeding. However, recent advances in genome editing have offered breeders a powerful tool for targeted improvements in these crops. Genome editing technologies enable scientists to make precise changes to an organism's DNA, potentially adding a trait or eliminating one. CRISPR-Cas9 is a revolutionary technology that allows geneticists to edit parts of the genome by removing, adding, or altering sections of the DNA sequence.

Recent advancements have been made in sequencing the genomes of several underutilized grain legumes. For grasspea, researchers have developed a draft genome sequence and are working to resequence diverse germplasm collections of this crop (Das et al. 2021). This foundational genomic information will enable more in-depth studies and genetic improvement efforts in grasspea. In addition, the first genome assembly of a *Lathyrus sativus* cultivar, Pusa-24, has been completed. Access to the assembled Pusa-24 genome can help accelerate genetics and breeding research in this nutritionally important legume (Rajarammohan et al. 2023).

For white lupin, scientists have adapted a highly efficient CRISPR/Cas9-based multiplex editing system using endogenous promoters. This optimization allows for the introduction of desired mutations at target sites in the lupin genome. Such capacity for precision editing through CRISPR holds the potential to rapidly improve agronomic traits in white lupin (Zhu et al. 2023). The seeds of the Lupin, specifically *Lupinus albus*, contain quinolizidine alkaloids (QA) naturally. Researchers are currently working on precision breeding to enhance the nutritional value of white lupin, which could provide high-quality plant-based protein for food and feed. They have identified potential genes for genome editing, which is a promising development in this area of research (Osorio and Till 2022).

A major genomic resource has also been developed for horsegram with the generation of a chromosome-scale draft genome sequence. The availability of the assembled horsegram genome provides a foundation to enable in-depth genetics, genomics, and breeding studies in this climate-resilient legume crop (Shirasawa et al. 2021). Successful application of precision gene editing like this demonstrates the potential of CRISPR-Cas9 for genetic improvement in understudied legumes. Looking ahead, advanced genome editing technologies such as base editors and prime editing could be harnessed to aid domestication efforts in wild crop relatives of neglected legumes. These techniques may also help in re-engineering plant metabolism in targeted ways to increase resilience to stresses and boost nutritional value in orphan legume crops. While progress is still in the early stages, emerging gene editing tools hold exciting promises for unlocking genetic potential in legumes that have so far received little research attention (Gasparini et al. 2021).

8.5 Unleashing the Potential of *de Novo* Domestication in Underutilized Legumes

Crop wild relatives represent a rich source of genetic diversity crucial for enhancing resistance to various stresses in crop plants. They can be transformed into new crops through processes known as "de novo domestication" or "redomestication" (Fernie and Yan 2019). The domestication of underutilized legume crops from their wild relatives plays a vital role in strengthening crop diversity and contributing to sustainable agriculture (Zhang et al. 2018).

Among the underutilized grain legume species, *Vigna stipulaceae* stands out as a promising candidate for de novo domestication due to its natural resilience to drought and salinity stress, along with reduced pod-shattering (Takahashi et al. 2019). Similarly, hairy vetch (*Vicia villosa*), classified as an "incompletely domesticated species," possesses inherent stress tolerance against both biotic and abiotic stresses, making it an ideal legume crop for de novo domestication (Renzi et al. 2020).

In the pursuit of introducing domestication-related traits in these wild relatives, key techniques include mutagenesis, forward screening, and CRISPR/Cas9-based gene editing (Li et al. 2018). Ethyl methanesulfonate mutagenesis, coupled with forward screening, played a pivotal role in the successful domestication of *Vigna stipulacea* Kuntze. The process involved selecting mutants with reduced pod shattering and diminished seed dormancy, contributing to the adaptation of this wild relative (Takahashi et al. 2019). Similarly, the revolutionary CRISPR/Cas9 genome editing technology has been applied to eliminate the g-glutamyl-b-cyano-alanine (GBCA) toxin from seeds of common vetch (*Vicia sativa*). This breakthrough offers the possibility of developing a zero-toxin vetch variety, addressing the increasing global demand for protein (Nguyen et al. 2020). The exploration of de novo domestication from crop wild relatives, such as *Vigna stipulaceae* and hairy vetch, represents an innovative approach to agricultural advancement. The application of mutagenesis, forward screening, and CRISPR/Cas9 technologies underscores the multifaceted strategies employed in enhancing the adaptability and productivity of new legume crops.

9 Emerging Challenges

In order to ensure the success of minor underutilized grain legume breeding programs at both national and international levels, it is crucial to address challenges. These challenges include limited awareness and funding, lack of germplasm diversity, climate change-induced variability, knowledge gaps, and limited collaboration. By adequately addressing these challenges, we can pave the way for the sustainable development of underutilized grain legume breeding programs.

- One of the primary challenges faced by these programs is the limited awareness and funding they receive. Often, the importance and potential of underutilized grain legumes are underestimated, resulting in insufficient resources and funding for breeding initiatives. To overcome this challenge, it is necessary to raise awareness about the significance of these legumes and highlight their potential benefits.
- The lack of diversity in germplasm is one of the major constraints. The availability of diverse and well-characterized germplasm is crucial for the genetic base of breeding programs. By enhancing efforts to collect and preserve germplasm, we can expand the genetic pool and enable the development of improved varieties of underutilized grain legumes.
- Climate change poses a significant threat to crop productivity and stability, which directly impacts underutilized grain legumes as well. To tackle this challenge, it is imperative to develop climate-resilient cultivars that can withstand extreme weather events and changing environmental conditions.
- In addition, there is a pressing need for increased knowledge sharing and collaboration among stakeholders involved in minor grain legume breeding programs. We can tap into collective expertise by fostering stronger networks and research partnerships and finding innovative solutions to common challenges. Collaboration across national and international borders can also lead to exchanging valuable insights and resources.

To ensure the sustainable development of minor grain legume breeding programs, it is essential to establish global networks, promote germplasm exchange, and address policy barriers. These efforts will facilitate the sharing of genetic resources, knowledge, and best practices, ultimately contributing to the advancement of breeding programs worldwide (Dhaliwal et al. 2020).

10 Breeding Progress/Varietal Development

Diverse breeding initiatives, coupled with the application of biotechnological techniques like micro-propagation, molecular markers research, and genetic transformation, are being employed to promote the cultivation of less familiar legume crops (Samal et al. 2023). These efforts aim to encourage farmers to diversify their agricultural practices, enhancing the profitability and sustainability of their systems. Utilizing cutting-edge methods such as micro-propagation facilitates the rapid multiplication of plant material, ensuring a more efficient and scalable production process. Molecular markers research plays a pivotal role in identifying and selecting desirable traits, contributing to the development of improved varieties with enhanced characteristics. Genetic transformation techniques further enable the introduction of specific genes, augmenting the resilience and productivity of underutilized crops. Collectively, these strategies empower farmers to explore a wider array of legume crops, fostering agricultural diversity and economic benefits. Recent reports highlighted a significant challenge in the development of effective phenotyping and breeding methods for underutilized legumes (Takahashi et al. 2016). The limited availability of genetic diversity accessible to breeding programs poses a hurdle for contemporary efforts aimed at enhancing disease resistance, improving quality, and increasing yields. While gene banks hold a substantial amount of genetic variation within grain legume seeds, this diversity remains underutilized in ongoing breeding initiatives. The untapped potential within gene bank resources underscores the need for more comprehensive integration of genetic diversity into active breeding operations for underutilized legumes.

A pioneering adzuki bean map was established by utilizing an F2 population resulting from the cross between adzuki bean and *Vigna nakashimae*, featuring a modest 132 mapped markers (Kaga et al. 1996). Genes responsible for starch synthesis (Peterbauer et al. 1999), encoding abscisic acid-specific glucosyltransferase (Xu et al. 2002), and conferring resistance to bruchids (Chen et al. 2005) or mosaic virus (Chen et al. 2009), have been successfully cloned and examined in adzuki bean. Research has explored the induction of mutations in cluster beans through the application of both physical and chemical mutagens. A cultivar developed through the utilization of these mutagens has been subsequently released (Chopra 2005). The induction of heterophylly in cluster bean was observed through the application of gamma rays and sodium azide treatment (Badami and Bhalla 1992). Achieving viable mutations in cluster bean has been explored through the use of gamma rays in lower doses (up to 50 KR) and lower concentrations of EMS (<0.2%) (Dube et al. 2011). In 2013, Souleymane and colleagues substantiated the aphid tolerance of the enhanced line IT97K-556-6, wherein two resistance loci were delineated, as previously documented by Huynh et al. in 2015. Additionally, a novel source of aphid resistance was pinpointed from a wild relative of cowpea, denoted as TVNu 1158. In grasspea, several varieties and lines have been developed by combining low β-ODAP (<0.1%) with high-yield potential and resistance to biotic and abiotic stresses (Kumar et al. 2013).

Researchers are actively working to broaden the application of the CRISPR/Cas9 system within the grasspea improvement program. Their focus is on engineering signaling pathways and regulatory mechanisms associated with both ODAP biosynthesis and responses to biotic and abiotic stresses (Emmrich 2017). Recent advances in technologies allow sequencing of the large genome of fababean. Under the NORFAB project, a comprehensive annotated reference transcriptome for fababean has been established (Escobar-Herrera et al. 2020). This valuable resource is helpful in the creation of gene models for the reference assembly process. Progress was also made in the field of gene discovery and development of a diagnostic molecular marker for the terminal inflorescence gene (ti) in fababean besides the efficient gene discovery for quality traits (Avila et al. 2006, 2007). Leveraging a diverse set of molecular markers including RAPD, ISSR, SSR, and conserved orthologous sequences, researchers successfully constructed the initial genetic map of horsegram. This milestone was achieved using an intra-specific mapping population derived from the cross between HPK4 and HPKM249. The resulting genetic map spanned a length of 1423.4 cM, showcasing a mean marker interval of 9.6 cM (Chahota et al. 2020). The following table (Table 4) depicts the list of underutilized grain legume varieties released in India.

11 Maintenance Breeding

Maintenance breeding is the specialized field within plant breeding that focuses on the principles and methods associated with nucleus seed production and the preservation of specific plant varieties (Yadav et al. 2022). Belonging to the Papilionaceae family, underutilized grain legume exhibits floral biology that strongly promotes self-pollination. Nevertheless, the likelihood of outcrossing in some crops like grasspea varies based on the flower color, and it can reach up to 30% (Rahman et al. 1995; Almeida et al. 2015). Maintaining the genetic purity of this frequently cross-pollinated crop is essential. This can be achieved through the implementation of adequate isolation distances or employing protective measures such as enclosing unopened buds in cages to prevent pollination by insects. Different approaches can be considered in variety maintenance for self-pollinated underutilized grain legumes depending on the reproduction system of the crop (self-pollinating, often cross-pollinating). For self-pollinated grain legumes, the different procedures of variety maintenance and nucleus seed production are (1) positive or negative mass selection without progeny test and (2) positive mass selection with progeny test. The primary goals of this breeding are preservation of traits, avoiding inbreeding, and population stability (Yadav et al. 2022). During the final stage of inspection, the allowable percentage of off-types is limited to 0.10% for plots involved in foundation seed production and 0.20% for those engaged in certified seed production. Seed or seed pieces with an intact seed coat are classified as pure seed, and the stipulated physical purity requirement stands at 98% for both foundation and certified seed. It is noteworthy that in numerous countries where grain legumes are still cultivated, an informal seed system persists. In this system, economically disadvantaged farmers rely on landraces and saved seeds from their own farms.

12 Coordinated System of Testing

The enhancement of crops demands a comprehensive, interdisciplinary, and multisectoral approach that influences recent advances in science and technology. Acknowledging the importance of a systematic research initiative focused on less common and underutilized crops, the Indian Council of Agricultural Research (ICAR) launched the All India Coordinated Research Project on Underutilized and Under

Table 4 List of underutilized grain legume varieties released in India

Crop	Varieties released	Salient features	Breeding methods	References
Grasspea	Pusa 24	High yielding, with low ODAP (0.2%), suitable for upland situation	Germplasm selection	Kumar et al. (2011)
	Nirmal	High yielding (1.1 t/ha), ODAP content (0.3%)	Germplasm selection	
	BioL-212 (Ratan)	Yield potential is 1.5–1.6 t/ha, low ODAP (0.1%), bold seeded, blue flower, Powdery mildew resistant	Somaclonal variant of Pusa 24	
	Prateek	Yield potential is 1.5–1.6 t/ha, low ODAP (0.10%), blue flower, tolerant to Powdery mildew	LS 82046 × A 60	
	Mahateora	Yield potential is 1.5–1.6 t/ha, low ODAP (0.07%), pink flower, creamy white seeded, early maturing	Ratan × JRL 2	
	Bidhan Khesari-1	Yield potential is 1.5–1.6 t/ha, short duration (108–116 days), low ODAP (0.10%), white flowers with white seed coat	Prateek × Ratan	
Ricebean	RBL-1	Normal duration with high yield	Pureline selection from material collected from Rajasthan	Pattanayak et al. (2019), Raiger et al. (2009)
	RBL-6	Early type	Pureline selection from material collected from Nagaur, Rajasthan	
	RBL-35	Normal duration with green color grain	Pureline selection from material collected from Nagaur, Rajasthan	
	BRS-1	High-yielding variety suitable for hilly region	Germplasm selection	
Cowpea	Kohinoor	Red and bold seeds	Single plant selection from Itan (IL-68-786)	AICRP Annual Report (2022–23), Tripathi et al. (2019)
	Arka Garima	Photo insensitive, suitable for vegetable purpose, tolerant to heat, drought, and low moisture stress	Derivative of the cross TUV 762 x *V. unguiculata* subsp. *sesquipedalis*	
	TC-901	Early maturity (70 days) with semi-determinate habit and field resistance to mosaic & root rot	Mutant of EC394763	
	Kashi Nidhi	Golden mosaic virus-resistant	Pedigree selection (Kashi Unnati x Cowpea-263)	
	UPC- 9202	Medium late variety which matures in 80–85 days	Pedigree selection from intervarietal cross	
	Kashi Shyamal	Early maturity (40–42 (IVREP-1) (Kala Jhamla) days); average yield under normal conditions: 79 qt/ha (fruit)	Selection from local collection (Kala Jhamla)	
	GC-7 (BanasTejas)	Less infection of root rot, yellow mosaic virus and cercospora leaf spot diseases	GC 2 x PGCP 1	
	PhuleSonali	Moderate resistance to YMV and completely free from dry root rot	RC 101 x PhuleVithai	
	Shalimar Cowpea 2 (SKUA-WCP-149)	Resistant to cowpea mosaic virus and Ascochyta blight and tolerant to drought stress		
	VBN 4 (VCP 14-001)	Resistant to cowpea mosaic virus and Ascochyta blight and tolerant to drought stress	VBN 1 x VCP 10-001	

(continued)

Table 4 (continued)

Crop	Varieties released	Salient features	Breeding methods	References
Cluster bean	Agaita Guara 111	Forage type, non-branching type, medium-sized grains, suitable for inter-cropping	Pedigree method of selection from the cross G 325 × FS 277	Roy et al. (2018), Jindal et al. (2021), Directorate of Research, SKNAU, Jobner (2023), Seednet (2023)
	Bundel Guar 1 (IGFRI 212-1)	Dual purpose type, suitable for drier areas with moderate rainfall, moderately resistant to bacterial blight and major pests	Single plant selection from an indigenous material of Rajasthan (Acc. no. B.5/54)	
	Durgajay	Dual purpose type, suitable for late sowing, grain rich in gum and protein content	Single plant selection from local germplasm collected from Nagaur, Rajasthan	
	Durgapura Safed	Dual purpose type, preferred for grain production, grain rich in gum and protein content	Single plant selection from local germplasm of Rajasthan	
	Guar Kranti (RGC 1031)	Suitable for late sowing, dual purpose, highly nutritive fodder, high-quality grain for protein and gum	A derivative of the cross RGC 936 × RGC 986/P 10 through pedigree method of selection	
	Guar Lathi (RGC 1066)	Unbranched type, early maturity, suitable for inter- and mixed cropping	Selection from the cross RGC 936 × RGC 197	
	Gujarat Guar 1(GAUG 34)	Late maturing, branched, medium bold seeds, moderately resistant to bacterial blight disease	Mutant of Kutch 8	
	Karan Guar 1 (RGr 12-1)	Early maturity, drought tolerant, lodging and shattering resistant, moderately resistant to bacterial blight, Alternaria blight, and leaf spot and root rot diseases, low incidence of jassids/ leaf fly	Selection from the cross RGC 1002 × GAUG 04	
	Karan Guar 14 (RGR 18-1)	Lesser infestation of bacterial leaf blight, root rot, and *Alternaria* blight and lower incidence of white fly, leaf hopper, and aphids	RGC 1038 x RGC 1017	AICRP Annual Report (2022–23)
Fababean	Vikrant	Good yield, tolerant to insect pests, diseases, lodging		Arya et al. (2022)
	Swarna Gaurav	High yielding and nutritionally rich variety	Selection made from the germplasm collected from Bihar	Singh et al. (2015)

Exploited Plants in 1982. Subsequently, the project underwent a redesignation and is now known as the All India Coordinated Research Project on Underutilized Crops, recently renamed as the All India Coordinated Research Network on Potential Crops (AICRN). Presently, this network is engaged in comprehensive research on 16 crops with significance in terms of food, fodder, and industrial applications. The research activities span various agro-climatic regions across the country. In 1992, a comprehensive research initiative focusing on arid legumes, namely mothbean, cluster bean, horsegram, and cowpea, commenced with the establishment of the All India Coordinated Research Project on Arid Legumes.

The project's initial location was at ICAR-NBPGR, New Delhi, but in 1995, a committee chaired by Dr. P.N. Bahl, former ADG, ICAR, recommended its relocation to RAU, Bikaner. Following the suggestions of the Jain Committee, the AICRP on Arid Legumes transitioned into the All-India Network Research Project on Arid Legumes in the latter part of 1999, with its headquarters subsequently moving to CAZRI, Jodhpur. Upon the approval of the EFC Memo, a significant development occurred during the XII Plan period starting from April 1, 2015. The headquarters of this network project underwent another shift, this time to ICAR-Indian Institute of Pulses Research, Kanpur. This relocation aimed at enhancing the coordination of research activities related to arid legumes within the framework of the broader national initiative. The ongoing All India Network Project on Arid Legumes, now recognized as AICRP on Kharif Pulses, is dedicated to addressing practical and region-specific challenges. This program places a significant focus on activities such as the collection, assessment, characterization, and utilization of germplasm, as well as the development of varieties. Moreover, it aims to create pertinent production and protection technologies. The project actively engages in the formulation of productive and sustainable cropping systems and conducts on-farm testing, along with the seamless transfer of technology. The multifaceted approach of the program ensures a comprehensive exploration of solutions to enhance agricultural productivity and sustainability in the context of

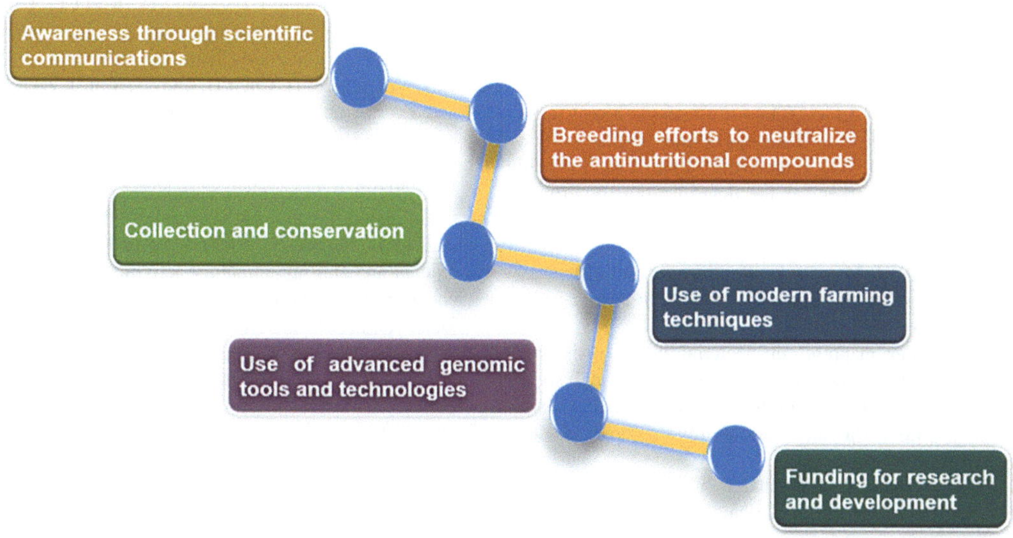

Fig. 6 Future thrust areas for the underutilized grain legumes

arid legumes and kharif pulses which includes many underutilized grain legumes.

13 Conclusion and Future Thrust Area

In the area of underutilized grain legume breeding, the future thrust lies in a multifaceted approach that encompasses various crucial aspects. Firstly, heightened awareness through scientific communications will play a pivotal role in disseminating knowledge about the significance of underutilized grain legumes and their potential contributions to global food security. Concurrently, there is a need for robust efforts in the collection and conservation of diverse genetic resources to ensure the preservation of valuable traits within these legumes. The integration of advanced genomic tools and technologies is imperative for accelerating breeding programs, enabling precise trait selection, and enhancing the overall efficiency of the breeding process. Furthermore, dedicated breeding efforts aimed at neutralizing anti-nutritional compounds in these legumes will be instrumental in enhancing their nutritional value. In Fig. 6, a detailed view of future thrust areas for the underutilized legumes is given. Adoption of modern farming techniques will be essential to optimize productivity, making the cultivation of underutilized grain legumes more attractive to farmers. Adequate funding for research and development will be crucial to sustain and propel these initiatives forward, facilitating breakthroughs in breeding methodologies and ensuring the successful integration of underutilized grain legumes into mainstream agricultural practices.

References

Adebo JA (2023) A review on the potential food application of lima beans (Phaseolus lunatus L.) an underutilized crop. Appl Sci 13(3):1996

Adedayo BC, Anyasi TA, Taylor MJ, Rautenbauch F, Le Roes-Hill M, Jideani VA (2021) Phytochemical composition and antioxidant properties of methanolic extracts of whole and dehulled Bambara groundnut (Vigna subterranea) seeds. Sci Rep 11(1):14116

Aditya JP, Bhartiya A, Chahota RK, Joshi D, Chandra N, Kant L, Pattanayak A (2019) Ancient orphan legume horse gram: a potential food and forage crop of future. Planta 250:891–909

Agarwal S, Chauhan ES (2019) Adzuki beans-physical and nutritional characteristics of beans and its health benefits. Int J Health Sci Res 9(4):304–310

Akpapunam MA, Sefa-Dedeh S (1997) Jack bean (Canavalia ensiformis): nutrition related aspects and needed nutrition research. Plant Foods Hum Nutr 50:93–99

All India Coordinated Research Project on Kharif Pulses (Indian Council of Agricultural Research)- Annual Report (Cowpea, Cluster bean, Horse gram&Moth bean) 2022–23

Almeida NF, Rubiales D, Vaz Patto MC (2015) Grass pea. Grain Legumes:251–265

Alsamman AM, Mousa KH, Nassar AE, Faheem MM, Radwan KH, Adly MH, El Allali A (2023) Identification, characterization, and validation of NBS-encoding genes in grass pea. Front Genet 14:1187597

Amkul K, Wang L, Somta P, Wang S, Cheng X (2019) Construction of a high density linkage map and genome dissection of bruchid resistance in zombi pea (Vigna vexillata (L.) A. Rich). Sci Rep 9(1):11719

Arbaoui M, Link W, Satovic Z, Torres AM (2008) Quantitative trait loci of frost tolerance and physiologically related trait in faba bean (Vicia faba L.). Euphytica 164:93–104

Arya RK, Hooda JS, Kaushik SK, Kumar S, Raigar HL, Gill RK, Kumar P (2022) Development of new promising varieties of faba bean through traditional pedigree method for commercial cultivation in plain zone of India. Indian J Tradit Knowl 21(4):895–904

Avila CM, Atienza SG, Moreno MT, Torres AM (2007) Development of a new diagnostic marker for growth habit selection in faba bean (Vicia faba L.) breeding. Theor Appl Genet 115:1075–1082

Avila CM, Satovic Z, Sillero JC, Rubiales D, Moreno MT, Torres AM (2004) Isolate and organ-specific QTLs for ascochyta blight resistance in faba bean (*Vicia faba* L). Theor Appl Genet 108:1071–1078

Avila CM, Nadal S, Moreno MT, Torres AM (2006) Development of a simple PCR-based marker for the determination of growth habit in Vicia faba L. using a candidate gene approach. Mol Breed 17:185–190

Badami PS, Bhalla JK (1992) Mutagenic effectiveness and efficiency of gamma rays, magnetic fields and sodium azide in cluster bean. Adv Pl Sci 5:534–541

Bajaj A (2011) Molecular mapping of bacterial blight resistance gene, Drought Tolerant QTL (s) and Genetic Diversity Analysis in Cluster bean {Cyamopsis Tetragonoloba (L) Taub}. Ph.D. Thesis, Chaudhary Charan Singh Haryana Agricultural University, Hisar, India

Bamshaiye OM, Adegbola JA, Bamishaiye EI (2011) Bambara groundnut: an under-utilized nut in Africa. Adv Agric Biotechnol 1(1):60–72

Barpete S, Parmar D, Sharma NC, Kumar S (2012) Karyotype analysis in grass pea (Lathyrus sativus L.). J Food Legumes 25(1):14–17

Basu S, Roberts JA, Azam-Ali SN, Mayes S (2007) Bambara groundnut. In: Pulses, sugar and tuber crops. Springer Berlin Heidelberg, Berlin, Heidelberg, pp 159–173

Bhadkaria A, Srivastava N, Bhagyawant SS (2021) A prospective of underutilized legume moth bean (Vigna aconitifolia (Jacq.) Marechàl): phytochemical profiling, bioactive compounds and in vitro pharmacological studies. Food Biosci 42:101088

Bhagyawant SS, Bhadkaria A, Narvekar DT, Srivastava N (2019) Multivariate biochemical characterization of rice bean (Vigna umbellata) seeds for nutritional enhancement. Biocatal Agric Biotechnol 20:101193

Bhardwaj J, Chauhan R, Swarnkar MK, Chahota RK, Singh AK, Shankar R, Yadav SK (2013) Comprehensive transcriptomic study on horse gram (Macrotyloma uniflorum): De novo assembly, functional characterization and comparative analysis in relation to drought stress. BMC Genomics 14:1–18

Bohra A, Kilian B, Sivasankar S, Caccamo M, Mba C, McCouch SR, Varshney RK (2022) Reap the crop wild relatives for breeding future crops. Trends Biotechnol 40(4):412–431

Bonifácio EM, Fonsêca A, Almeida C, Dos Santos KG, Pedrosa-Harand A (2012) Comparative cytogenetic mapping between the lima bean (Phaseolus lunatus L.) and the common bean (P. vulgaris L.). Theor Appl Genet 124:1513–1520

Boukar O, Fatokun CA, Huynh BL, Roberts PA, Close TJ (2016) Genomic tools in cowpea breeding programs: status and perspectives. Front Plant Sci 7:757

Burbano-Erazo E, León-Pacheco RI, Cordero-Cordero CC, López-Hernández F, Cortés AJ, Tofiño-Rivera AP (2021) Multi-environment yield components in advanced common bean (Phaseolus vulgaris L.)× tepary bean (P. acutifolius A. Gray) interspecific lines for heat and drought tolerance. Agronomy 11(10):1978

Caracuta V, Weinstein-Evron M, Kaufman D, Yeshurun R, Silvent J, Boaretto E (2016) 14,000-year-old seeds indicate the Levantine origin of the lost progenitor of faba bean. Sci Rep 6(1):37399

Chahota RK, Sharma V, Rana M, Sharma R, Choudhary S, Sharma TR, Isobe SN (2020) Construction of a framework linkage map and genetic dissection of drought-and yield-related QTLs in horsegram (Macrotyloma uniflorum). Euphytica 216:1–11

Chankaew S, Sriwichai S, Rakvong T, Monkham T, Sanitchon J, Tangphatsornruang S, Somta P (2022) The first genetic linkage map of Winged bean [Psophocarpus tetragonolobus (L.) DC.] and QTL mapping for flower-, pod-, and seed-related traits. Plan Theory 11(4):500

Chen GH, Hsu MP, Tan CH, Sung HY, Kuo CG, Fan MJ, Chen CS (2005) Cloning and characterization of a plant defensin VaD1 from azuki bean. J Agric Food Chem 53(4):982–988

Chen X, Yi J, Zhang H, Chen H, Yang Y, Wan J, Zhai H (2009) Clone and expression analysis of pathogenesis-related gene VaPR3 in azuki bean (Vigna angularis). Jiangsu J Agric Sci 25(5):1119–1123

Cheng A, Raai MN, Zain NAM, Massawe F, Singh A, Wan-Mohtar WAAQI (2019) In search of alternative proteins: unlocking the potential of underutilized tropical legumes. Food Secur 11:1205–1215

Chinnapun D, Sakorn N (2022) Structural characterization and antioxidant and anti-inflammatory activities of new chemical constituent from the seeds of bambara groundnut (Vigna subterranea (L.) Verdc.). CyTA-J Food 20(1):93–101

Choi HW, Kim MY, Lee SH, Sultana S, Bang JW (2013) Molecular cytogenetic analysis of the Vigna species distributed in Korea. Genes Genom 35:257–264

Chopra VL (2005) Mutagenesis: investigating the process and processing the outcome for crop improvement. Curr Sci 89:353–359

Chowdhury MA, Sllnkard AE (1997) Natural outcrossing in grasspea. J Hered 88(2):154–156

Chu L, Zhao P, Wang K, Zhao B, Li Y, Yang K, Wan P (2021) VaSDC1 is involved in modulation of flavonoid metabolic pathways in black and red seed coats in adzuki bean (Vigna angularis L.). Front Plant Sci 12:679892

Cobb JN, Biswas PS, Platten JD (2019) Back to the future: revisiting MAS as a tool for modern plant breeding. Theor Appl Genet 132:647–667

Cowling WA, Buirchell BJ, Tapia ME (1998) *Lupin. Lupinus L.* International Plant Genetic Resources Institute (IPGRI)

Crépon K, Marget P, Peyronnet C, Carrouée B, Arese P, Duc G (2010) Nutritional value of faba bean (Vicia faba L.) seeds for feed and food. Field Crop Res 115(3):329–339

Cui XH, Hao FS, Chen H, Chen J, Wang XC (2008) Expression of the Vicia faba VfPIP1 gene in Arabidopsis thaliana plants improves their drought resistance. J Plant Res 121:207–214

Das A, Parihar AK, Barpete S, Kumar S, Gupta S (2021) Current perspectives on reducing the β-ODAP content and improving potential agronomic traits in grass pea (Lathyrus sativus L.). Front Plant Sci 12:703275

Deepika DD, Padhi SR, Gore PG, Tripathi K, Katral A, Chandora R, Abhishek GJ, Kondal V, Singh R, Bharadwaj R, Bhatt KC, Rana JC, Riar A (2022) Nutritional potential of Adzuki Bean germplasm and mining nutri-dense accessions through multivariate analysis. Food Secur 12(22):4159

Desta KT, Yoon H, Shin MJ, Lee S, Wang XH, Choi YM, Yi JY (2022) Variability of anthocyanin concentrations, total metabolite contents and antioxidant activities in adzuki bean cultivars. Antioxidants 11(6):1134

Dhaliwal SK, Talukdar A, Gautam A, Sharma P, Sharma V, Kaushik P (2020) Developments and prospects in imperative underexploited vegetable legumes breeding: a review. Int J Mol Sci 21(24):9615

Díaz-Ruiz R, Torres AM, Satovic Z, Gutierrez MV, Cubero JL, Román B (2010) Validation of QTLs for Orobanche crenata resistance in faba bean (Vicia faba L.) across environments and generations. Theor Appl Genet 120:909–919

Directorate of Research, SKNAU, Jobner, Rajasthan. Accessed on July 10, 2023. http://www.sknau.ac.in

Doi K, Kaga A, Tomooka N, Vaughan DA (2002) Molecular phylogeny of genus Vigna subgenus Ceratotropis based on rDNA ITS and atpB-rbcL intergenic spacer for cpDNA sequences. Genetica 114:129–145. https://doi.org/10.1016/B978-0-12-397935-3.00011-6

Drayner JM (1959) Self-and cross-fertility in field beans (Vicia faba Linn.). J Agric Sci 53(3):387–403

Dube KG, Bajaj AS, Gawande AM (2011) Mutagenic efficiency and effectiveness of gamma rays and EMS in Cyamopsis tetragonoloba (L.) var. Sharada. Asiatic J Biotechnol Resour 2(4):436–440

Dwivedi SL, Ortiz R (2021) Food crops low in seed-phytate improve human and other non-ruminant animals' health. Crop Breed Genet Genom 3(4):e210009

El-Adawy TA, Rahma EH, El-Bedawey AA, Gafar AF (2001) Nutritional potential and functional properties of sweet and bitter lupin seed protein isolates. Food Chem 74(4):455–462

Ellwood SR, Phan HT, Jordan M, Hane J, Torres AM, Avila CM, Oliver RP (2008) Construction of a comparative genetic map in faba bean (Vicia faba L.); conservation of genome structure with Lens culinaris. BMC Genomics 9(1):1–11

Emmrich P (2017) Unlocking the potential of grass pea (Lathyrus sativus) for food security. Agric Dev 31:67–70

Escobar-Herrera L, Kreplak J, Nadzieja M, Terezol M, Desmet C, Aubert G, Andersen SU (2020) The faba bean pan-transcriptome. In: PAG XXVIII, plant and animal genome

Ezeagu IE, Ibegbu MD (2010) Biochemical composition and nutritional potential of UKPA: a variety of tropical lima beans (*Phaseolus lunatus*) from Nigeria-a short report. Polish J Food Nutr Sci 60(3):231

FAO (2009) How to feed the world in 2050. Insights from an Expert Meeting at FAO, 2050 (1), 1–35

FAO (2018) Food and Agriculture Organization of the United Nations, Rome

Fan W, Lou HQ, Gong YL, Liu MY, Cao MJ, Liu Y, Zheng SJ (2015) Characterization of an inducible C2H2-type zinc finger transcription factor Vu STOP 1 in rice bean (Vigna umbellata) reveals differential regulation between low pH and aluminum tolerance mechanisms. New Phytol 208(2):456–468

Fang Y, Wang L, Sapey E, Fu S, Wu T, Zeng H, Han T (2021) Speed-breeding system in soybean: integrating off-site generation advancement, fresh seeding, and marker-assisted selection. Front Plant Sci 12:717077

Fernie AR, Yan J (2019) De novo domestication: an alternative route toward new crops for the future. Mol Plant 12(5):615–631

Flint F (2011) Mapping quantitative traits and strategies to find quantitative trait genes. Methods 53:163–174. https://doi.org/10.1016/j.ymeth.2010.07.007

Fuchs J, Strehl S, Brandes A, Schweizer D, Schubert I (1998) Molecular—cytogenetic characterization of the Vicia faba genome—heterochromatin differentiation, replication patterns and sequence localization. Chromosom Res 6:219–230

Fuller DQ, Murphy C (2018) The origins and early dispersal of horse gram (Macrotyloma uniflorum), a major crop of ancient India. Genet Resour Crop Evol 65:285–305

Garcia T, Duitama J, Zullo SS, Gil J, Ariani A, Dohle S, Chacón-Sánchez MI (2021) Comprehensive genomic resources related to domestication and crop improvement traits in Lima bean. Nat Commun 12(1):702

Gasparini K, dos Reis MJ, Peres LEP, Zsögön A (2021) De novo domestication of wild species to create crops with increased resilience and nutritional value. Curr Opin Plant Biol 60:102006

Gautam RK, Kumari J, Kuldeep Tripathi RK, Pamarthi SC, Mamta Singh KS, Singh H (2023) PGR management: theory and practices. ICAR-National Bureau of Plant Genetic Resources, New Delhi

Ghosh B, Mitra J, Chakraborty S, Bhattacharyya J, Chakraborty A, Sen SK, Neerathilingam M (2015) Simple detection methods for antinutritive factor β-ODAP present in Lathyrus sativus L. by high pressure liquid chromatography and thin layer chromatography. PLoS One 10(11):e0140649

Gore PG, Das A, Tripathi K, Pratap A, Nair RM (2021) Understanding G× E interaction for nutritional and antinutritional factors in a diverse panel of Vigna stipulacea (Lam.) Kuntz germplasm tested over the locations. Front Plant Sci 12:766645

Gore PG, Gupta V, Singh R, Tripathi K, Kumar R, Kumari G, Pratap A (2022) Insights into the genetic diversity of an underutilized Indian legume, Vigna stipulacea (Lam.) Kuntz., using morphological traits and microsatellite markers. PLoS One 17(1):e0262634

Grela ER, Rybiński W, Matras J, Sobolewska S (2012) Variability of phenotypic and morphological characteristics of some Lathyrus sativus L. and Lathyrus cicera L. accessions and nutritional traits of their seeds. Genet Resour Crop Evol 59:1687–1703

Gu J, Bk A, Wu H, Lu P, Nawaz M, Barrow C et al (2022) Impact of processing and storage on protein digestibility and bioavailability of legumes. Food Rev Int 39:1–28. https://doi.org/10.1080/87559129.2022.2039690

Gujaria-Verma N, Ramsay L, Sharpe AG, Sanderson LA, Debouck DG, Tar'an B, Bett KE (2016) Gene-based SNP discovery in tepary bean (Phaseolus acutifolius) and common bean (P. vulgaris) for diversity analysis and comparative mapping. BMC Genomics 17:1–16

Gupta N, Shrivastava N, Singh PK, Bhagyawant SS (2016) Phytochemical evaluation of moth bean (Vigna aconitifolia L.) seeds and their divergence. Biochem Res Int 2016:1

Gupta PK (1900) Genetics classical to modern. Rastogi Publications

Han DG, Bae MJ, An BJ (2022) Anti-Inflammatory Activity of Velvet Bean (*Mucuna pruriens*) Substances in LPS-Stimulated RAW 264.7 Macrophages. Molecules (Basel, Switzerland) 27(24):8797

Ho WK, Chai HH, Kendabie P, Ahmad NS, Jani J, Massawe F, Mayes S (2017) Integrating genetic maps in bambara groundnut [Vigna subterranea (L) Verdc.] and their syntenic relationships among closely related legumes. BMC Genomics 18:1–9

https://www.pulsedb.org/organism/988181

https://www.pulsedb.org/organism/983374

https://www.nbpgr.ernetin/PGR_Databases.aspx

https://www.genesys-pgr.org/

https://www.gene.affrc.go.jp/databases-plant_search_en.php

https://seedvault.nordgen.org

Huynh BL, Ehlers JD, Ndeve A, Wanamaker S, Lucas MR, Close TJ, Roberts PA (2015) Genetic mapping and legume synteny of aphid resistance in African cowpea (*Vigna unguiculata* L. Walp.) grown in California. Mol Breed: new strategies in plant improvement 35(1):36

Iwata-Otsubo A, Lin JY, Gill N, Jackson SA (2016) Highly distinct chromosomal structures in cowpea (Vigna unguiculata), as revealed by molecular cytogenetic analysis. Chromosom Res 24:197–216

Jiang J, Gill BS (2006) Current status and the future of fluorescence in situ hybridization (FISH) in plant genome research. Genome 49(9):1057–1068

Jindal Y, Sehrawat SK, Chhabra AK, Kumar N, Kumar S, Kumar S, Yadav SS, Dahia M, Niwas R (2021) Varieties of CCS HAU: continued efforts towards food security (PUB ##21-058). Directorate of Research, CCS Haryana Agricultural University, Hisar, Haryana, p 152

Johnson JB, Neupane P, Bhattarai SP, Trotter T, Naiker M (2022) Partitioning of nutritional and phytochemical constituents in nine Adzuki bean genotypes from Australia. J Agric Food Res 10:100398

Kaga A, Ohnishi M, Ishii T, Kamijima O (1996) A genetic linkage map of azuki bean constructed with molecular and morphological markers using an interspecific population (Vigna angularis x V. nakashimae). Theor Appl Genet 93:658–663

Kang YJ, Satyawan D, Shim S, Lee T, Lee J, Hwang WJ, Lee SH (2015) Draft genome sequence of adzuki bean, Vigna angularis. Sci Rep 5(1):8069

Katoch R (2013) Nutritional potential of rice bean (Vigna umbellata): an underutilized legume. J Food Sci 78(1):C8–C16

Kaul T, Eswaran M, Thangaraj A, Meyyazhagan A, Nehra M, Raman NM, Balamurali B (2019) Rice Bean (Vigna umbellata) draft genome sequence: unravelling the late flowering and unpalatability related genomic resources for efficient domestication of this underutilized crop. bioRxiv: 816595

Kearney L (2001) Molecular cytogenetics. Best Pract Res Clin Haematol 14(3):645–668

Khazaei H, Purves RW, Hughes J, Link W, O'Sullivan DM, Schulman AH, Stoddard FL (2019) Eliminating vicine and convicine, the main

anti-nutritional factors restricting faba bean usage. Trends Food Sci Technol 91:549–556
Kiranmai K, Lokanadha Rao G, Pandurangaiah M, Nareshkumar A, Amaranatha Reddy V, Lokesh U, Sudhakar C (2018) A novel WRKY transcription factor, MuWRKY3 (Macrotyloma uniflorum Lam. Verdc.) enhances drought stress tolerance in transgenic groundnut (Arachis hypogaea L.) plants. Front Plant Sci 9:346
Kitano-Okada T, Ito A, Koide A, Nakamura Y, Han KH, Shimada K, Fukushima M (2012) Anti-obesity role of adzuki bean extract containing polyphenols: in vivo and in vitro effects. J Sci Food Agric 92(13):2644–2651
Kumar S, Bejiga G, Ahmed S, Nakkoul H, Sarker A (2011) Genetic improvement of grass pea for low neurotoxin (β-ODAP) content. Food Chem Toxicol 49(3):589–600
Kumar S, Gupta P, Barpete S, Sarker A, Amri A, Mathur PN et al (2013) Grass pea. In: Singh M, Upadhyaya HD, Bisht IS (eds) Genetic and genomic resources for grain legume improvement. Elsevier, London, pp 269–293
Kumar S, Gupta P, Choukri H, Siddique KH (2020) Efficient breeding of pulse crops. In: Accelerated plant breeding, volume 3: food legumes, pp 1–30
Kumari M, Naresh P, Acharya GC, Laxminarayana K, Singh HS, Raghu BR, Aghora TS (2022) Nutritional diversity of Indian lablab bean (Lablab purpureus (L.) Sweet): an approach towards biofortification. S Afr J Bot 149:189–195
Lampart-Szczapa E (1996) Preparation of protein from lupin seeds. Food Nahrung 40(2):71–74
Li T, Yang X, Yu Y, Si X, Zhai X, Zhang H, Xu C (2018) Domestication of wild tomato is accelerated by genome editing. Nat Biotechnol 36(12):1160–1163
Li X, Yang H, Buirchell B, Yan G (2011) Development of a DNA marker tightly linked to low-alkaloid gene iucundus in narrow-leafed lupin (Lupinus angustifolius L.) for marker-assisted selection. Crop Pasture Sci 62(3):218–224
Lin R, Renshaw D, Luckett D, Clements J, Yan G, Adhikari K, Yang H (2009) Development of a sequence-specific PCR marker linked to the gene "pauper" conferring low-alkaloids in white lupin (Lupinus albus L.) for marker assisted selection. Mol Breed 23:153–161
Link T, Seibel C, Voegele RT (2014) Early insights into the genomesequence of Uromyces Fabae. Front Plant Sci 5:587
Liu C, Wu J, Wang L, Fan B, Cao Z, Su Q, Wang S (2017) Quantitative trait locus mapping under irrigated and drought treatments based on a novel genetic linkage map in mungbean (Vigna radiata L.). Theor Appl Genet 130:2375–2393
Maass BL, Jamnadass RH, Hanson J, Pengelly BC (2005) Determining sources of diversity in cultivated and wild Lablab purpureus related to provenance of germplasm by using amplified fragment length polymorphism. Genet Resour Crop Evol 52:683–695
Madesis P, Ganopoulos I, Ralli P, Tsaftaris A (2012) Barcoding the major Mediterranean leguminous crops by combining universal chloroplast and nuclear DNA sequence targets. Genet Mol Res 11(3):2548–2558
Majidian P (2022) Legume breeding: from conventional method to modern technique. In: Legumes research-volume 1. IntechOpen
Mangena P (ed) (2020) Advances in legume research: physiological responses and genetic improvement for stress resistance, vol 1. Bentham Science Publishers
Masand S, Yadav SK (2016) Overexpression of MuHSP70 gene from Macrotyloma uniflorum confers multiple abiotic stress tolerance in transgenic Arabidopsis thaliana. Mol Biol Rep 43:53–64
Massawe FJ, Dickinson M, Roberts JA, Azam-Ali SN (2002) Genetic diversity in bambara groundnut (Vigna subterranea (L.) Verdc) landraces revealed by AFLP markers. Genome 45(6):1175–1180
Maquet A, Vekemans X, Baudoin JP (1999) Phylogenetic study on wild allies of Lima bean, Phaseolus lunatus (Fabaceae), and implications on its origin. Plant Syst Evol 218:43–54
Merga B, Egigu MC, Wakgari M (2019) Reconsidering the economic and nutritional importance of faba bean in Ethiopian context. Cogent Food Agric 5(1):1683938
Minde JJ, Venkataramana PB, Matemu AO (2021) Dolichos Lablab-an underutilized crop with future potentials for food and nutrition security: a review. Crit Rev Food Sci Nutr 61(13):2249–2261
Mishra B, Gautam M, Thakur N, Chahota R (2023) Phylogenetic relationship among cultivated and wild species of genus Macrotyloma based on cytology and molecular markers
Missanga JS, Venkataramana PB, Ndakidemi PA (2021) Recent developments in Lablab purpureus genomics: a focus on drought stress tolerance and use of genomic resources to develop stress-resilient varieties. Legume Sc 3(3):e99
Mobini SH, Warkentin TD (2016) A simple and efficient method of in vivo rapid generation technology in pea (Pisum sativum L.). In Vitro Cell Dev Biol Plant 52:530–536
Mohanty CS, Singh V, Chapman MA (2020) Winged bean: an underutilized tropical legume on the path of improvement, to help mitigate food and nutrition security. Sci Hortic 260:108789
Muñoz LC, Duque MC, Debouck DG, Blair MW (2006) Taxonomy of tepary bean and wild relatives as determined by amplified fragment length polymorphism (AFLP) markers. Crop Sci 46(4): 1744–1754
Nagaraj NC, Muniyappa V, Satyan BA (1981) Resistance source for mung bean yellow mosaic virus
National Academy of Sciences (1979) In Tropical legumes: resources for the future, Report of an Ad hoc panel of the advisory committee on technology innovation board on science and technology for international development, Washington, D.C
National Research Council (2006) Lost crops of Africa. Volome11: vegetable. The National Academies Press, Washington, D.C.. https://www.nap.edu/read/11763/chapter/12
Nayak SP, Lone RA, Fakhrah S, Chauhan A, Sarvendra K, Mohanty CS (2022) Mainstreaming underutilized legumes for providing nutritional security. In: Future Foods. Academic Press, pp 151–163
Nedumaran S, Abinaya P, Jyosthnaa P, Shraavya B, Rao P, Bantilan C (2015) Grain legumes production, consumption and trade trends in developing countries; Working Paper Series No. 60
Nguyen V, Riley S, Nagel S, Fisk I, Searle IR (2020) Common vetch: a drought tolerant, high protein neglected leguminous crop with potential as a sustainable food source. Front Plant Sci 11:818
Obiakor PN (2015) Effects of processing techniques on chemical, functional and microbial properties of two varieties of Lima Bean (Phaseolus lunatus) and maize (Zea mays) flours and sensory properties of their products (Doctoral dissertation)
Ojuederie OB, Balogun MO, Fawole I, Igwe DO, Olowolafe MO (2014) Assessment of the genetic diversity of African yam bean (SphenostylisstenocarpaHochst ex. A Rich. Harms) accessions using amplified fragment length polymorphism (AFLP) markers. Afr J Biotechnol 13(18):1850
Okii D, Tukamuhabwa P, Odong T, Namayanja A, Mukabaranga J, Paparu P, Gepts P (2014) Morphological diversity of tropical common bean germplasm. Afr Crop Sci J 22(1):59–68
Oliveira APD, Nomura CS, Silva SGD, Naozuka J (2018) Iron and selenium speciation in enriched adzuki bean sprouts after fractionation procedures by graphite furnace atomic absorption spectrometry. J Anal Pharm Res 7(2):104–112
Oliveira ARDS, Martins LDV, Bustamante FDO, Muñoz-Amatriaín M, Close T, da Costa AF, Brasileiro-Vidal AC (2020) Breaks of macrosynteny and collinearity among moth bean (Vigna aconitifolia), cowpea (V. unguiculata), and common bean (Phaseolus vulgaris). Chromosom Res 28(3–4):293–306
Omondi OS (2011) The potential for Njahi (Lablab Purpureus L.) in improving consumption adequacy for protein, iron and zinc in households: a case for Nandi south District, Kenya (Doctoral dissertation). University of Nairobi

Osorio CE, Till BJ (2022) A bitter-sweet story: unraveling the genes involved in quinolizidine alkaloid synthesis in Lupinus albus. Front Plant Sci 12:795091

Pandurangaiah M, Lokanadha Rao G, Sudhakarbabu O, Nareshkumar A, Kiranmai K, Lokesh U, Sudhakar C (2014) Overexpression of horsegram (Macrotyloma uniflorum Lam. Verdc.) NAC transcriptional factor (Mu NAC4) in groundnut confers enhanced drought tolerance. Mol Biotechnol 56:758–769

Parihar AK, Barpete S, Das A, Lamichaney A, Gupta S (2022) Lathyrus breeding. In: Fundamentals of field crop breeding. Springer Nature Singapore, Singapore, pp 1323–1389

Pathak R, Singh M, Henry A (2011) Genetic diversity and interrelationship among cluster bean (Cyamopsis tetragonoloba) genotypes for qualitative traits. Indian J Agric Sci 81(5):402

Pathak R, Roy MM (2015) Climatic responses, environmental indices and interrelationships between qualitative and quantitative traits in cluster bean Cyamopsis tetragonoloba (L) Taub. under arid conditions. Proc Natl Acad Sci Sect Biol Sci 85:147–154

Pathak R (2023) Genetics, physiology and cultivation of Moth Bean, Cowpea and Horse Gram. Springer Nature

Pathirana R, Carimi F (2022) Management and utilization of plant genetic resources for a sustainable agriculture. Plan Theory 11(15):2038

Pattanayak A, Roy S, Sood S, Iangrai B, Banerjee A, Gupta S, Joshi DC (2019) Rice bean: a lesser known pulse with well-recognized potential. Planta 250:873–890

Peterbauer T, Mucha J, Mayer U, Popp M, Glössl J, Richter A (1999) Stachyose synthesis in seeds of adzuki bean (Vigna angularis): molecular cloning and functional expression of stachyose synthase. Plant J 20(5):509–518

Petterson DS, Sipsas S, Mackintosh JB (1997) The chemical composition and nutritive value of Australian pulses, 2nd edn. Grains Research and Development Corporation

Petterson DS (1998) Composition and food uses of lupins. In: Lupins as crop plants: biology, production and utilization, pp 353–384

Piepho HP, Möhring J (2007) Computing heritability and selection response from unbalanced plant breeding trials. Genetics 177(3):1881–1888

Plewiński P, Książkiewicz M, Rychel-Bielska S, Rudy E, Wolko B (2019) Candidate domestication-related genes revealed by expression quantitative trait loci mapping of narrow-leafed lupin (Lupinus angustifolius L.). Int J Mol Sci 20(22):5670

Popoola JO, Adebambo A, Ejoh S, Agre P, Adegbite AE, Omonhinmin CA (2017) Morphological diversity and cytological studies in some accessions of Vigna vexillata (L.) A. Richard. Ann Res Rev Biol 19:1–12

Porch TG, Cichy K, Wang W, Brick M, Beaver JS, Santana-Morant D, Grusak MA (2017) Nutritional composition and cooking characteristics of tepary bean (Phaseolus acutifolius Gray) in comparison with common bean (Phaseolus vulgaris L.). Genet Resour Crop Evol 64:935–953

Priyadarshan PM (2019) Plant breeding: classical to modern

Rahate KA, Madhumita M, Prabhakar PK (2021) Nutritional composition, anti-nutritional factors, pretreatments-cum-processing impact and food formulation potential of faba bean (Vicia faba L.): a comprehensive review. Lwt 138:110796

Rahman MM, Kumar J, Rahman MA, Afzal MA (1995) Natural outcrossing in Lathyrus satlvus L. Indian J Genet Plant Breed 55(02):204–207

Raiger HL, Phogat BS, Dua RP & Sharma SK (2009) Improved varieties and cultivation practices of Rice bean

Rajarammohan S, Kaur L, Verma A, Singh D, Mantri S, Roy JK, Kandoth PK (2023) Genome sequencing and assembly of Lathyrus sativus-a nutrient-rich hardy legume crop. Scientific Data 10(1):32

Rajendran K, Sarker A, Singh M, Abd El-Moneim AM, Nakkoul H (2019) Variation for seed protein and ODAP content in grass pea (Lathyrus sativus L.) germplasm collections. Indian J Genet Plant Breed 79(02):438–443

Ramtekey V, Bhuriya A, Ayer D, Parekh V, Modha K, Kale B, Patel R (2019) Molecular tagging of photoperiod responsive flowering in Indian bean [Lablab purpureus (L.) Sweet]. Indian J Genet Plant Breed 79(Sup-01):264–269

Ramya KR, Tripathi K, Pandey A, Barpete S, Gore PG (2021) Rediscovering the potential of multifaceted orphan legume grasspea-a sustainable resource with high nutritional values. Front Nutr 8(2022):826208

Reddy SM (2007) University botany-iii:(plant taxonomy, plant embryology, plant physiology), vol 3. New Age International

Renzi JP, Chantre GR, Smýkal P, Presotto AD, Zubiaga L, Garayalde AF, Cantamutto MA (2020) Diversity of naturalized hairy vetch (Vicia villosa Roth) populations in central Argentina as a source of potential adaptive traits for breeding. Front Plant Sci 11:189

Rizvi AH, Sarker A, Dogra A (2016) Enhancing grasspea (Lathyrus sativus L.) production in problematic soils of South Asia for nutritional security. Indian J Genet Plant Breed 76(04):583–592

Roy AK, Agrawal RK, Shahid Ahmad RV, Kumar AK, Mall NR, Bhardwaj RM, Singh DN, Kantwa SR, Faruqui SA (2018) Database of forage crop varieties: 2018. AICRP on Forage crops and utilization. ICAR-IGFRI, Jhansi, p 344

Rychel-Bielska S, Nazzicari N, Plewiński P, Bielski W, Annicchiarico P, Książkiewicz M (2020) Development of PCR-based markers and whole-genome selection model for anthracnose resistance in white lupin (Lupinus albus L.). J Appl Genet 61(4):531–545

Saharan MS, Saharan GS, Singh JV (2004) Inheritance of leaf margin and branching behaviour in cluster bean (Cyamopsis tetragonoloba)

Salimath PM, Toker C, Sandhu JS, Kumar J, Suma B, Yadav SS, Bahl PN (2007) Conventional breeding methods. In: Chickpea breeding and management. CABI, Wallingford, pp 369–390

Samal I, Bhoi TK, Raj MN, Majhi PK, Murmu S, Pradhan AK, Guru PN (2023) Underutilized legumes: nutrient status and advanced breeding approaches for qualitative and quantitative enhancement. Front Nutr 10:1110750

Samineni S, Sen M, Sajja SB, Gaur PM (2019) Rapid Generation advance (RGA) in Chickpea to Produce up to Seven Generations Per Year and Enable Speed Breeding. Crop J 8:164–169

Saxena KB, Saxena RK, Hickey LT, Varshney RK (2019) Can a speed breeding approach accelerate genetic gain in pigeonpea? Euphytica 215:1–7

Seednet (2023) Accessed on December 6–8, 2023. https://seednetgov.in

Shamurailatpam A, Madhavan L, Yadav SR, Bhat KV, Rao SR (2015) Heterochromatin distribution and comparative karyo-morphological studies in Vigna umbellata Thunberg, 1969 and V. aconitifolia Jacquin, 1969 (Fabaceae) accessions. Compar Cytogen 9(1):119

Shirasawa K, Chahota R, Hirakawa H, Nagano S, Nagasaki H, Sharma T, Isobe S (2021) A chromosome-scale draft genome sequence of horsegram (Macrotyloma uniflorum). Gigabyte 2021:1

Singh AK, Singh IS, Sundaram PK (2015) Swarna Gaurav: high yielding nutritionally rich faba bean variety. J AgriSearch 2(3):179–182

Singh A, Dubey PK, Chaurasia R, Dubey RK, Pandey KK, Singh GS, Abhilash PC (2019) Domesticating the undomesticated for global food and nutritional security: four steps. Agronomy 9(9):491

Singh RK, Prasad A, Muthamilarasan M, Parida SK, Prasad M (2020) Breeding and biotechnological interventions for trait improvement: status and prospects. Planta 252:1–18

Siriwardhane D, Egawa Y, Tomooka N (1991) Cross-compatibility of cultivated Adzuki Bean (Vigna angularis) and Rice Bean (V. umbellata) with their wild relatives. Plant Breed 107(4):320–325

Somta P, Jomsangawong A, Yundaeng C, Yuan X, Chen J, Tomooka N, Chen X (2018) Genetic dissection of Azuki bean weevil (Callosobruchus chinensis L.) resistance in moth bean (Vigna aconitifolia [Jaqc.] Maréchal). Genes 9(11):555

Souleymane A, Aken'ova ME, Fatokun CA, Alabi OY (2013) Screening for resistance to cowpea aphid (*Aphis craccivora* Koch) in wild and cultivated cowpea (*Vigna unguiculata* L. Walp.) accessions. Int J Sci Environ Technol 2:611–621

Srinivasan K (2020) Cluster beans. In: Nutritional composition and antioxidant properties of fruits and vegetables. Academic Press, pp 301–311

Suranjika S, Pradhan S, Kalia RK, Dey N (2023) De novo assembly of the whole genome of Moth bean (Vigna aconitifolia), an underutilized Vigna species of India. bioRxiv: 2023–05

Suso MJ, Bebeli PJ, Palmer RG (2015) Reproductive biology of grain legumes. In: Grain legumes, pp 365–399

Tahir F, Hassani A, Kouadria M, Rezzoug W (2019) Study of morpho-physiological and biochemical behavior of cultivated legume (Lens culinaris MedikSsp culinaris) in Dry Area of Algeria. Ukrainian J Ecol 9(4):535–541

Takahashi Y, Somta P, Muto C, Iseki K, Naito K, Pandiyan M, Tomooka N (2016) Novel genetic resources in the genus Vigna unveiled from gene bank accessions. PLoS One 11(1):e0147568

Takahashi Y, Sakai H, Yoshitsu Y, Muto C, Anai T, Pandiyan M, Naito K (2019) Domesticating Vigna stipulacea: a potential legume crop with broad resistance to biotic stresses. Front Plant Sci 10:1607

Talukdar D (2011) Fluorescent-banded karyotype analysis and identification of chromosomes in three improved Indian varieties of grass pea (Lathyrus sativus L.). Chromo Sci 13(1+ 2):3–10

Talukdar D (2013) Comparative morpho-physiological and biochemical responses of lentil and grass pea genotypes under water stress. J Nat Sci Biol Med 4(2):396

Toker C (2004) Estimates of broad-sense heritability for seed yield and yield criteria in faba bean (Vicia faba L.). Hereditas 140(3):222–225

Tresina PS, Paulpriya K, Mohan VR, Jeeva S (2017) Effect of gamma irradiation on the nutritional and antinutritional qualities of Vigna aconitifolia (Jacq.) Marechal: an underutilized food legume. Biocatal Agric Biotechnol 10:30–37

Tripathi K, Gore PG, Ahlawat SP, Tyagi V, Semwal DP, Gautam NK, Kumar A (2019) Cowpea genetic resources and its utilization: Indian perspective-A review. Legume Res Int J 42(4):437–446

Tripathi K, Gore PG, Rajpoot SK, Singh N, Gupta V (2021) Minni Payaru [Vigna stipulacea (Lam.) Kuntz.]: an underutilized ancient legume of India. Indian J Tradit Knowl 20(4):1084–1087

Tripathi K, Sadhukhan R, Das A, Jana K, Semwal DP, Kumar A, Ahlawat SP (2022a) Khesari (Lathyrus sativus L.), an ancient legume for future gain: an expedition collection from parts of West Bengal state of Eastern India. Indian J Tradit Knowl 21(2):395–403

Tripathi K, Pamarthi RK, Semwal DP, Tyagi V, Brahmi P, Pandey A (2022b) Efforts for the Revival of Vigna subterranea (L.) Verdc. (Bambara groundnut) through Germplasm Introduction in India. Indian J Plant Genet Resour 36(1):116–121

Tsamo AT, Ndibewu PP, Dakora FD (2018) Phytochemical profile of seeds from 21 Bambara groundnut landraces via UPLC-qTOF-MS. Food Res Int 112:160–168

Tyagi A, Sharma P, Saxena S, Sharma R, Mithra SA, Solanke AU, Gaikwad K (2019) The genome size of cluster bean (Cyamopsis tetragonoloba) is significantly smaller compared to its wild relatives as estimated by flow cytometry. Gene 707:205–211

Uzun BÜLENT, Arslan C, Karhan M, Toker C (2007) Fat and fatty acids of white lupin (Lupinus albus L.) in comparison to sesame (Sesamum indicum L.). Food Chem 102(1):45–49

Varshney RK, Mohan SM, Gaur PM, Gangarao NVPR, Pandey MK, Bohra A, Gowda CLL (2013) Achievements and prospects of genomics-assisted breeding in three legume crops of the semi-arid tropics. Biotechnol Adv 31(8):1120–1134

Vasconcelos MW, Grusak MA, Pinto E, Gomes A, Ferreira H, Balázs B, Iannetta P (2020) The biology of legumes and their agronomic, economic, and social impact. In: The plant family Fabaceae: biology and physiological responses to environmental stresses, pp 3–25

Vatanparast M, Shetty P, Chopra R, Doyle JJ, Sathyanarayana N, Egan AN (2016) Transcriptome sequencing and marker development in winged bean (Psophocarpus tetragonolobus; Leguminosae). Sci Rep 6(1):29070

Waines JG (1978) Protein contents, grain weights, and breeding potential of wild and domesticated Tepary Beans 1. Crop Sci 18(4):587–589

Wang P, Zhou G, Jian J, Yang H, Renshaw D, Aubert MK, Li C (2021a) Whole-genome assembly and resequencing reveal genomic imprint and key genes of rapid domestication in narrow-leafed lupin. Plant J 105(5):1192–1210

Wang LX, Jie W, Luo GL, Yuan XX, Dan G, Hu LL, Cheng XZ (2021b) Construction of a high-density adzuki bean genetic map and evaluation of its utility based on a QTL analysis of seed size. J Integr Agric 20(7):1753–1761

Watson A, Ghosh S, Williams MJ, Cuddy WS, Simmonds J, Rey MD, Hickey LT (2018) Speed breeding is a powerful tool to accelerate crop research and breeding. Nat Plants 4(1):23–29

Wojciechowski MF, Lavin M, Sanderson MJ (2004) A phylogeny of legumes (Leguminosae) based on analysis of the plastid matK gene resolves many well-supported subclades within the family. Am J Bot 91(11):1846–1862

Xu ZJ, Nakajima M, Suzuki Y, Yamaguchi I (2002) Cloning and characterization of the abscisic acid-specific glucosyltransferase gene from adzuki bean seedlings. Plant Physiol 129(3):1285–1295

Yadav U, Singh N, Kaur A, Thakur S (2018) Physico-chemical, hydration, cooking, textural and pasting properties of different adzuki bean (Vigna angularis) accessions. J Food Sci Technol 55:802–810

Yadav RN, Kumar PR, Hussain Z, Yadav S, Lal SK, Kumar A, Yadav SK (2022) Maintenance breeding 13. In: Fundamentals of field crop breeding, p 703

Yang H, Tao Y, Zheng Z, Shao D, Li Z, Sweetingham MW, Li C (2013) Rapid development of molecular markers by next-generation sequencing linked to a gene conferring phomopsis stem blight disease resistance for marker-assisted selection in lupin (Lupinus angustifolius L.) breeding. Theor Appl Genet 126:511–522

Yang S, Grall A, Chapman MA (2018) Origin and diversification of winged bean (Psophocarpus tetragonolobus (L.) DC.), a multipurpose underutilized legume. Am J Bot 105(5):888–897

Yao L, Jiang Y, Lu X, Wang B, Zhou P, Wu T (2016) A R2R3-MYB transcription factor from Lablab purpureus induced by drought increases tolerance to abiotic stress in Arabidopsis. Mol Biol Rep 43:1089–1100

Yuan J, Yang R, Wu T (2009) Bayesian mapping QTL for fruit and growth phenological traits in Lablab purpureus (L.) Sweet. Afr J Biotechnol 8(2):167

Yuan J, Wang B, Wu TL (2011) Quantitative trait loci (QTL) mapping for inflorescence length traits in Lablab purpureus (L.) sweet. Afr J Biotechnol 10(18):3558–3566

Zhang H, Li Y, Zhu JK (2018) Developing naturally stress-resistant crops for a sustainable agriculture. Nat Plants 4(12):989–996

Zhao P, Chu L, Wang K, Zhao B, Li Y, Yang K, Wan P (2022) Analyses on the pigment composition of different seed coat colors in adzuki bean. Food Sci Nutr 10(8):2611–2619

Zhou G, Yang H, Renshaw D, Zou M, Thomas G, Li C (2021) Identification of gray leaf spot disease candidate gene in narrow-leafed lupin (Lupinus angustifolius L.). Front Genet 12:695791

Zhu X, Xu W, Liu B, Zhan Y, Xia T (2023) Adaptation of high-efficiency CRISPR/Cas9-based multiplex genome editing system in white lupin by using endogenous promoters. Physiol Plant 175(4):e13976

Soybean [*Glycine max* (L.) Merr.] Breeding

K. H. Singh, Sanjay Gupta, M. Shivakumar, and V. Nataraj

Abstract

A major crop in oilseed production, soybeans are an essential source of animal feed and a high value health food in the modern day due to their unique combination of leguminous, oilseed, and nutraceutical qualities. Although increasing yield is the main objective of soybean breeding, there is a lot of room for genetic advancement to satisfy the various needs of the soy sector. This chapter covers the global history of soybean and looks at the morphological characteristics that are important for breeding. The study delves into the primary methods of breeding soybeans, the genetic foundation of significant qualitative and quantitative features, and the essential measures to enhance yield components, resistance to major diseases, protein and oil augmentation, and oil quality improvement. It also discusses the necessity of lowering or getting rid of unfavorable characteristics like lipoxygenase that generate off-flavors and anti-nutritional elements like the Kunitz trypsin inhibitor. This chapter also discusses transgenic, molecular, and conventional methods for accomplishing these breeding goals.

Keywords

Glycine max · Breeding methods · Maturity · Speed breeding · Disease resistance · Yield components · Off-flavor · Kunitz trypsin inhibitor · Oleic acid · Marker-assisted breeding · Transgenic

1 Introduction

Soybeans, often referred to as the "miracle bean," possess a unique combination of approximately 20% oil and 40% protein, while also enhancing soil fertility by adding nitrogen to the soil. Globally, this crop plays a significant role, contributing 25% to the production of edible oil and 65% to animal feed. Despite belonging to the Leguminosae family, the use of soybeans as a human protein source is limited, with a staggering 94% of the world's soybean production being crushed for oil and animal feed. According to FAO estimates, 348.8 million metric tons of soybeans were produced from 133.79 million hectares of acreage worldwide in 2022. Major soybean producers worldwide include Brazil, the USA, Argentina, China, and India, with production figures of 120.7, 116.4, 43.9, 20.3, and 13.0 million metric tons, respectively, and acreage of 40.9, 34.9, 10.2, and 12.1 million hectares, respectively, in 2022, at varying productivity rates (source: FAOSTAT 2024; [FAO](https://www.fao.org/faostat/en/#data/QCL). In India, soybean cultivation in the hilly regions of the Himalayan states has a rich history dating back centuries. The journey of soybeans in India over the last five decades has been nothing short of remarkable. It was in the 1970s that soybeans truly began to make significant strides in the country. The growth in both the area under cultivation and production has been nothing short of spectacular. In 1970, soybeans covered a mere 30,000 hectares, producing 10,000 quintals. Fast forward to 2022, soybean cultivation has expanded to a staggering 12 million hectares, yielding 12.9 million metric tons—an unprecedented growth trajectory that few crops can rival. In the fiscal year 2021–22, India earned Rs 2207 crores through the export of soybean oil meal (source: [SEAI] (https://seaofindia.com/export-of-oil-meals-april-2021/)). Soybean cultivation in India is concentrated in three states: Madhya Pradesh, Maharashtra, and Rajasthan, which contribute significantly to both area and production. The continuous increase in global soybean production is closely tied to the development of new cultivars, which have played a pivotal role in enhancing yields. Research has shown a positive correlation between yield gains and the release year of soybean varieties, with newer varieties consistently producing higher yields (de Felipe et al. 2016; Fox et al. 2013; Jin et al. 2010; Morrison et al. 2000; Ramteke et al. 2011; de Toledo et al. 1990). Key driv-

K. H. Singh (✉) · S. Gupta · M. Shivakumar · V. Nataraj
ICAR-National Soybean Research Institute,
Indore, Madhya Pradesh, India

G. Prasad Dixit et al. (eds.), *Fundamentals of Legume Breeding*, https://doi.org/10.1007/978-981-96-7838-9_13

ers for yield improvements in soybean varieties include increased seed density, pod production, and lodging resistance. In India, Selection Cycle 1 cultivars (developed through exotic material selection) demonstrated four times higher yield and harvest index compared to the local variety Kalitur (Karmakar and Bhatnagar 1996). Selection Cycle 2 cultivars, developed through hybridization between exotic and local material, exhibited 19% and 16% higher yield and harvest index, respectively, over Selection Cycle 1 cultivars. An assessment of yield gains in Indian varieties from 1969 to 2008 revealed a remarkable 103% improvement (23 kg/year) at a rate of approximately 2.6% per year.

2 Origin, Evolution, and Distribution of Species and Forms: Wild Relatives

Soybean is believed to have originated from the eastern part of North China as the primary center and Manchuria as the secondary center of origin (Prakash Tiwari 2006). The *Glycine* genus comprises two subgenera, *Glycine Willd* and *Soja* (Moench) F.J. Hermann, with a total of 28 species classified under these two subgenera. *Soja* contains two annual species, *Glycine soja* Sieb & Zucc (wild) and *Glycine max* (cultivated), while subgenus *Glycine* contains 26 perennial species: *Glycine canescens, Glycine clandestine, Glycine latrobeana, Glycine tabacina, Glycine tomentella, Glycine falcate, Glycine latifolia, Glycine argyria, Glycine cyrtoloba, Glycine curvata, Glycine arenaria, Glycine microphylla, Glycine albicans, Glycine hirticaulis, Glycine lactovirens, Glycine dolichocarpa, Glycine pindanica, Glycine stenophita, Glycine peratosa, Glycine rubiginosa, Glycine aphyonota, Glycine pullenii, Glycine gracei, Glycine montis-douglas, Glycine pescadrensis,* and *Glycine syndetika.* The annual species were used for food or feed purposes (Nawaz et al. 2018a; b; Sherman-Broyles et al. 2014; Singh et al. 2007). The wild species *Glycine soja* is considered the progenitor of the cultivated species *Glycine max* based on cytological, genomic, and proteomic evidence. Soybean's origin has been historically associated with Eastern Asia, with later insights suggesting its roots in northern and central China. This crop emerged as a domesticated staple around the eleventh century B.C. in northern China, spreading to southern China, Korea, Japan, and Southeast Asia between the eleventh and thirteenth centuries B.C. (Anderson et al. 2019). Although soybean has been cultivated in Europe since the seventeenth century, its introduction to India is estimated to have occurred between 1870 and 1880. Population genomics methods help us trace the origins of crop plants and identify the genetic factors underlying the domestication syndrome, including reductions in genetic diversity, selection sweeps, and the specific genomic regions and genes involved in domestication. Research has shown that cultivated soybean was domesticated from its wild relative, *Glycine soja* (Sieb. and Zucc.), about 5000 years ago in the temperate regions of China (Zhou et al. 2015). Chung et al. (2014) cataloged genome variation in an annual soybean population by high-depth resequencing of 10 cultivated and 6 wild accessions, identifying 3.87 million high-quality single-nucleotide polymorphisms (SNPs). Phylogenetic analysis of the nuclear genome supported the idea of a single origin for cultivated soybeans. Other studies also propose a single domestication origin for soybean, a view supported by recent genome resequencing research (Zhou et al. 2015; Han et al. 2016). Resequencing of 302 wild, landrace, or improved soybean accessions suggests that all domesticated soybeans stem from a single cluster of *G. soja* wild soybeans, reinforcing the single-origin theory. Several genetic bottlenecks occurred during the domestication and diversification of soybean, particularly in the domestication of Asian landraces and the introduction of a limited number of landraces to North America (Hyten et al. 2006). Resequencing of both wild and cultivated soybean accessions revealed the impact of artificial selection during domestication, showing a significant reduction in genetic diversity (Lam et al. 2010). However, naturally occurring introgression helped mitigate the genetic bottlenecks during soybean domestication (Wang et al. 2019).

Biological Properties of Soybean

The taxonomy of wild annual and cultivated soybean is as follows:

Order	Fabales
Family	Fabaceae (Leguminosae)
Subfamily	Papilionoideae
Tribe	Phaseoleae
Subtribe	Glycininae
Genus	*Glycine* Willd.
Subgenus	*Soja* (Moench) F.J. Herm.
Species	*Glycine soja* Sieb. & Zucc.
Species	*Glycine max* (L.) Merr.

Botanical Description of the Soybean Plant

From a taxonomic perspective, the genus *Glycine* Willd. belongs to the family Leguminosae, also known as Fabaceae, under the subfamily Papilionoideae, tribe Phaseoleae, and subtribe Glycineae (Singh 2010; Maranna et al. 2021). The genus *Glycine* Willd. is subdivided into two distinct subgenera: *Glycine* and *Soja* (Moench) F.J. Hern. *Glycine max* is a palaeopolyploid (2n = 20) that has undergone multiple genome duplication and subsequent rediploidization events with simultaneous rearrangements among chromosomes, resulting in up to 12 occurrences of a given genome region in *G. max*. The *Glycine* subgenus comprises numerous (approximately 26) perennial inbreeding species that are native to regions including Australia, West and Central Pacific islands,

Papua New Guinea, Indonesia, the Philippines, and Taiwan. The Soja subgenus consists of two annual, diploid (2n = 40) species: *Glycine max* (cultivated soybean) and *Glycine soja* (wild soybean), both of which are prone to easy hybridization. Wild soybeans (*Glycine soja* Sieb. &Zucc), considered the direct ancestor of cultivated soybean (*G. max*) constitute the primary gene pool for *G. max*. While a secondary gene pool has not been reported for soybeans, 26 wild species make up the tertiary gene pool. Novel sources of resistance to various stresses, including soybean cyst nematode, sap-sucking insects, rust, sclerotinia stem rot, and sudden death syndrome, as well as salt tolerance, have been identified in *G. soja* and wild perennial species. Several genes contributing to salt tolerance have been identified using annual wild relatives. In India, *Glycine soja* has been instrumental in introducing resistance to Yellow Mosaic Virus (YMV) in *Glycine max*, with the associated genomic region tagged (Rani et al. 2018).

3 Reproductive Biology: Emasculation – Pollination Techniques

The flowers of the soybean are very small and are located in the axils of the leaves. They are arranged in clusters, with each cluster containing 20–24 flowers. The size of each flower is around 5–6 mm. The flowers are self-pollinating, with the upper petals (called the standard or banner) darker and larger than the lateral petals (wings) and the two lower petals (keel or boat). The color of the petals can be white or purple. Soybean plants start flowering after the formation of 6–10 leaves, and the flowering period can vary depending on the variety, lasting 20–30 days in early varieties and 40–80 days in late varieties. Pods develop after flowering and can be light yellow, grayish-yellow, light brown, dark gray, or dark brown when ripe. Soybeans themselves can vary in color from yellow, light green, light brown, to black. The size and shape of the seeds can also vary, with an average weight of 1000 seeds ranging from 80 to 250 grams. The seeds consist of the seed coat, seed pod, and cotyledons. The optimal temperature for soybean growth is between 20 and 24 °C, with seeds germinating in 2–3 days at a rate of 90–100%. The flowering and podding phase begins 30–65 days after germination, and this phase is the longest in the soybean plant's development. During the flowering phase, the plant grows rapidly, with an average daily growth of 1.5–2.5 cm. Each cluster of flowers blooms for 6–11 days, with the first flower usually opening from the second or third leaf axil.

Crossing Affinity

Intra- and interspecific crossability is a valuable approach for understanding species relationships. Crosses between species with similar genomes typically result in normal pod formation and fertile F1 seeds, with the hybrids being fertile. In contrast, crosses between species with significantly different genomes often lead to seed abortion, and any resulting hybrids are usually sterile (Newell and Hymowitz 1983; Grant et al. 1984; Singh and Hymowitz 1985a, b, 1988, 1989; Singh et al. 1992; Kollipara et al. 1993, 1994). Normal pod set and fertile F1 hybrids are expected when the parental species are both morphologically and genomically similar. Soybean (*Glycine max*) is an annual plant from the Fabaceae family and one of the world's most widely cultivated ancient crops. Native to Southeast Asia, soybeans have been grown in China for over 6000 years, and there are currently more than 2500 varieties. The height of the soybean plant can vary from low to tall, depending on soil and climatic conditions. The plant is mainly upright, with both horizontal and creeping forms. Stem thickness can vary greatly among varieties, ranging from 3–5 mm to 20–22 mm. The color of the hypocotyls of newly grown soybean seedlings can be green or purple. Stem color often correlates with flower color, with white flowers typically associated with dark green stems and purple flowers with light purple stems. Stem feathers, which serve as protection, can vary in density. The number and location of lateral branches depend on the genotype and growing conditions. The first-order side branches often form many pods, most of which are shed before the top pods ripen. Stem hairs can be scattered, semi-scattered, or absent, depending on the variety. Soybean pods are bushy or scattered, depending on the location of the side branches. Stem growth can cease after the top pods form, or it can continue until the pods begin to ripen. Stem forms can be incomplete, intermediate, or finished, based on the growth of the neck. Leaves are intricate, with three petals arranged on a leaf band, alternately at the base. Leaf shape can vary from lanceolate to heart-shaped, oval, or round, depending on factors like temperature, light, humidity, and nutrition. Leaves turn yellow and shed naturally as pods ripen. The root system is well developed, with the taproot penetrating the soil to a depth of 1–1.5 meters. Root development mainly occurs in the tillage layer, with many side roots forming. The amount and size of tubers depend on various factors, including plant genotype, soil microflora, and soil fertility. Soybean leaves can range from 5 to 10 cm in length and 3–10 cm in width, with a single plant having 20–40 or more leaves. Leaf surface characteristics depend largely on the plant's genotype. Soybean flowers are small, unsightly, and do not attract insects.

4 Genetic Resources

Soybean germplasm, comprising approximately 170,000 accessions worldwide, is a valuable genetic resource for improving yield and specific traits. China holds the largest collection with over 23,000 cultivated and 7000 wild accessions, followed by the USDA-Agricultural Research Service (ARS) with more than 20,000 accessions, and Japan with 11,300 accessions (Carter Jr et al. 2004). India also maintains a significant collection of 8025 soybean germplasm accessions, with 50% originating from exotic sources. The ICAR-Indian Institute of Soybean Research (IISR) in Indore is a National Active Germplasm Site (NAGS) for soybean and serves as the national repository of soybean varieties and germplasm. The augmentation, conservation, characterization, evaluation, and documentation of soybean germplasm are carried out by ICAR-IISR in association with the ICAR-National Bureau of Plant Genetic Resources (NBPGR) in New Delhi. The ICAR-NBPGR in New Delhi holds 5021 soybean germplasm accessions in Long-Term Storage (LTS). Exotic germplasm has been collected from various countries including the USA, Canada, South Africa, Australia, Germany, China, Nepal, Italy, UK, Argentina, Hungary, Taiwan, Morocco, Nigeria, Thailand, Romania, Sri Lanka, Trinidad, Philippines, Brazil, Yugoslavia, Papua New Guinea, Ghana, Israel, Korea, Rhodesia, Indonesia, Fiji, and Japan. Indigenous germplasm has been collected from states like Uttar Pradesh, Punjab, Himachal Pradesh, Sikkim, West Bengal, Assam, Meghalaya, Madhya Pradesh, Manipur, Bihar, Maharashtra, Gujarat, Jammu & Kashmir, Mizoram, Arunachal Pradesh, Rajasthan, Delhi, Karnataka, and Tamil Nadu (Verma and Thomas 1991). In India, the early growth of soybeans can be attributed to landraces or indigenous varieties like Kalitur, JS 2, Type 49, and Punjab White/Punjab 1 (Selection from Nanking). Introduced varieties such as Bragg, Clark 63, Davis, Hardee, Hill, Improved Pelican, KM 1 (AVRDC), Lee, and Monetta (EC 2587) have also contributed to the growth of soybean cultivation in India (Prakash Tiwari 2006). Cultivated soybean (*Glycine max*) and wild *Glycine soja* have the same chromosome number (2n = 40) and share a common gene pool (primary gene pool GP1). Soybean lacks a secondary gene pool (GP II), but 26 wild perennial species in the subgenus *Glycine* are included in the tertiary gene pool (GP III). The genetic base of modern soybean cultivars is narrow, and soybean breeders are limited to crossing within the primary gene pool (GP-1). Wild soybeans are crucial genetic resources for understanding genetic diversity, genomic variations, and environmental adaptations in soybean (Nawaz et al. 2020). Pre-breeding efforts utilizing landraces and crop wild relatives are being conducted to broaden the genetic base of soybean. Pre-breeding for soybean is carried out at various institutes including ICAR-IISR in Indore, ICAR-IARI in New Delhi, ICAR-VPKAS in Almora, PAU in Ludhiana, and GBPUAT in Pantnagar. The trait-specific germplasm is listed below (Table 1).

Table 1 List of trait-specific germplasm at ICAR- NSRI, Indore

S. No.	Trait	Germplasm
1	Photoperiod insensitivity	MACS 330, EC 325097, EC 333894, EC 333897, EC 34101, EC 325118, EC 390977, EC 538822
2	Drought resistance	EC 538828, EC 602288, JS 97-52, Kaeri 651-6, Young
3	Rust resistance	EC 241778, EC 241780
4	Rust differential	Wyne
5	Yellow mosaic virus resistance	UPSM 534, DS 3105, PLSO 84, EC24660, B463
6	Rhizoctonia root rot resistance	AGS48 and EC34117
7	Long juvenile	AGS 25, CM 60
8	Salinity tolerance	EC 34115
9	Mechanical damage	DOKO
10	High temperature	NRC 146
11	Water logging tolerance	JS 20-38, TGX 317-37E, Hardee, JS 20-76
12	Seed longevity	JS 335
13	Vegetable	Harasoya, Karune, MACS 1508, AGS 186, AGS 190, GC 84058-18-4, NRC 188
14	Null KTI	NRC 127, NRC 142, NRC 197
15	Lox 2 free	NRC 132, NRC 142, NRC 150, NRC 197
16	High protein	G 11, G 15 A
17	High oleic acid	IC 210
18	Antibiosis	PS 1572, PS 1556, F4P21, P3P18
19	Anthracnose	EC 457254
20	Charcoal rot	JS 20-98
21	Early maturity	EC 771186, EC 572154

5 Breeding and Genetics

Soybean, a major oilseed crop in the legume family, is a crucial source of nutrients and food worldwide. It was domesticated along with its wild relative, *Glycine soja*, originating from East China. Genetic variation has further adapted soybean crops to different environments in various countries. In China, soybean was domesticated 6000–9000 years ago, with a wide distribution of wild soybeans. However, Indian soybean cultivars have an extremely narrow genetic base (Shivakumar et al. 2019; Maranna et al. 2021). Studies by Delannay et al. (1983) and Gizlice et al. (1994) have shown that 80% of the gene pool of Northern American soybean originated from introductions from both northern and southern sources. This narrow genetic diversity makes the crop vulnerable to various stresses, leading to crop losses (Amrate et al. 2023). Extensive use of a single genetic male sterility has been identified as a cause of loss in several crops (Ramlal et al. 2022). Increasing genetic diversity through breeding and the use of wild relatives is crucial for improving the resilience of soybean crops to various stresses.

5.1 Breeding Objectives in Soybean

Productivity of soybean in India is relatively low at 1.0 ton/ha compared to leading soybean-producing countries such as the USA, Brazil, Argentina, and China, where productivity reaches 2.5 ton/ha. Therefore, enhancing soybean productivity and seed yield per unit area is a critical breeding objective in India. Breeding for early-maturing soybean varieties is necessary to improve cropping intensity and farmer income (Maranna et al. 2023). Developing varieties resistant to key diseases such as anthracnose, pod blight, yellow mosaic virus, and *Rhizoctonia* aerial blight is essential to mitigate productivity losses caused by disease outbreaks in recent years. Varieties with tolerance to abiotic stresses such as waterlogging, drought, high temperature, and salinity are also needed to enhance climate resilience in soybean. Other breeding objectives include developing genotypes suitable for mechanical harvesting, high oil content, and high oleic acid content, as well as varieties resistant to pod shattering and suitable for intercropping.

In India, varieties like JS 335 (100 days' duration) and JS 95-60 (90 days' duration), bred by JNKVV, Jabalpur, have been widely cultivated. However, these varieties have become susceptible to disease. The susceptibility of soybean seed coats to mechanical forces during harvesting and threshing, leading to seed germination losses, is attributed to low lignin content. Breeding efforts are underway to develop lines with higher lignin content and grain yield. Improved seed treatment techniques involving polymer coating with biocontrol agents like *Trichoderma*, as well as the application of salicylic acid, potassium phosphate, and alpha-tocopherol at pod formation or seed development stages, have been recommended to address seed quality issues and field weathering. Traits of importance in soybean breeding include maturity duration, plant height for mechanical harvesting, seed yield, oil content, protein content, shattering resistance, seed weight, number of pods, number of seeds per pod, resistance/tolerance against diseases and insect-pests, and tolerance to drought, waterlogging, and photo insensitivity. Conventional plant breeding has increased the allelic combinations that boost seed yield per unit area, but Indian cultivars still lag behind other major soybean countries. Studies have identified positive alleles for yield from wild accessions, suggesting that higher yields may come from more seeds per unit area or from a longer seed-filling period. Grain yield in soybean is mainly determined by the number of seeds per pod and 100-seed weight. Marker-assisted breeding is proving to be a valuable tool for addressing these complex genetic traits and enhancing productivity in soybean.

6 Photoperiod and Long Juvenility Traits in Soybean

Photoperiodism, first thoroughly described by Garner and Allard (1920), refers to the phenomenon in which many plants flower in response to changes in day length. They demonstrated that some species flower when the day length falls below a critical threshold (short-day plants), while others flower when the day length exceeds a critical threshold (long-day plants). Obligate photosensitive plants require exposure to an inductive day length to flower, whereas quantitative photosensitive plants delay flowering under non-inductive conditions. Day length varies significantly with latitude and season; while it remains constant at the equator, it increases as one moves toward higher latitudes. Although photoperiodism is generally a survival mechanism, it can become a form of abiotic stress when plant species are introduced to new geographic areas or climates. For example, a short-day species from higher latitudes introduced to the shorter day lengths of lower latitudes may flower too early, resulting in minimal biomass and low economic yield. Conversely, introducing this genotype to even higher latitudes may cause delayed flowering, excessive vegetative growth, and potential plant death due to extreme frost later in the season. Soybean is a quantitative photosensitive short-day crop. Its flowering is induced after 5–7 days of exposure to day lengths shorter than its critical day length (CDL). A soybean genotype that flowers after 12 hours of daylight under inductive conditions may require up to 16 hours of daylight to flower under non-inductive conditions or may not flower at all. Soybean, originating from its wild ancestor *G. soja* in China's longer photoperiodic and temperate climate, has adapted to latitudes ranging from 50°N to 40°S. However, individual soybean genotypes are adapted to a narrow latitudinal range of about 200 km. Genotypes adapted to northern regions, when grown in southern regions, flower earlier, with reduced vegetative growth and lower yields. Conversely, genotypes from higher latitudes, when grown in northern regions, often flower late in the growing season and fail to mature before the onset of killing frost.

7 Maturity Classification System in Soybean

Each soybean genotype requires a specific critical day length and should be grown in regions that provide the optimal conditions for its growth and yield potential. The influence of photoperiod on soybean growth and reproduction is so significant that a separate maturity classification system has been established in North America. This system categorizes soybean lines into one of thirteen maturity groups (MGs) (000-X), based on their adaptation to different latitudes. Photoperiodic genes were originally identified as genes influencing flowering and maturity. In soybeans, 12 loci have been reported to control the timing of flowering and maturity: E1, E2, E3, E4, E5, E6, E7, E8, E9, E10, E11, and J. Notably, E7 is considered an allele of E1, rather than a distinct locus, and E6 has recently been identified as an allele of the J locus. Photoperiod directly affects the E1, E3, and E4 loci, which interact with E2, E9, and J to regulate flowering and maturity responses.

7.1 Long Juvenility Trait

The limiting effect of short day lengths prevented the soybean crop from expanding into lower latitudes, prompting the search for genotypes with an extended vegetative period under highly inductive conditions (i.e., short photoperiods and/or high temperatures). Miyasaka and Medina (1981) planted the maturity group VIII cultivar *Santa Maria* at 23°S latitude in Brazil, where it flowered later and exhibited better growth compared to other cultivars of the same maturity group. In a glasshouse with a photoperiod of less than 13 hours and 30 minutes, *Santa Maria* flowered in 50–55 days, a significantly longer flowering duration than that reported by Garner and Allard (1930) for *Biloxi* grown in a photoperiod shorter than 14 hours. Hartwig and Edwards Jr (1970) identified another genotype, PI 274454, which flowered in 65 days when grown at 5° latitude (with a photoperiod of less than 13 hours) in Colombia. At 23° latitude, however, it flowered so late that it could not survive the killing frost. Hartwig and Kiihl (1979) noted that lines PI 159925 and *Santa Maria* had delayed flowering (approximately 60 days) under short days, compared to "Bragg" and "Tracy," which flowered in about 30 days. Based on various studies, it has been concluded that recessive alleles at two loci are necessary for the expression of the LJ trait. The combination of all recessive alleles results in an even greater level of LJ by further delaying flowering under short-day conditions. In crosses showing single-gene segregation, it was found that both parents were homozygous recessive for another gene. While many studies used PI 159925 as the LJ parent, varying inheritance results were attributed to common non-segregating genes and the involvement of two genes.

7.2 Utilization and Impact of Long Juvenility on World Soybean Production

Currently, the USA, Brazil, Argentina, China, and India contribute 40%, 38%, 12%, 4%, and 4%, respectively, to the global soybean production, which totals 123.6 million tons annually. Soybean production began modestly in 1961 with just 25 million tons, largely dominated by two countries: the USA (18.5 million tons) and China (6.3 million tons). While the USA remains the leading producer with 123.6 million tons, Brazil (117.8 million tons), Argentina (37.7 million tons), China (14.1 million tons), and India (13.7 million tons) are also significant producers. Among these, India lies at lower latitudes (8°04' N to 37°06' N), whereas Brazil spans both high and low latitudes (5°01' N to 33°04' S). Soybean cultivation in both India and Brazil began with the introduction of soybean varieties from the USA, initially in relatively high-latitude regions. In India, these temperate and high-latitude varieties flowered and matured early under the subtropical lower-latitude conditions, resulting in much lower yields than in the USA. However, these varieties proved advantageous in areas with limited rainfall in India, and they formed the foundation of the country's soybean breeding program.

Soybean was introduced to Brazil in 1882 in the tropical lower latitudinal regions (12°S – 13°S), but it could not adapt due to extremely low yields. As a result, soybean cultivation remained confined to Brazil's higher latitudes. Commercial soybean farming in Brazil remained limited to these higher latitudes until the 1980s. Technological advancements, such as the development of varieties suited for lower latitudes, the use of *Rhizobium* to reduce reliance on mineral nitrogen, and the improvement of acidic Cerrado soils through lime, enabled soybean cultivation to expand into Brazil's lower latitudes. The identification of natural mutants in popular soybean varieties from higher latitudes led to the creation of varieties capable of thriving in lower latitudes, facilitating the crop's spread to Brazil's Central Northern region. While varieties for lower latitudes had been developed, their genetic basis remained unclear and was often attributed to long juvenility. The history of soybean development in Brazil is divided into two phases: the developmental phase in the subtropics (up to 1980) and the expansion phase in the tropics (from the 1980s onward). In the 1980s, soybean production in the tropics and subtropics was 5 million and 11 million tons, respectively. By 2017, these figures had risen to 72.9 million tons and 41.3 million tons, respectively. Currently, the tropical lower latitudes account for 61.8% of Brazil's total soybean area. The area devoted to soybean cultivation in the tropical region has consistently grown, becoming a significant environmental concern in Brazil. Brazil's success in soybean production is largely attributed to long juvenility, with 72.9 million tons of production from the tropics representing 23.7% of global soybean production (307.3 million tons).

The lower latitudes of the USA have also benefited from the long juvenile (LJ) trait, with the developed varieties such as Padre and Vernal. The use of the LJ trait in countries with short growing seasons is still in its early stages. The LJ trait leads to delayed flowering and maturity, and the relatively good rainfall from late September to March/April in Brazil's lower latitudes is conducive to the growth of full-season LJ varieties (120–140 days). However, many tropical regions experience limited rainfall or require soybean varieties that can fit into the short duration between successive rice crops. In Asian countries, short-duration varieties (70–85 days) are commonly used in rice-based farming systems as they fit within the rotation windows between rice crops. To develop short-duration, high-yielding varieties, Australian researchers, as part of ACIAR Project 9040, introduced several LJ near-isogenic pairs with different genetic backgrounds from the southern USA. These were evaluated under various photothermal conditions in controlled environments and at tropical and subtropical locations in Australia and Thailand. Their modeling studies showed that the LJ trait delays the onset of flowering by approximately 200 day-degrees (with a base temperature of 9 °C), without significantly altering the genotype's sensitivity to temperature or photoperiod. Under the warm conditions typical of the tropics and summer season in the subtropics, the LJ trait delayed flowering by about 10–14 days, with no significant impact on the genotype's photothermal sensitivity. Specifically, there was no evidence that the LJ trait introduced photoperiod sensitivity into otherwise short-duration germplasm, which, in the absence of the trait, was insensitive to day length variations in tropical or subtropical environments.

Previous studies indicated that high-yielding cultivars from the mid-west USA were comparable in yield to local Australian cultivars but had significantly lower biomass and reduced height, which was insufficient for commercial mechanized soybean cultivation. Mayers et al. (1991) extended the day length by using incandescent lighting over the canopy and observed a delay in flowering, along with an increase in plant height and seed yield. Despite this, these varieties offered an advantage in areas with limited rainfall in India, forming the foundation of the Indian soybean breeding program.

8 Breeding Methodologies and Approaches

For many decades, plant breeders have used conventional breeding methods to improve soybeans, resulting in the release of over 180 soybean varieties in India. To accelerate breeding progress and improve efficiency, researchers have explored various breeding strategies and integrated them with new technologies. Most of the soybean varieties developed in India were the result of hybridization and pedigree selection (Tiwari 2001b; Tiwari and Raut 2004; Tiwari and Tiwari 2023). Conventional breeding has led to the development of mega varieties such as Gaurav (JS 72-44), JS-335, JS 93-05, and JS 95-60. From an Indian perspective, the time intervals of soybean improvement are as follows:

- Before 1970: Kalitur in MP and other local soybeans in foot hills of Himalaya
- Selection cycle 1 (1970–90): Introduction of soybean varieties: Bragg, Lee, Improved Pelican, Hardee. During this era, four-times higher yield over Kalitur (local cultivar) was evidenced.
- Selection cycle 2 (1990–2015): Hybridization of breeding lines and developed cultivars. This era evidenced 19% higher yield than cycle 1cultivars.
- Integrated solutions (2015 onward): Use of marker-assisted selection. NRC 127 first variety developed through MAS in soybean.

Karmakar and Bhatnagar (1996) evaluated forty-three soybean varieties from two different selection cycles, representing most of the varieties developed in India's domestic breeding program from 1969 to 1993. Varieties from selection cycle 1 exhibited four times higher seed yield and harvest index compared to the traditional farm variety, Kalitur. This yield increase was attributed to higher biomass, more pods per plant, greater mean seed weight, and a longer seed-filling duration, while also showing reduced plant height, improved lodging tolerance, fewer seeds per pod, and earlier flowering and maturity. Varieties from selection cycle 2 showed a 19% higher seed yield and a 16% increase in harvest index compared to cycle 1, with longer seed-filling duration, more seeds per pod, and shorter plant height. The annual genetic gain in seed yield for soybean varieties was approximately 22 kg per hectare.

A recent study indicated that the annual genetic gain in seed yield for soybean varieties released in India was approximately 23 kg/ha from 1969 to 2008. This resulted in a total seed yield increase of 103.5%, or 2.6% per year, over the 39-year period (Ramteke et al. 2011). The yield increase was attributed to a rise in the number of pods and seeds per plant, with seeds per plant increasing by 1.56% per year. Tiwari et al. (1999) summarized that current Indian soybean varieties fall into different groups based on their breeding history, which include (1) indigenous soybean varieties, (2) varieties from direct introductions and/or selections made from them, (3) varieties developed through hybridization and primarily pedigree selection, and (4) varieties developed through mutation breeding.

The genetics and inheritance of various soybean traits have been reviewed by Prabhakar and Tiwari (1991), Tiwari (2001b), and Tiwari and Raut (2004). Indigenous soybean varieties, which are landraces or selections derived from them, have been known for a long time. These include (a) a pool of black-seeded indigenous varieties, such as Bhat/

Bhatmas, which are native to the northern hill regions but were also grown in scattered areas of central India under names like Kalitur, (b) a yellow-seeded pool from the northern hill/Tehri-Garhwal region, currently represented by JS 2, and (c) a group of indigenous varieties with small yellow seeds, exemplified by Type 49. The traditional black-seeded indigenous variety Kalitur played a key role in the soybean revolution. Promising soybean varieties were identified as early as the 1970s by Saxena and Pandey (1971), Lal and Mehta (1972), Lal et al. (1974), and Singh and Saxena (1975). There are 20 research centers across 18 states in India under the All India Coordinated Research Project on Soybean.

Genomic selection and marker-assisted recurrent selection have been used to enhance quantitative traits in cross-pollinated species (Massman et al. 2013; Beyene et al. 2015, 2016). However, this selection approach is not feasible for self-pollinated crops due to the labor-intensive nature of crossing (Bernardo 2010).

Research efforts have been focused on maximizing realized yield by optimizing maturity and enhancing disease resistance, to cater to the needs of stakeholders. Multi-parent advanced generation inter-cross (MAGIC) populations, derived from breeder-relevant germplasm, offer platforms for increased recombination and high-resolution or fine mapping of quantitative traits in crop species (Cavanagh et al. 2008). Plant height is a key trait in soybean breeding, as it is closely linked to plant structure and yield. Using a high-density genetic linkage map constructed in a recombinant inbred line (RIL) population derived from DongnongL13 × Henong60, Wang et al. (2022) identified 33 quantitative trait loci (QTL) associated with plant height, five of which were consistently detected across multiple environments. Two candidate genes linked to plant height, Glyma.02G133000 and Glyma.05G240600, were predicted and validated (Wang et al. 2022).

Priyanatha et al. (2022) also conducted a GWAS for traits such as plant height, days to maturity, lodging score, 100-seed weight, and seed yield. Seed number per pod (SNPP) is a key yield component and a practical target for direct selection in field breeding trials, although progress in soybean breeding for this trait has been limited. Another important yield-related trait in soybean is the main stem node number (MSNN). Li et al. (2021) used 144 four-way recombinant inbred lines (RILs) to identify QTL for MSNN at two planting densities across five environments through linkage and association studies. They identified 40 QTL at one density and 28 QTL at the other, along with four candidate genes associated with MSNN. Early leaf senescence in soybean can help allocate dry matter to seeds, shorten the maturation period, and prevent green stem disorder. Soybean pubescence plays a crucial role in insect resistance and tolerance to drought and other stresses (Li et al. 2022). Shook et al. (2021) introduced the "Parental Allele Tracing, Recombination Identification, and Optimal Prediction (PATRIOT)" approach, which uses marker data to quickly identify lines carrying specific alleles, enhancing the accuracy of genomic relatedness and diversity estimates, and improving genomic prediction (GP). Developing resistant cultivars is the most cost-effective way to prevent yield loss, and resistance to diseases and pests remains one of the top breeding objectives in soybean.

The QTL mapping strategy involves several steps, including QTL detection, localization, and fine mapping. It helps to identify markers linked to the gene of interest (GOI). Once the GOI is tagged with a marker, this linked marker can be used for marker-assisted selection (MAS), which is less influenced by environmental factors. Phenotyping facilities for traits of interest in plants, segregants, lines, or germplasm are also critical and must be established. Novel approaches for discovering new QTLs or genes include (i) association mapping, including genome-wide association studies (GWAS), (ii) nested association mapping, and (iii) next-generation sequencing-based methods for mapping genes and QTLs. Several QTLs related to yield and associated traits have been mapped in soybean (Mansur et al. 1993, 1996; Maughan et al. 1996; Mian et al. 1996; Panthee et al. 2007; Li et al. 2008b; Soybase 2012). Bobby et al. (2008) identified four QTLs for pod number (qPN001-qPN004) on chromosomes 2, 6, and 8; two QTLs for seed number (qSN001 and qSN002) on chromosomes 5b and 11b; five QTLs for 100-seed weight (qSW001 to qSW005) on chromosomes 5a, 6, 8, 9, and 11c; and two QTLs for total seed weight (qTSW001 and qTSW002) on chromosomes 5b and 17c. A more comprehensive list of agronomically important QTLs in soybean is provided by Hu et al. (2011). Additionally, meta-QTL analysis is ongoing in soybean, which offers further opportunities. Meta-analysis is a valuable tool in linkage analysis to optimize QTLs, narrow the confidence interval (CI), and enhance the accuracy and validity of QTL positioning (Löffler 2009).

In soybean, Hu et al. (2011) conducted pioneering work in this area by (i) compiling published QTLs, (ii) creating a consensus map that integrated published maps with a reference map, (iii) obtaining consensus QTLs through meta-analysis, (iv) mining genes using bioinformatics tools, and (v) providing markers for consensus QTLs with high effects and narrow confidence intervals (CIs) suitable for marker-assisted selection (MAS).

In India, QTLs have been identified for various traits, including high seed longevity (linked to markers Satt 538, Satt 285, Satt 600, and Satt 434) as well as high oil and protein content. One of the two genes responsible for resistance to Yellow Mosaic Virus (YMV) in soybean has been found to be associated with markers Satt 322 and GMAC 7, located on the C2 linkage group. Basic and strategic research on diseases, including the integration of resistance and comprehensive disease management strategies using bioagents, is essential for future focus. Soybean cultivation during the rabi/summer season should be avoided. The management of insect pests, such as the girdle beetle, must also be prioritized. In Central India's major soybean-growing areas, significant insect pests include green semi-loopers (*Gessonia gemma*, *Chrysodeixis acuta*, and *Diachrysia orichalcea*), the girdle beetle (*Obereopsis brevis*), stemfly (*Melanagromyza soja*), and blue beetle (*Cneorane* sp.). Additionally, pests such as the leaf miner (*Bilobata subsecivella*) in Marathwada, Bihar hairy caterpillar (*Spilosoma obliqua*) in parts of the northern plains and Terai regions, and others are becoming increasingly problematic. PI416937 has been reported to be drought-resistant, with no impact on seed yield (Singh et al. 1990).

Several studies suggest that higher water use efficiency (WUE) is typically achieved through a reduction in stomatal conductance, which, however, can negatively impact accumulation. There is a growing market-driven demand for the breeding and development of specialty soybeans. Unlike other food crops, soybean is complex as it represents an agro-industrial venture, requiring both agricultural production and industrial involvement for value addition. Soybeans and their products cater to both direct consumers and industrial users. The Directorate of Soybean Research (DSR) in Indore has made significant strides in identifying and developing soybean varieties suited for specific food products. For example, varieties such as MAUS71, Punjab1, and LSb1 are excellent for soymilk production, while PK1029, Ankur, and Punjab 1 are ideal for tofu production. Indian soybean varieties have been found to contain trypsin inhibitors ranging from 35 to 115 mg/g of soy meal (Kumar et al. 2001). Lines such as NRC 101 and NRC 102, developed using PI 542044 (a source of the null allele), are devoid of Kunitz trypsin inhibitor (KTI). Additionally, the Satt 409 marker, which is closely linked with the Ti locus, has been validated and is now used for marker-assisted selection to develop KTI-free soybean varieties (Rani et al. 2011).

Genotypes carrying null alleles for lipoxygenases have been obtained from the USDA, and efforts to introgress these null alleles into Indian soybean varieties are currently underway. A new simple sequence repeat (SSR) marker closely linked to the lipoxygenase 1 gene (Rani et al. 2013) and lipoxygenase 2 gene (Kumar, personal communication) has recently been identified. Two lines, 'NRC 102', which is free from Kunitz trypsin inhibitor, and 'IC 210', an indigenous line with high oleic acid content (42%), are now ready for commercialization. This advancement represents a significant step toward promoting specialty soybeans and establishing a new commercial niche market in India. Vegetable soybean, harvested at the near R7 growth stage while the pod is still green and the seed fill is about 80–90%, is used as a fresh or frozen vegetable. The variety 'Harit Soya' (Himso 1563) was released in 2001 for culinary use in northern India. Additionally, some vegetable-type soybean lines were sourced from AVRDC, and both the Ranchi and Bangalore centers have utilized AVRDC germplasm. Varieties such as 'Swarna Vasundhara', 'Karune,' and 'NRC 188' have been released from these efforts.

Recent advances in marker-assisted selection have been employed, particularly for developing varieties free from KTI and lox 2 genes (Kumar et al. 2013, 2015; Rani et al. 2015; Maranna et al. 2016). As soybean production continues to grow to meet adding demand, there's a growing need for the application of ultramodern tools such as genomic selection, genome editing, speed breeding, and genome-wide association analysis to ameliorate agronomic traits in soybeans. Integrating conventional breeding with recent tools in genome mapping, molecular breeding, genomics, sequencing, and bioinformatics can enhance breeding effectiveness and dock the duration of variety development (Massman et al. 2013; Beyene et al. 2015, 2016). Economically important traits such as growth habit, stress tolerance, and seed yield are targeted when opting elite germplasm with desirable alleles; still, it should be noted that the genes governing these traits frequently suffer post-transcriptional variations (Ku et al. 2022). Multi-parent advanced generation inter-cross (MAGIC) populations deduced from elite panels give platforms for increased recombination and high-resolution mapping (Cavanagh et al. 2008).

The soybean MAGIC population has been employed as an inheritable resource for varietal development programs, performing in the establishment of different parentage lines with high yields and resistance to biotic and abiotic stresses (Maranna et al. 2024). Likewise, the soybean MAGIC population has been used for mapping quality traits (Hashemi et al. 2022). Successful transfer of qualitative and quantitative traits through conventional and marker-assisted breeding has been achieved in numerous crops. While conventional methods are typically sufficient for qualitative breeding, where the transfer of traits can be visually observed, molecular breeding tools and techniques are invaluable for quantitative trait improvement. Molecular markers aid in identifying new sources of variation and locating genomic regions or loci that control inherited quantitative traits (QTLs). Integrating genetics, genomics, and phenomics of qualitative and quantitative traits can enhance the understanding of soybean improvement for yield and component traits through genome editing approaches. In conclusion, the studies featured in this Research Topic will enhance our understanding of key traits and their underlying molecular mechanisms, while also developing effective strategies for trait improvement. These findings will be valuable for future efforts in QTL mapping, marker-assisted selection, map-based cloning, GWAS, genomic selection, gene editing, and breeding by design in soybeans. It is expected that molecular breeding in soybeans will advance significantly, bridging laboratory research with field selection, ultimately resulting in the development of new cultivars with desirable traits through an integrated approach.

9 Genomics Resources

A wealth of genomic, transcriptomic, proteomic, and metabolomic resources and databases has been developed for soybeans, which are essential for advancing soybean improvement. The soybean genome sequencing project, conducted by the DOE-JGI-Community Sequencing Program, resulted in the Glyma-1.0 genome sequence assembly. This assembly revealed a genome size of approximately 950 Mb, compared to the expected 1115 Mb, distributed across 20 chromosomes (Schmutz et al. 2010). The predicted protein-coding regions were estimated to include 66,153 genes, with over 46,000 genes predicted with high confidence. These genome sequence data and gene annotations are accessible through The Phytozome database (http://www.phytozome.net/). Furthermore, several research initiatives have generated extensive datasets for population, comparative, and functional genomics studies (Kim et al. 2019; Ratnaparkhe et al. 2020; Kajiya-Kanegae et al. 2021; Liu et al. 2020, 2024; Huang et al. 2024). The soybean physical map is publicly available through the Soybean Breeders Toolbox (SBT), hosted on the Soybase website (http://www.soybase.org), for the benefit of the research community.

Whole transcriptome sequencing and analysis of differentially expressed genes offer valuable functional genomic markers for population genomics and transcriptomic studies. Gene expression patterns in soybeans have been explored using global expression analysis methods, including high-density expression arrays, Serial Analysis of Gene Expression (SAGE), and other functional genomics approaches. Ratnaparkhe et al. (2020) conducted a whole-genome analysis of the soybean genotype EC 241780 to explore the genomic landscape associated with rust resistance. By comparing the sequence of EC 241780 with rust-susceptible cultivars like Williams 82 and JS-335, they identified sequence variations and prioritized candidate genes. Their study revealed three genes encoding NBS-LRR family proteins (Glyma18G51715, Glyma18G51741, and Glyma18G51765) as promising candidates for the Rpp1 locus, offering insights into genome-wide sequence variation at the Rpp1 locus. This can assist in developing rust-resistant soybean cultivars by effectively utilizing genomic resources. Molecular genetic diversity in soybeans and their wild relatives has been studied using various markers, including random amplified polymorphic DNA (RAPD), amplified fragment length polymorphisms (AFLPs), simple sequence repeats (SSRs), and single nucleotide polymorphisms (SNPs). Recent advancements in genome sequencing have significantly contributed to genome-wide association studies (GWAS).

GWAS is commonly used in soybeans and other plant species to identify markers linked to abiotic and biotic stresses, as well as other agronomic traits. The introgression of alleles from wild soybeans (Glycine soja Siebold and Zucc.) offers a valuable strategy to improve genetic diversity and germplasm for important traits. To explore the genetic architecture underlying seed protein and oil content, Yang (2021) performed linkage mapping using a recombinant inbred line (RIL) population created from a cross between the *G. max* cultivar 'Osage' and the *G. soja* accession PI 593983. They identified seven significant QTLs for seed protein on chromosomes 14 and 20, and one for seed oil on chromosome 8. Within these associated genomic regions, eight genes were proposed as candidate genes. Using a restricted two-stage multi-locus GWAS with 15,501 SNP linkage-disequilibrium block markers in a population of 361 germplasm accessions from Northeast China, Feng et al. identified 73 QTLs linked to seed protein, explaining 71.70% of the phenotypic variation, 28 of which were new. They also annotated 120 candidate genes, classified into 13 functional categories.

A team of Canadian researchers has collaboratively focused on soybean improvement, including conducting GWAS on multiple traits to identify novel alleles related to seed yield, seed quality, and agronomic traits from modern Chinese parents (Priyanatha et al. 2022). They developed a genomic panel that included modern Canadian (CD) and Chinese (CH) cultivars, as well as progeny from CD × CH crosses. This panel was field-tested and genotyped using 32 K high-quality SNPs developed through genotyping-by-sequencing. A putative gene was identified for each of the traits related to seed yield, seed oil, and protein concentration (Priyanatha et al. 2022).

10 Biotic Stresses

Soybeans are vulnerable to numerous biotic and abiotic stresses that can lead to significant yield losses. Among the common biotic stresses are anthracnose, charcoal rot, Rhizoctonia aerial blight, yellow mosaic disease (YMD), pod blight, bacterial pustules, frog eye, rust, and leaf spot, each prevalent in various regions. The cultivation of soybeans in northern India became possible with the development of YMV-resistant varieties like PK 416 and PS 564, derived from the resistance source PI 171443 (UPSM 534) (Agarwal et al. 2013; Singh et al. 1974). *Glycine soja* PI 393551 is the donor for MYMIV, with duplicate dominant inheritance of MYMIV resistance genes revealed (Anita

Rani et al. 2018). UPSM 534 and *Glycine soja* have been utilized as sources of YMV resistance, with well-understood inheritance of resistance and closely linked SSR markers identified (Rani et al. 2017, 2018). The inheritance of resistance to pod blight has also been elucidated (Nataraj et al. 2020). Continuous efforts in resistance breeding have led to the development of several disease-resistant soybean varieties in India. Additionally, the endophytic fungus *Mucor cie-cinelloides* and the bacterium *Paenibacillus macerans* have been recognized for their effectiveness in controlling diseases (anthracnose and charcoal rot) and insects (stem fly). The Bt 127SC strain has also proven to be an effective biocontrol agent against lepidopteran insects affecting soybeans.

10.1 Anthracnose

Anthracnose, caused by *Colletotrichum truncatum*, has become increasingly devastating in recent years, especially in Madhya Pradesh. *C. truncatum* is a hemi-biotrophic fungal pathogen with a dual mode of pathogenicity, acting through both biotrophic and necrotrophic modes. In India, resistant genotypes have been identified under both field conditions and controlled conditions. Under high disease pressure in field conditions, genotypes EC 538828, EC 34372, EC 457254, AKSS 67, and Karune were found to be resistant. It has also been discovered that resistance against anthracnose disease is governed by two genes interacting in a complementary manner (Nataraj et al. 2020).

10.2 Charcoal Rot

Charcoal rot disease, caused by *Macrophomina phaseolina*, is a necrotrophic fungal pathogen that affects approximately 500 plant species. In soybeans, symptoms of above-ground charcoal rot start to appear from the R4 stage. Host plant genetic resistance is considered the most effective and sustainable approach for combating losses due to this disease. Coser et al. (2017) deciphered the genetic architecture of charcoal rot resistance and identified genes that confer resistance. Both field and in vitro screenings were conducted on a diverse panel of 459 plant introductions from the USDA soybean germplasm core collection. Under field conditions using the RSS disease index, four significant SNPs (ss715588228, ss715618004, ss715631906, and ss715631726) and putative candidate genes (Glyma.04g053100, Glyma.14g002000, Glyma.18g248100, and Glyma.18 g228600) governing biotic and abiotic stress responses were identified.

10.3 Rhizoctonia Aerial Blight

Rhizoctonia root rot and aerial blight disease, caused by *Rhizoctonia solani* Kuhn (teleomorph *Thanatephorus cucumeris* (Frank) Donk), is one of the most economically significant soil-borne diseases of soybeans in India. The disease appears in severe form in the states of Madhya Pradesh, Chhattisgarh, and Uttarakhand. Genotypes such as JS 20-30, JS 20-57, JSM 222, MACS 1407, PS1611 Cat 2126 B, and PS 1583 are resistant to this disease (Amrate and Shrivastava 2021; Surbhi et al. 2021). Given its economic significance, there is a pressing need to develop soybean varieties resistant to this disease and to identify genes/loci conferring resistance.

10.4 Yellow Mosaic Disease (YMD)

Yellow mosaic disease in soybeans is caused by *Mungbean yellow mosaic virus* (MYMV) (Morinaga et al. 1990) and *Mungbean yellow mosaic India virus* (MYMIV) (Mandal et al. 1997). The nucleotide sequence of the virus causing YMD in soybeans across northern plains and central India shows 89% sequence similarity to MYMIV and has been designated as MYMIV-[Sb] (Usharani et al. 2004). MYMIV is a single-stranded DNA virus of the *Geminiviridae* family, transmitted by the whitefly, *Bemisia tabaci* (Govindan et al. 2014), and can also infect other legumes such as black gram and green gram (Govindan et al. 2014). Soybean genotype SL 525 has been found to be resistant to MYMIV. Genetic analysis of MYMIV resistance in SL 525 revealed that the MYMIV resistance gene is located on chromosome 6 (LG C2) within a 3.5-cM genome region between two SSR markers, GMAC7L and Satt322 (Rani et al. 2017). *Glycine soja* Siebold &Zucc. (PI 393551) has also been found to be resistant to MYMIV. Genetic mapping studies on this *Glycine soja* accession revealed tight linkages of MYMIV resistance with SSR marker BARCSOYSSR_08_0867 (chromosome 8) and with BARCSOYSSR_14_1416 (chromosome 14) and BARCSOYSSR_14_1417 (chromosome 14) (Rani et al. 2018).

10.5 Rust

Asian soybean rust, caused by *Phakopsora pachyrhizi* Syd. & P. Syd., was first reported in Japan in 1903 (Hennings 1903). In India, southern parts such as Dharwad, Sangli, and Pune are prone to rust, where this disease can cause significant yield losses. A total of seven rust-resistant loci have been identified to date. Furthermore, three genes,

Glyma18G51715, Glyma18G51741, and Glyma18G51765, were found to be candidate genes within the RPP1 locus (Ratnaparkhe et al. 2022).

Details of rust resistance loci and the resistant donors

Locus	Chromosome	Resistant donor	Reference
Rpp1	18	PI 200492	Hyten et al. (2007)
Rpp2	16	PI 224270	Garcia et al. (2008)
Rpp3	6	PI 200456	Garcia et al. (2008)
Rpp4	18	PI 459025	Garcia et al. (2008)
Rpp5	3	PI 200456	Garcia et al. (2008)
Rpp6	18	PI 567102B	Li et al. (2012)
Rpp7	19	PI 605823, PI 605823	Childs et al. (2018)

11 Insect Pests

A total of 22 species of insect pests, including 7 sucking pests, 11 defoliators, 2 stem borers, 1 pod borer, and 1 root feeder have been reported to damage the soybean crop. Stem fly (*Melanagromyza sojae*), girdle beetle (*Oberea brevis*), white fly (*Bemisia tabaci*), Bihar hairy caterpillar *(Spilarctia obliqua)*, gram pod borer (*Helicoverpa armigera*), and Thrips (*Thrips tabaci*) are the major insects causing severe damage to the crop. Studies have been conducted to develop host resistance against these insects. Few genotypes expressed antixenosis/antibiosis against *Spodoptera litura*, which can be used in insect-resistant breeding programs. Wild soybean (*Glycine soja*) is reported to be highly resistant to stem fly and girdle beetle and being used to introgress resistance against these insects in *G. max*.

12 Abiotic Stress Resistance Including Climate Change

Drought and waterlogging represent major abiotic stresses affecting soybean production globally, especially in India where the crop relies heavily on the monsoon. Two released varieties, NRC 7 and JS 97-52, have demonstrated drought tolerance. Drought tolerance is a complex trait, with 37 QTLs related to various traits such as transpiration rate, plant height, seed weight, relative root and shoot length identified (Ren et al. 2020). Broad bed furrow technology has significantly benefited soybean cultivation, offering advantages in both drought and excess water conditions. An impact assessment of this technology has estimated yield increases of 7.5% and 22.2% over flatbed sowing in improved and farmer's practices, respectively.

13 Agronomic and Yield Traits

Plant height is a crucial trait in soybean breeding, as it is closely linked to plant shape and yield. Wang et al. (2022) identified 33 QTLs associated with plant height, with five being consistently detected across multiple environments. Additionally, 62 plant height QTNs were discovered, with 26 repeatedly identified using various methods through multi-locus association analysis in a population of 455 accessions genotyped with 63,306 SNP markers. Two candidate genes associated with plant height, Glyma.02G133000 and Glyma.05G240600, were predicted and validated (Wang et al. 2018). Priyanatha et al. (2022) conducted GWAS on plant height, days to maturity, lodging score, 100-seed weight, and seed yield, offering insights into potentially valuable genetic resources from Chinese modern cultivars for further soybean improvement. Seed number per pod (SNPP) is an important yield component. Liu et al. (2021) investigated the molecular mechanisms behind SNPP differences between soybean and cowpea, identifying the PIN1 and CKX gene families and exploring differences in auxin and cytokinin pathways. Additionally, Li et al. (2021) identified QTLs for main stem node number (MSNN) in soybean, predicting four candidate genes associated with MSNN.

Early leaf senescence in soybean can aid in directing dry matter to seeds and reduce the maturation period. Yamatani et al. identified two soybean mutant lines, els1-1 and els1-2, which are involved in chlorophyll biosynthesis during leaf development and chlorophyll degradation during leaf senescence. Tokumitsu et al. (2022) investigated the relationship between the G gene and its homolog G-like, focusing on their role in chlorophyll accumulation in seed coats of mature seeds. Soybean pubescence plays an important role in insect resistance, drought tolerance, and other stress responses. Li et al. (2022) mapped QTLs for pubescence traits, identifying ten QTLs for pubescence length and nine for pubescence density, along with candidate genes associated with these traits.

14 Value Addition

The utilization of soybean in India has primarily focused on oil extraction, with the seed meal used as feed for cattle and poultry. However, its potential in human food has been limited due to traditional taste preferences. Despite its high protein content of 40–45% and oil content of approximately 20%, challenges such as anti-nutritional factors like Kunitz trypsin inhibitor (KTI) and undesirable attributes like lipoxygenase (LOX2) have restricted its use in food. To address this, varieties with desirable traits such as KTI-free, LOX2-free, and medium oleic content have been developed, with associated genes identified and markers created to aid breeding efforts (Anita and Vineet 2015; Rani and Kumar 2022). The future of soybean in India lies in value addition, particularly in addressing protein malnutrition in a predominantly vegetarian population. Soybean can serve as a rich protein source, benefiting a significant portion of the population. There is a range of soybean-derived ingredients like soy concentrate and soy protein isolates, which can be used in various food products like soy nuts, soy chap, and soy sticks, gaining acceptance in the Indian diet. Soybean flour, when combined with cereals and millets, can enrich them with high-quality protein. Efforts are underway to popularize soybean as a vegetable, similar to sweet peas, with specific varieties released for this purpose. Notably, food-grade soybean varieties free from anti-nutritional factors have been developed, leading to wider acceptance and usage in the food industry. However, industry support is crucial to fully unlock soybean's potential in India. By promoting its diverse uses and ensuring profitability for farmers and the industry, soybean can become a transformative crop, addressing nutritional challenges and contributing significantly to India's agricultural landscape. Continued research, innovation, and collaboration are essential for realizing soybean's role in shaping the future of Indian agriculture.

15 Future Thrust

Currently, soybean cultivation in India is primarily concentrated in three states: Madhya Pradesh, Maharashtra, and Rajasthan. However, promising results from multi-location yield evaluation trials under AICRP-Soybean suggest opportunities for expanding soybean cultivation to states such as Punjab, Haryana, and Uttar Pradesh. These regions, traditionally associated with paddy cultivation, could potentially transition to high-yielding soybean varieties, achieving average productivity ranging from 2.5 to 3.0 t/ha. Bihar and Orissa have also demonstrated potential for higher soybean productivity, particularly with short-maturity duration varieties. Efforts should focus on developing photo-insensitive varieties, resistant to post-harvest sprouting (PHS), with high oleic acid content in the oil, and food-grade quality, tailored to the specific cropping systems and consumer preferences in various regions. Collaborative support from industry, state agricultural departments, seed production agencies, and policy planning is essential to ensure higher prices and procurement, ultimately contributing to increased soybean productivity and production in India. Conventional breeding methods, which played a pivotal role in the early phase of soybean improvement, needs to be reoriented with the use of genetically diverse parents and pre-bred diverse material in the crosses, large F_2 populations, marker-assisted breeding methods, development of hot-spots and screening facilities for prevalent abiotic and biotic stresses, genome editing.

Development and use of core collection and mini core collection. Identifying novel sources for environmental adaptation is crucial to support breeding strategies. Genome-environment association (GEA), combined with GWAS, has been used to predict drought and heat tolerance in related legumes, such as the common bean, which has been proposed as a diploid model for soybean. With the availability of extensive soybean marker datasets, GEA will be a valuable tool for studying environmental adaptation in soybean in the future. Moreover, the availability of genomic datasets, along with linked phenotypic and marker databases, lays the foundation for next-generation breeding technologies such as genomic prediction, genome-wide selection scans, machine learning, and speed breeding. The high-quality soybean datasets also facilitate genomic-assisted backcrossing, replacing traditional marker-assisted backcrossing and accelerating future breeding efforts. New technologies like long-read sequencing are being used to generate high-quality reference genomes and de novo assemblies of over 20 accessions, contributing to pangenomes. Long-read sequencing is set to become the gold standard for high-fidelity variation mapping across populations and will enable the construction of larger de novo assembled pangenomes. Pangenomes are now expanding to higher-level taxa, as demonstrated by Zhuang et al. (2022), Wang et al. (2019), Li et al. (2014), Huang et al. (2024), Fang et al. (2017), Landis and Doyle (2023), Cannon and Shoemaker (2012), Schmutz et al. (2010), Doyle et al. (2004), and Zhang et al. (2008), and will soon be applied to functional genomics, advancing super-pangenomics-guided breeding. Development of extra early maturing and early maturing (95 days) biotic stress resistant/tolerant varieties of soybean with high yield potentials. Extra early varieties are required for cropping systems of soybean/potato, onion, garlic/wheat. Presently two extra early varieties JS 95-60 and JS 20-34 are available which were released in 2007 and 2014, respectively. New extra early varieties, full maturity varieties (>95 days) with AICRP yield average > 2.5 t/ha, and varieties with drought, heat, and waterlogging tolerance need to be developed. Seed coat strengthening to reduce germination losses during field operations, storage, and transport of soybean seed. Development of soybean varieties with enhanced seed longevity for storage under ambient room conditions.

References

[SEAI](https://seaofindia.com/export-of-oil-meals-april-2021/)

Agarwal DK, Billore SD, Sharma AN et al (2013) Soybean: introduction, improvement, and utilization in India—problems and prospects. Agric Res 2:293–300. https://doi.org/10.1007/s40003-013-0088-0

Amrate PK, Shrivastava MK (2021) Yield response and pathological characterization of promising genotypes of soybean against major diseases in Madhya Pradesh. J Oilseeds Res 38(4):380–384

Amrate PK, Shrivastava MK, Bhale MS, Agrawal N, Kumawat G, Shivakumar M, Nataraj V (2023) Identification and genetic diversity analysis of high-yielding charcoal rot resistant soybean genotypes. Sci Rep 13(1):8905

Anderson EJ, Ali ML, Beavis WD, Chen P, Clemente TE, Diers BW, Graef GL, Grassini P, Hyten DL, McHale LK, Nelson RL (2019) Soybean [Glycine max (L.) Merr.] breeding: history, improvement,

production and future opportunities. In: Advances in plant breeding strategies: legumes: volume 7, pp 431–516
Anita R, Vineet K (2015) Development and commercialization of first ever kunitz trypsin inhibitor free soybean genotype. Curr Sci 109:1–3
Bernardo R (2010) Genomewide selection with minimal crossing in self-pollinated crops. Crop Sci 50:624–627. https://doi.org/10.2135/cropsci2009.05.0250
Beyene Y, Semagn K, Mugo S, Tarekegne A, Babu R, Meisel B et al (2015) Genetic gains in grain yield through genomic selection in eight biparental maize populations under drought stress. Crop Sci 55:154–163. https://doi.org/10.2135/cropsci2014.07.0460
Beyene Y, Semagn K, Crossa J, Mugo S, Atlin GN, Tarekegne A et al (2016) Improving maize grain yield under drought stress and non-stress environments in subsaharanafrica using marker-assisted recurrent selection. Crop Sci 56:344–353. https://doi.org/10.2135/cropsci2015.02.0135
Bobby R, Richard B, Willsheana C, Stella KK, Khalid M, Masum A, My Abdelmajid K (2008) Genetic Analysis of Yield Components in the PI 438489B by ‗Hamilton' Recombinant Inbred Line (RIL) Population of Soybean [Glycine max (L.) Merr.] Journal of Agricultural Science 4:98–105
Cannon SB, Shoemaker RC (2012) Evolutionary and comparative analyses of the soybean genome. Breed Sci 61(5):437–444
Carter TE Jr, Nelson RL, Sneller CH, Cui Z (2004) Genetic diversity in soybean. In: Soybeans: improvement, production, and uses 16:303–416. Book Editor(s):Richard M. Shibles, James E. Harper, Richard F. Wilson, Randy C. Shoemaker First published: 01 January 2004. https://doi.org/10.2134/agronmonogr16.3ed.c8
Cavanagh C, Morell M, Mackay I, Powell W (2008) From mutations to MAGIC: resources for gene discovery, validation and delivery in crop plants. Curr Opin Plant Biol 11:215–221. https://doi.org/10.1016/j.pbi.2008.01.002
Childs SP, King ZR, Walker DR, Harris DK, Pedley KF, Buck JW, Boerma HR, Li Z (2018) Discovery of a seventh Rpp soybean rust resistance locus in soybean accession PI 605823. Theor Appl Genet 131:27–41
Chung WH, Jeong N, Kim J, Lee WK, Lee YG, Lee SH, Yoon W, Kim JH, Choi IY, Choi HK, Moon JK (2014) Population structure and domestication revealed by high-depth resequencing of Korean cultivated and wild soybean genomes. DNA Res 21(2):153–167
Coser SM, Chowda Reddy RV, Zhang J, Mueller DS, Mengistu A, Wise KA, Allen TW, Singh A, Singh AK (2017) Genetic architecture of charcoal rot (Macrophominaphaseolina) resistance in soybean revealed using a diverse panel. Front Plant Sci 8:1626
de Felipe M, Gerde JA, Rotundo JL (2016) Soybean genetic gain in maturity groups III to V in Argentina from 1980 to 2015. Crop Sci 56(6):3066–3077
de Toledo JFF, de Almeida LA, de Kiihl RA, Menosso OG (1990) Ganhogenéticoem soja no Estado do Paraná, via melhoramento. Pesquisa Agropecuária Brasileira, Brasília 25(1):8994
Delannay X, Rodgers DM, Palmer RG (1983) Relative genetic contributions among ancestral lines to North American soybean cultivars 1. Crop Sci 23(5):944–949
Doyle JJ, Doyle JL, Rauscher JT, Brown AHD (2004) Evolution of the perennial soybean polyploid complex (Glycine subgenus Glycine): a study of contrasts. Biol J Linn Soc 82(4):583–597
Fang C, Ma Y, Wu S, Liu Z, Wang Z, Yang R, Hu G, Zhou Z, Yu H, Zhang M, Pan Y (2017) Genome-wide association studies dissect the genetic networks underlying agronomical traits in soybean. Genome Biol 18:1–14
FAOSTAT (2024) [FAO] (https://www.fao.org/faostat/en/#data/QCL)
Fox CM, Cary TR, Colgrove AL, Nafziger ED, Haudenshield JS, Hartman GL, Specht JE, Diers BW (2013) Estimating soybean genetic gain for yield in the northern United States—influence of cropping history. Crop Sci 53(6):2473–2482
Garcia A, Calvo ÉS, de Souza Kiihl RA, Harada A, Hiromoto DM, Vieira LGE (2008) Molecular mapping of soybean rust (Phakopsorapachyrhizi) resistance genes: discovery of a novel locus and alleles. Theor Appl Genet 117(4):545–553
Garner WW, Allard HA (1920) Effect of the relative length of day and night and other factors of the environment on growth and reproduction in plants. Mon Weather Rev 48(7):415–415
Garner WW, Allard HA (1930) Photoperiodic response of soybeans in relation to temperature and other environmental factors. United States Department of Agriculture, Washington, DC
Gizlice Z, Carter TE Jr, Burton JW (1994) Genetic base for North American public soybean cultivars released between 1947 and 1988. Crop Sci 34:1143–1151
Govindan K, Nagarajan P, Angappan K (2014) Molecular studies on transmission of mung bean yellow mosaic virus (MYMV) by Bemisiatabaci Genn. in mungbean. Afr J Agril Res 9(38):2874–2879
Grant JE, Grace JP, Brown AHD, Putievsky E (1984) Interspecific hybridization in Glycine Willd. subgenus Glycine (Leguminosae). Aust J Bot 32(6):655–663
Han Y, Zhao X, Liu D, Li Y, Lightfoot DA, Yang Z, Zhao L, Zhou G, Wang Z, Huang L, Zhang Z (2016) Domestication footprints anchor genomic regions of agronomic importance in soybeans. New Phytol 209(2):871–884
Hartwig EE, Edwards CJ Jr (1970) Effects of morphological characteristics upon seed yield in soybeans 1. Agron J 62(1):64–65
Hartwig EE, Kiihl RA (1979) Identification and utilization of a delayed flowering character in soybeans for short-day conditions. Field Crop Res 2:145–151
Hashemi SM, Perry G, Rajcan I, Eskandari M (2022) SoyMAGIC: an unprecedented platform for genetic studies and breeding activities in soybean. Front Plant Sci 13:945471. https://doi.org/10.3389/fpls.2022.945471
Hennings VP (1903) A few new Japanese Uredinaceae. Hedwigia 42:S107–S108
http://pgrportal.nbpgr.ernet.in/(S(cjgkfi55dwalem55xrugfk3g))/SearchResult.aspx
Hu G, Chen Q, Liu C, Jiang H, Wang J, Zhaoming Q (2011). Integration of Major QTLs of Important Agronomic Traits in Soybean. In—Soybean - Molecular Aspects of Breeding||. Dr. Aleksandra Sudaric (Ed.), ISBN: 978-953-307-240-1, InTech. pp. 81–118
Huang Y, Koo DH, Mao Y, Herman EM, Zhang J, Schmidt MA (2024) A complete reference genome for the soybean cv. Jack. Plant Commun 5(2):100765
Hyten DL, Song Q, Zhu Y, Choi IY, Nelson RL, Costa JM, Specht JE, Shoemaker RC, Cregan PB (2006) Impacts of genetic bottlenecks on soybean genome diversity. Proc Natl Acad Sci 103(45):16666–16671
Hyten DL, Hartman GL, Nelson RL, Frederick RD, Concibido VC, Narvel JM, Cregan PB (2007) Map location of the Rpp1 locus that confers resistance to soybean rust in soybean. Crop Sci 47(2):837–838
Jin J, Liu X, Wang G, Mi L, Shen Z, Chen X, Herbert SJ (2010) Agronomic and physiological contributions to the yield improvement of soybean cultivars released from 1950 to 2006 in Northeast China. Field Crop Res 115(1):116–123
Kajiya-Kanegae H, Nagasaki H, Kaga A, Hirano K, Ogiso-Tanaka E, Matsuoka M, Ishimori M, Ishimoto M, Hashiguchi M, Tanaka H, Akashi R, Isobe S, Iwata H (2021) Whole-genome sequence diversity and association analysis of 198 soybean accessions in mini-core collections. DNA Res 28(1):dsaa032
Karmakar PG, Bhatnagar PS (1996) Genetic improvement of soybean varieties released in India from 1969 to 1993. Euphytica 90(1):95–103
Kim JY, Jeong S, Kim KH, Lim WJ, Lee HY, Jeong N, Moon JK, Kim N (2019) Dissection of soybean populations according to selection signatures based on whole-genome sequences. Gigascience 8(12)
Kollipara KP, Singh RJ, Hymowitz T (1993) Genomic diversity in aneudiploid (2n = 38) and diploid (2n = 40) Glycine tomentella revealed by cytogenetic and biochemical methods. Genome 36:391–396

Kollipara KP, Singh RJ, Hymowitz T (1994) Genomic diversity and multiple origins of tetraploid (2n = 78, 80) Glycine tomentella. Genome 37:448–459

Ku YS, Cheung MY, Cheng SS, Nadeem MA, Chung G, Lam HM (2022) Using the knowledge of post-transcriptional regulations to guide gene selections for molecular breeding in soybean. Front Plant Sci 13:867731. https://doi.org/10.3389/fpls.2022.867731. PMID: 35432392; PMCID: PMC9009170

Kumar V, Rani A and Tiwari SP (2001) Comparative Activities of trypsin inhibitor among released cultivars of soybean. Indian Journal of Nutrition and Dietetics 38:437–440

Kumar V, Rani A, Rawal R, Hussain SM (2013) Lipoxygenase-2 free Indian soybean genotypes. Curr Sci 104:586–587

Kumar V, Rani A, Rawal R, Mourya V (2015) Marker assisted accelerated introgression of null allele of kunitz trypsin inhibitor in soybean. Breed Sci 65(5):447–452

Lal MS, Mehta SK (1972) Performance of soybean varieties from the USA. Indian J. Agric. Sciences 42:461–463

Lal MS, Mehta SK, Pandya SC (1974) Stability of performance of soybean varieties in Madhya Pradesh. Indian J. Agric. Sciences 44:217–220

Lam HM, Xu X, Liu X, Chen W, Yang G, Wong FL, Li MW, He W, Qin N, Wang B, Li J, Jian M, Wang J, Shao G, Wang J, Sun SS, Zhang G (2010) Resequencing of 31 wild and cultivated soybean genomes identifies patterns of genetic diversity and selection (2010). Nat Genet 42(12):1053–1059

Landis JB, Doyle JJ (2023) Genomic relationships of Glycine remota, a recently discovered perennial relative of soybean, within Glycine. Syst Bot 48(1):78–87

Li W, Zheng DH, Van K, and Lee SH (2008b) QTL mapping for major agronomic traits across two years in soybean (*Glycine max* L. Merr.). J. Crop Sci. Biotechnol. 11:171–190

Li S, Smith JR, Ray JD, Frederick RD (2012) Identification of a new soybean rust resistance gene in PI 567102B. Theor Appl Genet 125:133–142

Li YH, Zhou G, Ma J, Jiang W, Jin LG (2014) De novo assembly of soybean wild relatives for pan-genome analysis of diversity and agronomic traits. Nat Biotechnol 32(10):1045–1052

Li WX, Wang P, Zhao H, Sun X, Yang T, Li H, Hou Y, Liu C, Siyal M, Raja Veesar R, Hu B (2021) QTL for main stem node number and its response to plant densities in 144 soybean FW-RILs. Front Plant Sci 12:666796

Li Y, Chu L, Liu X, Zhang N, Xu Y, Karikari B et al (2022) Genetic architecture and candidate genes for pubescence length and density and its relationship with resistance to common cutworm in soybean. Front Plant Sci 12:771850

Liu Y, Du H, Li P, Shen Y et al (2020) Pan-genome of wild and cultivated soybeans. Cell 182(1):162–176

Liu LM, Zhang HQ, Cheng K, Zhang YM (2021) Integrated bioinformatics analyses of PIN1, CKX, and yield-related genes reveals the molecular mechanisms for the difference of seed number per pod between soybean and cowpea. Front Plant Sci 12:749902

Liu Z, Yang Q, Liu B, Li C, Shi X, Wei Y, Guan Y, Yang C, Zhang M, Yan L (2024) De novo genome assembly of a high-protein soybean variety HJ117. BMC Genom Data 25(1):25

Löffler M (2009) Revealing the genetic architecture of FHB resistance in hexaploid wheat (Triticum aestivum L.) by QTL meta-analysis. Molecular Breeding 23: 473–488

Mandal B, Varma A, Malathi VG (1997) Systemic infection of Vigna mungo using the cloned DNAs of the blackgram isolate of mungbean yellow mosaic geminivirus through agroinoculation and transmission of the progeny virus by whiteflies. J. Phytopathol. 145, 505–510

Mansur LM, Lark KG, Kross H, Oliveira A (1993) Interval mapping of quantitative trait loci for reproductive, morphological and seed traits of soybean (Glycine max L.). Theor. Appl. Genet. 86:907–913

Mansur LM, Orf JH, Chase K, Jarvik T, Cregan PB, Lark KG (1996) Genetic mapping of agronomic traits using recombinant inbred lines of soybean. Crop Sci. 36:1327–1336

Maranna S, Verma K, Talukdar A, Lal SK, Kumar A, Mukherjee K (2016) Introgression of null allele of Kunitz trypsin inhibitor through marker-assisted backcross breeding in soybean (Glycine max L. Merr.). BMC Genet 17:1–9

Maranna S, Nataraj V, Kumawat G, Chandra S, Rajesh V, Ramteke R, Patel RM, Ratnaparkhe MB, Husain SM, Gupta S, Khandekar N (2021) Breeding for higher yield, early maturity, wider adaptability and waterlogging tolerance in soybean (Glycine max L.): a case study. Sci Rep 11(1):22853

Maranna S, Kumawat G, Nataraj V, Gill BS, Nargund R, Sharma A, Rajput LS, Ratnaparkhe MB, Gupta S (2023) Development of improved genotypes for extra early maturity, higher yield and Mungbean Yellow Mosaic India Virus (MYMIV) resistance in soybean (Glycine max). Crop Pasture Sci 74(12):1165–1179

Maranna S, Nataraj V, Kumawat G, Mehetre SP, Reddy R, Jaybhay S, PG S, Rathod S, Agarwal N, Rajesh V, Kumar S, Rajput LS, Talukdar ACG, Chandra S, Ratnaparkhe MB, Ramteke R, Satpute GK, Gupta S, Singh KH (2024) Understanding of G × E interactions of yield attributes in soybean MAGIC population and characterization for charcoal rot resistance. Agron J 1–12:1290. https://doi.org/10.1002/agj2.21572

Massman JM, Jung HJG, Bernardo R (2013) Genomewide selection versus marker-assisted recurrent selection to improve grain yield and stover-quality traits for cellulosic ethanol in maize. Crop Sci 53:58–66. https://doi.org/10.2135/crop.Sci.2012.02.0112

Maughan PJ, Saghai Maroof MA, Buss GR (1996) Molecular-marker analysis of seed weight: Genomic locations, gene action and evidence for orthologous evolution among three legume species, Theor. Appl. Genet. 93:574–579

Mayers JD, Lawn RJ, Byth DE (1991) Agronomic studies on soybean (Glycine max (L.) Merrill) in the dry season of the tropics. III. Effect of artificial photoperiod extension on phenology, growth and seed yield. Aust J Agric Res 42(7):1109–1121

Mian MAR, Bailey MA, Tamulonis JP, Shipe ER, Carter Jr. TE, Parrot WA, Ashley DA, Hussey RS, Boerma HR (1996) Molecular markers associated with seed weight in two soybean populations. Theor. Appl. Genet. 93:1011–1016

Miyasaka S, Medina JC (1981) Soybean in Brazil. ITAL, Campinas, SP

Morinaga T, Ikegami M, Miura K (1990) Physical mapping and molecular cloning of mung bean yellow mosaic virus DNA. Intervirology 31(1):50–56

Morrison MJ, Voldeng HD, Cober ER (2000) Agronomic changes from 58 years of genetic improvement of short-season soybean cultivars in Canada. Agron J 92(4):780–784

Nataraj V, Maranna S, Kumawat G, Gupta S, Rajput LS, Kumar S, Sharma AN, Bhatia VS (2020) Genetic inheritance and identification of germplasm sources for anthracnose resistance in soybean [Glycine max (L.) Merr.]. Genet Resour Crop Evol 67:1449–1456

Nawaz MA, Yang SH, Chung G (2018a) Wild soybeans: an opportunistic resource for soybean improvement. INTECH. https://doi.org/10.5772/intechopen.74973

Nawaz MA, Yang SH, Chung G (2018b) Wild soybeans: an opportunistic resource for soybean improvement. In: Rediscovery of landraces as a resource for the future, p 9

Nawaz MA, Lin X, Chan TF, Ham J, Shin TS, Ercisli S, Golokhvast KS, Lam HM, Chung G (2020) Korean wild soybeans (Glycine soja Sieb &Zucc.): geographic distribution and germplasm conservation. Agronomy 10(2):214

Newell CA, Hymowitz T (1983) Hybridization in the genus Glycine subgenus Glycine willd. (Leguminosae, Papilionoideae). Am J Bot 70(3):334–348

Panthee DR, Pantalone VR, Saxton AM, West DR, Sams CE (2007) Quantitative trait loci for agronomic traits in soybean. Plant Breeding 126:51–57

Prabhakar, Tiwari SP (1991) Genetic improvement of soybean. In Genetic Improvement Of Oilseed Crops. Ed. Jafar Nizam, S.A. Farook and I.A. Khan. Ukaaz Publi., Hyderabad. Chapter 3. pp. 43–60

Prakash Tiwari S (2006) Plant introductions in soybean – achievements and opportunities. Indian J Plant Genet Resour 19(3):353–365

Priyanatha C, Torkamaneh D, Rajcan I (2022) Genome-wide association study of soybean germplasm derived from Canadian× Chinese crosses to mine for novel alleles to improve seed yield and seed quality traits. Front Plant Sci 13:866300

Ramlal A, Nautiyal A, Baweja P, Kumar Mahto R, Mehta S, Pujari Mallikarunja B, Vijayan R, Saluja S, Kumar V, Kumar Dhiman S, Lal SK (2022) Harnessing heterosis and male sterility in soybean [Glycine max (L.) Merrill]: a critical revisit. Front Plant Sci 13:981768

Ramteke R, Gupta GK, Murlidharan P, Sharma SK (2011) Genetic progress of soybean varieties released during 1969 to 2008 in India. Indian J Genet Plant Breed 71(04):333–340

Rani A, Kumar V (2022) Soybean breeding. In: Fundamentals of field crop breeding. Singapore, Springer Nature Singapore, pp 907–944

Rani A, Kumar V, Mourya V, Singh RK, Husain SM (2011) Validation of SSR markers linked to null kunitz trypsin inhibitor allele in Indian soybean [Glycine max (L.) Merr.] population. J. Plant Biochem. Biotechnol. 20:258–261

Rani A, Kumar V, Rawal R (2013) Identification of simple sequence repeat markers linked to lipoxygenase-1 gene in soybean. J. Plant Biotechnol. 22:488–491

Rani A, Kumar V, Gill BS, Rathi P, Shukla S, Singh RK (2017) Linkage mapping of Mungbean yellow mosaic India virus (MYMIV) resistance gene in soybean. Breed Sci 67:95–100

Rani A, Vineet Kumar BS, Gill SS, Rathi P, Singh RK (2018) Mapping of duplicate dominant genes for Mungbean yellow mosaic India virus resistance in Glycine soja. Crop Sci 58(4):1566–1574

Ratnaparkhe MB, Marmat N, Kumawat G, Shivakumar M, Kamble VG, Nataraj V, Ramesh SV, Deshmukh MP, Singh AK, Sonah H, Deshmukh RK (2020) Whole genome re-sequencing of soybean accession EC241780 providing genomic landscape of candidate genes involved in rust resistance. Curr Genomics 21(7):504–511

Ratnaparkhe MB, Nataraj V, Shivakumar M, Chandra S, Ramesh SV, Kumawat G, Kamble V, Rajput LS, Kumar S, Rajesh V, Satpute GK (2022) Genomic design for biotic stresses in soybean. In: Genomic designing for biotic stress resistant oilseed crops. Springer International Publishing, Cham, pp 1–54

Ren H, Han J, Wang X, Zhang B, Yu L, Gao H, Hong H, Sun R, Tian Y, Qi X, Liu Z (2020) QTL mapping of drought tolerance traits in soybean with SLAF sequencing. Crop J 8(6):977–989

Saxena MC, Pandey RK (1971) Characeristics and performance of some promising varieties of soybean at Pantnagar. Indian J. Agric. Sciences 41:355–360

Schmutz J, Cannon SB, Schlueter J, Ma J, Mitros T, Nelson W, Hyten DL et al (2010) Genome sequence of the paleopolyploid soybean. Nature 463:178–183

Sherman-Broyles S, Bombarely A, Powell AF, Doyle JL, Eqan AN, Coate JE, Doyle JJ (2014) The wild side of a major crop: soybean's perennial cousins from down under. Am J Bot 101:1651–1665. https://doi.org/10.3732/ajb.1400121

Shivakumar M et al (2019) NAM population—a novel genetic resource for soybean improvement: development and characterization for yield and attributing traits. Plant Genet Resour Characteriz Utiliz 17(6):545–553. https://doi.org/10.1017/S1479262119000352

Shook JM, Lourenco D, Singh AK (2021) PATRIOT: a pipeline for tracing identity-by-descent for chromosome segments to improve genomic prediction in self-pollinating crop species. Front Plant Sci 12:676269

Singh G (2010) The soybean: botany, production and uses. CABI. ISBN 1845936450, 9781845936457. 494 pages

Singh RJ, Hymowitz T (1985a) The genomic relationships among six wild perennial species of the genus Glycine subgenus Glycine Willd. Theor Appl Genet 71:221–230

Singh RJ, Hymowitz T (1985b) Intra- and interspecific hybridization in the genus Glycine, subgenus Glycine Willd.: chromosome pairing and genome relationships. Z Pflanzenzüchtg 95:289–310

Singh RJ, Hymowitz T (1988) The genomic relationship between Glycine max (L.) Merr. and G. soja Sieb. and Zucc. as revealed by pachytene chromosome analysis. Theor Appl Genet 76:705–711

Singh RJ, Hymowitz T (1989) The genomic relationships among Glycine soja Sieb. and Zucc., G. max (L.) Merr. and 'G. gracilis' Skvortz. Plant Breed 103:171–173

Singh BB, Saxena MC (1975) Soybean varieties for different agro-climatic zones in India. Seed Technlology News 5:5–7

Singh BB, Gupta SC, Singh BD (1974) Sources of field resistance to rust and yellow mosaic diseases of soybeans. Indian J Genet Plant Breed 34:400–404

Singh RJ, Patterson RP, Carter TE Jr. (1990) Field drought tolerance of a soybean plant introduction. Crop Science 30:118–123

Singh RJ, Kollipara KP, Hymowitz T (1992) Genomic relationships among diploid wild perennial species of the genus Glycine Willd. subgenus Glycine revealed by crossability, meiotic chromosome pairing and seed protein electrophoresis. Theor Appl Genet 85:276–282

Singh RJ, Nelson RL, Chung G (2007) Soybean (Glycine max (L.) Merr.). Genetic resources, chromosome engineering, and crop improvement. Oilseed Crops 4:13–50

SoyBase (2012) https://www.soybase.org/

Surbhi K, Singh KP, Singh NK, Aravind T (2021) Assessment of genetic diversity among soybean genotypes differing in response to aerial blight (Rhizoctonia solani Kuhn) using SSR markers. J Phytopathol 169(1):37–44

Tiwari SP (2001b) Indian scenario and needs in soybean genetic resources. Proceedings of India Soy Forum, Indore 48–57

Tiwari SP, Raut VM (2004) Breeding and Crop Improvement. In — Soybean Production and Improvement in India||. Ed. NB Singh, GS Chauhan, AK Vyas, OP Joshi. NRCS, ICAR Publi 30–46

Tiwari SP, Tiwari SP (2023) The Indian soybean revolution – ascertaining the determinants and the tipping point an analytical review. J Oilseeds Res 40(1–2):1–12

Tiwari SP, Joshi OP, Sharma AN (1999) The saga of success – The advent and renaissance of soybean: A landmark in Indian agriculture. pp 1–54

Tokumitsu Y, Kozu T, Yamatani H, Ito T, Nakano H, Hase A, Sasada H, Takada Y, Kaga A, Ishimoto M, Kusaba M (2022) Functional divergence of G and its homologous genes for green pigmentation in soybean seeds. Front Plant Sci 12:796981

Usharani KS, Surendranath B, Haq QMR, Malathi VG (2004) Yellow mosaic virus infecting soybean in northern India is distinct from the species infecting soybean in southern and western India. Curr Sci:845–850

Verma VD, Thomas TA (1991) Soybean genetic resources and their evaluation. Indian J Pl Genet Resour 4(1):29–33

Wang YY, Li YQ, Wu HY, Hu B, Zheng JJ, Zhai H, Lv SX, Liu XL, Chen X, Qiu HM, Yang J, Zong CM, Han DZ, Wen ZX, Wang DC, Xia ZJ (2018) Genotyping of soybean cultivars with medium-density array reveals the population structure and QTNs underlying maturity and seed traits. Front Plant Sci 9:610

Wang J, Li X, Do Kim K, Scanlon MJ, Jackson SA, Springer NM, Yu J (2019) Genome-wide nucleotide patterns and potential mechanisms of genome divergence following domestication in maize and soybean. Genome Biol 20:1–16

Wang J, Hu B, Jing Y, Hu X, Guo Y, Chen J, Liu Y, Hao J, Li WX, Ning H (2022) Detecting QTL and candidate genes for plant height in soybean via linkage analysis and GWAS. Front Plant Sci 12:803820

Yang Y (2021) Fine mapping and candidate gene identification of a soybean seed protein and oil qtl from a wild soybean accession and linkage analysis for whole plant biomass, carbon, nitrogen, and seed composition using a RIL mapping population (Master's thesis, University of Missouri-Columbia)

Zhang B, Pan X, Stellwag EJ (2008) Identification of soybean microRNAs and their targets. Planta 229:161–182

Zhou Z, Jiang Y, Wang Z, Gou Z, Lyu J, Li W (2015) Resequencing 302 wild and cultivated accessions identifies genes related to domestication and improvement in soybean. Nat Biotechnol 33(4):408–414

Zhuang Y, Wang X, Li X et al (2022) Phylogenomics of the genus Glycine sheds light on polyploid evolution and life-strategy transition. Nat Plants 8:233–244

New Breeding Tools and Grain Legume Improvement

Abhishek Bohra, Ashok K. Parihar, and Girish Prasad Dixit

Abstract

Safeguarding nutrition security worldwide and strengthening global climate action have accelerated transition toward alternative protein sources. Grain legumes are the most affordable sources of plant-based proteins and have served as an integral part of the human diet for millennia. Transformative breeding methods must be embraced to fast-track yield increase and climate adaptation in these crops. State-of-the-art genetic technologies, statistical methods, and genomics-assisted breeding approaches hold great promise to address future breeding objectives. Detailed molecular profiles of gene bank collections, breeder's germplasm, and artificial populations pave the way for efficient utilization of allelic diversity for developing future-ready grain legumes. Efforts to identify and track novel genetic variations should leverage the growing wealth of genetic sequence information of tens to hundreds of individuals. Modern breeding tools including genomic selection (GS), speed breeding (SB), and haplotype-based breeding (HBB) augment rapid integration of the beneficial alleles in crop breeding programs. In the long term, increasing the rate of genetic gain in crop breeding programs calls for optimization of breeding strategies like optimal contribution selection (OCS) that could sustain genetic gain under constrained scenarios (e.g., permissible inbreeding rates/coancestry). Harnessing the full potential of cutting-edge genetic solutions to bring rapid and precise changes in the desired traits would be crucial for climate-proofing of future grain legume varieties.

Keywords

Domestication · gene editing · genomics · genomic selection · grain legumes · haplotype · pangenome

A. Bohra (✉) · A. K. Parihar · G. P. Dixit
ICAR-Indian Institute of Pulses Research (IIPR), Kanpur, India

1 Introduction

Grain legumes are crucial to global food security and safeguarding economic returns in small-holding farming systems (Bohra et al. 2014a, b, 2015, 2020a). Owing to their low water and carbon footprints and ability to fix atmospheric nitrogen, grain legume crops attract increasing attention in relation to sustainable agro-ecological development. Traditional breeding has delivered significant gains in grain legume crops by enhancing the yield and quality traits of modern crop varieties (Singh et al. 2016). However, the rate of genetic progress of these crops must accelerate in order to feed the world population amid declining inputs and resources for future agricultural production. The pursuit of producing more with less without compromising the environment has also brought grain legumes into the spotlight (Ferreira et al. 2021). Crop breeders need to tailor breeding methodologies to harness the full potential of modern genomic resources such as whole genome sequence, pangenomes, genes, and the haplotypes (Garg et al. 2024) to integrate them with the data developed through large-scale, accurate and precise phenotyping of plants (Bohra et al. 2021). We begin with the characterization of genetic diversity held in gene banks and experimental populations. The status of implementation of genomic selection (GS)/optimal cross designs, speed breeding (SB), and gene editing (GEd) is discussed while presenting the scope for haplotype-based breeding and de novo domestication.

2 High-Throughput Genetic Marker Systems for Breeding Applications

Advances in sequencing and genotyping technologies have made remarkable improvements in our ability to identify genetic markers and apply them in large samples in a cost-effective and time-saving manner (Bohra et al. 2020b). The sequence-based genotyping driven by the next-generation sequencing (NGS) methods has integrated the discovery and

G. Prasad Dixit et al. (eds.), *Fundamentals of Legume Breeding*, https://doi.org/10.1007/978-981-96-7838-9_14

Table 1 Genome-wide SNP arrays for high throughput genotyping of grain legumes

Crop	Chip name	SNPs	References
Soybean	SoySNP50K	52,041	Song et al. (2013)
Chickpea	Axiom®*Cicer*SNP Array	50,590	Roorkiwal et al. (2018)
Groundnut	Axiom®*Arachis*SNP Array	58,233	Pandey et al. (2017)
Pigeonpea	Axiom®*Cajanus* SNP Array	56,512	Saxena et al. (2018)
Cowpea	Illumina Cowpea iSelect Consortium Array	51,128	Muñoz-Amatriaín et al. (2017)
Pea	GenoPea 13.2K SNP Array	13,204	Tayeh et al. (2015)

mapping of genetic markers in genome-wide scale in hundreds to thousands of samples. In recent years, single nucleotide polymorphism (SNP) has been established as the preferred DNA marker system owing to genome abundance, cost-effectiveness, ease of assaying, and amenability to automation. The extensive use of high-throughput genotyping enabled by a variety of methods such as WGRS, skim sequencing, specific-locus amplified fragment sequencing (SLAF-seq), genotyping by sequencing (GBS), restriction-site associated DNA sequencing (RAD-Seq), and diversity arrays technology (DArT)/DArT-Seq and fixed SNP arrays has supported large-scale investigations on gene bank characterization, population genetics, trait discovery and introgression and selections in breeding programs of grain legumes (Varshney et al. 2019a). Table 1 provides a list of SNP chips developed in grain legume crops to support genetic research and crop breeding.

3 Bridging the Gap Between Germplasm Diversity and Crop Breeding

3.1 Comprehensive Characterization to Unleash the Potential of Gene Bank Collections

Gene banks conserve crop diversity that is essential for sustainable plant breeding and for safeguarding future food security (Bohra et al. 2022a). Globally, 1.1 million accessions of grain legumes are held in various gene banks. The evolutionary and breeding potential of the enormous genetic diversity archived in large gene bank holdings is severely constrained by the limited germplasm characterization. Declining costs and growing throughput of genotyping technologies have facilitated in-depth genetic profiling of large germplasm collections. Bandillo et al. (2015) elucidated the genetic structure of 14,000 soybean accessions by genotyping them with a 50K-SNP chip. Such comprehensive molecular characterization provides researchers with new information on the genome-wide patterns and distribution of the genetic variation in the gene bank collections. The molecular passport data could assist gene bank managers in correcting the mislabeled biological status of the germplasm collections in the historical records besides facilitating identification of the conservation gaps and the duplicate accessions. For example, a wild-specific allele of the *SHATTERPROOF2* homolog identified through the genetic analysis of the WGRS data of 3366 chickpea accessions assisted in correcting the mislabeling of the chickpea accession ICC 16369 (Varshney et al. 2021a).

In the presence of the phenotyping data, the genome-wide genetic marker information on the germplasm helps link the genetic variation with the phenotyping diversity, thus making beneficial alleles available for use in breeding programs (Varshney et al. 2019b). Focused identification of germplasm strategy (FIGS) has also been implemented in grain legume to mine useful variation rapidly and precisely from large germplasm collection according to the breeding objectives. FIGS has led to the creation of trait-specific subsets of germplasm collections by using the eco-geographic and trait data of the collection sites that experience high selection pressure. FIGS identified a set of accessions in faba bean (Khazaei et al. 2013) and soybean (Haupt and Schmid 2020) that hold great potential to improve adaptive traits.

3.2 Pangenomics for Capturing Untapped Diversity

A growing body of literature on reference genome assemblies in different crops has illuminated the stark limitation of a single genome sequence to capture the entire spectrum of genetic diversity within a species (Aggarwal et al. 2022; Khan et al. 2024). The sequencing of the whole genome of multiple accessions has revealed genes unique to one (Private) or a few individuals (Dispensable) besides those present across all genomes (Core). The difference in genetic variations, particularly the large structural variations (SVs) that the diverse germplasm resource may contain, became quite evident in a recent study that investigated the genomic composition of the three soybean genotypes Williams 82 (Wm82), Zhonghuang 13 (ZH13), and a wild soybean W05. The detection of these large SVs such as copy number variation (CNV) and presence-absence variation (PAV) is largely neglected in the era of short-read sequencing technologies because of the technical limitations.

With the evolution of long-read sequencing technologies [e.g., Pacific Biosciences (PacBio) and Oxford Nanopore Technologies], current research on plant genome sequencing provides evidence in support of the important role that these large SVs play in crop adaptation and evolutionary transitions. In soybean, Liu et al. (2020) built de novo genome assembly of 26 diverse accessions by using a combination of multiple sequencing platforms (e.g., single molecule real-time (SMRT), chromosome conformation capture (Hi-C), and Illumina sequencing (HiSeq)). The 26 high-quality genome assemblies were then used to develop a graph-based pangenome which provided new insights into gene families from different categories, i.e., core (20,623), soft core (8163), dispensable (28,679), and private (27). The study highlighted the functional importance of the PAV in related to seed luster variation in soybean.

Recently, a pea pan-genome based on 116 diverse accessions shed light on the domestication history, resolving *Pisum abyssinicum* as a separate species from *P. fulvum* and *P. sativum* (Yang et al. 2022). With the support from third-generation sequencing technologies and advanced genome assembly software solutions, the pangenomics has unleashed new functional diversity for improving stress response, quality, and productivity traits of grain legumes.

4 Elucidating the Genetic Architecture of Complex Traits

4.1 Building Next-Generation Population for Genetic Research and Breeding

Genetic analysis of the biparental populations derived from crosses between two inbred lines has elucidated the genetic architecture of important traits in grain legumes (Bohra et al. 2014a, b; Varshney et al. 2013). Genomic regions underlying complex traits have been dissected by genetic analysis of various kinds of artificial populations including F_2, backcross, and recombinant inbred lines (RILs). However, the power and precision of biparental population for QTL detection is limited by low genetic recombination and limited alleles. In this context, the development of multi-parental population has gained increasing popularity due to their greater power for mapping complex traits. Two standard multi-parent designs are multi-parent advanced generation inter-cross (MAGIC) and nested association mapping (NAM).

In MAGIC, multiple founders are crossed in pairs following a funnel breeding design, and the resulting progenies are selfed to generate stable RILs. The NAM design is relatively simple and involves crossing diverse founders to a common parent, thus yielding a series of interconnected RILs. NAM populations have improved genetic linkage analysis for different traits in grain legumes, such as seed-flooding tolerance and seed and seedling rot in soybean (Ali et al. 2020; Song et al. 2017; Bolanos-Carriel et al. 2023), and seed and pod weight in groundnut (Gangurde et al. 2020), 100-seed weight, flowering time, Ascochyta blight and seed quality in chickpea (Lakmes et al. 2022, 2023; Kalve et al. 2022), and protein content, disease resistance, growth habit, flowering and maturity times in pigeonpea (Saxena et al. 2020). Similarly, the development and application of MAGIC design have been reported in soybean, faba bean, cowpea, pigeonpea, and chickpea (Arrones et al. 2020; Hashemi et al. 2022).

4.2 Identifying Candidate Genes for Important Traits

The genetic linkage analysis has embraced innovations in methods of discovery and mapping of genetic variations in the artificial population and natural collections. These methods have been greatly scaled up due to the availability of the reference genome sequence, thus guiding easy, rapid, and accurate mapping of the sequence reads from mapping individuals from the whole population or the selected bulks. The new trait mapping methods employed in grain legume crops include QTL-Sequencing (QTL-Seq), Mut-Map etc., which have narrowed down the candidate genomics regions to few kbs. In parallel, the genome-wide genotyping of the natural collections using GBS or WGRS has also augmented trait mapping efforts in grain legumes through genome-wide association studies (GWAS). Table 2 presents recent examples of WGRS-based GWAS that have been reported in soybean, common bean, chickpea, pigeonpea, groundnut, mungbean, and adzuki bean. Besides QTL mapping, high-throughput sequencing of RNAs collected from diverse tissue samples spanning a range of growth stages from germination to senescence augments trait mapping efforts by illustrating specific and differential patterns of gene expression profiles. RNA-Seq has facilitated the development of comprehensive gene expression atlases in grain legume crops that catalog candidate genes for relevance to crop breeding (Table 3). These modern genomic resources delineate target genomic regions for functional validation and genetic manipulation.

5 Genomic Selection for Improving Genetic Gain

Genomic selection (GS) has been extremely popular in livestock breeding since Meuwissen et al. (2001) propounded the concept of predicting individual's genetic value using genome-wide marker information. Genomic selection improves the rate of genetic gain in a breeding program by increasing the selection intensity and accuracy, and reducing the cycle time (Sinha et al. 2021). The triumph of GS in ani-

Table 2 GWAS in grain legumes based on WGRS datasets of natural collections

Crop	Traits	References
Soybean	Oil and fatty acid biosynthesis	Zhou et al. (2015)
Soybean	Agronomic traits and fatty acid content	Fang et al. (2017)
Common bean	Common bacterial blight and angular leaf spot	Lobaton et al. (2018)
Common bean	Disease defense-related traits	de Souza et al. (2023)
Common bean	Yield-related traits	Wu et al. (2020)
Chickpea	Yield-related traits	Varshney et al. (2021a)
Pigeonpea	Agronomically important traits	Varshney et al. (2017)
Groundnut	28 agronomic traits	Lu et al. (2024)
Mungbean	Agronomic traits	Han et al. (2022)
Adzuki bean	Domestication traits	Chu et al. (2024)

Table 3 RNA sequencing-based comprehensive gene expression atlas in grain legumes

Crop	Number of genes identified	References
Common bean	26,964	O'Rourke et al. (2014)
Chickpea	15,947	Kudapa et al. (2018)
Pigeonpea	28,793	Pazhamala et al. (2017)
Cowpea	24,866	Yao et al. (2016)
Pea	46,099	Alves-Carvalho et al. (2015)

mal breeding has inspired GS implementation in plant breeding programs, especially for improving the quantitative traits with low heritability (Saxena et al. 2021). The intrinsic nature of GS to account for the small effect QTL regions makes it highly suitable for the prediction of quantitative traits. GS facilitates rapid selection of worthy individuals from the "untested" breeding population that has been genotyped using genome-wide markers but lacks phenotypic data.

Training population or reference population represents an essential prerequisite for GS that contains individuals having both genotypic and phenotypic records. The genomics-estimated breeding values (GEBVs) are estimated from the genotyping data of the selection candidates based on the marker effects summed across the whole genome (Hayes et al. 2009). The marker effects are calculated from the genotyping and phenotyping data of the training population. The prediction accuracy is the correlation between the GEBVs and observed phenotypes. In this way, GS bypasses the need for recurrent field-testing in plant-breeding programs.

Jarquín et al. (2014) reported up to 0.64 prediction accuracy for grain yield in soybean. The authors used 301 experimental lines as training population and genotyped them with 52,349 SNPs. Similarly, Bandillo et al. (2023) also made genomic prediction for grain yield in soybean based on a population 1500 lines assayed by over 20,000 SNPs. In a recent study, Kaler et al. (2022) tested 11 prediction models for canopy wilting and carbon isotope ratio in soybean. The training population comprised 346 accessions genotyped with 42,509 SNPs. The authors attributed differences in prediction accuracies among different crops to different levels of LD/LD decay. Inclusion of significant trait-associated markers in models contributed to increased prediction accuracies. In pea, Carpenter et al. (2018) obtained 0.56 prediction accuracies for *Ascochyta* blight disease resistance, whereas the accuracies ranged between 0.19 and 0.65 yield, flowering time, seed weight, and lodging susceptibility (Annicchiarico et al. 2019).

GS for grain yield, protein content, and biotic and abiotic stress response in lentils revealed prediction varying between 0.34 and 0.83 (Gebremedhin et al. 2024). The authors analyzed 2081 breeding lines using 64,781 SNP markers and reported suitability of BayesR for genomic prediction. Recent research on 305 MAGIC-RILs scored with 15,124 SNPs demonstrated the potential of genomic prediction for improving different traits (seed size, flowering time, and nutrient concentrations) in cowpea (Huynh et al. 2024). The author implemented Bayesian Ridge Regression (BRR) to train the prediction model, yielding prediction accuracies in the range of 0.4–0.58. Saradadevi et al. (2021) estimated GEBVs for grain yield, long cooking time, iron, and zinc from the data of the African bean panel of 358 and obtained moderate to high genomic heritability for these traits. The authors applied economic weights on GEBVs to build an optimum selection index and proposed a crossing design based on optimal contribution selection (OCS). OCS balanced the genetic gain derived from the weighted index in a scenario constrained by the maximum permissible target inbreeding rate. In the long term, strategies like OCS help to sustain genetic gain in a breeding program while minimizing the loss of genetic diversity.

6 Accelerating Generation Turnover by Speeding Breeding

Length of the crop breeding cycle presents a major stumbling block against the pace of crop improvement programs (Aggarwal et al. 2025; Samantara et al. 2022). Research efforts to increase generation turnover in plant breeding programs have led researchers to learn new methods that stimulate plant growth and development. The earlier approaches (e.g., shuttle breeding) enabling rapid cycling of crossing, selection, and fixation relied on reducing generation time by growing off-season crops in greenhouse or nurseries or by using short-season environments. For example, in India, two additional generations of chickpea are obtained in greenhouses during the off-season (February to September) besides the normal chickpea crop grown in fields during the crop season (October to January), resulting in three generations in a year (Gaur et al. 2007). Exposing chickpea plants to a 24-h day length in the greenhouse using incandescent lights could reduce the time to anthesis. The genotypes grown under 24-h photoperiod flowered 25 to 35 days after sowing.

The artificial light may be stopped once early flowering has occurred. Alternatively, two chickpea crops can be obtained in winter (end of September/first week of October to mid-January) and spring (mid-January/first week of February to April) seasons, while crop sown in June in rain-out shelter facility may be harvested in September, thus leading to one more generation during the rainy season. Supplementary light should overcome the sunlight limitations during the rainy season, thus ensuring the optimum sunshine hours for better crop growth. However, high cost and logistics complications put a serious limit on the frequent use of the nurseries for generation advancement, and also, safeguarding sufficient quantity of the seeds in these nurseries is a big challenge (Jähne et al. 2020).

Researchers have been studying the impact of light on plant growth and development for a long time. In 1968, Russian botanist Andrei Famintsyn made the earliest report on the influence of artificial light on plant growth and development. Since then, the evolving landscape of lighting systems encompasses incandescent lamps, tungsten halogen lamps, fluorescent lamps, high-pressure mercury lamps, high- and low-pressure sodium lamps, metal-halide lamps, and light-emitting diode (LED) (Loi et al. 2021). Among the modern lighting systems, LEDs have attracted widespread attention owing to their energy-efficiency, greater longevity, narrow spectrum, small size, low maintenance, and low temperature. Apart from the duration of the light period, spectral composition, intensity, and light direction constitute the key characteristics of the light that should be considered in studies aimed at assessing the impact of light on plant growth and development (Bayat et al. 2018).

Production of double haploids (DHs) is another important technique that allows rapid progression to homozygosity in plant genetic research and breeding. However, a demand for specialized facilities, skill, and funding resources put serious limitations on its widespread use in plant breeding. To accelerate basic and applied research, in vitro protocols have been optimized in different grain legume crops including common bean, chickpea, pea, lentil, and faba bean (Mobini and Warkentin 2016).

The term "speed breeding" refers to a collection of methods that trigger plant growth and development through manipulation of light characteristics and its duration, control of temperature and humidity, adjustment of plant densities, and immature seed harvest while circumventing the need for any labor-intensive tissue culture practices such as embryo rescue. Extended photoperiod has been reported to trigger the development rate in plants, especially the long-day or day-neutral plants. Studies on grain legume crops have shown a significant genetic variability for photoperiod response in diverse genotypes. More recently, lentil landraces of diverse latitudinal origin uncovered considerable variability for traits [e.g., days to flowering (40–120) and maturity (64–150)], when investigated under 22-h day length at 25 °C (Mitache et al. 2024).

Watson et al. (2018) obtained 6 generations per year of chickpea using SB conditions. Chickpea plants flowered uniformly in 33 ± 2 days when grown under an extended photoperiod (22 hours) provided by high-pressure sodium lamps and a controlled temperature regime (22/17 °C). Research has shown a positive impact of the application of growth hormones on plant development in controlled conditions. For example, Mobini and Warkentin (2016) used flurprimidol to reduce the internode length of pea plants, thus maximizing plant densities (266 plants/m^2) in the experimental settings. By adjusting the day length to 20 hours and temperatures at 21 °C/16 °C light/dark, the authors demonstrated the RGT method to be 30–45 days faster than the traditional single seed descent (SSD) method. In soybean (a short-day plant), Jähne et al. (2020) demonstrated the possibility of adjusting SB conditions, and the authors obtained five generations per year by supplying blue-light enriched far-red-deprived light spectrum for 10 hours.

7 Tailoring Grain Legumes with Modern Gene-Editing Tools

Three decades ago, research by Rouet et al. (1994) created the major breakthrough in the field of genome engineering and revolutionized molecular biology. Their research opened exciting opportunities to examine cellular DNA repair machinery that is crucial to preserve genome integrity. Rouet et al. (1994) introduced specific double-strand breaks (DSBs) to human genome by using a rare-cutting endonuclease from *Saccharomyces cerevisiae* (I-*Sce*I). Since then, zinc finger nucleases (ZFNs), transcription activator-like effector nucleases (TALENs), and clustered regularly interspaced palindromic repeat (CRISPR)-Cas9 have evolved as advanced programmable tools for bringing desired mutations in the target genetic sequences (Kumar et al. 2023). Both ZFN and TALEN act as dimers and are based on the fusion of a Fok1 nuclease domain, respectively, with the DNA binding domains namely zinc fingers and TALE repeats (Malzahn et al. 2017).

Thus, protein-DNA binding remains at the core of the target specificity of both ZFNs and TALENs; however, the designing of TALENs is easier than ZFNs. The CRISPR-Cas9 system is the most simple, cost-effective, flexible, scalable, and rapid. Because of these features, the CRISPR-Cas9 system has been the most popular among the current range of gene-editing platforms. CRISPR-Cas9 engages two components: CRISPR-associated endonuclease (Cas9) and a guide RNA (gRNA). A protospacer adjacent motif (PAM) located in the vicinity of the target DNA is required for the binding of Cas9. The breaks are induced by three base pairs upstream of the PAM upon binding of gRNA with the target DNA (Malzahn et al. 2017). PAM sequence thus reserves a binding signal and differentiates between the non-self and self DNA. Professor Emmanuelle Charpentier (Max Planck Unit for the Science of Pathogens, Berlin, Germany) and Professor Jennifer Doudna (University of California, Berkeley, USA) jointly won the Nobel Prize in Chemistry in 2020 for their ground-breaking research on CRISPR-Cas9 system.

In recent years, base editor (BE) and primer editor (PE) representing the latest versions of CRISPR-Cas9, can precisely alter the target DNA sequence without DSBs and HDR (Anzalone et al. 2020). The gene-editing products based on BE or PE may be considered SDN1 type (Tripathi et al. 2024). Anzalone et al. (2020) have reviewed different types of genome editing strategies and CRISPR-derived editing agents, such as nucleases, transposes/recombinases, BE, and PE. The availability of efficient transformation and regeneration protocols sets an essential prerequisite for undertaking gene-editing experiments in any crop. The gene editing has met with success in some of the legume crops including model legumes *Medicago truncatula* and *Lotus japonicas* and other crops like soybean, cowpea, chickpea, and peanut (Baloglu et al. 2022).

The recalcitrant nature of legumes to transformation owing to the low efficiency and genotype dependence of protocols poses a big hurdle to undertake gene-editing experiments in these crops. Alternative systems like hairy root transformation (referred to as hairy CRISPR) can be applied to overcome their inherent problems in relation to transformation recalcitrance. A recent study by de Koning et al. (2023) in common bean assessed the sgRNA efficiency through hairy root induction by *Rhizobium rhizogenes* K599 and reported a transformation frequency of 42–48%.

8 Genomics-Assisted Breeding for Varietal Improvement: Notable Examples in Grain Legumes

The use of DNA marker technologies in breeding programs has accelerated breeding decisions by facilitating accurate, early, and large-scale selections. Genomics-assisted breeding (GAB) has been successful in transferring the gene(s)/QTL(s) into diverse genetic backgrounds in a rapid and cost-effective manner. Marker-assisted selection (MAS) and marker-assisted backcrossing (MABC) have been the standard techniques for improving traits, particularly involving QTLs that explain a substantially high proportion of phenotypic variations (PVs) for the target trait (Varshney et al. 2014, 2021b, c). For example, in chickpea, genomic regions controlling drought tolerance, *Fusarium* wilt, and *Ascochyta* blight were transferred into different genetic backgrounds by harnessing marker-trait associations (MTAs).

In India and Africa, GAB has led to the release and commercial cultivation of nine improved chickpea varieties over the last five years. The GAB-bred improved Indian chickpea varieties include Pusa Chickpea 10,216 and Super Annigeri (released in 2019), Pusa Chickpea Manav (released in year 2020), Pusa Chickpea 4005 (BG 4005), IPCL 4-14 (IPCMAS 1) and IPCMB 19-3 (released in year 2021), and Pusa Chickpea Shyam (released in year 2022). The two chickpeas that were developed in Ethiopia following GAB include Geletu and Girar, which were released in the years 2019 and 2020, respectively. Similarly, the GAB efforts in groundnut met with success with the release of two improved varieties Girnar 4 and Girnar 5 in India, which had 78% oleic acid content and an oleic to linoleic ratio of 17 (Kamdar et al. 2022). The early selection for *ahFAD2* mutant allele was made using cleaved amplified polymorphic sequence (CAPS) markers. CAPS markers were developed by Chu et al. (2011) and the authors used MAS to combine genes for nematode resistance and O/L trait, resulting in the development of an improved line 'Tifguard High O/L' carrying genes for both target traits. A new groundnut variety YH61 was bred in China by GAB by using Kompetitive allele-specific PCR (KASP) markers and Axiom_*Arachis* 58 K SNP array (Tang et al. 2022).

9 Building Haplotype Catalogs and Haplotype-Based Breeding

Large genetic sequence datasets have been generated in grain legume crops, thanks to the growing affordability, accuracy, and throughput of DNA sequencing systems (Bohra and Singh 2015). High-quality reference genome sequence is now available for several grain legume crops and the latest additions to this are: pea (Kreplak et al. 2019; Yang et al. 2022), faba bean (Jayakodi et al. 2023), and adzuki bean (Chu et al. 2024). The genome resequencing attempts have also made the large WGRS datasets available for applied and basic research in these crops (Table 2). The WGRS datasets offer new avenues to catalog genome-wide genetic variations and identify haplotypes or haplotype blocks and associate them with the traits of agronomic significance (e.g., grain yield, stress response, adaptation, and quality).

Recent research has demonstrated the superiority of selecting haplotypes over individual genetic markers because of the better ability of haplotypes to capture historical recombination events, linkage disequilibrium (LD), and genetic relatedness and to detect QTL with greater statistical power while offering possibilities to identify high-order allelic interactions (Sallam et al. 2020). The application and challenges of using haplotype-based approaches for functional gene discovery and genomic prediction have been reviewed elsewhere (Bhat et al. 2021). Bevan et al. (2017) proposed a haplotype assembly approach focusing on a set of haplotype-defined markers and using them for assembly superiors haplotypes/haplotype combinations in future cultivars. In grain legumes, WGRS datasets have facilitated the identification of superior haplotypes for future improvement of breeding relevant traits such as plant height in soybean (Bhat et al. 2022), seed size and development in chickpea (Varshney et al. 2021a), and drought tolerance in pigeonpea (Sinha et al. 2020). In chickpea, haplo-pheno analysis identified 24 superior haplotypes by targeting a set of 20 candidate genes underlying seed size and development, which were then successfully validated in historical data on 129 chickpea varieties (Varshney et al. 2021a).

10 *De Novo* Domestication: Designing Future Grain Legumes

Grain legumes, domesticated around 12,000 years ago, are referred to as "founder crops" that played an important role in shifting human's nomadic lifestyle from hunter-gatherer to permanent settlement and farming. Pea, faba bean, lentil, and chickpea are some of the world's oldest crops that underwent human-mediated cultivation in the Fertile Crescent in the Neolithic Age (Bohra et al. 2022b). Domestication is a slow and complex evolutionary process that has shaped morphologies and physiologies of the current cultivated crops. However, the process of domesticating wild species has caused an accumulation of deleterious genetic variations in the domesticated species. The resulting genetic load referred to as "cost of domestication," can greatly reduce the fitness of the individuals carrying them. It is evident from research on domesticated crops that the domestication-related loci are limited in number and exert strong effects on the phenotypes. The genetic architecture of domestication loci makes them easier to manipulate "de novo" (in the wild genetic background itself) by using state-of-the art genetic tools (e.g., gene editing or introgression).

As an alternative route to transferring adaptation and stress response traits from wild to cultivars, de novo domestication introduces desirable genetic changes into a few domestication genes in the wild background, which could deliver predictable breeding outcomes. In the absence of the genome sequence and transformation system, Takahashi et al. (2019) proposed a combination of mutagenesis and forward phenotype screening as an alternative solution to facilitate de novo domestication of wild species. The screening of EMS mutagenized population derived from *Vigna stipulacea* Kuntze (known as Minnipayaru, which is an underutilized legume species) led to the recovery of the mutant plants showing reduced seed dormancy and reduced pod shattering while retaining resistance to pests and diseases. Development of legume crops that match the breeding objectives in the low-input agriculture settings is the need of the hour.

11 Conclusion and Prospects

In-depth characterization of germplasm resources of grain legumes has provided breeders with the new allelic variation for crop improvement. New generation of experimental populations and trait mapping methods elucidate the functional diversity for precise genetic manipulation. Application of new breeding tools such as GS and SB has helped increase selection efficiency and generations per unit time, leading to a rise in the rate of genetic gain in breeding programs of grain legume crops. Desirable and rapid alterations in target DNA sequence obtained through GEd hold great promise for designing future legume crops. The superior haplotypes mined from growing genetic sequence data would facilitate new approaches for tailoring future cultivars for better yield and stress response.

References

Aggarwal SK, Singh A, Choudhary M, Kumar A, Rakshit S, Kumar P, Bohra A, Varshney RK (2022) Pangenomics in Microbial and Crop Research: Progress, Applications, and Perspectives. Genes (Basel) 13(4):598

Aggarwal A, Jeena AS, Mehra K, Kumar B, Kashyap S, Yadav DK, Maurya AK, Venkatesh SC, Singla P, Bohra A (2025) Speed-bred crops for food security and sustainable agriculture. Planta (In press)

Ali MJ, Xing G, He J, Zhao T, Gai J (2020) Detecting the QTL-allele system controlling seed-flooding tolerance in a nested association mapping population of soybean. Crop J 8:781–792

Alves-Carvalho S, Aubert G, Carrère S et al (2015) Full-length de novo assembly of RNA-seq data in pea (Pisumsativum L.) provides a gene expression atlas and gives insights into root nodulation in this species. Plant J84:1–19

Annicchiarico P, Nazzicari N, Pecetti L, Romani M, Russi L (2019) Pea genomic selection for Italian environments. BMC Genomics 20:603

Anzalone AV, Koblan LW, Liu DR (2020) Genome editing with CRISPR–Cas nucleases, base editors, transposases and prime editors. Nat Biotechnol 38:824–844

Arrones A, Vilanova S, Plazas M et al (2020) The Dawn of the age of multi-parent MAGIC populations in plant breeding: novel powerful next-generation resources for genetic analysis and selection of recombinant elite material. Biology (Basel) 9:229

Baloglu MC, Celik Altunoglu Y, Baloglu P, Yildiz AB, Türkölmez N, Özden Çiftçi Y (2022) Gene-Editing Technologies and Applications in Legumes: Progress, Evolution, and Future Prospects. Front Genet 13:859437

Bolanos-Carriel C, Balk C, Wickramasinghe D, Acharya B, Dorrance AE (2023) Screening the soybean nested association mapping (SoyNAM) parents for resistance towards isolates of *Phytophthora sojae, Fusarium graminearum*, and species of *Globisporangium*. Plant Health Prog 24:336–344. https://doi.org/10.1094/php-12-22-0126-rs

Bandillo NB, Jarquin D, Posadas LG, Lorenz AJ, Graef GL (2023) Genomic selection performs as effectively as phenotypic selection for increasing seed yield in soybean. Plant Genome 16(1):e20285

Bandillo N, Jarquin D, Song Q, Nelson R, Cregan P, Specht J, Lorenz A (2015) A population structure and genome-wide association analysis on the USDA soybean germplasm collection. Plant Genome 8:1–13

Bayat L, Arab M, Aliniaeifard S, Seif M, Lastochkina O, Li T (2018) Effects of growth under different light spectra on the subsequent high light tolerance in rose plants. AoB Plants 10(5):ply052

Bevan MW, Uauy C, Wulff BB, Zhou J, Krasileva K, Clark MD (2017) Genomic innovation for crop improvement. Nat 543(7645): 346–354

Bhat JA, YuD BA, Ganie SA, Varshney RK (2021) Features and applications of haplotypes in crop breeding. Commun Biol 4:1266

Bhat JA, Karikari B, Adeboye KA, Ganie SA, Barmukh R, Hu D, Varshney RK, Yu D (2022) Identification of superior haplotypes in a diverse natural population for breeding desirable plant height in soybean. Theor Appl Genet 135(7):2407–2422

Bohra A, Singh NP (2015) Whole genome sequences in pulse crops: a global community resource to expedite translational genomics and knowledge-based crop improvement. Biotechnol Lett 37: 1529–1539

Bohra A, Pandey MK, Jha UC et al (2014a) Genomics-assisted breeding in four major pulse crops of developing countries: present status and prospects. Theor Appl Genet 127:1263–1291

Bohra A, Jha UC, Kishor PB, Pandey S, Singh NP (2014b) Genomics and molecular breeding in lesser explored pulse crops: current trends and future opportunities. Biotechnol Adv 32:1410–1428

Bohra A, Sahrawat KL, Kumar S et al (2015) Genetics- and genomics-based interventions for nutritional enhancement of grain legume crops: status and outlook. J Appl Genet 56:151–161

Bohra A, Saxena KB, Varshney RK et al (2020a) Genomics-assisted breeding for pigeonpea improvement. Theor Appl Genet 133:1721–1737

Bohra A, Jha UC, Godwin I, Varshney RK (2020b) Genomic interventions for sustainable agriculture. Plant Biotechnol J 18:2388–2405

Bohra A, Satheesh Naik SJ, Kumari A, Tiwari A, Joshi R (2021) Integrating phenomics with breeding for climate-smart agriculture. In: Omics Technologies for Sustainable Agriculture and Global Food Security (Vol II) (Kumar A, Kumar R, Shukla P, Patel HK Eds). Springer, Singapore. Pp 1–24

Bohra A, Kilian B, Sivasankar S, Caccamo M, Mba C, McCouch SR, Varshney RK (2022a) Reap the crop wild relatives for breeding future crops. Trends Biotechnol 40:412–431

Bohra A, Tiwari A, Kaur P et al (2022b) The key to the future lies in the past: insights from grain legume domestication and improvement should inform future breeding strategies. Plant Cell Physiol 263:1554–1572

Carpenter MA, Goulden DS, Woods CJ, Thomson SJ, Kenel F, Frew TJ, Cooper RD, Timmerman-Vaughan GM (2018) Genomic Selection for Ascochyta Blight Resistance in Pea. Front Plant Sci 9:1878

Chu Y, Wu CL, Holbrook CC, Tillman BL, Person G, OziasAkins P (2011) Marker-assisted selection to pyramid nematode resistance and the high oleic trait in peanut. Plant Genome 4:110–117

Chu L, Yang K, Chen C, Zhao B, Hou Y, Wang W, Zhao P, Wang K, Wang B, Xiao Y, Li Y, Li Y, Song Q, Liu B, Fan R, Bohra A, Yu J, Sonnenschein EC, Varshney RK, Tian Z, Jian J, Wan P (2024) Chromosome-level reference genome and resequencing of 322 accessions reveal evolution, genomic imprint and key agronomic traits in adzuki bean. Plant Biotechnol J 22:2173–2185

de Koning R, Daryanavard H, Garmyn J, Kiekens R, Toili MEM, Angenon G (2023) Fine-tuning CRISPR/Cas9 gene editing in common bean (*Phaseolus vulgaris* L.) using a hairy root transformation system and *in silico* prediction models. Front Plant Sci 14:1233418

de Souza IP, de Azevedo BR, Coelho ASG et al (2023) Whole-genome resequencing of common bean elite breeding lines. Sci Rep 13:12721

Fang C, Ma Y, Wu S et al (2017) Genome-wide association studies dissect the genetic networks underlying agronomical traits in soybean. Genome Biol 18:161

Ferreira H, Pinto E, Vasconcelos MW (2021) Legumes as a cornerstone of the transition toward more sustainable agri-food systems and diets in Europe. Front Sustain Food Syst 5:694121

Gangurde SS, Wang H, Yaduru S et al (2020) Nested-association mapping (NAM)-based genetic dissection uncovers candidate genes for seed and pod weights in peanut (*Arachishypogaea*). Plant Biotechnol J 18:1457–1471

Garg V, Bohra A, Mascher M, Spannagl M, Xu X, Bevan MW, Bennetzen JL, Varshney RK (2024) Unlocking plant genetics with telomere-to-telomere genome assemblies. Nature Genetics 56: 1788–1799

Gaur PM, Samineni S, Gowda CLL, Rao BV (2007) Rapid generation advancement in chickpea. SAT eJournal 3:1–3

Gebremedhin A, Li Y, Shunmugam ASK, Sudheesh S, Valipour-Kahrood H, Hayden MJ, Rosewarne GM, Kaur S (2024) Genomic selection for target traits in the Australian lentil breeding program. Front Plant Sci 14:1284781

Han X, Li L, Chen H, Liu L, Sun L, Wang X, Xiang Y, Wan Z, Liu C (2022) Resequencing of 558 Chinese mungbean landraces identifies genetic loci associated with key agronomic traits. Front Plant Sci 13:1043784

Hashemi SM, Perry G, Rajcan I, Eskandari M (2022) SoyMAGIC: an unprecedented platform for genetic studies and breeding activities in soybean. Front Plant Sci 13:945471

Haupt M, Schmid K (2020) Combining focused identification of germplasm and core collection strategies to identify genebank accessions for central European soybean breeding. Plant Cell Environ 43:1421–1436

Hayes BJ, Bowman PJ, Chamberlain AC et al (2009) Accuracy of genomic breeding values in multi-breed dairy cattle populations. Genet Sel Evol 41:51

Huynh BL, Stangoulis JCR, Vuong TD et al (2024) Quantitative trait loci and genomic prediction for grain sugar and mineral concentrations of cowpea [*Vigna unguiculata* (L.) Walp.]. Sci Rep 14:4567

Jähne F, Hahn V, Würschum T, Leiser WL (2020) Speed breeding short-day crops by LED-controlled light schemes. Theor Appl Genet 133:2335–2342

Jarquín D, Kocak K, Posadas L, Hyma K, Jedlicka J, Graef G, Lorenz A (2014) Genotyping by sequencing for genomic prediction in a soybean breeding population. BMC Genomics 15(1):740

Jayakodi M, Golicz AA, Kreplak J et al (2023) The giant diploid f aba genome unlocks variation in a global protein crop. Nat 615:652–659

Kaler AS, Purcell LC, Beissinger T et al (2022) Genomic prediction models for traits differing in heritability for soybean, rice, and maize. BMC Plant Biol 22(1):87

Kalve S, Gali KK, Tar'an B (2022) Genome-wide association analysis of stress tolerance indices in an interspecific population of chickpea. Front Plant Sci 13:933277

Kamdar JH, Jasani MD, Chandrashekar AB et al (2022) Does improved oleic acid content due to marker-assisted introgression of *ahFAD2* mutant alleles in peanuts alter its mineral and vitamin composition? Front Plant Sci 13:942617

Khan AW, Garg V, Sun S et al (2024) Cicer super-pangenome provides insights into species evolution and agronomic trait loci for crop improvement in chickpea. Nat Genet 56:1320

Khazaei H, Street K, Bari A, Mackay M, Stoddard FL (2013) The FIGS (focused identification of germplasm strategy) approach identifies traits related to drought adaptation in *Viciafaba* genetic resources. PLoS One 8:e63107

Kreplak J, Madoui MA, Cápal P et al (2019) A reference genome for pea provides insight into legume genome evolution. Nat Genet 51:1411–1422

Kudapa H, Garg V, Chitikineni A, Varshney RK (2018) The RNA-Seq-based high resolution gene expression atlas of chickpea (*Cicer arietinum* L.) reveals dynamic spatio-temporal changes associated with growth and development. Plant Cell Environ 41(9):2209–2225

Kumar M, Prusty MR, Pandey MK, Singh PK, Bohra A, Guo B, Varshney RK (2023) Application of CRISPR/Cas9-mediated gene editing for abiotic stress management in crop plants. Front Plant Sci 14:1157678

Lakmes A, Jhar A, Penmetsa RV, Wei W, Brennan AC, Kahriman A (2022) The Quantitative Genetics of Flowering Traits in Wide Crosses of Chickpea. Agriculture 12(4):486

Lakmes A, Jhar A, Penmetsa RV, Wei W, Brennan AC, Kahriman A (2023) Inheritance of seed weight and growth habit in 10 intercross chickpea (*Cicer arietinum*) nested association mapping populations. Plant Breed 142(1):86–96

Liu Y, Du H, Li P, Shen Y, Peng H, Liu S, Zhou GA, Zhang H, Liu Z, Shi M, Huang X, Li Y, Zhang M, Wang Z, Zhu B, Han B, Liang C, Tian Z (2020) Pan-Genome of Wild and Cultivated Soybeans. Cell 182(1):162–176.e13

Lobaton JD, Miller T, Gil J, Ariza D, de la Hoz JF, Soler A, Beebe S, Duitama J, Gepts P, Raatz B (2018) Resequencing of Common Bean Identifies Regions of Inter-Gene Pool Introgression and Provides Comprehensive Resources for Molecular Breeding. Plant Genome 11(2). https://doi.org/10.3835/plantgenome2017.08.0068

Loi M, Villani A, Paciolla F, Mulè G, Paciolla C (2021) Challenges and opportunities of light-emitting diode (LED) as key to modulate antioxidant compounds in plants. A review. Antioxidants 10:42

Lu Q, Huang L, Liu H et al (2024) A genomic variation map provides insights into peanut diversity in China and associations with 28 agronomic traits. Nat Genet 56:530–540

Malzahn A, Lowder L, Qi Y (2017) Plant genome editing with TALEN and CRISPR. Cell Biosci 7:21

Meuwissen TH, Hayes BJ, Goddard ME (2001) Prediction of total genetic value using genome-wide dense marker maps. Genetics 157:1819–1829

Mobini SH, Warkentin TD (2016) A simple and efficient method of in vivo rapid generation technology in pea (*Pisum sativum* L.). In Vitro Cell Dev Biol Plant 52:530–536

Mitache M, Baidani A, Bencharki B, Idrissi O (2024) Exploring genetic variability under extended photoperiod in lentil (Lens Culinaris Medik): vegetative and phenological differentiation according to genetic material's origins. Plant Methods 20(1):9

Muñoz-Amatriaín M, Mirebrahim H, Xu P et al (2017) Genome resources for climate-resilient cowpea, an essential crop for food security. Plant J 89:1042–1054

O'Rourke JA, Iniguez LP, Fu F et al (2014) An RNA-Seq based gene expression atlas of the common bean. BMC Genomics 15:866

Pandey M, Agarwal G, Kale S et al (2017) Development and evaluation of a high density genotyping 'Axiom_*Arachis*' Array with 58 K SNPs for accelerating genetics and breeding in groundnut. Sci Rep 7:40577

Pazhamala LT, Purohit S, Saxena RK, Garg V, Krishnamurthy L, Verdier J, Varshney RK (2017) Gene expression atlas of pigeonpea and its application to gain insights into genes associated with pollen fertility implicated in seed formation. J Exp Bot 68:2037–2054

Roorkiwal M, Jain A, Kale SM, Doddamani D, Chitikineni A, Thudi M, Varshney RK (2018) Development and evaluation of high-density Axiom® CicerSNP Array for high-resolution genetic mapping and breeding applications in chickpea. Plant Biotechnol J 16:890–901

Rouet P, Smih F, Jasin M (1994) Introduction of double-strand breaks into the genome of mouse cells by expression of a rare-cutting endonuclease. Mol Cell Biol 14:8096–8106

Sallam AH, Conley E, Prakapenka D, Da Y, Anderson JA (2020) Improving Prediction Accuracy Using Multi-allelic Haplotype Prediction and Training Population Optimization in Wheat. G3 (Bethesda) 10(7):2265–2273

Samantara K, Bohra A, Mohapatra SR et al (2022) Breeding more crops in less time: a perspective on speed breeding. Biology 11:275

Saradadevi R, Mukankusi C, Li L et al (2021) Multivariate genomic analysis and optimal contributions selection predicts high genetic gains in cooking time, iron, zinc, and grain yield in common beans in East Africa. Plant Genome 14:e20156

Saxena RK, Rathore A, Bohra A et al (2018) Development and application of high-density Axiom *Cajanus* SNP Array with 56K SNPs to understand the genome architecture of released cultivars and founder genotypes. Plant Genome 11:3

Saxena RK, Jiang Y, Khan A, Zhao Y, Singh VK, Bohra A, Sonappa M, Rathore A, Sameerkumar CV, Saxena CV, Reif J, Varshney RK (2021) Characterization of heterosis and genomic prediction based establishment of heterotic pattern for developing better hybrids in pigeonpea. Plant Genome 14:e20125

Saxena RK, Hake A, Hingane A, Sameer Kumar CV, Bohra A, Sonnappa M, Rathore A, Kumar A, Mishra A, Tikle AN, Sudhakar C, Rajamani S, Patil DK, Singh IP, Singh NP, Varshney RK (2020) Translational Pigeonpea Genomics Consortium for accelerating genetic gains in pigeonpea (*Cajanus cajan* L.). Agronomy 10:1289

Sinha P, Singh VK, Bohra A et al (2021) Genomics and breeding innovations for enhancing genetic gain for climate resilience and nutrition traits. Theor Appl Genet 134:1829–1843

Sinha P, Singh VK, Saxena RK, Khan AW, Abbai R, Chitikineni A, Desai A, Molla J, Upadhyaya HD, Kumar A, Varshney RK (2020) Superior haplotypes for haplotype-based breeding for drought tolerance in pigeonpea (*Cajanus cajan* L.). Plant Biotechnol J 18(12):2482–2490

Singh IP, Bohra A, Singh F (2016) An overview of varietal development program of pigeonpea in India. Legum Perspect 11:39–42

Song Q, Hyten DL, Jia G, Quigley CV, Fickus EW, Nelson RL, Cregan PB (2013) Development and evaluation of SoySNP50K, a high-density genotyping array for soybean. PLoS One 8:e54985

Song Q, Yan L, Quigley C, Jordan BD, Fickus E, Schroeder S, Song BH, Charles An YQ, Hyten D, Nelson R, Rainey K, Beavis WD, Specht J, Diers B, Cregan P (2017) Genetic characterization of the soybean nested association mapping population. Plant Genome 10:1–14. https://doi.org/10.3835/plantgenome2016.10.0109

Takahashi Y, Sakai H, Yoshitsu Y, Muto C, Anai T, Pandiyan M, Senthil N, Tomooka N, Naito K (2019) Domesticating Vigna Stipulacea: A Potential Legume Crop With Broad Resistance to Biotic Stresses. Front Plant Sci 10:1607

Tang Y, Qiu X, Hu C et al (2022) Breeding of a new variety of peanut with high-oleic-acid content and high-yield by marker-assisted backcrossing. Mol Breed 42:42

Tayeh N, Aluome C, Falque M et al (2015) Development of two major resources for pea genomics: the GenoPea 13.2K SNP Array and a high-density, high-resolution consensus genetic map. Plant J 84:1257–1273

Tripathi L, Ntui VO, Tripathi JN (2024) Application of CRISPR/Cas-based gene-editing for developing better banana. Front Bioeng Biotechnol 12:1395772

Varshney RK, Mohan SM, Gaur PM et al (2013) Achievements and prospects of genomics-assisted breeding in three legume crops of the semi-arid tropics. Biotechnol Adv 31:1120–1134

Varshney RK, Thudi M, Nayak SN et al (2014) Genetic dissection of drought tolerance in chickpea (*Cicer arietinum* L.). Theor Appl Genet 127:445–462

Varshney RK, Pandey MK, Bohra A, Singh VK, Thudi M, Saxena RK (2019a) Toward the sequence-based breeding in legumes in the post-genome sequencing era. Theor Appl Genet 132:797–816

Varshney RK, Thudi M, Roorkiwal M et al (2019b) Resequencing of 429 chickpea accessions from 45 countries provides insights into genome diversity, domestication and agronomic traits. Nat Genet 51:857–864

Varshney RK, Roorkiwal M, Sun S et al (2021a) A chickpea genetic variation map based on the sequencing of 3,366 genomes. Nature 599:622–627

Varshney RK, Bohra A, Yu J, Graner A, Zhang Q, Sorrells ME (2021b) Designing future crops: genomics-assisted breeding comes of age. Trends Plant Sci 26:631–649

Varshney RK, Bohra A, Roorkiwal M et al (2021c) Fast-forward breeding for a food-secure world. Trends Genet 37:1124–1136

Varshney RK, Saxena RK, Upadhyaya HD, Khan AW, Yu Y, Kim C, Rathore A, Kim D, Kim J, An S, Kumar V, Anuradha G, Yamini KN, Zhang W, Muniswamy S, Kim JS, Penmetsa RV, von Wettberg E, Datta SK (2017) Whole-genome resequencing of 292 pigeonpea accessions identifies genomic regions associated with domestication and agronomic traits. Nat Genet 49(7):1082–1088

Watson A, Ghosh S, Williams MJ et al (2018) Speed breeding is a powerful tool to accelerate crop research and breeding. Nat Plants 4:23–29

Wu J, Wang L, Fu J et al (2020) Resequencing of 683 common bean genotypes identifies yield component trait associations across a north-south cline. Nat Genet 52:118–125

Yang T, Liu R, Luo Y, Hu S, Wang D, Wang C, Pandey MK, Ge S, Xu Q, Li N, Li G, Huang Y, Saxena RK, Ji Y, Li M, Yan X, He Y, Liu Y, Wang X, Xiang C, Varshney RK, Ding H, Gao S, Zong X (2022) Improved pea reference genome and pan-genome highlight genomic features and evolutionary characteristics. Nat Genet 54(10):1553–1563

Yao S, Jiang C, Huang Z et al (2016) The *Vigna unguiculata* Gene Expression Atlas (VuGEA) from de novo assembly and quantification of RNA-seq data provides insights into seed maturation mechanisms. Plant J 88:318–327

Zhou Z, Jiang Y, Wang Z et al (2015) Resequencing 302 wild and cultivated accessions identifies genes related to domestication and improvement in soybean. Nat Biotechnol 33:408–414

Nutritional Quality Enhancement in Legumes

Neeraj Kumar, Chirag Maheshwari, C. Bharadwaj, Rahul Kaldate, and T. Vinutha

Abstract

The rising global population and widespread deficiencies in micronutrients pose ongoing challenges to the quantity and quality of food security, exacerbating issues of malnutrition and associated health concerns termed "hidden hunger," notably in developing nations. Despite considerable efforts in the agriculture sector to enhance nutrient content in cereals, successes have been insufficient to reduce malnutrition significantly. Legumes, integral to the food systems in sub-Saharan Africa and South Asia, have seen limited efforts to enhance their nutrient content. However, the germplasm pool of legume crops has genetic variation for a wide range of nutritional properties, which can be used to improve the nutrient content of seeds by selective breeding. The targeted identification and utilization of precise nutritional traits in breeding programs can help mitigate malnutrition. Several challenges, including the intricate balance between yield and protein content, the interplay of various micronutrients, and the need to reduce anti-nutritional factors, hinder the nutritional enhancement of legumes. Conventional breeding methods, while effective, could be faster and improved by limited genetic variation. However, modern plant breeding tools offer promising solutions. Genetic engineering, marker-assisted selection (MAS), and genome editing enable precise manipulation of plant genomes, accelerating breeding and expanding the pool of genetic resources. Advances in gene editing, such as CRISPR/Cas9 technology, offer novel avenues for improving nutritional content while addressing regulatory and consumer concerns. Integrating these modern breeding approaches is essential for developing bio-fortified legume crops. This chapter aims to understand the nutritional merit of legumes and the associated breeding techniques to breed nutritionally superior legumes, combining knowledge of conventional and modern breeding tools to develop micronutrient-dense, anti-nutrient-free varieties to address malnutrition and hidden hunger.

Keywords

Legumes · Breeding methods · Nutrition · Quality breeding · modern breeding tools

1 Introduction

Food legumes are essential for human and animal nutrition as they provide high-quality proteins necessary for a balanced diet. These crops also play a significant role in sustainable agriculture by fixing atmospheric nitrogen and improving soil fertility in low-input farming systems. In addition to proteins, carbohydrates, and dietary fiber, legume seeds are rich in micronutrients such as minerals, fatty acids, vitamins, and folic acid, which contribute to overall health (Singh et al. 2011). The Leguminosae or Fabaceae family, commonly known as legumes, peas, or beans, consists of over 20,000 species (Stagnari et al. 2017) that thrive in various habitats around the world (Black et al. 2006). While the terms legumes, pulses, and beans are sometimes used interchangeably, there are distinctions among them. Any plant belonging to the Fabaceae family, including its leaves, stems, and pods, is referred to as a legume. The edible seeds of legume plants, such as beans, lentils, and peas, are known as pulses. For example, a pea pod is classified as a legume, while the pea inside the pod is referred to as a pulse. The entire legume plant is used in agriculture as cover crops and animal feed. Various members of the Leguminosae family, including clover, alfalfa, faba beans, chickpeas, cowpeas, peas, peanuts, pigeon peas, lentils, and sweet lupin, have diverse applications in human and animal nutrition as well as in medicine. Legumes have a unique functional property

N. Kumar · C. Maheshwari · C. Bharadwaj (✉) · T. Vinutha
ICAR-Indian Agricultural Research Institute, Pusa, New Delhi, India

R. Kaldate
Assam Agricultural University, Jorhat, Assam, India

G. Prasad Dixit et al. (eds.), *Fundamentals of Legume Breeding*, https://doi.org/10.1007/978-981-96-7838-9_15

characterized by their ability to form symbiotic associations with rhizobia bacteria, which fix atmospheric nitrogen to benefit the plant. In return, the plant provides carbon to the bacteria (Peix et al. 2015). This symbiotic relationship reduces production costs and minimizes environmental pollution risks associated with synthetic nitrogen fertilizers. As a result, agricultural systems fix 50–70 million tons of biological nitrogen annually, including 16.4 million tons in soybeans and 12–25 million tons in fodder legumes (Herridge et al. 2008).

Legumes are an affordable source of protein, minerals (Fe, Mg, P, and Zn), and vitamins (B_1, B_2, B_3, B_6, and B_9). They contain 8–27% dietary fiber with 3.3–13.8% of it being soluble fiber and have a modest glycemic index (Sánchez-Chino et al. 2015). The storage proteins constitute the most important protein fraction in legumes, with globulins making up almost 70% of the total protein, followed by albumins and glutelins at 10–20% each (Sharif et al. 2018). Legumes can be categorized based on their lipid content into two groups: (1) low-fat legumes (1–6%) such as chickpeas, lentils, beans, and broad beans; (2) high-fat legumes such as peanuts and soybeans. The nutritional benefits and significance of legumes were broadly reviewed by Jukanti et al. (2012) and Foyer et al. (2016). Conversely, these pulses lack tryptophan and sulfur-containing amino acids (AA) but are abundant in lysine which is lacking in cereal grains. The low biological value of pulses is mostly caused by their low methionine content, as well as their low digestibility due to the presence of anti-nutritional factors (ANFs) and protease inhibitors. The combined number of monosaccharides and oligosaccharides, known as total sugars, make up a minor fraction of the overall carbohydrate content in pulses.

The concentration of amylase in the starch directly affects the solubility of the starch, its ability to bind with lipids, and other functional characteristics. Amylose and amylopectin contribute to the structural composition of starch granules. Consuming a significant number of pulses has been found to induce flatulence in humans. Excessive accumulation of gas in the digestive system leads to discomfort, spasms, pain, and loose bowel movements. Human mucosa lacks the hydrolytic enzyme α-1,6-galactosidase, which makes raffinose family sugars difficult to digest. These oligosaccharides generate significant quantities of CO_2, H, and CH_4 resulting in the production of flatus. Whereas genetic manipulation and selective breeding can be used to generate pulse cultivars with decreased levels of raffinose family oligosaccharides. The most prevalent soluble sugars in most pulses are raffinose, ajugose, verbascose, and stachyose which account for 31.1–76% of the total. Verbascose is a prominent oligosaccharide found in *Vigna mungo, Vigna radiata*, and *Cajanus cajan*, whereas stachyose is the predominant oligosaccharide found in *Cicer arietinum*, *Lens culinaris*, and *Vigna unguiculata*.

Legumes are a good source of dietary micronutrients and essential AA, which makes them potentially important in addressing the increasing problem related to malnutrition. Improving grain nutritional quality in legume by a plant breeding strategy is a financially feasible choice. An extensive characterization of the worldwide legume germplasm, involving both cultivated and wild species, has the potential to significantly assist in the identification of genotypes with elevated micronutrient content. Pre-breeding and breeding initiatives are necessary to cultivate breeding lines that are abundant in micronutrients and evaluate their performance at multiple sites to analyze the influence of the environment on the assimilation of micronutrients. Moreover, by enhancing legume genomic resources, we can effectively identify the different genomic regions that regulate micronutrient content using biparental quantitative trait loci (QTL) mapping and genome-wide association studies (GWAS). Similarly, remarkable progress in functional genomics has offered insights into the specific candidate gene(s) that regulate the synthesis and accumulation of numerous micronutrients with specific activity. Adopting an integrated breeding method that incorporates all these technologies can certainly assist us in reducing or mitigating the nutritional portfolio of legume crops for different applications in agriculture and industry. This chapter attempts to understand the overall development in genetic augmentation of nutritional quality in legumes and advancements in these directions.

2 Characterization of Storage Proteins in Legumes

Emerging demands for plant-based foods with enhanced health benefits are driven by shifting lifestyles, limited availability, and increasing costs of animal proteins, as well as a growing consumer preference for market trends and healthier diets. Leguminous crops stand out in addressing these needs because of their abundance of phytochemicals and rich protein content, and vital nutrients crucial for promoting a healthier life (Halliwell 2024). Particularly, pulses within the legume family serve as exceptional and cost-effective sources of protein, fiber, beneficial carbohydrates, and minerals. They have a low-fat content, apart from oilseeds that are mostly grown because of their lipid content. The content of carbohydrates, fatty acids, protein, and micronutrients in grain legumes differs based on the type of legume, environmental circumstances, and farming methods. Major grain legumes typically contain about 18–50% protein, 0.8–21% oil, 60–69% carbohydrates, and 0.9–7.2% fiber (Vollmann 2016; González and Arellano 2009) as summarized in Table 1. Proteins in legumes are commonly categorized into three groups: structural, storage, and biologically active. Storage proteins, which are crucial nitrogen reserves for

Table 1 Proximate composition of important legume crops

Legume species	Composition (g/100 g)					Reference
	Protein	Carb	Crude fiber	Fat	Ash	
Dry bean (*Phaseolus* spp)	16–25	39–56	6.3–18.9	1.5	2.9–4.3	Celmeli et al. (2018)
Pigeon peas (*Cajanus cajan*)	16.8–29.2	62.7–62.8	3.2–8.6	1.2–1.3	3.2–4.0	Wu et al. (2024)
Black gram (*Vigna mungo*)	24	51–59	5.5	1.1–2.4	1.4	Girish et al. (2012)
Lentils (*Lens culinaris*)	24.6	63.4	13.87–24.91	0.8	2.7	Dhull et al. (2023), Liberal et al. (2024)
French bean (*Phaseolus vulgaris*)	24.84	45.66	16.91	1.03	4.69	Azmah et al. (2023)
Chickpea (*Cicer arietnum*)	22.75–23.44	62.24–63.24	5.62–5.84	4.99–5.4	2.92–3.16	Tania et al. (2023)
Green gram (*Vigna radiata*)	26–27	62–70	4.6	1.6–1.8	1–3.7	Mubarak (2005)
Long beans (*Vigna unguiculata* subsp. Sesquipedalis)	27.89	49.63	8.59	1.43	3.93	Azmah et al. (2023)
Soybean (*Glycine max*)	41.22	30–34	4.5	17.02	8.14	Miao et al. (2024)

germination, are the most abundant. Biologically active proteins consist of enzymes, enzyme inhibitors, and lectins, whereas structural proteins contain membrane ribosomal and chromosomal proteins. Storage proteins can be classified into different categories depending on their solubility in different solvents.

These categories include albumins, globulins, glutelins, and prolamins. Albumins are soluble in water, globulins are soluble in a weak salt solution, prolamins dissolve in 75% ethanol, whereas glutelins are difficult to dissolve but can be dissolved in a weak alkaline solution. They also vary in frequency between monocotyledonous grains and dicotyledonous legumes. Legumes are mostly composed of albumins and globulins, whereas cereals are mainly composed of prolamins and glutelins. Albumins typically have low molecular weights ranging from 5 to 80 kDa and may include lectins, inhibitors like trypsin, and amylase. Asparagine, arginine, and glutamine are common nitrogen-containing AA found in storage proteins, serving as nitrogen sources for germination. Legumes contain globulin proteins, such as legumins and vicilins, which are also referred to as 7S and 11S globulins depending on their Svedberg sedimentation values (S). Legume proteins have a low content of S-containing AA like cysteine, methionine, and tryptophan except for soybeans, which contain all essential AA necessary for human diets and exhibit good digestibility. Soybean protein is of high quality, nearly comparable to animal protein sources in nutritional value. Overall, legumes are nutritionally valuable because of their proteins along with a rich assortment of macro and micronutrients.

3 Metabolic Pathways of Various Nutrient Acquisition in Legumes

Understanding the acquisition, transportation, and accumulation of minerals in legume seeds is pivotal for maximizing nutrient availability. Despite the identification of certain genes involved, many others remain elusive. Research has been primarily focused on genes facilitating mineral translocation to vegetative tissues and seeds, yet there's a dearth of knowledge concerning phloem-expressed genes linked to mineral loading and mobilization to various sink tissues. While transporter studies have offered insights into their functions, comprehensive whole-plant investigations are imperative to pinpoint the most relevant transporters for delivering minerals to seeds. Although extensive research has been conducted on mineral acquisition and mobilization in plants, various stresses can impede nutrient availability, thereby hindering crop growth.

Legumes, categorized as "strategy I" plants, employ specific mechanisms to enhance iron (Fe) uptake and transport. They acidify the rhizosphere by employing an H^+-ATPase enzyme, which enhances the solubility of Fe^{3+} (Santi and Schmidt 2009). Subsequently, chelate reductase, ferric reduction oxidase (FRO2), aids in reducing Fe^{3+} to Fe^{2+}, which is then absorbed by the root's PM through Fe^{2+} iron-regulated transporter (IRT1) or its homologs. Rhizosphere acidification involves proton release and subsequent absorption of excess cations, particularly Fe^+, in N_2 fixation (Sinclair and Krämer 2012). In legumes, proteins such as HA2, FRO2, and IRT1 mainly facilitate the absorption and delivery of iron to the roots. Homologs involved in transporting Fe from leaves to roots, like bZIP23, FIT1, OPT3, and IRT1 have been detected in different legumes crops (Xiong et al. 2012).

In legume shoots, there's relatively limited understanding of iron uptake and regulation. The IRT-like transporters facilitate the uptake of Fe in soybean plants' shoots, and they transport it through the xylem as complexes of ferric citrate.

Xylem unloading is an important stage in distributing Fe to various tissues and sink cells. Expression patterns indicate the involvement of YSL and ZIP transporters in unloading metals from the xylem. The oligopeptide transporter (OPT) is believed to have a substantial impact on the precise long-distance communication of iron from shoots to roots in Arabidopsis (Kumar et al. 2017). Further, the significant concentrations of nicotianamine (NA) found in shoot tissues, along with its potent affinity for various ions, suggest that YSL transporters play a vital role in the movement of metals from xylem to leaves and seeds. The NRAMP family genes are important for maintaining iron balance, while the YSL and OPTs genes have key functions in transporting iron-NA complexes in/out of the phloem tissue (Palmer and Guerinot 2009).

Zinc (Zn) is mostly absorbed through the PM of root cells in legumes in the form of Zn^{2+}. The ZIP (zinc-regulated, iron-regulated transporter-like proteins) transporters have a crucial function in the absorption and movement of Zn from the roots to the seeds (Palmgren et al. 2008). Various proteins like ZIP, MTP (metal tolerance protein), ZIF (zinc-induced facilitator), and HMA (heavy metal ATPase) are involved in Zn transport. MTPs facilitate the mobilization of multiple metal ions, including iron, within the cytoplasm. In barrel medic plants, several ZIP genes are upregulated under Zn deficiency, indicating their active role in Zn transport. The ZIP genes were found in various tissues like seeds, leaves, and roots of many dicots (Lin et al. 2009; Astudillo et al. 2013). Manganese (Mn) is essential for plants as it acts as a cofactor in critical activities like lipid production and photosynthesis. Plants take up Mn from the soil primarily as Mn^{2+}. Although few transporters exclusively for Mn have been identified in plants, various transporters like CDF/MTP, IRT1, NRAMP, P-Type-ATPase, VIT, and YSL assist in Mn transport (Xia et al. 2010; Socha and Guerinot 2014).

Plants primarily absorb phosphorus (P) from the soil in the form of phosphate (Pi). This uptake can happen through root epidermal cells driven by a proton gradient produced by PM H^+-ATPases or with the assistance of arbuscular mycorrhizal fungi (AMF), commonly found in legumes. The *pht (Phosphate transporter)* genes are active in the roots, seeds, and other aerial parts, indicating their possible role in the movement of internal Pi. In *M. truncatula*, the Pi-transporter genes, such as *MtPT1* and *MtPT2*, belonging to the *Pht1* family, exhibit significant expression levels in Pi-deprived roots (Liu et al. 2008).

4 Anti-nutrient Factors in Legumes

In food and nutrition literature, plant defense metabolites are often called anti-nutrient factors (ANFs) or natural toxicants. The biological effects of molecules ranging from high-molecular-weight proteins to simple AA and oligosaccharides depend on their shapes. The factors destroyed by heat treatment are the α-amylase and protease inhibitors, goitrogens, lectins, and antivitamins. Non-protein amino acids are heat-stable ANFs. Depending on their physiochemical properties, these ANFs can be categorized into groups (Enneking and Wink 2000). Plants have ANFs due to evolutionary adaptation, allowing them to live and complete their life cycle. Due to their toxicity, unpalatability, or indigestibility, it becomes necessary to eliminate them from food sources either by choosing plant genotypes that have less content of these compounds or through post-harvest treatment like germination, leaching, boiling, extraction, or fermentation before consumption. Some important ANFs and compounds in legumes are saponins, oxalates, phytic acid, tannins, and raffinose family oligosaccharides. Saponins are a class of molecules that consist of lipophilic triterpene and hydrophilic sugar components, and they have both hydrophilic and lipophilic properties. In pea flour, saponin-B and DDMP cause a bitter "off-flavor" (Heng et al. 2006).

Similarly, oxalates can form soluble or insoluble salts with Ca, Fe, Mg, Na, and K. Water-soluble oxalates are $K_2C_2O_4$, $Na_2C_2O_4$, and $(NH_4)2C_2O_4$, while water insoluble oxalate is CaC_2O_4 (Noonan and Savage 1999). The water-soluble oxalates reduce mineral absorption, while the water-insoluble oxalates lead to the formation of urinary stones. Phytic acid (PA) is crucial in reducing micronutrient bioavailability, and consuming foods rich in PA results in mineral deficiencies, specifically iron and zinc, in humans (Pramitha et al. 2021). Legumes contain myoinositol (dihydrogen phosphate), which limits phosphate use because it is primarily present in the organic form of inositol hexakisphosphate (IP6). Therefore, a decrease in IP6 levels can enhance the availability of phosphates and minerals in crops and grains. Both pigeon peas and chickpeas have modest levels of phytic acid. To increase the intake of pulses and nutritional content, breeding programs should choose genotypes with low levels of phytic acid (Chitra et al. 1995).

Raffinose family oligosaccharides (RFOs) are a prominent group of ANFs with notable adverse effects on monogastric animals and humans. Unlike PA, which binds to minerals and reduces their availability for absorption by humans and other animals, RFOs contribute to the production of flatulence (Konsam et al. 2014; Zhawar et al. 2011). Proanthocyanidins or catechins are widely found as secondary plant metabolites in many plants, such as cocoa beans, nuts, tea, wines, fruits, and legumes (Petroski and Minich 2020). Tannins function as antioxidants by effectively neutralizing free radicals. Additionally, they can bind with dietary elements such as Fe2+, Cu2+, and Zn2+ through a process known as chelation (Karamac 2009). This anti-nutritional factor varies greatly among legume species and their possible impact on human health (Table 2).

Table 2 Major anti-nutrient factors (ANFs) and their effects

Anti-nutritional factor	Effect on human body	Legume and reference
Saponins	Astringent and bitter off-flavor	Pea seeds (Reim and Rohn 2015)
Oxalates	Kidney stones	Soybeans (370 mg/100 g dry weight (DW)), lentils and peas (168–293 mg/100 g DW), chickpeas (192 mg/100 g DW), and common beans (98–117 mg/100 g DW) (Petroski and Minich 2020)
Phytic acid (PA)	Deficiencies of minerals such as Fe and Zn The reduced availability of essential minerals in legumes	11.33–14 mg/g in chickpea,15.64–18.82 mg/g in common beans, 19.65–22.85 mg/g in faba bean, 8.56–17.1 mg/g in lentil, and 9.93–12.40 mg/g in peas (Shi et al. 2018)
Raffinose family oligosaccharides	RFOs are responsible for flatulence	Chickpea, raffinose sugars were found: 0.38–0.99 g/100 g (Konsam et al. 2014; Roorkiwal et al. 2021)
Tannins	Chelate dietary elements like Fe^{2+},Cu^{2+}, and Zn^{2+}	Kidney bean (*Phaseolus vulgaris* L.) contained a significant amount of catechins (20.1 mg/kg; Arts et al. 2000)

5 Major Objective in Nutritional Quality Traits in Legumes

Breeding better varieties in conventional breeding method is challenging. Plant breeding techniques have been practiced for thousands of years and involve selecting plants with desirable traits and inter-crossing them to produce offspring with improved characteristics. These methods rely heavily on the natural variation within plant populations and the phenotypic selection of traits like drought tolerance, yield, and disease resistance. While conventional breeding has been successful in developing numerous high-yielding and resilient crop varieties, it is inherently slow, often taking many years to produce significant improvements. Additionally, it faces challenges such as limited genetic variation, difficulty in transferring desirable traits from wild relatives, and the inability to precisely control which genes are passed on. New improved tools in plant breeding offer methods to overcome the constraints of conventional breeding techniques (Fig. 1).

Techniques such as genetic engineering, MAS, and genome editing (e.g., CRISPR/Cas9) allow precise manipulation of plant genomes. Genetic engineering enables the introduction of novel traits from unrelated species, expanding the pool of available genetic resources. MAS and genomic selection accelerate the breeding process by using molecular markers to identify desirable traits at the seedling stage, reducing the reliance on phenotypic selection. Genome editing allows for targeted modifications, providing breeders with the ability to fine-tune plant genomes with unprecedented accuracy. The challenges inherent in conventional breeding, such as long development times and limited genetic diversity, highlight the need for modern plant breeding tools. These advanced techniques offer faster, more precise, and more efficient ways to develop crop varieties that can meet the growing demands of a rapidly increasing global population, adapt to changing climatic conditions, and contribute to sustainable agricultural practices. Adapting modern breeding tools is vital for ensuring food security and addressing the agricultural challenges of the twenty-first century.

The major breeding objective and challenge in legumes breeding is dealing with negative correlation between yield and protein content, as well as protein and sulfur-containing AA. Further, augmenting Fe, Zn, and other micronutrients with yield is also the major objective. Reducing ant nutritional content in various legume cros is also an important breeding objective. Traditional breeding programs needs evaluation of a genetically diverse panel and detection of potential donor parents for desirable nutritional profile. Enhancing legume nutritional profile requires the identification of potential donors to transfer the desired trait or gene to the elite genotype to obtain a biofortified variety (Guindon et al. 2021). However, biofortification, achieved by implementation of plant breeding techniques, is considered the most dependable, sustainable, and economical approach. The process of generating a biofortified product is time-consuming and hindered by the absence of efficient screening and evaluation methods. The utilization of new molecular markers can provide an additional benefit by significantly reducing time by enabling fast identification of the superior recombinants possessing the desired characteristic.

5.1 Exploiting Germplasm and Genetic Diversity for Quality Enhancement

Genetic biofortification involves using plant breeding methods to develop crops that have elevated amounts of essential micronutrients and reduced levels of harmful ANFs (Bouis 2003). Before attempting to improve the availability of a specific element through breeding, it is essential to screen a diverse germplasm to understand the genetic variation associated with the desired trait (McCouch et al. 2013). Breeders can use heterosis, transgressive segregation, and additive gene effects to boost micronutrients content by utilizing genetic variability. The transgenic techniques can offer additional sources of vari-

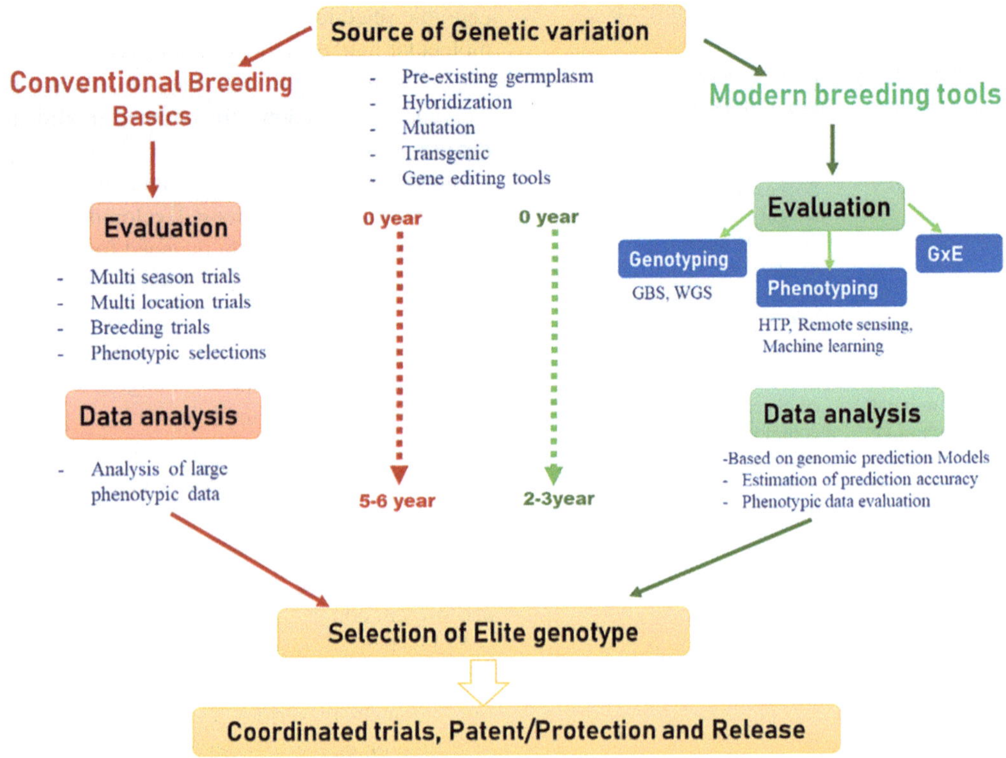

Fig. 1 Comparison of conventional and modern breeding schemes

ability when the requisite genetic variation is unavailable (Francis et al. 2017). Studies on legume crops have shown substantial genetic variations in vital micronutrients in chickpea, common bean, cowpea, lentil, mung bean, pea, and soybean (Table 3). Biofortification programs can use the identified promising germplasms to map QTLs or genes related to high nutrient content. Further, the molecular markers associated with high micronutrient concentrations will be employed for MAS. Considering its sustainability and lack of legal and political restrictions, breeding appears to be the ideal bio-fortification approach (Saltzman et al. 2017).

The initial stage of establishing a legume genetic biofortification program involves identifying a certain micronutrient level to be achieved for each crop. Genetic biofortification is influenced by factors such as the presence of genetic variation and knowledge about genes that regulate the absorption, movement, distribution, and accumulation of nutrients in different plant parts (Bouis and Welch 2010). Micronutrients make up a small proportion of the overall grain weight. Therefore, accurate assessment is necessary to effectively evaluate genetic diversity for breeding with consistent and high nutrient concentrations. Nevertheless, there is still a need to discover rapid, precise, and cost-effective techniques for finding genotypes with high nutrient concentrations. Pre-breeding techniques can generate introgression lines and breeding materials from wild micronutrient donor cultivars for genetic biofortification of crops.

The phenomenon of transgressive segregation can be utilized to generate genetic diversity for a certain nutritional attribute, especially when it is challenging to accomplish using traditional selection methods (Bouis and Welch 2010). To choose the parental line and breeding strategy, it is essential to have a comprehensive understanding of the genetics of the desired trait in every breeding program. Additionally, studying the relationship between nutritional attributes and yield might assist in identifying nutrient-rich cultivars with higher yields and desired features (Bouis and Welch 2010).

A positive association among iron and zinc levels in peanut, common bean, pearl millet, sesame seeds, wheat, and maize suggests that these two nutrients can be improved together (Pixley et al. 2011; Velu et al. 2012). While in chickpea, Diapari et al. (2014) observed that the grain yield was negatively correlated with Zn and Fe levels. Rapid phenotyping of nutritional traits is always a challenging task for a breeder as it requires multidisciplinary collaborative approach and sophisticated instrumentation facilities. Typically, inductively coupled plasma-mass spectrometry (ICP-MS) is employed for mineral measurement. However, this method needs costly apparatus, experienced manpower, and extensive sample preparation. The colorimetric method utilized for mineral measurement is semi-quantitative and

Table 3 Micronutrient contents in major grain legume crops

S. No.	Legumes	Iron (Fe)	Zinc (Zn)	Carotenoids	Folate	References
1	Chickpea	60–70 mg/kg	40–45 mg/kg	22–44 μg/g	351–589 μg	Karaca et al. (2020)
2	Lentil	75.6–100 mg/kg	36.7–50.6 mg/kg	–	216–290 μg	Ray et al. (2014)
		42–132 ppm	23–78 ppm	–	–	Kumar et al. (2016)
3	Common bean	55.4–176.1 mg/kg	5.1–77.7 mg/kg	–	–	Murube et al. (2021)
4	Cowpea	69.77–134.16 mg/kg	33.79–40.53 mg/kg	–	–	Gerrano et al. (2022)
		–	–	β-carotene (13.2 ± 2.9 mg/100 g; Lutein (58 ± 12.8 mg/100 g) Zeaxanthin (14.7 ± 3.1 mg/100 g)	–	Sodedji et al. (2022)
5	Pea	47.7–58.1 mg/kg	27.4 34 mg/kg	–	–	Dissanayaka (2019)
		29.22–90.53 μg/g	12.83–51.47 μg/g	–	23–30 μg/ 100 g	Klein et al. (2020)
		22–490 mg/kg		–	–	Jha et al. (2015)
6	Mung bean	3.4–4.4 mg/100 g	1.2–2.1 mg/100 g	–	–	Kumar and Pandey (2020)

requires significant effort when employed for wide screening. Atomic absorption spectrometry (AAS) is a technique used to quantitatively detect elements in a sample by measuring the absorption of optical radiation by free atoms. X-ray fluorescence spectroscopy (XRF) is a reliable and efficient method for determining the concentrations of elements in samples (Singh et al. 2013).

5.2 QTLs for Nutrient Accumulation

Integrating genetic data with nutritional concentration data is a precise method for identifying QTLs that are linked with traits of interest. The process of analyzing the nutritional content of genetically characterized germplasm (composite set, core collection, mini-core collection, and reference set) can establish a connection between genetic information and worldwide mineral nutrition (Norton et al. 2010). The region that is associated with genetic markers through QTL mapping can subsequently be separated or cloned to identify and analyze the relevant genes. Genomic regions responsible for controlling nutritional content/concentration have been identified through molecular mapping in several legumes (Table 4). Understanding the genetic basis of qualitative traits in legumes is challenging due to the influence of many loci. The development of legume genetic resources has changed significantly in the past decade. This change has helped researchers understand the genetic factors that regulate crucial features, such as the micronutrient content of grains.

The biparental mapping approach was primarily employed to examine genetic regions that regulate micronutrient content in different types of grain legumes. Studies on deciphering genes and mechanisms to enhance the nutritional composition of seeds by QTL mapping was restricted to a small number of nutrients found in legumes. QTLs for seed nutrient concentrations have been found in various plant species such as *M. truncatula* (Sankaran et al. 2009), *L. japonicus* (Klein and Grusak 2009), common bean (Casañas et al. 2013), soybean (Ramamurthy et al. 2014), lentil (Aldemir et al. 2017), and chickpea (Sab et al. 2020). Most of the research performed to map and identify the QTL(s)/gene(s) responsible for micronutrient content in legumes have predominantly shown a quantitative mode of inheritance. For example, in the case of *P. vulgaris*, two genes (*PvIRT1* and *PvIRT2*) located on chromosome 3 and two genes (*PvbZIP2* and *PvbZIP3*) located on chromosome 11 coincide with QTLs for iron and zinc levels (Jiang et al. 2008).

5.3 Genomic Approaches for Nutritional Breeding in Pulses

Conventional breeding approaches showed some success in improving the quality of legumes by enhancing their nutrient level, but not to the desired level. Hence, it is crucial to harness the potential of genomics to speed up the advancement of nutritionally enhanced cultivars. New developments in next-generation sequencing (NGS) technologies have led to a significant decrease in the cost of sequencing services. As a result, genome sequences for key legumes such as *Medicago*, chickpea, and pigeonpea have become more accessible. While there are affordable genotyping platforms for using genomics-assisted breeding (GAB) in legume crops, the cost of accurately and efficiently estimating nutrient levels causes a major hurdle. Plant genome sequencing provides an opportunity to analyze and comprehend the genes involved in nutrient assimilation process.

The genome sequencing of several legume crops such as soybean (Schmutz et al. 2010), pigeon pea (Varshney et al. 2012), chickpea (Varshney et al. 2013), common bean (Schmutz et al. 2014), peanut (Bertioli et al. 2016), and lupin (Hane et al. 2017) was completed. These sequences can be used to detect the genes responsible for nutritional traits and serve as a basis for applying genomics in legume breeding. NGS-based genotyping technologies can be utilized to gain insights into the genetic basis of nutritional attributes using accurate marker-trait association (MTA), gene identification, and the development of functional markers. Their efficacy has been demonstrated in many agronomic characteristics for the purpose of genetic mapping, MAS, and genomic selection (GS). GAB techniques, such as marker-assisted backcrossing (MABC) and marker-assisted recurrent selection (MARS), can be employed to enhance the quality of one or more nutritional characteristics. However, advanced genotyping techniques such as genotyping by sequencing (GBS) and whole genome re-sequencing (WGRS) utilize NGS to identify a significant number of single nucleotide polymorphisms (SNPs) marker across the entire genome.

This allows for precise genetic and association mapping at a high resolution. Furthermore, recently popularized sequencing-based mapping methods, such as Bulked segregant RNA-Seq (BSR-Seq), Seq-BSA, QTL-Seq, MutMap, and Indel-Seq, can be used to map nutritional traits. The study conducted by Diapari et al. (2014) found unique functional allelic variants linked with Fe and Zn content in chickpea seeds. Similarly, three potential genes associated with the storage of Fe and Zn in mature soybean seeds have been discovered by Liu et al. (2011). Further researchers have discovered two SNP markers in lentils that are strongly linked to the amounts of Fe and Zn in the seeds (Khazaei et al. 2017). Transcriptome sequencing has become an alternative to genome sequencing for targeted expressed gene sequenc-

Table 4 QTLs/genomics region identified/reported for various micronutrient content in legumes

Crop	Trait	Parental crosses and Populations	QTL identified	Reference
Chickpea	Iron content in seed	ICC 4958 × ICC 8261	*CaqFe1.1; CaqFe3.1;CaqFe4.1*	Upadhyaya et al. (2016)
	Zinc content in seed	ICC 4958 × ICC 8261	*CaqZn2.1; CaqZn3.1;CaqFZ4.1;CaqFZ7.1*	Upadhyaya et al. (2016)
	Protein content	187 accessions	4 QTLs	Jadhav et al. (2015)
Common bean	Iron content in seed	MAGIC population	*SdFe6.1; SdFe6.2*	Diaz et al. (2020)
	Zinc content in seed	MAGIC population	*SdZn8.1*	Diaz et al. (2020)
	Iron content (mg/seed)	G14519 x G4825	*QFe_contDaAA6.1 QFe_contPaAA1.1QFe_contPaAA7.1*	Blair et al. (2010)
	Zinc content (mg/seed)	G14519 x G4825	*QZn_contDaAA1.1*	Blair et al. (2010)
	Calcium content in seed coat	Xana × Cornell49242	$Ca1^{xc}$; $Ca7^{xc}$; $Ca9^{xc}$	Casañas et al. (2013)
	Magnesium content in seed coat	Xana × Cornell49242	$Mg7^{xc}$	Casañas et al. (2013)
Lentil	Fe and Zn content in seed	ILL 8006 × CDC Milestone	*FeQTL1.1–1.3; FeQTL2.1–2.3; FeQTL4.1–4.6; FeQTL5.1–5.4; FeQTL6.1–6.2; 7.1–7.3*	Aldemir et al. (2017)
	Zinc content Iron content	96 accessions	Four MTAs; Three MTAs	Singh et al. (2017)
Mung bean	P compounds in the seed	V1725BG × AusTRCF321925	2 for phytic acid; 4 for inorganic P; 1 for total P	Sompong et al. (2012)
Field pea	Protein content	Wt10245 × Wt11238	*prot1, prot2, prot3, prot4, prot5*	Irzykowska and Wolko (2004)
		Carneval × MP1401	PC.LGIII.cccc18.E_2000	Tar'an et al. (2004)
	Raffinose and glucose concentration	Champagne × Terese	RafCleS2.c; GlcT2.b	Dumont et al. (2009)
Peanut	Linoleic acid, oil content, oleic acid, O/L ratio, protein content	TG26 × GPBD4	Seed linoleic 1-1; Seed oleic 1-13; Seed oleic/linoleic 1-1; Seed protein 1-1	Sarvamangala et al. (2011)
Soybean	Seed mineral, cysteine, and methionine concentrations	82 × DSR-173; Williams 82 × Vinton 81 Williams82 × NKS19-90		Ramamurthy et al. (2014)

ing, in addition to trait mapping. It elucidates gene activity and the molecular mechanisms behind nutrient mobilization in crops. It is possible to identify potential genes related to nutritional characteristics by analyzing gene expression profile data from transcriptome assemblies (Pandey et al. 2016).

5.4 Gene Editing for Nutritional Enhancement in Legumes

The development of transgenic biofortified pulses has not yet yielded successful reports due to the lack of reliable and repeatable plant regeneration techniques and biosafety regulations. However, a study by Tan et al. (2018) successfully developed genetically engineered chickpea with high iron content by introducing soybean *ferritin* (*GmFER*) genes and chickpea *nicotianamine synthase 2* (*CaNAS2*) gene. Similarly, Khandal et al. (2020) expressed the *cytokinin oxidase/dehydrogenase 6* (*CaCKX6*) gene in the roots of chickpea lines, which resulted in elevated levels of minerals, including Zn (27–62%), Fe (22–48%), Cu (26–61%), Mg (13–22%), K (11–27%), and P (5–19%) in transgenic lines. The NAS genes obtained from chickpea (*CaNas2*) and rice (*OsNas2*) facilitate the production of NA and are responsible for the absorption and movement of Fe in plants. Das Bhowmik et al. (2019) have successfully created stable genetically modified chickpea plants containing the genes for GUS (*uidA*) and Fe-biofortification (*OsNAS2* and *CaNAS2*).

An earlier study found that the introduction of a *methionine-rich storage albumin* gene from Brazil nut into common beans resulted in an increase in the levels of methionine and cysteine (Aragao et al. 1999). Similarly, the overexpression of the *sunflower seed albumin* gene resulted in a significant increase in the methionine content of lupin (Molvig et al. 1997). In soybean, seed protein content, oleic acid, and pro-vitamin were found to be increased by expressing the bacterial *PSY* (phytoene synthase) gene (Schmidt et al. 2015). Similarly, Kim et al. (2012) observed enhanced beta-carotene in soybean via overexpression of *PSY* and *carotene desaturase* genes. Soybean protein is deficient in S-containing essential AA, *viz.*, cysteine, and methionine. There are reports showing enhanced cysteine and methionine content in soybean due to the overexpression of several genes, such as the *O-acetylserine sulfhydrylase* gene (Kim et al. 2012), *maize zein protein* gene (Dinkins et al. 2001), and *cystathionine γ-synthase* gene (Song et al. 2013).

When cultivated and wild relatives lack the desired nutrient variation or when biofortified crop varieties are difficult to produce, transgenic breeding and genome editing can produce bio-enriched crops with desirable nutritional traits. This method can also aim to boost the bioavailability of micronutrients by removing ANFs or ingesting promoters, in addition to enriching the concentration of micronutrients (Garg et al. 2018). This strategy not only enhances the concentration of micronutrients but also specifically focuses on eliminating ANFs that can improve the availability of micronutrients (Garg et al. 2018). Genome editing is a highly effective technique used to disable, reduce, or modify the expression of genes without causing any changes to the genetic composition of the organism. The CRISPR/Cas9 genome editing technique can be employed to improve the micronutrient content of seeds and decrease the presence of ANFs in plants.

Progress in molecular biology has greatly transformed the methods used for altering DNA in a more precise and focused manner. This is achieved using transcription activator-like effector nucleases (TALENs), zinc-finger nucleases (ZFNs), and short palindromic repeat (CRISPR)–associated protein (Cas9) in legumes and other crops (Michno et al. 2015; Tang et al. 2016). Among several genetic engineering methods, RNA interference (RNAi) stands out as a superior method due to its increased capability to specifically silence genes. This significant breakthrough has the potential to contribute to the development of crops that are both nutritionally rich and have reduced levels of ANFs (Tang and Galili 2004). The RNAi technology has been employed to decrease the amount of BOAA in grass peas and to lower the level of Arah2, an allergen, by 25% in crude peanut extract (Dodo et al. 2008).

It is necessary to employ advanced technologies like TALENs, ZFNs, RNAi, and CRISPR/Cas9 to enhance the nutritional profile of legumes. CRISPR-edited plants are devoid of any exogenous DNA, making them potentially more acceptable than conventional genetically modified crops. The utilization of artificial nucleases (ZFNs and TALENs) enables precise and rapid modification of genomes, which has the potential to expedite both fundamental and practical research. This approach offers the ability to create modified genomes in a predictable manner. CRISPR/Cas9 genome editing is a more advanced and efficient technology compared to first-generation tools like ZFNs and TALENs. It involves the use of Cas9 protein and guide RNAs to target numerous places in the genome. This method is characterized by its simplicity in designing and cloning processes (Jaganathan et al. 2018). Utilizing these recent genome editing tools will assist in the creation of non-GMO plants with the required characteristics, which can enhance their potential for higher yields in challenging environments and enhance their nutritional value.

CRISPR-Cas9 technology has been successfully implemented in legume crops such as *M. truncatula*, soybean, cowpea, and chickpea, which already have proven transformation protocols. Various GmU6 promoters were evaluated to determine their effectiveness in regulating gRNA expression in soybean hairy roots (Di et al. 2019). The GmU6-8 and GmU6-10 promoters were discovered to have significantly higher activity in improving editing efficiency compared to the other nine promoters. CRISPR/Cas9 technology has effectively performed gene editing in soybean to modify seed-related properties, including seed oil profile (Do et al. 2019), undesirable beany flavor of seed product (Wang et al. 2020), and isoflavone content and resistance to soybean mosaic virus (Zhang et al. 2020). The CRISPR-Cas9 gene editing technique has been used to create grass pea seeds with extremely low or non-existent levels of β-ODAP by targeting the *BAHD-AT3* gene, which is responsible for producing a crucial enzyme in the β-ODAP biosynthesis pathway.

This approach aims to develop genetically modified grass peas that are free of transgenes and have significantly reduced levels of β-ODAP in their seeds, thus ensuring safer food in the future (Saha et al. 2023). The utilization of CRISPR/Cas9 technology has opened new avenues for functional genomics in grain legume crops. To achieve genetic transformation, such as CRISPR/Cas9 gene editing, it is essential to have the capability to transport the DNA/RNA components and regenerate a complete plant. Most food legumes exhibit resistance to the absorption and incorporation of foreign DNA, and some of them are recalcitrant (Ochatt et al. 2018). Therefore, for genome editing to be successful in improving legumes, it is essential to have efficient techniques for modifying plants and regenerating complete plants, as well as a regulatory framework that supports these methods and consumer approval of genetically modified crops.

5.5 Genomic Selection for Nutritional Traits

Genomic selection (GS) predicts breeding values using marker-phenotype associations. In the early breeding cycle, this can reduce cross-to-cross generation interval and increase genetic gain per time unit. Plant breeding operations have been transformed by cost-effective, high-throughput genotyping systems that enable GS at scale. Nowadays GS is being routinely used in breeding programs to enhance the genetic gains. Genomic selection works on the assumption that the genetic variance for a trait is controlled by the combined effects of numerous genes. These genes often consist of hundreds to thousands of variants, each having minor individual effects (Crossa et al. 2017). Most of these variations may not be able to be accurately mapped or identified due to their minor effect. GS initially utilizes genome-wide markers and observed phenotype data from populations to establish connections between markers and phenotypes, rather than finding linked loci for the trait. Subsequently, it employs genome-wide markers to estimate the genetic gain of the populations being tested.

The efficiency of genomic selection relies on multiple factors, but the most crucial one is the training population. This population comprises individuals who have undergone genotyping and phenotyping for the specific trait of interest. It is utilized to construct the predictive regression model or to train prediction models employing machine learning techniques. It is possible to estimate the performance of genotyped individuals who have not been phenotyped (referred to as the "test population" or "selection candidates") without the need for field trials or phenotyping. The best advantage of GS is that it shortens the breeding cycle and enables the quick selection of superior genotypes by using genome-wide high-density markers as predictors for evaluating and selecting early generations in breeding programs.

Soybean was among the first crops where GS was employed to enhance yield and agronomic characteristics using GBS (Jarquín et al. 2014). Additional experiments have shown that GS can rapidly produce improved lines and increase the rate of genetic improvement in various legume crops such as chickpea, pea, pigeon pea, soybean, and groundnut (Varshney et al. 2018). GS is becoming more popular in crop breeding programs, but it has some limitations and challenges. Genomic selection can greatly enhance the precision of selection for phenotypic traits with low heritability, such as grain yield, by utilizing both genotype and phenotype information. Nevertheless, for qualities that exhibit high heritability, such as grain weight, disease resistance, and plant height, GS may not offer as many advantages compared to phenotype-based selection. This is because the cost-to-genetic-gain ratio of marker profiling may not be as competitive as direct phenotypic selection for such traits. GS has been utilized in lentils to accelerate the process of selecting desired traits such as protein content, seed size index, and 100-grain weight. This was achieved by developing genomic prediction models using Bayesian methodology, which were based on 64,781 genome-wide SNPs markers (Gebremedhin et al. 2024). GS holds the possibility of early generation selection and breeding decision with possibly omitting rigorous phenotyping for nutritional traits, which again is expensive in bio-fortification program in legumes.

6 Conclusions

In conclusion, legumes are indispensable contributors to human and animal nutrition, offering high-quality proteins crucial for balanced diets. Their role extends beyond nutrition, with their ability to fix atmospheric nitrogen, enhancing soil fertility in sustainable agricultural systems. Legumes are rich in proteins, carbohydrates, fibers, and micronutrients and offer significant health benefits. Despite their nutritional value, challenges like deficiencies in sulfur-containing AA persist, along with anti-nutritional factors causing digestive discomfort. However, ongoing research aims to address these issues through genetic and breeding interventions. By leveraging global germplasm resources and advancing genomic technologies, we can develop micronutrient-rich legume varieties, paving the way for improved agricultural and industrial applications. This not only enhances food security and nutrition worldwide but also underscores our collective responsibility and the urgency of the task at hand.

References

Aldemir S, Ates D, Temel HY, Yagmur B, Alsaleh A, Kahriman A et al (2017) QTLs for iron concentration in seeds of the cultivated lentil (Lens culinaris Medic.) via genotyping by sequencing. Turk J Agric For 41:243–255. https://doi.org/10.3906/tar-1610-33

Aragao FJL, Barros LMG, De Sousa MV, Grossi de Sa MF, Almeida ERP, Gander ES et al (1999) Expression of a methionine-rich storage albumin from the Brazil nut (Bertholletiaexcelsa H.B.K., Lecythidaceae) in transgenic bean plants (Phaseolus vulgaris L., Fabaceae). Genet Mol Biol 22:445–449. https://doi.org/10.1590/S1415-47571999000300026

Arts ICW, van de Putte B, Hollman PCH (2000) Catechin contents of foods commonly consumed in The Netherlands. 1. Fruits, vegetables, staple foods, and processed foods. J Agric Food Chem 48:1746–1751. https://doi.org/10.1021/jf000025h

Astudillo C, Fernandez AC, Blair MW, Cichy KA (2013) The Phaseolus vulgaris ZIP gene family: identification, characterization, mapping, and gene expression. Front Plant Sci 4:286

Azmah UN, Makeri MU, Bagirei SY, Shehu AB (2023) Compositional characterization of starch, proteins and lipids of long bean, dwarf long bean, mung bean and French bean seed flours. Measurement: Food 12:100111

Bertioli DJ, Cannon SB, Froenicke L, Huang G, Farmer AD, Cannon EK et al (2016) The genome sequences of Arachis duranensis and Arachis ipaensis, the diploid ancestors of cultivated peanut. Nat Genet 48:438–446

Black M, Bewley JD, Hlamer P (2006) The encyclopedia of seeds: science, technology and uses. CAB International, Wallingford, p 828. https://doi.org/10.1093/aob/mcm225

Blair MW, Medina JI, Astudillo C, Rengifo J, Beebe SE, Machado G et al (2010) QTL for seed iron and zinc concentration and content in a Mesoamerican common bean (Phaseolus vulgaris L.) population. Theor Appl Genet 121:1059–1070. https://doi.org/10.1007/s00122-010-1371-0

Bouis HE (2003) Micronutrient fortification of plants through plant breeding: can it improve nutrition in man at low cost? Proc Nutr Soc 62:403–411. https://doi.org/10.1079/pns2003262

Bouis HE, Welch RM (2010) Biofortification - A sustainable agricultural strategy for reducing micronutrient malnutrition in the global south. Crop Sci 50:S20–S32

Casañas F, Pérez-Vega E, Almirall A, Plans M, Sabaté J, Ferreira JJ (2013) Mapping of QTL associated with seed chemical content in a RIL population of common bean (Phaseolus vulgaris L.). Euphytica 192:279–288. https://doi.org/10.1007/s10681-013-0880-8

Celmeli T, Sari H, Canci H, Sari D, Adak A, Eker T et al (2018) The nutritional content of common bean (Phaseolus vulgaris L.) landraces in comparison to modern varieties. Agronomy 8(9):166. https://doi.org/10.3390/agronomy8090166

Chitra U, Vimala V, Singh U, Geervani P (1995) Variability in phytic acid content and protein digestibility of grain legumes. Plant Foods Hum Nutr 47:163–172. https://doi.org/10.1007/bf01089266

Crossa J, Pérez-Rodríguez P, Cuevas J, Montesinos-López O, Jarquín D, de los Campos G, Burgueño J, González-Camacho JM, Pérez-Elizalde S, Beyene Y, Dreisigacker S (2017) Genomic selection in plant breeding: methods, models, and perspectives. Trend Plant Sci 22:961–975

Das Bhowmik SS, Cheng AY, Long H, Tan GZH, Hoang TML, Karbaschi MR et al (2019) Robust genetic transformation system to obtain non-chimeric transgenic chickpea. Front Plant Sci 10:524. https://doi.org/10.3389/fpls.2019.00524

Dhull SB, Kinabo J, Uebersax MA (2023) Nutrient profile and effect of processing methods on the composition and functional properties of lentils (Lens culinaris Medik): a review. Legume Sci 5(1):e156

Di Y-H, Sun X-J, Hu Z, Jiang Q-Y, Song G-H, Zhang B et al (2019) Enhancing the CRISPR/Cas9 system based on multiple GmU6 promoters in soybean. Biochem Biophys Res Commun 519:819–823. https://doi.org/10.1016/j.bbrc.2019.09.074

Diapari M, Sindhu A, Bett K, Deokar A, Warkentin TD, Tar'an, B. (2014) Genetic diversity and association mapping of iron and zinc concentrations in chickpea (Cicer arietinum L.). Genome 57:459–468. https://doi.org/10.1139/gen-2014-0108

Diaz S, Ariza Suarez D, Izquierdo P, Lobaton JD, de la Hoz JF, Acevedo F et al (2020) Genetic mapping for agronomic traits in a MAGIC population of common bean (Phaseolus vulgaris L.) under drought conditions. BMC Genomics 21:1–20. https://doi.org/10.1186/s12864-020-07213-6

Dinkins RD, Reddy MSS, Meurer CA, Yan B, Trick H, Thibaud-Nissen F et al (2001) Increased Sulphur amino acids in soybean plants overexpressing the maize 15 kDazein protein. In Vitro Cell Dev Biol Plant 37:742–747. https://doi.org/10.1007/s11627-001-0123-x

Dissanayaka D (2019) Genome wide association study to identify single nucleotide polymorphism markers for Fe, Zn, and se concentration in field pea seeds. master's thesis. University of Saskatchewan, Saskatoon, SK, Canada

Do PT, Nguyen CX, Bui HT, Tran LT, Stacey G, Gillman JD et al (2019) Demonstration of highly efficient dual gRNA CRISPR/Cas9 editing of the homeologous GmFAD2–1A and GmFAD2–1B genes to yield a high oleic, low linoleic and α-linolenic acid phenotype in soybean. BMC Plant Biol 19:311. https://doi.org/10.1186/s12870-019-1906-8

Dodo HW, Konan KN, Chen FC, Egnin M, Viquez OM (2008) Alleviating peanut allergy using genetic engineering: the silencing of the immunodominant allergen Ara h 2 leads to its significant reduction and a decrease in peanut allergenicity. Plant Biotechnol J 6:135–145. https://doi.org/10.1111/j.1467-7652.2007.00292.x

Dumont E, Fontaine V, Vuylsteker C, Sellier H, Bodèle S, Voedts N et al (2009) Association of sugar content QTL and PQL with physiological traits relevant to frost damage resistance in pea under field and controlled conditions. Theor Appl Genet 118:1561–1571. https://doi.org/10.1007/s00122-009-1004-7

Enneking D, Wink M (2000) Towards the elimination of anti-nutritional factors in grain legumes. In: Linking research and marketing opportunities for pulses in the 21st century: Proceedings of the third international food legumes research conference. Springer Netherlands, Dordrecht, pp 671–683

Foyer CH, Lam HM, Nguyen HT, Siddique KH, Varshney RK, Colmer TD et al (2016) Neglecting legumes has compromised human health and sustainable food production. Nat Plants 2:1–10. https://doi.org/10.1007/978-981-13-0253-4_1

Francis D, Finer JJ, Grotewold E (2017) Challenges and opportunities for improving food quality and nutrition through plant biotechnology. Curr Opin Biotechnol 44:124–129. https://doi.org/10.1016/j.copbio.2016.11.009

Garg M, Sharma N, Sharma S, Kapoor P, Kumar A, Chunduri V et al (2018) Biofortified crops generated by breeding, agronomy, and transgenic approaches are improving lives of millions of people around the world. Front Nutr 5. https://doi.org/10.3389/fnut.2018.00012

Gebremedhin A, Li Y, Shunmugam AS, Sudheesh S, Valipour-Kahrood H, Hayden MJ, Rosewarne GM, Kaur S (2024) Genomic selection for target traits in the Australian lentil breeding program. Front Plant Sci 14:1284781

Gerrano AS, Mbuma NW, Mumm RH (2022) Expression of nutritional traits in vegetable cowpea grown under various south African agro-ecological conditions. Plan Theory 11:1422. https://doi.org/10.3390/plants11111422

Girish TK, Pratape V, Rao UP (2012) Nutrient distribution, phenolic acid composition, antioxidant and alpha-glucosidase inhibitory potentials of black gram (Vigna mungo L.) and its milled by-products. Food Res Int 46:370–377. https://doi.org/10.1016/j.foodres.2011.12.026

González-Pérez S, Arellano JB (2009) Vegetable protein isolates. In: Phillips GO, Williams PA (eds) Handbook of Hydrocolloids, 2nd edn. Woodhead Publishing, Oxford, pp 383–419

Guindon MF, Cazzola F, Palacios T, Gatti I, Bermejo C, Cointry E (2021) Biofortification of pea (Pisum sativum L): a review. J Sci Food Agric 101:3551–3563. https://doi.org/10.1002/jsfa.11059

Halliwell B (2024) Understanding mechanisms of antioxidant action in health and disease. Nat Rev Mol Cell Biol 25(1):13–33

Hane JK, Ming Y, Kamphuis LG, Nelson MN, Garg G, Atkins CA et al (2017) A comprehensive draft genome sequence for lupin (Lupinus angustifolius), an emerging health food: insights into plant–microbe interactions and legume evolution. Plant Biotechnol J 15:318–330. https://doi.org/10.1111/pbi.12615

Heng L, Vincken JP, van Koningsveld G, Legger A, Gruppen H, van Boekel T et al (2006) Bitterness of saponins and their content in dry peas. J Sci Food Agric 86:1225–1231. https://doi.org/10.1002/jsfa.2473

Herridge DF, Peoples MB, Boddey RM (2008) Global inputs of biological nitrogen fixation. Plant Soil 311:1–18. https://doi.org/10.1007/s11104-008-9668-3

Irzykowska L, Wolko B (2004) Interval mapping of QTLs controlling yield related traits and seed protein content in Pisum sativum. J Appl Genet 45:297–306

Jadhav AA, Rayate SJ, Mhase LB, Thudi M, Chitikineni A, Harer PN et al (2015) Marker-trait association study for protein content in chickpea (Cicer arietinum L.). J Genet 94:279–286. https://doi.org/10.1007/s12041-015-0529-6

Jaganathan D, Ramasamy K, Sellamuthu G, Jayabalan S, Venkataraman G (2018) CRISPR for crop improvement: an update review. Front Plant Sci 9:985. https://doi.org/10.3389/fpls.2018.00985

Jarquín D, Kocak K, Posadas L, Hyma K, Jedlicka J, Graef G, Lorenz A (2014) Genotyping by sequencing for genomic prediction in a soybean breeding population. BMC Genomics 15:1–10

Jha AB, Ashokkuma K, Diapari M, Ambrose SJ, Zhang H, Taran B et al (2015) Genetic diversity of folate profiles in seeds of common bean, lentil, chickpea and pea. J Food Compos Anal 42:134–140. https://doi.org/10.1016/j.jfca.2015.03.006

Jiang W, Struik PC, van Keulen H, Zhao M, Jin LN, Stomph TJ (2008) Does increased zinc uptake enhance grain zinc mass concentration in rice? Ann Appl Biol 153:135–147. https://doi.org/10.1111/j.1744-7348.2008.00243.x

Jukanti AK, Gaur PM, Gowda CL, Chibbar RN (2012) Nutritional quality and health benefits of chickpea (Cicer arietinum L.): a review. Br J Nutr 108:11–26

Karaca N, Ates D, Nemli S, Ozkuru E, Yilmaz H, Yagmur B et al (2020) Identification of SNP markers associated with iron and zinc concentrations in CicerSeeds. Curr Genomics 21:212–223. https://doi.org/10.1016/j.ygeno.2019.09.012

Karamac M (2009) Chelation of cu(II), Zn(II), and Fe(II) by tannin constituents of selected edible nuts. Int J Mol Sci 10:5485–5497. https://doi.org/10.3390/ijms10125485

Khandal H, Gupta SK, Dwivedi V, Mandal D, Sharma NK, Vishwakarma NK et al (2020) Root specific expression of chickpea cytokinin oxidase/dehydrogenase 6 leads to enhanced root growth, drought tolerance and yield without compromising nodulation. Plant Biotechnol J 18:2225–2240. https://doi.org/10.1111/pbi.13378

Khazaei H, Podder R, Caron CT, Kundu SS, Diapari M, Vandenberg A, Bett KE (2017) Marker–Trait Association Analysis of Iron and Zinc Concentration in Lentil (Medik.) Seeds. Plant Genome 7:10. https://doi.org/10.3835/plantgenome2017.02.0007

Kim MJ, Kim JK, Kim HJ, Pak JH, Lee JH, Kim DH et al (2012) Genetic modifica-tion of the soybean to enhance the β-carotene content through seed-specific expression. PLoS One 7:e48287. https://doi.org/10.1371/journal.pone.004828

Klein MA, Grusak MA (2009) Identification of nutrient and physical seed trait QTL in the model legume Lotus japonicus. Genome 52:677–691. https://doi.org/10.1139/g09-039

Klein A, Houtin H, Rond-Coissieux C, Naudet-Huart M, Touratier M, Marget P et al (2020) Meta-analysis of QTL reveals the genetic control of yield-related traits and seed protein content in pea. Sci Rep 10:15925. https://doi.org/10.1038/s41598-020-72548-9

Konsam S, Chellapilla B, Ram G, Chellapilla TS, Jain PK (2014) Molecular diversity of chickpea ('Cicer arietinum'L.) genotypes differing in their Raffinose family oligosaccharides viz., raffinose and stachyose content as revealed through SSR markers. Aust J Crop Sci 8(8):1175–1185

Kumar S, Pandey G (2020) Biofortification of pulses and legumes to enhance nutrition. Heliyon 6:e03682. https://doi.org/10.1016/j.heliyon.2020.e03682

Kumar J, Gupta DS, Kumar S, Gupta S, Singh NP (2016) Current knowledge on genetic biofortification in lentil. J Agric Food Chem 64:6383–6396. https://doi.org/10.1021/acs.jafc.6b02171

Kumar RK, Chu HH, Abundis C, Vasques K, Rodriguez DC, Chia JC et al (2017) Iron-nicotianamine transporters are required for proper long distance iron signaling. Plant Physiol 175:1254–1268. https://doi.org/10.1104/pp.17.00821

Liberal Â, Almeida D, Fernandes Â, Pereira C, Ferreira IC, Vivar-Quintana AM, Barros L (2024) Nutritional, chemical, and antioxidant screening of selected varieties of lentils (Lens culinaris spp.) from organic and conventional agriculture. J Sci Food Agric 104(1):104–115

Lin YF, Liang HM, Yan SY, Boch A, Clemens S, Chen CC et al (2009) Arabidopsis IRT3 is a zinc-regulated and plasma membrane localized zinc/iron transporter. New Phytol 182:392–404. https://doi.org/10.1111/j.1469-8137.2009.02766.x

Liu J, Versaw WK, Pumplin N, Gomez SK, Blaylock LA, Harrison MJ (2008) Closely related members of the Medicago truncatula PHT1 phosphate transporter gene family encode phosphate transporters with distinct biochemical activities. J Biol Chem 283:24673–24681. https://doi.org/10.1074/jbc.m802695200

Liu G, Xu H, Zhang L, Zheng Y (2011) Fe binding properties of two soybean (Glycine max L.) LEA4 proteins associated with antioxidant activity. Plant Cell Physiol 52:994–1002. https://doi.org/10.1093/pcp/pcr052

McCouch S, Baute GJ, Bradeen J, Bramel P, Bretting PK, Buckler E et al (2013) Agriculture: feeding the future. Nature 499:23–24

Miao X, Niu H, Sun M, Dong X, Hua M, Su Y et al (2024) A comparative study on the nutritional composition, protein structure and effects on gut microbiota of 5 fermented soybean products (FSPs). Food Res Int 183:114199

Michno JM, Wang X, Liu J, Curtin SJ, Kono TJ, Stupar RM (2015) CRISPR/Cas mutagenesis of soybean and Medicago truncatula using a new web-tool and a modified Cas9 enzyme. GM Crops Food 6:243–252. https://doi.org/10.1080/21645698.2015.1106063

Molvig L, Tabe LM, Eggum BO, Moore AE, Craig S, Spencer D et al (1997) Enhanced methionine levels and increased nutritive value of seeds of trans-genic lupins (Lupinus angustifolius L.) expressing a sunflower seed albumin gene. Proc Natl Acad Sci USA 94:8393–8398. https://doi.org/10.1073/pnas.94.16.8393

Mubarak AE (2005) Nutritional composition and antinutritional factors of Mung bean seeds (Phaseolus aureus) as affected by some home traditional processes. Food Chem 89:489–495. https://doi.org/10.1016/j.foodchem.2004.01.007

Murube E, Beleggia R, Pacetti D, Nartea A, Frascarelli G, Lanzavecchia G et al (2021) Characterization of nutritional quality traits of a common bean Germplasm collection. Foods 10:1572. https://doi.org/10.3390/foods10071572

Noonan SC, Savage GP (1999) Oxalate content of foods and its effect on humans. Asia Pac J Clin Nutr 8:64–74. https://doi.org/10.1046/j.1440-6047.1999.00038.x

Norton GJ, Deacon CM, Xion L, Huang S, Meharg AA, Price AH (2010) Genetic mapping of the rice ionome in leaves and grain: identification of QTLs for 17 elements including arsenic, cadmium, iron and selenium. Plant Soil 329:139–153. https://doi.org/10.1007/s11104-009-0141-8

Ochatt S, Conreux C, Moussa Mcolo R, Despierre G, Magnin-Robert J-B, Raffiot B (2018) Phytosulfokine-alpha, an enhancer of in vitro regeneration competence in recalcitrant legumes. Plant Cell Tissue Organ Cult 135:189–201. https://doi.org/10.1007/s11240-018-1455-0

Palmer CM, Guerinot ML (2009) Facing the challenges of CuFe and Zn homeostasis in plants. Nat Chem Biol 5:333–340. https://doi.org/10.1038/nchembio.166

Palmgren MG, Clemens S, Williams LE, Krämer U, Borg S, Schjørring JK et al (2008) Zinc biofortification of cereals: problems and solutions. Trends Plant Sci 13:464–473. https://doi.org/10.1016/j.tplants.2008.06.005

Pandey MK, Roorkiwal M, Singh VK, Ramalingam A, Kudapa H, Thudi M et al (2016) Emerging genomic tools for legume breeding: current status and future prospects. Front Plant Sci 7:455

Peix A, Ramírez-Bahena MH, Velázquez E, Bedmar EJ (2015) Bacterial associations with legumes. Crit Rev Plant Sci 34(1–3):17–42

Petroski W, Minich DM (2020) Is there such a thing as "anti-nutrients"? A narrative review of perceived problematic plant compounds. Nutrients 12:2929. https://doi.org/10.3390/nu12102929

Pixley KV, Palacios-Rojas N, Glahn RP (2011) The usefulness of iron bioavailability as a target trait for breeding maize (Zea mays L.) with enhanced nutritional value. Field Crop Res 123:153–160. https://doi.org/10.1016/j.fcr.2011.05.011

Pramitha JL, Rana S, Aggarwal PR, Ravikesavan R, Joel AJ, Muthamilarasan M (2021) In: Kumar G (ed) Chapter Three-Diverse role of phytic acid in Plants and Approaches to Develop low-Phytate Grains to Enhance Bioavailability of Micronutrients, vol 107. Elsevier, United States: Academic Press, pp 89–120

Ramamurthy RK, Jedlicka J, Graef GL, Waters BM (2014) Identification of new QTLs for seed mineral, cysteine, and methionine concentrations in soybean [Glycine max (L.) Merr.]. Mol Breed 34:431–445. https://doi.org/10.1007/s11032-014-0045-z

Ray H, Bett K, Ta'ran B, Vandenberg A, Thavarajah D, Warkentin T (2014) Mineral micronutrient content of cultivars of field pea, chickpea, common bean, and lentil grown in Saskatchewan, Canada. Crop Sci 54:1698–1708. https://doi.org/10.2135/cropsci2013.08.0568

Reim V, Rohn S (2015) Characterization of saponins in peas (Pisum sativum L.) by HPTLC coupled to mass spectrometry and a hemolysis assay. Food Res Int 76:3–10. https://doi.org/10.1016/j.foodres.2014.06.043

Roorkiwal M, Pandey S, Thavarajah D, Hemalatha R, Varshney RK (2021) Molecular mechanisms and biochemical pathways for micronutrient acquisition and storage in legumes to support biofortification for nutritional security. Front Plant Sci 12. https://doi.org/10.3389/fpls.2021.682842

Sab S, Lokesha R, Mannur DM, Somasekhar, Jadhav K, Mallikarjuna BP et al (2020) Genome wide SNP discovery and mapping QTLs for seed ironand zinc concentrations in chickpea (Cicer arietinum L.). Front Nutr 7:559120. https://doi.org/10.3389/fnut.2020.559120

Saha T, Shee R, Sahid S, Shee D, Roy C, Sharma R, Pandey A, Paul S, Datta R (2023) Designer grass pea for transgene-free minimal neurotoxin-containing seeds with CRISPR-Cas9. bioRxiv, pp 2023–03

Saltzman A, Birol E, Oparinde A, Andersson MS, Asare-Marfo D, Diressie MT et al (2017) Availability, production, and consumption of crops biofortified by plant breeding: current evidence and future potential. Ann N Y Acad Sci 1390:104–114. https://doi.org/10.1111/nyas.13314

Sánchez-Chino X, Jiménez-Martínez C, Dávila-Ortiz G, Álvarez-González I, Madrigal-Bujaidar E (2015) Nutrient and nonnutrient components of legumes, and its chemopreventive activity: a review. Nutr Cancer 67:401–410. https://doi.org/10.1080/01635581.2015.1004729

Sankaran RP, Huguet T, Grusak MA (2009) Identification of QTL affecting seed mineral concentration and content in the model legume Medicago truncatula. Theor Appl Genet 119:241–253. https://doi.org/10.1007/s00122-009-1033-2

Santi S, Schmidt W (2009) Dissecting iron deficiency-induced proton extrusion in Arabidopsis roots. New Phytol 183:1072–1084. https://doi.org/10.1111/j.1469-8137.2009.02908.x

Sarvamangala C, Gowda MVC, Varshney RK (2011) Identification of quantitative trait locifor protein content, oil content and oil quality in groundnut (Arachis hypogaea. L). Field Crop Res 122:49–59. https://doi.org/10.1016/j.fcr.2011.02.010

Schmidt MA, Parrott WA, Hildebrand DF, Berg RH, Cooksey A, Pendarvis K et al (2015) Transgenic soya bean seeds accumulating β-carotene exhibit the col-lateral enhancements of oleate and protein content traits. Plant Biotechnol J 13:590–600. https://doi.org/10.1111/pbi.12286

Schmutz J, Cannon SB, Schlueter J, Ma J, Mitros T, Nelson W et al (2010) Genome sequence of the palaeopolyploid soybean. Nature 463:178–183

Schmutz J, McClea PE, Mamidi S, Wu GA, Cannon SB, Grimwood J et al (2014) A reference genome for common bean and genome-wide analysis of dual domestications. Nat Genet 46:707–713

Sharif HR, Williams PA, Sharif MK, Abbas S, Majeed H, Masamba KG et al (2018) Current progress in the utilization of native and modified legume proteins as emulsifiers and encapsulants–A review. Food Hydrocoll 76:2–16. https://doi.org/10.1016/j.foodhyd.2017.01.002

Shi L, Arntfield SD, Nickerson M (2018) Changes in levels of phytic acid, lectins and oxalates during soaking and cooking of Canadian pulses. Food Res Int 107:660–668. https://doi.org/10.1016/j.foodres.2018.02.056

Sinclair SA, Krämer U (2012) The zinc homeostasis network of land plants. Biochim Biophys Acta 1823:1553–1567. https://doi.org/10.1016/j.bbamcr.2012.05.016

Singh A, Singh S, Babu JDP (2011) Heritability, character associationvand path analysis studies in early segregating population of field pea (Pisum sativum L. var. arvense). Int J Plant Breed Genet 5:86–92. https://doi.org/10.3923/ijpbg.2011.86.92

Singh UM, Sareen P, Sengar RS, Kumar A (2013) Plant ionomics: a newer approach to study mineral transport and its regulation. Acta Physiol Plant 35:2641–2653. https://doi.org/10.1007/s11738-013-1316-8

Singh A, Sharma V, Dikshit HK, Aski M, Kumar H, Thirunavukkarasu N et al (2017) Association mapping unveils favorable alleles for grain iron and zinc concentrations in lentil (Lens culinaris subsp. culinaris). PLoS One 21:e0188296. https://doi.org/10.1371/journal.pone.018829

Socha AL, Guerinot ML (2014) Mn-euvering manganese: the role of transporter gene family members in manganese uptake and mobilization in plants. Front Plant Sci 5:106

Sodedji FAK, Ryu D, Choi J, Agbahoungba S, Assogbadjo AE, N'Guetta S-PA et al (2022) Genetic diversity and association analysis for carotenoid content among sprouts of cowpea (Vigna unguiculata L. Walp). Int J Mol Sci 23:3696. https://doi.org/10.3390/ijms23073696

Sompong U, Somta P, Raboy V, Srinives P (2012) Mapping of quantitative trait loci for phytic acid and phosphorus contents in seed and seedling of mungbean (Vigna radiata (L.) Wilczek). Breed Sci 62:87–92. https://doi.org/10.1270/jsbbs.62.87

Song S, Hou W, Godo I, Wu C, Yu Y, Matityahu I et al (2013) Soybean seeds expressing feedback-insensitive cystathionine -synthase exhibit a higher content of methionine. J Exp Bot 64:1917–1926. https://doi.org/10.1093/jxb/ert053

Stagnari F, Maggio A, Galieni A, Pisante M (2017) Multiple benefits of legumes for agriculture sustainability: an overview. Chem Biol Technol Agric 4(1):2

Tan GZH, Das Bhowmik SS, Hoang TML, Karbaschi MR, Long H, Cheng A et al (2018) Investigation of baseline iron levels in Australian chickpea and evaluation of a transgenic biofortification approach. Front Plant Sci 9. https://doi.org/10.3389/fpls.2018.00788

Tang G, Galili G (2004) Using RNAi to improve plant nutritional value: from mechanism to application. Trends Biotechnol 22:463–469. https://doi.org/10.1016/j.tibtech.2004.07.009

Tang F, Yang S, Liu J, Zhu H (2016) Rj4, a gene controlling nodulation specificity in soybeans, encodes a thaumatin-Like protein but not the onepreviously reported. Plant Physiol 170:26–32. https://doi.org/10.1104/pp.15.01661

Tania A, Akram A, Hanif NQ, Ajmal M, Seerat W, Nijabat A, Mehak A (2023) Proximate composition, fungal isolation and contamination of aflatoxin B1 in chickpea seeds from the Punjab, Pakistan. Nat Product Res 37(19):3314–3322

Upadhyaya HD, Bajaj D, Das S, Kumar V, Gowda CL, Sharma S et al (2016) Genetic dissection of seed-iron and zinc concentrations in chickpea. Sci Rep 6:24050

Varshney RK, Chen W, Li Y, Bharti AK, Saxena RK, Schlueter JA et al (2012) Draft genome sequence of pigeonpea (Cajanus cajan), an orphan legume crop of resource-poor farmers. Nat Biotechnol 30:83–89. https://doi.org/10.1038/nbt.2022

Varshney RK, Song C, Saxena RK, Azam S, Yu S, Sharpe AG et al (2013) Draft genome sequence of chickpea (Cicer arietinum) provides a resource for trait improvement. Nat Biotechnol 31:240–246

Varshney RK, Thudi M, Pandey MK, Tardieu F, Ojiewo C, Vadez V, Whitbread AM, Siddique KH, Nguyen HT, Carberry PS, Bergvinson D (2018) Accelerating genetic gains in legumes for the development of prosperous smallholder agriculture: integrating genomics, phenotyping, systems modelling and agronomy. J Exp Bot 69(13):3293–3312

Velu G, Singh RP, Huerta-Espino J, Peña RJ, Arun B, Mahendru-Singh A et al (2012) Performance of biofortified spring wheat genotypes in target environments for grain zinc and iron concentrations. Field Crop Res 137:261–267. https://doi.org/10.1016/j.fcr.2012.07.018

Vollmann J (2016) Soybean versus other food grain legumes: a critical appraisal of the United Nations International Year of Pulses 2016. J Land Manage Food Environ 67(1):17–24. https://doi.org/10.1515/boku-2016-0002

Wang J, Kuang H, Zhang Z, Yang Y, Yan L, Zhang M et al (2020) Generation of seed lipoxygenase-free soybean using CRISPR-Cas9. Crop J 8:432–439. https://doi.org/10.1016/j.cj.2019.08.008

Wu J, Zhou Q, Zhou C, Cheng KW, Wang M (2024) Strategies to promote the dietary use of pigeon pea (Cajanus cajan L.) for human nutrition and health. Food Frontiers

Xia J, Yamaji N, Kasai T, Ma JF (2010) Plasma membrane-localized transporter for aluminum in rice. Proc Natl Acad Sci USA 107:18381–18385. https://doi.org/10.1073/pnas.1004949107

Xiong H, Kobayashi T, Kakei Y, Senoura T, Nakazono M, Takahashi H et al (2012) AhNRAMP1 iron transporter is involved in iron acquisition in peanut. J Exp Bot 63:4437–4446. https://doi.org/10.1093/jxb/ers117

Zhang P, Du H, Wang J, Pu Y, Yang C, Yan R et al (2020) Multiplex CRISPR/Cas9-mediated metabolic engineer-ing increases soya bean isoflavone content and resistance to soya bean mosaic virus. Plant Biotechnol J 18:1384–1395. https://doi.org/10.1111/pbi.13302

Zhawar VK, Kaur N, Gupta AK (2011) Phytic acid and raffinose series oligosaccharides metabolism in developing chickpea seeds. Physiol Mol Biol Plants 17:355–362. https://doi.org/10.1007/s12298-011-0080-8

Quality Seed Production in Legumes

P. K. Katiyar, Abhishek Bohra, Esha Jaiswal, S. J. Satheesh Naik, and Girish Prasad Dixit

Abstract

Pulses are affordable sources of protein and minerals, with the pulse-based systems having the least environmental footprint. Pulses are vital to safeguard the nutritional security and livelihood generation of the marginalized section of the community. The productivity and sustainability of crop production system rely heavily on the timely supply of the quality seed of the improved varieties. Continuous replacement of old varieties in farmers' fields with the latest one expands climate adaptation capacity of the production systems. Unavailability and poor access to the quality seed of pulses has been a land-standing bottleneck in realizing the productivity gains in these crops. However, recent efforts on improving the production of quality seed of pulses in India have been remarkable. The synergies among different Indian Council of Agricultural Research (ICAR) institutes, State Agricultural Universities (SAUs), National Seeds Corporation (NSC), State Seeds Corporations (SSCs), and other stakeholders with adequate support from government policies have led to a significant increase in production and accessibility of the quality seed of pulse crops. The improved rate of seed and varietal replacement resulting from these new initiatives is contributing to overcome the gaps in crop yields between experimental plots and farmers' fields. This chapter discusses the latest trends and prospects of quality seed production of pulse crops in India.

Keywords

Pulse · Breeder seed · Genetic gain · Varietal replacement · Seed replacement · Yield

P. K. Katiyar (✉) · A. Bohra · E. Jaiswal · S. J. Satheesh Naik · G. P. Dixit
ICAR-Indian Institute of Pulses Research, Kanpur, India

1 Introduction

Pulses being an affordable source of plant-based dietary protein and calories for vegetarian population remain vital to food and nutritional security worldwide (Bohra et al. 2015). These crops are usually grown with minimum input in stress-prone areas and marginal lands. Owing to their unique property of fixing atmospheric nitrogen symbiotically in their root nodules and improving soil physical property, these crops contribute significantly to improving soil health and maintaining soil fertility. Besides the higher protein content of pulses, the rich amino-acid profile of these crops helps them to complement cereal-based diets, thus leading to superior biological value than either of the components alone. Therefore, pulses have a profound significance in alleviating protein hunger and malnutrition that is prevalent among the poorer section of the society, especially in the developing and underdeveloped countries (Bohra et al. 2014). India's rising population imposes high pressure on land and other nature reserves to meet its nutritional and development needs. Inevitably, the natural resource inventory of land, water, and bio-diversity faces the threat of depletion. Challenges ahead for meeting food demands are formidable considering the non-availability of propitious factors of growth, plummeting factor productivity in key cropping systems, and precipitously diminishing resource base (Singh et al. 2020). A more inclusive, comprehensive, and sustainable green revolution is the need of the hour which can reinforce and strengthen crop production in our nation without exacerbating the depletion of our natural resources.

Indian agriculture has improved remarkably post-independence, transforming the country from being a food-scarce to food-exporting country (Pathak et al. 2022). Since the foundation of solid agricultural systems rests on the availability of quality seed, the Indian seed industry has emerged as a key player in making available quality seed of improved varieties and taking Indian agriculture to new heights (Parihar and Dixit 2016). Farmers must have access to a sufficient supply of quality planting materials at a rea-

G. Prasad Dixit et al. (eds.), *Fundamentals of Legume Breeding*, https://doi.org/10.1007/978-981-96-7838-9_16

sonable price and at the right time if they are to effect significant changes in the agricultural landscape of our nation. Notably, this is still one of the biggest issues facing the seed industry today. Sustained production is heavily reliant on the presence of a robust seed supply chain which ensures the availability of improved seeds to the farmers (Louwaars and Manicad 2022). A plethora of Indian farmers are reliant on local sources for procuring seed for crop production as they lack access to quality seed (Pattanaik 2013). The private seed sector is in need of reinforcement to fulfill the seed requirement of pulse crops.

Seed is essential for crop production, environmental protection, and food security (Chauhan et al. 2015; Varshney et al. 2021). Desired growth rates in agriculture can only be achieved if the input of agriculture, i.e., "seed," gets its rightful paramountcy. Seed offers to integrate production, protection, and quality enhancement technologies in a single entity, in a cost-effective way. Seed security is having a strong positive correlation to food security (Nabuuma et al. 2022), and this relation has been clearly evident from increased food grain production vis-à-vis growth in quality seed production. In context to significant advances registered in Indian Agriculture, the progress in the seed sector has been phenomenal with contributions from researchers, policymakers, governmental agencies, and private companies (Vijay et al. 2023). In India, three generation system of downstream seed multiplication system is followed encompassing breeder seed which is multiplied into foundation and certified seed in subsequent generations. Downstream seed production chain is entirely dependent on the regular supply of seed stock which is genetically pure, i.e., breeder seed, which lays down the foundation for all subsequent generations. To ascertain the accessibility of quality seed of superior varieties to farming community, it is imminent for crop breeding institutes, viz., ICAR institutes and State Agricultural Universities (SAUs), to produce breeder seed exhibiting the highest genetic purity.

2 Generation System of Seed Multiplication

As shown in Fig. 1, India practices three generation system of seed multiplication. Breeder seed is the progeny of nucleus seed and exclusively produced by the originating plant breeder or crop breeding institute or a sponsored institute, while the production of foundation and certified seed is primarily executed by National Seeds Corporation (NSC), state seeds corporations (SSCs), various seed cooperatives, NGOs, private seed companies, farmers produce organizations, etc. (Vijay et al. 2023).

2.1 Seed Supply System in India

With the aim to produce seeds of highest quality, very strong linkages need to exist between various stakeholders of seed producers and supply chain in India. Indian Council of Agricultural Research (ICAR) along with various SAUs under NARES system shoulders the responsibility of breeder seed production (Agarwal et al. 2020; Vijay et al. 2023), while National Seeds Corporation (NSC), State Seeds Corporations (SSCs), and other stakeholders of the seed industry are involved in the production and distribution of foundation as well as certified seed at large (Fig. 2).

2.2 Breeder Seed Production

Production of breeder seed in adequate quantity is necessary to establish an effective seed supply chain. The production of breeder seed is demand-driven and is executed as per the indent of the Department of Agriculture and Farmers' Welfare, Government of India. The responsibility of breeder seed production is entrusted to the National Agriculture Research System (NARS) which comprises ICAR institutes, Central / State Agricultural Universities, along with other public sector entities. This seed is subsequently utilized by stakeholders in the public and private domains to produce foundation, certified, and truthfully labeled seed to sustain an efficient seed chain that ensures farmers have timely access to quality seeds. Breeder seed indent for pulses has remained very inconsistent; for instance, the indent of breeder seed in 2016–17, 2017–18, 2018–19, 2019–20, 2020–21, 2021–22, and 2022–23 was 17670.5 q, 18595.6q, 16470.4 q, 15284.5 q, 15448.26 q, 11700.6 q, and 9550.73 q, respectively (Fig. 3). Producing sufficient quantity of breeder seed is vital for making available a sufficient quantity of quality seed of pulse crops and bolstering the seed chain. The production of breeder seed has consistently remained higher than the respective indents with a record production of 263998q during 2017–2018 (Fig. 4). Individual crop-wise indent and production in case of breeder seed are shown in Table 1.

2.3 Quality Seed

Limited availability of quality seeds has remained one of the principal obstacles in enhancing the productivity of pulse crops (Gowda et al. 2013; Varshney et al. 2018). The requirement of quality seed of major pulse crops as per 100% seed replacement rate (SRR) and actual SRR is mentioned in Table 2.

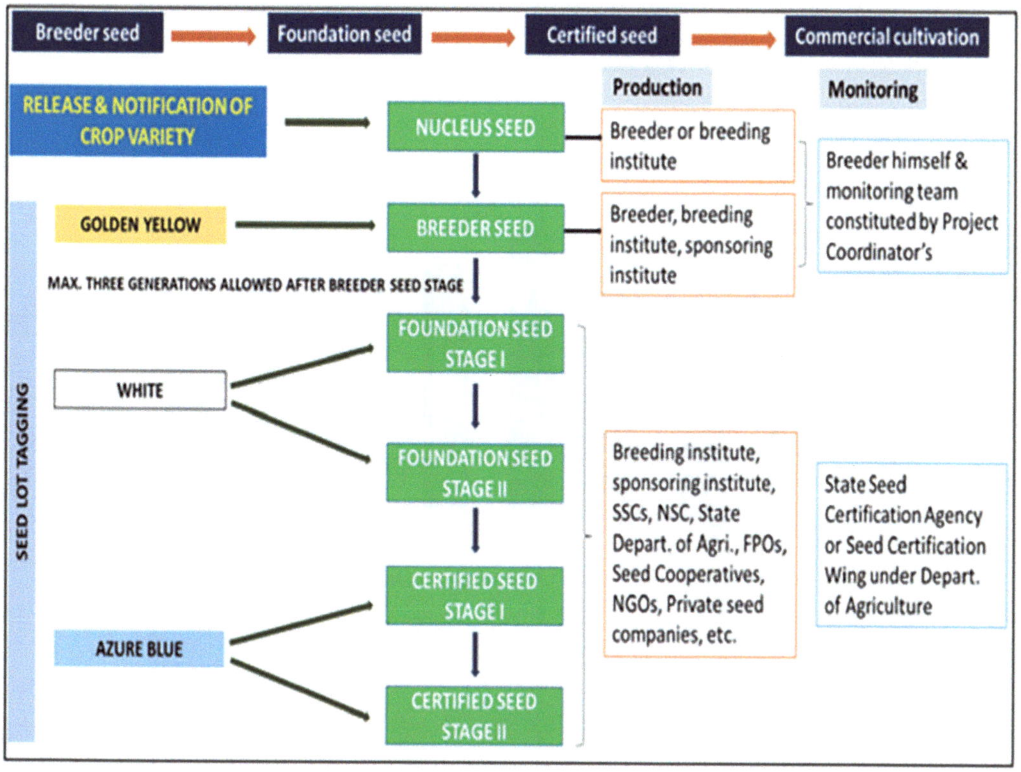

Fig. 1 Seed production and monitoring system in India

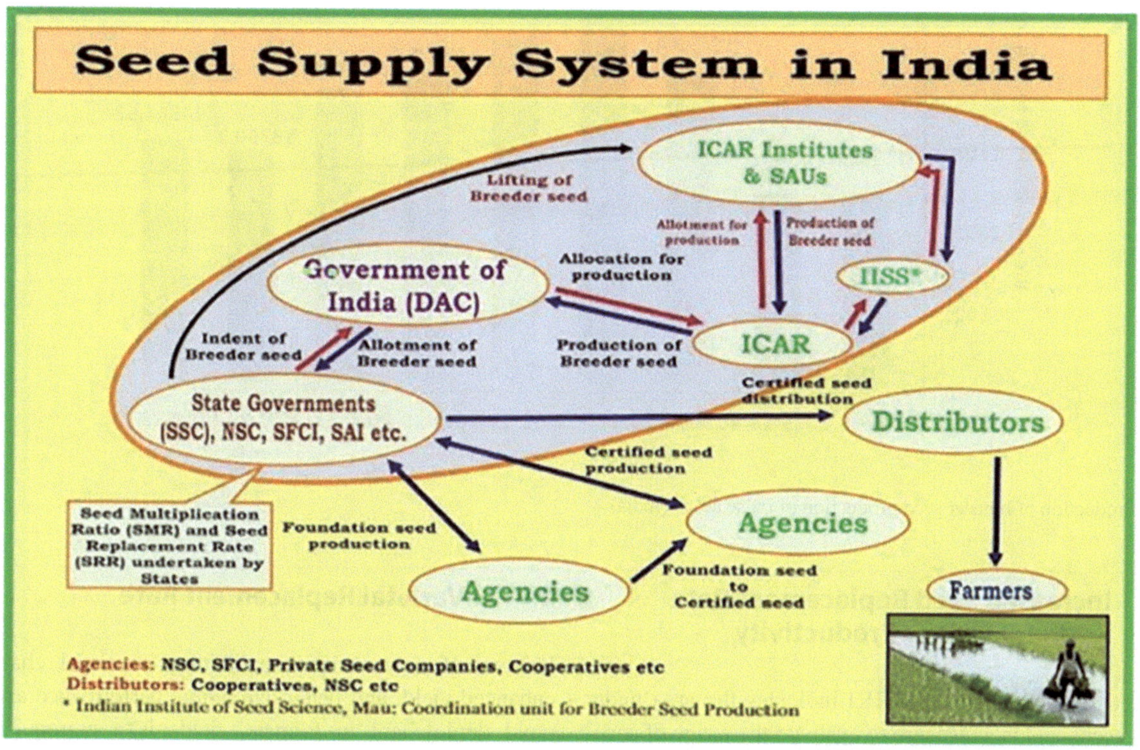

Fig. 2 Seed supply system in India. (Source: www.seednet.gov.in. Ministry of Agriculture & Farmers Welfare, GOI)

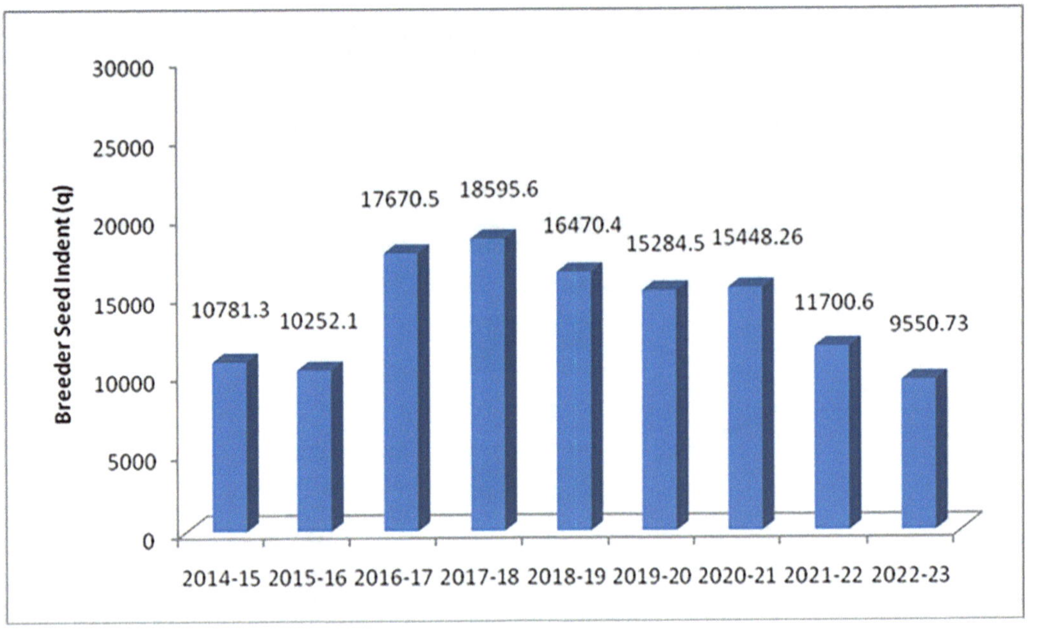

Fig. 3 Indent of breeder seed production of pulse crops (India)

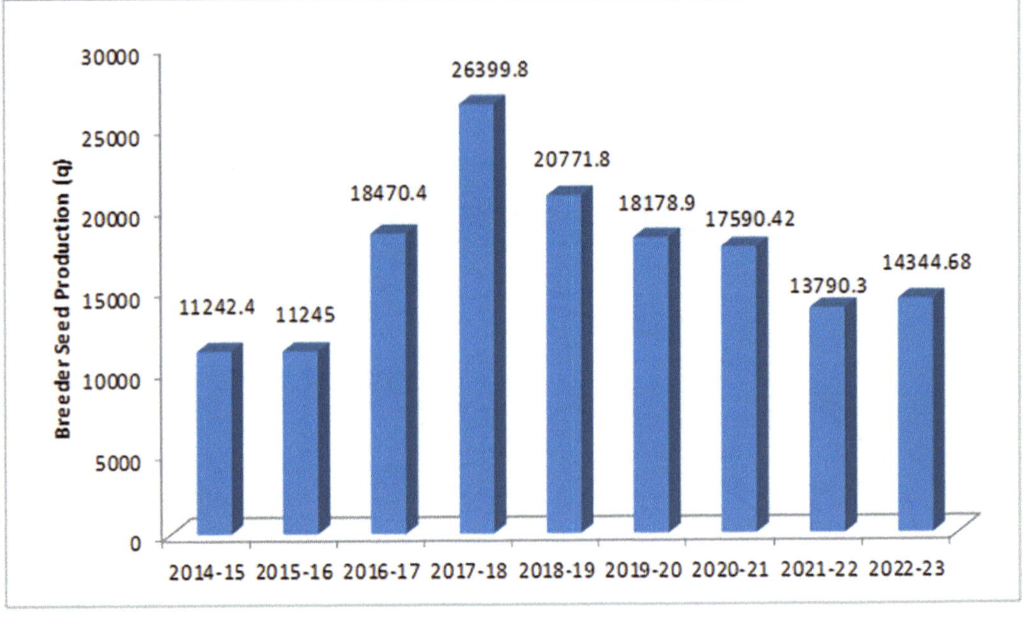

Fig. 4 Production of breeder seed production of pulse crops (India)

3 Increased Seed Replacement Rate (SRR) for Higher Productivity

The seed replacement rate (SRR) indicates the area under quality seeds and is a measure to evaluate the usage of certified and/or quality seed in a crop (Chauhan et al. 2016). Improving SRR contributes to enhanced productivity of the crop. As shown in Fig. 5, the SRR of all pulse crops demonstrated an increasing trend during the last decade.

4 Varietal Replacement Rate

Release of new varieties exhibiting improved characters, enhanced yield, and better agronomic performance and their rapid deployment in farmers' fields is a prerequisite for increasing crop production and climate adaptation of farming systems (Varshney et al. 2021; Bohra et al. 2020; Atlin et al. 2017). Further, the inclusion of new improved varieties in seed chain and its rapid conversion from breeder seed to certified

Table 1 Breeder seed production of pulse crops (India)

Crops	2016–17		2017–18		2018–19		2019–20		2020–21		2021–22		2022–23	
	Indent	Prod.	Indent	Prod.	Indent	Prod.	Indent	Prod.	Indent	Prod.	Indent	Prod.	Indent	Prod.
Chickpea	12477.8	12396.3	13537.6	17061.1	11795.6	14162.8	11686.9	13291.6	11396.30	12972.08	8379.92	9515.69	6809.59	10524.63
Pigeonpea	1257.0	1531.4	1464.1	5123.7	661.3	942.3	302.7	954.2	376.19	513.22	423.37	724.76	255.52	388.36
Field pea	959.0	1127.1	735.6	787.2	719.3	1079.1	846.1	1058.5	846.10	1058.46	805.69	637.73	670.13	869.22
Mung	1254.1	1495.6	1406.8	1534.4	1372.2	1670.0	808.5	935.6	736.35	872.68	712.82	1309.71	458.30	634.91
Urd	816.4	773.8	624.7	862.8	772.2	1362.4	397.2	528.9	816.84	748.07	553.42	735.08	445.06	659.83
Lentil	565.4	648.2	561.2	677.3	673.4	972.1	813.2	1036.7	813.19	1036.71	584.14	598.98	622.90	905.36
Rajmash	44.9	69.8	33.1	35.6	21.7	28.2	23.7	35.9	4.35	5.66	16.05	23.72	15.98	27.18
Horse gram	34.1	33.6	26.2	30.1	14.4	21.2	21.6	29.7	22.20	25.87	13.25	17.98	20.84	23.78
Indian bean	12.7	13.0	1.0	2.7	10.0	15.5	5.0	5.4	1.80	3.50	3.8	5.53	4.15	6.25
Cowpea	67.6	135.8	27.0	57.0	68.8	97.8	44.6	66.2	34.10	40.03	26.3	32.79	17.44	43.95
Lathyrus	114.6	166.0	146.5	178.8	165.3	170.3	198.5	57.2	198.50	57.22	158.5	131.1	155.19	117.20
Moth Bean	62.2	66.3	26.3	43.5	30.7	39.3	16.5	22.3	38.15	42.24	20.9	41.11	15.00	28.60
Guar Cluster bean	4.7	13.7	5.6	5.6	165.6	210.9	120.2	156.8	164.19	214.68	2.4	16.16	60.63	115.41
Total	17670.5	18470.4	18595.6	26399.8	16470.4	20771.8	15284.5	18178.9	15448.26	17590.42	11700.6	13790.3	9550.73	14344.68

Table 2 Requirement of quality seed of major pulse crops

	Seed requirement @ 100% SRR			Seed Requirement @ actual SRR		
Crops	Certified seed (000′ q)	Foundation seed (000′ q)	BS (q)	Certified seed (000′ q)	Foundation seed (000′ q)	BS (q)
Chickpea	4437.0	295.8	19720.0	1372.8	91.5	6101.4
Pigeonpea	339.0	3.4	33.9	207.0	2.1	20.7
Mung	459.0	11.5	286.9	166.9	4.2	104.3
Urd	422.5	10.6	264.1	137.9	3.4	86.2
Lentil	367.5	12.3	408.3	148.8	5.0	165.3
Total	**6025.0**	**333.6**	**20713.2**	**2033.4**	**106.2**	**6477.9**

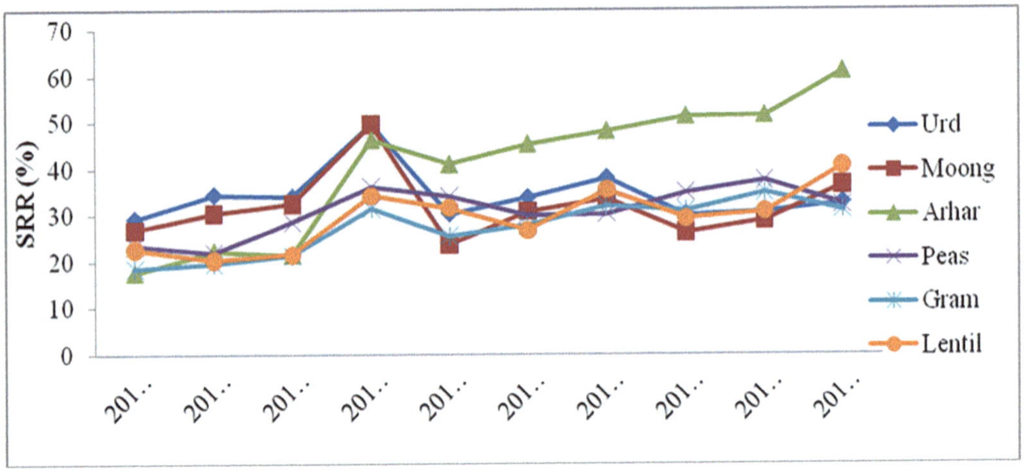

Fig. 5 The trend in SRR of the major pulse crops in India

seed are crucial for improving varietal replacement rate and seed replacement rate (Chauhan et al. 2016; Singh et al. 2020). Research evidence supports that accelerating the rate of genetic gain relies on VRR (Spielman and Smale 2017). Tables 3 and 4 depict the percent share of newly released varieties (<5 and <10 years) toward the total breeder seed indent.

In general, the share of newly released varieties has increased considerably over the years. For instance, in pigeonpea, chickpea, mungbean, urd bean, field pea, and lentil, the share of <5 years old varieties increased from 7.2% to 39.4%, 5.1% to 49.4%, 9.5% to 70.0%, 4.0% to 46.2%, 4.5% to 59.8%, and 4.7% to 76.7%, respectively, from 2016–17 to 2022–23 (Table 3). Likewise, the share of <10 years old varieties in pigeonpea, chickpea, mungbean, urd bean, field pea, and lentil has increased from 43% to 75.4%, 56.9% to 85.4%, 57.8% to 95.6%, 70.2% to 79.4%, 19.4% to 93.5%, 44.9% to 96.4%, respectively over the same period of time (Table 4).

5 Enhancing Breeder Seed Production

Realizing the significance of quality seed in enhancing indigenous pulse production and ensuring self-sufficiency in the country, the Government of India conceived and executed a noble program by strengthening/reinforcing the existing breeder seed production program in India. Aimed at scaling up quality seed supply in major pulse crop growing regions of the country, the Department of Agriculture and Farmers' Welfare (DA&FW), Government of India, initiated a mega program on "Enhancing breeder seed production for increasing indigenous production of pulses in India" during 2016 (https://www.nfsm.gov.in/ICAR.aspx). Since the commencement of this program in 2016, substantial quantity of additional breeder seed (21713.3 q) of different pulse crops was produced. During 2016–2023, the quantity of additional breeder seed produced by different national agricultural research centers across the country in different pulse crops ranged from 2087.5 q (fieldpea) to 9318.7 q (chickpea) (Fig. 6). This has resulted in raising substantially quality seed supply in the country to boost both productivity and production in pulses, which in turn has made the country self-reliance in the pulses seed sector.

6 Establishing Seed Hubs to Improve Accessibility and Availability of Quality Seed

Unavailability and poor access to quality seed have remained a crucial bottleneck in overcoming the gaps in crop yields at experimental settings and yields obtained at farmers' fields (Wimalasekera 2015). It is demonstrated

Table 3 Percent share of varieties <5 years old in total breeder seed indents

Crop	2016–17	2017–18	2018–19	2019–20	2020–21	2021–22	2022–23
Pigeonpea	7.2	6.7	3.5	4.5	22.0	26.8	39.4
Chickpea	5.1	9.2	41.4	34.5	50.3	37.7	49.4
Mung	9.5	1.6	21.1	28.0	46.4	59.2	70.0
Urd	4.0	7.1	6.3	9.1	17.2	25.9	46.2
Fieldpea	4.5	7.2	24.1	26.9	54.8	53.3	59.8
Lentil	4.7	5.6	20.4	24.6	70.2	64.4	76.7

Table 4 Percent share of varieties <10 years old in total breeder seed indents

Crop	2016–17	2017–18	2018–19	2019–20	2020–21	2021–22	2022–23
Pigeonpea	43.0	36.6	36.5	44.1	53.5	66.7	75.4
Chickpea	56.9	55.7	54.5	70.2	81.7	80.0	85.4
Mung	57.8	53.3	68.8	71.2	79.5	89.4	95.6
Urd	70.2	65.6	32.9	40.6	45.2	61.3	79.4
Fieldpea	19.4	46.4	62.0	64.6	85.5	85.4	93.5
Lentil	44.9	48.5	53.8	69.6	98.9	90.8	96.4

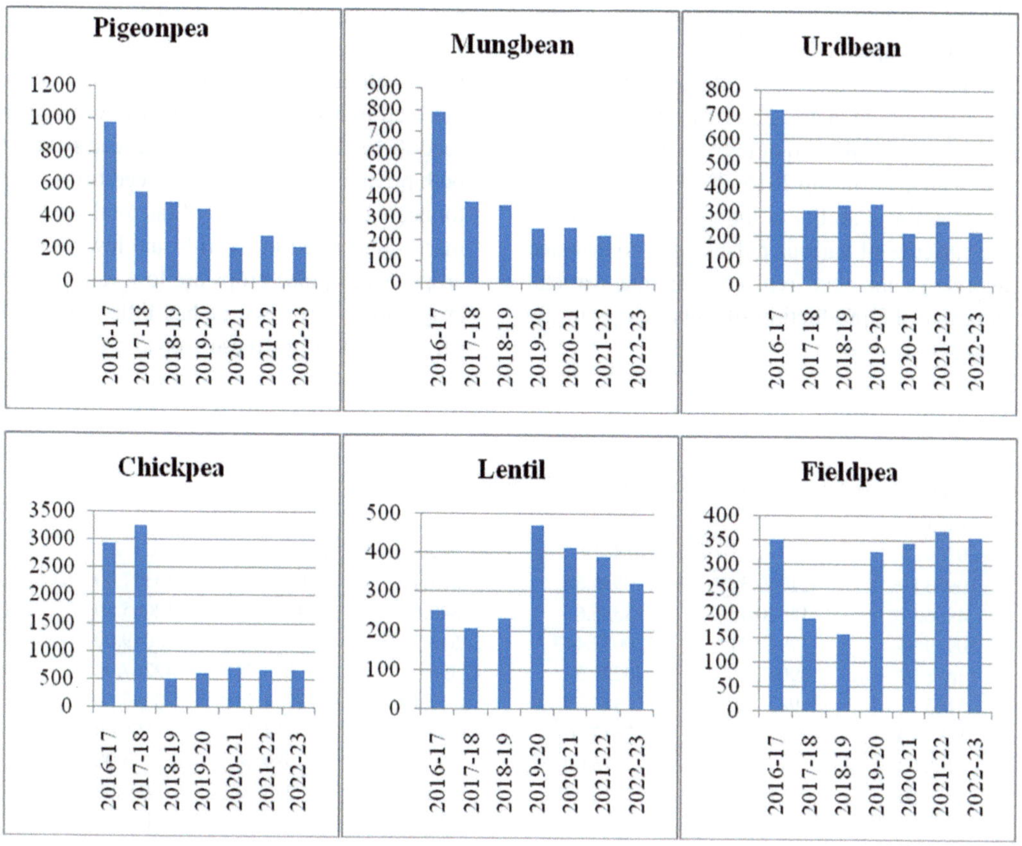

Fig. 6 Production of additional breeder seed during 2016–2023

that pulse productivity can be increased many-folds by simply making available the quality seeds of right crop/variety at right time/place. Analyzing the significance of quality seeds in agricultural production, the Department of Agriculture and Farmers' Welfare (DA&FW), Government of India, has taken a novel initiative to ensure quality seed availability locally through the creation of seed hubs through a dedicated project "Creation of seed hubs for increasing indigenous production of pulses in India" with a total outlay of INR 2253.108 million (https://www.nfsm.gov.in/ICAR.aspx). The project is aimed at establishing 150 seed hubs located mainly in ICAR Institutes/State Agricultural Universities/Krishi Vigyan Kendras (Agriculture Science Centers) belonging to non-governmental organizations and public-funded institutions in different pulses-producing areas covering 24 states of the country.

A total of 52421.6 q, 111,801 q, 123784.3 q, 109614.7 q, 110706.5 q, 988,949 q, and 102156.3 q of quality seed of various pulses were produced through the years 2016–17, 2017–18, 2018–19, 2019–20, 2020–21, 2021–22, and 2022–23, respectively (Table 5). Overall, quality seed to the tune of 110190.18q of mungbean, 92179.64q of urd bean, 126228.68q of pigeonpea, 266020.29q of chickpea, 61209.96q of lentil, 35398.09q of fieldpea, 2452.86q of mothbean, 3589.96q of cowpea, 2266.4q of horsegram, 8211.36q of rajmash, and 1586.11q of lathyrus was produced through different seed hubs during 2016–2023. The increased production and greater availability of the quality seed of different improved varieties of various pulses have contributed to a significant rise in the seed replacement rate, varietal replacement rate, and the production of pulse crops in the nation.

7 Opportunities for Seed System in India

The Indian agriculture offers numerous competitive advantages which are as follows:

(a) Large irrigated farmlands
(b) Ideal agro-climatic zones
(c) Gap between current productivity and potential productivity
(d) Accessibility to the skilled, educated, and technical and scientific workforce.

Indian agriculture requires market access, technological advancement, and policy intervention in order to capitalize and establish competitive advantage in the global market through:

1. *Public sector:* The Indian seed business benefits greatly from the extensive network of public sector institutions, such as NSC, SSCs, SCAs, STLs, SAUs, and ICAR institutes, which are involved in seed production, maintenance, and quality assessment.
2. *Private Sector:* The presence of over 600 seed companies, including multinational corporations, operating in the seed industry is yet another advantage for the Indian seed sector. India boasts a strong seed system that includes both private and public sector seed firms. The private sector accounting for 58.8% of the total seed available is primarily interested in hybrids and low-volume, high-value crops. Since pulse crops are low-value and large-volume crops, the private sector has little contribution over the seeds produced for these crops.

Table 5 Quantity of quality seed produced under seed hub during 2016–2023

	Production (q)						
Crop	2016–17	2017–18	2018–19	2019–20	2020–21	2021–22	2022–23
Mungbean	10953.80	21050.34	20535.62	13561.41	16117.51	13337.00	14634.50
Urdbean	4956.92	15520.42	18336.67	12840.46	14875.27	12665.50	12984.40
Pigeonpea	10919.94	17927.21	20032.44	24848.52	22634.21	13649.00	16217.36
Chickpea	17628.97	39057.56	40225.57	42336.01	41658.78	42267.00	42846.40
Lentil	5165.91	9821.20	12884.78	8717.44	7754.53	8946.50	7919.60
Field pea	1687.50	4825.99	7517.70	4662.64	5712.2	5655.00	5337.06
Mothbean	150.00	475.00	615.00	86.86	389.00	325.00	412.00
Cowpea	379.00	1315.50	1014.27	190.19	278.00	196.00	217.00
Horsegram	64.40	425.00	1015.00	192.00	152.00	169.00	249.00
Rajmash	320.26	851.10	1257.26	2002.74	1004.00	1535.00	1241.00
Lathyrus	194.91	531.70	350.00	176.50	131.00	104.00	98.00
Total	**52421.61**	**111801.02**	**123784.31**	**109614.77**	**110706.5**	**98849.00**	**102156.32**

3. *Public-Private Partnership:* In India, there is an open, transparent, and reliable public-private partnership (PPP) that has greatly advanced the seed sector and will continue to do so by exploring and capturing new horizons. The liberal policies of the government and the ideal atmosphere offered to private seed firms have resulted in the establishment of trust, which has been propagated effortlessly. Licensing crop varieties to private seed businesses and to small-scale entrepreneurs who are essentially novices, for nominal rates and with easily agreeable terms and conditions, has strengthened this partnership further. More than 300 small and large seed companies have been drawn to associate with the various ICAR Institutes and SAUs as a result, and this has proved to be highly beneficial in quickly distributing quality seeds of recently released varieties.

For instance, IARI has non-exclusive agreements in place for the production of seeds for various IARI-released field and vegetable crop varieties with over 200 private seed companies and one non-governmental organization. With funding from the National Agricultural Innovation Project, the Business Planning and Development (BPD) unit of IARI promotes Public-Private Partnerships (PPP) and provides industries and entrepreneurs with technological know-hows, consulting services, project planning, and business expansion training. Start-up companies and aspiring entrepreneurs can make use of BPD's assistance to launch their agricultural businesses. They are registered as Corporate Members of IARI-BPD, and BPD offers them a moderately equipped office space with internet access, meeting rooms, and sitting areas and working laboratories. In addition, support from laboratories and instrumentation facilities spread across many divisions, as well as mentorship from IARI scientists are offered. This has aided in the formation of numerous alliances with various private businesses in order to commercialize its technology. In order to facilitate the fastest possible distribution of new technology to farmers, IARI has also established BEEJ INDIA, a farmers' producer company.

8 Desirable Goals for Self-Sufficiency in Seed Systems

8.1 Use of Molecular Tools for Varietal Identity and Genetic Purity

All seed testing labs in India operate under the procedures and protocols established by ISTA for varietal identification and genetic purity. The Grow-Out Test and other conventional tests, which are among the older methods for determining varietal purity, are influenced by the environment, take a lot of time, and are often not very effective. Molecular markers have emerged as an indispensable tool for the majority of the crop improvement programs. Their use for the establishment of genetic purity and varietal identification must be promoted and popularized.

8.2 Farmers' Participatory Seed Production

To ensure timely and easy availability of quality seed to farmers, the farmers' participatory seed production needs to be reinforced. It has to be scientifically backed in terms of contaminants, admixtures, and seed production protocols during the seed production process and further advancement of the planting material to the subsequent generation in the production process.

8.3 Poor Conversion Ratio

Meeting the demand for breeder seeds of different varieties was once challenging. However, with the help of various initiatives, such as the Mega Seed Project, the production of breeder seed as per the indented quantity is carried out successfully with very few exceptions. It has been noted in many crops that the breeder seed indents are so exceptionally high that only a fraction of that can cover the whole area of a particular crop if it is actually converted to foundation and certified seeds. The breeder seed-producing centers and the breeders are burdened further when breeder seed is over-indented and used directly without being added to the seed chain. Maintaining the availability of high-quality breeder seed for various crops requires management over this procedure.

8.4 Public Sector Seed Corporations

State Farms Corporation of India which is currently merged with National Seed Corporation has access to a huge land area and decent infrastructure. Despite this, their performance has remained below satisfactory. Strengthening the public sector seed corporations can be a great asset to the seed industry. Once the public seed sector is reinforced, it can increase its share in the seed production program and can even overtake the private sector in the long run. Few farms like those at Suratgarh, Sardargarh, and Jetsar in Sriganganagar district of Rajasthan which are spread out over 16,000 ha can prove to be a great asset to the seed production program.

8.5 Awareness on Seed Trade

Owing to the limited availability and accessibility of quality seeds of new varieties in particular states/region, there lies ample opportunity for the seed companies to invest in quality seed production of pulses crops.

8.6 Reaching the Unreached

Seed corporations and seed certification agencies are not present in seven northeastern states of India. Additionally, vast expanses of rice fallows lying vacant can be exploited for the production of *rabi* pulses once quality seeds are made available to these states. To significantly alter the pulse production scenario of our country, additional efforts must be put into producing seeds of different varieties under various pulse crops. Self-help groups, voluntary organizations, and small farmers can take initiatives to make quality seed available to remote or untapped areas as well help small-scale entrepreneurs.

8.7 Climate Smart Seed Production and Storage Facilities

Erratic monsoons and fluctuating temperatures are an impediment to the production of seeds. Temperature swings that occur suddenly throughout the crop season in traditional seed-producing regions for different crops are a severe concern that should serve as a reminder to look for safe alternative locations for the production of climate-smart seed production. Similarly, climate-smart storage facilities are necessary to maintain the quality of the seeds.

9 Conclusion

Ensuring seed security in pulses is crucial for sustainable agricultural systems and food security. Initiatives that focus on improving the availability and accessibility of quality seed for pulses in India have been ensured through Pulses Seed Hubs. This program aims to digitalize seed hubs and enhance breeder seed production data for pulses. The goal is to provide farmers with access to quality seed production information. A web-based information system has been developed to analyze and retrieve data related to seed hub centers, variety-wise production, seed lifting, and expenditure. Agricultural growth, especially crop production, plays a significant role in achieving food security. Enhancing seed availability improves smallholder farmers' access to improved seed varieties and encourages adoption. Seeds are the fundamental input for sustainable agriculture. The quality of seeds significantly impacts crop production, with direct contributions estimated at 15–20% (and potentially up to 45% with efficient management of other inputs). Over the last 30 years, India has witnessed significant developments in its seed industry.

Key milestones include the National Seed Project Phases I, II, and III, which restructured the seed infrastructure, and the New Seed Development Policy (1988–1989), which opened access to high-quality seeds for Indian farmers. The policy changes encouraged private sector investments, leading to a wide product choice for farmers and a market-driven seed industry. The Seeds Act, 1966, was enacted to regulate seed quality and trade in India. National Seeds Programme (1975–85), a World Bank-aided initiative helped in the establishment of State Seed Corporations, Seed Certification Agencies, and testing laboratories for ensuring quality and quantity of the seed. Production as well as distribution subsidies for seed production program and its distribution further strengthened seed delivery system in India. India follows a limited generation system for seed multiplication, recognizing three generations: breeder, foundation, and certified seed. Seed certification is voluntary in India, with labeling being mandatory. Quality seed production of pulse crops through pulse seed hub focuses on local seed quality enhancement, safe storage, and seed development to ensure sustainable and profitable seed supply for farmers.

References

Agarwal DK, Sripathy KV, Govind P, Udaya B, Kamble U (2020) Breeder seed production and supply chain in pulses: Indian perspective. In: Abstracts: international conference on pulses the climate smart crops: challenges and oppurtunities. ICAR-Indian Institute of Pulses Research, Kanpur, pp 79–87

Atlin GN, Cairns JE, Das B (2017) Rapid breeding and varietal replacement are critical to adaption of cropping systems in the developing world to climate change. Glob Food Sec 12:31–37

Bohra A, Jha UC, Kavi Kishor PB, Pandey S, Singh NP (2014) Genomics and molecular breeding in lesser explored pulse crops: current trends and future opportunities. Biotechnol Adv 32:1410–1428

Bohra A, Sahrawat KL, Kumar S, Joshi R, Parihar AK, Singh U, Singh D, Singh NP (2015) Genetics- and genomics-based interventions for nutritional enhancement of grain legume crops: status and outlook. J Appl Genet 56:151–161

Bohra A, Jha UC, Godwin I, Varshney RK (2020) Genomic interventions for sustainable agriculture. Plant Biotechnol J 18:2388–2405

Chauhan JS, Singh AL, Rajendra Prasad S, Pal S (2015) Quality seed: a mega factor in enhancing crop productivity. In: Singh AL (ed) Recent advances in crop physiology, vol II. Daya Publishing House, New Delhi, pp 357–426

Chauhan JS, Singh BB, Gupta S (2016) Enhancing pulses production in India through improving seed and variety replacement rates. Indian J Genet Plant Breed 76(4):410–419

Gowda CLL, Srinivasan S, Gaur PM, Saxena KB (2013) Enhancing the productivity and production of pulses in India. In: Shetty PK, Ayyappan S, Swaminathan MS (eds) Climate change and sustainable food security. National Institute of Advanced Studies, Bengalaru and Indian Council of Agricultural Research, New Delhi, pp 145–159

Louwaars NP, Manicad G (2022) Seed systems resilience—an overview. Seeds 1(4):340–356

Nabuuma D, Reimers C, Hoang KT et al (2022) Impact of seed system interventions on food and nutrition security in low- and middle-income countries: a scoping review. Glob Food Sec 33:100638

Parihar AK, Dixit GP (2016) Varietal spectrum of seed production of pulses in India: an updated approach. Proc Natl Acad Sci India B: Biol Sci 86(2):247–252

Pathak H, Mishra JP, Mohapatra T (2022) Indian agriculture after independence. Indian Council of Agricultural Research, New Delhi, p 426

Pattanaik BB (2013) Quality seeds: contribution of national seeds corporation. In: Singh AK (ed) Golden Jubilee Souvenir 2012–13. New Delhi National Seeds Corporation Ltd., pp 14–18

Singh RP, Chintagunta AD, Agarwal DK, Kureel RS, Kumar SPJ (2020) Varietal replacement rate: prospects and challenges for global food security. Global Food Secur 25:100324

Spielman DJ, Smale M (2017) Policy options to accelerate variety change among smallholder farmers in South Asia and Africa South of the Sahara. Environment and Production Technology Division, p 68, IFPRI Discussion Paper 01666

Varshney RK, Thudi M, Pandey MK, Tardieu F, Ojiewo C, Vadez V, Whitbread AM, Siddique KHM, Nguyen HT, Carberry PS, Bergvinson D (2018) Accelerating genetic gains in legumes for the development of prosperous smallholder agriculture: integrating genomics, phenotyping, systems modelling and agronomy. J Exp Bot 69(13):3293–3312

Varshney RK, Bohra A, Roorkiwal M, Barmukh R, Cowling W, Chitikineni A, Lam HM, Hickey LT, Croser J, Edwards D, Farooq M, Crossa J, Weckwerth W, Harvey M, Kumar A, Bevan MW, Siddique KHM (2021) Rapid delivery systems for a food-secure future. Nat Biotechnol 39:1179–1181

Vijay D, Vijayakumar HP, Manjunath Prasad CT, Choudhury PR (2023) Indian seed sector. In: Ghosh PK, Das A, Saxena R, Banerjee K, Kar G, Vijay D (eds) Trajectory of 75 years of Indian agriculture after independence. Springer, Singapore

Wimalasekera R (2015) Role of seed quality in improving crop yields. In: Hakeem K (ed) Crop production and global environmental issues. Springer, Cham

MIX
Papier aus verantwortungsvollen Quellen
Paper from responsible sources
FSC® C105338

If you have any concerns about our products,
you can contact us on
ProductSafety@springernature.com

In case Publisher is established outside the EU,
the EU authorized representative is:
Springer Nature Customer Service Center GmbH
Europaplatz 3, 69115 Heidelberg, Germany

Printed by Libri Plureos GmbH
in Hamburg, Germany